Introduction to Computer Numerical Control (CNC)

Third Edition

James V. Valentino
Joseph Goldenberg

Upper Saddle River, New Jersey
Columbus, Ohio

Library of Congress Cataloging-in-Publication Data
Valentino, James.
Introduction to computer numerical control (CNC) / James V. Valentino and Joseph Goldenberg. —3rd ed.
p. cm.
ISBN 0-13-094424-6
1. Machine-tools—Numerical control. I. Goldenberg, Joseph. II. Title.

TJ1189 .V32 2003
621.9'023—dc21 2002074965

Editor in Chief: Stephen Helba
Executive Editor: Ed Francis
Production Editor: Christine M. Buckendahl
Production Coordination: Kelly Mulligan, Carlisle Publishers Services
Design Coordinator: Diane Ernsberger
Cover Designer: Jason Moore
Production Manager: Brian Fox
Marketing Manager: Mark Marsden

This book was set in New Century Schoolbook by Carlisle Communications, Ltd., and was printed and bound by Courier Kendallville, Inc. The cover was printed by Phoenix Color Corp.

Pearson Education Ltd.
Pearson Education Australia Pty. Limited
Pearson Education Singapore Pte. Ltd.
Pearson Education North Asia Ltd.
Pearson Education Canada, Ltd.
Pearson Educación de Mexico, S.A. de C.V.
Pearson Education—Japan
Pearson Education Malaysia Pte. Ltd.
Pearson Education, *Upper Saddle River, New Jersey*

10 9 8 7 6
ISBN: 0-13-094424-6

To my wife Barbara, and to our delightful children, Sarah and Andrew

James V. Valentino

To my wife Erica, and to our wonderful children, Janet, Simon, and Victoria, for their endless love, support, and inspiration

Joseph Goldenberg

PREFACE

One of the greatest challenges facing the United States today is in the area of manufacturing. To a large extent the computer has revolutionized this technology. It has virtually transformed the process of product design, analysis, and manufacture. Industries are finding that the new manufacturing technology demands well-trained personnel. Education is now being viewed as a continuous and long-term investment.

The third edition of *Introduction to Computer Numerical Control (CNC)* has been expanded and improved. The blueprint reading material has been separated as follows: Chapter 5—*Review of Basic Blueprint Reading for CNC Programmers* and Chapter 6—*Review of Basic Geometric Dimensioning and Tolerancing for CNC Programmers.*

Chapter 18 now includes a presentation on creating and simulating a complete part program using *Mastercam* CNC software. The third edition introduces the use of CNC software for writing, verifying, and simulating the milling word address programs in this text. To this end, a new Chapter 20, titled *Verifying Part Programs,* has been added. Included with this edition is a bound CD-ROM disk containing powerful, industrial quality CNC verification and simulation software.

The software displays real-time solid model animation of the machining that results from a part program. Additionally, it has an inspection mode that enables students to section as well as verify the dimensions of the machined part.

The milling part programs in the text have been edited so they will work properly with the verification and simulation software.

- Each chapter begins with a brief listing of objectives and ends with a chapter summary.
- Illustrations and photographs are used liberally throughout to reinforce pictorially what is being discussed.
- Students are frequently directed to boxed-in key terms and concepts.
- Flowcharts are used to teach CNC process planning and program planning.
- The important topic of job setup is discussed in the many solved programming examples.
- Fundamental word address (G and M code) programming is stressed.
- Industrial standard practices and terms are emphasized in the solved programming examples.
- Needless cross-referencing has been eliminated. Each program is listed with all explanations appearing on the same page.
- Pattern recognition is emphasized. The student is taught to recognize a certain group of programming commands as a programming pattern. For example, pattern A commands start up the CNC machine, whereas pattern B commands cause a tool change to take place.
- An excellent assortment of review exercises is provided at the end of each chapter. These exercises supply the student such important information as the operation to be performed, tooling, tool speed, tool feed, and job setup data.

- The industry standard Fanuc controller is emphasized throughout the text.
- Important mathematical principles are reviewed before programming is presented. A special chapter on right-triangle trigonometry provides the student with the critical mathematical information needed to understand programming.
- The student is exposed to the big picture of CNC shop activities. A special chapter explains the most important operations to be carried out in manufacturing a part.
- Appendixes contain information useful to the CNC student. They include a list of important safety precautions; summaries of G and M codes for milling and turning operations; recommended speeds and feeds for different materials with respect to drilling, milling, and turning operations; and important and easy-to-use machining formulas.
- A comprehensive glossary of key CNC terms is provided at the end of the book.
- Verification and simulation software enables students to visualize the effects of a written part program.

Introduction to Computer Numerical Control (CNC), Third Edition, can be used as an entry-level text for many different types of training applications. These include:

- Undergraduate one-semester or two-semester CNC courses
- Manual component of a CNC programming course
- Industry training course
- Seminar on CNC programming
- Adult education course
- Reference text for self-study

This textbook is designed to be used in many types of educational institutions:

- Four-year engineering schools
- Four-year technology schools
- Community colleges
- Trade schools
- Industrial training centers

This work is the result of several years of experience in running CNC courses for both industrial personnel and the students at Queensborough Community College. We found that many existing texts were either too general or too advanced for direct application. As a result, we drafted supplementary notes containing step-by-step information. The notes were enhanced and tested extensively in the classroom. Several colleagues, both in industry as well as in education, were called upon for their input. A thorough market survey also influenced the final content. It should be noted that all the programs presented have been thoroughly tested. The student is advised to take the appropriate safety precautions when running them on a CNC machine.

ACKNOWLEDGMENTS

The authors are indebted to many persons and industrial organizations for their assistance in preparing this manuscript. They are listed as follows:

- Allen-Bradley Company
- Amatrol, Inc.
- American SIP Corporation
- Autodesk Corporation
- Boston Digital Corporation
- Bridgeport Machines, Inc.
- Chick Machine Tool Company
- Cincinnati Millicron, Inc.
- Cleveland Twist Drill Company
- CNC Software, Inc.
- Command Corporation International
- Dapra Corporation
- Dell Computer Corporation
- DoAll Company
- EDO Corporation
- EMCO MAIER Corporation
- Gateway Computer Corporation
- GE Fanuc Automation Corporation
- Gibbs and Associates
- GN Telematic, Inc.
- Greco Systems
- Greenleaf Corporation
- GTE Valenite Corporation
- Handsvent Industries, Inc.
- Index Corporation
- Ingersoll Cutting Tools
- Institute of Advanced Machining Sciences
- Intergraph Corporation
- International Manufacturing Computer Services, Inc.
- Kennametal, Inc.
- KT-Swazey, Milwaukee, Wisconsin
- Lasercut, Inc.
- Laserdyne Division, Lumonics Corporation
- Macro Machine Tool Company
- Maho Machine Tool Corporation
- Mitsui Machine Technology, Inc.
- Monarch Machine Tool Company
- Niagara Cutter, Inc.
- Predator Software, Inc.
- SMW Systems, Inc.
- Solidworks Corporation
- Stripit, Inc.

- Surfware, Inc.
- Tri Star Computer Corporation
- TSD Universal/DeVlieg Ballard Tooling Systems Division
- Unigraphics Solutions, Inc.
- Visionary Design Systems, Inc.
- Waukesha Cutting Tools, Inc.
- Robert Brumm, State University of New York and Alfred State College
- Kurt Carlson, Cleveland Industrial Training Center
- Dan Krier, Moraine Park Technical College
- Zhongming Liang, Purdue University at Fort Wayne
- Mr. Martin Powell, senior CLT, Queensborough Community College

The authors would also like to thank the reviewers of this edition for their helpful comments and suggestions: Keith Green, Mineral Area College; Brian J. Hill, Corning Community College; and Kent Kohkonen, Brigham Young University.

CONTENTS

CHAPTER 9 *Word Address Programming* 174

CHAPTER 10 *Programming Hole Operations* 186

CHAPTER 11 *Programming Linear Profiles* 205

CHAPTER 12 *Programming Circular Profiles* 232

CHAPTER 16 *Fundamental Concepts of CNC Lathe Programming* *331*

CHAPTER 17 *Techniques and Fixed Cycles for CNC Lathe Programming* *357*

CHAPTER

1

INTRODUCTION TO COMPUTER NUMERICAL CONTROL MANUFACTURING

1-1 CHAPTER OBJECTIVES

At the conclusion of this chapter you will be able to

1. Explain what computer numerical control (CNC) is and what basic components comprise CNC systems.
2. State the objectives, advantages, and special requirements concerning CNC use.
3. Identify the different media used to input and store CNC programs.
4. Describe the two different punched-tape formats used with CNC machines.

1-2 INTRODUCTION

The basic concepts of numerical control (NC) and computer numerical control (CNC) technology are discussed. Traditional NC and contemporary CNC hardware configurations are described. The important benefits to be derived from CNC operations are listed and explained. The different types of media used for storage and input of CNC programs are then explored. The reader is introduced to different formats for punched tape. Machining centers with automatic tool changers, the latest development in CNC, are considered.

1-3 NUMERICAL CONTROL DEFINITION, ITS CONCEPTS AND ADVANTAGES

Numerical control has been used in industry for more than 40 years. Simply put, numerical control is a method of automatically operating a manufacturing machine based on a code of letters, numbers, and special characters. A complete set of coded instructions for executing an operation is called a program. The program is translated into corresponding electrical signals for input to motors that run the machine. Numerical control machines can be programmed manually. If a computer is used to create a program, the process is known as computer-aided programming. The approach taken in this text will be in the form of manual programming.

Traditionally, numerical control systems have been composed of the following components:

Tape punch: converts written instructions into a corresponding hole pattern. The hole pattern is punched into tape, which passes through this

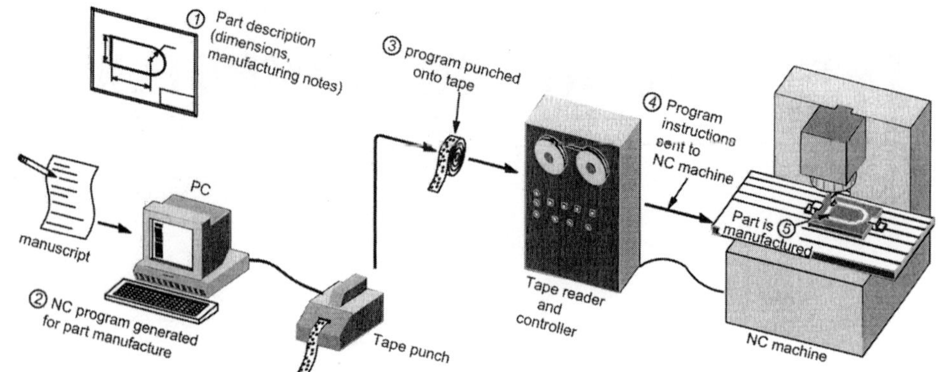

FIGURE 1–1 Components of traditional NC systems.

device. Much older units used a typewriter device called a Flexowriter. Later devices included a microcomputer coupled with a tape punch unit.

Tape reader: reads the hole pattern on the tape and converts the pattern to a corresponding electrical signal code.

Controller: receives the electrical signal code from the tape reader and subsequently causes the NC machine to respond.

NC machine: responds to programmed signals from the controller. Accordingly, the machine executes the required motions to manufacture a part (spindle rotation on/off, table and or spindle movement along programmed axis directions, etc.). See Figure 1–1.

NC systems offer some of the following advantages over manual methods of production:

1. Better control of tool motions under optimum cutting conditions.
2. Improved part quality and repeatability.
3. Reduced tooling costs, tool wear, and job setup time.
4. Reduced time to manufacture parts.
5. Reduced scrap.
6. Better production planning and placement of machining operations in the hands of engineering.

1–4 DEFINITION OF COMPUTER NUMERICAL CONTROL AND ITS COMPONENTS

A computer numerical control (CNC) machine is an NC machine with the added feature of an on-board computer. The on-board computer is often referred to as the machine control unit or MCU. Control units for NC machines are usually hard wired. This means that all machine functions are controlled by the physical electronic elements that are built into the controller. The on-board computer, on the other hand, is "soft" wired. Thus, the machine functions are encoded into the computer at the time of manufacture. They will not be erased when the CNC machine is turned off. Computer memory that holds such information is known as ROM or read-only memory. The MCU usually has an alphanumeric keyboard for direct or manual data input (MDI) of part programs. Such programs are stored in RAM or the random-access memory portion of the

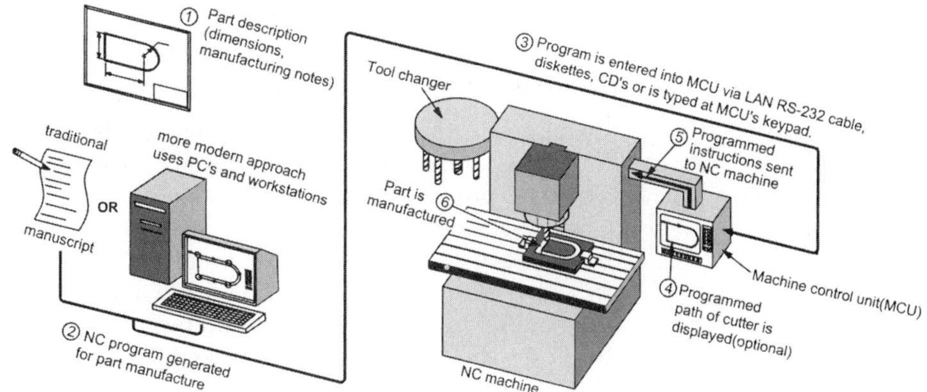

FIGURE 1–2 Components of modern CNC systems.

computer. They can be played back, edited, and processed by the control. All programs residing in RAM, however, are lost when the CNC machine is turned off. These programs can be saved on auxiliary storage devices such as punched tape, magnetic tape, or magnetic disk. Newer MCU units have graphics screens that can display not only the CNC program but the cutter paths generated and any errors in the program.

The components found in many CNC systems are shown in Figure 1–2.

Machine control unit: generates, stores, and processes CNC programs. The machine control unit also contains the machine motion controller in the form of an executive software program. See Figure 1–3.

NC machine: responds to programmed signals from the machine control unit and manufactures the part.

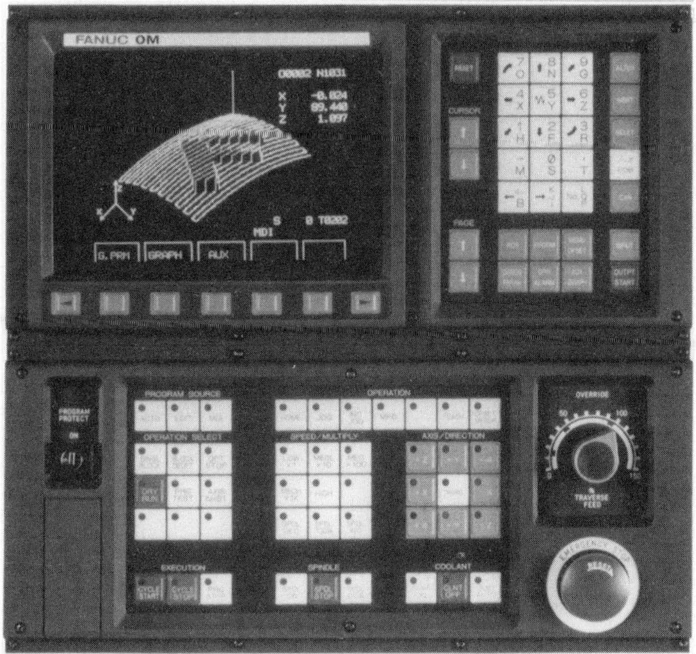

FIGURE 1–3 A modern machine control unit.

1-5 ADVANTAGES OF CNC COMPARED WITH NC

Computer numerical control opens up new possibilities and advantages not offered by older NC machines. Some of these are

1. Reduction in the hardware necessary to add a machine function. New functions can be programmed into the MCU as software.
2. The CNC program can be written, stored, and executed directly at the CNC machine.
3. Any portion of an entered CNC program can be played back and edited at will. Tool motions can be electronically displayed upon playback.
4. Many different CNC programs can be stored in the MCU.
5. Several CNC machines can be linked together to a main computer. Programs written via the main computer can be downloaded to any CNC machine in the network. This is known as direct numerical control or DNC. See Figure 1–4.
6. Several DNC systems can also be networked to form a large distributive numerical control system. Refer to Figure 1–5.
7. The CNC program can be input from zip or floppy disks or downloaded from local area networks.

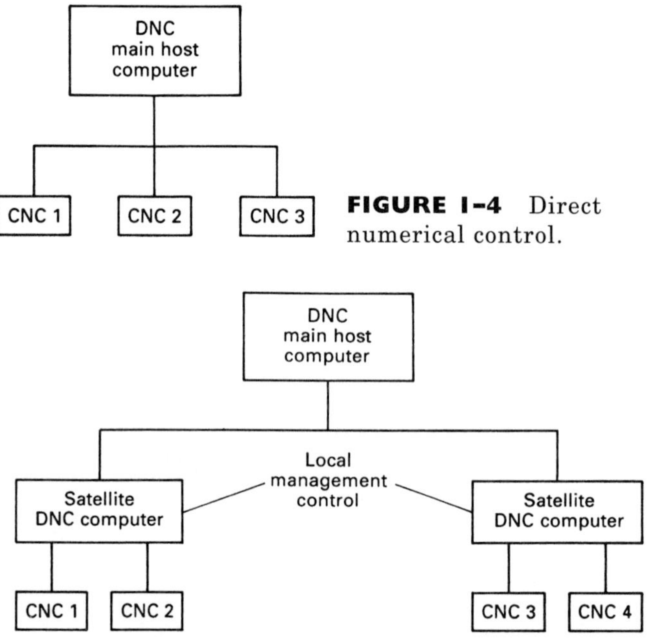

FIGURE 1–4 Direct numerical control.

FIGURE 1–5 Distributive numerical control.

1-6 SPECIAL REQUIREMENTS FOR UTILIZING CNC

Computer numerical control machines can dramatically boost productivity. The CNC manager, however, can only ensure such gains by first addressing several critical issues.
Among these are

1. Sufficient capital must be allocated for purchasing quality CNC equipment.
2. CNC equipment must be maintained on a regular basis. This can be accomplished by obtaining a full-service contract or by hiring an in-house technician.

3. Personnel must be thoroughly trained in the setup and operation of CNC machines.
4. Careful production planning must be studied. This is because the hourly cost of operating a CNC machine is usually higher than that for conventional machines.

1–7 FINANCIAL REWARDS OF CNC INVESTMENT

Investors are encouraged to look to the CNC machine tool as a production solution with the following savings benefits:

1. Savings in direct labor. One CNC machine's output is commonly equalivent to several conventional machines.
2. Savings in operator training expenses.
3. Savings in shop supervisory costs.
4. Savings due to tighter, more predictable production scheduling.
5. Savings in real estate since fewer CNC machines are needed.
6. Savings in power consumption since CNC machines produce parts with a minimum of motor idle time.
7. Savings from improved cost estimation and pricing.
8. Savings due to the elimination of construction of precision jigs, and the reduced need for special fixtures. Maintenance and storage costs of these items are also reduced.
9. Savings in tool engineering/design and documentation. The CNC's machining capability eliminates the need for special form tools, special boring bars, special thread cutters, etc.
10. Reduced inspection time due to the CNC machine's ability to produce parts with superior accuracy and repeatibility. In many cases, only spot checking of critical areas is necessary without loss of machine time.

Using Payback Period to Estimate Investment Efficiency

The Payback Period calculation estimates the number of years required to recover the net cost of the CNC machine tool.

$$\text{Payback Period} = \frac{\text{Net cost of CNC} - \text{Net cost of CNC} \times \text{Tax Credit}}{\text{Savings} - \text{Savings} \times \text{Tax Rate} + \text{Yearly Depreciation of CNC} \times \text{Tax Rate}}$$

Using Return on Investment (ROI) to Estimate Investment Efficiency

The ROI calculation predicts what percent of the net cost of the CNC will be recovered each year. The ROI calculation accounts for the useful life of the CNC machine tool.

$$\text{ROI} = \frac{\text{Average Yearly Savings} - \text{Net cost of CNC/years of life}}{\text{Net cost of CNC}}$$

■ EXAMPLE 1–1

Given the investment figures in Table 1–1 for implementing a new CNC machine tool, determine the payback period and the annual return on investment. The CNC is conservatively estimated to have a useful life of 12 years.

TABLE 1–1

Initial Investment ($)	One-time savings in tooling ($)	Net Cost of CNC ($)	Average yearly savings ($)	Tax credit (10%)	Tax rate (46%)	Yearly depreciation of CNC ($)
130,250	35,000	95,250	63,100	0.1	0.46	10,900

$$\text{Payback Period} = \frac{95{,}250 - 95{,}250 \times 0.1}{63{,}100 - 63{,}100 \times 0.46 + 10{,}900 \times 0.46}$$

$$\text{Payback Period} = 2.19 \text{ years}$$

This calculation estimates that the net cost of the CNC will be recovered in 2.19 years.

$$\text{ROI} = \frac{63{,}100 - 95{,}250/12}{95{,}250}$$

$$\text{ROI} = .57$$

This calculation estimates that the investor can expect 57% of the net cost of the CNC or (.57 × \$95,250) = \$54,293 to be recovered each year if the CNC machine's useful life is 12 years. ■■

1–8 CNC MACHINING CENTERS AND TURNING CENTERS

Machining centers are the latest development in CNC technology. These systems come equipped with automatic tool changers capable of changing 90 or more tools. Many are also fitted with movable rectangular worktables called pallets. The pallets are used to automatically load and unload workpieces. At a single setup, machining centers can perform such operations as milling, drilling, tapping, boring, counterboring, and so on. Additionally, by utilizing indexing heads, some centers are capable of executing these tasks on many different faces of a part and at specified angles. Machining centers save production time and cost by reducing the need for moving a part from one machine to another. Two types of machining centers are shown in Figures 1–6 and 1–7.

A more complete discussion of the important features found on machining centers is deferred until Chapter 4.

Turning centers with increased capacity tool changers are also making a strong appearance in modern production shops. These CNC machines are capable of executing many different types of lathe cutting operations simultaneously on a rotating part. A modern turning center is shown in Figure 1–8.

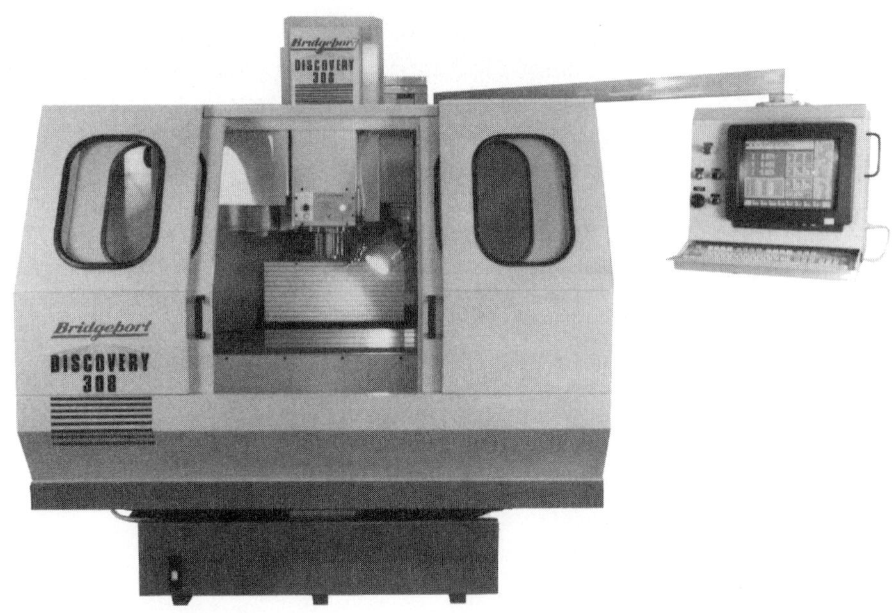

FIGURE 1–6 A vertical spindle machining center. (Photo courtesy of Bridgeport Machines, Inc.)

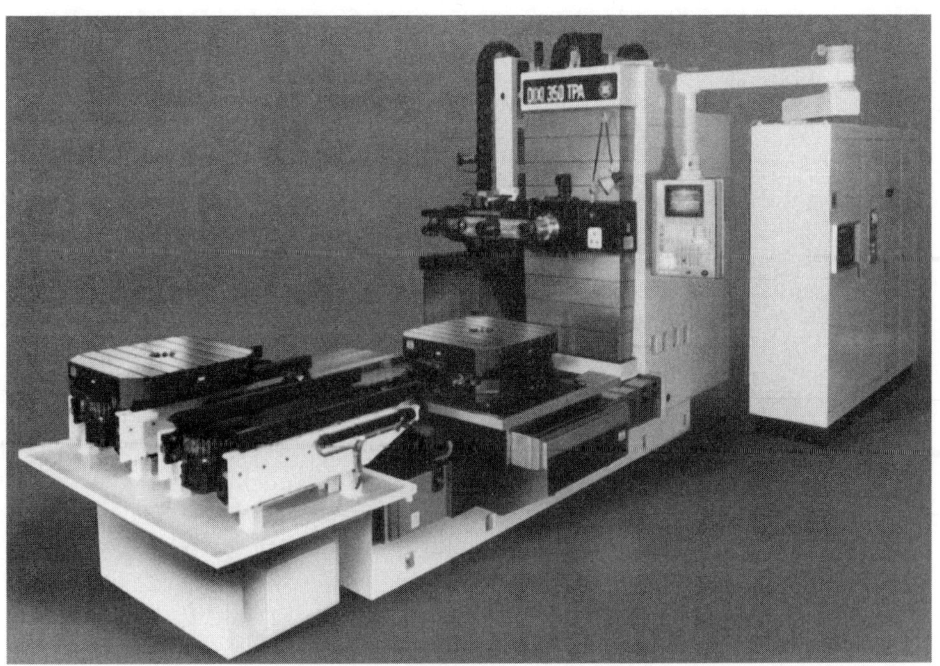

FIGURE 1–7 A horizontal spindle machining center with an automatic tool changer and two-pallet work changers. (Photo courtesy of American SIP.)

FIGURE 1–8 A modern CNC turning center. (Photo courtesy of The Monarch Machine Tool Company.)

1–9 OTHER TYPES OF CNC EQUIPMENT

In addition to machining centers and turning centers, CNC technology has also been applied to many other types of manufacturing equipment. Among these are wire electrical discharge machines (wire EDM) and laser cutting machines.

Wire EDM machines utilize a very thin wire (0.0008 to 0.012 in.) as an electrode. The wire is stretched between diamond guides and carbide that conduct current to the wire and cuts the part like a bandsaw. Material is removed by the erosion caused by a spark that moves horizontally with the wire. CNC is used to control horizontal table movements. Wire EDM machines are very useful for producing mold inserts, extrusion and trim dies, as well as form tools. See Figure 1–9.

Laser cutting CNC machines utilize an intense beam of focused laser light to cut the part. Material under the laser beam undergoes a rapid rise in temperature and is vaporized. If the beam power is high enough, it will penetrate through the material. Because no mechanical cutting forces are involved, lasers cut parts with a minimum of distortion. They have been very effective in machining slots and drilling holes. See Figure 1–10.

1–10 CNC INPUT AND STORAGE MEDIA

The information necessary to perform CNC operations could be entered manually into the control unit. This is a long and inefficient process. The machine is also prevented from making parts while this is being done. Thus, several devices have been developed for storing and loading programs written with the aid of a microcomputer or larger mainframe computer. These are shown in Figure 1–11.

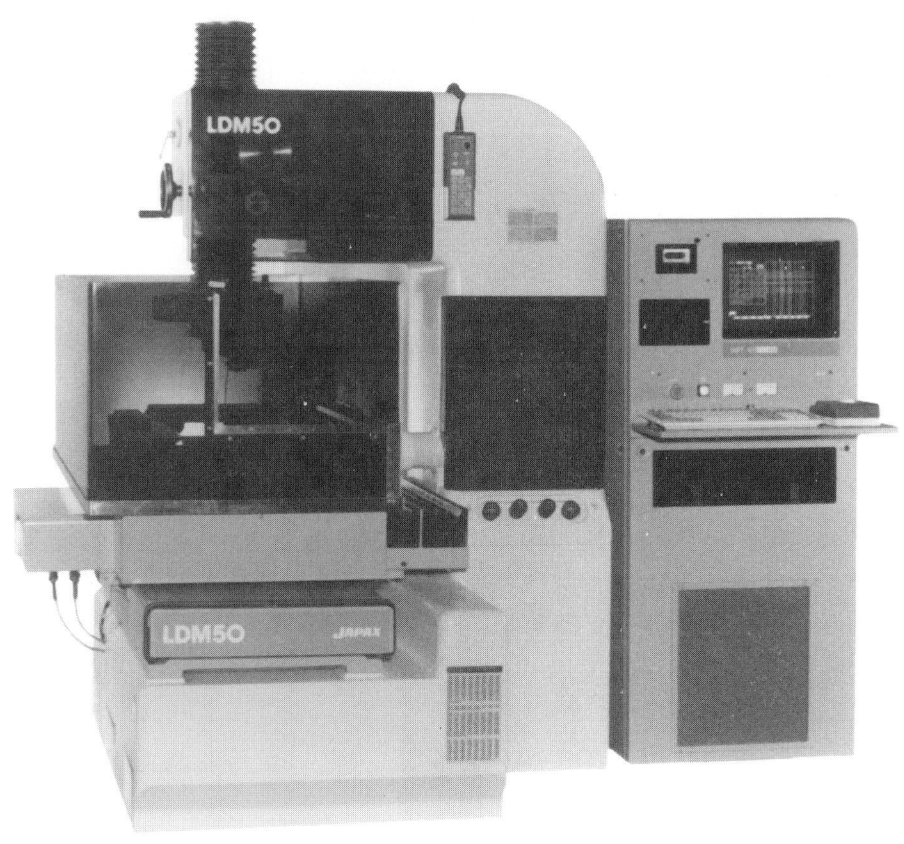

FIGURE 1-9 A CNC-controlled wire cutting EDM machine. (Photo courtesy of Mitsui Machine Tool Technology, Inc.)

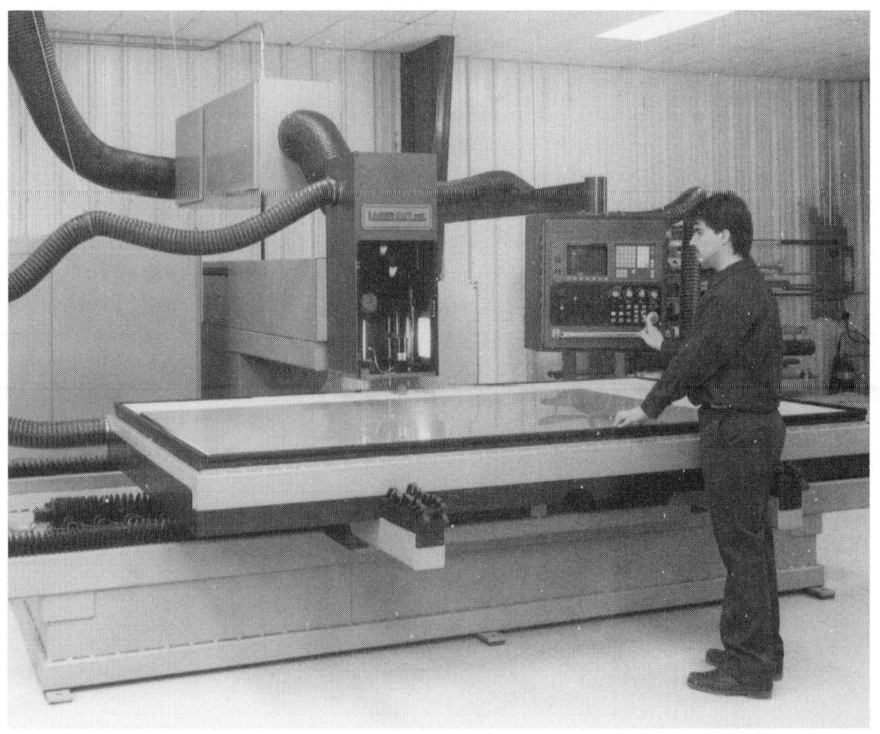

FIGURE 1-10 A laser cutting CNC machine. (Photo courtesy of Lasercut, Inc.)

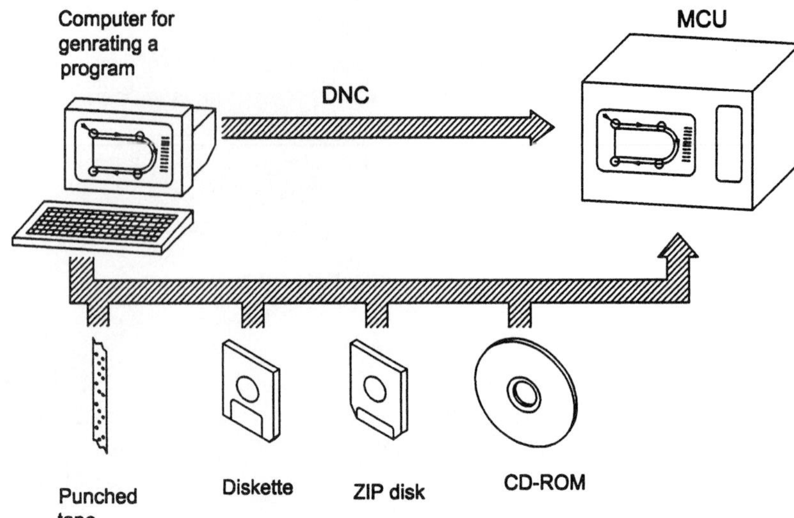

FIGURE 1-11 CNC input and storage media.

Punched Tape

It will be useful for the reader to become acquainted with the basic concept of binary number code prior to discussing tape formats.

Internally, computers and similar electronic control devices operate by a system of electrical switches. A 1 (one) is processed as an open switch and a 0 (zero) is processed as a closed switch. All numbers, letters, and special characters are represented in terms of unique sets of zeros and ones. The only code the computer can work on is binary. All code generated external to the computer must first be translated into binary before the computer can act on it. The binary code from the computer must also be translated back into a code operators can understand. The translation process is automatically executed by devices inside the computer.

Punched tape is one method of translating code into binary. Each hole in the tape is electronically read as a one and each blank as a zero. Tape is typically 1" wide and is made of paper, paper-Mylar (Mylar is a tough plastic), or aluminum-Mylar laminates. Paper tape is the least expensive. It is treated to resist water and oil and is the most popular. Mylar tape is much more expensive but is very durable. It is still used by CNC manufacturers to store information as executive tapes. The two types of tape formats currently in use are the EIA RS-244B and the EIA RS-358B. Different types of punching devices are used to translate the programmed instructions into a corresponding hole pattern on the tape. The hole pattern is usually read by a photoelectric light reader. Other electrical and mechanical methods have also been used.

> ***Note:*** Punched tape is being phased out and replaced by more modern storage devices such as disks, zip disks and CD-ROMs.

Disks and Zip Disks

These devices store a program in the form of a magnetic pattern on a plastic disk. During operation, the disk is made to spin and the pattern is read by recording heads in the disk drive unit. Disks, also known as "floppy" disks, can store up to 1.44 megabytes (MB) of information.

Zip disks have a larger storage capacity and can typically hold up to 100 MB of information. High capacity Zip disks capable of storing 250MB are also available.

CD-ROM

The compact disk (CD) is a relatively new and popular method of storing information in the form of a pattern of etched pits. An optical laser is used to read the pit pattern on the spinning disk. CDs offer many advantages over other types of storage devices. They are a very stable and durable medium ensuring almost indefinite storage life. Additionally, they are capable of storing large amounts of information. A typical CD has a storage capacity of 680MB.

Disks, Zip disks and CD-ROMs are used with personal computers (PCs) and workstations. They are referred to as random access mediums. This means that any information on the disks can be found and retrieved almost instantaneously.

1–11 CHAPTER SUMMARY

The following key concepts were discussed in this chapter:

1. Numerical control is a method of automatically operating a manufacturing machine based on a programmed set of instructions.
2. A CNC machine is similar to an NC machine except that an on-board computer is used to store, edit, and execute programmed instructions.
3. CNC use involves substantial investments in equipment, production planning, and personnel training.
4. CNC machine tools greatly boost productivity.
5. Different methods of input include manual data input (MDI), punched tape, floppy disks, Zip disks and CD-ROMs and direct transmission from a remote computer (DNC).
6. The machine control unit operates in binary code only. All inputted programs must be translated into binary.
7. The two punched-tape formats in use today are RS-244B and RS-358B. Punched tape stores CNC code in binary form.
8. Storage devices include punched tape, disks, Zip disks and CD-ROMs.

REVIEW EXERCISES

1.1. What is numerical control and what components comprise a traditional NC machine?

1.2. What are four objectives of numerical control?

1.3. What advantages does numerical control offer over manual methods?

1.4. What is a computer numerical control (CNC) machine?

1.5. What improvements do CNC machines offer over traditional NC machines?

1.6. What is meant by the terms direct numerical control and distributive numerical control?

1.7. Name four requirements that must be satisfied prior to using CNC in a shop.

1.8. Describe four devices for storing and inputting CNC programs.

1.9. What is binary code?

1.10. Name three advantages offered by machining centers.

1.11. Describe the financial rewards of CNC investment.

1.12. (a) What is estimated by payback period?
(b) What is estimated by return on investment (ROI)?

CHAPTER 2

MODERN MACHINE TOOL CONTROLS

2–1 CHAPTER OBJECTIVES

At the conclusion of this chapter you will be able to

1. Describe the two types of control systems used to output tool movement.
2. Explain the two types of loop systems used with CNC controllers.
3. Identify the four different types of motors used to control tool movements.
4. Explain how points are located using the Cartesian coordinate system.
5. Understand CNC machine axis of motion.
6. Explain the significant difference between incremental and absolute positioning.
7. State what delta dimensioning is.
8. Explain what datum dimensioning is.

2–2 INTRODUCTION

The two major types of CNC control systems are discussed. The Cartesian coordinate system for locating points is described. Conventions for assigning machine axes and various types of CNC machine movements are studied. The important issue of controlling tool movements is presented. The chapter ends with a consideration of absolute and incremental positioning modes for specifying tool movements.

2–3 DIFFERENT TYPES OF SYSTEM CONTROL

As was stated in Section 1–4, the MCU contains a machine motion controller for controlling tool movement. Many different types of controllers are available today. They include Fanuc, Allen-Bradley, GE, Okuma, Bendix, Mazak, and others. The physical appearance of these controllers is somewhat similar and each responds to a slightly different set of programmed codes. See Figure 2–1. All control systems, however, fall into two major categories: point to point and continuous path.

Point-to-Point Tool Movements

Point-to-point control systems cause the tool to move to a point on the part and execute an operation at that point only. The tool is not in continuous contact

General Safety Rules for Operating Any CNC Machine

➡ Prior to machine operations:
- Read the safety instructions label posted on the CNC machine.

- Confirm safety of personnel near the CNC machine.
- Remove all obstacles from the machine's working area.
- Make sure protection fence is closed.

➡ During machining operations:
- Do not enter the working area.
- Make sure oil or water does not splash power supply units.
- Wear protective goggles and clothing. Put long hair up.

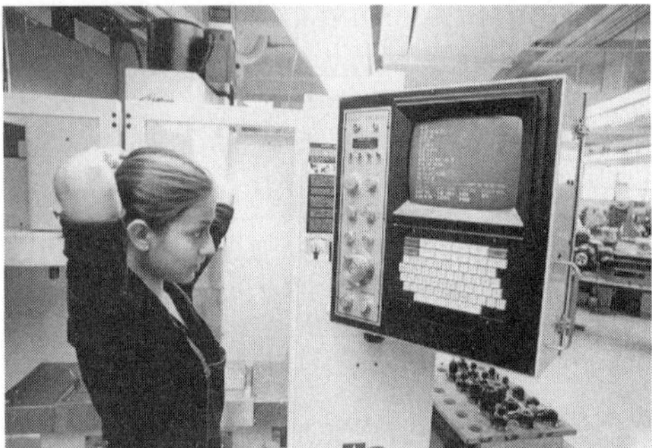

➡ Use the EMERGENCY STOP control when there is a possibility of personal injury or any machine/part damage.

➡ During troubleshooting and maintenance:
- Allow only qualified personnel to work on the machine.
- First turn off the power supply.
- Be aware that power is still present in the primary side power cables to the breaker box.

with the part while it is being moved to a working location. Some point-to-point operations are drilling, reaming, boring, tapping, and punching. See Figure 2–2.

Continuous-Path Tool Movements

Continuous-path controllers are so named because they cause the tool to maintain continuous contact with the part as the tool cuts a contour shape. Continuous-path operations include milling along lines at any angle, milling arcs, and lathe turning. See Figure 2–3.

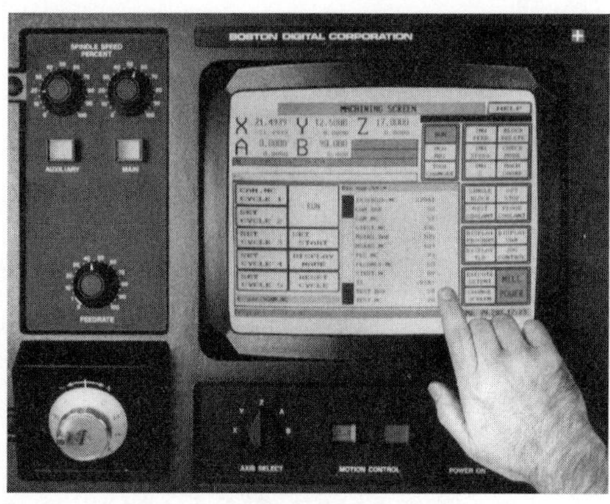

FIGURE 2–1 Machine control unit types. [(a) Photo courtesy of Boston Digital Corp. (b) Photo courtesy of Allen-Bradley Company Inc.]

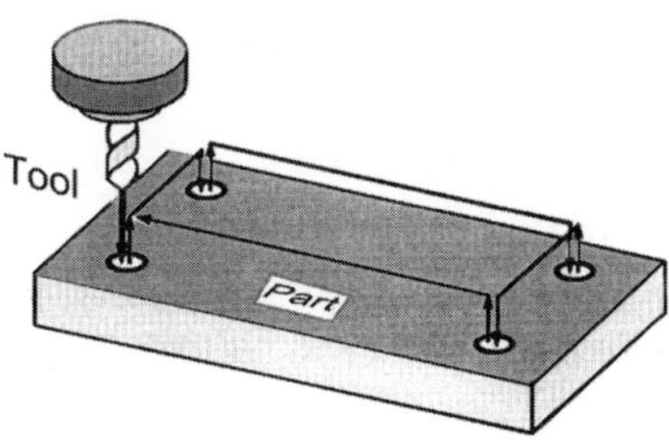

FIGURE 2–2 Point-to-point tool movement.

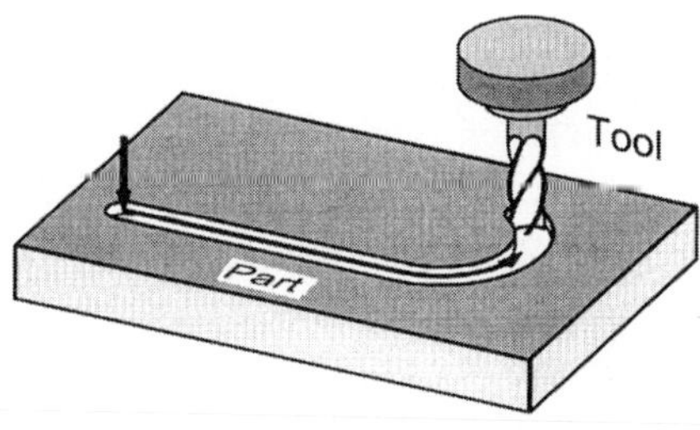

FIGURE 2–3 Continuous-path tool movement.

It should be noted that continuous-path controllers output motion by interpolating each position of the tool. Interpolation is a mathematical method of approximating the true or exact positions required to follow a precalculated path. The interpolated positions are determined such that they differ from the exact positions within an acceptable tolerance. Many continuous-path controllers interpolate curves as a series of straight-line segments. Very high accuracy can be achieved by making the line segments smaller and smaller. These concepts are illustrated in Figure 2–4.

Traditionally, continuous-path controllers were more expensive than point-to-point systems. Advancements in microelectronics, however, have reduced the cost and today most CNC machines come equipped with continuous-path controllers.

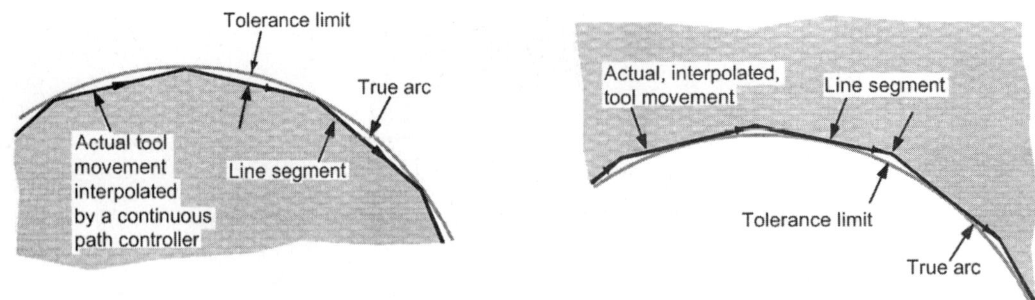

FIGURE 2–4 Interpolation used for continuous-path movement.

2–4 LOOP SYSTEMS FOR CONTROLLING TOOL MOVEMENT

A loop system sends electrical signals to drive motor controllers and receives some form of electrical feedback from the motor controllers. One of the important factors determining the tolerance to which a part can be cut is the loop system type.

There are two main systems in use today for controlling CNC machine movements: the open loop system and the closed loop system.

Open Loop Systems

An open loop system utilizes stepping motors to create machine movements. These motors rotate a fixed amount, usually 1.8°, for each pulse received. Stepping motors are driven by electrical signals coming from the MCU. The motors are connected to the machine table ball-nut lead screw and spindle. Upon receiving a signal, they move the table and/or spindle a fixed amount. The motor controller sends signals back indicating the motors have completed the motion. The feedback, however, is not used to check how close the actual machine movement comes to the exact movement programmed. See Figure 2–5.

Closed Loop Systems

Special motors called servos are used for executing machine movements in closed loop systems. Motor types include AC servos, DC servos, and hydraulic servos. Hydraulic servos, being the most powerful, are used on large CNC machines. AC servos are next in strength and are found on many machining centers.

A servo does not operate like a pulse counting stepping motor. The speed of an AC or DC servo is variable and depends upon the amount of current passing through it. The speed of a hydraulic servo depends upon the amount of fluid passing through it. The strength of current coming from the MCU determines the speed at which a servo rotates.

Servos are connected to the spindle. They are also connected to the machine table through the ball-nut lead screw. A device called a resolver continuously monitors the distance by which the table and/or spindle has moved and sends this information back to the MCU. The MCU can then adjust its signal as the actual table and/or spindle position approaches the programmed position. Systems that provide feedback signals of this type are called servo systems or servomechanisms. They can position tools with a very high degree of accuracy even when driving motors with high-horsepower ranges. See Figures 2–6 and 2–7.

Open loop systems have recently gained renewed interest for CNC applications. Improvements in stepping motor accuracy and power have, in some

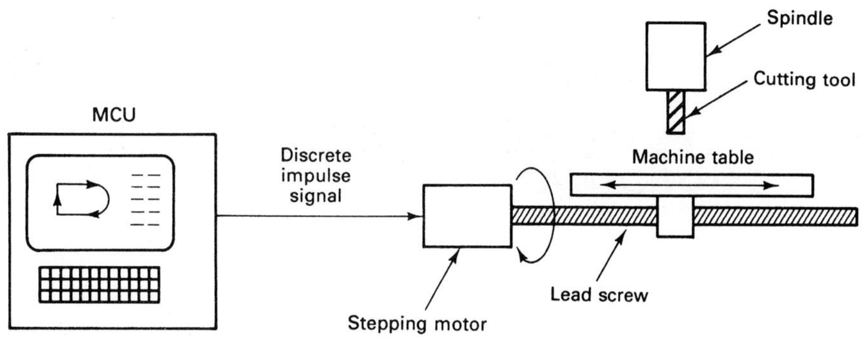

FIGURE 2-5 Configuration for an open loop system.

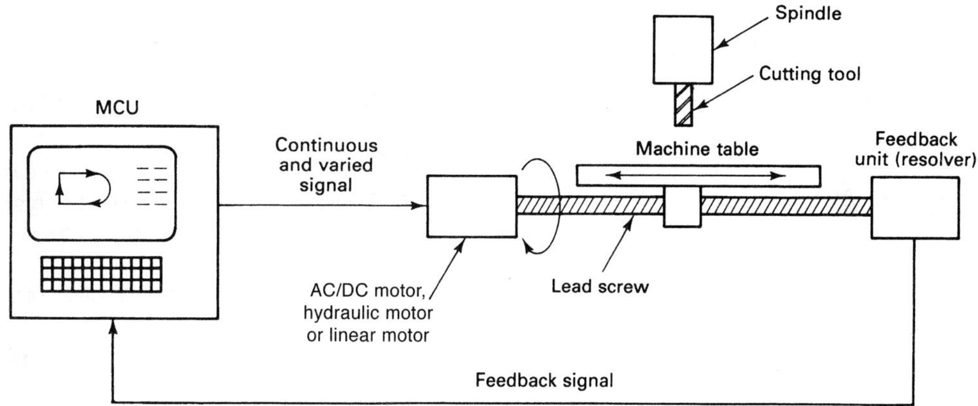

FIGURE 2-6 Configuration of a closed loop system.

FIGURE 2-7 AC servo motors. (Photo courtesy of French and Rodgers, Inc.)

cases, eliminated the need for expensive feedback system hardware and its associated circuitry. These newer systems represent substantial savings in machine and maintenance costs.

2–5 ESTABLISHING LOCATIONS VIA CARTESIAN COORDINATES

The location of a CNC tool at any time is controlled by a system of *XYZ* coordinates called Cartesian coordinates. This system is composed of three directional lines, called axes, mutually intersecting at an angle of 90°. The point of intersection is known as the origin. The *XY* coordinate plane is broken up into four quadrants. The value and sign of an ordered pair of (*X*, *Y*) coordinates determines the *X* and *Y* distances of a point from the origin and the quadrant in which the point is located. These concepts are illustrated in Figure 2–8.

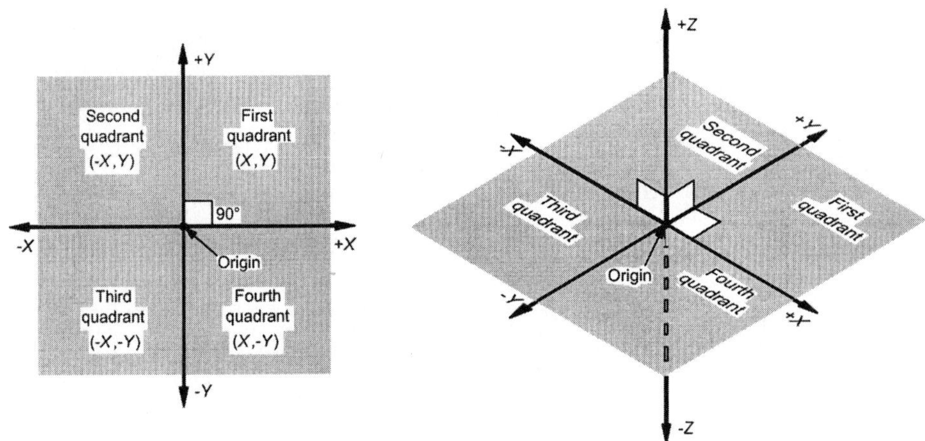

FIGURE 2–8 The Cartesian coordinate system.

■ EXAMPLE 2–1

Graphically indicate the locations of the points *A* (3, −2), *B* (1,4), *C* (−2, −3), and *D* (−3, 4) in the *XY* plane. See Figure 2–9.

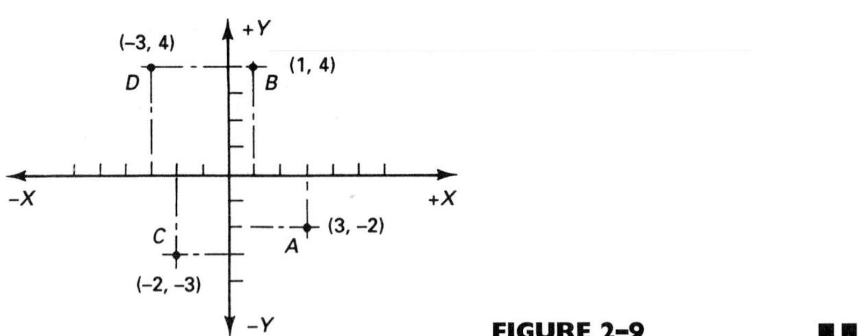

FIGURE 2–9 ■■

■ EXAMPLE 2–2

Graphically indicate the locations of the points *A* (3, 2, 1) and *B* (4, −1, −1) in *XYZ* space. These points are shown in Figure 2–10.

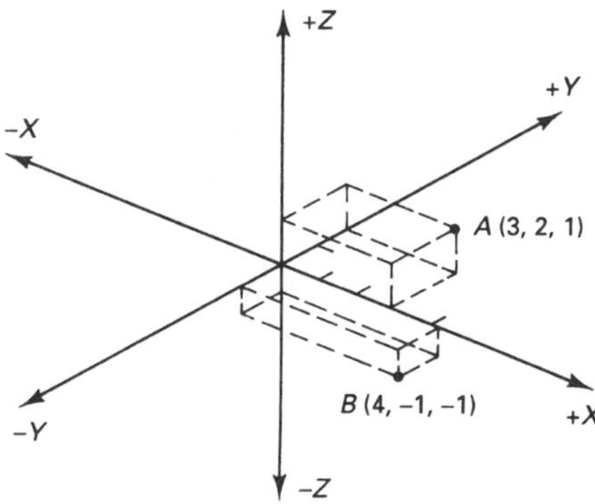

FIGURE 2–10

2–6 CNC MACHINE AXES OF MOTION

CNC equipment executes machining operations by performing some form of sliding linear motion and rotary motion. The actual method of movement is designed by the manufacturer and can vary from machine to machine. For example, the table can move in the horizontal plane (*XY*-axis motion) and the spindle in the vertical plane (*Z*-axis motion). The system will respond to a command to move the spindle (tool) along the $+X$ or $+Y$ axis by moving the table in the opposite direction, $-X$ or $-Y$. Because the machine automatically knows how to move in response to an axis command, the programmer need not be concerned whether it is the spindle or table that moves. The main point is that, in the end, the tool arrives at the programmed location. Thus, the machine axis will be defined in terms of spindle movement.

> ***Note:*** For programming purposes, programmers should consider the CNC machine table as locked with only the tool in the spindle moving along axes. $\pm X$, $\pm Y$, and $\pm Z$ axes.

Present standards for machine axes are established according to the industry standard report EIA RS-267A. These standards include:

1. Primary machine axes of movement should follow the right-hand rule. See Figure 2–11.
2. Spindle movement is primarily along the *Z* axis. Movement into the work is along the $-Z$ axis and movement away from the work is along the $+Z$ axis.
3. In a majority of milling machines, motion along the *X* axis is the longest travel perpendicular to *Z*. Motion indicated by $-X$ is directly opposite to that indicated by $+X$. The *X* axis is parallel to the work holding and is in the horizontal plane. The *X* axis moves to the right along a plane of the work as the operator looks at that plane.
4. With regard to milling machines, motion along the *Y* axis is the shortest travel perpendicular to *Z*. Motion indicated by $-Y$ is directly opposite to that indicated by $+Y$. The *Y* axis is in the same plane as the *X* axis. Looking at the plane, the operator will note that the *Y* axis is perpendicular to the *X* axis.

Most CNC milling machines can perform simultaneous motions along the *X*, *Y*, and *Z* axes and are called three-axis machines. See Figures 2–12 and 2–13.

More complex CNC machines have the capability of executing additional rotary motions as follows:

- Rotation about an axis parallel to the *X* axis or *A*-axis rotation
- Rotation about an axis parallel to the *Y* axis or *B*-axis rotation
- Rotation about an axis parallel to the *Z* axis or *C*-axis rotation

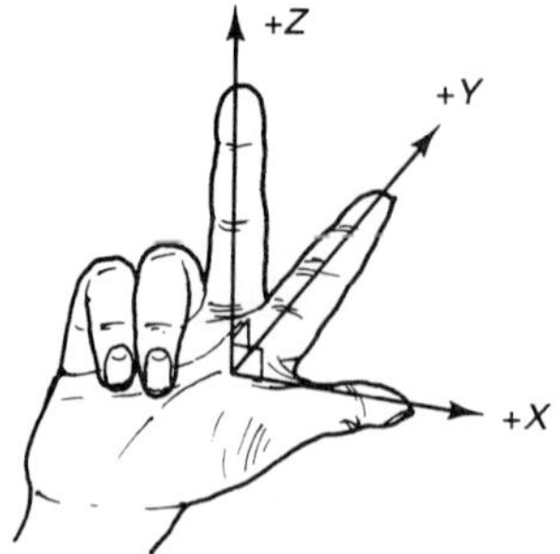

FIGURE 2–11 The right-hand rule for linear motion.

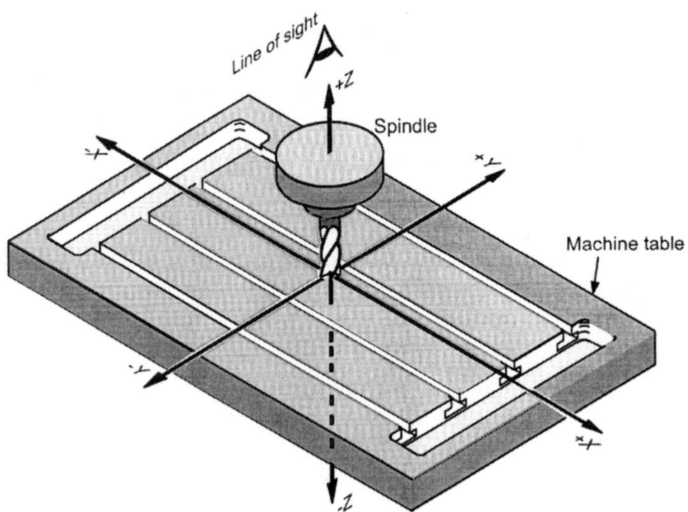

FIGURE 2–12 Machine axis for a three-axis vertical CNC machine (machine axis defined as spindle movement).

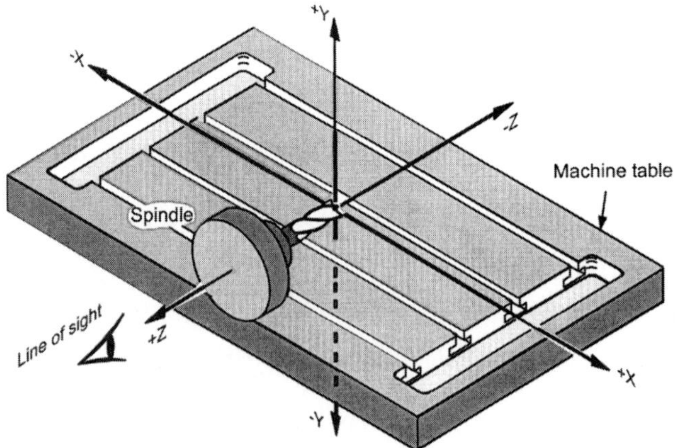

FIGURE 2–13 Machine axis for a three-axis horizontal CNC machine (machine axis defined as spindle movement).

For example, a three-axis horizontal machining center that is equipped with a rotary table will be capable of motion about a fourth axis or C-axis rotation. If the machining center has the additional capability of tilting the machine table on spindle about the A or B axis, a fourth and fifth axis are added. Such a machine is then capable of five simultaneous motions: three linear and two rotational. See Figure 2–14.

FIGURE 2–14 A rotary index table for adding a fourth axis to a CNC machine. (Photo courtesy of Dapra Corp.)

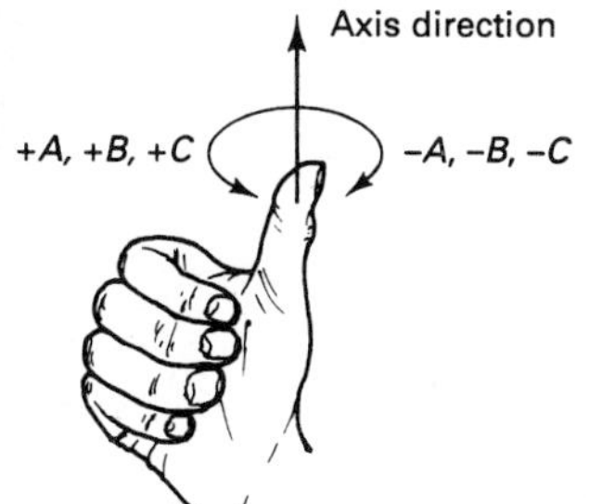

FIGURE 2–15 The right-hand rule for rotary motion.

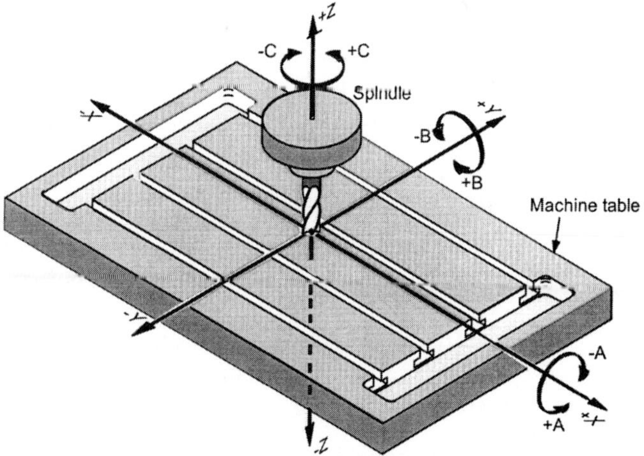

FIGURE 2–16 The six possible machine axes for a vertical CNC machine (machine axis defined as spindle movement).

Rotary motion directions also follow the right-hand rule. See Figure 2–15.

Four- and five-axis machines are used to machine parts with complex surfaces. See Figures 2–16 and 2–17.

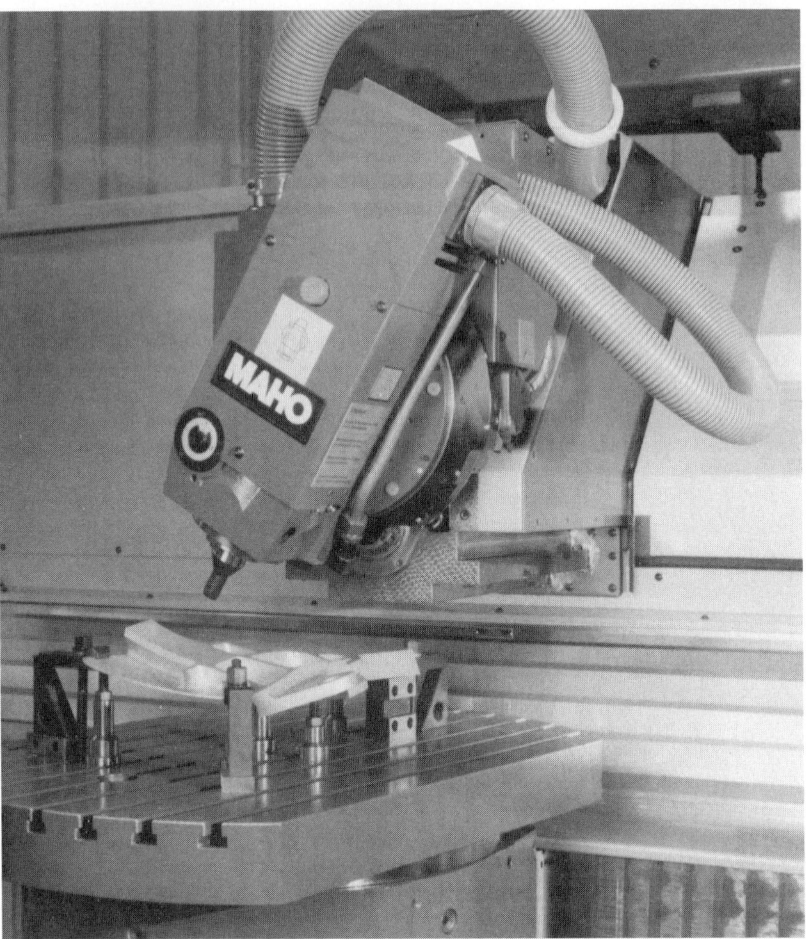

FIGURE 2–17 A five-axis machining operation. (Photo courtesy of Maho Machine Tool Corp.)

2–7 TYPES OF TOOL POSITIONING MODES

Within a given machine axes coordinate system, a CNC can be programmed to locate tool positions in the following modes: incremental, absolute, or mixed (incremental and absolute).

Incremental Positioning

Machines operating in this mode locate each new tool position by measuring from the last tool position established. See Figure 2–18 for an illustration of incremental positioning.

Incremental positioning has some drawbacks. The most notable is that if one incremental movement is in error, all other subsequent movements will also be incorrect.

> ***Note:*** With delta dimensioning each new dimension is specified by measuring it relative to the dimension previously entered. Delta dimensioning is well suited to incremental position programming.

Absolute Positioning

When operating in this mode, the machine determines each new tool position from a fixed home or specified origin (0, 0). Refer to Figure 2–19.

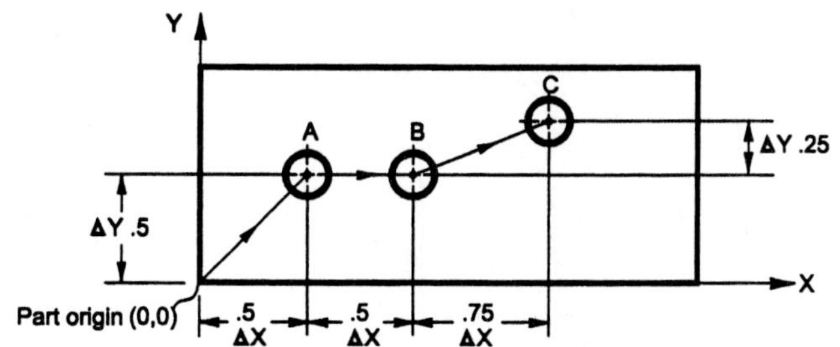

Tool position	Location	
	ΔX	ΔY
A	.5	.5
B	.5	0
C	.75	.25

FIGURE 2–18 Delta dimensioning for incremental mode positioning.

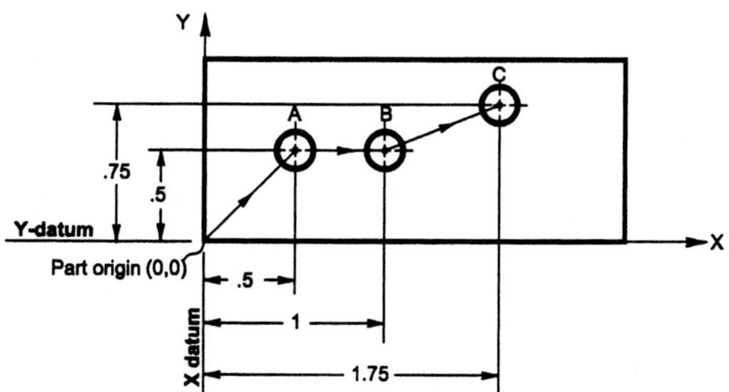

Tool position	Location	
	ΔX	ΔY
A	.5	.5
B	1	.5
C	1.75	.75

FIGURE 2–19 Datum dimensioning for absolute mode positioning.

Many modern controllers are capable of operating in either positioning mode. The programmer can switch from one to the other by simply inputting a single code.

> ***Note:*** With datum dimensioning, a datum or zero reference line is established. All linear dimensions are then taken relative to the datum line. Datum dimensioning is well suited to absolute position programming.

2–8 UNITS USED FOR POSITIONING COORDINATES

In the United States positioning coordinates are specified in either the English system (e.g., inch) or the metric system (e.g., millimeter). Most manufacturing outside the United States is done using the metric system. Up-to-date CNC machines are built to accept programs written with either unit. To use the metric system, parts must be dimensioned in metric. Conversion is easy if metric tools are also used. If inch tools are used, their dimensions must first be converted to metric to ensure that the proper coordinates are inputted for milling. It is safe to say that programmers will be working with both systems during their careers so both systems should be understood.

The millimeter and inch are related as follows:

$$1 \text{ mm} \approx 0.[illegible] \text{ in.} \quad .03937$$

$$1 \text{ in.} \approx 25.4 \text{ mm}$$

See Figure 2–20 for an illustration of a part dimensioned in both units.

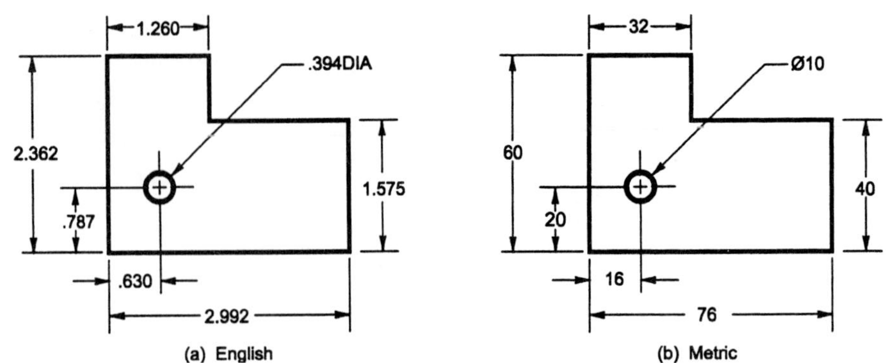

FIGURE 2–20 A part dimensioned in English and metric units.

2–9 CHAPTER SUMMARY

The following key concepts were discussed in this chapter:

1. Two different types of control systems are used to specify tool positions: point to point and continuous path. These modes can be mixed in one program.
2. The four types of drive motors used to position tools are stepping motors, AC motors, DC servos, and hydraulic servos.
3. Open loop systems cause tool movement by executing a discrete number of displacements. Feedback does not include information for checking true versus programmed position. Closed loop systems cause tool movement by executing continuous but varied displacements. Feedback does include information for checking true versus programmed position.
4. The Cartesian coordinate system is used to specify the direction and location of tool movement. Any point in space is located by specifying its *X, Y,* and *Z* coordinates.
5. CNC machine movements are specified by the machine axes.
6. Programmers should consider the CNC machine table as locked, with only the tool in the spindle moving along the machine axes.
7. In incremental positioning each new tool location is taken relative to the previous tool location.
8. In absolute positioning, each new tool location is taken relative to a fixed origin.
9. Delta dimensioning is well suited to incremental position programming.
10. Datum dimensioning is well suited to absolute position programming.

REVIEW EXERCISES

2.1. Identify the two types of control systems used to output tool movement. What is the difference between them?
2.2. What is interpolation? How is it used to cut curves?
2.3. What is a loop system?
2.4. Explain the difference between an open loop system and a closed loop system.
2.5. Name the factors of importance when selecting a loop system.
2.6. What are machine axes?
2.7. Spindle movement is primarily along the _____axis.
2.8. Table movements for most milling machines are along the _____ and_____axes.
2.9. What two types of programming modes can be used to specify tool position? How do they differ?

2.10. What is the difference between delta dimensioning and datum dimensioning?

2.11. Write the absolute X and Y coordinates of the points shown in Figure 2–21.

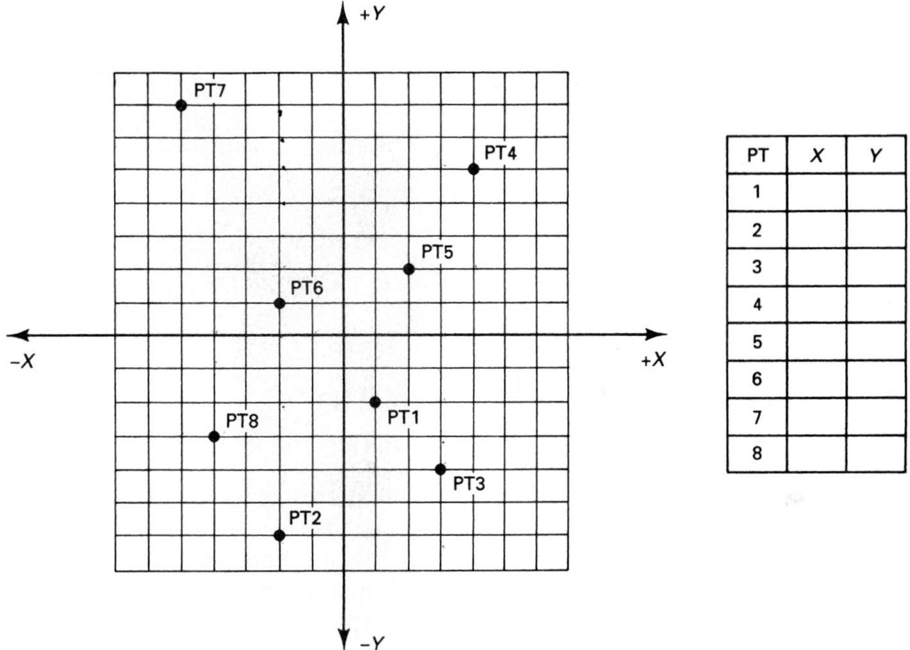

PT	*X*	*Y*
1		
2		
3		
4		
5		
6		
7		
8		

FIGURE 2–21

2.12. Write the incremental X and Y coordinates of the points in Figure 2–21. Use the following order: origin to PT1, from PT1 to PT2, from PT2 to PT3 . . . finish with PT8.

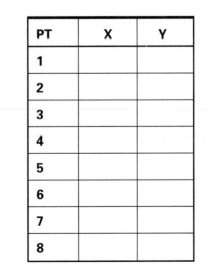

PT	X	Y
1		
2		
3		
4		
5		
6		
7		
8		

CHAPTER

3

Tooling for Hole and Milling Operations

3-1 CHAPTER OBJECTIVES

At the conclusion of this chapter you will be able to

1. State the types and applications of drills used with CNC equipment.
2. Describe tooling used for such hole operations as boring, reaming, tapping, counterboring, and countersinking.
3. Explain the meaning of tool speeds and feeds for hole operations.
4. Identify the types and applications of milling tools used with CNC equipment.
5. Explain the advantages and disadvantages of using special tooling such as spade drills, carbide insert drills, and carbide insert mills.
6. Explain the meaning of tool speeds and feeds for milling operations.
7. State the difference between climb milling and conventional milling.
8. Explain the need for using cutting fluid for CNC operations.

3-2 INTRODUCTION

The CNC programmer must be thoroughly knowledgeable in the machining operations pertaining to the CNC machine to be programmed. CNC tooling is tied to machining operations planning. This chapter presents a review of important tools and tool usage in CNC machining operations. Drilling accounts for a majority of hole operations is discussed first. Other hole operations considered are boring, reaming, tapping, counterboring, and countersinking. Tooling for milling operations is also considered. The reader is introduced to profile and face milling operations.

Carbide tooling is an important aspect of modern CNC tooling. Carbide insert tools for hole operations are discussed and illustrated. Tool speeds and feeds determine, to a large extent, the performance of a cutting tool. They are defined for drilling as well as milling operations. The chapter ends with a discussion of the use of cutting fluids with tooling.

3-3 TOOLING FOR DRILLING OPERATIONS

Drilling, in most cases, is the first machining operation performed in the total production of a hole. Drills usually do not produce holes of high accuracy. To give the hole its required characteristics of precision of size, shape, location, finish, and internal configurations, other tools must be used. These include reamers, borers, and taps.

Twist Drills

The most important tool used in drilling is the twist drill. This end cutting tool has two helical grooves or flutes cut around a center called a web. The flutes act as cutting edges for feeding the tool into the material and as channels for admitting lubricant and carrying away the cut chips. The web gives the drill strength in resisting deflections. Metal cutting twist drills are made from a wide range of materials ranging from carbon tool steel to solid carbide. They are sized by diameter as follows.

In the English system:

- Number sizing—from 1 (0.228 in.) to 97 (0.0059 in.)
- Letter sizing—from A (0.234 in.) to Z (0.413 in.)
- Fractional sizing—from 1/64 in. to 63/64 in.

In the metric system:

- From 0.2 mm to 50.5 mm

Twist drills are illustrated in Figure 3–1.

Straight shanks are common for drills up to 1/2 in. Larger drills can have straight or tapered shanks. The tangs of the tapered shank prevent slipping of the drill while cutting larger holes.

Drilling accuracy tends to decrease when either drill length or drill size is increased. Longer drills exhibit less stiffness and more torsional deflection. A good rule of thumb is to select the shortest drill possible for any hole operation. Drills must be sharpened by grinding the tip and the flute edges at precise angles.

When drilling through a material, good practice is to allow one-third of the diameter of the drill plus 100 thousandths to extend beyond the material as shown in Figure 3–2.

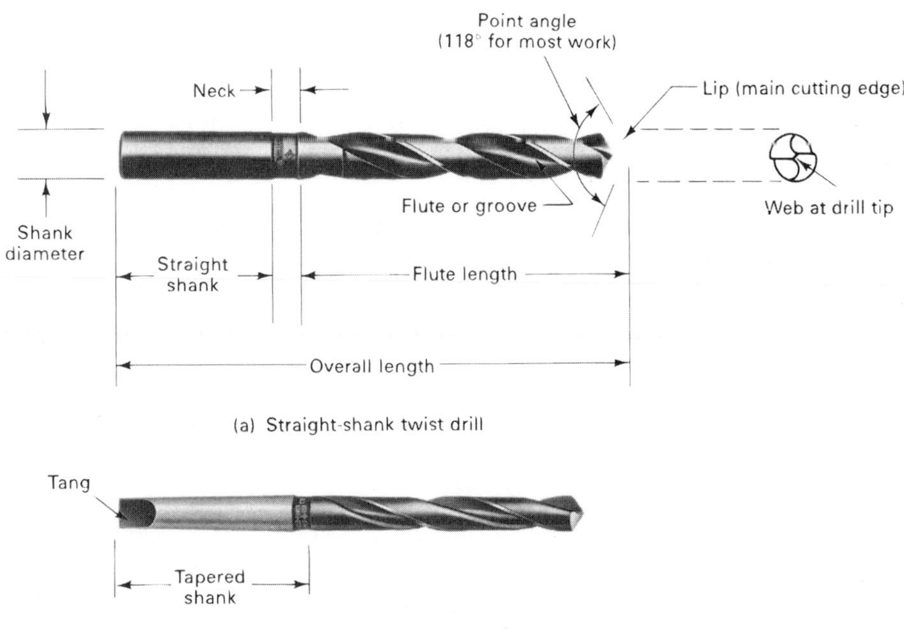

FIGURE 3–1 Parts of straight and tapered shank twist drills. (Photo courtesy of Cleveland Twist Drill Company.)

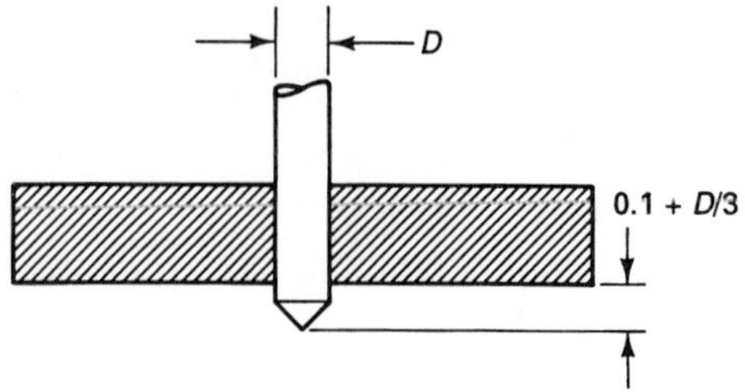

FIGURE 3–2 Drill clearance for through holes.

FIGURE 3–3 Center drill types. (Courtesy of Cleveland Twist Drill Company.)

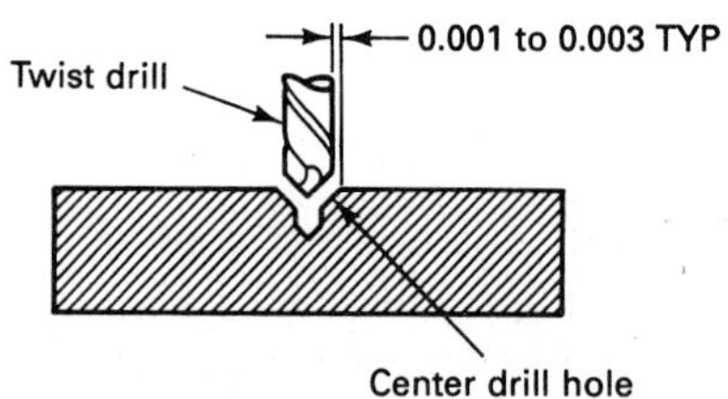

FIGURE 3–4 Drill depth for center drilling.

Center Drills

As was stated previously, twist drills are not capable of locating programmed hole centers with sufficient accuracy due to several factors including flute length, drill diameter, drill flexibility, cutting edge preparation, and material hardness. To better locate a hole center, a short stubby drill called a center drill is used first. The resulting starter hole is used to guide the twist drill into the material with a minimum of inaccuracy. Figure 3–3 illustrates a bell-type center drill as well as a spotting and centering drill.

Good practice is to create a center drill hole such that the countersunk portion is approximately 0.003 to 0.006 in. larger than the corresponding twist drill diameter. This is shown in Figure 3–4.

Coolant-Fed Drills

Coolant-fed drills have one or two holes passing from the shank to the cutting point. Compressed air, oil, or cutting fluid is passed through the drill as it operates. This system enables the cutting point and work to be cooled as chips are flushed out. These drills are especially useful for drilling deep holes. See Figure 3–5.

Spade Drills

A spade drill consists of a blade holder to which is bolted several different size drill point blades. See Figure 3–6. Spade drills offer several advantages over twist drills for drilling holes 1 in. in diameter and up. The larger web of the spade drill ensures that during penetration less flexturing occurs and thus a more accurate hole results. Tooling costs are lower with spade drills because eight standard blade holders will accommodate all blade widths, normally ranging from 5/8 to 6 in. Worn blades can be either resharpened or simply replaced

FIGURE 3–5 A coolant-fed drill. (Photo courtesy of Cleveland Twist Drill Company.)

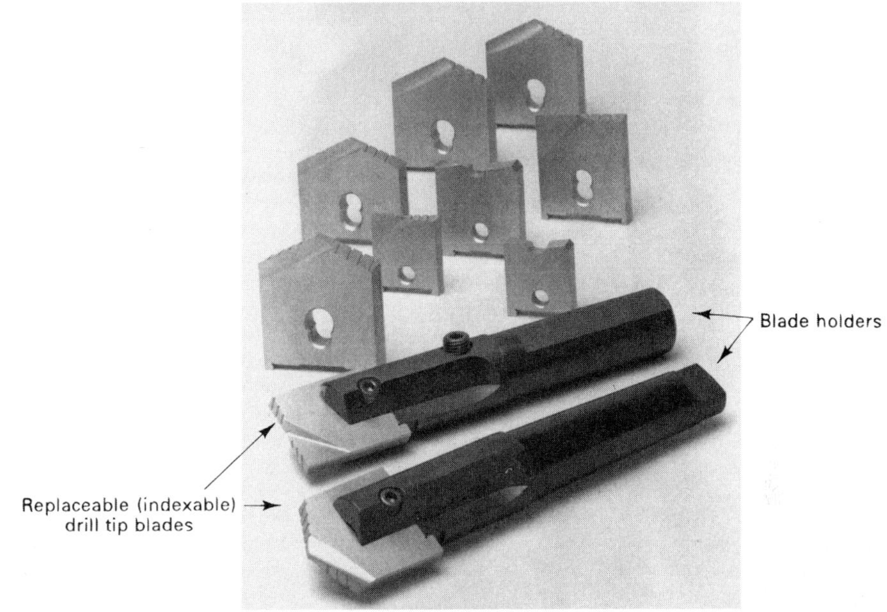

FIGURE 3–6 Spade drills. (Photo courtesy of Kennametal, Inc.)

with new ones. Job setup for CNC hole operations is also reduced. Spade drills are designed to machine a hole from the solid in one pass, eliminating the need for center drilling or multiple-pass drilling to gradually enlarge the hole size.

In order to utilize a spade drill, a 50% or greater torque machine is needed beyond that normally used for drilling with a standard twist drill. The machine and the setup must also have increased rigidity.

Most spade drills operate with coolant flowing through the drill for heat dissipation and flushing out the chips. Thus, a high-pressure coolant system is usually installed. The depth of holes is also limited with spade drills because flutes do not exist to help carry out the cut chips. The cutting edges of the blade incorporate chip splitting and breaking action to reduce chip size and to facilitate chip removal.

Indexable Insert Drills

These drills represent the latest state-of-the-art advancements in CNC hole drilling for hole diameters ranging from 5/8 to 3 in. They offer all the advantages of spade drills including replaceable (indexable) inserts. They are capable of drilling from the solid at penetration rates of 5 to 10 times that of twist drills or spade drills. The higher-strength carbide inserts allow the tool to be driven into the hardest of materials. See Figure 3–7. Feed forces are no higher than those required by twist drills. However, higher machining forces demand substantial spindle horsepower. A high-pressure coolant system is needed along with sufficient machine rigidity. The main disadvantages of using carbide inserts are their brittleness and their sensitivity to shock. If not used properly inserts tend to chip or crack.

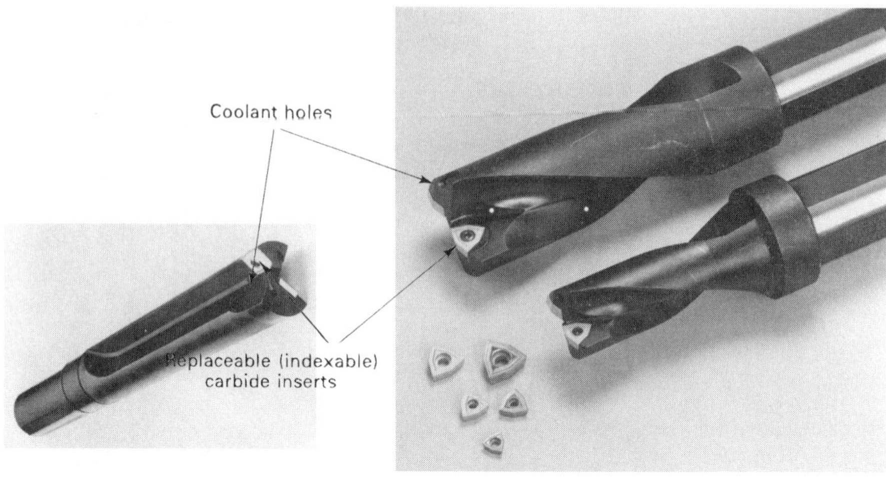

FIGURE 3–7 Carbide insert drills. [(a) Photo courtesy of Waukesha Cutting Tools, Inc. (b) Photo courtesy of Kennametal, Inc.]

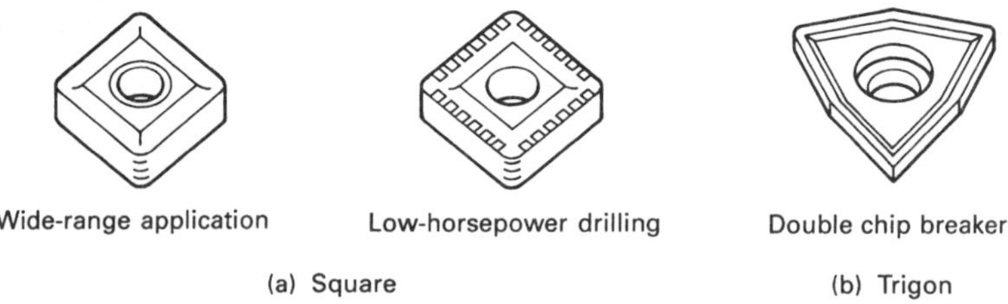

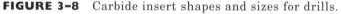

FIGURE 3–8 Carbide insert shapes and sizes for drills.

Carbide inserts used in drilling tools are manufactured in various sizes and shapes as illustrated in Figure 3–8.

3–4 CARBIDE INSERT TECHNOLOGY

A general system for identifying the shapes and sizes of inserts is normally given in toolmaker catalogs. The system created by Kennametal is shown in Figure 3–9.

Carbide inserts are graded by the American National Standards Institute (ANSI) or USA class according to the C system. Grades C1 to C4 are composed of tungsten carbide with cobalt binder. They are used to machine cast iron, work-hardened stainless steels, titanium, and other nonferrous metals. Grades C5 to C8 include titanium-carbide additions and are known as steel cutting grades. The ISO system is used in Europe and Japan. It employs the P group for cutting ferrous metals that form long chips, the M group for cutting ferrous and nonferrous metals that form long or short chips, and the K group for short chipping ferrous and nonferrous metals. It should be noted that the carbide composition for an ANSI class will not necessarily be the same for the corresponding ISO classification. A table listing ANSI and ISO classifications is given in Figure 3–10.

The general practice is to select the hardest and longest-wearing grade that will withstand a machining operation. The operator can then back off to a tougher but softer grade only if forced by insert chipping or fracturing.

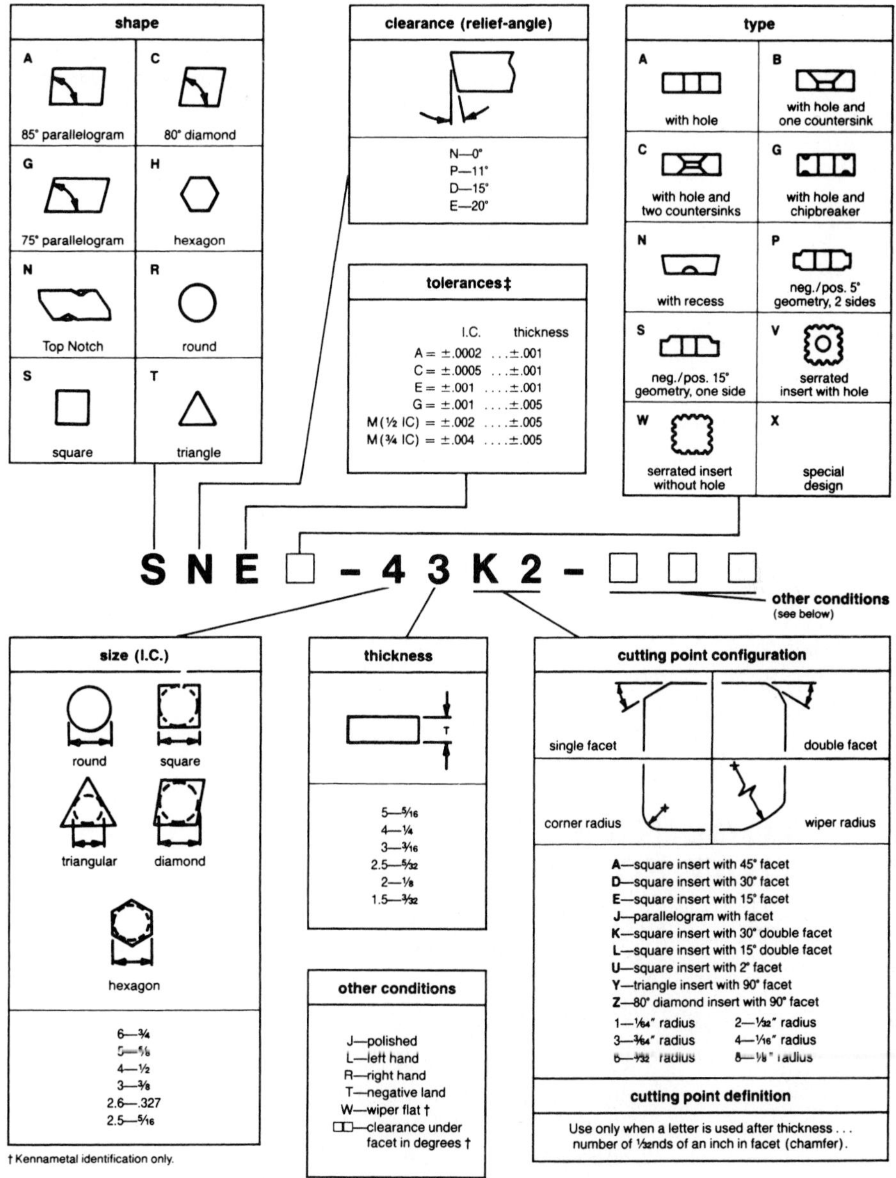

FIGURE 3–9 Kennametal insert identification system. (Courtesy of Kennametal, Inc.)

Today many vendors manufacture carbide inserts. Each has its own grading system. For example, Kennametal uses the K system, GTE Valenite the VC system, Sandvick the GC, SMA, S, H, and HM system, and Greenleaf the G (uncoated grades), GA (coated grades) system. Vendors also supply supplementary grades within a particular ANSI class. Vendor tool catalogs usually give a complete listing of the relationship between their grading system and the standard ANSI or ISO systems. The reader is advised to consult these catalogs for more detailed information.

ANSI (USA) class	ISO class	Material suitable for machining	Machining operation
C1	K30 K20	Cast iron, work-hardened stainless steels, and nonferrous metals	Rough machining
C2	K20 K25		General machining
C3	K10 K15		Finish machining
C4	K01 K05		Fine-finish machining
C5	K40 K50	Stainless steels, tool steel, and general alloy steels	Rough machining
C6	K25 K35		General machining
C7	K10 K25		Finish machining
C8	K01 K05		Fine-finish machining

FIGURE 3–10 Carbide insert grades and machining applications.

3–5 TOOLING FOR HOLE OPERATIONS THAT FOLLOW DRILLING

Additional hole operations following drilling include boring, reaming, tapping, counterboring, spotfacing, and countersinking.

Boring

Boring is used for two main purposes: enlargement of an existing hole and accurate readjustment of the center location of the enlarged hole. Better hole straightness and surface finish can also be achieved by boring. Some typical boring tools are shown in Figure 3–11.

As a rule the shortest boring bar should be selected for any operation. As with drills, the greater the length-to-diameter ratio the more flexible and error prone the boring bar will be. The finish of the surface inside the hole will also be affected because long bars tend to chatter.

Reaming

As was stated previously, a twist drill will not consistently cut holes to the exact size and surface finish. If a high degree of accuracy for these results is required, an additional operation called reaming must be included.

A reamer is a cylindrical tool with straight or helical cutting edges. The majority of reamers are made of high-speed steel. Carbide-tipped and solid reamers are also available as are cobalt high-speed-steel tools. Reamers cut on the sides as well as the ends. Hole type, size, and the number of holes to be reamed in a production run will determine the type of reamer used. As a rule of thumb, a straight-flute reamer is used as a starting tool. In reaming blind holes, if chips cause trouble, a right-hand spiral reamer may be used to pull up the chips.

FIGURE 3–11 Carbide insert boring tools. (Photo courtesy of Waukesha Cutting Tools, Inc.)

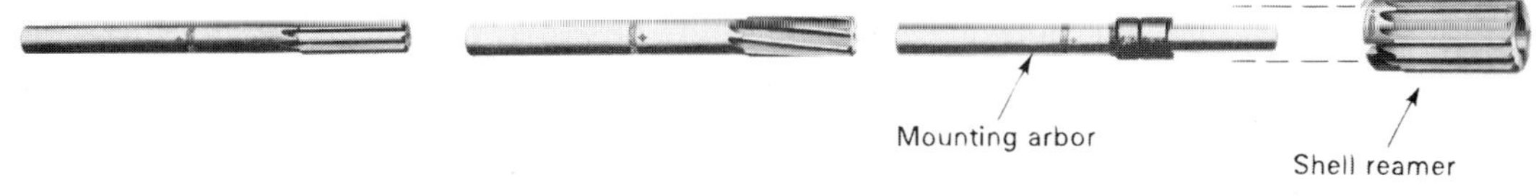

(a) Straight fluted (b) Spiral fluted (c) Shell reamer

FIGURE 3–12 Reaming tools. (Photo courtesy of Cleveland Twist Drill Company.)

Shell reamers are a very economical way of machining holes larger than 3/4 in. One shank fits several size ream cartridges. Some typical reaming tools are shown in Figure 3–12.

It should be noted that the reamer is guided by the existing hole. Therefore, it will not correct errors in hole location or its straightness. If these problems exist it is advisable to first bore, then ream.

Tapping

The process of cutting threads on the inside of a hole by a tap is called tapping. This can be a delicate and sometimes troublesome process depending upon the material type and the thread depth. The main problem with tapping is clearing

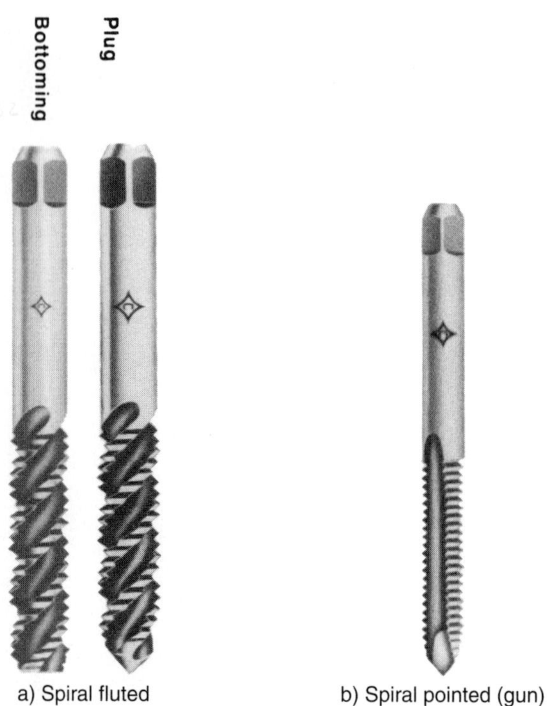

FIGURE 3–13 Tapping tools. [(a, b) Photos courtesy of Cleveland Twist Drill Company.]

chips from the hole. Taps are made from a variety of materials such as carbon tool steel, high-speed steel, and carbide. Titanium nitride (TiN) is often used to create a hard wear resistant outer coating or case on taps. Hand taps are not recommended for use in CNC machines unless they are driven by a special tapping attachment. The attachment contains a clutch that slips when the tap experiences too much torque or suddenly jams in the hole. For *blind threads*, normal practice is to use a spiral fluted tap. Plug types are used to cut threads to a specific depth while bottoming types cut threads at the bottom of the hole.

The flutes are designed to admit lubricant as well as force chips to flow back out of the hole.

Sprial pointed or "gun" taps are used in CNC machines to cut *through threads.* They are designed to admit lubricant as well as force the chips to flow through the hole ahead of the tap. The term "gun" comes from the fact that the tap "shoots" the chips ahead as it cuts the threads.

Some typical tapping tools used in CNC machines are shown in Figure 3–13.

Standard tables for selecting appropriate tap drills for tap sizes can be found in such references as Machinery's Handbook. Selecting the proper size tap drill insures that the tap will work properly in the hole when cutting threads.

Counterboring

It is often desirable to enlarge a hole to a depth slightly larger than the head of a specific bolt or pin. This allows for the head to be buried below the machined surface. For CNC operations, counterboring is usually done with an end mill or flat bottom drill.

Countersinking

Countersinking involves enlarging the top end of a hole in the form of cone-shaped depression. This will allow a flat or oval head machine screw to be flush or slightly below the surface when inserted. The cone angle is usually 82° or

 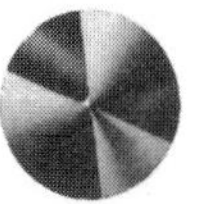

FIGURE 3–14 A three-flute countersinking tool. (Photo courtesy of Cleveland Twist Drill Company.)

90°, respectively, and is made with a countersunk tool. Threaded holes should be countersunk slightly larger than the tap diameter to protect the starting threads unless the production part drawing calls for no chamfer. A typical countersinking tool is shown in Figure 3–14.

Problems can arise concerning the difference between the theoretical vertex point on the tool and its actual existing value. An optical indicator should be used to determine the actual vertex height of the tool. This measurement will avoid countersinking too deep.

3–6 TOOL SPEEDS AND FEEDS FOR HOLE OPERATIONS

For a certain-diameter tool and tool type, the two most important parameters that must be specified when cutting a particular material are its speed and feed. These values influence tool life and cutting performance.

Tool Speed (Hole Operations)

Tool speed is defined as the speed of any point on the circumference of the cutter. It is usually expressed in surface feet per minute or sfpm. The tool speed and cutter diameter selected will determine the spindle revolutions per minute or rpm. These concepts are illustrated in Figure 3–15.

$$\frac{\pi \times \text{Tool Diameter}}{12} = \text{Tool Circumference (ft)}$$

$$\frac{\pi \times \text{Tool Diameter} \times \text{rpm}}{12} = \text{Tool Cutting Speed (sfpm)}$$

Because

$$\pi = 3.142 \quad \frac{\pi}{12} \approx \frac{1}{4}$$

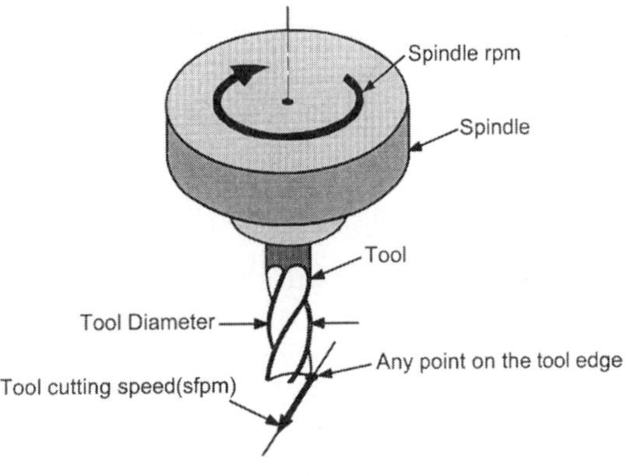

FIGURE 3–15 Tool speed.

Thus

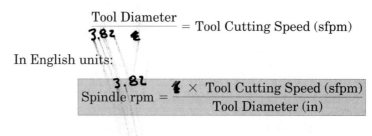

$$\frac{\text{Tool Diameter}}{4} = \text{Tool Cutting Speed (sfpm)}$$

In English units:

$$\text{Spindle rpm} = \frac{4 \times \text{Tool Cutting Speed (sfpm)}}{\text{Tool Diameter (in)}}$$

In metric units:

$$\text{Spindle rpm} = \frac{318.31 \times \text{Tool Cutting Speed (mmpm)}}{\text{Tool Diameter (mm)}}$$

where the tool diameter is expressed in millimeters (mm), and the tool cutting speed is expressed in millimeters per minute (mmpm).

The speed selected for a particular hole operation tool will depend on several factors, some of which are

- Type of hole operation, material hardness, hole depth
- Type of tool used and type of lubricant or coolant used
- Type of hold-down fixture and CNC machine used

Selecting a speed that is too high can lead to excessive dulling and burning of the tool's cutting edges. On the other hand, speeds that are too low may cause the tool to wear excessively or break under operation. Refer to Appendix C for a comprehensive list of drilling speeds.

For reaming, the cutting speeds should be approximately 1/2 to 2/3 that used for drilling the material.

Spindle rpm values for tapping operations are determined by a number of parameters and usually range from 50 to 300 rpm.

Newer machines with rigid tapping or tapping heads can tap at higher rpm values. Aluminum, for example, can be tappped at 3000 rpm with this type of eqiupment.

Counterboring and countersinking speeds should be in the neighborhood of 1/4 that used for drilling the material.

Tool Feed (Hole Operations)

Tool feed for hole operations is defined as the rate at which the tool advances into the work per revolution. It is expressed in inches per revolution or ipr.

By multiplying the tool feed in ipr by the spindle rpm, one arrives at the tool feed rate in inches per minute or ipm. The tool feed rate is the value to be entered into the CNC program.

In English units:

$$\text{Tool feed rate (ipm)} = \text{ipr} \times \text{rpm} \qquad \text{(Hole operations)}$$

where the penetration rate, ipr, is expressed in inches per revolution, and rpm is the spindle speed.

In metric units:

$$\text{Tool feed rate (mmpm)} = \text{s} \times \text{rpm}$$

where the penetration rate, s, is expressed in millimeters per revolution, rpm is the spindle speed and tool feed rate is given in millimeters per minute (mmpm).

Approximate values of tool feed rates for various hole operations are listed in Appendix C.

Reaming feed rates should be approximately two to three times greater than needed for drilling the same material.

Tapping feed rates can depend on the number of threads per inch for the tap and the spindle rpm.

■ EXAMPLE 3–1

Determine the feed rate for a tap having 18 threads per inch and rotated at 300 rpm.

$$\text{Actual feed rate} = \frac{1 \text{ in.}}{18 \text{ rev}} \times 300 \frac{\text{rev}}{\text{min}}$$

$$= 16.7 \text{ ipm}$$

Tapping is usually done with a floating clutch tap holder. For the holder to be effective, the programmed feed rate should be slightly less than the actual value. Thus,

Programmed feed rate = 16 ipm ■■

3–7 TOOLING FOR PROFILE MILLING AND FACING OPERATIONS

Profiling is the process of creating a contour by removing material with a cutter having teeth on its periphery. Refer to Figure 3–16 for an illustration of profile milling. Face milling involves the creation of a flat surface by making successive passes with the bottom of a face mill or end mill cutter. Face milling is illustrated in Figure 3–22. In profiling, the axis of rotation of the cutter is parallel to the machined surface. In face milling, the axis of rotation of the cutter is perpendicular to the machined surface. Both of these operations also involve continuous-path tool motion in which the tool remains in contact with the part.

End Mills

The most often used tool for profiling operations is the end mill cutter. It is particularly suited to CNC operations involving a minimum of setup when short or medium production runs are to be made. End mills are also used for facing, slotting, plunging, and cavity cutting. End mill cutters range in size from 0.032 to 2 in. in diameter. The parts of a single end mill cutter are shown in Figure 3–17.

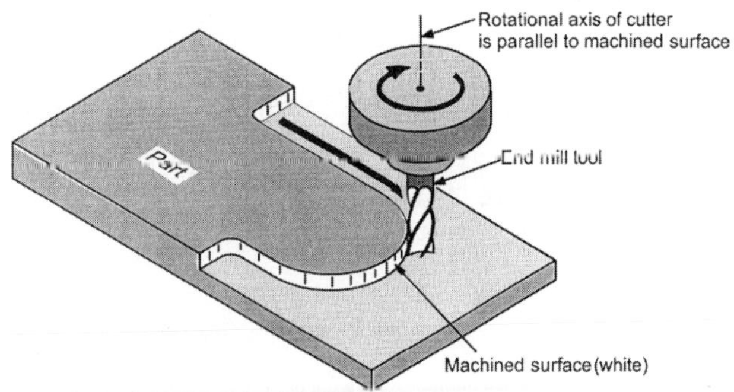

FIGURE 3–16 Profile milling.

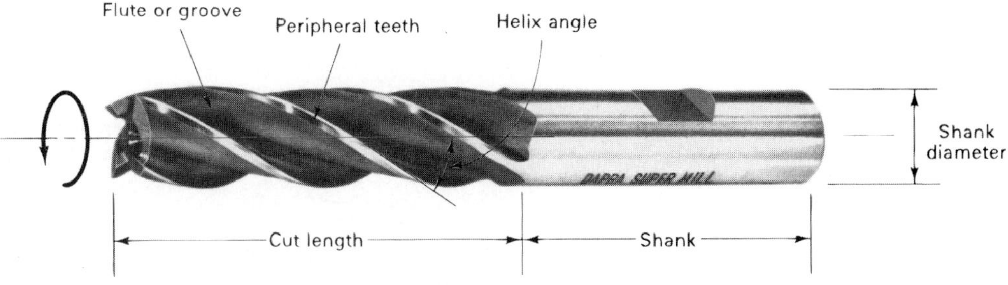

FIGURE 3–17 Parts of a four-flute end mill cutter. (Photo courtesy of Dapra Corp., Bloomfield, Ct.)

End mill tools are manufactured with two, three, four, or more flutes. The flutes are helically grooved portions of the tool used for carrying away cut chips and admitting lubricant or coolant. The four-flute end mill shown in Figure 3–17 does not have end teeth that go all the way to its center. Thus, it cannot be used for plunging directly into the center of a solid material. The end teeth on the two-flute end mill shown in Figure 3–18, however, do pass through its center. This tool is also called a center cutting mill. It can be used for rough drilling holes from solid, counterboring, and boring as well as slotting and cavity cutting. Holes that have been drilled with a drill can be sized to a high degree of accuracy by using an end mill. Center cutting mills are also manufactured with three or more flutes.

Roughing end mills are referred to as hogging cutters. They have grooves or scallops around the body and produce a broken-up chip. They generate lower side loads than those with smooth helical teeth. These factors enable roughing end mills to remove metal at a rate three times that of the usual end mills.

The versatility of end mills allows a shop to cut down on tooling costs, tooling inventory, and job setup time. Various types of end mill cutters and their uses are shown in Figures 3–18 through 3–20.

Shell End Mills

These types of milling cutters range in size from 1 1/4 to 6 in. in diameter. They represent a savings to most shops because several cutter sizes can be fitted to one mounting arbor. Facing cuts are usually made with these tools. A helical fluted shell mill and its mounting arbor are shown in Figure 3–21.

Carbide Indexable Insert End Mills

Modern tool technology has produced milling cutters with replaceable or indexable carbide inserts. Initial investment in these tools is somewhat high but they yield substantial long-term savings in terms of job setup times and tool regrinding costs. Worn inserts are simply replaced with factory-ordered prefaced

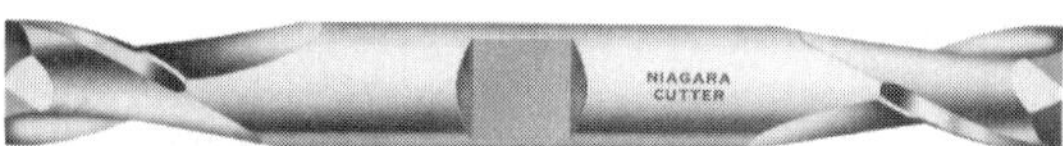

FIGURE 3–18 Double-end two-flute center cutting—plunging, profile, milling, slotting, and drilling. (Photo courtesy of Niagara Cutter.)

FIGURE 3–19 Ball end mill—milling radii in slots and holes. (Photo courtesy of DoAll Company.)

FIGURE 3–20 Roughing end mill—removing large amounts of metal. (Photo courtesy of Niagara Cutter.)

inserts. Tool inventory is reduced because the same tool can be used to cut different materials by switching to a different insert grade. Milling inserts are available in various shapes and sizes. The shape determines the strength and the number of cutting edges. Circular inserts are the strongest, followed by hexagonal, square, parallelogram shaped, diamond shaped, and triangular. See Figure 3–23. For a given size those with the largest included angles use less carbide than those with smaller angles. The CNC operator should select the strongest insert shape consistent with the profile to be cut.

Because these tools are very rigid, a high degree of accuracy can be achieved with their use. More exact diameters can be milled when a center cutting carbide tool is used for boring. Milling time is also shorter with carbide cutters. Depending upon the material being cut, tool speed can often be doubled or tripled and tool feed can be increased by 25% or more.

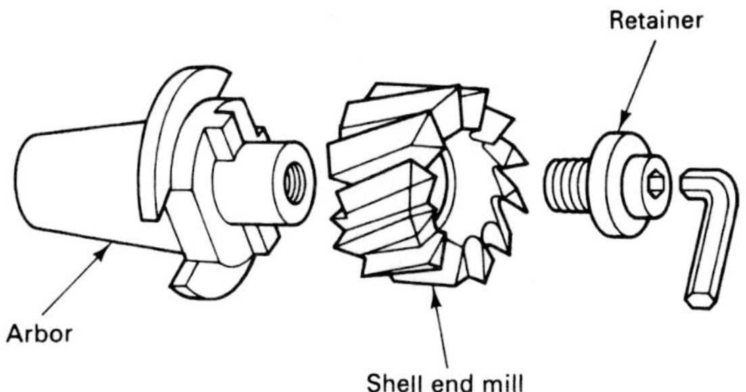

FIGURE 3–21 A shell end mill and arbor.

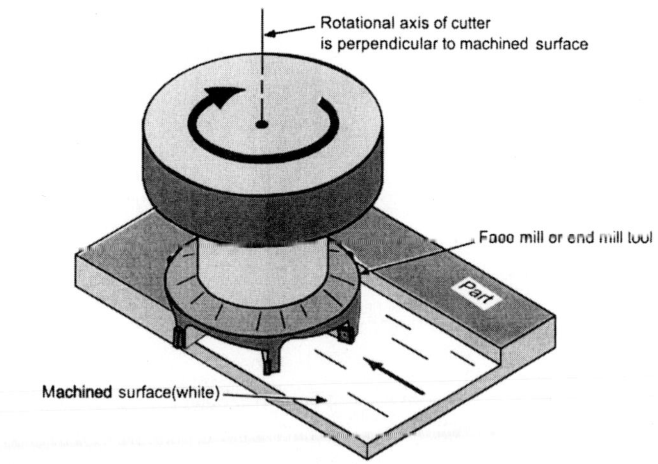

FIGURE 3–22 Face milling.

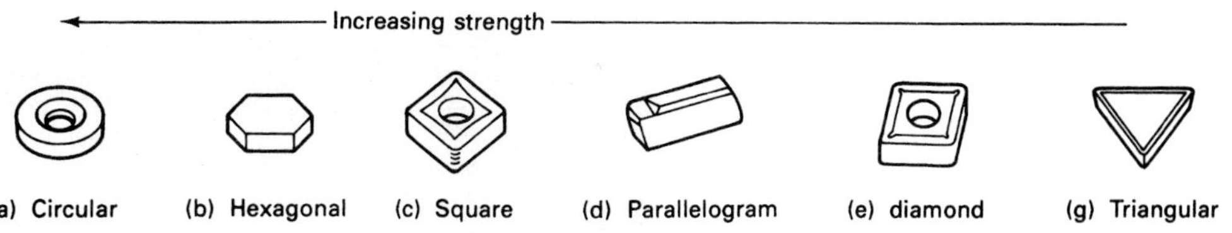

FIGURE 3–23 Carbide insert shapes and sizes used for milling tools.

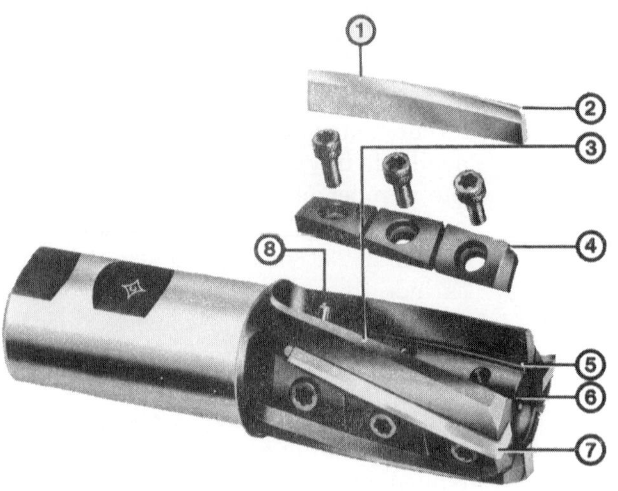

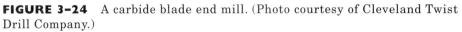

FIGURE 3-24 A carbide blade end mill. (Photo courtesy of Cleveland Twist Drill Company.)

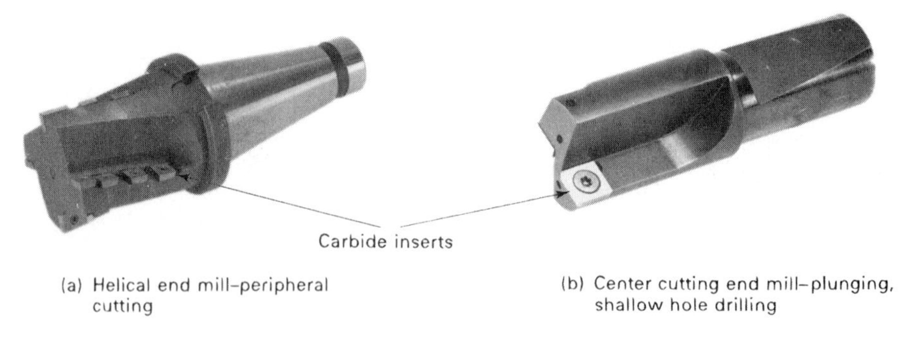

(a) Helical end mill–peripheral cutting

(b) Center cutting end mill–plunging, shallow hole drilling

(c) Ball nose end mill–plunging, blended radius cutting

FIGURE 3-25 Carbide insert end mills. (Photo courtesy of Kennametal, Inc.)

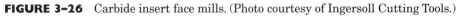

FIGURE 3-26 Carbide insert face mills. (Photo courtesy of Ingersoll Cutting Tools.)

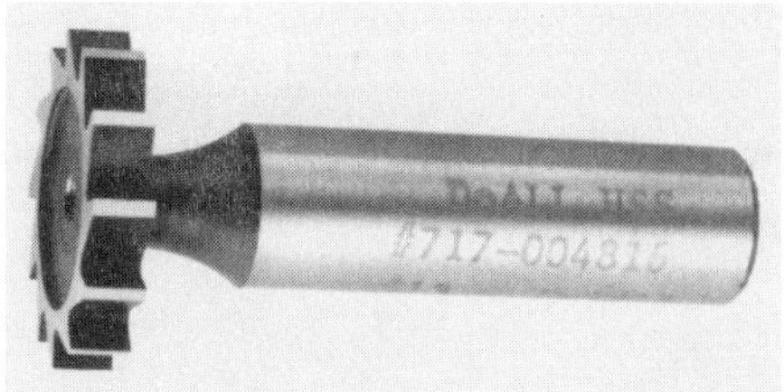

(a) Woodruff cutter for cutting slots

(b) Staggered tooth side milling cutter

FIGURE 3–27 (Photo courtesy of DoAll Company.)

Various carbide insert tools for milling are shown in Figures 3–24 through 3–26. Two typical cutters are shown in Figure 3–27.

In order to use these tools, the CNC milling machine must have sufficient horsepower and rigidity. It should also be capable of producing very steady and controllable feed rates so that table surge is kept to a minimum.

Miscellaneous Milling Cutters

Other types of milling tools include woodruff cutters, side cutters and plain end mills. Examples of these tools are shown in Figure 3–27.

3–8 COATED TOOLING

Many CNC cutting tools can be ordered with special coatings. These coatings are applied directly to high-speed-steel tools or to carbide inserts. One of the most popular coatings is titanium nitride (TiN). High-speed-steel tools with this coating can be operated at the higher cutting speeds of carbides. They have reduced sensitivity to shock, brittleness, and cracking—problems that often arise when using carbides. Titanium-nitride coatings tend to reduce friction and wear at the cutting edges. Coated tools have up to 20 times the tool life of uncoated carbide grades. Complex tool shapes not available with carbide tooling can be manufactured in high-speed steel and coated for best results.

3–9 TOOL SPEEDS AND FEEDS FOR MILLING OPERATIONS

As was stated previously, tool speed and tool feed are the two most important parameters to be specified for a cutting tool. This holds true especially for milling processes. The success of an operation in terms of accuracy of the cut, surface finish, and tool wear depends upon the proper specification of tool speeds and feeds.

Tool Speed

Tool speed was defined for hole operations and carries the same meaning for milling. Refer to Appendix C for a comprehensive list of suggested milling speeds for various materials.

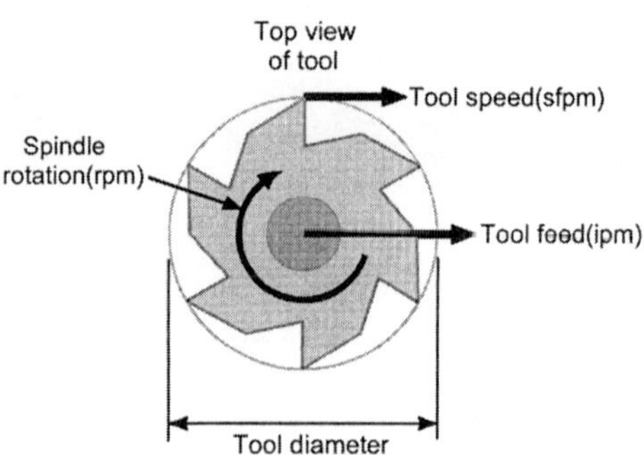

FIGURE 3–28 Tool feed.

Tool Feed

Tool feed is defined as the rate at which the cutter advances into the work. It is expressed in inches per minute (ipm). See Figure 3–28.

Tool feed for profiling depends upon the inches of material cut per tooth (ipt) or chip load, the number of teeth, and the spindle rpm. The theoretical tool feed rate is given by the following formulas.

In English units:

$$\text{Tool feed rate (ipm)} = \text{ipt} \times \text{Number of cutter teeth} \times \text{rpm} \qquad \text{(Profiling)}$$

In Metric units:

$$\text{Tool feed rate (mmpm)} = \text{cl} \times \text{Number of cutter teeth} \times \text{rpm} \qquad \text{(Profiling)}$$

where cl is the chip load expressed in millimeters per tooth and the feedrate is expresed in millimeters per minute (mmpm)

Approximate chip load recommendations for various materials are given in tool vendor catalogs.

Correct feed in milling may also depend upon another parameter called chip thickness. This is not the chip load (feed per tooth) but the actual thickness of the chips resulting from a given feed rate. The chip thickness will depend upon such factors as the geometry of the cutter, the lead angle used, and the position of the cutter on the workpiece. For general-purpose milling, the chip thickness should be between 0.004 and 0.008 in. Feed rates that produce chip sizes outside of this range will cause premature wear and breakdown of the carbide insert cutters.

The book formula for average chip thickness is given as

$$\text{Average chip thickness (in)} = \text{ipt} \times \sqrt{\frac{\text{Width of cut (in)}}{\text{Cutter diameter (in)}}}$$

where ipt is the chip load in inches per tooth and $0.004\text{in} \leq$ average chip thickness $\leq 0.008\text{in}$

■ EXAMPLE 3–2

A 3-in.-diameter end mill with five teeth is to machine a 1.125 width of cut to a square shoulder. See Figure 3–29. A chip load of 0.006 ipt and a spindle rpm

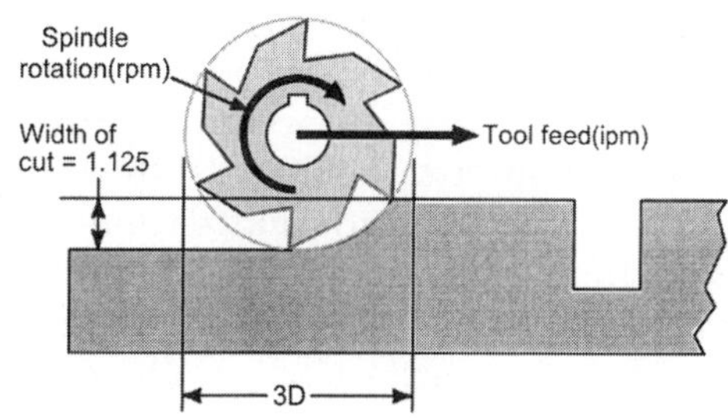

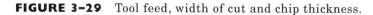

FIGURE 3–29 Tool feed, width of cut and chip thickness.

of 400 is initially recommended. Determine if the corresponding feed rate needs to be adjusted to maintain a chip thickness of 0.008 in.

$$\text{Tool feed rate} = 0.006 \times 5 \times 400$$
$$= 12 \text{ ipm}$$

$$\text{Average chip thickness} = \sqrt{\frac{1.125}{3}} \times 0.006$$
$$= 0.0037 \text{ in.}$$

This is out of the optimum range. Using 0.008 in. as the desired chip thickness gives a chip load of

$$\text{ipt} = \sqrt{\frac{\text{D}}{\text{Width of cut}}} \times \text{Average chip thickness}$$
$$= \sqrt{\frac{3}{1.125}} \times 0.008$$
$$= 0.013$$

The corresponding feed rate is then adjusted to

$$\text{Tool feed rate} = 0.013 \times 5 \times 400$$
$$= 26 \text{ ipm}$$

Thus, to achieve a chip thickness of 0.008 in. with a spindle speed of 400 rpm when executing a 1.125 width of cut, the feed rate must be adjusted to 26 ipm. It should be noted that these are given "book" values of chip thickness and adjusted feed rates. They should not be applied directly without taking into account the actual physical conditions. These include workpiece stability, fixturing, available horsepower, and required surface finish. ■■

3–10 FEED DIRECTIONS FOR MILLING OPERATIONS

Basically, there are two directions in which a milling cutter can be fed into the work—one with the feed in the direction of cutter rotation or climb milling and the other with the feed opposite the direction of cutter rotation or conventional milling.

Climb Milling

Climb or down milling causes the tool to make a chip of maximum thickness at the start of the cut at and near the part surface and minimum thickness at the end of the cut inside the material. The work is pushed down and into the cutter. Thus, less clamping and machining horsepower are required. The extra force on the table, however, means that the milling machine must have a backlash eliminator for eliminating play between the nut and the table screw. Most modern CNC machines come equipped with backlash eliminators. More accurate roughing cuts with a minimum of machining marks or surface breakouts are attainable with this method of milling. Chips are pushed behind and away from the cutter and do not tend to dull the tool by remaining on its front cutting edges. Climb milling is also recommended for machining thin parts or parts that are hard to hold down. Manufacturers of carbide-tipped tools suggest climb milling especially when cutting work-hardened materials. See Figure 3–30.

Conventional Milling

When conventional milling is practiced, the chip has no thickness at the start of the cut inside the material and attains maximum thickness at the end of the cut near and at the surface of the part. The chip is pushed up from the inside of the part and in front of the cutter.

Conventional or up milling requires more fixture hold-down force and is recommended for materials with hard outer scales such as castings, forgings, and hot rolled steel and for finishing CNC operations. Recall climb milling causes the cutter to cut the outer scale first, which can create high tool impact loads and excessive tool wear on hard-scale materials. Conventional milling is also recommended for cases where a tool length used may cause unacceptable chattering of the cutter with climb milling. See Figure 3–31.

A test part should be machined and inspected to verify if the decision to climb mill or conventional mill is appropriate.

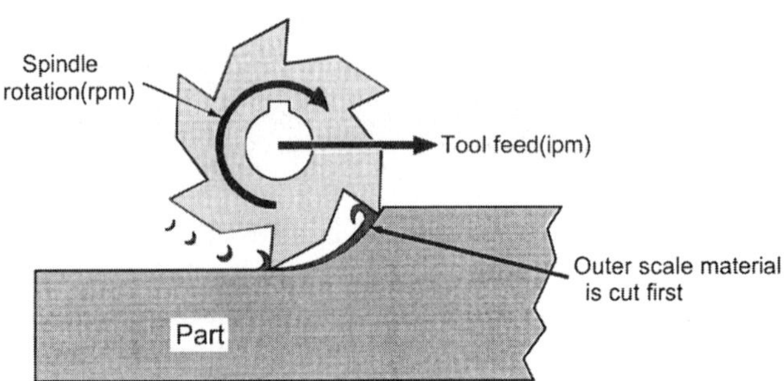

FIGURE 3–30 Climb milling.

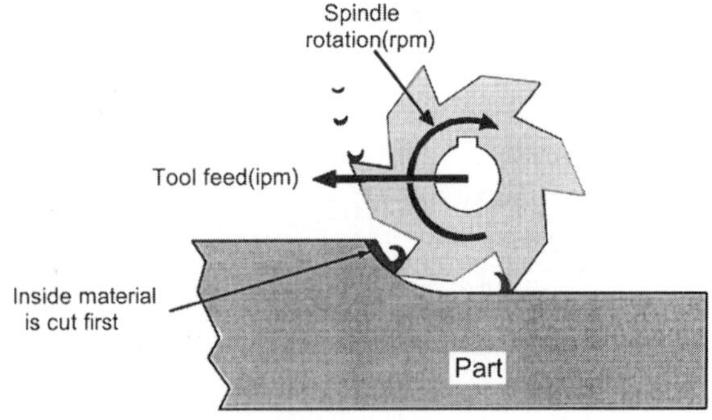

FIGURE 3–31 Conventional milling.

3–11 CUTTING FLUIDS FOR CNC OPERATIONS

Cutting fluids are used when executing most metal cutting operations with CNC machines. During the cutting process pieces of metal are removed by the shearing action of the tool. Heat is built up due to plastic deformation of the metal and sliding friction against the tool. The pieces of cut metal tend to weld themselves to the tool's cutting edges. This phenomenon is known as built-up edge and results in the tool's edges being dulled.

In addition to a substantial decrease in tool performance, built-up edge causes excessive tool wear, poor surface finish, and even cratering of the tool face. Cutting fluids attack these problems by offering such benefits as

- Cooling and lubricating the tool and work
- Controlling built-up edge and tool wear
- Flushing away chips
- Ensuring a good surface finish
- Enabling the tool to be operated at increased cutting speeds and depths of cut

Water-based solutions or cutting oils are the two most commonly used cutting fluids. They can be divided into three categories: cutting oils, emulsifiable oils, and synthetic cutting fluids.

Cutting fluids are applied in a variety of ways on CNC equipment. The most common method is to flood both the tool and work. See Figure 3–32. A pump is used to deliver the coolant. The used fluid is drained into the chip pan and back into the sump of the coolant pump for reuse. Two or more nozzles are used: one for directing the fluid into the cutting area and another for auxiliary cooling and flushing away chips.

FIGURE 3–32 Flood cooling during a milling operation. (Photo courtesy of EDO Corp.)

Fluids can also be delivered by a mist system. A small jet is used to spray soluble oil or synthetic water-miscible cutting fluids in very fine droplets. Water-miscible fluids are preferred over oil, which may present health hazards and tends to clog. Mist cooling is especially applicable to end milling operations in which the cutting speed is high and the areas cut are low.

3-12 CHAPTER SUMMARY

The following key concepts were discussed in this chapter:

1. The majority of hole operations involve drilling.
2. Because twist drills tend to deflect under loads, the shortest drill should be used for an operation.
3. A center drill should be used to accurately locate the center of a hole prior to drilling.
4. Spade drills and carbide insert drills can drill holes from the solid at higher feed rates than that normally used for twist drills.
5. To use spade drills and carbide insert drills, the CNC machine must have sufficient rigidity, spindle horsepower, and a high-pressure coolant system.
6. End mills can be used for profiling, rough drilling, boring, and counterboring operations.
7. Tool speed is the speed at which the cutting edge of the tool rotates.
8. Tool feed is the rate at which the tool advances into the material in inches per minute (ipm).
9. Feeds and speeds are important factors in determining tool life and cutting performance.
10. Climb milling involves feeding the cutter in the direction of its rotation. It is recommended for rough cuts on CNC equipment and hard-to-hold pieces.
11. Conventional milling involves feeding the cutter in a direction opposite to rotation. It is recommended for finishing cuts and for machining metals with hard outer scales.
12. Cutting fluids are commonly used for CNC metal cutting operations to decrease tool wear and improve cutting performance.

REVIEW EXERCISES

3.1. Explain the function of each of the parts of the following twist drills:
  (a) Flutes
  (b) Web
  (c) Lip

3.2. **(a)** What is the principal use of a center drill?
  (b) Why is a center drill needed?

3.3. **(a)** What type of attachment is recommended for use with taps in CNC machines.
  (b) What type of tap is used to cut blind threads in CNC machines.
  (c) What type of tap is used to cut through threads in CNC machines.

3.4. State three advantages and three disadvantages of using a spade drill.

3.5. How does the length-to-diameter ratio influence flexing errors in hole-making tools?

3.6. The two most important parameters for controlling tool wear and tool cutting performance for a particular material are cutting _____ and cutting _____ .

3.7. Describe the following types of milling operations:
(a) Profile milling
(b) Face milling

3.8. **(a)** What characteristic with regard to end teeth is present in a center-type end mill tool?
(b) List three kinds of machining operations center-type end mills can perform that regular end mills cannot.

3.9. Describe three advantages and three disadvantages of using carbide insert milling tools.

3.10. What are three advantages of using titanium-nitride-coated tools?

3.11. **(a)** What is climb milling?
(b) Give two cases in which it is preferred.

3.12. **(a)** What is conventional milling?
(b) Give two cases in which it is preferred.

3.13. Give four reasons for using cutting fluids in CNC operations.

CHAPTER 4

Exploring Features of CNC Machining Centers

4-1 CHAPTER OBJECTIVES

At the conclusion of this chapter you will be able to

1. Explain the four important functions of tool holders.
2. Describe the two most important tool holder designs commonly used in CNC machining centers.
3. Know how tool holders are captured in the spindles of machining centers.
4. Identify the three most common types of tool storage and tool changer systems.
5. Understand the importance of using pallet loading mechanisms.

4-2 INTRODUCTION

This chapter begins with a description of the most significant features found in machining centers. Tool holders are an integral part of interfacing various cutting tools to the CNC spindle. The types of tool holders used and the methods of securing tools to holders and holders to the spindle are considered. The programmer should have a good understanding of how tool changers work. Thus, the operation of tool changing systems is discussed and illustrated in detail. Pallet work changers reduce the idle time of machining centers and are considered at the end of the chapter.

4-3 BACKGROUND ON CNC MACHINING CENTERS

CNC machining centers were described briefly in Chapter 1. By definition, a machining center is a CNC machine that incorporates some form of automatic tool changing and is capable of performing multiple machining operations. In the development of machining centers, the horizontal spindle center came from the milling machine and the vertical spindle from the drilling machine. With the addition of automatic tool changers, each still tends to favor its original function. To favor drilling, vertical spindles tend to be lower in horsepower, relatively small in diameter, and higher in speed. Horizontal machining centers are heavier and slower to favor milling. There are about two verticals for every horizontal in the industry. Verticals are ideal for three-axis work on a single-part face with little or no part indexing required.

Safety Rules in Tooling Setup

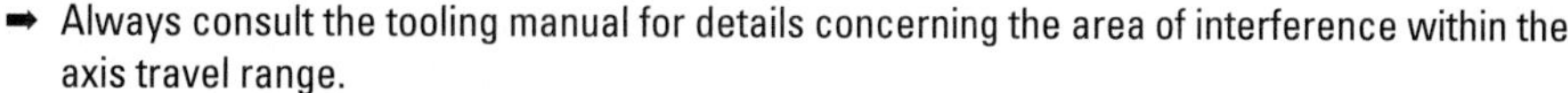

➡ Always consult the tooling manual for details concerning the area of interference within the axis travel range.

➡ Always allow for adequate clearance as follows:
- Between the cutter, workpiece, and work-holding devices during machining operations including tool changing and table cross-travel movements.
- Between the cutter and workpiece when machining multilevel surfaces.
- Always check the cutter and workpiece clearance before lowering the quill.

➡ Reduce the overhang of the quill and each cutting tool to assure maximum rigidity.

➡ After an emergency stop has been executed, always carry out inspection.

➡ Practice thorough cleaning after operations:
- Wipe the mating surfaces with a clean wiping cloth.
- Carefully move the palm and fingers over the part to feel if there are any foreign particles left.
- Apply a drop of oil on the taper or thread of the spindle nose and the shoulder.
- Wear a face mask for dust levels above OSHA standards.

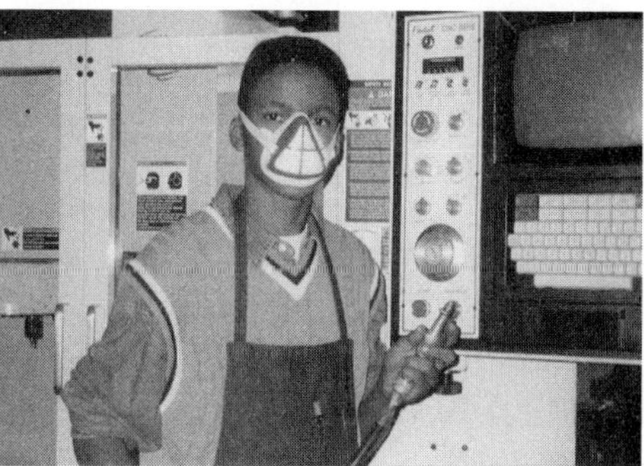

Long flat parts are much easier to fixture on a vertical and spindle thrusts are absorbed by the table. An important factor favoring verticals is that they are less expensive than horizontals. Horizontal spindle machines work at right angles to their tables and thus input torque to the machined parts. Better work holding or heavier fixturing is required although larger parts can benefit from their own weight. Most horizontals have rotary table options so that all four sides of the part can be easily accessed. Horizontals also benefit from the effect of gravity on the chips. See Figures 4–1 and 4–2.

FIGURE 4–1 A vertical machining center with carousel-type tool storage and spindle direct tool changer. (Photo courtesy of Cincinnati Milacron.)

A variety of new head designs is bringing horizontal and vertical machining centers closer together.

4–4 TOOLING SYSTEMS USED WITH AUTOMATIC TOOL CHANGERS

Machining centers use many different types of tools to execute multiple operations on a part. This means the machine must be capable of manipulating tools of various sizes and shapes. Tooling systems have been designed to mate different tools to the same spindle. A simple three-piece system would typically involve a tool holder, an intermediate component such as a boring bar or insert-type milling cutter body, and an indexable insert cutter. More complex systems may be composed of many additional pieces such as extension units, tool body adapters, and arbors. See Figure 4–3 for an illustration of modular tool systems commonly used with machining centers.

Automatic tool changer tool holders are multipurpose devices that are designed to:

1. Minimize tool inventory by interfacing different size cutting tools to the same spindle.
2. Be easily manipulated by the tool changing mechanism.

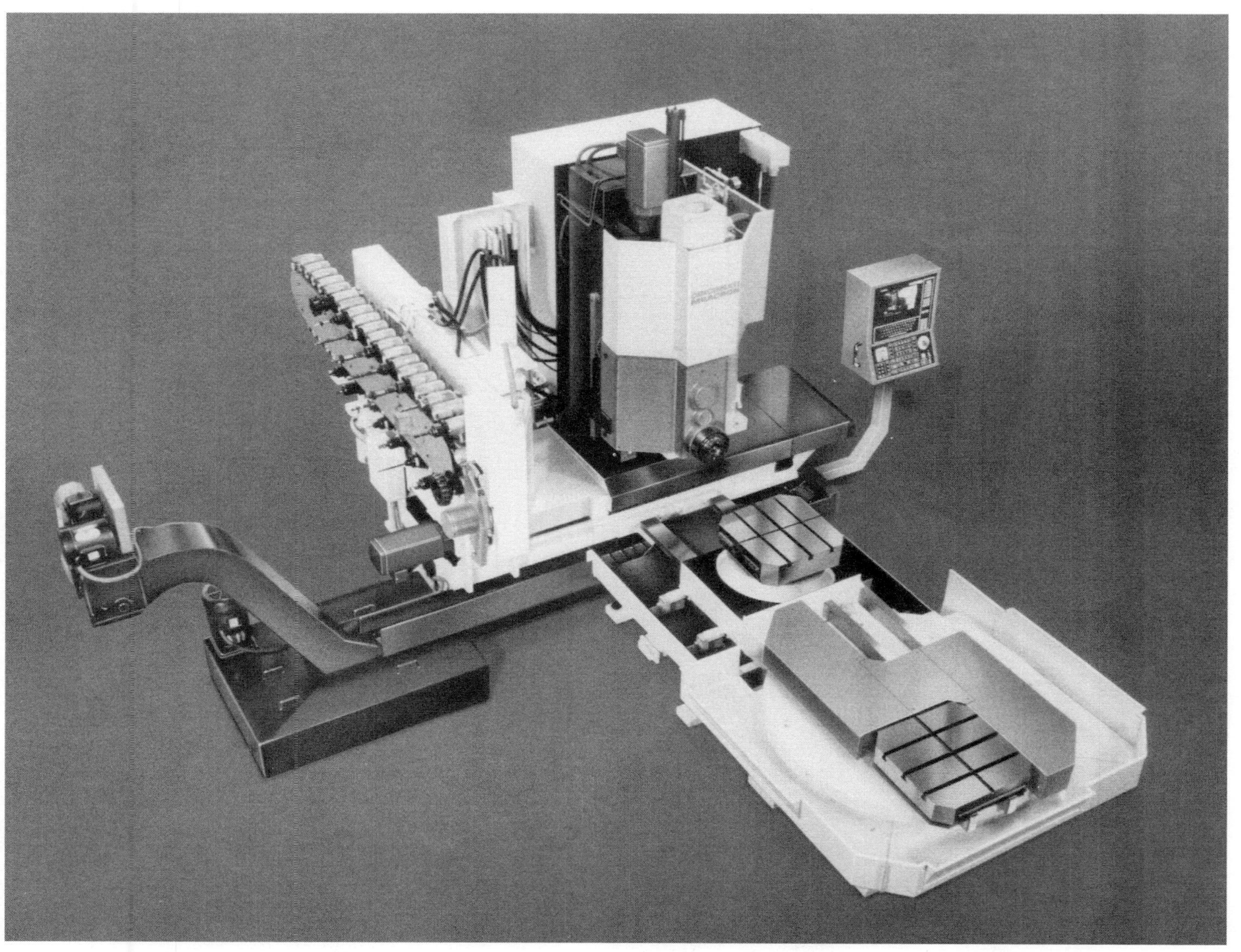

FIGURE 4–2 A horizontal spindle machining center with matrix magazine tool storage and pivot insert tool changer. (Photo courtesy of Cincinnati Milacron.)

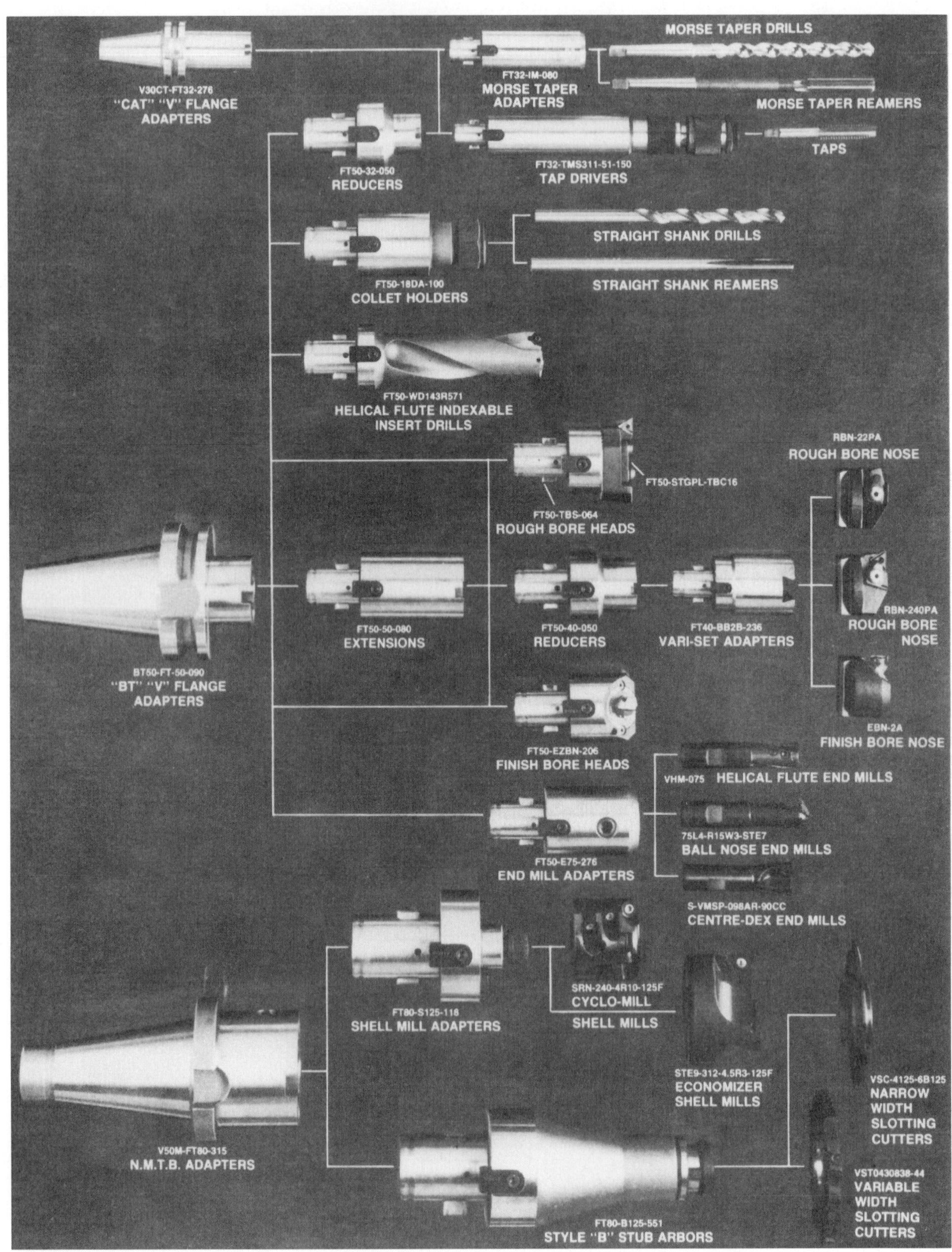

FIGURE 4–3 An assortment of CNC tooling systems based on the V-flange tool holder. Note the use of interchangeable extensions, adapters, arbors, boring heads, and cutting tools. (Photo courtesy of GTE Valenite Corp.)

3. Ensure repeatability of a tool—center the tool in the spindle such that the tool's relation to the work is repeated (within tolerance) every time the tool is used.
4. Provide fast and easy off-line tool assembly.

The CNC shop can choose from a broad spectrum of tool holding systems. By far, the V-flange or caterpillar shank tool holder is the most popular. A majority of machine tool builders are designing holders to conform to the American National Standard Institute (ANSI) standard B5.50 for tapered shank. This standard includes six basic sizes ranging from 30 through 60. Each machining center will accept only one size. Machining centers that have 10 to 40 horsepower commonly use size 50.

4–5 METHODS OF SECURING TOOLS IN TOOL HOLDERS

Tool holders are designed to be assembled off line. Once assembled, the system is loaded into the machining center's tool storage mechanism in preparation for running a machining program. Modular tooling systems are a key factor in boosting CNC machining time to approximately 70% compared to 30% for conventional machines. The two most important V-flange tool holder designs for gripping and centering tools are the end mill holder and the collet-and-chuck holder.

End Mill Tool Holders

End mill holders lock and center the cutting tool by way of one or two set screws. The screws press on the flats of the tool shank. Because the screws are both located on one side, a high degree of concentricity is not guaranteed. These tool holders are relatively inexpensive in comparison to collet-and-chuck holders. They are used to hold end mills, drills, boring bars, spade drills, and other tools with straight shanks of standard dimensions. They find wide application in many milling and boring operations not requiring very close tolerances. See Figure 4–4.

Collet-and-Chuck Tool Holders

This system involves the use of a collet for gripping and centering the tool in the tool holder. See Figure 4–5. Collet-and-chuck tool holders are more expensive than end mill holders but generally provide better tool holding capability and more accurate tool centering than end mill types. Collets come in two designs: double angle and single angle (slow taper). Double-angle collets can be used to hold many different tool shank shapes including tapers. Single-angle collets are especially useful for applying very high clamping forces on cutting tools. Approximately 60% to 70% of tooling sold for machining centers is either

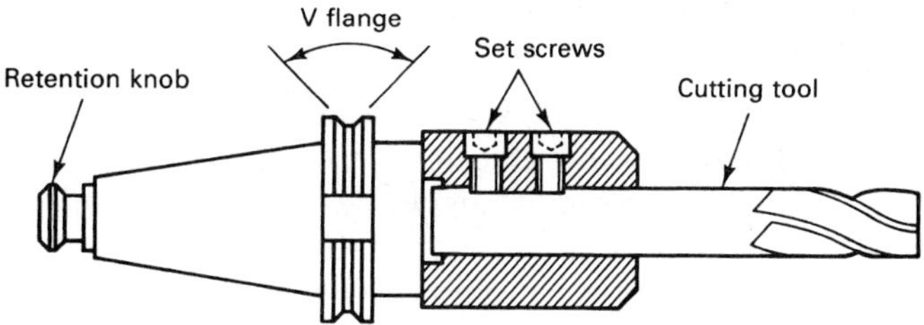

FIGURE 4–4 An end mill tool holder.

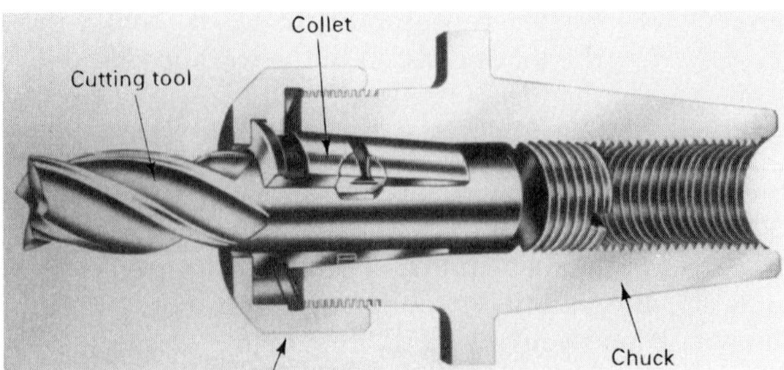

FIGURE 4–5 A collet-and-chuck tool holder. (Photo courtesy of TSD Universal/DeVlieg Ballard Tooling Systems Division.)

FIGURE 4–6 Collet and chuck tool holding systems. (Photo courtesy of Command Corporation International.)

part of a collet-and-chuck system or is made to be used with a collet-and-chuck system.

Various types of collet and chuck systems used with CNC machining centers including cutting tools, tool holders, adapters, collets, and retention knobs are shown in Figure 4–6.

4–6 METHODS OF SECURING TOOLING SYSTEMS TO THE CNC SPINDLE

In one system the tool interchange arm grips the tool holder and aligns it with the spindle. The spindle may descend on the tool holder or the arm may insert the holder into the spindle. As the spindle and tool holder mate, a split bushing retainer inside the spindle locks onto a retention knob on the top of the tool holder. The retainer then draws the tool holder up and into the spindle, creating positive clamping and centering action. See Figure 4–7.

Grooved flange tool holders are used with side-gripping interchange arms. For these configurations the holder is gripped from one side and placed into the spindle. The interchange arm disengages the holder by moving off to the side.

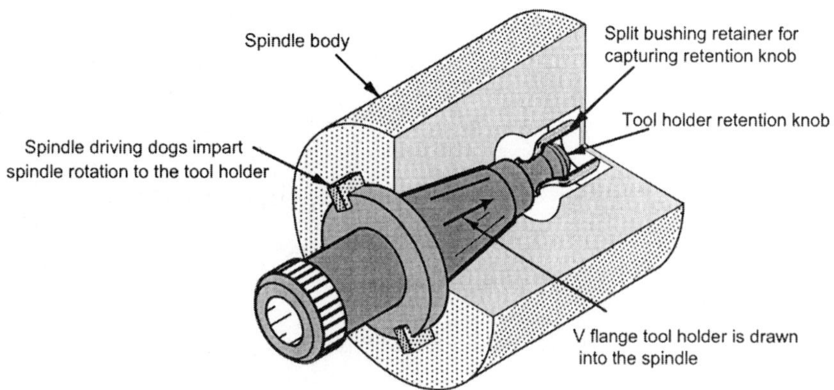

FIGURE 4–7 A tool holder with retention knob secured in the spindle by split bushing retainer.

4–7 AUTOMATIC TOOL CHANGER SYSTEMS

Many different types of mechanisms have been designed for storing and changing tools. The three most important are turret head, carousel storage with spindle direct changing, and matrix magazine storage with pivot insertion tool changer. Tool storage magazines may be horizontal or vertical.

Turret Head

This type of system is found on older NC drilling machines. The tools are stored in the spindles of a device called a turret head. When a tool is called by the program, the turret rotates (indexes) it into position. The tool can be used immediately without having to be inserted into a spindle. Thus, turret head designs provide for very fast tool changes. The main disadvantage of turret head changers is the limit on the number of tool spindles that can be used. See Figure 4–8.

Carousel Storage with Spindle Direct Tool Changer

Systems of this type are usually found in vertical machining centers. Tools are stored in a coded drum called a carousel. The drum rotates to the space where the current tool is to be stored. It moves up and removes the current tool, then rotates the new tool into position and places it into the spindle. On larger systems, the spindle moves to the carousel during a tool change. See Figure 4–9.

Horizontal Storage Matrix Magazine with Pivot Insertion Tool Changer

Chain-type storage matrix magazines have been popular in machining centers since early 1972. This type of system permits an operator to load many tools in a relatively small space. The chain may be located on the side or the top of the CNC machine. These positions enable tools to be stored away from the spindle and work. This will ensure a minimum of chip interference with the storage mechanism and a maximum of tool protection.

FIGURE 4–8 A turret head tool changing system. (Photo by James V. Valentino.)

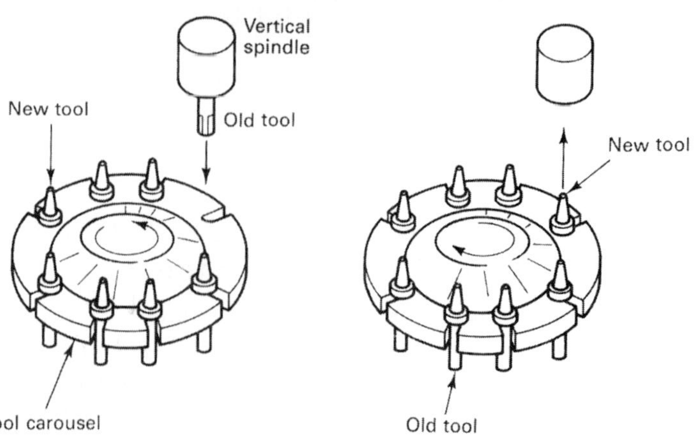

FIGURE 4–9 A carousel storage system with spindle direct tool changer. (Photo courtesy of Index Corp.)

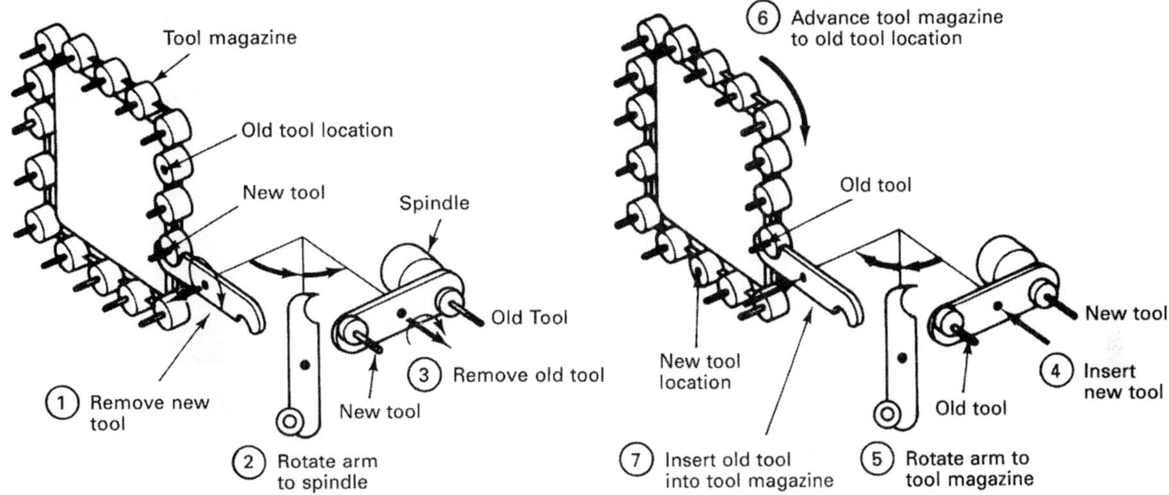

FIGURE 4–10 A storage matrix magazine and pivot insert tool changer. (Photo courtesy of Cincinnati Milacron.)

Upon entering a programmed tool change, the system advances to the proper tool via the chain mechanism. The pivot arm rotates and picks up both the new tool in the magazine and the old tool in the spindle. The magazine then advances to the space where the old tool is to be stored. The arm executes a rotation again and inserts the new tool into the spindle and the old tool into the magazine. A final rotation returns the arm back to its parked position. These steps are illustrated in Figure 4–10.

Two methods of tool identification are currently in use. One is the bar code designation. The code is imprinted and fastened to the tool. When the program calls for a specific tool, the controller looks for a particular tool code, not a specific location. Another tool identification system uses a computer microchip that is part of the tool or tool holder. The microchip contains the tool identification number and information related to the parameters of the tool. A special sensor reads the data and transfers it to the machine controller.

4–8 PALLET LOADING SYSTEMS

Pallet loading systems represent another means of cutting down on machine idle time due to setup operations. One of the main goals with pallets is to keep the machine tool running. A setup person can set up a job on an idle pallet. It can then be automatically loaded into the machine as soon as the part that is running is finished. See Figures 4–11 and 4–12.

FIGURE 4–11 A two-pallet work loading system used on a horizontal machining center. (Photo courtesy of KT-Swazey, Milwaukee, Wis.)

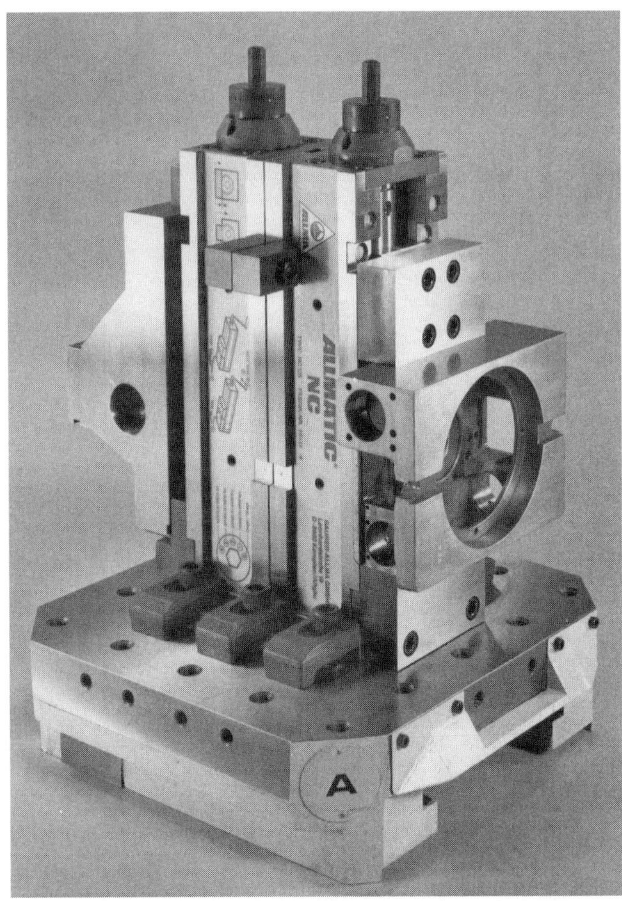

FIGURE 4–12 Details of a job setup clamped to a pallet. (Photo courtesy of Dapra Corp.)

A basic pallet system is normally composed of the following elements:

The pallet: holds the work directly or by means of a fixture.
The pallet loader: moves the pallet from the load/unload station to the pallet receiver/holder.
The pallet receiver/holder: grips and positions the pallet in the CNC machine.

Pallets are manufactured in a variety of shapes and sizes. These include round, square, rectangular, and three- and four-jaw chuck holding. The size and shape of the part to be manufactured will determine the type of pallet used.

A typical two-pallet operation is described in the following discussion. Refer to Figure 4–13 for a corresponding illustration.

Step 1: Assume pallet 2 is on the loader that is currently off line. The operator sets up a job by mounting the work to a fixture bolted to pallet 2. The operator then enters information into the MCU indicating that this pallet is ready.
Step 2: After completing the part program for the part on pallet 1, the MCU directs the machine to pull it from the receiver onto the loader. Pallet 1 is then removed and placed at an off-line holding location.
Step 3: Pallet 2 is then loaded into the receiver and locked into position. If the new job is identical to the last job run, the same part program is retained. If, however, a different type of job is to be run, the machine must

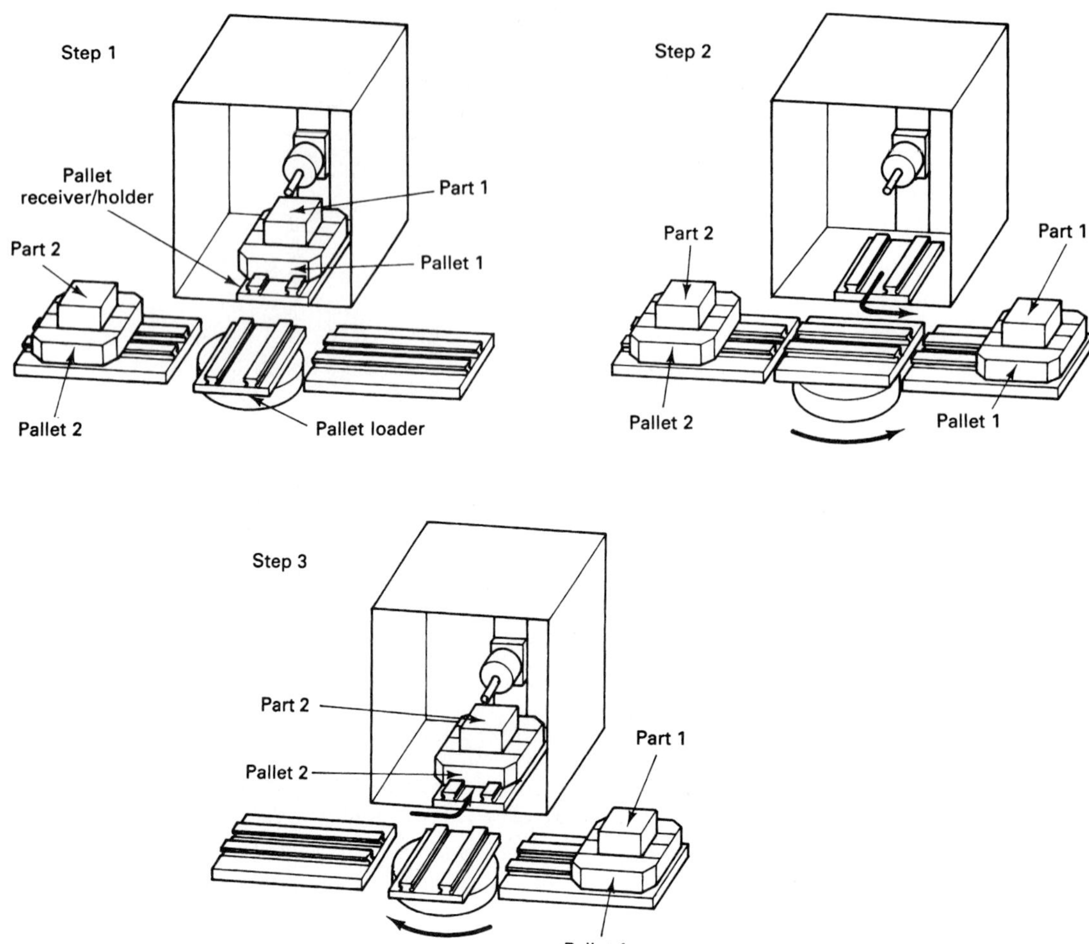

FIGURE 4–13 The operations involved in a two-pallet work changing system.

be signaled to load the corresponding part program. This information can be entered manually or automatically read from a microchip or bar code encoded on pallet 2. The part program is run for the work on pallet 2. The completed part on pallet 1 is removed. A new job is set up on pallet 1, thereby maintaining a continuous work cycle.

The two pallets are rotated in and out of the machining center until the entire job run for all parts is completed.

Pallet loading systems have been adapted to all types of CNC machines including vertical and horizontal machining centers.

4–9 CHAPTER SUMMARY

The following key concepts were discussed in this chapter:

1. A machining center is a CNC machine that incorporates some form of automatic tool changing.
2. A tool holder is used to mate several different size tools to the CNC spindle.
3. Tool holders are designed to center the tool in the spindle such that the tool's relation to the work is repeated every time the tool is used.

4. The two most important types of V-flange tool holders are the end mill holder and the collet-and-chuck holder.
5. Automatic tool storage and tool changer systems include turret head, carousel storage with spindle direct tool changer, and horizontal storage matrix magazine with pivot insertion tool changer.
6. Pallet loading systems are used to reduce machine idle time. They automatically rotate in a setup part and rotate out a completed part.

REVIEW EXERCISES

4.1. An automatic tool changer is found on what type of CNC equipment?

4.2. Name four important functions of a tool holder.

4.3. Describe the methods of capturing the tool for the following types of tool holders:
- (a) End mill holder
- (b) Collet-and-chuck holder

4.4. Explain an advantage and a disadvantage of using a collet-and-chuck holder as opposed to an end mill holder.

4.5. Describe how the following tool storage and tool changer mechanisms work:
- **(a)** Turret head
- **(b)** Carousel storage with spindle direct tool changer
- **(c)** Horizontal storage matrix magazine and pivot insertion tool changer

4.6. What is an advantage of using a storage matrix magazine?

4.7. What advantage does a pallet loading mechanism offer to a machining center?

CHAPTER

5

Review of Basic Blueprint Reading for CNC Programmers

5-1 CHAPTER OBJECTIVES

At the conclusion of this chapter you will be able to

1. Know the sizes and format for CNC prints.
2. State the alphabet of lines used in prints.
3. Understand the terms: othographic projection, first angle projection, third angle projection, and auxiliary projection.
4. Describe a section view and the types of section views used in prints.
5. Know how to read conventional tolerances.
6. Identify the various types of dimensioning systems and practices used in CNC prints.
7. Understand thread nomenclature, thread descriptions in CNC prints.
8. Know how to read surface texture symbols and notes.
9. Understand how to read material notes in drawings.
10. Read and understand part heat treatment notes.
11. Interpret coating and plating notes in drawings.

5-2 INTRODUCTION

The starting point for CNC programs is the CNC print. The print provides key information describing the shape and size of the part, tolerances and accuracy of production, the surface finish, and other important production notes. The CNC programmer must have a fundamental knowledge of how to read and interpret prints (see Figure 5–1).

A basic review of prints and print reading is presented in this chapter. The reader is also encouraged to consult other sources of information on blueprint reading listed at the conclusion of this chapter.

5-3 SHEET SIZES

The American National Standards Institute (ANSI) document Y14.1-1987 specifies the sizes of flat cut sheets used in the United States and Canada. A lettering system identifies the sheet sizes. A drawing form normally has a border that is .03″ thick. The lettering heights range from .120″ to .29″. These parameters vary according to sheet size. The European and international com-

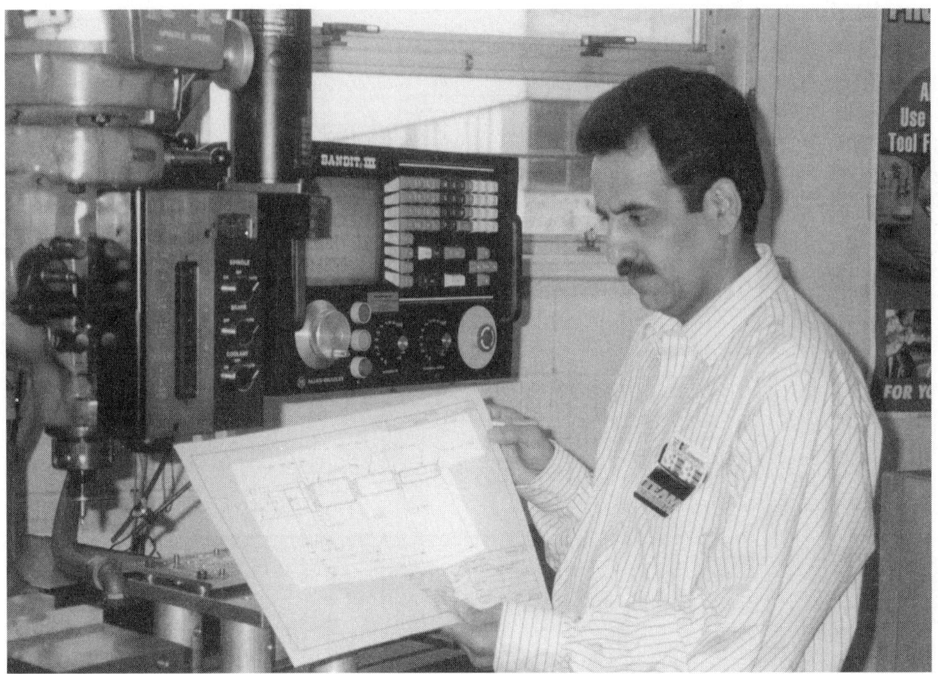

FIGURE 5–1 The part print serves as a starting point for writing the CNC program.

munity uses metric sheet sizes. Refer to Figure 5–2 for an illustration of standard drawing sheet sizes.

5–4 DRAWING FORMATS

A typical blank drawing form as recommended by ANSI Y14.1 for mechanical drawings is shown in Figure 5–3.

The important areas of information contained in this form will be described in detail in the sections to follow.

Title Block

A standard title block is shown in Figure 5–4. It is normally placed in the lower right hand corner of the drawing form and contains the following information:

Company name and address
Drawing title: The part name such as SHEAVE COVER, etc.
Size: A standard sheet size (A, B, C, etc.) is entered.
FSCM NO.: The Federal Supply Code Number is entered for manufacturers who make parts for the US government.
DWG NO.: The drawing number is assigned for a drawing log and is used to identify the part.
Rev: A letter (A, B, C, etc.) is used to indicate the latest revision or version of the drawing.

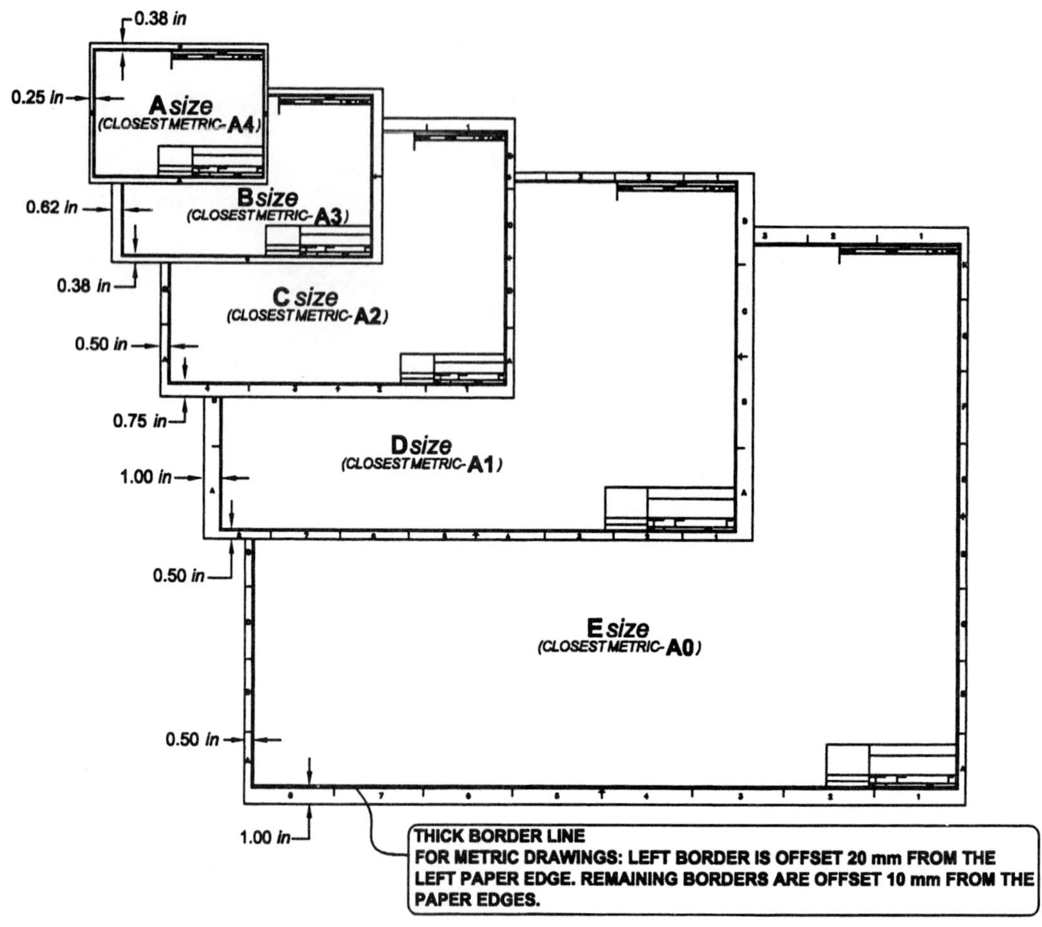

DRAWING SIZES - USA				
DRAWING SIZE	PAPER SIZE		BORDER SIZE	
	WIDTH (in)	LENGTH (in)	WIDTH (in)	LENGTH (in)
A	8.50	11.00	7.75	10.50
B	11.00	17.00	10.25	15.26
C	17.00	22.00	15.50	21.00
D	22.00	34.00	21.00	32.00
E	34.00	44.00	32.00	43.00

DRAWING SIZES - INTERNATIONAL				
DRAWING SIZE	PAPER SIZE		BORDER SIZE	
	WIDTH (mm)	LENGTH (mm)	WIDTH (mm)	LENGTH (mm)
A4	210	297	190	267
A3	297	420	277	390
A2	420	594	400	564
A1	594	841	574	811
A0	841	1189	821	1159

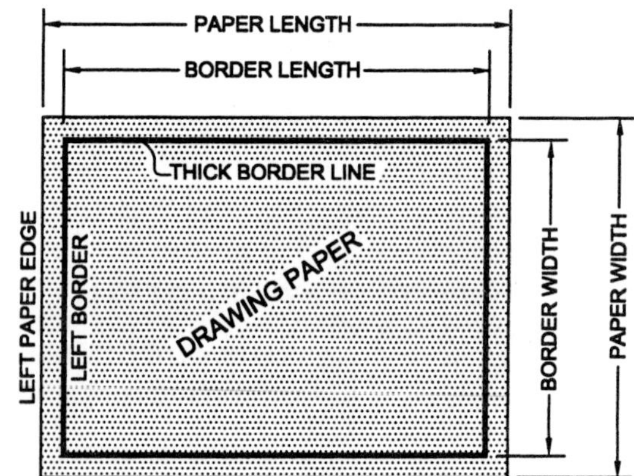

FIGURE 5–2 Standard drawing sheet sizes.

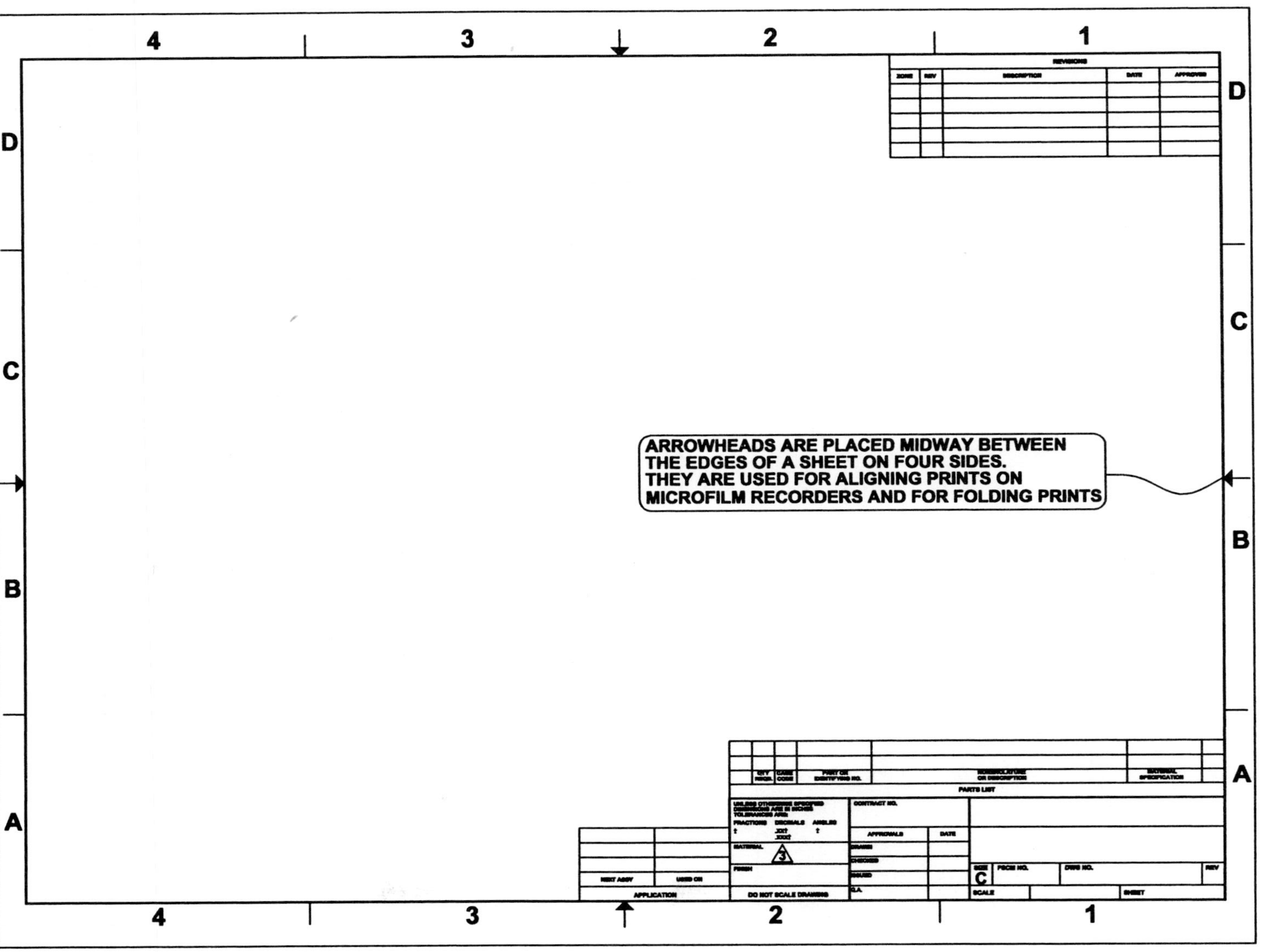

ARROWHEADS ARE PLACED MIDWAY BETWEEN
THE EDGES OF A SHEET ON FOUR SIDES.
THEY ARE USED FOR ALIGNING PRINTS ON
MICROFILM RECORDERS AND FOR FOLDING PRINTS

FIGURE 5–3 Typical drawing form as recommended by ANSI Y 14.1.

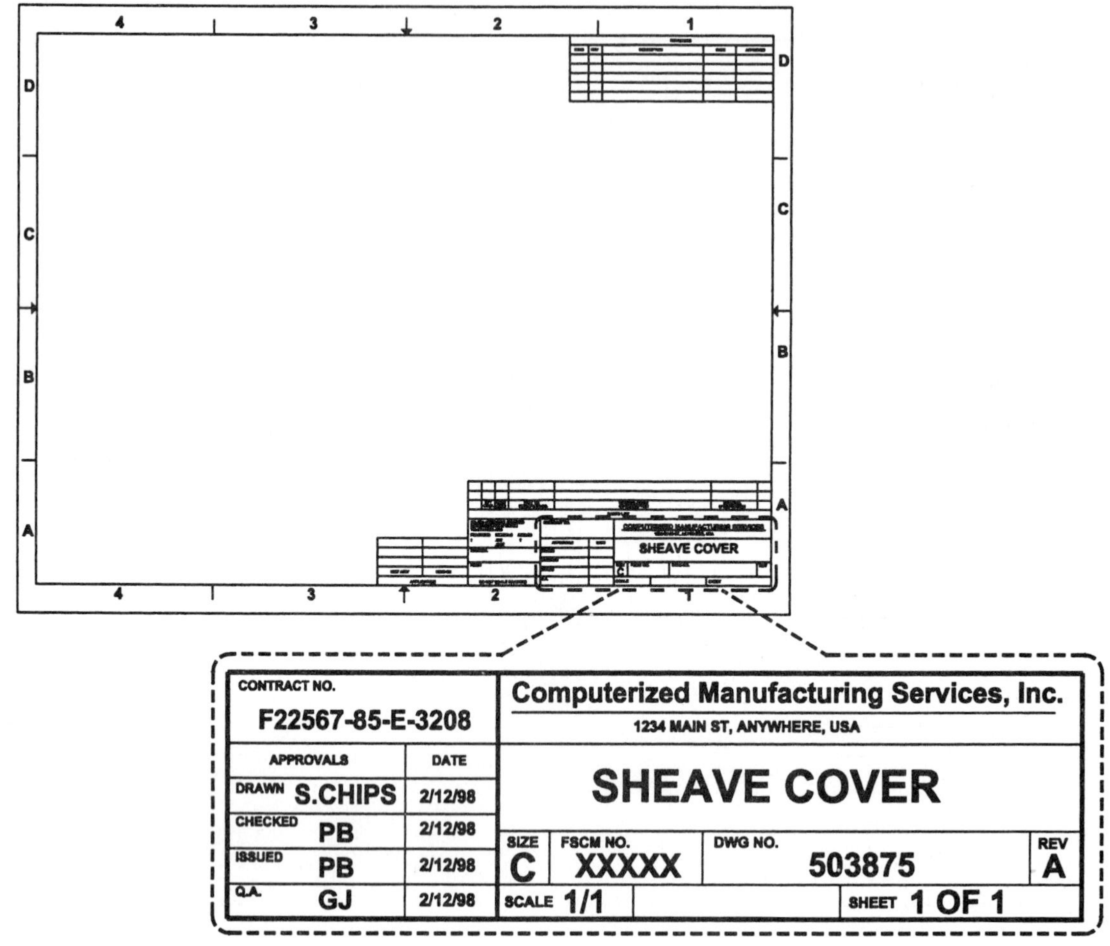

FIGURE 5–4 Details of the title block area of the drawing form.

Scale: The drawing scale is indicated. Formats include

English	**Metric**
Full or 1=1	Full or 1:1
Half or 1=2	Half or 1:2
Quarter or 1=4	Quarter or 1:4
Eighth or 1=8	Eighth or 1:8
Tenth or 1=10	Tenth or 1:10
Double or 2=1	Double or 2:1

Sheet: Each sheet describing the part is numbered in this box. The first sheet in a three sheet set, for example, would be labeled as 1 of 3.

Contract No: This number identifies the contract under which the part is being produced. The number is used for such accounting functions as charging time and cost.

Approvals: All individuals responsible for generating the drawing must register their signature and date.

Drawn: Refers to the drafter.

Checked: Refers to the drawing checker.

Issue: Refers to the person who gives final approval.

QA: Refers to the individual responsible for quality assurance.

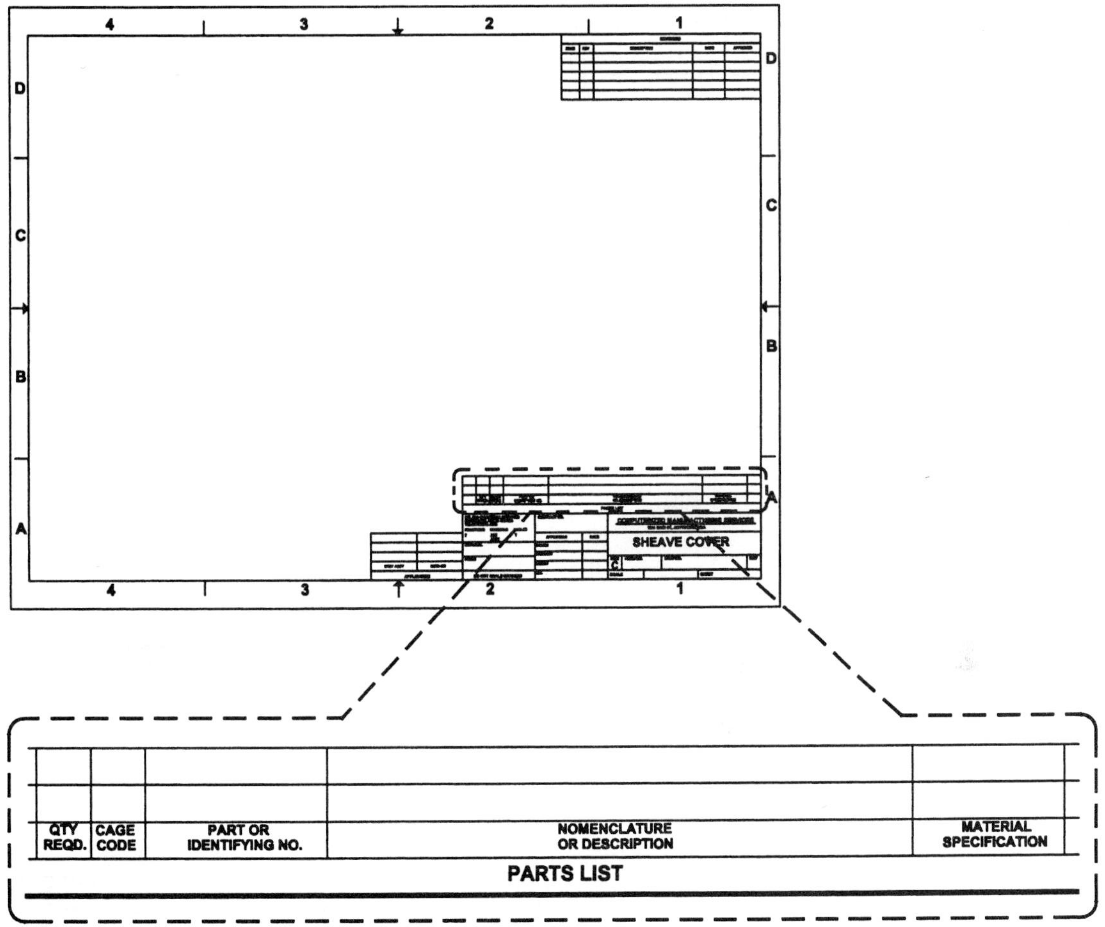

FIGURE 5–5 Details of the parts list area of the drawing form.

Parts List

The parts list, also known as the bill of materials, is located above the title block as shown in Figure 5–5. It can also be placed on a separate sheet and is used for drawings involving an assembly of parts.

The parts list block contains the following information:

Qty reqd.: The quantity required specifies the number of times each part in the assembly is used.

Cage code: The cage code indicates the inventory cage number where each part is stored.

Nomenclature or description: A brief description of every part type used in the assembly is written here.

Material specification: The material of every part type used in the assembly is entered here.

Revision Block

The revision block is placed in the upper right hand corner of the drawing as shown in Figure 5–6.

All changes to the drawing due to design, manufacturing, customer requirements and errors are recorded in the revision block according to the following format:

Zone: A zone is a particular area in the drawing to which a note or drawing description applies. The zone is located by using a number to indicate the

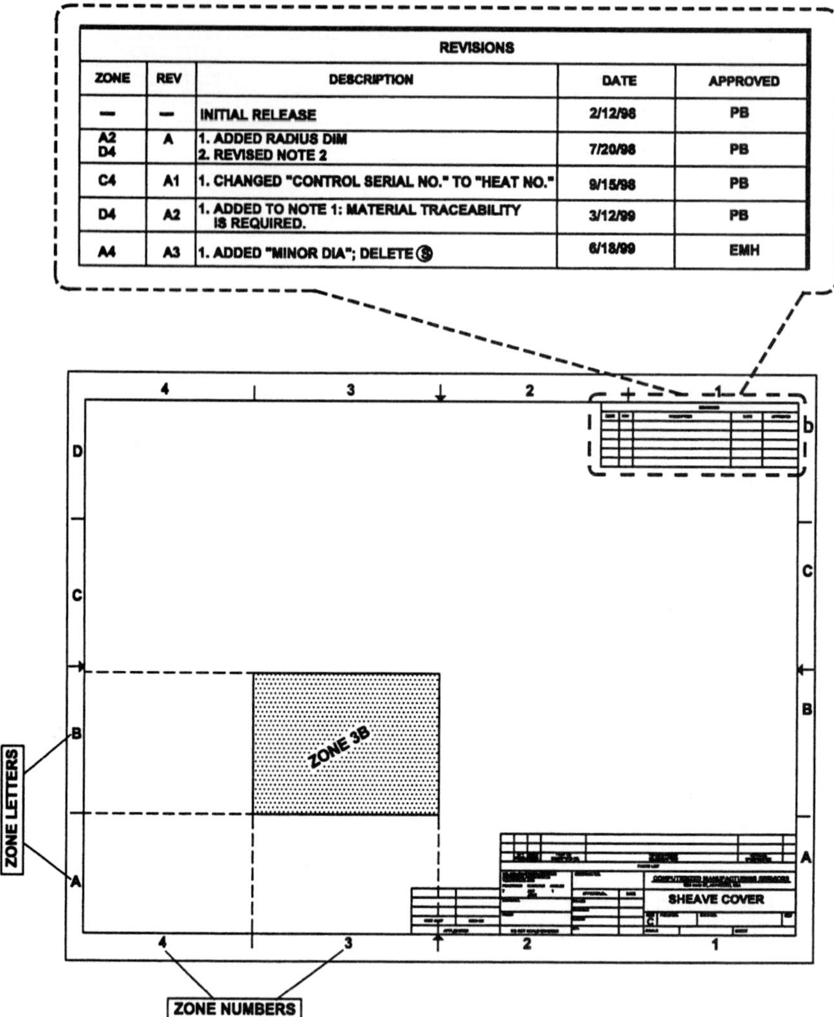

FIGURE 5–6 Details of the revision block area of the drawing form.

horizontal position range and a letter to specify the vertical position range. The zone 3B is shown in Figure 5–6.

Rev: A revision change is identified by a number attached to the revision letter, e.g., A1. Successive revision changes are placed in sequential and alphabetic order below the first revision note as shown in Figure 5–6.

Description: A brief description of the revision change is recorded in this area.

Date: This area contains the date of the revision.

Approved: The individual responsible for the change must register an approval.

Notes Block

The notes block is located to the left of the title block as shown in Figure 5–7. The following information is contained in the notes block.

Tolerances: The amount by which the dimensions in the drawing may vary is indicated in this section.

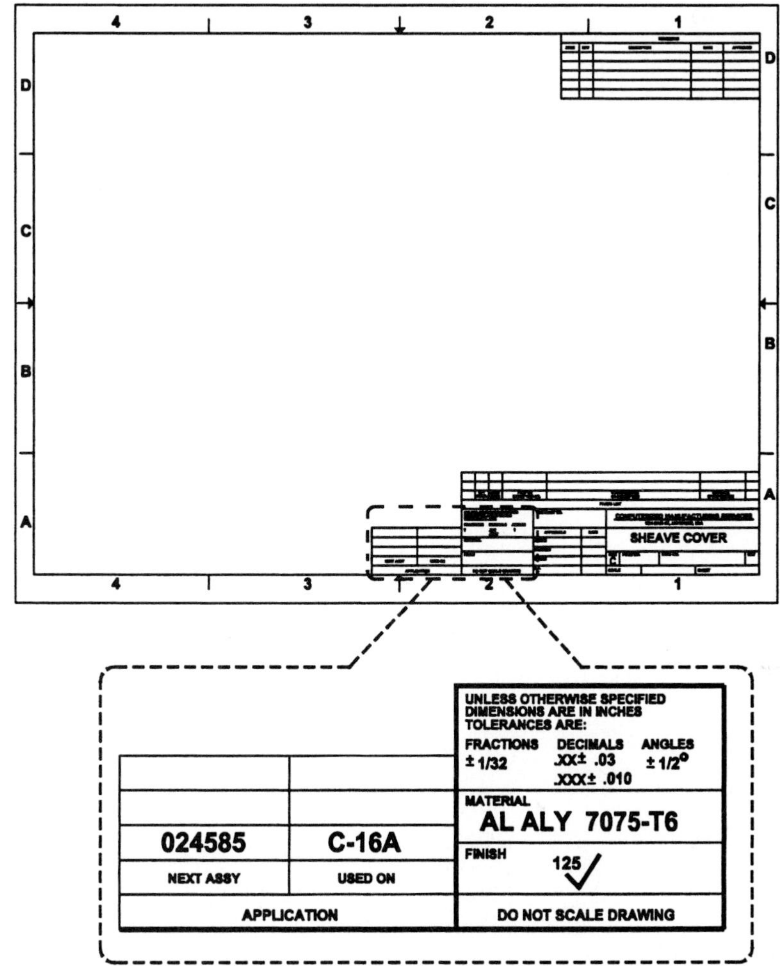

FIGURE 5–7 Details of the notes block area of the drawing form.

Material: The material from which the part is to be made is recorded in this area.

Finish: Final surface finish or degree of surface smoothness to be produced for the part is entered in this section. Refer to Section 5-13 for surface finish nomenclature and definitions.

Do not scale drawing: This note indicates that the print reader is not to use a scale or any other means to measure a dimension on the drawing.

Next assy: The drawing number of the next assembly that accepts this part is entered here.

Used on: The serial number of the final assembly this part belongs to is entered in this box.

If more extensive notes are required to describe the manufacturing processes and parameters for the part, they may be written to the left of the notes block as shown in Figure 5–8.

5–5 INTERPRETING LINES IN DRAWINGS

Different types of lines are used in mechanical drawings to convey important information describing a part's shape, dimensions and internal features. Conventions for lines in prints as set forth by the American National Standards Institute document ANSI Y14.2M-1987 are shown in Table 5–1.

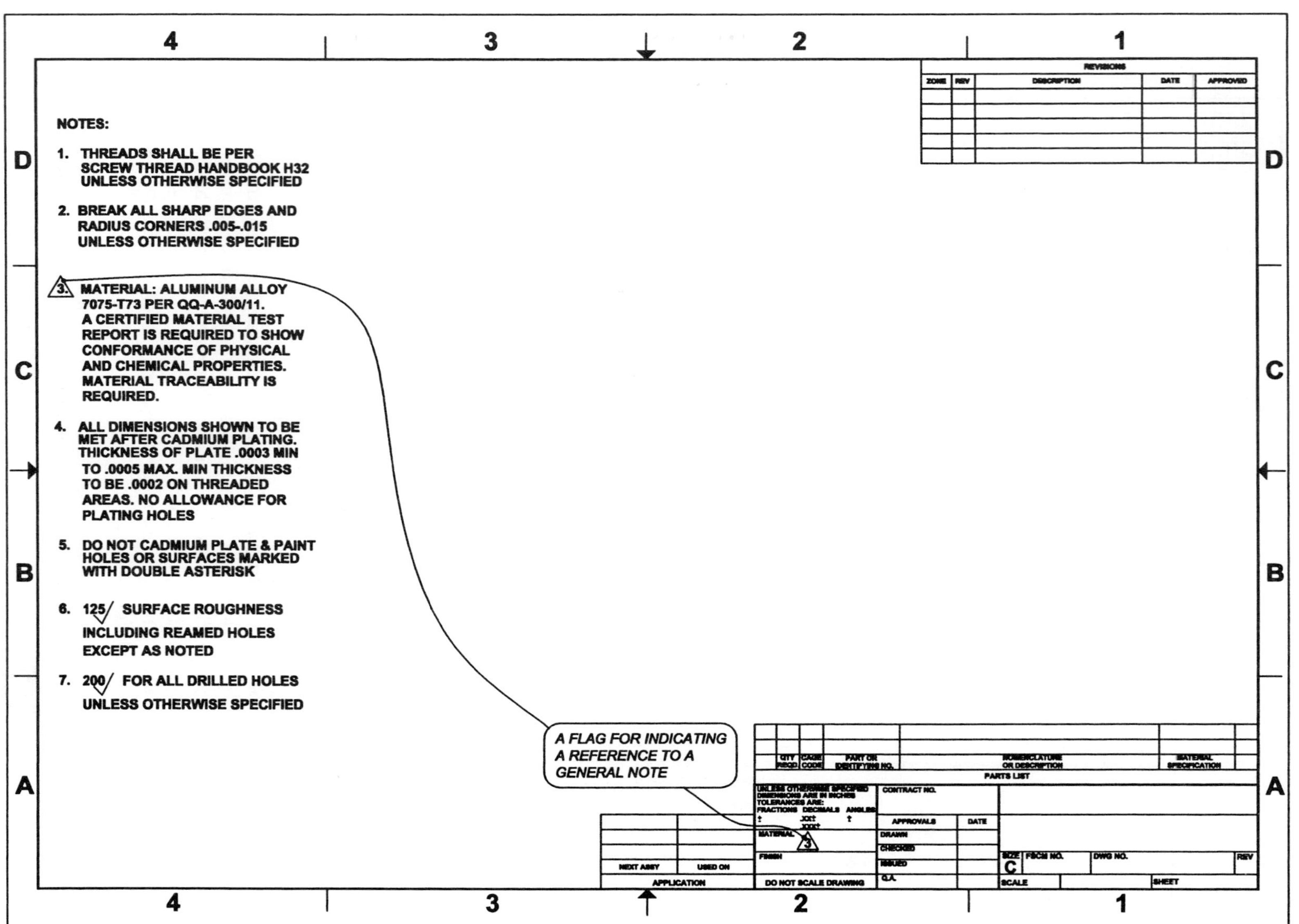

NOTES:
1. THREADS SHALL BE PER SCREW THREAD HANDBOOK H32 UNLESS OTHERWISE SPECIFIED
2. BREAK ALL SHARP EDGES AND RADIUS CORNERS .005-.015 UNLESS OTHERWISE SPECIFIED
3. MATERIAL: ALUMINUM ALLOY 7075-T73 PER QQ-A-300/11. A CERTIFIED MATERIAL TEST REPORT IS REQUIRED TO SHOW CONFORMANCE OF PHYSICAL AND CHEMICAL PROPERTIES. MATERIAL TRACEABILITY IS REQUIRED.
4. ALL DIMENSIONS SHOWN TO BE MET AFTER CADMIUM PLATING. THICKNESS OF PLATE .0003 MIN TO .0005 MAX. MIN THICKNESS TO BE .0002 ON THREADED AREAS. NO ALLOWANCE FOR PLATING HOLES
5. DO NOT CADMIUM PLATE & PAINT HOLES OR SURFACES MARKED WITH DOUBLE ASTERISK
6. 125 SURFACE ROUGHNESS INCLUDING REAMED HOLES EXCEPT AS NOTED
7. 200 FOR ALL DRILLED HOLES UNLESS OTHERWISE SPECIFIED
A FLAG FOR INDICATING A REFERENCE TO A GENERAL NOTE
REVISIONS
ZONE
REV
DESCRIPTION
DATE
APPROVED
QTY REQD
CAGE CODE
PART OR IDENTIFYING NO.
NOMENCLATURE OR DESCRIPTION
MATERIAL SPECIFICATION
PARTS LIST
UNLESS OTHERWISE SPECIFIED DIMENSIONS ARE IN INCHES TOLERANCES ARE:
FRACTIONS DECIMALS ANGLES
MATERIAL
FINISH
DO NOT SCALE DRAWING
CONTRACT NO.
APPROVALS
DATE
DRAWN
CHECKED
ISSUED
Q.A.
SIZE
C
FSCM NO.
DWG NO.
REV
SCALE
SHEET
NEXT ASSY
USED ON
APPLICATION

FIGURE 5–8 Use of additional notes in production drawings.

TABLE 5–1

THE ALPHABET OF LINES

Thick = .032 *in* OR 0.7 *mm*
Thin = .016 *in* OR 0.35 *mm*

LINE NAME	USE IN DRAWINGS	APPEARANCE	EXAMPLE
VISIBLE OR OBJECT	SHOWS VISIBLE EDGES OF AN OBJECT	*Thick*	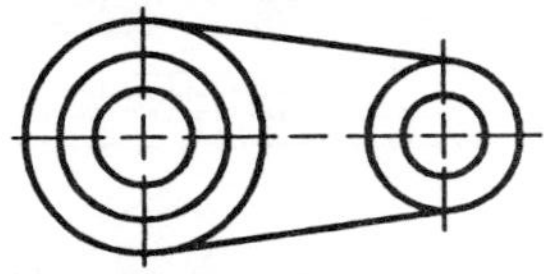
HIDDEN	INDICATES EDGES, SURFACES OR CORNERS OF AN OBJECT THAT ARE HIDDEN FROM VIEW	*Thin*	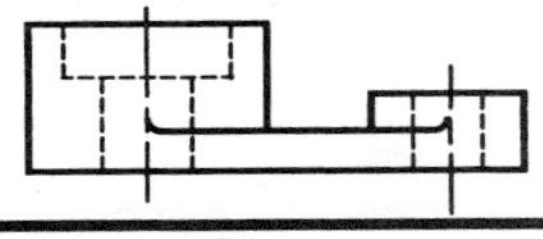
CENTER	INDICATES SYMMETRICAL FEATURES IN A PART. IT MARKS THE CENTERS OF HOLES AND CENTERS OF CYLINDERS AND BOLTS.	*Thin*	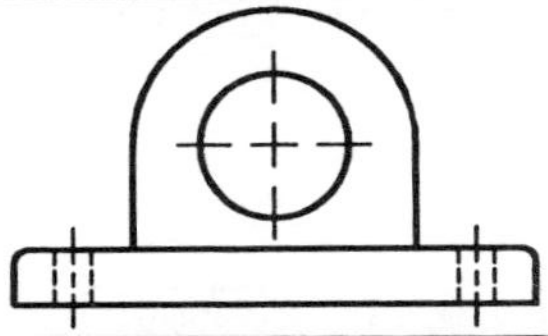
SYMMETRY	INDICATES PART SYMMETRY . IT IS USED FOR PARTIAL VIEWS OF SYMMETRICAL PARTS.	*Thick* *Thin*	
DIMENSION	INDICATES SIZE OR LOCATION	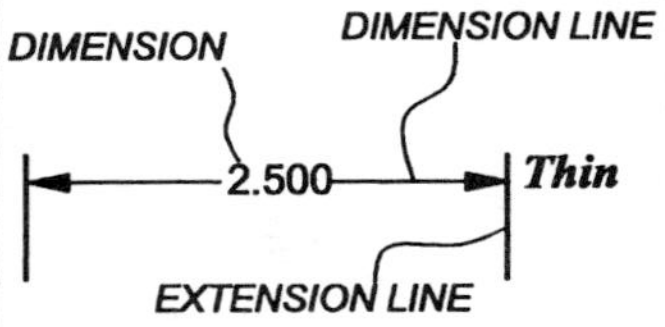 DIMENSION; DIMENSION LINE; 2.500; *Thin*; EXTENSION LINE	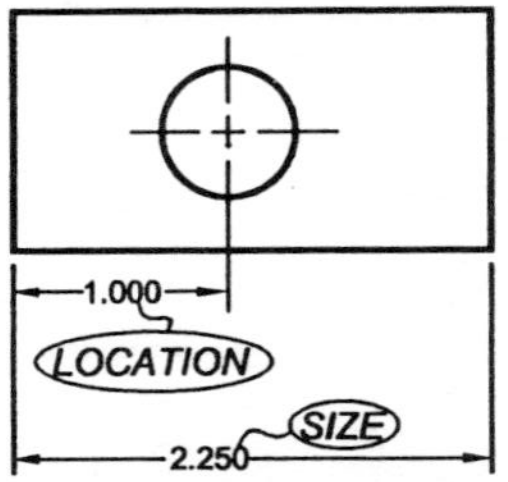 1.000; LOCATION; SIZE; 2.250
EXTENSION	INDICATES THE BEGINNING AND END OF A DIMENSION LINE		

TABLE 5-1 *(continued)*

CUTTING PLANE	INDICATES LOCATION OF CUTTING PLANE . OBJECT IS IMAGINED TO BE CUT BY THIS PLANE TO EXPOSE INTERNAL FEATURES.	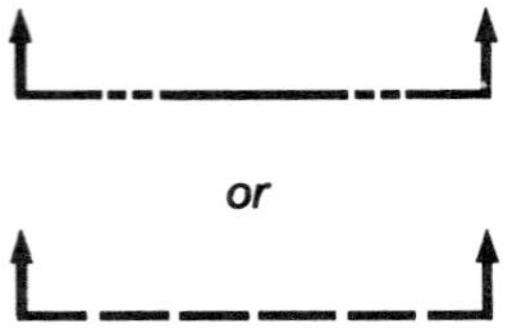 *Thick* or *Thick*	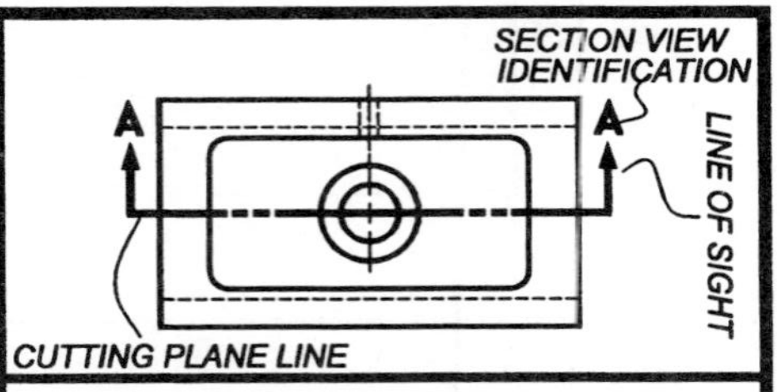 SECTION VIEW IDENTIFICATION A A LINE OF SIGHT CUTTING PLANE LINE
SECTION	INDICATES SOLID MATERIAL CUT BY THE CUTTING PLANE .	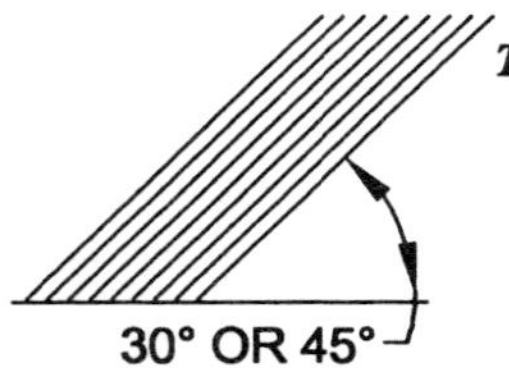 *Thin* 30° OR 45°	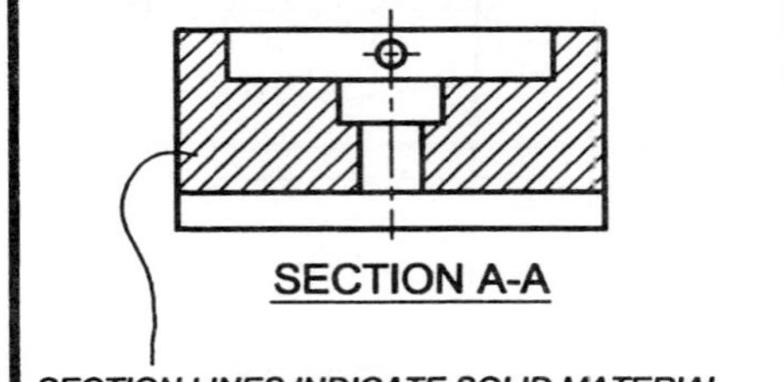 SECTION A-A SECTION LINES INDICATE SOLID MATERIAL CUT BY THE CUTTING PLANE
LEADER	CONNECTS A DIMENSION OR NOTE TO A PART FEATURE USED FOR HOLES,CHAMFERS, FILLETS ,SMALL CIRCLES AND ASSEMBLIES.	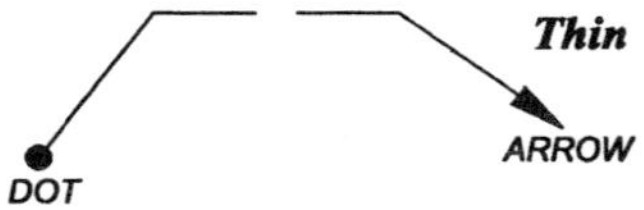 *Thin* DOT ARROW	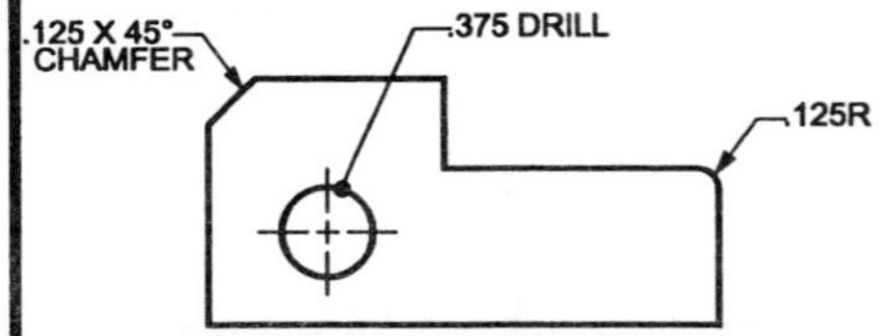 .125 X 45° CHAMFER .375 DRILL .125R
BREAK	INDICATES WHERE AN OBJECT IS BROKEN. USED TO SAVE DRAWING SPACE BY CUTTING OUT UNIMPORTANT REPETITIVE OR CONTINUOUS FEATURES. ALSO USED TO REVEAL INTERNAL FEATURES.	 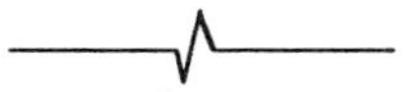 *Thick* SHORT BREAK LINE INDICATES SHORT BREAKS	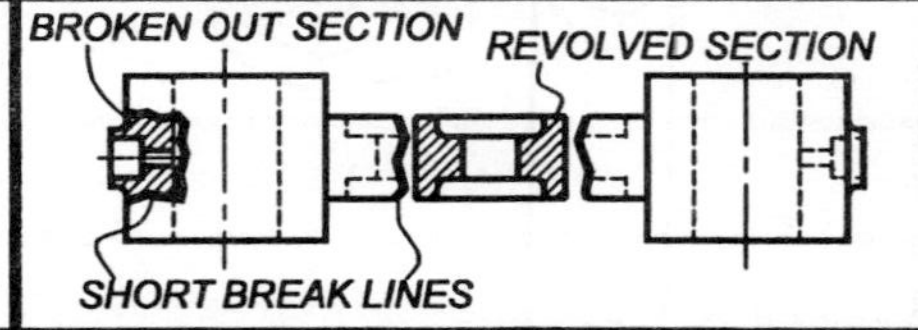 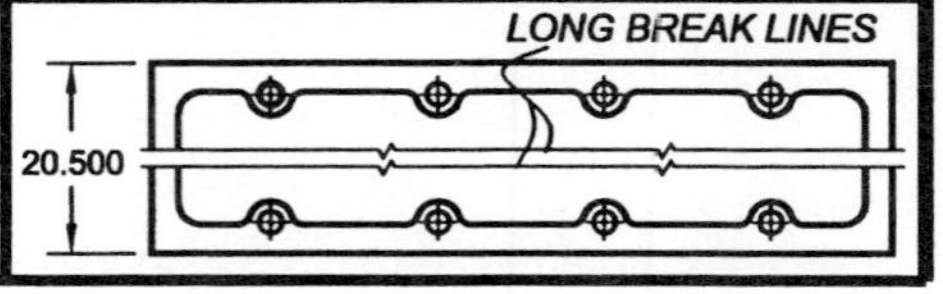 BROKEN OUT SECTION REVOLVED SECTION SHORT BREAK LINES
		Thin LONG BREAK LINE INDICATES LONG BREAKS	LONG BREAK LINES 20.500

5–6 PROJECTION CONVENTIONS USED IN DRAWINGS

An othographic projection is a technique for displaying an object according to the following conventions:

- The features of an object fall onto a projection plane.
- The projection plane is oriented perpendicular to the parallel lines of sight.

All objects possess three dimensions: length, height, and depth. Mechanical drawings utilize the multiview othographic system of showing an object. In this system, the object is surrounded by a box of six mutually perpendicular projection planes. The object is oriented such that each plane shows the true size of two dimensions as well as true angles and any hidden lines. A view is defined as the projection of an object's features onto a projection plane. The six principal othographic views are: front, top, right side, left side, rear and bottom. These concepts are illustrated in Figure 5–9.

The multiview othographic system of projection used in the United States and Canada is called third angle projection. Third angle projection involves placing the object in the third quadrant behind the projection planes as shown in Figure 5–10

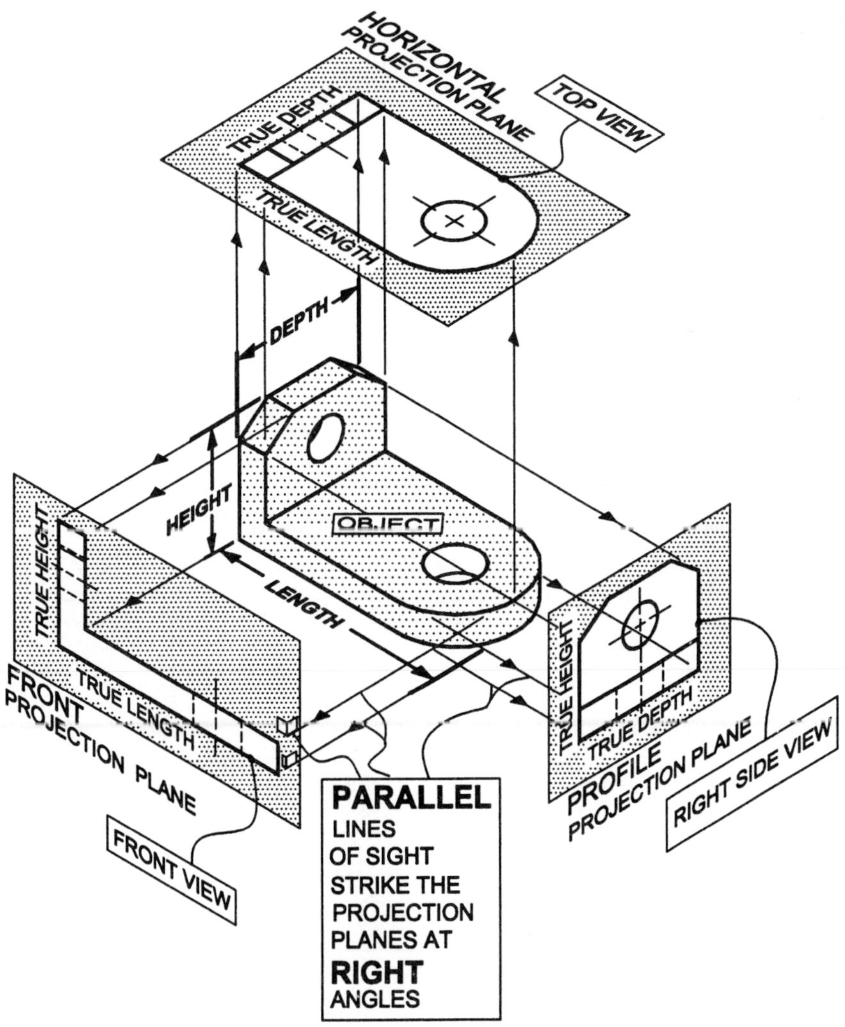

FIGURE 5–9 The multiview projection system with horizontal, front, and profile projection planes and corresponding top, front, and right side views.

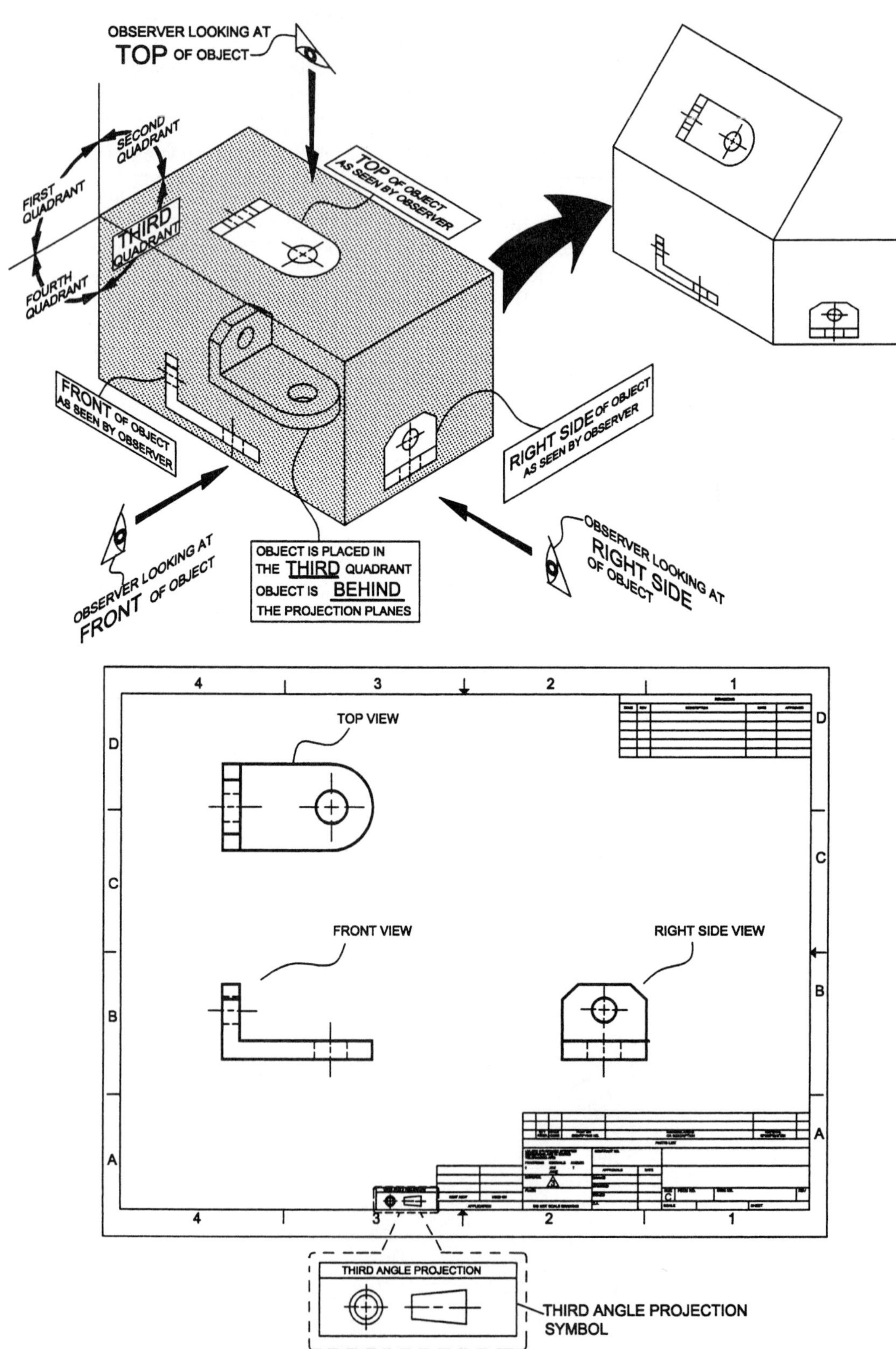

FIGURE 5–10 Third angle projection as used in the United States.

(top). The drawing form contains the note **THIRD ANGLE PROJECTION** and shows the truncated cone symbol as illustrated in Figure 5–10 (bottom).

European countries use another system of othographic projection known as first angle projection. With first angle projection, the object is placed in the first quadrant in front of the projection planes as Figure 5–11 (top) illustrates. First angle projection drawings contain the note **FIRST ANGLE PROJECTION** and display the truncated cone symbol shown in Figure 5–11 (bottom).

5–7 VISUALIZING 3D OBJECTS FROM 2D OTHOGRAPHIC VIEWS

Multiview othographic drawings present the minimum of views needed to completely describe a three-dimensional object. With practice the reader can develop the skills of fusing othographic views into a unified 3D image of the corresponding object. Important visualization principles to be followed when reading othographic drawings are given in Table 5–2.

5–8 AUXILIARY VIEWS

Some objects have inclined faces that do not appear true shape in any of the six principal othographic views. In these cases additional or auxiliary views are needed to show their true shape. An auxiliary view is folded out from the view in which the inclined face appears on edge. Refer to Figure 5–12 for an illustration of these concepts.

Figure 5–13 illustrates the common practice of omitting those portions of faces that do not appear true shape in an auxiliary view. This technique saves drafting time and makes the print easier to understand.

Viewplane lines like those shown in Figure 5–14 are used in cases involving auxiliary views of large or complex parts where drawing space is a problem. They indicate that the auxiliary view is to be found on another sheet or in a specific location of the same drawing sheet.

5–9 SECTIONAL VIEWS

A section view is used to expose the internal features of an object. It may also be used as an additional othographic view of surfaces that appear as hidden lines in the principal othographic views. The section view is created by passing an imaginary cutting plane through the object. Material that is cut is indicated by section lines as shown in Table 5–1. Figure 5–15 illustrates the principle of a section view.

Important sectioning conventions utilized for section views are shown in Table 5–3.

The different types of section views that appear in mechanical drawings are given in Table 5–4.

5–10 READING DIMENSIONS

A print's graphics describe the shape of an object and the dimensions indicate the size, location, and orientation of its individual geometric features. Tolerances specify the accuracy to which geometric features are to be held. The document entitled Dimensioning and Tolerancing ASME Y14.5M-1994 published by the American Society of Mechanical Engineers was adopted as the dimensioning standard for the United States by the American National Standards Institute (ANSI). It is available from the ASME at 345 East 47th St, N.Y. 10007. This section will discuss the most important aspects of the document.

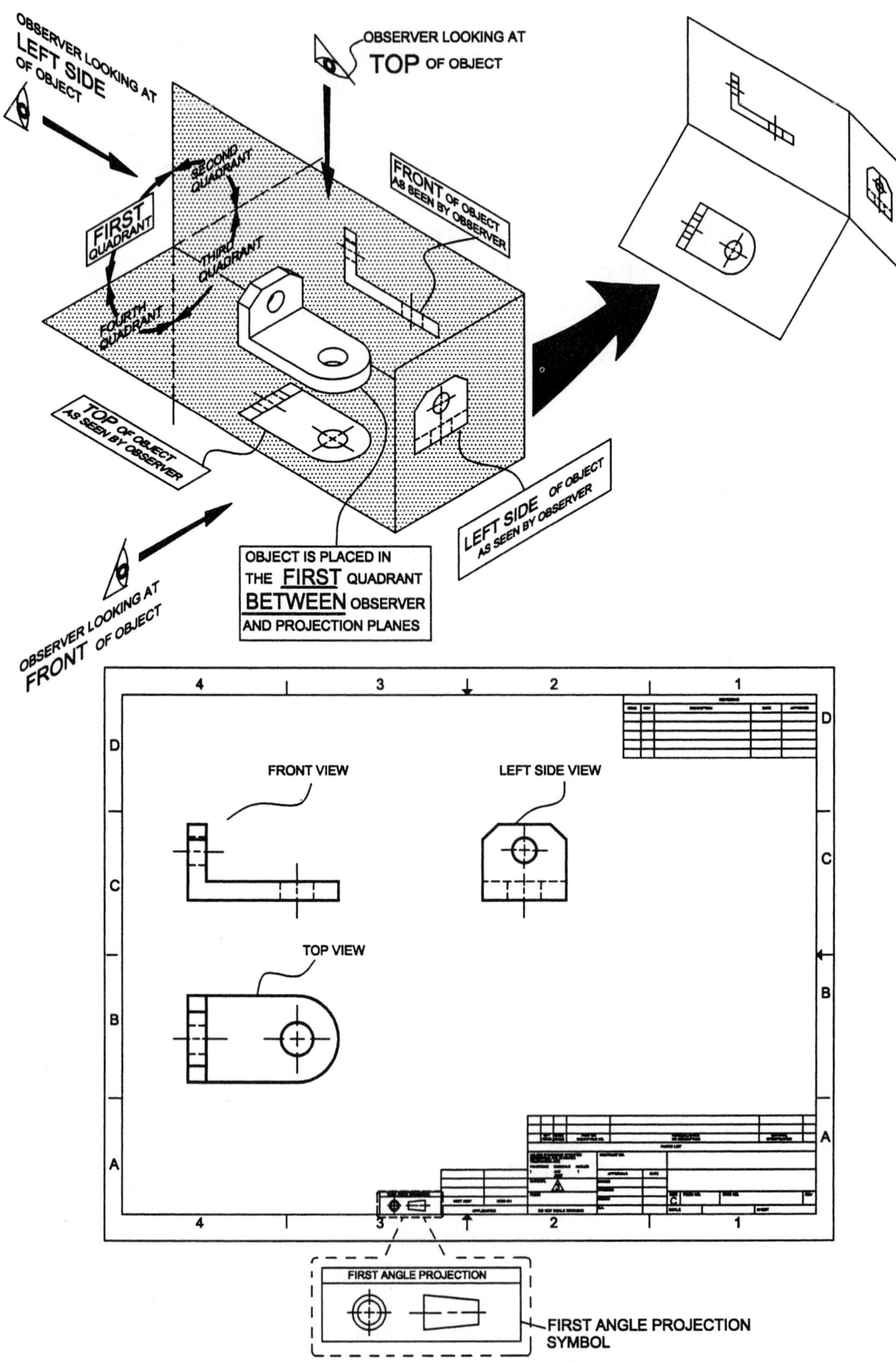

FIGURE 5-11 First angle projection as used in the European countries.

TABLE 5-2

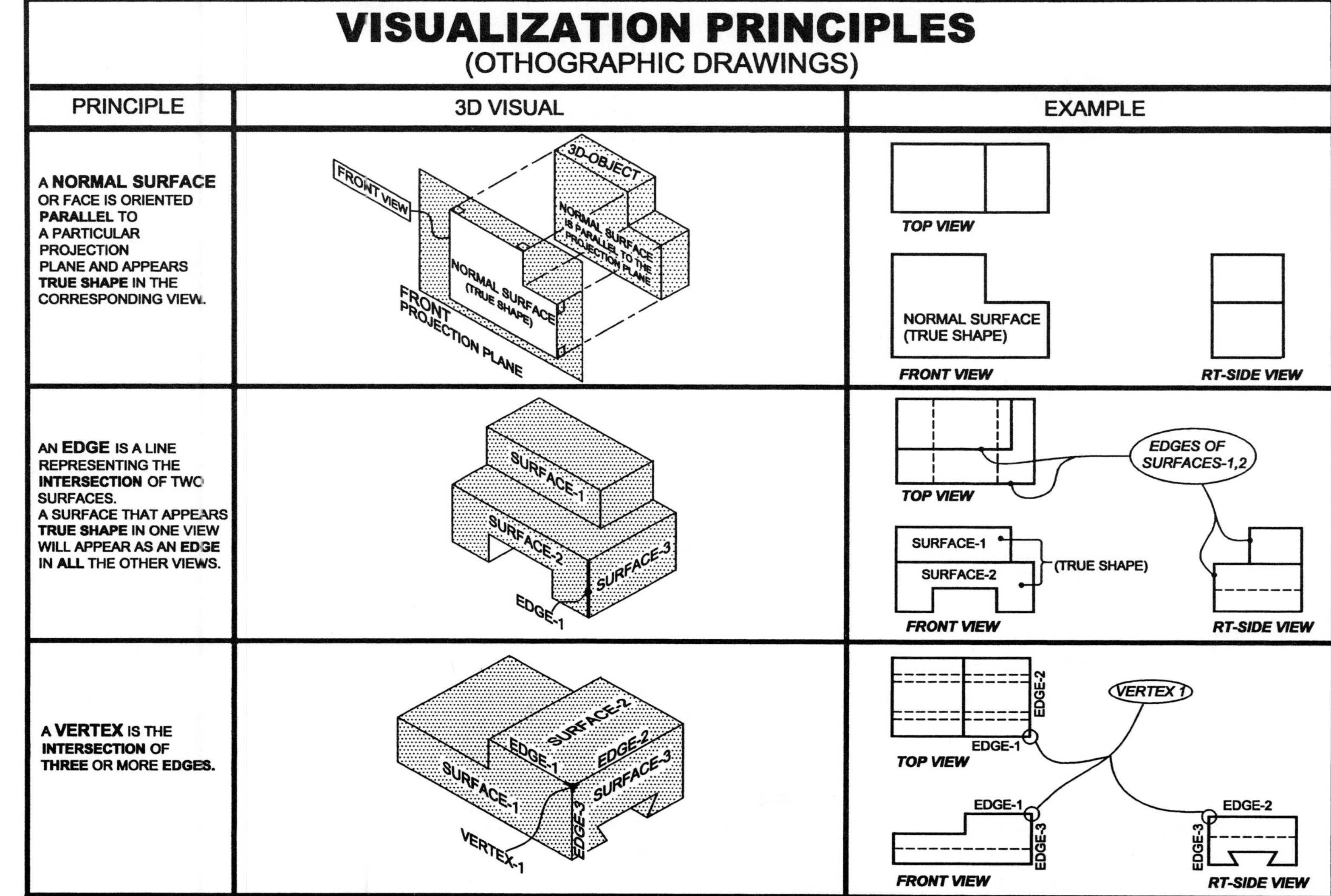

VISUALIZATION PRINCIPLES
(OTHOGRAPHIC DRAWINGS)

PRINCIPLE	3D VISUAL	EXAMPLE
A **NORMAL SURFACE** OR FACE IS ORIENTED **PARALLEL** TO A PARTICULAR PROJECTION PLANE AND APPEARS **TRUE SHAPE** IN THE CORRESPONDING VIEW.	FRONT VIEW 3D-OBJECT NORMAL SURFACE IS PARALLEL TO THE PROJECTION PLANE NORMAL SURFACE (TRUE SHAPE) FRONT PROJECTION PLANE	TOP VIEW NORMAL SURFACE (TRUE SHAPE) FRONT VIEW RT-SIDE VIEW
AN **EDGE** IS A LINE REPRESENTING THE **INTERSECTION** OF TWO SURFACES. A SURFACE THAT APPEARS **TRUE SHAPE** IN ONE VIEW WILL APPEAR AS AN **EDGE** IN **ALL** THE OTHER VIEWS.	SURFACE-1 SURFACE-2 SURFACE-3 EDGE-1	TOP VIEW EDGES OF SURFACES-1,2 SURFACE-1 SURFACE-2 (TRUE SHAPE) FRONT VIEW RT-SIDE VIEW
A **VERTEX** IS THE **INTERSECTION** OF **THREE** OR MORE **EDGES.**	SURFACE-2 EDGE-1 EDGE-2 SURFACE-3 SURFACE-1 EDGE-3 VERTEX-1	EDGE-2 VERTEX 1 EDGE-1 TOP VIEW EDGE-1 EDGE-3 EDGE-2 EDGE-3 FRONT VIEW RT-SIDE VIEW

TABLE 5–2 *(continued)*

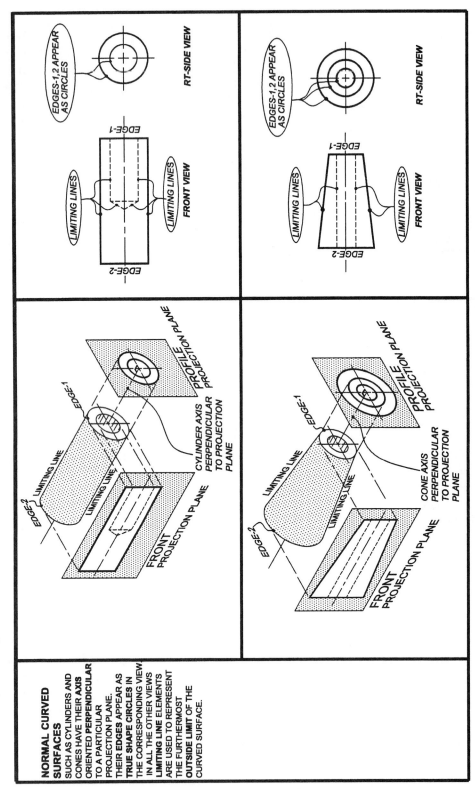

NORMAL CURVED SURFACES SUCH AS CYLINDERS AND CONES HAVE THEIR **AXIS** ORIENTED **PERPENDICULAR** TO A PARTICULAR PROJECTION PLANE. THEIR **EDGES** APPEAR AS **TRUE SHAPE CIRCLES** IN THE CORRESPONDING VIEW. IN ALL THE OTHER VIEWS **LIMITING LINE** ELEMENTS ARE USED TO REPRESENT THE FURTHERMOST **OUTSIDE LIMIT** OF THE CURVED SURFACE.	EDGE-2 LIMITING LINE EDGE-1 LIMITING LINE FRONT PROJECTION PLANE CYLINDER AXIS PERPENDICULAR TO PROJECTION PLANE PROFILE PROJECTION PLANE	LIMITING LINES EDGE-2 EDGE-1 LIMITING LINES FRONT VIEW EDGES-1,2 APPEAR AS CIRCLES RT-SIDE VIEW
	EDGE-2 LIMITING LINE EDGE-1 LIMITING LINE FRONT PROJECTION PLANE CONE AXIS PERPENDICULAR TO PROJECTION PLANE PROFILE PROJECTION PLANE	LIMITING LINES EDGE-2 EDGE-1 LIMITING LINES FRONT VIEW EDGES-1,2 APPEAR AS CIRCLES RT-SIDE VIEW

TABLE 5-2 *(continued)*

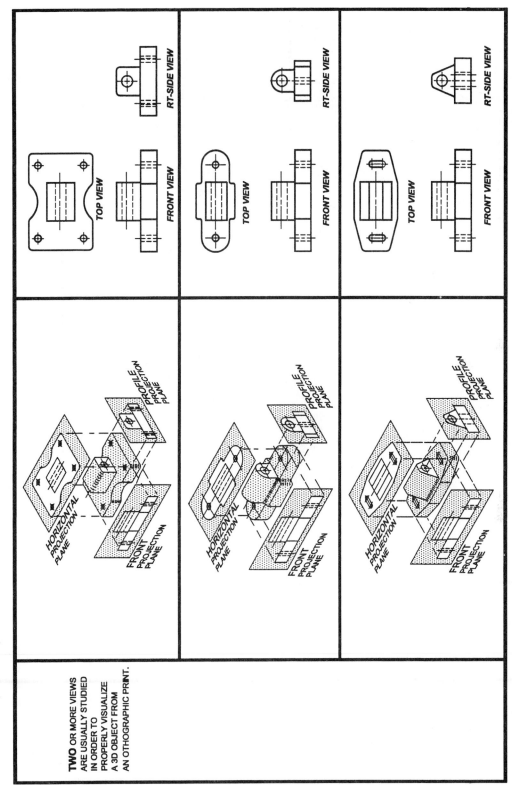

TWO OR MORE VIEWS ARE USUALLY STUDIED IN ORDER TO PROPERLY VISUALIZE A 3D OBJECT FROM AN OTHOGRAPHIC PRINT.	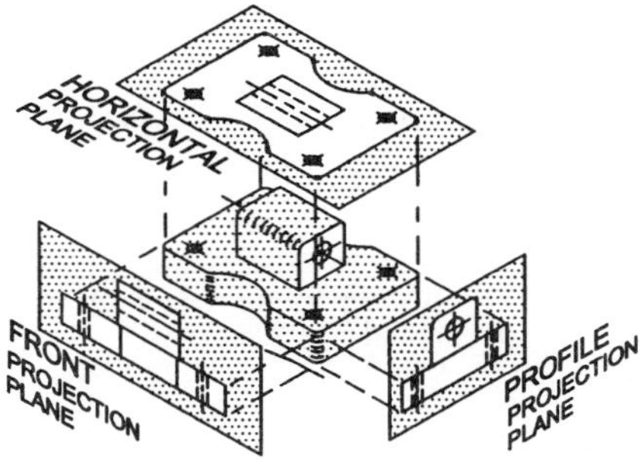 HORIZONTAL PROJECTION PLANE FRONT PROJECTION PLANE PROFILE PROJECTION PLANE	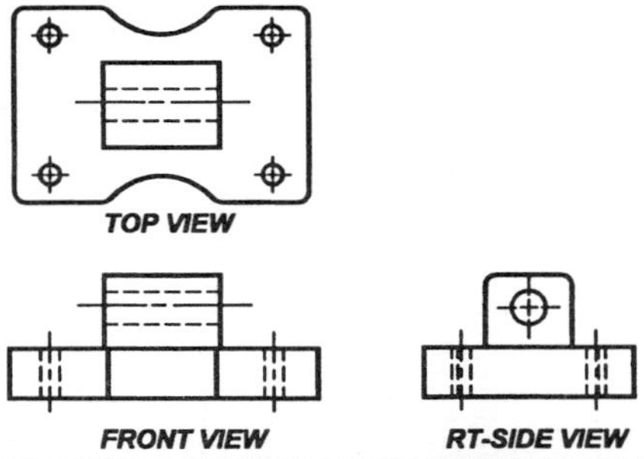 TOP VIEW FRONT VIEW RT-SIDE VIEW
	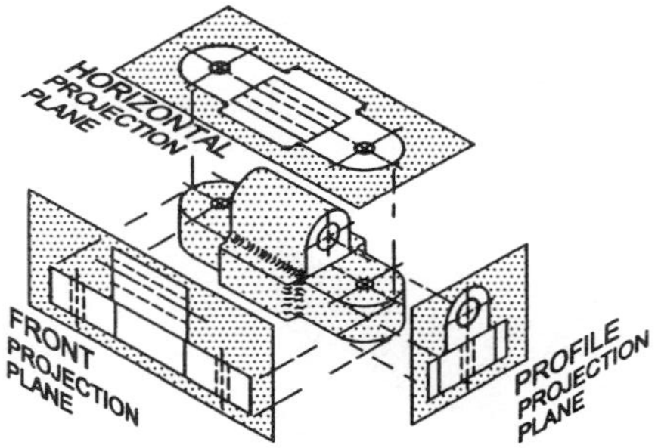 HORIZONTAL PROJECTION PLANE FRONT PROJECTION PLANE PROFILE PROJECTION PLANE	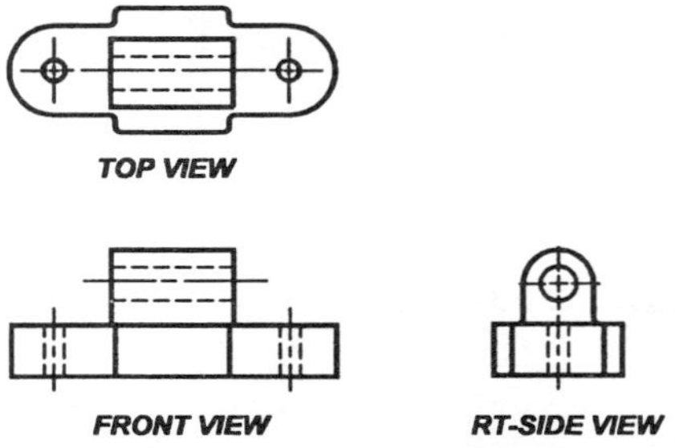 TOP VIEW FRONT VIEW RT-SIDE VIEW
	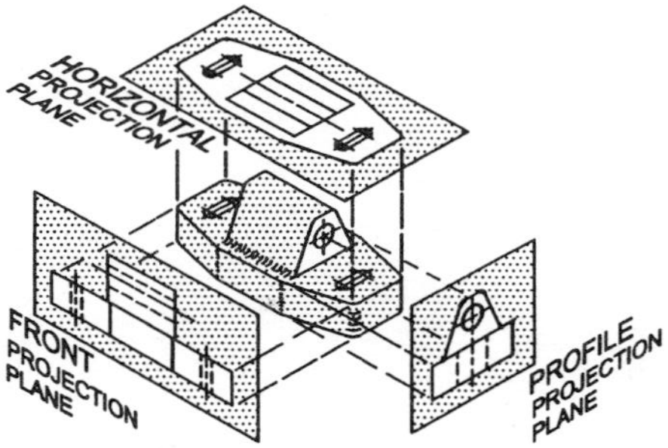 HORIZONTAL PROJECTION PLANE FRONT PROJECTION PLANE PROFILE PROJECTION PLANE	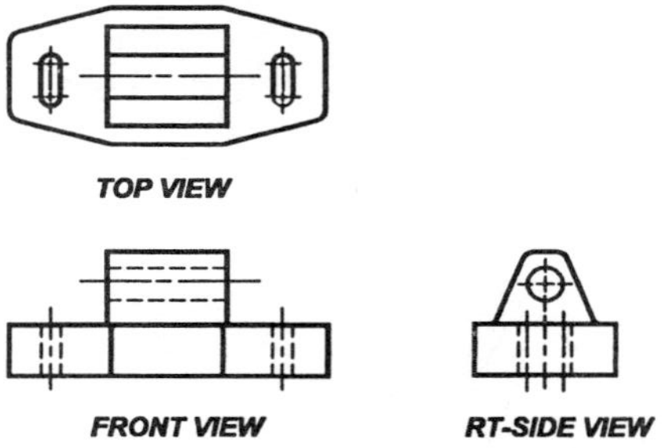 TOP VIEW FRONT VIEW RT-SIDE VIEW

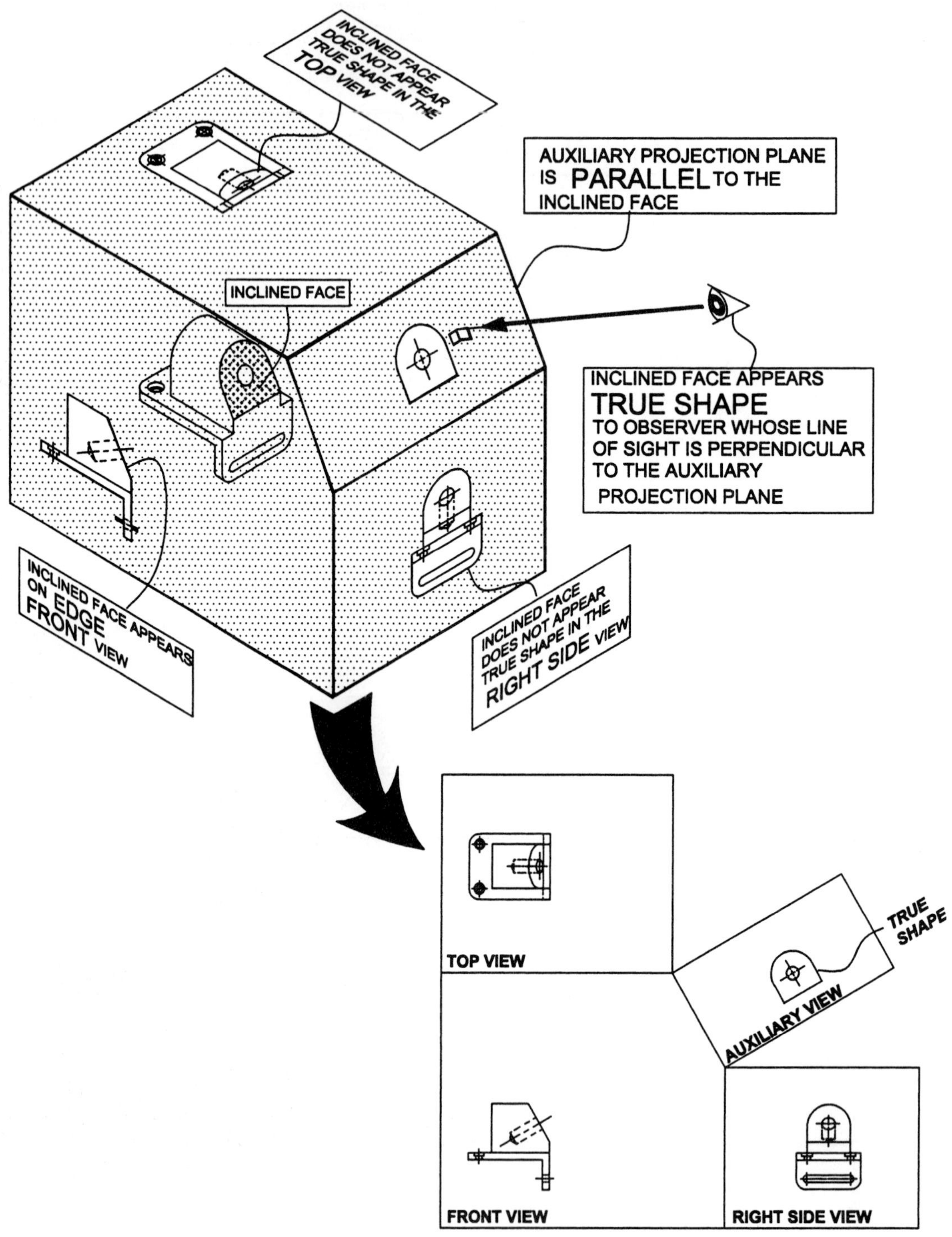

INCLINED FACE DOES NOT APPEAR TRUE SHAPE IN THE TOP VIEW
AUXILIARY PROJECTION PLANE IS PARALLEL TO THE INCLINED FACE
INCLINED FACE
INCLINED FACE APPEARS TRUE SHAPE TO OBSERVER WHOSE LINE OF SIGHT IS PERPENDICULAR TO THE AUXILIARY PROJECTION PLANE
INCLINED FACE APPEARS ON EDGE FRONT VIEW
INCLINED FACE DOES NOT APPEAR TRUE SHAPE IN THE RIGHT SIDE VIEW
TOP VIEW
TRUE SHAPE
AUXILIARY VIEW
FRONT VIEW
RIGHT SIDE VIEW

FIGURE 5–12 An auxiliary view shows the true shape of an inclined face.

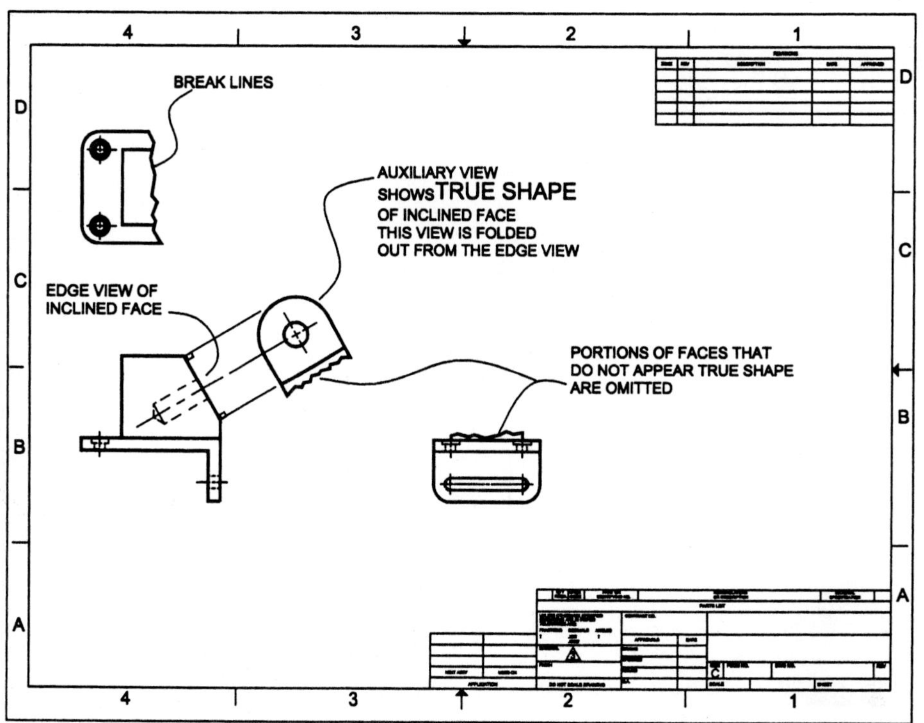

FIGURE 5–13 Practice of showing an auxiliary view in a mechanical drawing.

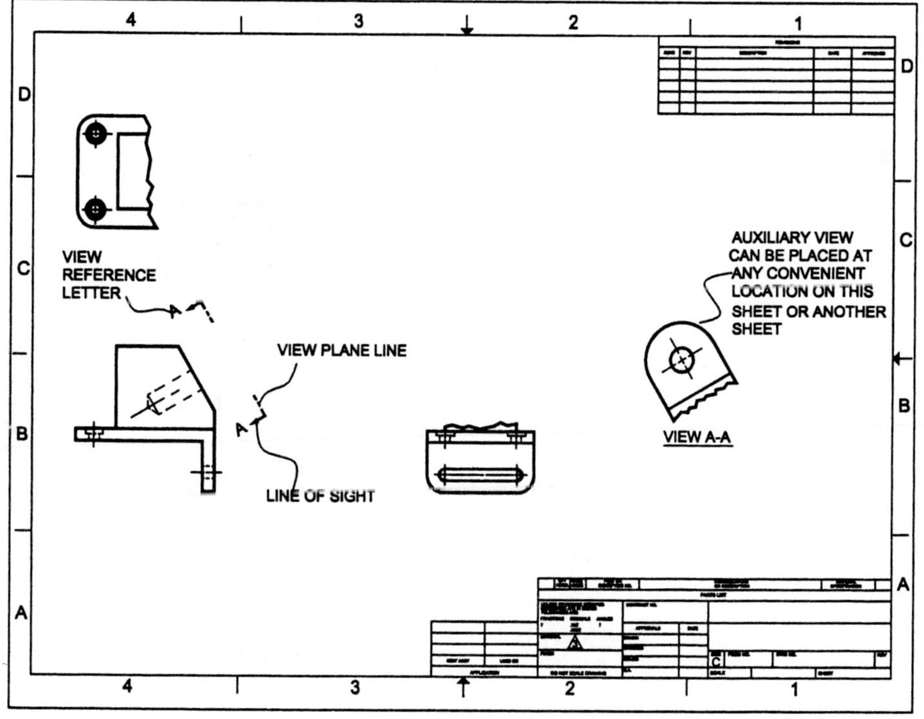

FIGURE 5–14 Using viewplane lines to place an auxiliary view at a convenient location.

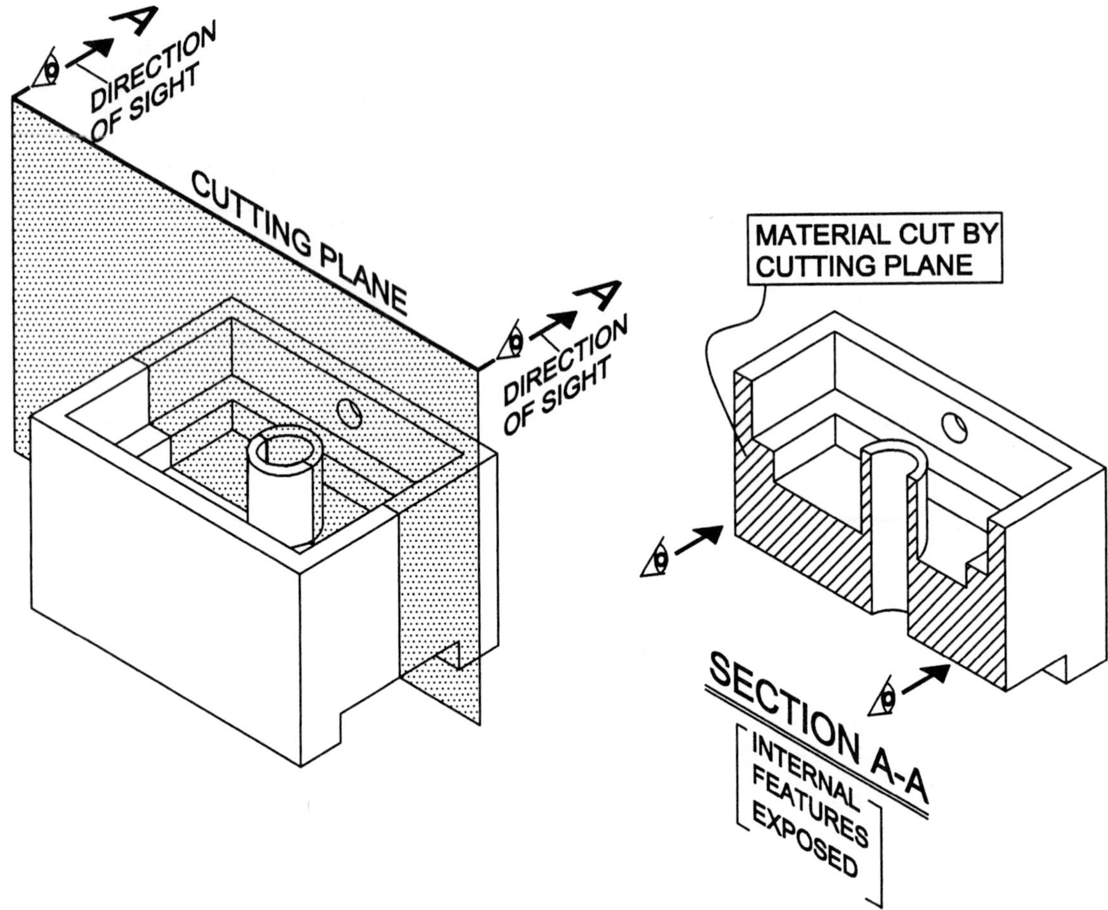

FIGURE 5–15 Exposing the internal features of an object via a section view.

Dimensioning Terminology

Units: Dimensioning units follow either the USA standards or the International system of units (SI).

USA

Units are expressed in inches and fractions of an inch. For numbers less than 1, a leading zero is used.

■ EXAMPLE 5–1

All dimension units are written with tolerances applied. The permitted variation must have the same number of significant digits following the decimal point as the specified dimension. ■■

TABLE 5–3

SECTIONING CONVENTIONS		
CONVENTION	3D-VISUAL	SECTIONING EXAMPLE
HIDDEN LINES ARE NOT SHOWN IN SECTION VIEWS		
SECTION VIEWS ARE ALWAYS DRAWN WITH CONTINUOUS BOUNDARIES.		CONTINUOUS SECTION BOUNDARY
TO AVOID GIVING A FALSE IMPRESSION OF THICKNESS, RIBS, WEBS AND GEAR TEETH ARE NOT SECTIONED	RIB	RIB NOT SECTIONED
	WEB OR GUSSET	WEB NOT SECTIONED
	SPOKE GEAR TOOTH	TOOTH AND SPOKE NOT SECTIONED REVOLVED SECTION OF SPOKE

■ EXAMPLE 5–2

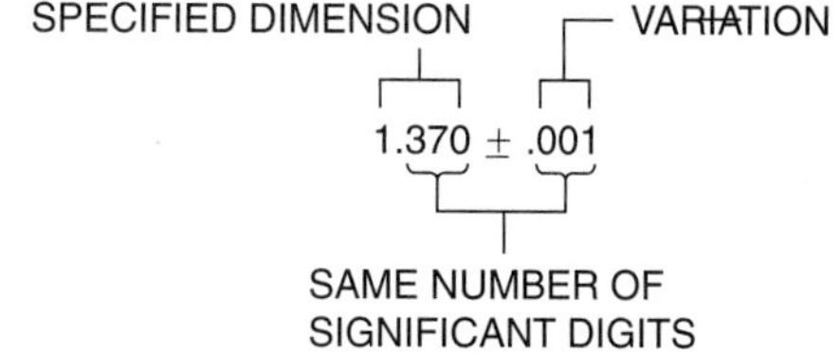

SI

Units are expressed in millimeters and fractions of a millimeter. For numbers less than 1, a leading zero is used. ■■

TABLE 5–4

LIST OF SECTION VIEWS			
SECTION TYPE	DESCRIPTION	3D-VISUAL	SECTIONING EXAMPLE
FULL	PART IS COMPLETELY CUT IN TWO BY A STRAIGHT CUTTING PLANE.	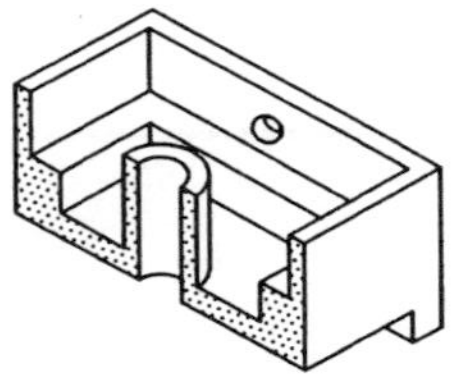	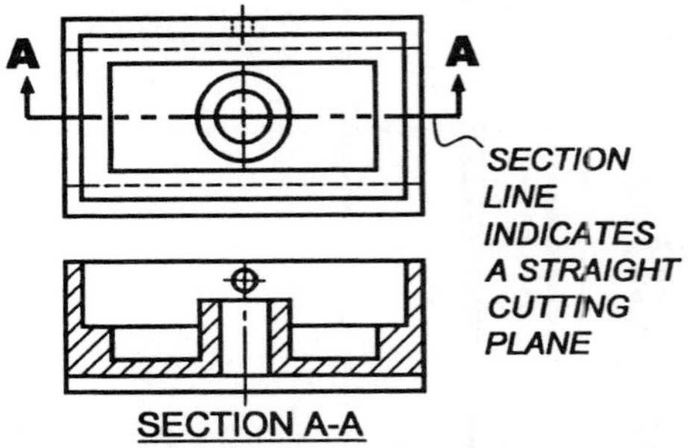 A A SECTION LINE INDICATES A STRAIGHT CUTTING PLANE SECTION A-A
OFFSET	PART IS COMPLETELY CUT IN HALF BY A BROKEN OR STAGGERED CUTTING PLANE.	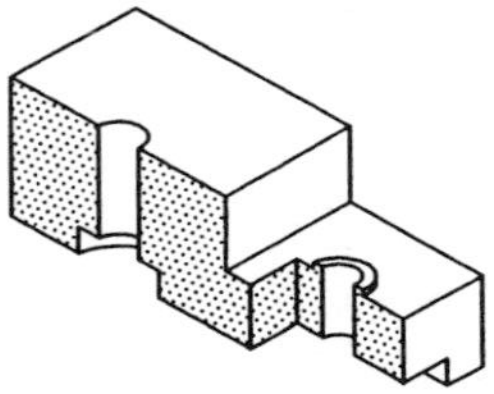	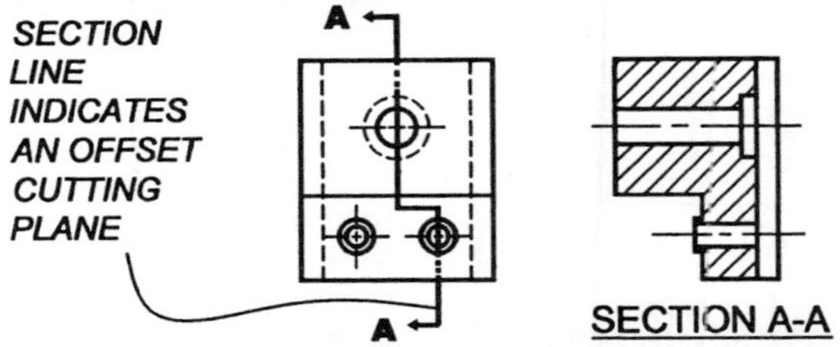 SECTION LINE INDICATES AN OFFSET CUTTING PLANE A A SECTION A-A
ALIGNED	PART IS CUT BY A BENT CUTTING PLANE. ALL SECTIONS ARE ALIGNED IN THE SAME VIEW SUCH THAT THEIR TRUE SHAPE IS SHOWN.	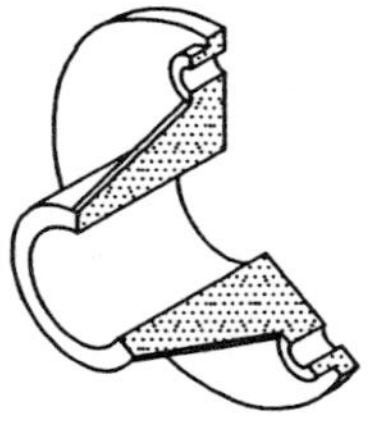	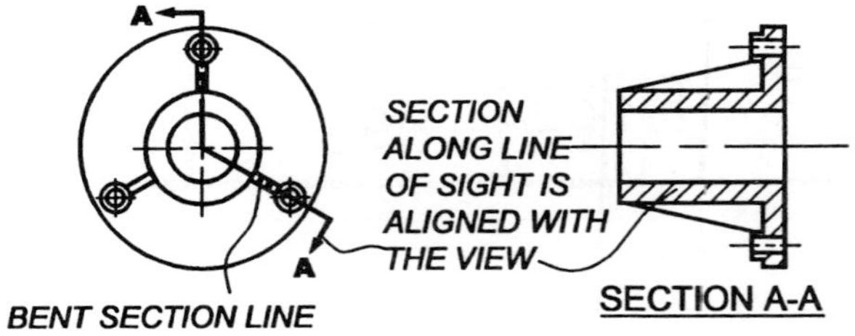 A A SECTION ALONG LINE OF SIGHT IS ALIGNED WITH THE VIEW BENT SECTION LINE SECTION A-A

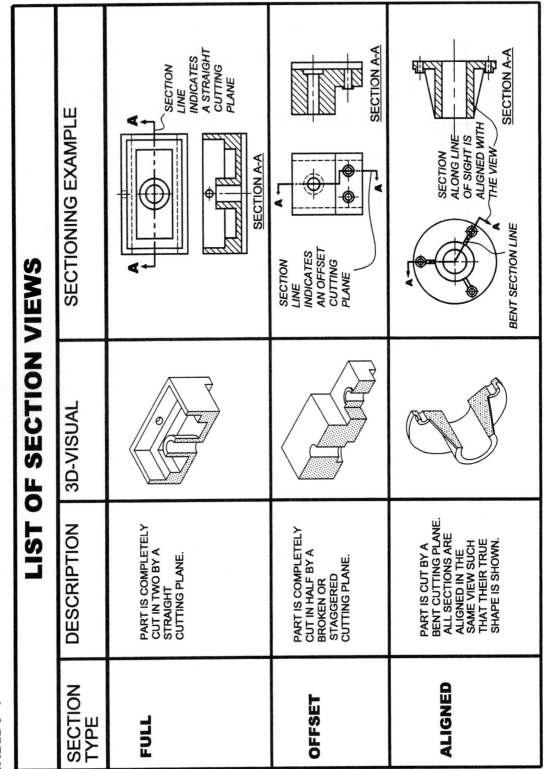

TABLE 5–4 *(continued)*

BROKEN OUT	A PORTION OF THE OBJECT IS BROKEN AWAY THUS EXPOSING THE INSIDE FEATURES. A SHORT BREAK LINE IS USED INSTEAD OF A SECTION LINE.	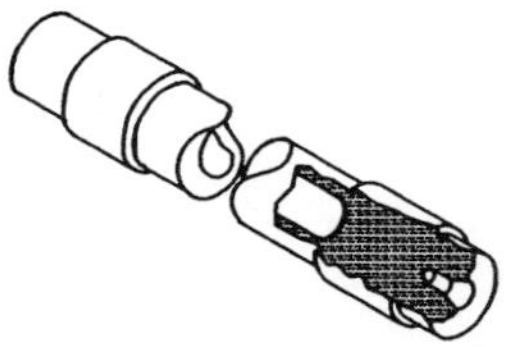	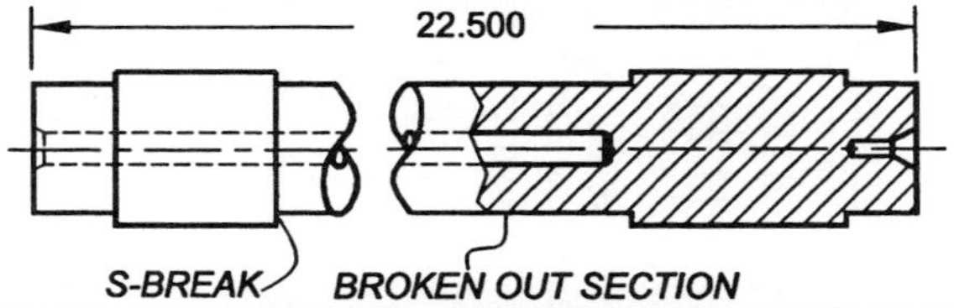 22.500 S-BREAK BROKEN OUT SECTION
PARTIAL	ONLY PART OF THE OBJECT IS CUT BY THE CUTTING PLANE.	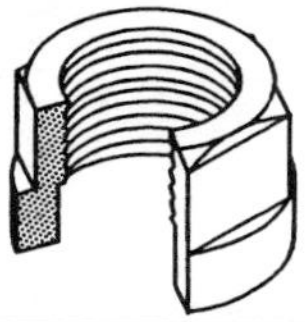	 A A SECTION A-A
REVOLVED	A RIB,WEB,GUSSET OR FLANGE THAT IS CONTINUOUS ALONG THE LENGTH OF THE OBJECT IS SECTIONED. THE TRUE SHAPE IS REVOLVED INTO THE VIEWING PLANE.	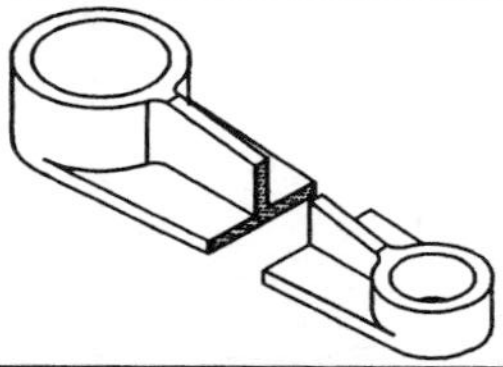	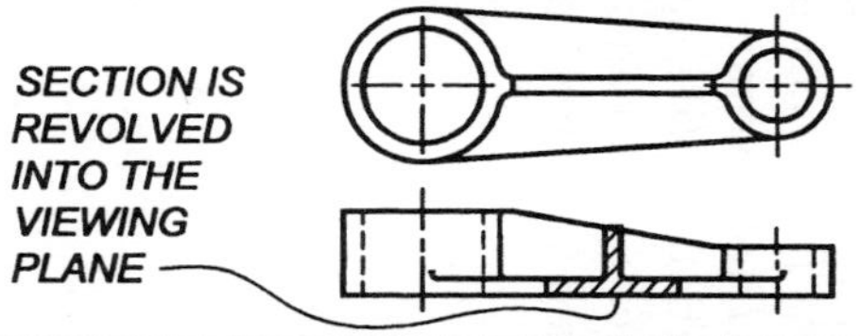 *SECTION IS REVOLVED INTO THE VIEWING PLANE*
REMOVED	THE SECTION IS DETACHED FROM THE PROJECTED VIEW AND PLACED ELSEWHERE. DETACHED SECTIONS CAN BE SHOWN AT A LARGER SCALE FOR CLARITY.	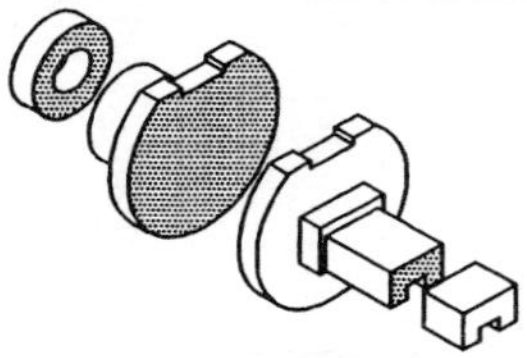	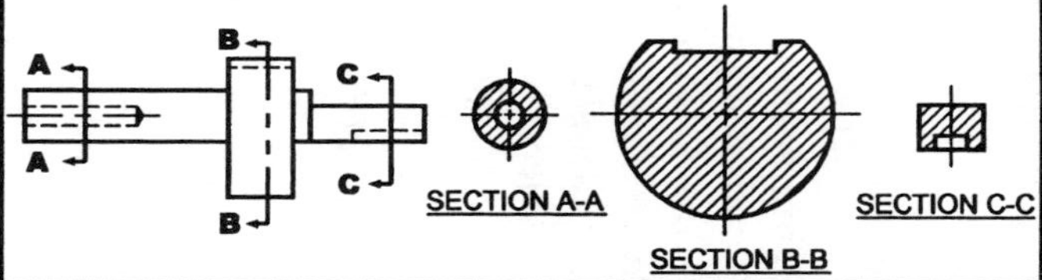 A A B B C C SECTION A-A SECTION B-B SECTION C-C
DETAIL	A PORTION OF THE OBJECT IS DISPLAYED AT A LARGER SCALE FOR CLARITY.		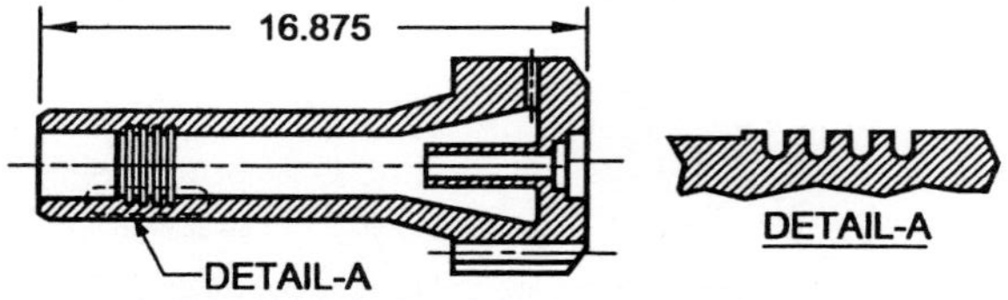 16.875 DETAIL-A DETAIL-A

■ EXAMPLE 5–3

0.7

SI (METRIC) UNITS HAVE LEADING ZEROS

All dimension units are written with tolerances applied. The permitted variation must have the same number of significant digits following the decimal point as the specified dimension.

Basic Size/Basic Dimension: A numerical value specifying the theoretical exact size, location, profile, or orientation of a geometrical feature of an object. Tolerance variations are applied to the basic dimension. A rectangle is used to identify a basic dimension. ■■

■ EXAMPLE 5–4

1.500——BASIC SIZE

Limits: The largest (upper limit) and smallest (lower limit) values permitted from the basic size. ■■

■ EXAMPLE 5–5

1.502——UPPER LIMIT
1.500——NOMINAL SIZE
1.498——LOWER LIMIT

Tolerance: The total permissible variation permitted in a dimension or the difference between the upper and lower limits. ■■

■ EXAMPLE 5–6

1.502——UPPER LIMIT
−1.498——LOWER LIMIT
.004——TOLERANCE

Bilateral/Unilateral: A way of expressing the limits in terms of + and − variations from the basic size ■■

■ EXAMPLE 5–7

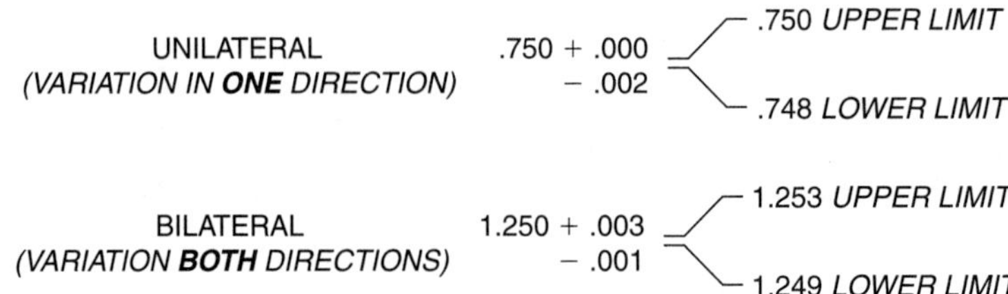

Allowance: Intentional minimum clearance (+allowance value) or maximum interference (−allowance value) assigned for mating parts.

Reference Dimension: In addition to the regular dimensions placed on a drawing, the draftsperson may also include important calculated dimensions. The idea is to reduce any additional size or location calculations that may be needed in the shop. Tolerances are not applied to reference dimensions. Parenthesis ( ) are used to indicate a reference dimension.

Nominal Size: The basic size of an object. Standard fractions are used when needed. Examples of nominal are stock size or thread diameter sizes.

Actual Size: The actual or final production size of a part.

Datum: A theoretically exact point, plane, or axis from which dimensions are taken. CNC part drawings utilize datums. This practice cuts down on tolerance error buildups and ensures that parts are machined with a maximum of accuracy.

Table 5–5 identifies the different types of dimensioning systems encountered in mechanical drawings.

A comprehensive listing of standard dimensioning practices is given in Table 5–6. The reader should be acquainted with these practices in order to properly interpret mechanical part drawings. ■■

TABLE 5–5

DIMENSIONING SYSTEMS

SYSTEM	DESCRIPTION	EXAMPLE
UNIDIRECTIONAL	ALL DIMENSION AND NOTES ARE ALIGNED HORIZONTALLY IN THE DRAWING. LOCATION AND SIZE DIMENSIONS ARE TAKEN FROM DATUMS.	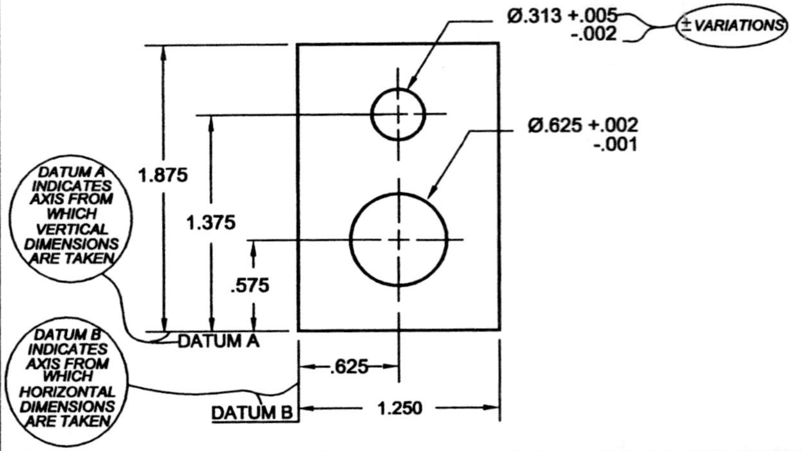 Ø.313 +.005 -.002; ± VARIATIONS; Ø.625 +.002 -.001; 1.875; 1.375; .575; DATUM A; .625; 1.250; DATUM B; DATUM A INDICATES AXIS FROM WHICH VERTICAL DIMENSIONS ARE TAKEN; DATUM B INDICATES AXIS FROM WHICH HORIZONTAL DIMENSIONS ARE TAKEN
COORDINATE	THE LOCATION OF FEATURES SUCH AS HOLES ARE GIVEN IN TERMS OF HORIZONTAL (X) AND VERTICAL (Y) COORDINATE DISTANCES FROM A DATUM ORIGIN. 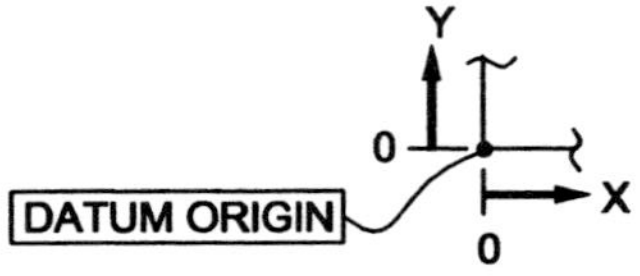 Y; X; 0; 0; DATUM ORIGIN	HOLE CHART (below); ± VARIATIONS; A; B; C; 0; 0

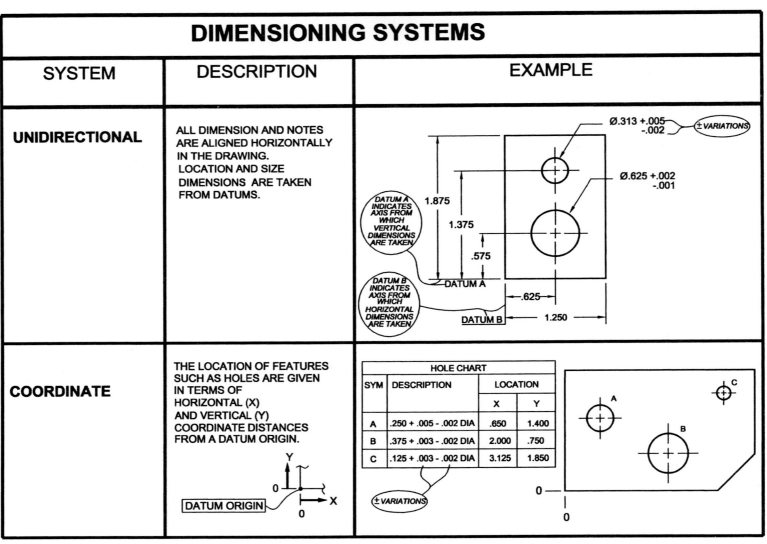

HOLE CHART

SYM	DESCRIPTION	LOCATION	
		X	Y
A	.250 + .005 - .002 DIA	.650	1.400
B	.375 + .003 - .002 DIA	2.000	.750
C	.125 + .003 - .002 DIA	3.125	1.850

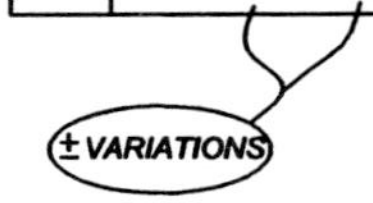

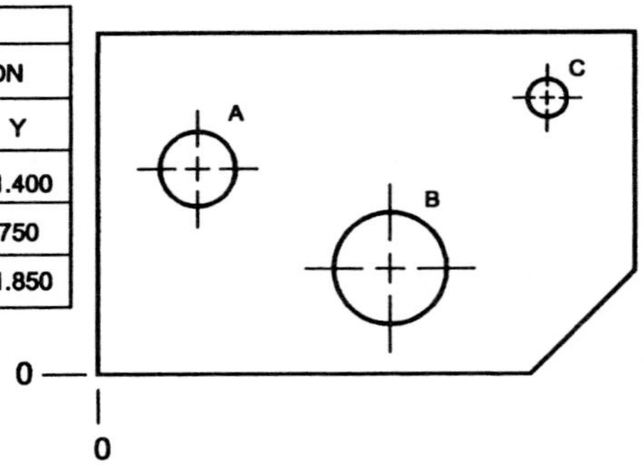

TABLE 5–5 *(continued)*

ARROWLESS

NO DIMENSION LINES ARE USED. EACH HORIZONTAL (X) AND VERTICAL(Y) LOCATION DIMENSION IS REFERENCED FROM A DATUM ORIGIN AND PLACED AT THE CORRESPONDING EXTENSION LINE.

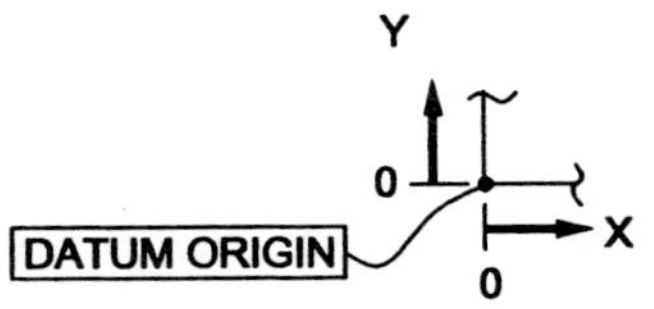

HOLE CHART		
SYM	DESCRIPTION	QNTY
A	.125 +.004 -.002 DIA	6
B	.187 +.003 -.002 DIA	3
C	.250+.005 -.002 DIA	2

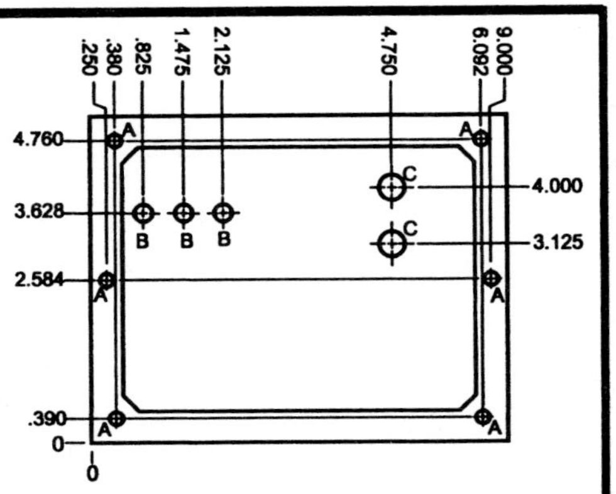

TABLE OR CHART

A TABLE OR CHART SYSTEM IS USED TO DESCRIBE A FAMILY OF PARTS HAVING THE SAME SHAPE BUT DIFFERENT DIMENSIONS.

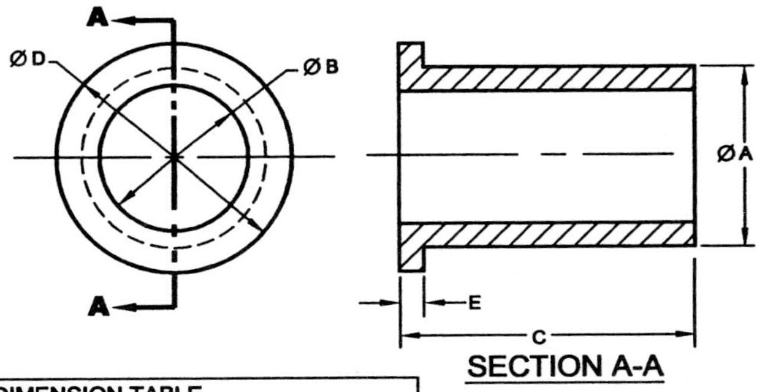

DIMENSION TABLE					
DASH NO.	A	B	C	D	E
-1	1.876 1.874	1.501 1.499	3.002 2.998	2.380 2.370	.255 (UPPER LIMIT) .245 (LOWER LIMIT)
-2	1.676 1.674	1.476 1.474	2.752 2.748	2.130 2.120	.255 .245
-3	1.676 1.674	1.476 1.474	2.502 2.498	2.125 2.125	.193 .183

TABLE 5–6

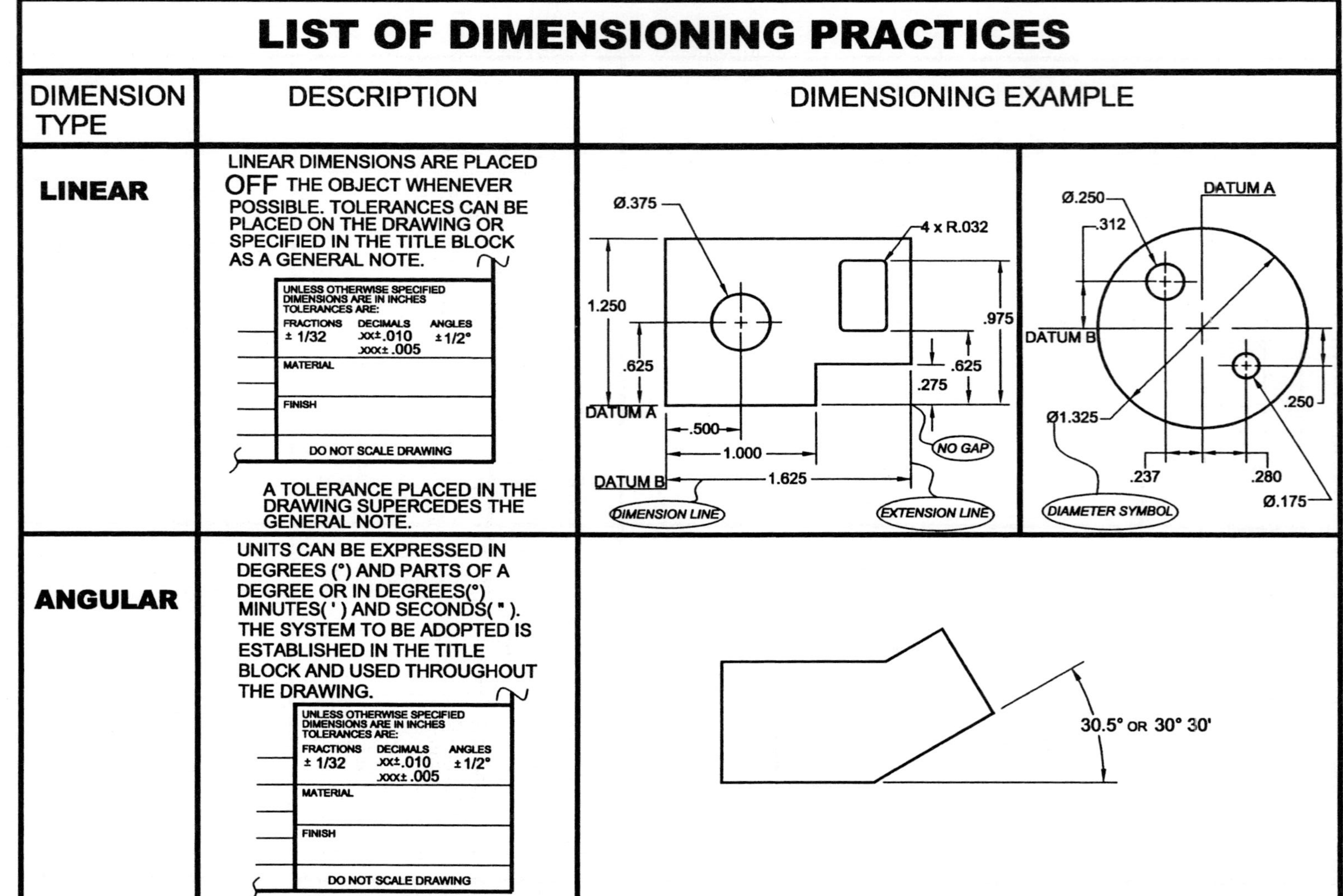

LIST OF DIMENSIONING PRACTICES		
DIMENSION TYPE	DESCRIPTION	DIMENSIONING EXAMPLE
LINEAR	LINEAR DIMENSIONS ARE PLACED OFF THE OBJECT WHENEVER POSSIBLE. TOLERANCES CAN BE PLACED ON THE DRAWING OR SPECIFIED IN THE TITLE BLOCK AS A GENERAL NOTE. UNLESS OTHERWISE SPECIFIED DIMENSIONS ARE IN INCHES TOLERANCES ARE: FRACTIONS ± 1/32 · DECIMALS .XX± .010, .XXX± .005 · ANGLES ± 1/2° MATERIAL FINISH DO NOT SCALE DRAWING A TOLERANCE PLACED IN THE DRAWING SUPERCEDES THE GENERAL NOTE.	
ANGULAR	UNITS CAN BE EXPRESSED IN DEGREES (°) AND PARTS OF A DEGREE OR IN DEGREES(°) MINUTES(') AND SECONDS("). THE SYSTEM TO BE ADOPTED IS ESTABLISHED IN THE TITLE BLOCK AND USED THROUGHOUT THE DRAWING. UNLESS OTHERWISE SPECIFIED DIMENSIONS ARE IN INCHES TOLERANCES ARE: FRACTIONS ± 1/32 · DECIMALS .XX± .010, .XXX± .005 · ANGLES ± 1/2° MATERIAL FINISH DO NOT SCALE DRAWING	

TABLE 5–6 *(continued)*

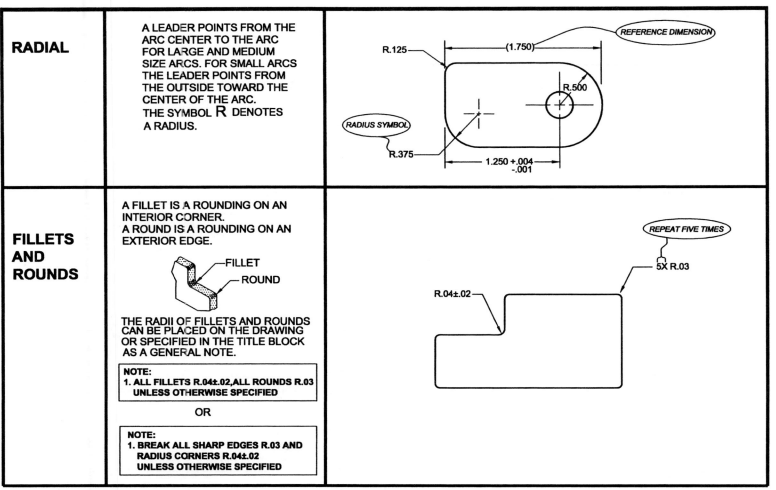

RADIAL	A LEADER POINTS FROM THE ARC CENTER TO THE ARC FOR LARGE AND MEDIUM SIZE ARCS. FOR SMALL ARCS THE LEADER POINTS FROM THE OUTSIDE TOWARD THE CENTER OF THE ARC. THE SYMBOL R DENOTES A RADIUS.	REFERENCE DIMENSION R.125 (1.750) R.500 RADIUS SYMBOL R.375 1.250 +.004 -.001
FILLETS AND ROUNDS	A FILLET IS A ROUNDING ON AN INTERIOR CORNER. A ROUND IS A ROUNDING ON AN EXTERIOR EDGE. FILLET ROUND THE RADII OF FILLETS AND ROUNDS CAN BE PLACED ON THE DRAWING OR SPECIFIED IN THE TITLE BLOCK AS A GENERAL NOTE. **NOTE:** **1. ALL FILLETS R.04±.02,ALL ROUNDS R.03 UNLESS OTHERWISE SPECIFIED** OR **NOTE:** **1. BREAK ALL SHARP EDGES R.03 AND RADIUS CORNERS R.04±.02 UNLESS OTHERWISE SPECIFIED**	REPEAT FIVE TIMES 5X R.03 R.04±.02

TABLE 5–6 *(continued)*

CHAMFERS	A CHAMFER IS A BEVELED EDGE MADE TO BREAK A SHARP CORNER.  CHAMFER METHOD-1: USE A LEADER POINTING TO THE CHAMFER. METHOD-2: USE SIDE/ANGLE DIMENSIONS. METHOD-3: USE SIDE/SIDE DIMENSIONS. METHOD-4: USE A GENERAL NOTE IN THE TITLE BLOCK. **REMOVE ALL BURRS AND BREAK SHARP EDGES 1/2 RADIUS OR 45° CHAMFER** *or* **TOLERANCE AND MACHINING NOTES** **(UNLESS OTHERWISE SPECIFIED)** **RADIUS OR CHAMFER ALL EDGES — .03 MAX** **FILLET RADII — .04±.02**	**METHOD-1** 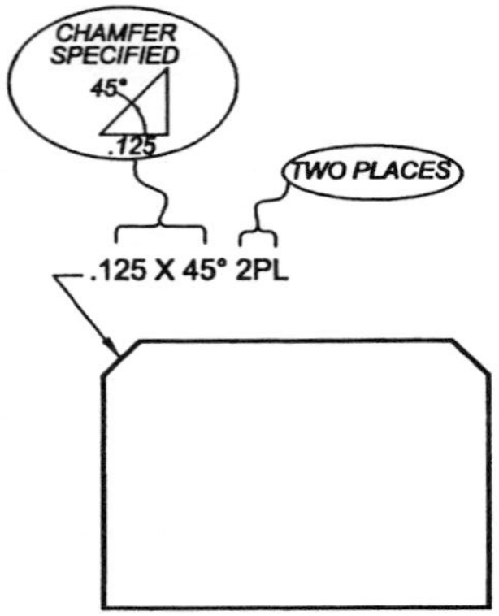 CHAMFER SPECIFIED 45° .125 TWO PLACES .125 X 45° 2PL	**METHODS-2 & 3** 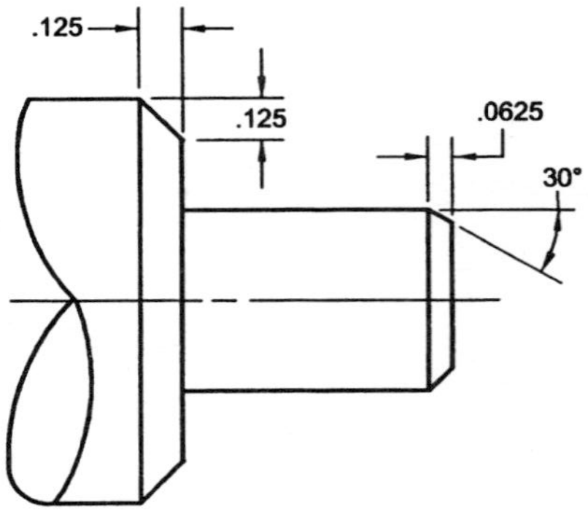 .125 .125 .0625 30°
FLAT TAPERS	A FLAT TAPER IS A MACHINED ANGLE OR BEVEL. METHOD-1: USE TRADITIONAL SIDE/ANGLE DIMENSIONS METHOD-2: USE THE CURRENT ASME Y14.5M-1994 FLAT TAPER SYMBOL. 	**METHOD-1** (TRADITIONAL) 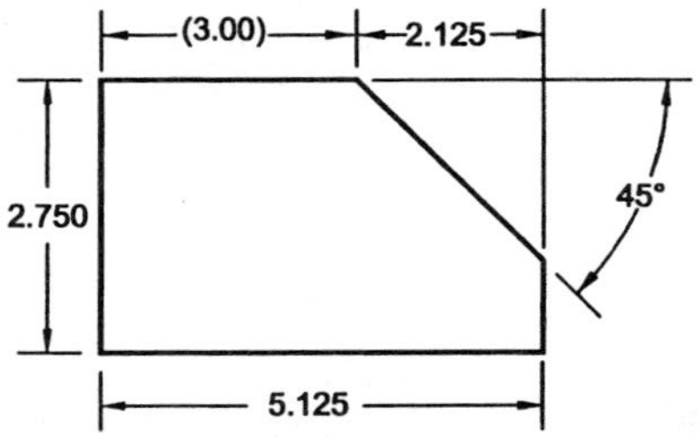 (3.00) 2.125 2.750 45° 5.125	**METHOD-2** (CURRENT ASME Y14.5M 1994) 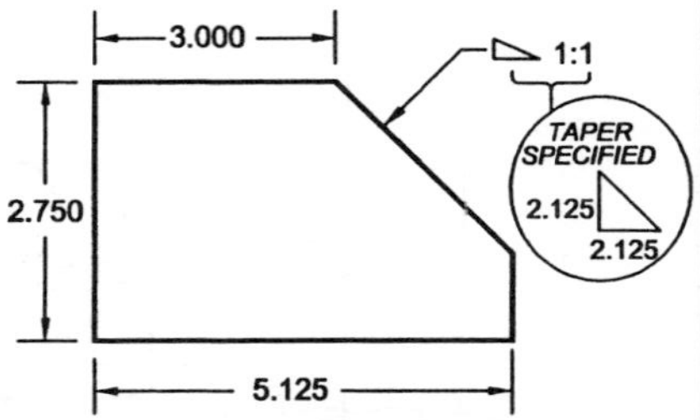 3.000 1:1 TAPER SPECIFIED 2.125 2.125 2.750 5.125

TABLE 5-6 *(continued)*

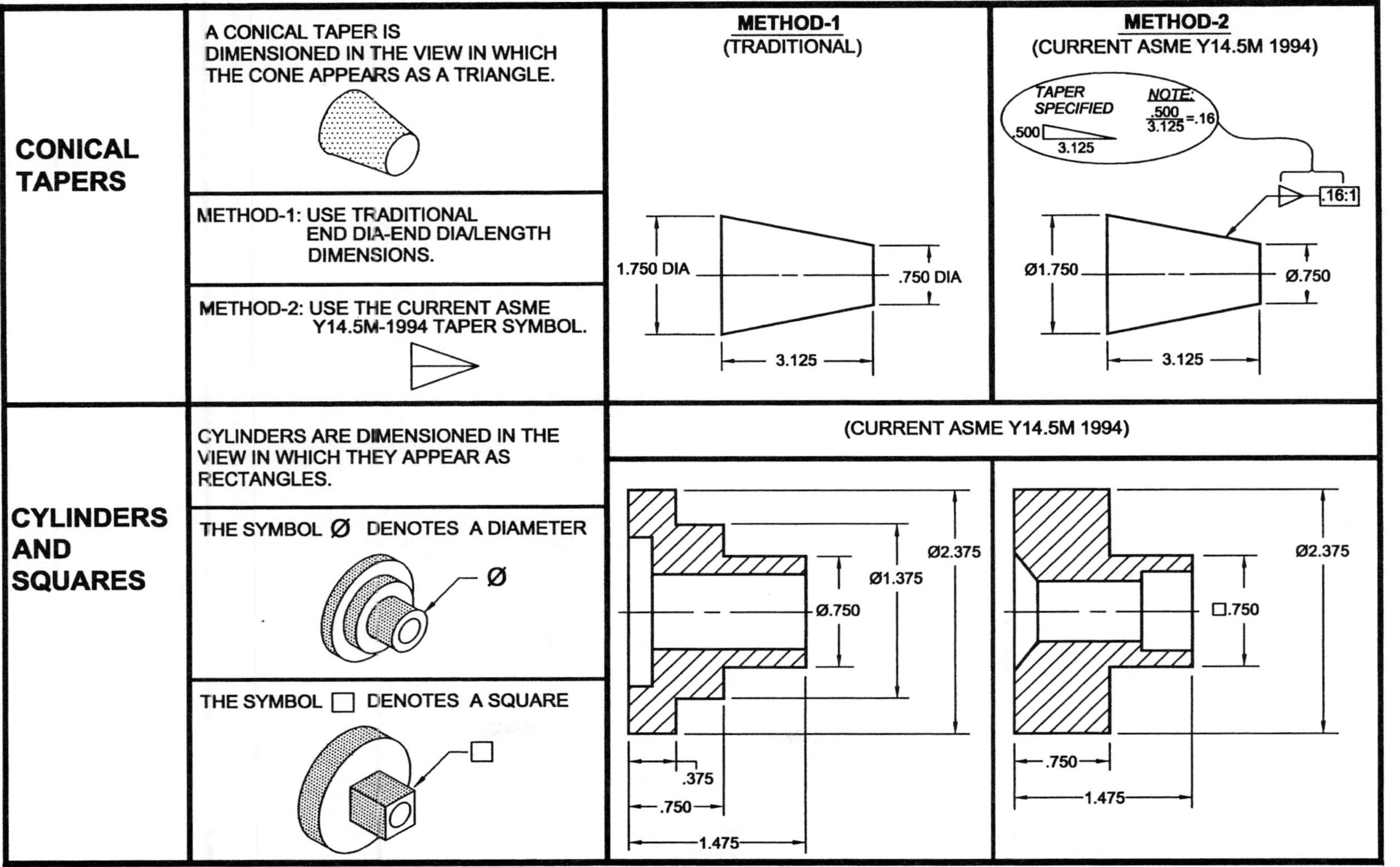

CONICAL TAPERS
A CONICAL TAPER IS DIMENSIONED IN THE VIEW IN WHICH THE CONE APPEARS AS A TRIANGLE.
METHOD-1: USE TRADITIONAL END DIA-END DIA/LENGTH DIMENSIONS.
METHOD-2: USE THE CURRENT ASME Y14.5M-1994 TAPER SYMBOL.
METHOD-1
(TRADITIONAL)
1.750 DIA
.750 DIA
3.125
METHOD-2
(CURRENT ASME Y14.5M 1994)
TAPER SPECIFIED
.500
3.125
NOTE:
.500 / 3.125 = .16
.16:1
Ø1.750
Ø.750
3.125
CYLINDERS AND SQUARES
CYLINDERS ARE DIMENSIONED IN THE VIEW IN WHICH THEY APPEAR AS RECTANGLES.
THE SYMBOL Ø DENOTES A DIAMETER
Ø
THE SYMBOL □ DENOTES A SQUARE
□
(CURRENT ASME Y14.5M 1994)
Ø2.375
Ø1.375
Ø.750
.375
.750
1.475
Ø2.375
□.750
.750
1.475

TABLE 5–6 *(continued)*

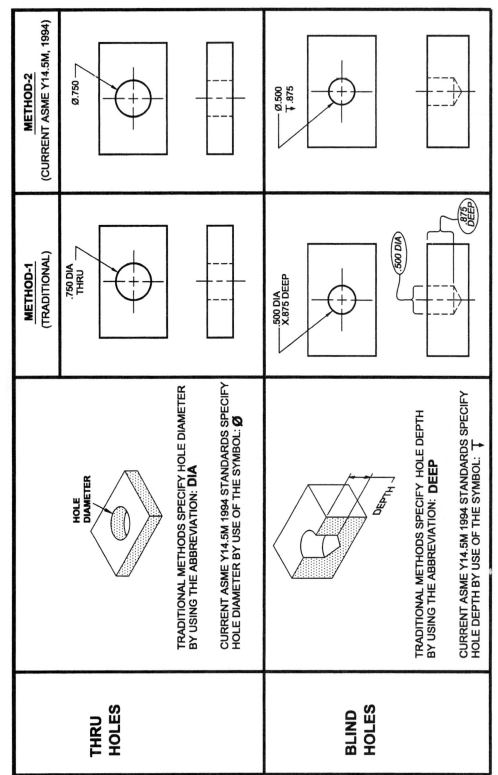

		METHOD-1 (TRADITIONAL)	**METHOD-2** (CURRENT ASME Y14.5M, 1994)
THRU HOLES	HOLE DIAMETER TRADITIONAL METHODS SPECIFY HOLE DIAMETER BY USING THE ABBREVIATION: **DIA** CURRENT ASME Y14.5M 1994 STANDARDS SPECIFY HOLE DIAMETER BY USE OF THE SYMBOL: **Ø**	.750 DIA THRU	Ø.750
BLIND HOLES	DEPTH TRADITIONAL METHODS SPECIFY HOLE DEPTH BY USING THE ABBREVIATION: **DEEP** CURRENT ASME Y14.5M 1994 STANDARDS SPECIFY HOLE DEPTH BY USE OF THE SYMBOL: ↧	.500 DIA X.875 DEEP .500 DIA .875 DEEP	Ø.500 ↧.875

TABLE 5–6 *(continued)*

		METHOD-1 (TRADITIONAL)	METHOD-2 (CURRENT ASME Y14.5M, 1994)
COUNTER-BORE & SPOTFACE HOLES	COUNTERBORE SPOTFACE TRADITIONAL METHODS SPECIFY A COUNTERBORE BY USE OF THE ABREVIATION: **C'BORE** A SPOTFACE IS INDICATED BY USING THE ABBREVIATION: **SP FACE** CURRENT ASME Y14.5M 1994 STANDARDS SPECIFY A COUNTERBORE OR SPOTFACE BY USE OF THE SYMBOL: ⌴	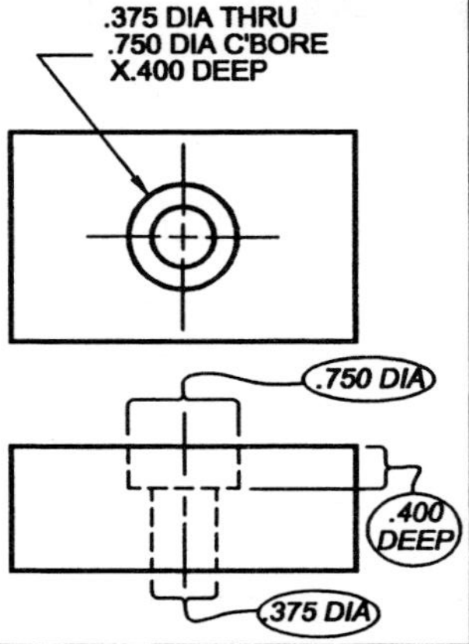 .375 DIA THRU .750 DIA C'BORE X.400 DEEP .750 DIA .400 DEEP .375 DIA	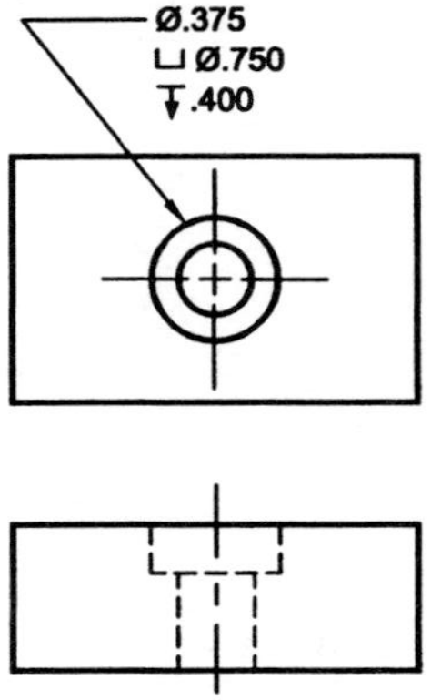 Ø.375 ⌴ Ø.750 ↧.400
COUNTER-SINK HOLES	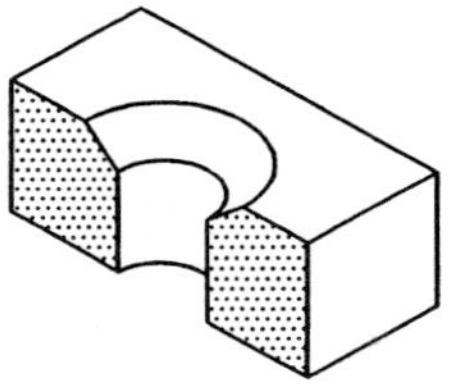 TRADITIONAL METHODS SPECIFY A COUNTERSINK BY USING THE ABBREVIATION: **C'SINK** CURRENT ASME Y14.5M 1994 STANDARDS SPECIFY A COUNTERSINK BY USE OF THE SYMBOL: ∨	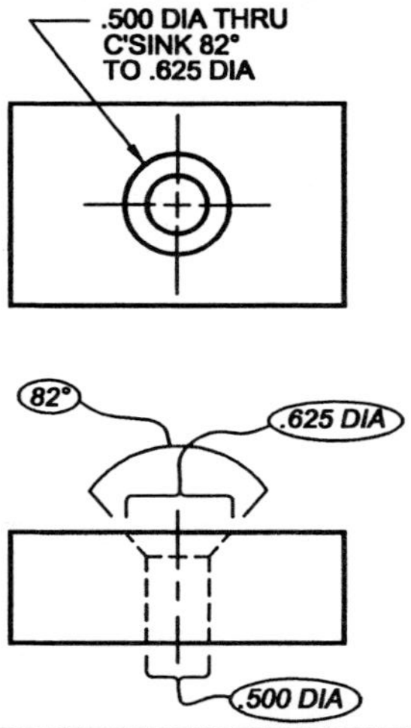 .500 DIA THRU C'SINK 82° TO .625 DIA 82° .625 DIA .500 DIA	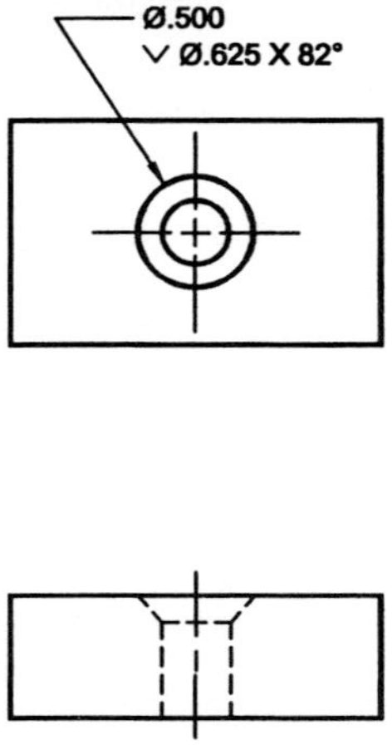 Ø.500 ∨ Ø.625 X 82°

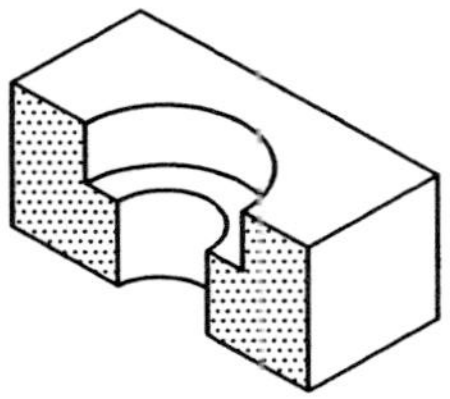

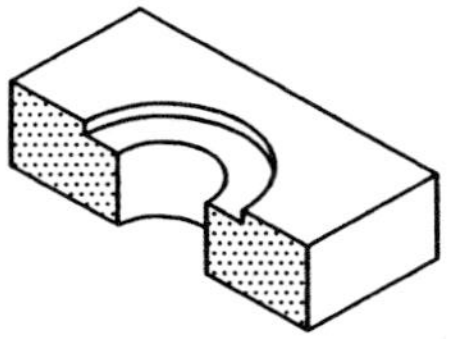

TABLE 5–6 *(continued)*

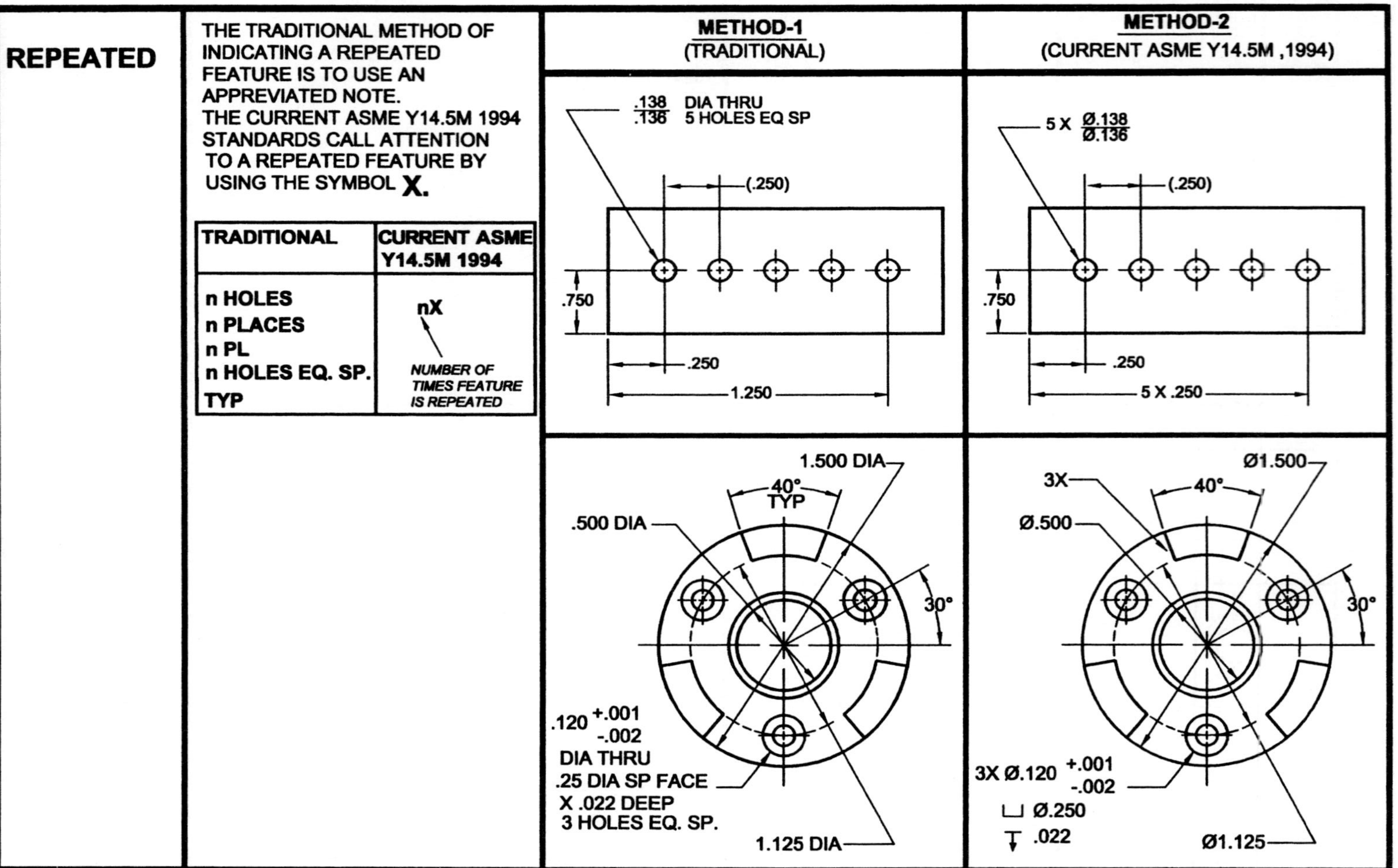
REPEATED
THE TRADITIONAL METHOD OF INDICATING A REPEATED FEATURE IS TO USE AN APPREVIATED NOTE.
THE CURRENT ASME Y14.5M 1994 STANDARDS CALL ATTENTION TO A REPEATED FEATURE BY USING THE SYMBOL X.
TRADITIONAL
CURRENT ASME Y14.5M 1994
n HOLES
n PLACES
n PL
n HOLES EQ. SP.
TYP
nX
NUMBER OF TIMES FEATURE IS REPEATED
METHOD-1
(TRADITIONAL)
.138 / .136 DIA THRU
5 HOLES EQ SP
(.250)
.750
.250
1.250
1.500 DIA
40° TYP
.500 DIA
30°
.120 +.001 -.002
DIA THRU
.25 DIA SP FACE
X .022 DEEP
3 HOLES EQ. SP.
1.125 DIA
METHOD-2
(CURRENT ASME Y14.5M ,1994)
5 X Ø.138 / Ø.136
(.250)
.750
.250
5 X .250
Ø1.500
3X
40°
Ø.500
30°
3X Ø.120 +.001 -.002
⌴ Ø.250
↧ .022
Ø1.125

TABLE 5–6 *(continued)*

SPHERICAL RADIUS	THE SYMBOL **SR** IS USED TO DENOTE A SPHERICAL RADIUS. 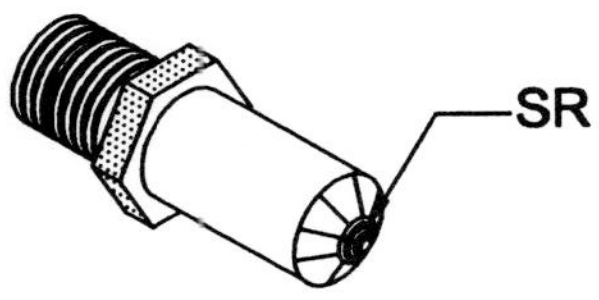 SR	(CURRENT ASME Y14.5M 1994) 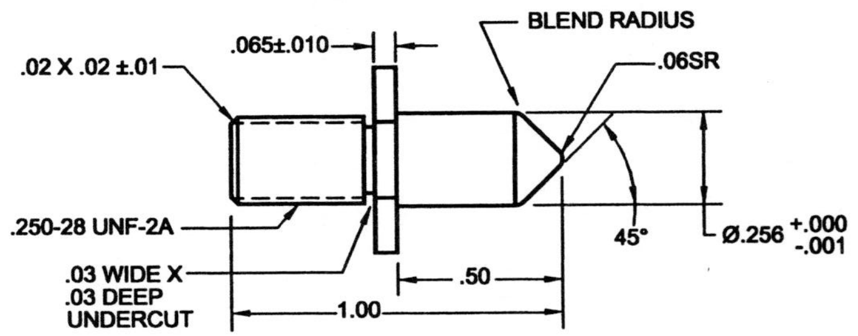 .02 X .02 ±.01 .065±.010 BLEND RADIUS .06SR .250-28 UNF-2A .03 WIDE X .03 DEEP UNDERCUT 45° Ø.256 +.000 -.001 .50 1.00
ORIGIN	THE ORIGIN SYMBOL ○ IS USED TO CLEARLY INDICATE THE ORIGIN FROM WHICH A DIMENSION IS TAKEN.	(CURRENT ASME Y14.5M 1994) 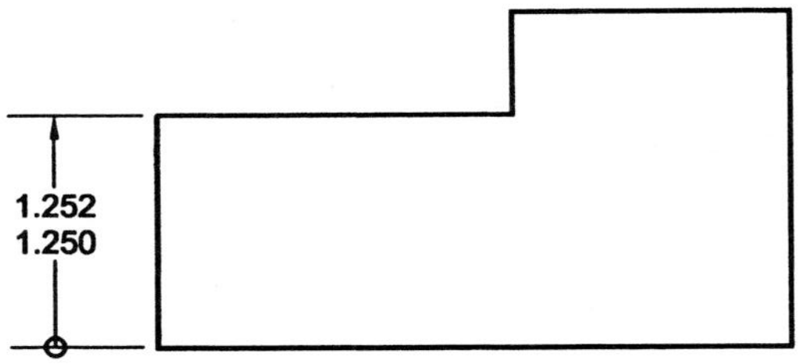 1.252 1.250

5–11 READING THREADS AND THREAD NOTES

Threads are used for joining two or more parts together, for adjusting the position of one part with respect to another and for transmitting power. The American National Standards Institute document ANSI Y14.6-1978 specifies the conventions to be followed when specifying threads in mechanical drawings.

Thread Terminology

The thread terminology presented here relates to the general illustration of internal and external threads as shown in Figure 5–16.

Axis: The centerline of the thread cylinder.
Body: The unthreaded portion of the screw shaft.
Chamfer: The angular relief machined on the last thread to allow easier engagement with the internal threads of the mating part.
Crest: The top of the thread teeth for external threads, and the bottom of the thread teeth for internal threads.

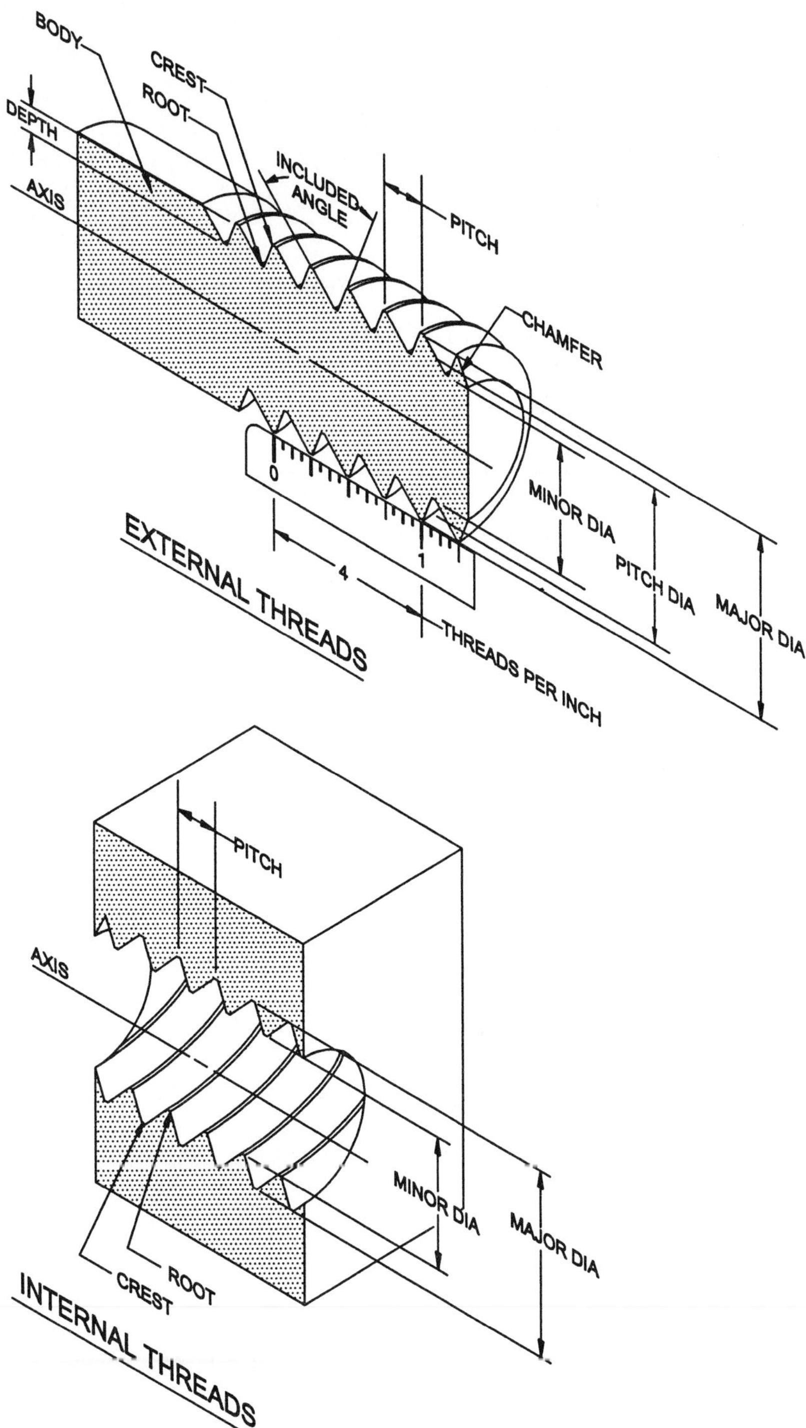

FIGURE 5–16 General illustration of internal and external threads.

Root: The bottom of the thread teeth for external threads, and the top of the thread teeth for internal threads.

Thread Depth: The perpendicular distance between the crest and root of the thread.

Die: The cutting tool used to cut external threads.

Included Angle: The angle made between threads by the thread cutting tool.

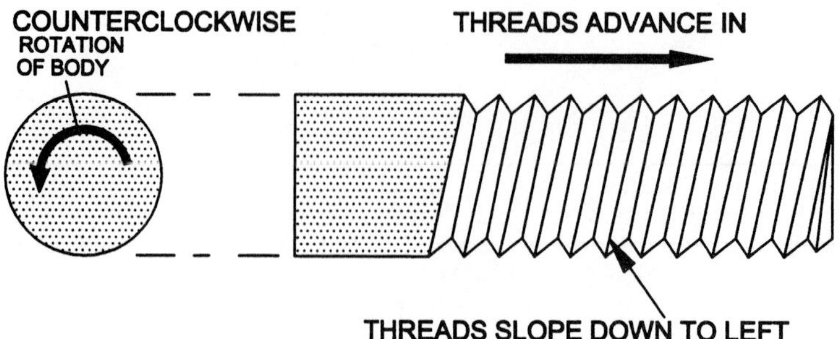

FIGURE 5–17 Left-handed threads.

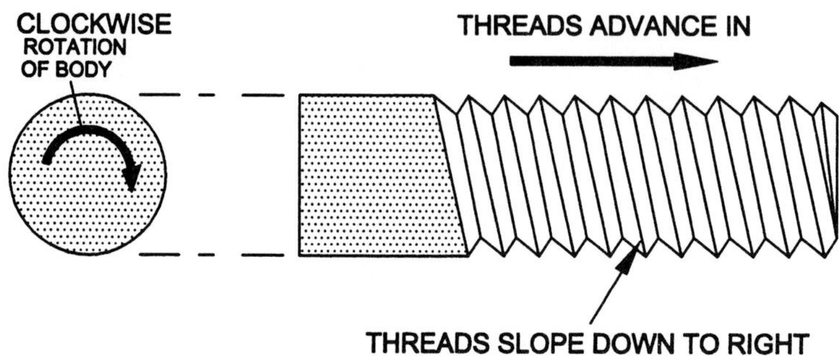

FIGURE 5–18 Right-handed threads.

Lead: The distance the thread travels along its axis during one complete revolution.

Left-Handed Threads: Threads cut so that they slope down to the left. They engage by rotating the thread shaft counterclockwise when viewed toward the mating thread. Left-handed threads are used in such applications as turnbuckles and are designated as LH in drawings. See Figure 5–17.

Right-handed Threads: Threads cut so that they slope down to the right. They engage by rotating the thread shaft clockwise when viewed toward the mating thread. If no specific designation is given in a drawing, the thread is assumed to be right-handed. Refer to Figure 5–18.

Major Diameter: The largest diameter on external or internal thread.

Minor Diameter: The smallest diameter on external or internal thread.

Pitch: The distance between two adjacent thread crests or roots. For single start threads pitch is given as:

PITCH = 1″/THREADS PER INCH

Pitch Diameter: The diameter of an imaginary cylinder that is located midway between the major and minor diameters.

Single Threads: Advance a distance of one pitch (1P) for every 360-degree revolution of the thread shaft. This thread is the most common and is used to transmit substantial pressure and power. See Figure 5–19.

Double Threads: Advance two pitches (2P) per 360-degree revolution of the thread shaft. Refer to Figure 5–20.

Triple Threads: Advance a distance of three pitches (3P) per 360-degree revolution of the thread shaft. Double and triple threads are used for cases where quick movement, not power, is needed. They appear in applications like valve stems and drafting instruments. See Figure 5–21.

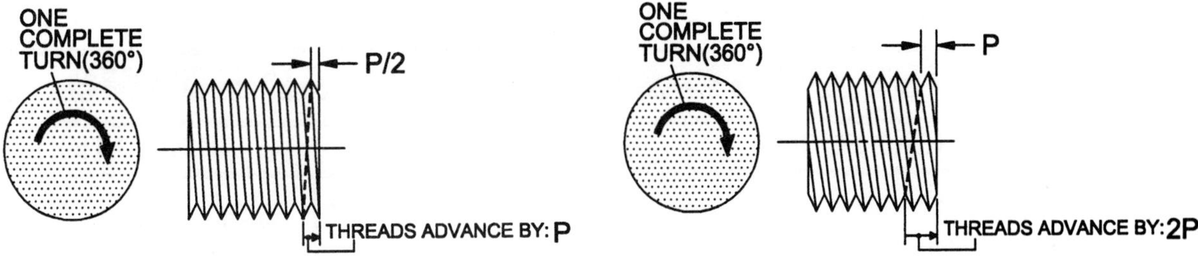

FIGURE 5–19 Single threads.

FIGURE 5–20 Double threads.

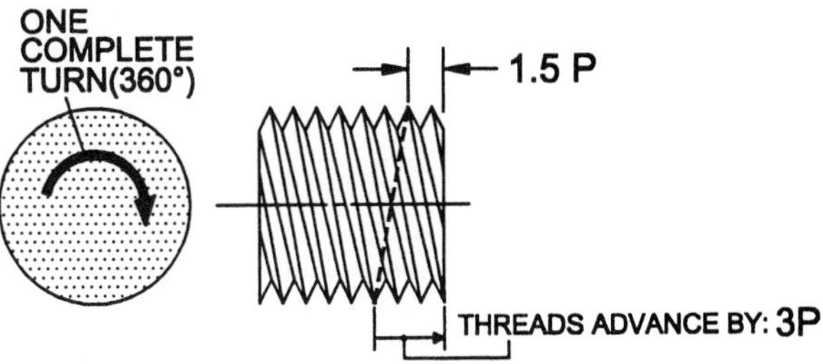

FIGURE 5–21 Triple threads.

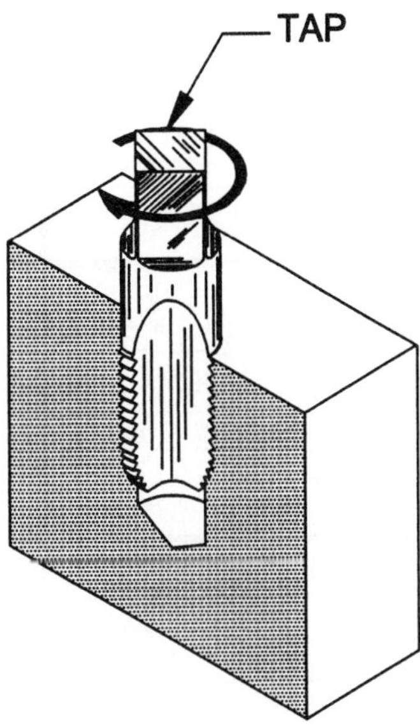

FIGURE 5–22 A tap cuts internal threads in a drilled hole.

Tap Drill: A drill used to machine a hole in a part prior to creating internal threads.

Tap: A tap is a thread cutting tool used to cut internal threads as shown in Figure 5–22. A complete table of tap and tap drills is available from such references as *Machinery's Handbook.*

Thread Form: The shape of the threads cut into the shaft or hole. The typical thread forms that appear on mechanical drawings for CNC operations are listed in Table 5–7.

Screw Thread Series: The different pitch versions of the same thread form comprise a thread series. A list of the screw thread series for the unified national thread form is given in Table 5–8.

Table 5–9 provides a listing of the dimensioning practices used to specify screw threads in mechanical drawings.

TABLE 5-7

LIST OF STANDARD THREAD FORMS

THREAD FORM	DESCRIPTION	APPLICATION
UNIFIED NATIONAL ROUND **(UNR)**	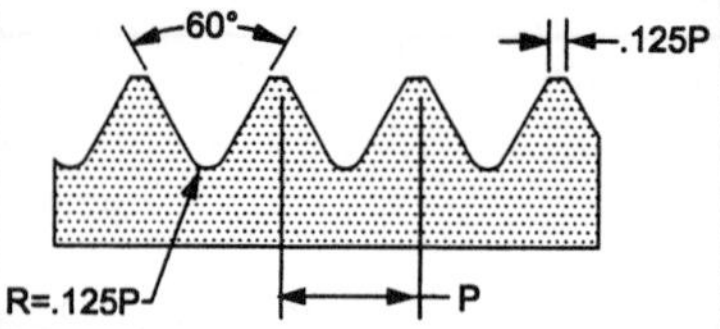 60° .125P R=.125P P	UNIFIED NATIONAL ROUND IS THE CURRENT STANDARD IN THE USA AND CANADA FOR FASTENING AND ADJUSTING PARTS.
AMERICAN NATIONAL **(N)**	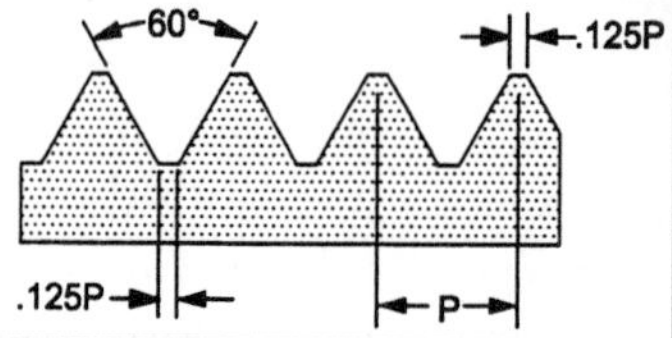 60° .125P .125P P	AMERICAN NATIONAL THREADS ARE STRONGER THAN SHARP V THREADS. THE AMERICAN NATIONAL SERIES IS USED FOR FASTENING AND JOINING.
METRIC **(M)**	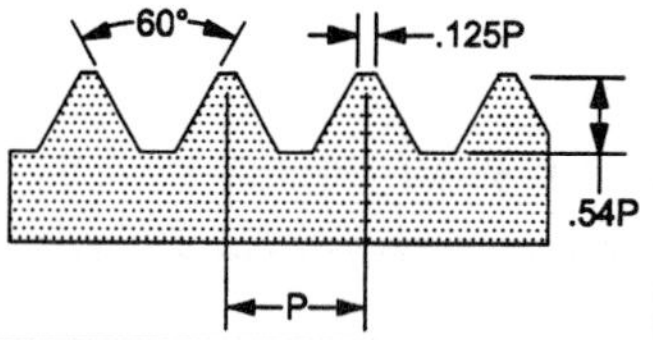 60° .125P .54P P	METRIC THREADS ARE USED OUTSIDE THE USA AND CANADA. THEY CONFORM TO THE ISO STANDARDS AND ARE USED FOR FASTENING AND ADJUSTING PARTS.
ACME	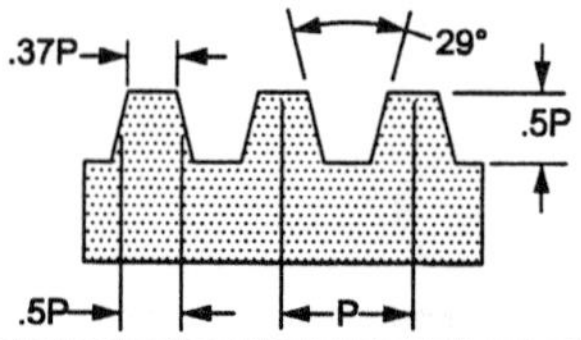 .37P 29° .5P .5P P	ACME THREADS HAVE REPLACED SQUARE THREADS FOR TRANSMITTING POWER. APPLICATIONS INCLUDE LEAD SCREWS, CLAMPS AND ENGINES.
BUTRESS	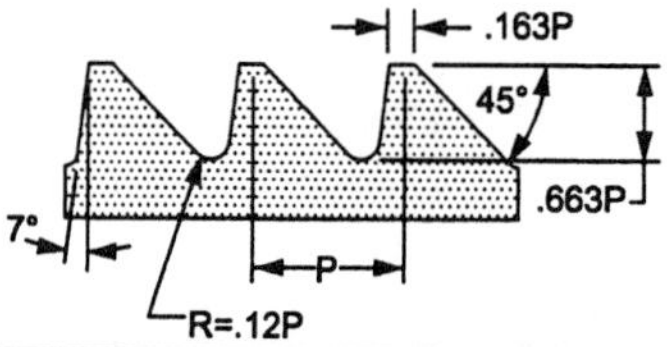 .163P 45° .663P 7° P R=.12P	BUTRESS THREADS ARE USED TO TRANSMIT POWER IN ONE DIRECTION ONLY. APPLICATIONS INCLUDE HYDRAULIC JACKS AND THE BREACH BLOCKS OF LARGE GUNS.
AMERICAN NATIONAL STANDARD TAPER	DWG TAPER NOTE: 1 IN 16 ON DIA; MEANING: 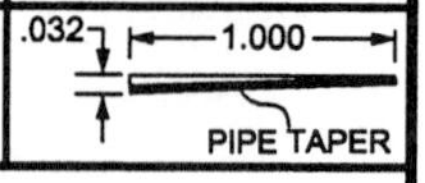 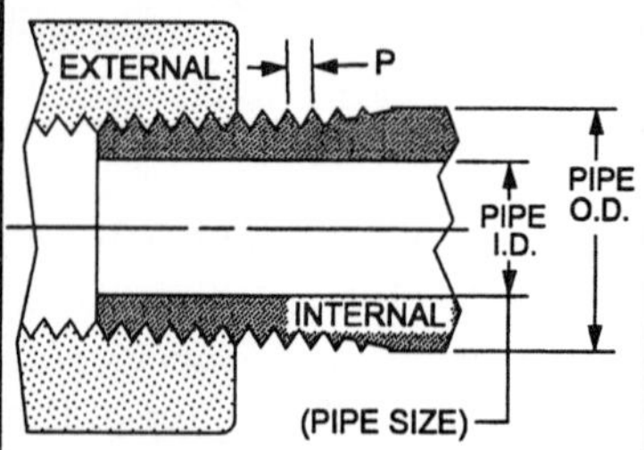 .032 1.000 PIPE TAPER; EXTERNAL P INTERNAL PIPE I.D. PIPE O.D. (PIPE SIZE)	AMERICAN NATIONAL STANDARD TAPER ARE USED ON PIPES AND PIPE FITTINGS. REGULAR THREADS ARE USED FOR STANDARD PLUMBING APPLICATIONS AND DO NOT CREATE PRESSURE TIGHT JOINTS. DRYSEAL THREADS CREATE PRESSURE TIGHT JOINTS AND ARE USED IN AUTOMOTIVE, HYDRAULIC AND REFRIGERATION PARTS.

TABLE 5-8

SCREW THREAD SERIES		
GRADE	DESIGNATION	APPLICATION
COARSE	**UNC**	USED FOR CAST IRON, SOFT METALS WHEN RAPID ASSEMBLY IS DESIRED.
FINE	**UNF** *OR* **NF**	USED FOR AEROSPACE AND AUTOMOTIVE PARTS WHEN A LARGE TIGHENTING FORCE IS REQUIRED.
FINE	**UNEF** *OR* **NEF**	FOR APPLICATIONS WHERE THE LENGTH OF THREAD ENGAGEMENT IS SHORT AND HIGH STRESSES MUST BE WITHSTOOD.
CONSTANT PITCH	**4UN,6UN ... , 32UN** (*THREADS PER INCH*)	CONSTANT PITCH THREADS HAVE THE SAME PITCH SIZE REGARDLESS OF THE THREAD DIAMETER. THEY ARE USED FOR LARGE DIAMETER AND HIGH PRESSURE APPLICATIONS.

TABLE 5-9

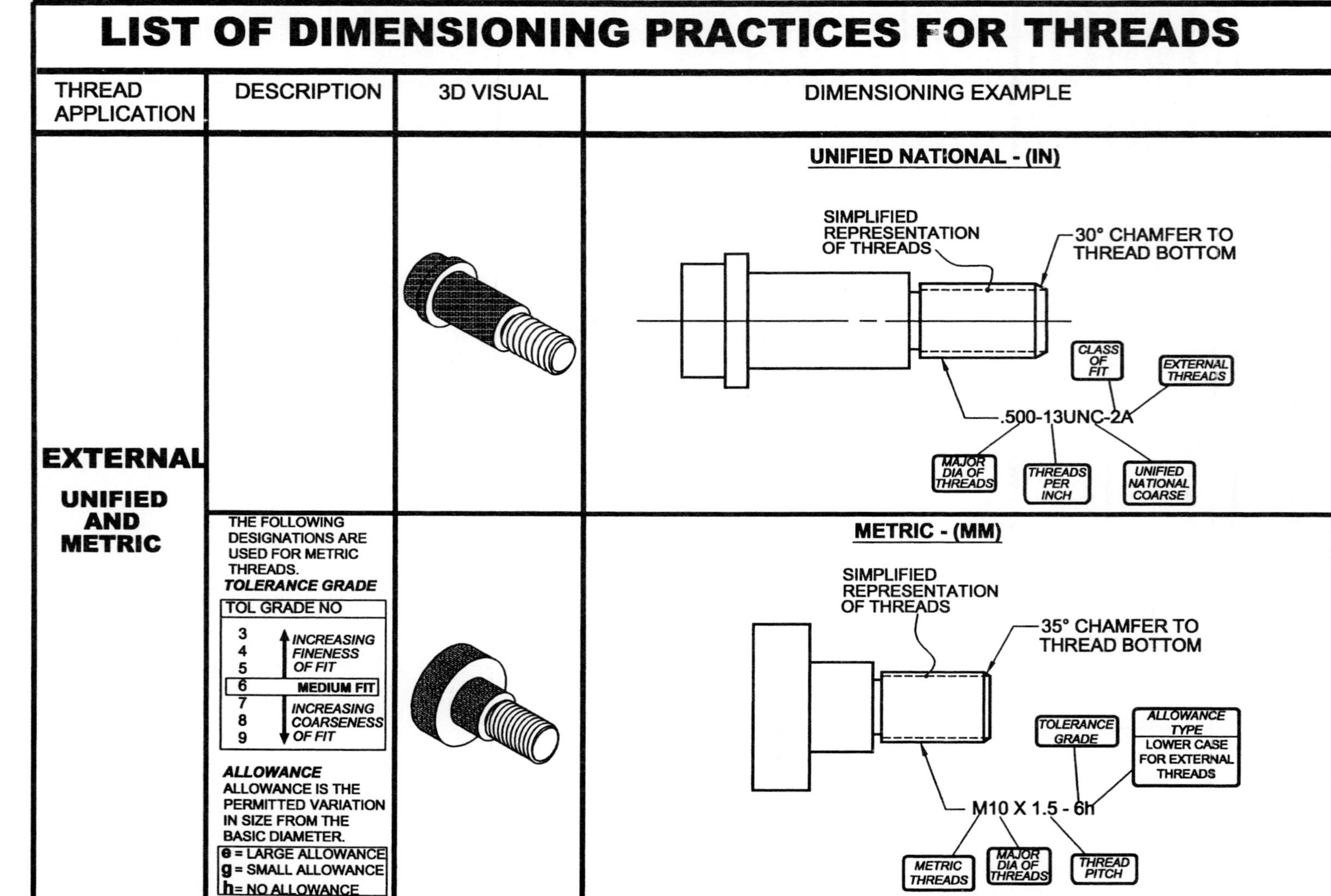

LIST OF DIMENSIONING PRACTICES FOR THREADS

THREAD APPLICATION	DESCRIPTION	3D VISUAL	DIMENSIONING EXAMPLE
EXTERNAL **UNIFIED AND METRIC**			**UNIFIED NATIONAL - (IN)** SIMPLIFIED REPRESENTATION OF THREADS 30° CHAMFER TO THREAD BOTTOM .500-13UNC-2A MAJOR DIA OF THREADS THREADS PER INCH UNIFIED NATIONAL COARSE CLASS OF FIT EXTERNAL THREADS
	THE FOLLOWING DESIGNATIONS ARE USED FOR METRIC THREADS. ***TOLERANCE GRADE*** TOL GRADE NO 3, 4, 5 — INCREASING FINENESS OF FIT 6 — MEDIUM FIT 7, 8, 9 — INCREASING COARSENESS OF FIT ***ALLOWANCE*** ALLOWANCE IS THE PERMITTED VARIATION IN SIZE FROM THE BASIC DIAMETER. **e** = LARGE ALLOWANCE **g** = SMALL ALLOWANCE **h** = NO ALLOWANCE		**METRIC - (MM)** SIMPLIFIED REPRESENTATION OF THREADS 35° CHAMFER TO THREAD BOTTOM M10 X 1.5 - 6h METRIC THREADS MAJOR DIA OF THREADS THREAD PITCH TOLERANCE GRADE ALLOWANCE TYPE LOWER CASE FOR EXTERNAL THREADS

TABLE 5–9 *(continued)*

EXTERNAL **ACME**	**ACME** -THREAD COMES IN TWO FORMS: **G** -GENERAL PURPOSE THE GRADES IN THIS CATEGORY ARE: **2G, 3G, 4G** **C** -CENTRALIZING THE GRADES IN THIS CATEGORY ARE: **2C, 3C, 4C, 5C, 6C** THESE THREADS ARE MACHINED TO HAVE LIMITED CLEARANCE IN ORDER TO MAINTAIN PROPER ALIGNMENT.	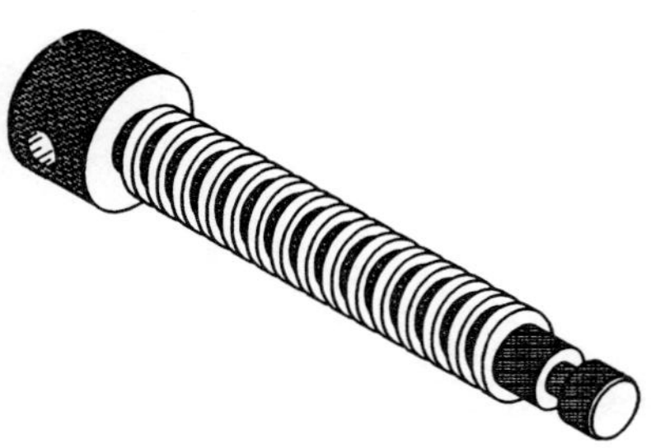	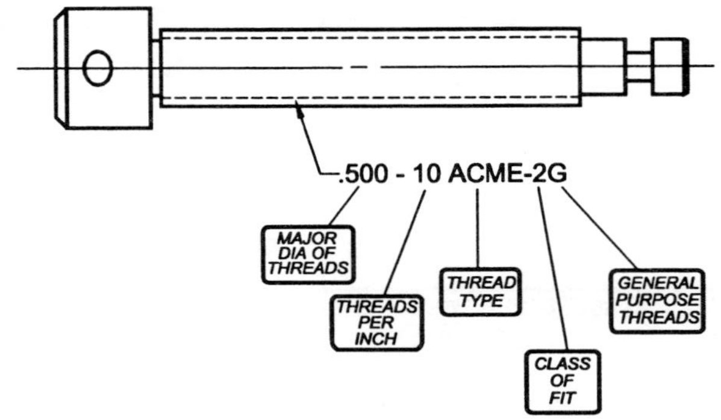 .500 - 10 ACME-2G MAJOR DIA OF THREADS THREADS PER INCH THREAD TYPE CLASS OF FIT GENERAL PURPOSE THREADS
EXTERNAL **&** **INTERNAL** **PIPE**	REGULAR PIPE THREAD COMES IN TWO FORMS: **NPS** -NATIONAL PIPE STRAIGHT **NPT** -NATIONAL PIPE TAPERED **DRYSEAL** PIPE THREAD ALSO COMES IN TWO FORMS: **NPSF** -NATIONAL PIPE STRAIGHT FINE **NPTF** -NATIONAL PIPE TAPERED FINE	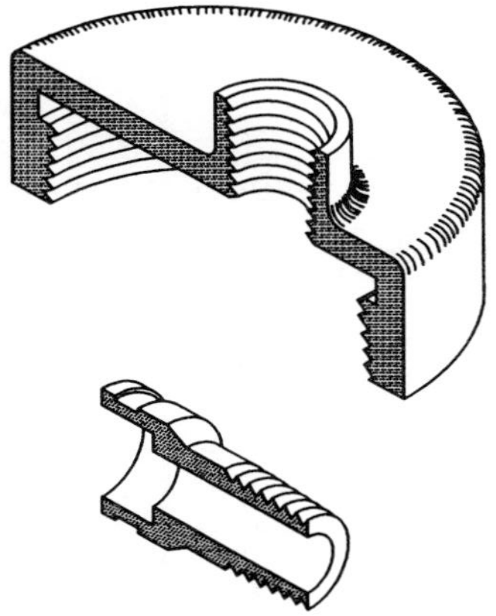	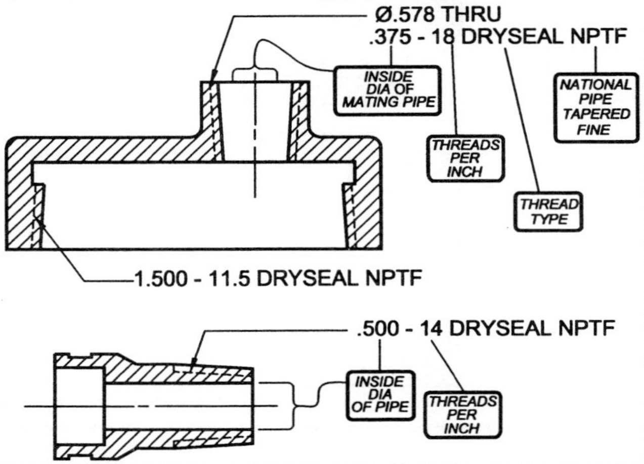 Ø.578 THRU .375 - 18 DRYSEAL NPTF INSIDE DIA OF MATING PIPE THREADS PER INCH THREAD TYPE NATIONAL PIPE TAPERED FINE 1.500 - 11.5 DRYSEAL NPTF .500 - 14 DRYSEAL NPTF INSIDE DIA OF PIPE THREADS PER INCH

TABLE 5–9 *(continued)*

		UNIFIED NATIONAL - (IN)	
		METHOD-1 (TRADITIONAL)	**METHOD-2** (CURRENT ASME Y14.5M,1994)
INTERNAL DRILL & TAP **UNIFIED AND METRIC**	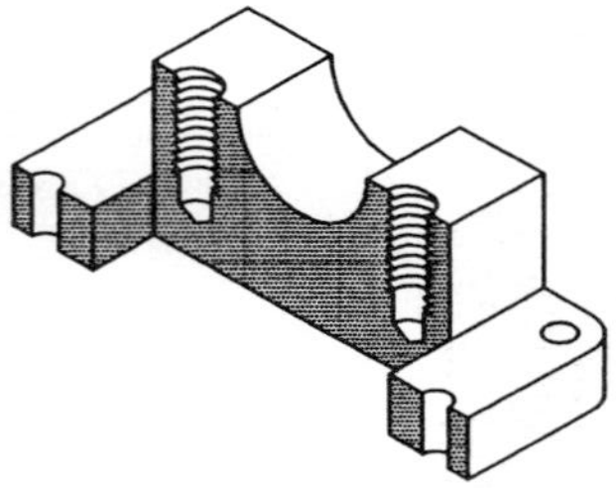	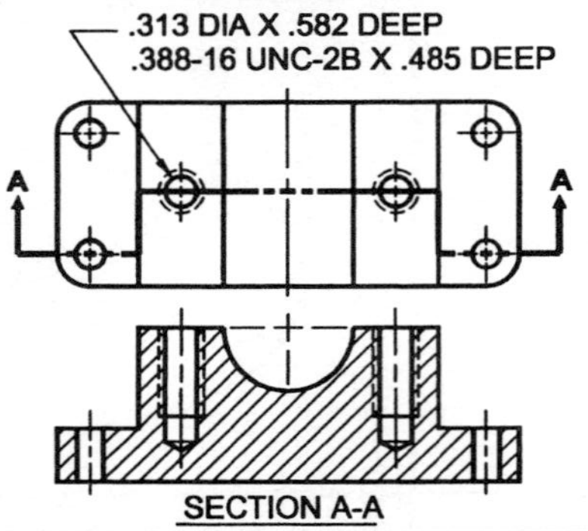 .313 DIA X .582 DEEP .388-16 UNC-2B X .485 DEEP A A SECTION A-A	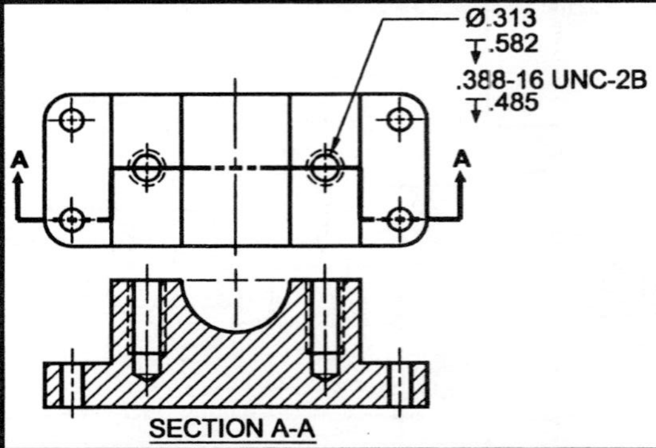 Ø.313 ↧.582 .388-16 UNC-2B ↧.485 A A SECTION A-A
		METRIC - (MM)	
		METHOD-1 (TRADITIONAL)	**METHOD-2** (CURRENT ASME Y14.5M,1994)
	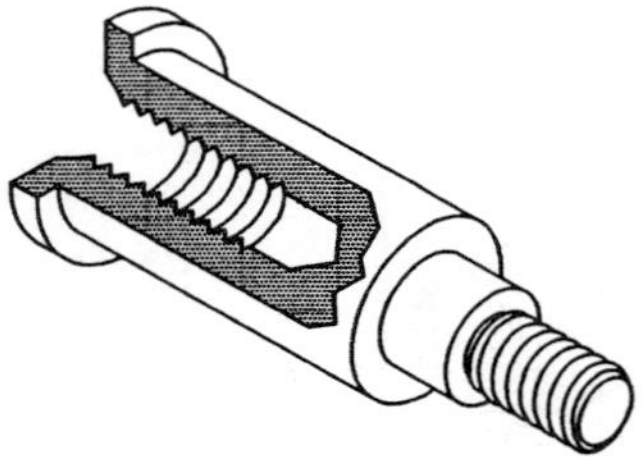	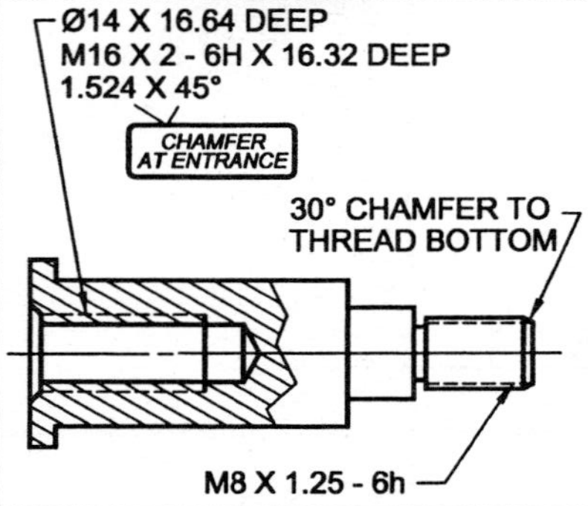 Ø14 X 16.64 DEEP M16 X 2 - 6H X 16.32 DEEP 1.524 X 45° CHAMFER AT ENTRANCE 30° CHAMFER TO THREAD BOTTOM M8 X 1.25 - 6h	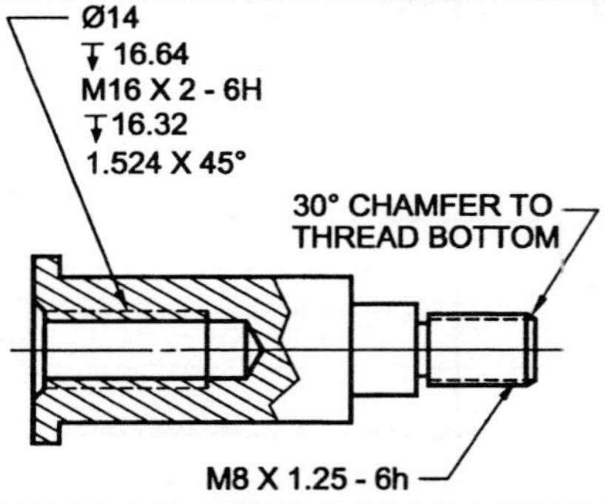 Ø14 ↧ 16.64 M16 X 2 - 6H ↧16.32 1.524 X 45° 30° CHAMFER TO THREAD BOTTOM M8 X 1.25 - 6h

TABLE 5–9 *(continued)*

THREAD APPLICATION	3D VISUAL	DIMENSIONING EXAMPLE	
INTERNAL THROUGH **UNIFIED AND METRIC**	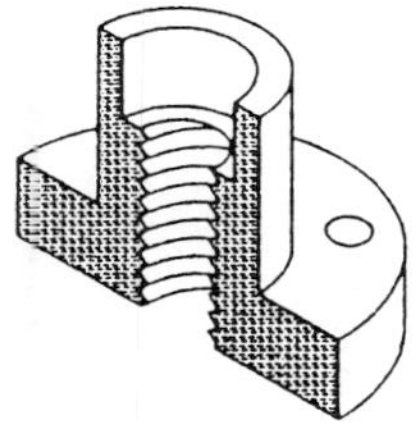	**UNIFIED NATIONAL - (IN)**	
		METHOD-1 (TRADITIONAL) 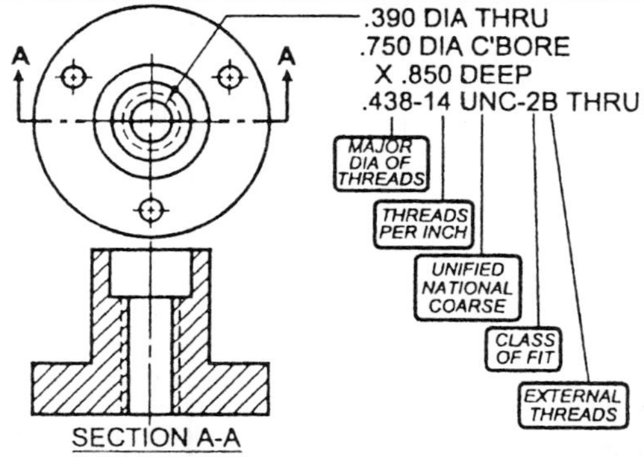	**METHOD-2** (CURRENT ASME Y14.5M,1994) 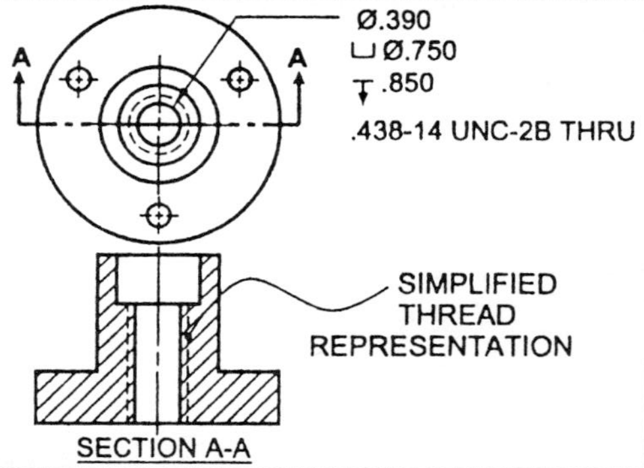
		A A .390 DIA THRU .750 DIA C'BORE X .850 DEEP .438-14 UNC-2B THRU MAJOR DIA OF THREADS THREADS PER INCH UNIFIED NATIONAL COARSE CLASS OF FIT EXTERNAL THREADS SECTION A-A	A A Ø.390 ⌴ Ø.750 ↧ .850 .438-14 UNC-2B THRU SIMPLIFIED THREAD REPRESENTATION SECTION A-A
		METRIC - (MM)	
		METHOD-1 (TRADITIONAL) 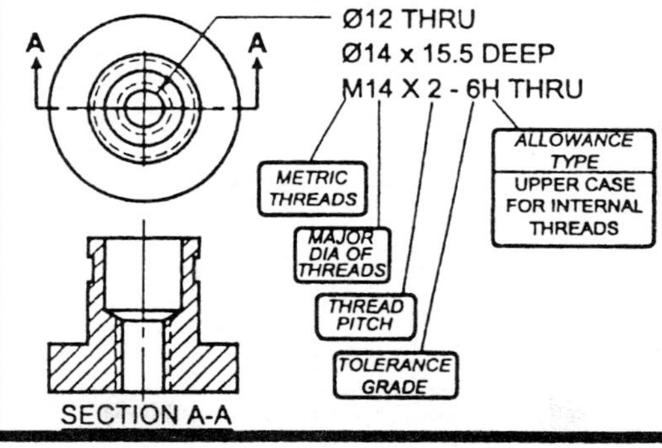	**METHOD-2** (CURRENT ASME Y14.5M,1994) 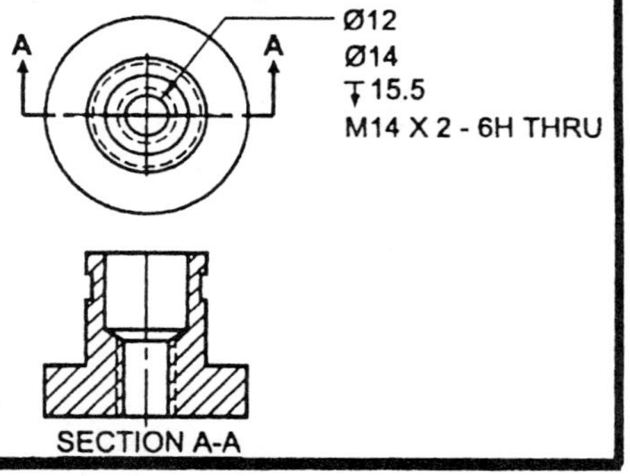
		A A Ø12 THRU Ø14 x 15.5 DEEP M14 X 2 - 6H THRU METRIC THREADS MAJOR DIA OF THREADS THREAD PITCH TOLERANCE GRADE ALLOWANCE TYPE UPPER CASE FOR INTERNAL THREADS SECTION A-A	A A Ø12 Ø14 ↧ 15.5 M14 X 2 - 6H THRU SECTION A-A

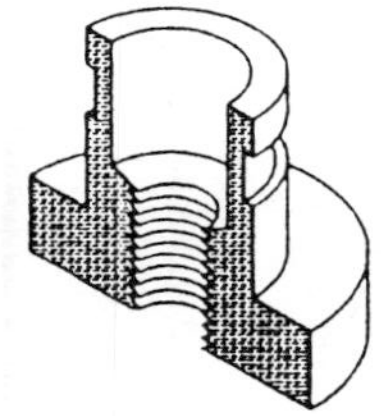

5-12 READING SURFACE FINISH SYMBOLS AND NOTES

The surface finish or degree of surface smoothness on a part has a direct influence on such operating factors as friction, fatigue life, corrosion resistance, contact stresses and vibration. Finish also controls the part's dimensional accuracy.

Surface Finish Terminology

Microinch (μ_{in}): One millionth of an inch or .000001 in.
Micrometer (μ_m): One millionth of a meter or .000001 m.
Roughness: The finer irregularities of surface due to the production process (tool feed marks, etc.).
Roughness Average (Ra): The average roughness expressed in microinches micrometers, or a corresponding system of grade numbers (N1 to N12).
A listing of surface roughness that results from different production operations is given in Table 5–10.

TABLE 5-10

SURFACE ROUGHNESS *vs* OPERATION

OPERATION	TYPE OF SURFACE PRODUCED	AVERAGE ROUGHNESS Ra USA (μ_{in})	AVERAGE ROUGHNESS Ra METRIC (μ_m)
SAND CASTING, ROUGH FORGING, TORCH OR SAW CUTTING.	VERY ROUGH, LOW GRADE SURFACE	1000 ✓	25 ✓
ROUGH MACHINING INVOLVING HEAVY CUTS AND COURSE FEEDS IN MILLING, TURNING, SHAPING, BORING AND ROUGH FILING. SAND CASTING AND ROUGH FORGING.	ROUGH, LOW GRADE SURFACE	500 ✓	12.5 ✓
COARSE SURFACE GRINDING, ROUGH FILING, DISK GRINDING, RAPID FEEDS IN TURNING, MILLING, DRILLING, BORING, AND SHAPING. FORGING AND CASTINGS.	COARSE PRODUCTION SURFACE	250 ✓	6.3 ✓
MACHINING OPERATIONS INVOLVING HIGH SPEEDS, FINE FEEDS AND LIGHT CUTS WITH SHARP TOOLS.	MEDIUM PRODUCTION SURFACE	125 ✓	3.2 ✓
MACHINING OPERATIONS INVOLVING HIGH SPEEDS, FINER FEEDS AND LIGHTER CUTS WITH SHARP TOOLS.	GOOD MACHINE FINISH	63 ✓	1.6 ✓
MACHINING OPERATIONS INVOLVING HIGH SPEEDS, EXTREMELY FINE FEEDS AND EXTRA LIGHT CUTS WITH SHARP TOOLS. THIS SURFACE IS EASILY PRODUCED BY CENTERLESS CYLINDRICAL AND SURFACE GRINDING	HIGH GRADE MACHINE FINISH	32 ✓	0.80 ✓
CYLINDRICAL GRINDING, EMERY BUFFING COARSE HONING OR LAPPING.	HIGH GRADE MACHINE FINISH	16 ✓	0.40 ✓
HONING, LAPPING OR BUFFING	VERY FINE MACHINE FINISH	8 ✓	0.20 ✓
FINE HONING, LAPPING OR BUFFING	EXTREMELY FINE MIRROR FINISH SURFACE	4 ✓	0.050 ✓
HIGHEST DEGREE OF HONING, LAPPING OR BUFFING	SATIN OR HIGHLY POLISHED MIRROR SURFACE	2 *to* 1 ✓	0.050 *to* 0.025 ✓

Roughness Width: The distance parallel to the nominal surface between successive peaks or ridges, expressed in microinches or micrometers.

Roughness Width Cutoff: The maximum spacing for repetitive surface irregularities to be included in the measurement of average roughness.

Waviness: Surface irregularities caused by machining or work deflections, vibration, chatter, heat treatment, or warping strains. Roughness is considered to be superimposed upon a wavy surface. Waviness is spaced further apart than roughness width cutoff, as shown in Figure 5–23.

Lay: The direction of the predominant surface pattern that is caused by the method of production.

Flaws: Random surface irregularities such as local cracks, blow holes, checks, scratches, pits, or burrs.

Table 5–11 provides a listing of the type of surface finish symbols and notes the reader will encounter in reading mechanical drawings.

5–13 READING MATERIAL SPECIFICATIONS

The part material heavily influences such important machining parameters as

- The cutting tool material.
- Cutting speed.
- Cutting feed.
- Depth of cut.
- CNC machine horsepower required to take a cut.
- Tool wear.
- Cutting temperature.
- Type of cutting fluid used.

Most materials that are machined are metals. Harder metals such as steels require the use of high speed steel (HSS) or carbide cutting tools. These materials are machined at low speeds and feeds and cutting fluids are used. Softer metals like aluminum can be machined with high speed steel tools operating at

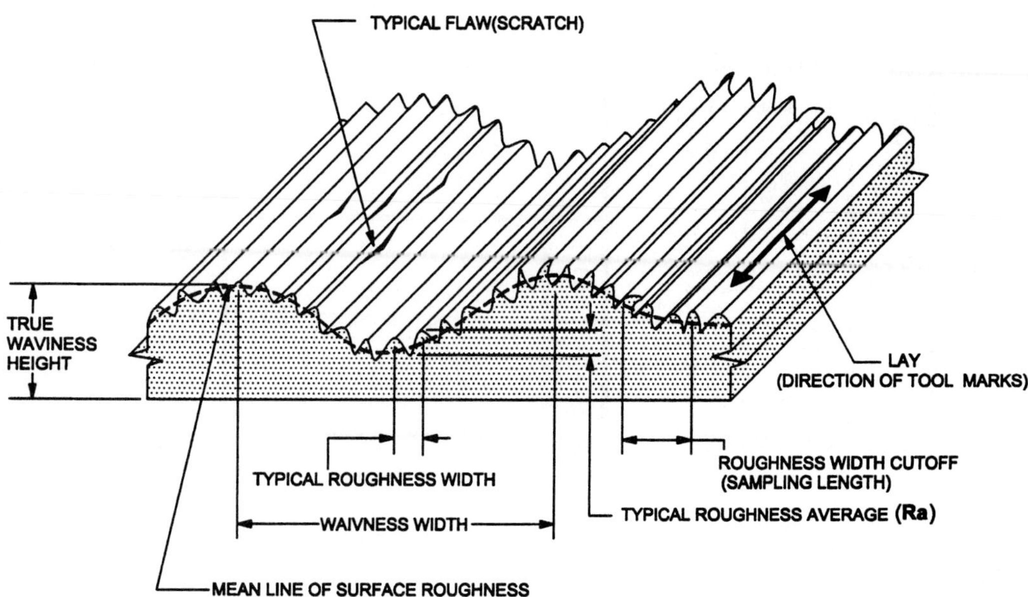

FIGURE 5–23 An exaggerated picture of a part's surface.

TABLE 5-11

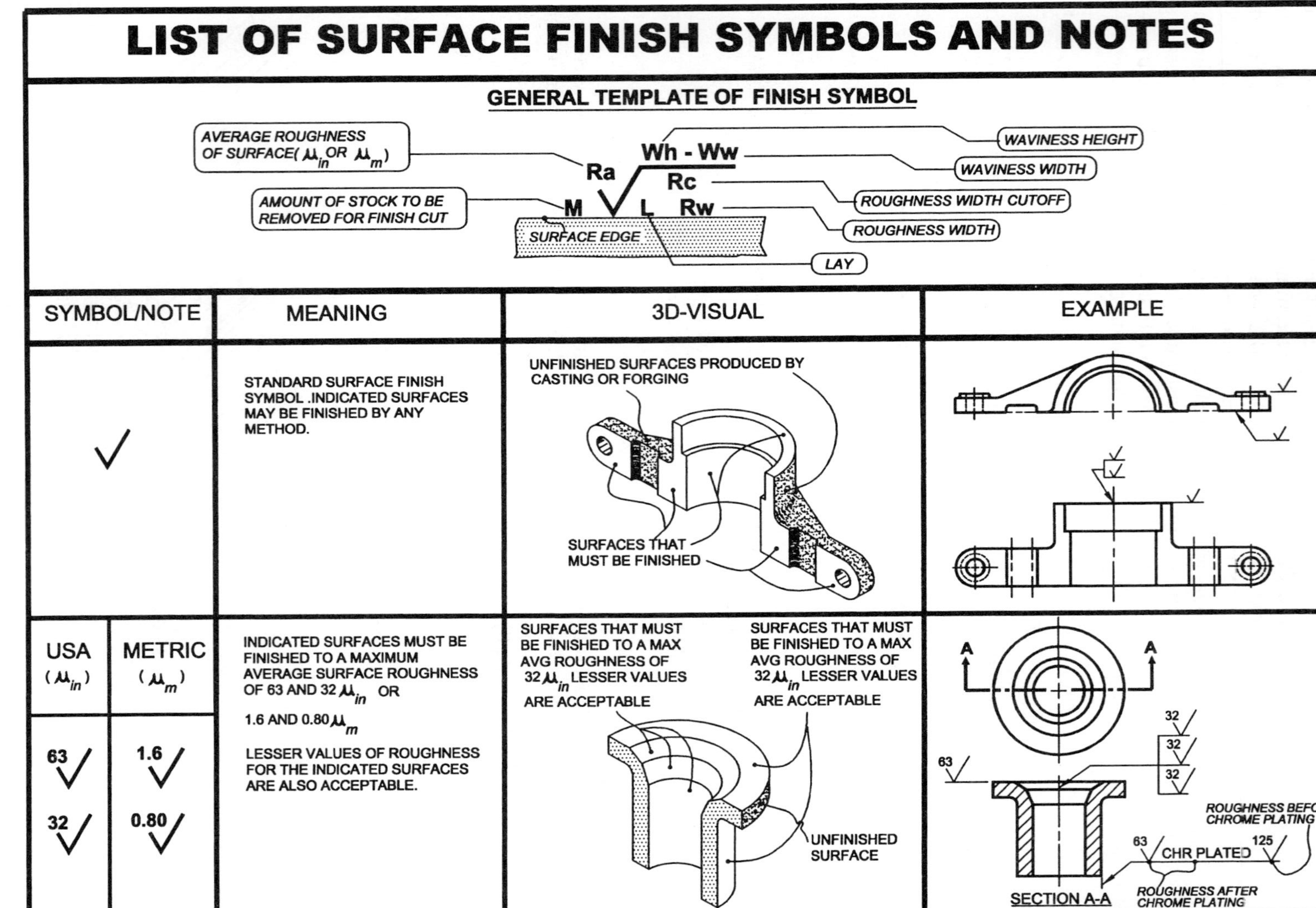
LIST OF SURFACE FINISH SYMBOLS AND NOTES
GENERAL TEMPLATE OF FINISH SYMBOL
AVERAGE ROUGHNESS OF SURFACE(µin OR µm)
AMOUNT OF STOCK TO BE REMOVED FOR FINISH CUT
Ra
M
Wh - Ww
Rc
L
Rw
SURFACE EDGE
WAVINESS HEIGHT
WAVINESS WIDTH
ROUGHNESS WIDTH CUTOFF
ROUGHNESS WIDTH
LAY
SYMBOL/NOTE
MEANING
3D-VISUAL
EXAMPLE
STANDARD SURFACE FINISH SYMBOL .INDICATED SURFACES MAY BE FINISHED BY ANY METHOD.
UNFINISHED SURFACES PRODUCED BY CASTING OR FORGING
SURFACES THAT MUST BE FINISHED
USA (µin)
METRIC (µm)
63
32
1.6
0.80
INDICATED SURFACES MUST BE FINISHED TO A MAXIMUM AVERAGE SURFACE ROUGHNESS OF 63 AND 32 µin OR
1.6 AND 0.80 µm
LESSER VALUES OF ROUGHNESS FOR THE INDICATED SURFACES ARE ALSO ACCEPTABLE.
SURFACES THAT MUST BE FINISHED TO A MAX AVG ROUGHNESS OF 32 µin LESSER VALUES ARE ACCEPTABLE
SURFACES THAT MUST BE FINISHED TO A MAX AVG ROUGHNESS OF 32 µin LESSER VALUES ARE ACCEPTABLE
UNFINISHED SURFACE
A
A
63
32
32
32
SECTION A-A
63
CHR PLATED
125
ROUGHNESS BEFORE CHROME PLATING
ROUGHNESS AFTER CHROME PLATING

TABLE 5-11 *(continued)*

UNLESS OTHERWISE SPECIFIED DIMENSIONS ARE IN INCHES TOLERANCES ARE: FRACTIONS ± 1/32 — DECIMALS .XX± .01, .XXX± .005 — ANGLES ± 1/2° MATERIAL FINISH 125 √ DO NOT SCALE DRAWING	FINISH ALL SURFACES TO AN AVERAGE ROUGHNESS VALUE OF 125 μ_{in} .		
NOTES: 1. 125 √ ALL OVER , UNLESS OTHERWISE SPECIFIED.			
NOTES : 1. REMOVE ALL BURRS BREAK ALL EDGES EXCEPT AS NOTED 2. 125 √ SURFACE ROUGHNESS INCLUDING REAMED HOLES, EXCEPT AS NOTED. 3. 200 √ FOR ALL DRILLED HOLES , UNLESS OTHERWISE SPECIFIED.	FINISH ALL SURFACES INCLUDING REAMED HOLES TO A MAXIMUM AVERAGE ROUGHNESS OF 125 μ_{in} . FINISH ALL DRILLED HOLES TO A MAXIMUM AVERAGE ROUGHNESS OF 200 μ_{in} . ANY LESSER VALUES OF ROUGHNESS FOR EACH OPERATION ARE ACCEPTABLE.		
USA (μ_{in}): 63 / 32 √ METRIC (μ_m): 1.6 / 0.80 √	INDICATED SURFACES MUST BE FINISHED TO WITHIN AN AVERAGE ROUGHNESS RANGE OF 63 TO 32 μ_{in} OR 1.6 TO 0.80 μ_m .	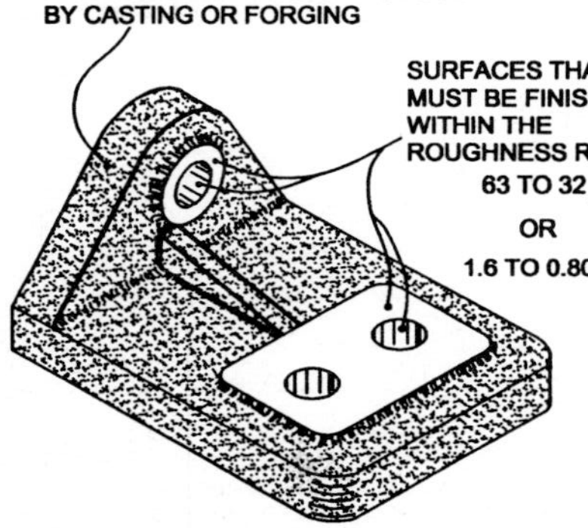	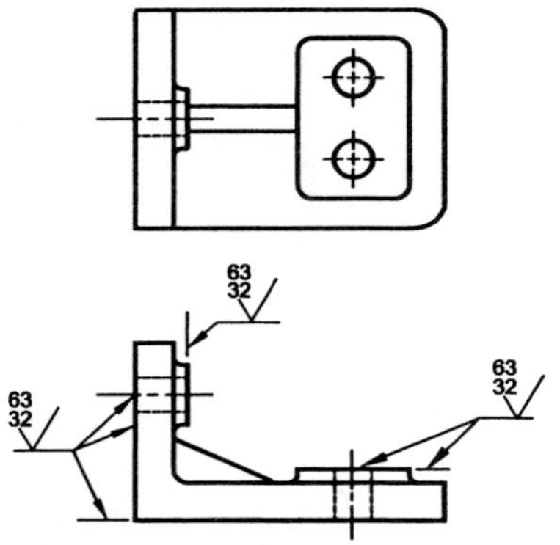

TABLE 5–11 *(continued)*

∀ (machining required)		SURFACE FINISH OF INDICATED SURFACES IS ACCOMPLISHED BY MACHINING. THE FORGED OR CAST PART MUST BE PRODUCED WITH EXTRA MATERIAL FOR THAT PURPOSE.	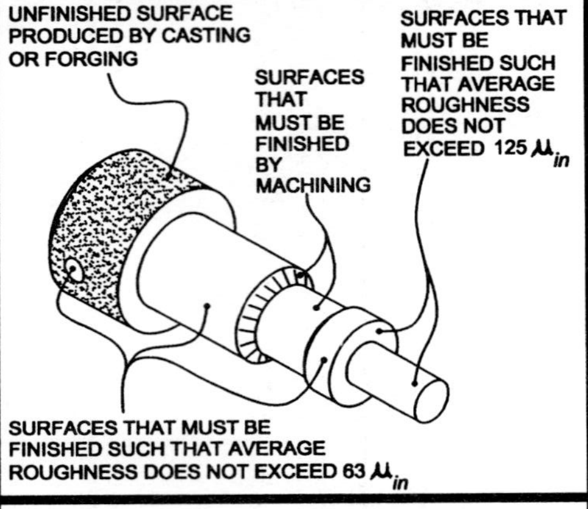 UNFINISHED SURFACE PRODUCED BY CASTING OR FORGING SURFACES THAT MUST BE FINISHED BY MACHINING SURFACES THAT MUST BE FINISHED SUCH THAT AVERAGE ROUGHNESS DOES NOT EXCEED 125 μ_{in} SURFACES THAT MUST BE FINISHED SUCH THAT AVERAGE ROUGHNESS DOES NOT EXCEED 63 μ_{in}	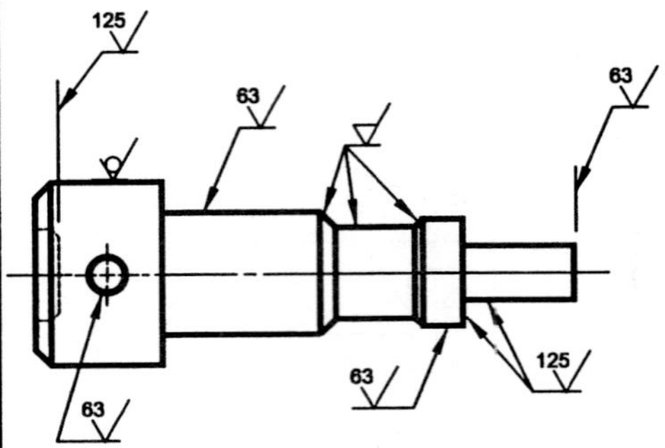 125 63 63 63 125 63
USA (μ_{in})	METRIC (μ_m)	SURFACE FINISH IS PRODUCED BY MACHINING. THE AMOUNT OF STOCK TO BE REMOVED IS .063 in OR 1.6mm 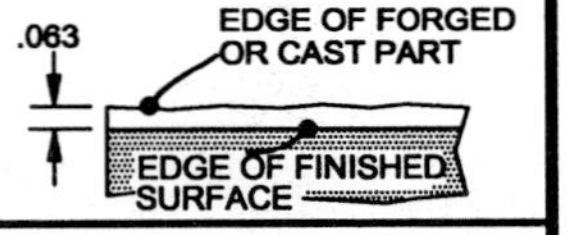 . .063 EDGE OF FORGED OR CAST PART EDGE OF FINISHED SURFACE		
.063	1.6			
(material removal prohibited symbol)		MATERIAL REMOVAL IS PROHIBITED		
√=		LAY IS **PARALLEL** TO THE SURFACE EDGE INDICATED.	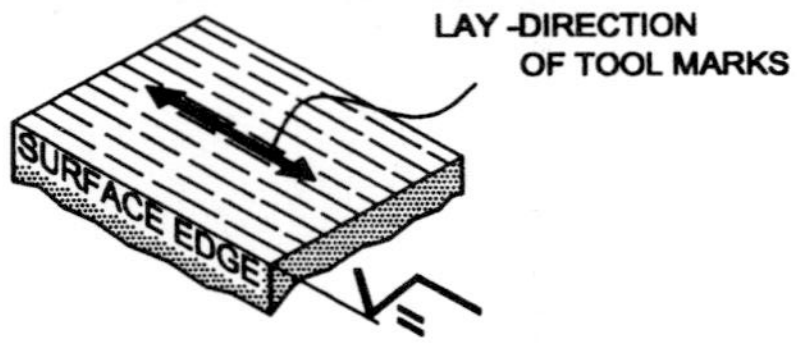 LAY -DIRECTION OF TOOL MARKS SURFACE EDGE	
√⊥		LAY IS **PERPENDICULAR** TO THE SURFACE EDGE INDICATED.	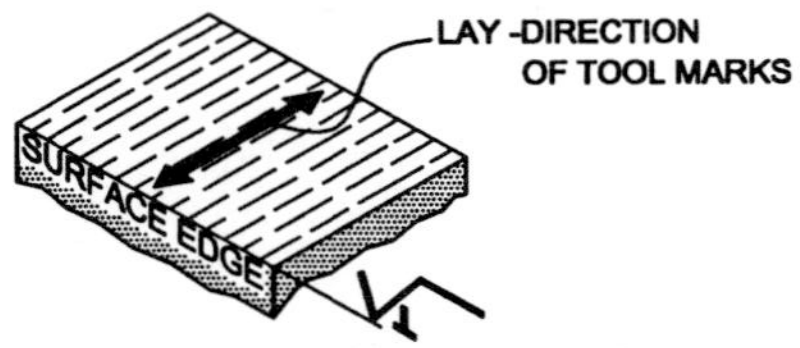 LAY -DIRECTION OF TOOL MARKS SURFACE EDGE	
√X		LAY IS **ANGULAR** IN BOTH DIRECTIONS TO THE SURFACE EDGE INDICATED.	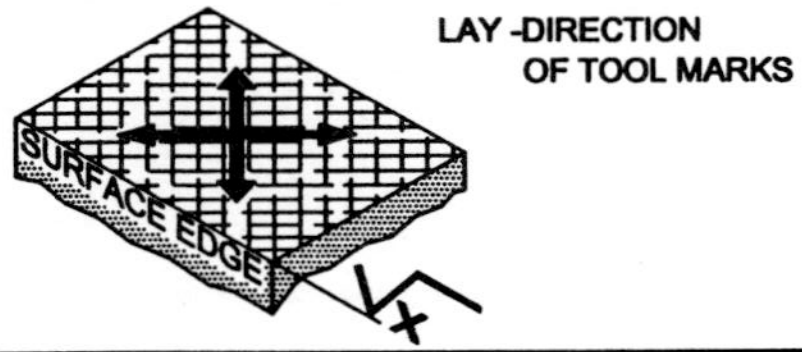 LAY -DIRECTION OF TOOL MARKS SURFACE EDGE	

TABLE 5-11 *(continued)*

Symbol	Description	Illustration	
$\sqrt{}$ M	LAY IS **MULTIDIRECTIONAL** WITH RESPECT TO THE SURFACE EDGE INDICATED.	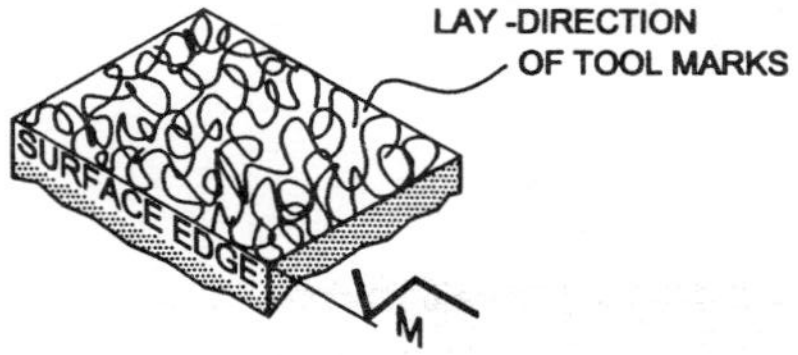 LAY -DIRECTION OF TOOL MARKS; SURFACE EDGE; $\sqrt{}$ M	
$\sqrt{}$ C	LAY IS **CIRCULAR** WITH RESPECT TO THE SURFACE EDGE INDICATED.	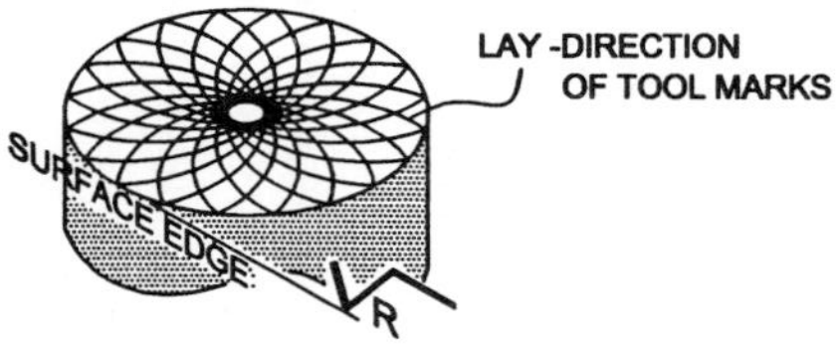 LAY -DIRECTION OF TOOL MARKS; SURFACE EDGE; $\sqrt{}$ R	
$\sqrt{}$ R	LAY IS **RADIAL** WITH RESPECT TO THE SURFACE EDGE INDICATED.	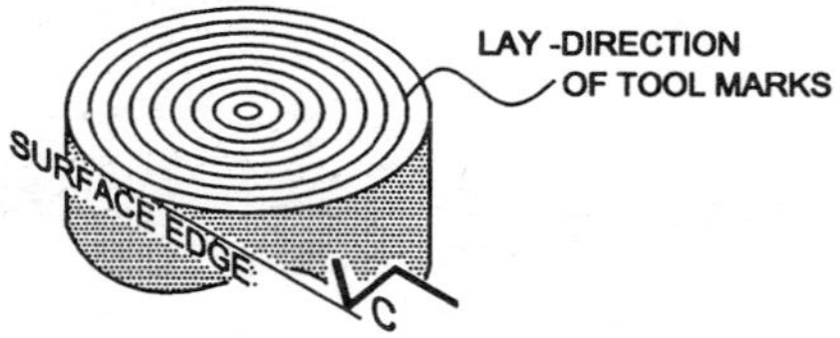 LAY -DIRECTION OF TOOL MARKS; SURFACE EDGE; $\sqrt{}$ C	
$\sqrt{}$ P	LAY IS **PARTICULATE NON-DIRECTIONAL** OR **PROTUBERANT** WITH RESPECT TO THE SURFACE EDGE INDICATED.	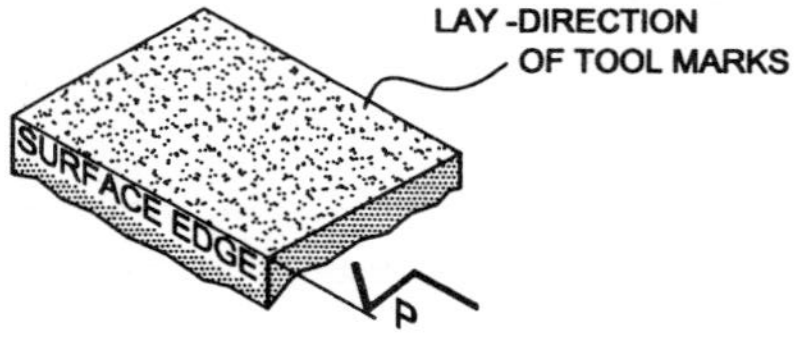 LAY -DIRECTION OF TOOL MARKS; SURFACE EDGE; $\sqrt{}$ P	
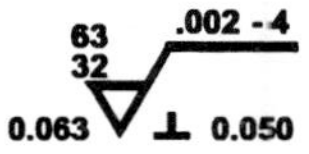 63 / 32 $\sqrt{}$ .002 - 4; 0.063; ⊥ 0.050	1. THE FINISHED SURFACE IS PRODUCED BY MACHINING. 2. LEAVE .063in STOCK FOR FINISH MACHINING. 3. AVERAGE SURFACE ROUGHNESS MUST BE IN THE RANGE OF 63 TO 32 μ_{in}.	4. ROUGHNESS SAMPLING WIDTH IS 0.050in. IF NO VALUE IS GIVEN, THE DEFAULT IS USUALLY 0.030in OR 0.80mm. 5. LAY IS PERPENDICULAR TO THE INDICATED SURFACE EDGE.	6. THE MAXIMUM WAVINESS HEIGHT MUST BE 0.002in. ANY LESSER VALUE IS ACCEPTABLE. 7. THE WAVINESS WIDTH MUST BE 4in, MAXIMUM. ANY LESSER VALUE IS ACCEPTABLE.

high speeds and feeds. Many times the use of a cutting fluid is not needed for cutting aluminum and its alloys.

Refer to Appendices C and D for a listing of tool speeds and feeds and machining fluids for various metals.

Metallic materials can be classified into two broad categories: ferrous and non-ferrous. Ferrous metals have iron as their principal element and are magnetic. These metals include cast irons and steels. Non-ferrous metals have little or no iron and are not magnetic. Aluminum, magnesium, copper, and zinc are examples of non-ferrous metals.

A metal consisting of one element is called a pure metal. Pure metals are soft and possess low strength. An alloy is a metal that is composed of one or more chemical elements. At least one of the elements in an alloy is a pure metal.

Ferrous Metals

Cast Iron

Cast iron is an alloy of iron, carbon, silicon, and various other elements existing in insignificant amounts. The American Society of Testing and Materials (ASTM) identification number system is used to specify cast irons in production drawings. Table 5–12 gives a listing of ASTM codes for commonly used cast irons.

Steel

Steel is an alloy of iron containing varying amounts of carbon (0.08 to 1.5%). It is the most important of all ferrous materials and the dominant material used in manufacturing and construction. The properties of steel can be changed by

TABLE 5–12

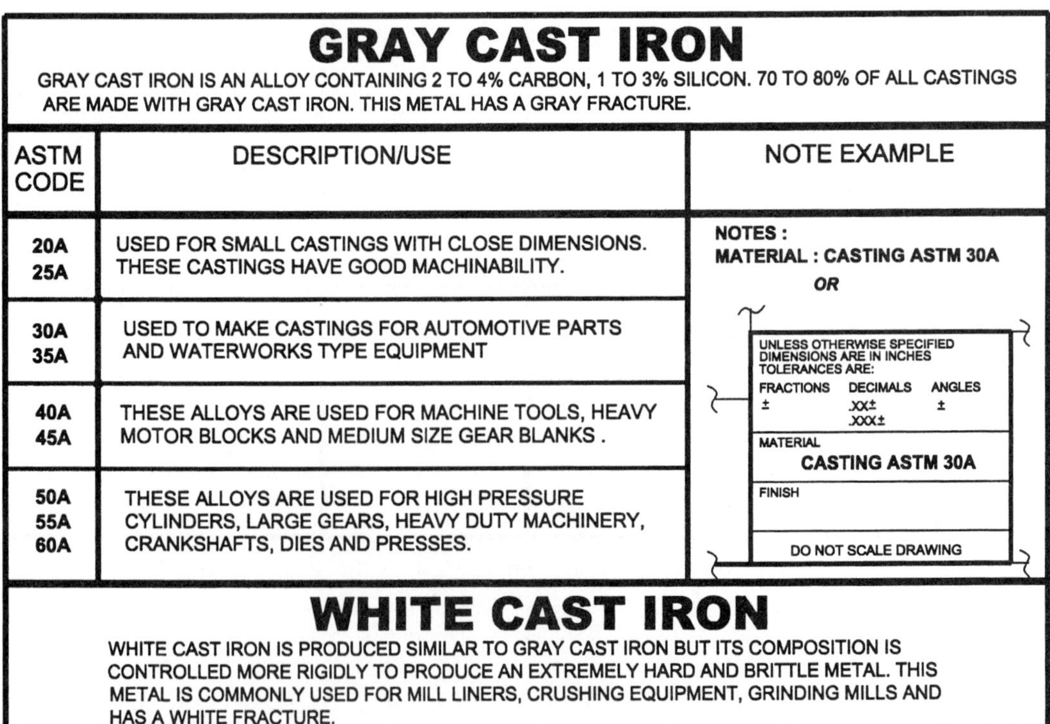

GRAY CAST IRON

GRAY CAST IRON IS AN ALLOY CONTAINING 2 TO 4% CARBON, 1 TO 3% SILICON. 70 TO 80% OF ALL CASTINGS ARE MADE WITH GRAY CAST IRON. THIS METAL HAS A GRAY FRACTURE.

ASTM CODE	DESCRIPTION/USE	NOTE EXAMPLE
20A 25A	USED FOR SMALL CASTINGS WITH CLOSE DIMENSIONS. THESE CASTINGS HAVE GOOD MACHINABILITY.	NOTES : MATERIAL : CASTING ASTM 30A *OR*
30A 35A	USED TO MAKE CASTINGS FOR AUTOMOTIVE PARTS AND WATERWORKS TYPE EQUIPMENT	UNLESS OTHERWISE SPECIFIED DIMENSIONS ARE IN INCHES TOLERANCES ARE: FRACTIONS ± DECIMALS .XX± .XXX± ANGLES ±
40A 45A	THESE ALLOYS ARE USED FOR MACHINE TOOLS, HEAVY MOTOR BLOCKS AND MEDIUM SIZE GEAR BLANKS .	MATERIAL CASTING ASTM 30A
50A 55A 60A	THESE ALLOYS ARE USED FOR HIGH PRESSURE CYLINDERS, LARGE GEARS, HEAVY DUTY MACHINERY, CRANKSHAFTS, DIES AND PRESSES.	FINISH DO NOT SCALE DRAWING

WHITE CAST IRON

WHITE CAST IRON IS PRODUCED SIMILAR TO GRAY CAST IRON BUT ITS COMPOSITION IS CONTROLLED MORE RIGIDLY TO PRODUCE AN EXTREMELY HARD AND BRITTLE METAL. THIS METAL IS COMMONLY USED FOR MILL LINERS, CRUSHING EQUIPMENT, GRINDING MILLS AND HAS A WHITE FRACTURE.

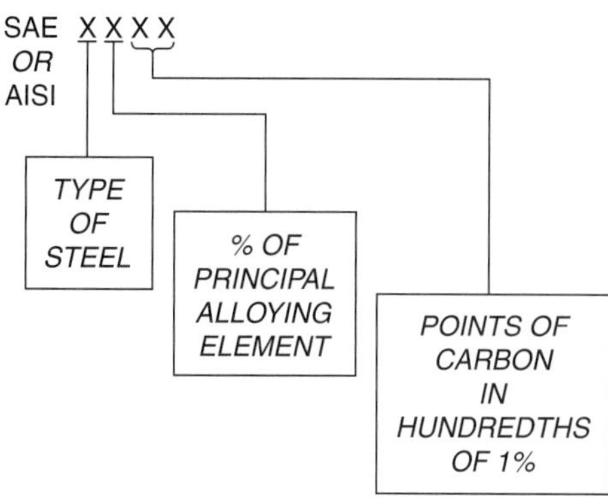

FIGURE 5–24 The SAE/AISI system for identifying steels.

adding other alloying elements such as nickel, molybdenum, chromium, and vanadium, and by heat treatment. Production prints utilize the American Iron and Steel Institute (AISI) and the Society of Automotive Engineers (SAE) numbering system for specifying steels as shown in Figure 5–24.

The SAE and the ASTM have developed a unified numbering system (UNS) to provide a method of correlating the different numbering systems of metals and alloys. UNS numbers do not contain information regarding the percentages of alloying elements and can only be used for identification purposes.

The UNS numbering system consists of an upper case letter followed by a five digit code. In most cases, the letter matches the first letter in the name of metal family (A-aluminum, S-steel, etc.).

Table 5–13 lists the SAE/AISI and UNS codes for important steels used in manufacturing.

Non-Ferrous Metals

Aluminum

Aluminum, the most heavily used non-ferrous metal, is mined from bauxite ore, which contains approximately 45% aluminum plus oxygen and other impurities. Its main properties include good conductivity, light weight, high resistance to corrosion, non-sparking, non toxic, non-magnetic, and easily formed and machined.

Production drawings utilize the Aluminum Association (AA) numbering system for wrought and cast alloys.

Wrought Aluminum Alloys

Wrought alloys are produced by running the coarse-grained metal between a set of rollers at an elevated temperature. The metal is compressed into specific shapes such as bars, rods, and tubes having smaller grain size and improved strength. Additional heat treatment or tempering of the metal further increases its strength.

Alclad aluminum possesses high strength and superior resistance to corrosion. It is produced by bonding a thin layer of corrosion resistant aluminum to the high strength alloy during the rolling process.

Most wrought alloys are easily machined but some may require special tools and production practices.

The AA system for wrought alloys consists of four digits followed by a temper designation as illustrated in Figure 5–25.

TABLE 5–13

CARBON STEEL

CARBON STEEL IS AN ALLOY OF IRON AND CARBON WITH INSIGNIFICANT AMOUNTS OF OTHER ELEMENTS. APPROXIMATELY 85% OF ALL STEEL PRODUCED IS CARBON STEEL.

C= Carbon

TYPE	SAE *OR* AISI CODE	UNS No.	ALLOY CONTENT	DESCRIPTION/USE	NOTE EXAMPLE
LOW CARBON	**1006** *TO* **1020**	**G10060** *TO* **G10200**	0.06 *TO* 0.20C	SOFT PLASTIC STEELS. USED FOR CHAIN, STAMPINGS MACHINE PARTS AND CASE HARDENED MACHINE PARTS	**NOTES** **MATERIAL : SAE 1020**
MEDIUM CARBON	**1020** *TO* **1050**	**G10200** *TO* **G10500**	0.20 *TO* 0.50C	MEDIUM HARD STEELS. USED FOR GEARS, SHAFTS, CONNECTING RODS, MACHINE PARTS, CRANKSHAFTS AND HEAT TREATED MACHINE PARTS.	*CARBON STEEL* *0% OR NO ADDITIONAL ALLOYING ELEMENT* *0.20% CARBON IN THE ALLOY*
HIGH CARBON	**1050** *AND UP*	**G10500** *AND UP*	0.50 *TO* 1.50C	TOUGH HARD STEELS KNOWN AS **TOOL STEELS**. USED FOR SCREWDRIVERS, DIES, LOCOMOTIVE WHEELS, DRILLS, TAPS, AND MILLING CUTTERS.	*OR*
RE-SULFERIZED	**1110** *TO* **1151**	**G11100** *TO* **G11510**	Sulphur is added	THESE STEELS ARE KNOWN AS **FREE MACHINING**. THE ELEMENT SULPHUR PRODUCES ALLOYS THAT CREATE SMALL CHIPS AND LESS TOOL WEAR WHEN CUT.	UNLESS OTHERWISE SPECIFIED DIMENSIONS ARE IN INCHES TOLERANCES ARE: FRACTIONS ± DECIMALS .XX± .XXX± ANGLES ±
RE-SULFERIZED AND RE-PHOSPHORIZED	**1211** *TO* **1215**	**G12110** *TO* **G12150**	Sulphur and Phosphorus is added	THESE **FREE MACHINING** STEELS HAVE INCREASED STRENGTH AND HARDNESS BUT LESS DUCTILITY	MATERIAL **SAE 1020** FINISH DO NOT SCALE DRAWING

TABLE 5-13 *(continued)*

ALLOY STEEL

IN ADDITION TO CARBON, ALLOY STEELS HAVE SIGNIFICANT AMOUNTS OF OTHER ELEMENTS SUCH AS MANGANESE, NICKEL, CHROMIUM, MOLYBDENUM, TUNGSTEN, ETC. BETTER STRENGTH-TO-WEIGHT RATIOS ARE ACHIEVEDWITH THESE STEELS.

C= Carbon, Mn= Manganese, Ni=Nickel, Cr=Chromium, Mo= Molybdenum, V= Vanadium, W= Tungsten

TYPE	SAE OR AISI CODE	UNS No.	ALLOY CONTENT	DESCRIPTION/USE	NOTE EXAMPLE
MANGANESE	1330 TO 1345	G13300 TO G13450	0.30C;1.75Mn TO 0.46C;1.75Mn	THESE STEELS PRODUCE BETTER SURFACE FINISH. USES INCLUDE CARBURIZED GEARS AND SPLINE SHAFTS.	**NOTES** **MATERIAL:** **NICKEL ALLOY STEEL** **SAE 2330** Nickel STEEL 3.5% Nickel ALLOYING ELEMENT 0.30% CARBON IN THE ALLOY *OR* UNLESS OTHERWISE SPECIFIED DIMENSIONS ARE IN INCHES TOLERANCES ARE: FRACTIONS ± DECIMALS .XX± .XXX± ANGLES ± MATERIAL **(NICKEL STEEL) SAE 2330** FINISH DO NOT SCALE DRAWING
NICKEL	23XX		3.5Ni	NICKEL STEELS EXHIBIT INCREASED TOUGHNESS AND STRENGTH. USES ARE TOOLS, PRECISION VISES, ARMOR, GEARS, AUTOMOTIVE, MACHINE PARTS AND BALL BEARINGS.	
	25XX		5Ni		
NICKEL-CHROMIUM	31XX TO 34XX		1.25Ni, 0.70Cr TO 3Ni, 0.77Cr	THESE STEELS HAVE TOUGHNESS, STRENGTH AND SUPERIOR WEAR RESISTANCE. USES INCLUDE BOLTS, STEERING ARMS, TRANSMISSION CHAINS AND GEARS.	
MOLYBDENUM	40XX	G40XXX	0.2 - 0.25Mo	GOOD STRENGTH AT HIGH TEMPERATURES AND RESISTANCE TO SHOCK IS EXHIBITED BY THESE STEELS. USES INCLUDE FORGING DIES, BALL BEARINGS, HIGH TEMP STEAMLINES, GEARS AND PROPELLER SHAFTS.	
	44XX	G44XXX	0.40 - 0.52Mo		
NICKEL-CHROMIUM-MOLYBDENUM	43XX	G43XXX	1.82Ni;0.50Cr;0.25Mo	THESE STEELS EXHIBIT HARDNESS, STRENGTH AND GOOD RESISTANCE TO RUST.SOME USES ARE TRUCK AND BUS GEARS,HEAVY DUTY SHAFTS AND TRANSMISSION CHAIN PINS.	
	86XX	G86XXX	0.55Ni;0.50Cr;0.20Mo		
	98XX		1.00Ni;0.80Cr;0.25Mo		
NICKEL-MOLYBDENUM	46XX	G46XXX	1.33Ni;0.20Mo	THIS FAMILY OF STEELS HAVE GOOD STRENGTH AT HIGH TEMPERATURES AND EXCELLENT RESISTANCE TO SHOCK. USES INCLUDE CAMS AND SHAFTS.	
	48XX	G48XXX	3.50Ni;0.25Mo		
CHROMIUM	51XX	G51XXX	0.93Cr (low Chromium)	THESE STEELS EXHIBIT HARDNESS, STRENGTH AND GOOD RESISTANCE TO RUST.SOME USES ARE CYANIDE HARDENED GEARS AND SHAFTS.	
	51100	G51986	1.02Cr (medium Chromium)		
	52100	G52986	1.45Cr (high Chromium)		
CHROMIUM-VANADIUM	61XX	G51XXX	0.95Cr;0.15V	INCREASED TOUGHNESS, HARDNESS AND RESISTANCE TO SHOCK IS EXHIBITED BY THESE STEELS. USES INCLUDE GEARS AND HIGH DUTY LEAF SPRINGS.	

TABLE 5–13 *(continued)*

TUNGSTEN-CHROMIUM	**72XX**		1.75W;0.75Cr	THIS FAMILY OF STEELS EXHIBITS GOOD STRENGTH, HARDNESS, WEAR RESISTANCE AND SHOCK RESISTANCE AT HIGH TEMPERATURES. USES ARE CUTTING TOOLS, TOOL BITS AND MILLING CUTTERS.	
SILICON-MANGANESE	**92XX**	**G92XXX**	0.76Mn;1.5Si	THESE STEELS ARE TOUGH AND RESISTANT TO WEAR AND ABRASION.THEY ARE DIFFICULT TO MACHINE AND ARE USUALLY MADE IN CAST FORM. USES INCLUDE ROCK CRUSHING EQUIPMENT, MINING TOOLS AND LEAF SPRINGS.	
BORON	**50BXX** **51BXX** **81BXX** **94BXX**	**G50XXX** **G51XXX** **G81XXX** **G94XXX**	**B indicates Boron as the alloying element**	SUPERIOR HARDNESS AND IMPROVED MACHINABILITY AND COLD FORMING ARE EXHIBITED . LOWER COSTS SINCE SMALLER AMOUNTS OF BORON CAN PRODUCE THE SAME HARDNESS IN STEEL AS LARGER AMOUNTS OF CARBON.	
LEAD	**XXLXX**		**L indicates Lead as the alloying element**	GOOD HARDNESS AND IMPROVED MACHINABILITY ARE EXHIBITED BY THESE STEELS.	

TABLE 5–13 *(continued)*

STAINLESS STEEL

CORROSION RESISTANT STAINLESS STEEL(**CRES**)IS DEFINED AS ANY STEEL HAVING 4% OR MORE CHROMIUM. IN REALITY STAINLESS STEELS USUALLY HAVE AT LEAST 10% CHROMIUM.

TYPE	SAE CODE	AISI CODE	UNS No.	DESCRIPTION/USE	NOTE EXAMPLE
CHROMIUM-MANGANESE-NICKEL	**302XX**	**2XX**	**S302XX**	THESE STEELS EXHIBIT GOOD RESISTANCE TO CORROSION. USES ARE AUTOMOTIVE PARTS, RAILROAD CAR BODIES AND STRUCTURAL MEMBERS.	**NOTES:** **MATERIAL:** **CRES SAE 30303F** *OR* UNLESS OTHERWISE SPECIFIED DIMENSIONS ARE IN INCHES TOLERANCES ARE: FRACTIONS ± DECIMALS .XX± .XXX± ANGLES ± MATERIAL **CRES SAE 30303F** FINISH DO NOT SCALE DRAWING
CHROMIUM-NICKEL	**303XX**	**3XX**	**S303XX**	THIS FAMILY OF STAINLESS STEELS INCLUDES THE **FREE MACHINING TYPE 30303F** THAT IS USED FOR PARTS MANUFACTURED ON CNC MACHINES.	
CHROMIUM	**514XX** **515XX**	**4XX** **5XX**	**S40XXX** **S50XXX**	CHROMIUM STAINLESS STEELS EXHIBIT SUPERIOR RESISTANCE TO CORROSION AT HIGHER TEMPERATURES AND INCLUDE THE **FREE MACHINING TYPES 51416F, 51420F, 51430F AND 51440F.** THESE STEELS ARE ESPECIALLY SUITED FOR PARTS MANUFACTURED ON CNC MACHINES.	

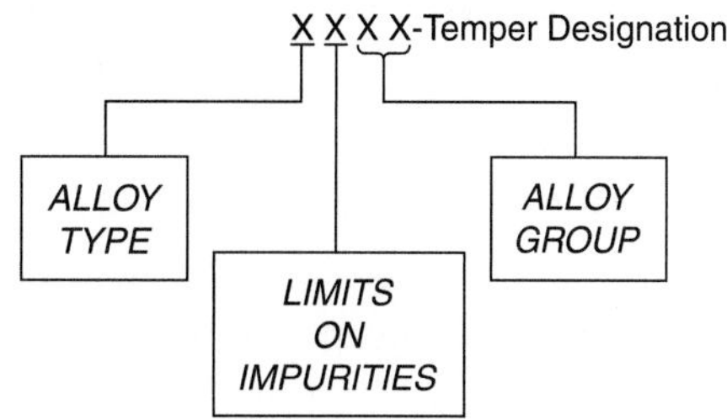

FIGURE 5–25 The AA system for identifying wrought aluminum alloys.

TABLE 5–14

ALLOY GROUP	
No.	MAJOR ALLOYING ELEMENT
1XX.X	Aluminum-95.0% or greater
2XX.X	Copper
3XX.X	Silicon + Copper or Manganese
4XX.X	Silicon
5XX.X	Magnesium
6XX.X	unused
7XX.X	Zinc
8XX.X	Tin
9XX.X	other elements

TABLE 5–15

TEMPER DESIGNATIONS	
DESIGNATION	MEANING
F	AS-FABRICATED
O	ANNEALED
H1	STRAIN HARDENED ONLY
H2	STRAIN HARDENED AND PARTIALLY ANNEALED
H3	STRAIN HARDENED AND THERMALLY STABILIZED
W	SOLUTION HEAT TREATED
T1	COOLED FROM AN ELEVATED TEMPERATURE SHAPING PROCESS AND NATURALLY AGED
T2	COOLED FROM AN ELEVATED TEMPERATURE SHAPING PROCESS, COLD WORKED AND NATURALLY AGED
T3	SOLUTION HEAT TREATED,COLD WORKED AND NATURALLY AGED
T4	SOLUTION HEAT TREATED AND NATURALLY AGED
T5	COOLED FROM AN ELEVATED TEMPERATURE SHAPING PROCESS ,COLD WORKED AND ARTIFICIALLY AGED
T6	SOLUTION HEAT TREATED AND ARTIFICIALLY AGED
T7	SOLUTION HEAT TREATED AND STABILIZED
T8	SOLUTION HEAT TREATED, COLD WORKED AND ARTIFICIALLY AGED
T9	SOLUTION HEAT TREATED,ARTIFICIALLY AGED AND COLDWORKED
T10	COOLED FROM AN ELEVATED TEMPERATURE SHAPING PROCESS, COLD WORKED AND ARTIFICIALLY AGED

First digit values for indicating the major alloying element in wrought aluminum alloys are in Table 5–14.

Commonly used temper designations and their meanings are given in Table 5–15.

A listing of AA/ANSI codes specifying wrought aluminum alloys in production drawings is given in Table 5–16.

Cast Aluminum Alloys

Cast alloys are made by pouring the molten metal into a mold to form a part. Increased strength can be accomplished by applying heat treatment after casting. Wrought alloys tend to have higher strength than castings. In most cases thicker sections are required to produce a part as a casting.

A part with intricate curved surfaces may be less expensive to produce in rough form as a casting rather than a machining from wrought stock.

The AA identification system for castings uses a four digit code with the last digit separated by a period as shown in Figure 5–26.

First digit values to indicate the major alloying element in cast aluminum alloys are given in Table 5–17.

Table 5–18 provides a listing of AA/ANSI codes for different types of cast aluminum alloys used in manufacturing.

TABLE 5–16

WROUGHT - ALUMINUM ALLOYS

AA/ANSI CODE	COMMON TEMPERS	DESCRIPTION/USE	NOTE EXAMPLE
1100	O, F, HXX	GOOD FORMING PROPERTIES ARE EXHIBITED BY THIS ALLOY. IT IS USED FOR MAKING DECORATIVE PARTS.	TEMPER DESIGNATION NOTES: MATERIAL: AL ALY 7075-T6 ALLOY TYPE ALLOY GROUP CONTROL ON IMPURITIES (0=NO CONTROL) OR UNLESS OTHERWISE SPECIFIED DIMENSIONS ARE IN INCHES TOLERANCES ARE: FRACTIONS ± DECIMALS .XX± .XXX± ANGLES ± MATERIAL 7075-T6 FINISH DO NOT SCALE DRAWING
3003	O, F, HXX	THIS ALLOY IS SLIGHTLY STRONGER AND LESS DUCTILE THAN 1100. USES INCLUDE COOKING UTENSILS, BUILDING HARDWARE, PRESSURE VESSELS AND STORAGE TANKS.	
Alclad 3003	O, F, HXX	THIS ALLOY IS USED FOR CHEMICAL EQUIPMENT, PRESSURE VESSELS AND STORAGE TANKS.	
2011	T3, T8	A **FREE MACHINING** ALLOY USED FOR AUTOMATIC PRODUCTION ON SCREW MACHINES.	
Alclad 2014	T3, T4, T6	AN ALLOY USED FOR HEAVY DUTY STRUCTURES AND AIRCRAFT STRUCTURES.	
2024	T3, T4	THIS ALLOY IS USED FOR AIRCRAFT STRUCTURES,TRUCK WHEELS AND SCREW MACHINE PRODUCTS.	
6061	T4, T6	STRONG CORROSION RESISTANCE IS EXHIBITED BY THIS ALLOY. IT IS USED FOR HEAVY DUTY STRUCTURES, TRUCK FRAMES , PIPE AND PIPE FLANGES, RAILROAD CARS AND MARINE APPLICATIONS.	
7075	T6	THIS ALLOY HAS LOWER DUCTILITY THAN 2024. IT IS USED FOR AIRCRAFT STRUCTURES AND KEYS.	
Alclad 7075	T6	THIS IS THE STRONGEST ALCLAD PRODUCT. IT FINDS WIDE USE IN AIRCRAFT STRUCTURES.	

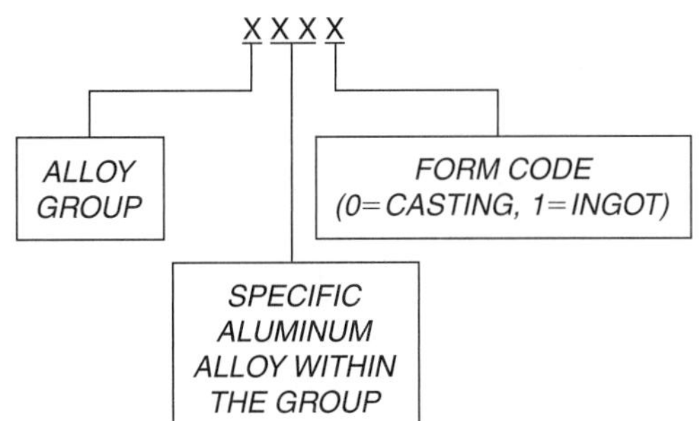

FIGURE 5-26 The AA system for identifying cast aluminum alloys.

TABLE 5-17

ALLOY GROUP	
No.	**MAJOR ALLOYING ELEMENT**
1XX.X	Aluminum-95.0% or greater
2XX.X	Copper
3XX.X	Silicon + Copper or Manganese
4XX.X	Silicon
5XX.X	Magnesium
6XX.X	unused
7XX.X	Zinc
8XX.X	Tin
9XX.X	other elements

TABLE 5-18

CAST - ALUMINUM ALLOYS			
AA/ANSI CODE	**COMMON TEMPERS**	**DESCRIPTION/USE**	**NOTE EXAMPLE**
208.0	F	A SAND CASTING THAT CAN BE HEAT TREATED AND IS USED FOR GENERAL PURPOSE.	NOTES: MATERIAL: AL ALY CASTING 535.0 ALLOY GROUP / CASTING OR UNLESS OTHERWISE SPECIFIED DIMENSIONS ARE IN INCHES TOLERANCES ARE: FRACTIONS ± DECIMALS .XX± .XXX± ANGLES ± MATERIAL CASTING 535.0 FINISH DO NOT SCALE DRAWING
355.0	T6, T7	THIS ALLOY IS FORMED BY EITHER SAND CASTING OR PERMANENT MOLD CASTING. USES INCLUDE HIGH STRENGTH AIRCRAFT PARTS AND PRESSURE VESSELS.	
356.0	T5, T6	FORMING PROCESSES INCLUDE SAND AND PERMANENT MOLD CASTINGS. IT IS USED FOR AUTO TRANSMISSIONS AND WHEELS.	
535.0	F, T5	A SAND CASTING THAT EXHIBITS GOOD CORROSON RESISTANGE AND MACHINABILITY.	
712.0	F, T5	A SAND CASTING THAT IS USED FOR MACHINE PARTS.	
850.0	T5	THIS IS A BEARING ALLOY FORMED BY SAND OR PERMANENT MOLD CASTING.	

5-14 UNDERSTANDING HEAT TREATMENT NOTES

Many parts may require heat treatments before or after rough machining and again after finish machining. These treatments involve controlled heating and cooling of the metal. They can be accompanied by diffusion processes in which the surface of the part is hardened by absorbing carbon, nitrogen, or a combination of both. Typical reasons for applying heat treatments are:

- Softening hard metals to improve their machinability
- Improving hardness and wear resistance after machining
- Easily machining regular carbon steels then heat treating to achieve same performance as alloy steels
- Preventing distortions and cracking resulting from internal stresses induced by rough machining

In larger companies the methods engineer specifies the sequence of operations such as rough machining, heat treatment, and finish machining, and the material allowances for each process. In medium to smaller companies, the CNC programmer must specify the amount of material that is to be left for these operations. Allowances are determined by applying reference data from publications such as the *Machining Data Handbook* to the following production factors:

- Size of the part
- Part material
- Type of machining operation
- Type of workholding fixture
- Type of heat treatment

It is important to note that a part's dimensions should be checked after heat treatments. In particular, stress relief and nitriding heat treatments will definitely affect dimensional accuracy.

Material Hardnesses

Hardness of a metal has been defined in terms of its resistance to local penetration, to scratching, to machining, to wear or abrasion, and to yielding. The most common type of testing measures resistance to penetration by a hardened steel ball of known dimensions or a conical shaped diamond brale penetrator. The hardness number is important to predicting the properties and performance of a metal.

Rockwell Hardness Tests

This is the most widely used test of hardness. A minor load is first applied to a steel ball or diamond brale penetrator to create a reference point. This is followed by the application of a major load. The major load is removed and the tester displays the Rockwell hardness number based upon the depth of penetration sensed. The tester has a reversed scale so that harder metals having smaller indentations produce higher Rockwell numbers and softer metals with larger indentations generate lower Rockwell numbers. A letter is commonly placed to the left of the hardness number to indicate the combination of major load and penetrator used. This is shown in Figure 5–27.

■ EXAMPLE 5–9

HRC 50-60
Specifies that a hardness range of 50 to 60 was obtained when a major load of 150kg was applied to a brale penetrator. ■■

ROCKWELL HARDNESS TEST SCALES			
SCALE	MAJOR LOAD(kg)	INDENTOR	TYPE OF MATERIAL TESTED
A	60	DIAMOND BRALE	EXTREMELY HARD STEEL AND CARBIDES
B	100	1/16" DIA STEEL BALL	MEDIUM-HARD METALS AND NON-FERROUS METALS
C	150	DIAMOND BRALE	THICK HARD METALS THAT ARE HARDER THAN B100

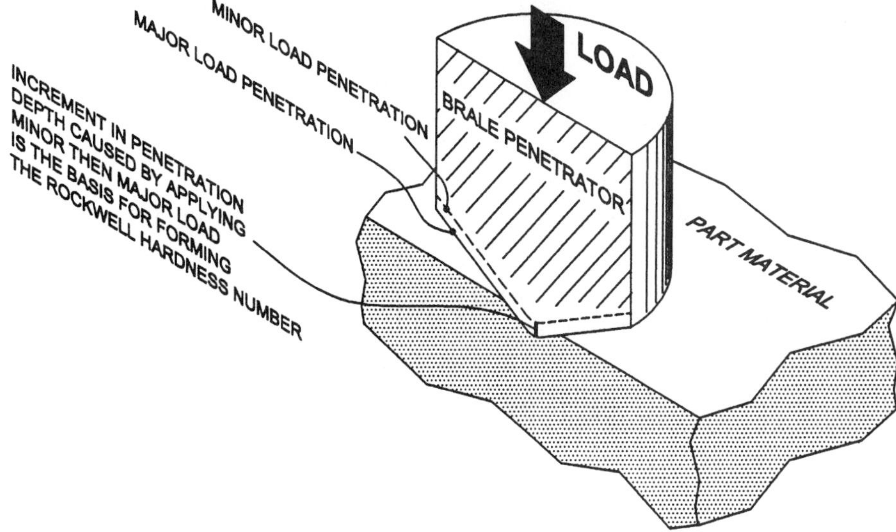

FIGURE 5–27 Common Rockwell Hardness Test Scales.

Rockwell hardness testing is easily adapted to mass production and causes minor marks on the surface of the tested part.

Brinell Hardness Tests

The Brinell test is an earlier method of measuring hardness that is still in use today. The test involves using a 500-, 1500-, or 3000-kg load to press a hardened steel ball 1 cm in diameter into the test metal. The resulting surface area of indentation is measured. The Brinell hardness number (BHN) is then determined as:

$$\text{BHN} = \frac{\text{LOAD}}{\text{SURFACE AREA OF INDENTATION}}$$

Harder metals have lower indentation surface areas and higher Brinell numbers. Softer metals have larger indentification areas and lower hardness numbers.

Brinell tests cannot be used in very hard or soft metals or metals only having hardness at their surfaces. The test may also produce unacceptable marks in the surface of tested parts. One advantage is that the Brinell number reflects hardness over a larger area of the material than does Rockwell. Conversion tables exist for relating Brinell hardness numbers to Rockwell hardness values. Refer to Table 5–19.

TABLE 5–19

HARDNESS CONVERSIONS FOR SAE CARBON AND ALLOY STEELS		
BRINELL(BHN) (10 mm carbide ball, 3000 kg load)	ROCKWELL(HR)	
	C	B
653	60	—
555	54.5	—
477	49.5	—
415	44.5	—
375	40.5	—
331	35.5	108.5
285	30	105.5
⋮	⋮	⋮

HARDER

Heat Treatments Affecting the Entire Part Material

The heat treatments listed in Table 5–20 affect the surface as well as the interior of the part material. They influence the mechanical properties of the metal but do not change its chemical composition.

Heat Treatments Affecting Selected Part Areas

The through heat treatments described in Table 5–21 cause some amount of distortion. In many cases it may be necessary to harden only selected areas of the part—the punch portion of a center punch or the contact surfaces of a gear. This approach has many benefits. It ensures greater dimensional stability as well as faster and less costly production time. Furthermore, it provides for certain areas of the part to be hard and wear resistant as needed and other areas to be flexible to prevent cracking during operation.

Surface heat treatments involve hardening the part at its surface and at some controlled depth below the surface. They fall into two broad categories: those that do not change the chemical composition of the metal (Table 5–21) and those that change the chemical composition of the surface region (Table 5–22).

5–15 INTERPRETING SURFACE COATING NOTES

It is common practice to treat parts with surface coatings after machining. The principal reasons for applying coatings include

- Enhancing the part's visual appearance
- Improving corrosion resistance
- Decreasing service wear, fatigue, and friction control
- Controlling or altering dimensions

The CNC programmer needs to understand that coatings alter part dimensions. Special attention must be paid to drawing notes that specify whether the dimensions are to apply before or after coating. If the dimensions are to be met after coating, the programmer must subtract the coating thickness from the given dimensions. The CNC program is then written for the corresponding before-coat dimensions. A listing of the most commonly applied surface coatings is given in Table 5–23.

TABLE 5–20

LIST OF THROUGH HEAT TREATMENTS
(ENTIRE MATERIAL OF PART IS AFFECTED)

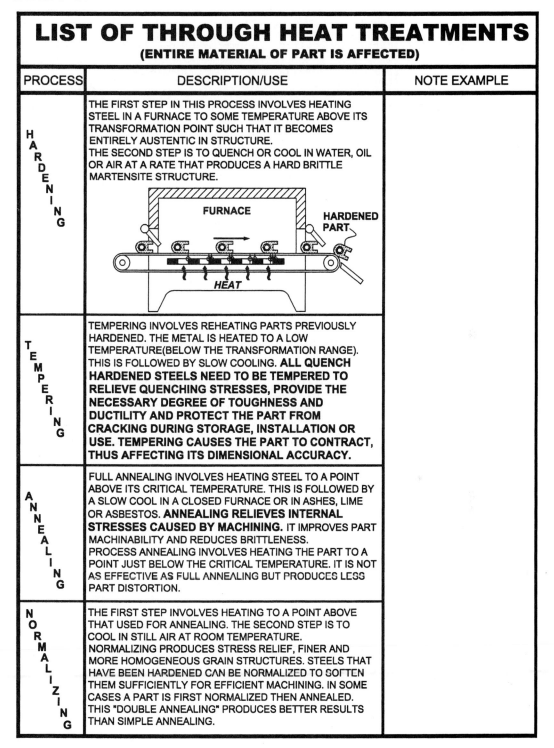

PROCESS	DESCRIPTION/USE	NOTE EXAMPLE
HARDENING	THE FIRST STEP IN THIS PROCESS INVOLVES HEATING STEEL IN A FURNACE TO SOME TEMPERATURE ABOVE ITS TRANSFORMATION POINT SUCH THAT IT BECOMES ENTIRELY AUSTENTIC IN STRUCTURE. THE SECOND STEP IS TO QUENCH OR COOL IN WATER, OIL OR AIR AT A RATE THAT PRODUCES A HARD BRITTLE MARTENSITE STRUCTURE. 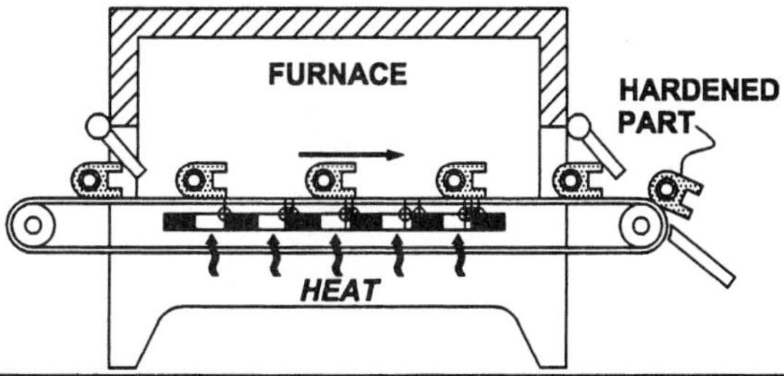 FURNACE HARDENED PART HEAT	
TEMPERING	TEMPERING INVOLVES REHEATING PARTS PREVIOUSLY HARDENED. THE METAL IS HEATED TO A LOW TEMPERATURE(BELOW THE TRANSFORMATION RANGE). THIS IS FOLLOWED BY SLOW COOLING. **ALL QUENCH HARDENED STEELS NEED TO BE TEMPERED TO RELIEVE QUENCHING STRESSES, PROVIDE THE NECESSARY DEGREE OF TOUGHNESS AND DUCTILITY AND PROTECT THE PART FROM CRACKING DURING STORAGE, INSTALLATION OR USE. TEMPERING CAUSES THE PART TO CONTRACT, THUS AFFECTING ITS DIMENSIONAL ACCURACY.**	
ANNEALING	FULL ANNEALING INVOLVES HEATING STEEL TO A POINT ABOVE ITS CRITICAL TEMPERATURE. THIS IS FOLLOWED BY A SLOW COOL IN A CLOSED FURNACE OR IN ASHES, LIME OR ASBESTOS. **ANNEALING RELIEVES INTERNAL STRESSES CAUSED BY MACHINING.** IT IMPROVES PART MACHINABILITY AND REDUCES BRITTLENESS. PROCESS ANNEALING INVOLVES HEATING THE PART TO A POINT JUST BELOW THE CRITICAL TEMPERATURE. IT IS NOT AS EFFECTIVE AS FULL ANNEALING BUT PRODUCES LESS PART DISTORTION.	
NORMALIZING	THE FIRST STEP INVOLVES HEATING TO A POINT ABOVE THAT USED FOR ANNEALING. THE SECOND STEP IS TO COOL IN STILL AIR AT ROOM TEMPERATURE. NORMALIZING PRODUCES STRESS RELIEF, FINER AND MORE HOMOGENEOUS GRAIN STRUCTURES. STEELS THAT HAVE BEEN HARDENED CAN BE NORMALIZED TO SOFTEN THEM SUFFICIENTLY FOR EFFICIENT MACHINING. IN SOME CASES A PART IS FIRST NORMALIZED THEN ANNEALED. THIS "DOUBLE ANNEALING" PRODUCES BETTER RESULTS THAN SIMPLE ANNEALING.	

TABLE 5–20 *(continued)*

SPHERODIZING	THIS PROCESS IS SIMILAR TO NORMALIZING EXCEPT THAT THE STEEL IS HEATED TO A POINT JUST BELOW THE CRITICAL TEMPERATURE. A ROUNDED OR GLOBULAR FORM OF CARBIDE IS PRODUCED. SPHERODIZING IS APPLIED TO STEEL TO IMPROVE DUCTILITY, FORMING QUALITIES AND MACHINABILITY.	**NOTES** **1. MATERIAL: SAE 1030** **2. ROUGH MACHINE** **3. STRESS RELIEVE** **4. FINISH MACHINE + .003 TO .005 GRIND ALLOWANCE** **5. HARDEN AND TEMPER TO 50 TO 60 HRC** **6. FINISH GRIND**
STRESS RELIEVING	**STRESS RELIEVING IS A COST EFFECTIVE METHOD OF LOWERING INTERNAL STRESSES CAUSED BY MACHINING.** IT IS A PROCESS THAT IS SIMILAR TO ANNEALING EXCEPT THAT THE PART IS HEATED TO A POINT 100 TO 200°F BELOW THE CRITICAL TEMPERATURE. WHILE THE PROCESS IS NOT AS EFFECTIVE AS ANNEALING IT OFFERS THE FOLLOWING ADVANTAGES: 1. LESS COSTLY AND TIME CONSUMING THAN ANNEALING 2. SUFFICIENT IN MOST CASES TO ELIMINATE CRACKING AND DISTORTION DUE TO MACHINING STRESSES 3. INVOLVES LESS DISTORTION ON COMPLEX PARTS THAN ANNEALING **IT IS COMMON PRACTICE TO APPLY STRESS RELIEF TO PARTS WITH CLOSE TOLERANCES.** FINISHED MACHINING AND GRINDING USUALLY FOLLOW STRESS RELIEVING OR ANNEALING TO ENSURE DIMENSIONAL ACCURACY.	

TABLE 5-21

LIST OF SELECTIVE HEAT TREATMENTS

(SURFACE OF PART AFFECTED-CHEMICAL COMPOSITION OF METAL *NOT* AFFECTED)

PROCESS	DESCRIPTION/USE	NOTE EXAMPLE
INDUCTION HARDENING	THIS TECHNIQUE INVOLVES PLACING THE PART INSIDE COPPER INDUCTION COILS AND INDUCING AN ELECTRIC CURRENT THROUGH THE METAL VIA A HIGH FREQUENCY CURRENT. THE RESISTANCE OF THE METAL TO THE FLOW OF CURRENT CAUSES ITS SURFACE TO HEAT UP TO A POINT ABOVE THE CRITICAL TEMPERATURE IN A MATTER OF SECONDS. AN AUTOMATIC SPRAY OF WATER, OIL OR COMPRESSED AIR IS USED TO QUENCH AND SUBSEQUENTLY HARDEN THE SURFACE. INDUCTION HEATING CAN ALSO BE USED FOR THROUGH HARDENING, ANNEALING, NORMALIZING AND STRESS RELIEVING IN SELECTED AREAS OF THE PART. 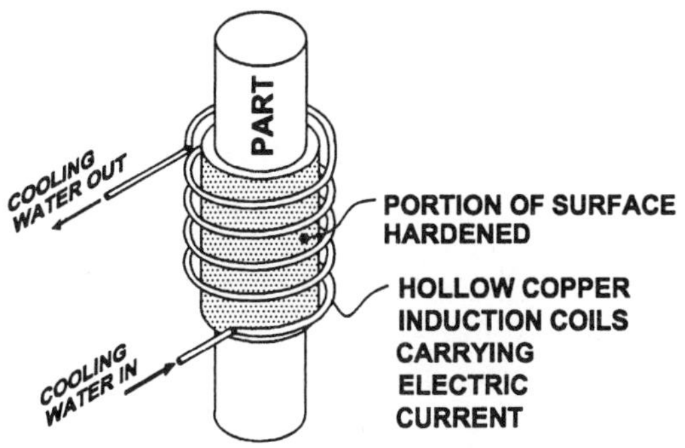 PART COOLING WATER OUT PORTION OF SURFACE HARDENED HOLLOW COPPER INDUCTION COILS CARRYING ELECTRIC CURRENT COOLING WATER IN	NOTES 1. MATERIAL: SAE 1040 2. ROUGH MACHINE 3. STRESS RELIEVE 4. FINISH MACHINE 5. INDUCTION HARDEN AND TEMPER AREAS MARKED WITH A DOUBLE ASTERISK TO 55 TO 60 HRC
FLAME HARDENING	THE FIRST STEP IN THIS PROCESS INVOLVES USING AN OXY-ACETYLENE TORCH TO RAPIDLY HEAT THE SURFACE OF THE PART TO A POINT ABOVE THE CRITICAL TEMPERATURE. THIS IS FOLLOWED BY A QUENCHING SPRAY TO GENERATE A HARDENED LAYER VARYING IN DEPTH FROM THE SURFACE TO AS MUCH AS 0.25 in INTO THE METAL. IN THE LAST STEP A LOW TEMPERATURE TORCH FOLLOWS THE QUENCHING NOZZLE TO TEMPER THE HARDENED AREAS. THIS METHOD IS USUALLY APPLIED TO PLAIN CARBON STEELS(0.40% - 0.95%C), ALLOY STEELS AND CAST IRONS. 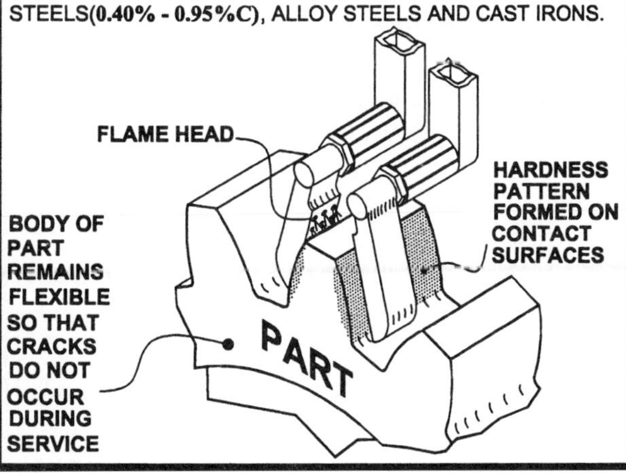 FLAME HEAD HARDNESS PATTERN FORMED ON CONTACT SURFACES BODY OF PART REMAINS FLEXIBLE SO THAT CRACKS DO NOT OCCUR DURING SERVICE PART	NOTES 1. MATERIAL: SAE 1050 2. ROUGH MACHINE 3. STRESS RELIEVE 4. FINISH MACHINE 5. FLAME HARDEN AREAS MARKED WITH A DOUBLE ASTERISK TO 0.10 IN DEEP MINIMUM. SURFACE HARDNESS AFTER TEMPERING 65 HRC MINIMUM

TABLE 5–22

LIST OF SELECTIVE HEAT TREATMENTS (SURFACE OF PART AND CHEMICAL COMPOSITION OF SURFACE *ARE* AFFECTED)		
PROCESS	DESCRIPTION/USE	NOTE EXAMPLE
CARBURIZING *or* CASE HARDENING	THE FIRST STEP IN THIS PROCESS INVOLVES HEATING LOW CARBON STEEL(LESS THAN 0.3%C) TO ITS CRITICAL TEMPERATURE WHILE IN A CARBON SATURATED ATMOSPHERE. THE CARBON WILL PENETRATE OR DIFFUSE INTO THE STEEL'S SURFACE CARBURIZING IT. IN THE SECOND STEP, THE METAL IS SLOW COOLED WHILE IN THE CARBON ATMOSPHERE SO THAT THE SURFACE AND A PORTION BELOW IS HARDENED. THIS REGION OF HARDNESS IS KNOWN AS A **CASE** .THE CORE REMAINS SOFT AND DUCTILE. 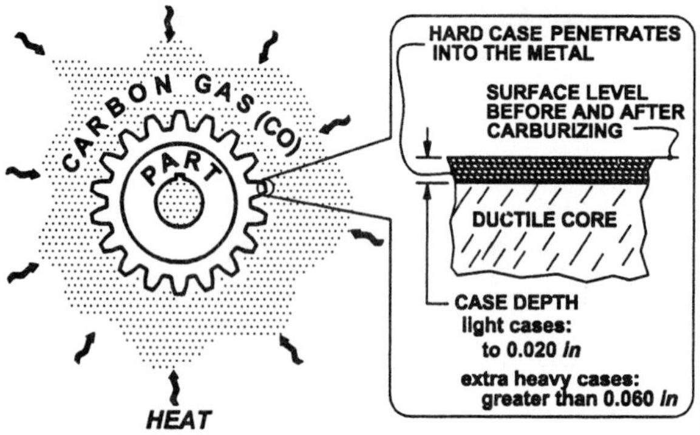 CARBON GAS (CO) PART HEAT HARD CASE PENETRATES INTO THE METAL SURFACE LEVEL BEFORE AND AFTER CARBURIZING DUCTILE CORE CASE DEPTH light cases: to 0.020 *in* extra heavy cases: greater than 0.060 *in*	NOTES 1. MATERIAL: SAE 1030 2. ROUGH MACHINE 3. STRESS RELIEVE 4. FINISH MACHINE 5. CARBURIZE TO 0.020 TO 0.025 IN DEEP 6. REMOVE CASE BY MACHINING IN AREAS MARKED WITH A DOUBLE ASTERISK 7. HARDEN AND TEMPER TO 50 TO 60 HRC
NITRIDING	THIS PROCESS INVOLVES PLACING CERTAIN ALLOY STEEL PARTS INTO AN ATMOSPHERE OF NITROGEN(AMMONIA GAS) AND HEATING TO A POINT BELOW THE CRITICAL TEMPERATURE. THE NITROGEN IS DIFFUSED INTO THE SURFACE OF THE STEEL. THE RESULTING CASE IS VERY HARD AND CONSISTS OF A WHITE BRITTLE OUTER LAYER, A HARD NITRIDE LAYER AND A DIFFUSION ZONE OF DECREASING HARDNESS. QUENCHING IS **NOT** REQUIRED AND THE PART MAY BE SLOW COOLED IN THE RETORT. THE WHITE LAYER IS USUALLY REMOVED. **NITRIDING CAUSES VERY MINIMAL PART DISTORTION BUT CREATES GROWTH IN DIMENSIONS ON TREATED SURFACES. FOR NORMAL CASES 0.005 TO 0.020 in GROWTH IS IN THE RANGE OF 0.0005 TO 0.001 in.** NITRIDED SURFACES ARE FINISHED GROUND AS NEEDED 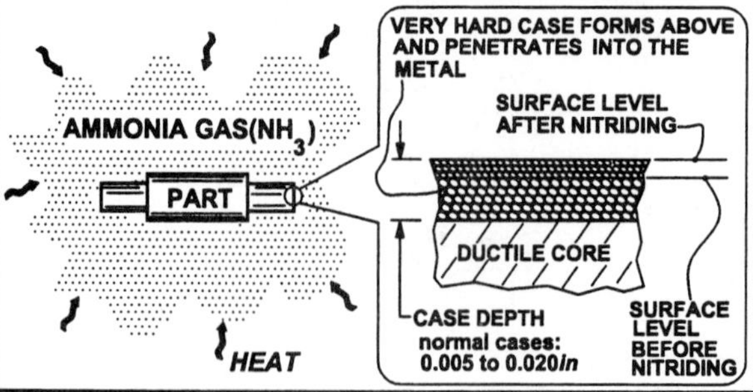 AMMONIA GAS(NH_3) PART HEAT VERY HARD CASE FORMS ABOVE AND PENETRATES INTO THE METAL SURFACE LEVEL AFTER NITRIDING DUCTILE CORE CASE DEPTH normal cases: 0.005 to 0.020*in* SURFACE LEVEL BEFORE NITRIDING	NOTES 1. MATERIAL: AISI 4140 2. ROUGH MACHINE 3. HARDEN AND TEMPER TO 30 TO 40 HRC 4. FINISH MACHINE 5. NITRIDE ALL OVER TO 0.010 TO 0.015 DEEP 6. REMOVE WHITE LAYER 0.001 TO 0.002

TABLE 5-23

LIST OF PROTECTIVE COATINGS

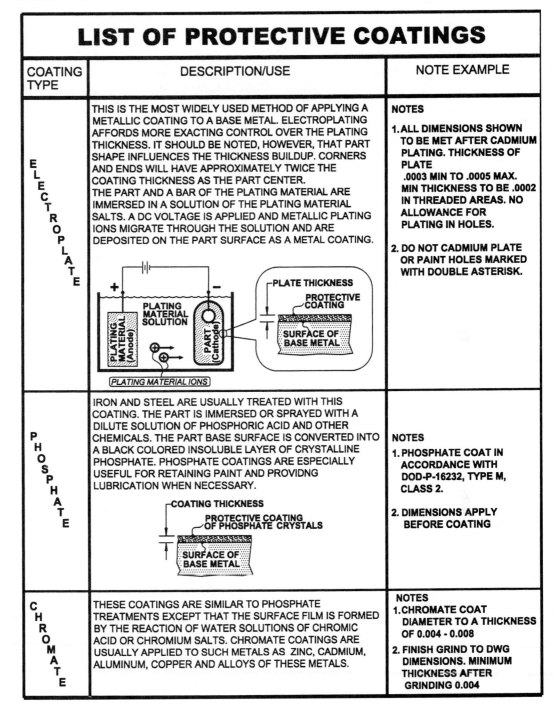

COATING TYPE	DESCRIPTION/USE	NOTE EXAMPLE
ELECTROPLATE	THIS IS THE MOST WIDELY USED METHOD OF APPLYING A METALLIC COATING TO A BASE METAL. ELECTROPLATING AFFORDS MORE EXACTING CONTROL OVER THE PLATING THICKNESS. IT SHOULD BE NOTED, HOWEVER, THAT PART SHAPE INFLUENCES THE THICKNESS BUILDUP. CORNERS AND ENDS WILL HAVE APPROXIMATELY TWICE THE COATING THICKNESS AS THE PART CENTER. THE PART AND A BAR OF THE PLATING MATERIAL ARE IMMERSED IN A SOLUTION OF THE PLATING MATERIAL SALTS. A DC VOLTAGE IS APPLIED AND METALLIC PLATING IONS MIGRATE THROUGH THE SOLUTION AND ARE DEPOSITED ON THE PART SURFACE AS A METAL COATING. 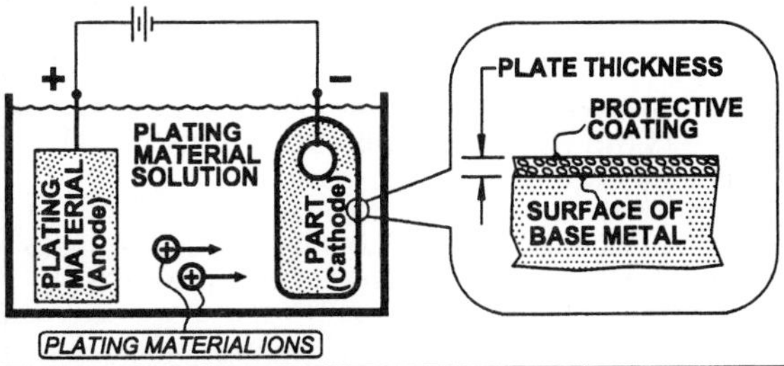 	NOTES 1. ALL DIMENSIONS SHOWN TO BE MET AFTER CADMIUM PLATING. THICKNESS OF PLATE .0003 MIN TO .0005 MAX. MIN THICKNESS TO BE .0002 IN THREADED AREAS. NO ALLOWANCE FOR PLATING IN HOLES. 2. DO NOT CADMIUM PLATE OR PAINT HOLES MARKED WITH DOUBLE ASTERISK.
PHOSPHATE	IRON AND STEEL ARE USUALLY TREATED WITH THIS COATING. THE PART IS IMMERSED OR SPRAYED WITH A DILUTE SOLUTION OF PHOSPHORIC ACID AND OTHER CHEMICALS. THE PART BASE SURFACE IS CONVERTED INTO A BLACK COLORED INSOLUBLE LAYER OF CRYSTALLINE PHOSPHATE. PHOSPHATE COATINGS ARE ESPECIALLY USEFUL FOR RETAINING PAINT AND PROVIDNG LUBRICATION WHEN NECESSARY. 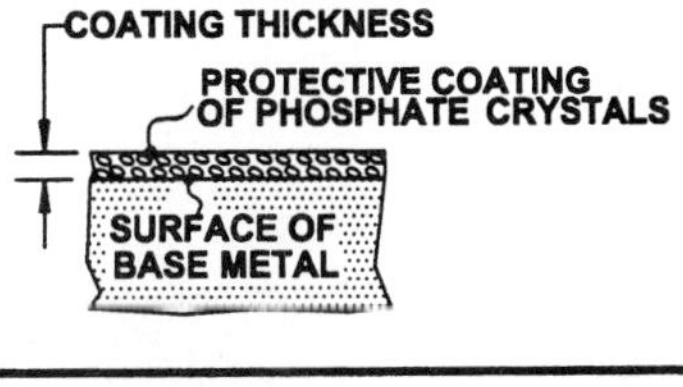 	NOTES 1. PHOSPHATE COAT IN ACCORDANCE WITH DOD-P-16232, TYPE M, CLASS 2. 2. DIMENSIONS APPLY BEFORE COATING
CHROMATE	THESE COATINGS ARE SIMILAR TO PHOSPHATE TREATMENTS EXCEPT THAT THE SURFACE FILM IS FORMED BY THE REACTION OF WATER SOLUTIONS OF CHROMIC ACID OR CHROMIUM SALTS. CHROMATE COATINGS ARE USUALLY APPLIED TO SUCH METALS AS ZINC, CADMIUM, ALUMINUM, COPPER AND ALLOYS OF THESE METALS.	NOTES 1. CHROMATE COAT DIAMETER TO A THICKNESS OF 0.004 - 0.008 2. FINISH GRIND TO DWG DIMENSIONS. MINIMUM THICKNESS AFTER GRINDING 0.004

TABLE 5-23 *(continued)*

ELECTROLESS	THIS TECHNIQUE INVOLVES DIPPING THE PART IN AN AQUEOUS SOLUTION OF SODIUM HYPOPHOSPHITE AND NICKEL SALTS. THE PART ACTS AS A CATALYST ATTRACTING THE NICKEL IONS TO ITS SURFACE. UNLIKE ELECTROPLATING, ELECTROLESS NICKEL PLATING DOES NOT TEND TO CREATE THICKNESS BUILDUPS AT PART ENDS AND IN CAVATIES. A UNIFORM THICKNESS COATING IS PRODUCED OVER ALL THE PART'S SURFACES.	**NOTES** **1. ELECTROLESS NICKEL PLATE 0.001 IN AREAS MARKED WITH DOUBLE ASTERISK.** **2. DIMENSIONS APPLY AFTER COATING**
HOT DIP	CORROSION RESISTANT COATINGS ARE APPLIED TO A GREAT NUMBER OF ALUMINUM AND STEEL PARTS BY DIPPING THEM INTO CERTAIN TYPES OF MOLTEN METALS. **GALVANIZING** OR DIPPING STEEL INTO A MOLTEN ZINC BATH IS ONE OF THE MOST COMMONLY USED METHODS TO PROTECT STEEL PARTS. DIPPING STEEL INTO A MOLTEN BATH OF TIN IS CALLED **TIN PLATING.** AN ADVANTAGE OF HOT DIPPING IS LOW COST. A DISADVANTAGE IS THE LACK OF THICKNESS CONTROL. HOT DIPPING TENDS TO CREATE UNEVEN BUILDUPS OVER THE SURFACES OF THE PART ESPECIALLY IN CAVATIES AND ON PROTRUSIONS.	
ANODIZE	ANODIZING IS THE MOST COMMONLY USED METHOD OF APPLYING PROTECTIVE COATINGS TO ALUMINUM METALS.AS WITH ELECTROPLATING, PART SHAPE INFLUENCES THE THICKNESS BUILDUP WITH CORNERS AND ENDS HAVING APPROXIMATELY TWICE THE THICKNESS AS PART CENTER AREAS. THE ALUMINUM PART IS PLACED INTO AN ELECTROLYTE ACID BATH.A DIRECT CURRENT IS APPLIED IN WHICH THE PART BECOMES THE ANODE(HENCE THE TERM ANODIZING). THE BATH YIELDS OXYGEN WHICH REACTS WITH THE ALUMINUM SURFACE TO FORM A THICKENED, HARD, CERAMIC FILM OF ALUMINUM OXIDE. THE FILM BUILDS UP ABOVE THE SURFACE OF THE PART AND PENETRATES INTO THE BASE METAL AS WELL. NORMAL PRACTICE IS THAT DRAWING DIMENSIONS ARE TO APPLY AFTER ANODIZING. 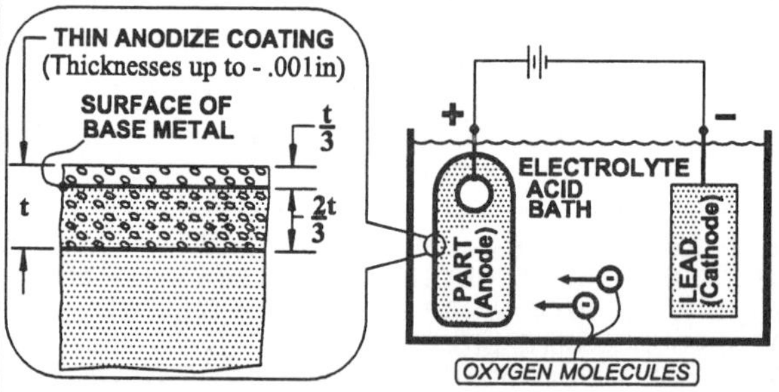 THIN ANODIZE COATING (Thicknesses up to - .001in) SURFACE OF BASE METAL t $\frac{t}{3}$ $\frac{2t}{3}$ + – ELECTROLYTE ACID BATH PART (Anode) LEAD (Cathode) OXYGEN MOLECULES	**NOTES** **1. ANODIZE ENTIRE PART PER MILL - A-8625, TYPE III, CLASS 1. THICKNESS RANGE .0005-.001 INCHES**

5–16 CHAPTER SUMMARY

The following key concepts were discussed in this chapter:

1. Standard sheet sizes in the United States are specified by the American National Standards Institute document (ANSI) Y14.1. The units used are inches (in.). European countries use standard sheets sized in millimeter (mm) units.
2. In addition to the drawing area, mechanical drawing forms also have four general areas of information: title block, parts list, revision block, and notes block.
3. Line types are used to convey different types of graphic and non-graphic information in drawings. Line types include visible or object, hidden, center, dimension, extension, cutting plane, section, leader, and break.
4. Mechanical drawings utilize the principle of othographic projection to show the true shape and size of part faces.
5. Third angle projection is an othographic projection wherein the object is placed in the third quadrant behind the projection planes. Third angle projection is used in the United States.
   First angle projection is an othographic projection that involves placing the object in the first quadrant in front of the projection planes. European countries use first angle projection.
6. Auxiliary views are needed to show the true shape and size of part faces that are inclined with respect to the principal othographic views.
7. A section view is used to expose the internal features of a part.
8. Dimension standards for mechanical drawings are outlined in the document ASME Y14-5M-1994.
   Inch (in.) units are used for dimensioning parts in the United States. The international communities use units of millimeters (mm) for dimensioning parts.
9. Tolerances control the accuracy with which a part is manufactured.
10. The different versions of pitch for the same thread form comprise a thread series.
11. Surface roughness influences such key operating factors as friction, fatigue life, corrosion resistance, contact stresses, and vibration.
12. The part material heavily influences such important machining parameters as cutting tool material, cutting speed, cutting feed, and depth of cut.
13. CNC programmers take into consideration the necessary machining operations and material allowances that are made when the part drawing notes call for heat treatments.
14. The CNC programmer reads the part drawing notes to determine if the part dimensions are to be met before or after specified coatings.

REVIEW EXERCISES

5.1. What is the width and length of a B-size drawing?
5.2. How much larger is a C-size drawing than a B-size?
5.3. What size metric drawing is the closest to C-size?
5.4. Title Block is located in the lower right hand corner. True or false?
5.5. Where is the Parts list or the Bill of Materials located?
5.6. Name the type of notes that applies to the entire drawing.
5.7. What does a symmetry line indicate?
5.8. Describe a section line.
5.9. Name the six principal othographic views.
5.10. What angle projection is used in the United States? Describe it.

5.11. What is an Edge on a part?
5.12. In what kind of geometric shape does a cylinder project on a profile projection plane? On a front projection plane?
5.13. Auxiliary projection plane is parallel to the inclined face. True or False?
5.14. How is a section view created?
5.15. In a section view, are ribs, webs, and gear teeth sectioned?
5.16. Can a detail section appear at a larger scale?
5.17. What is meant by referring to a dimension as Basic Size?
5.18. Describe the term Datum. Is it used for CNC part drawings?
5.19. Are fillets the same as rounds?
5.20. Specify a .25 diameter hole with a .375 diameter, .25 deep counterbore in ASME Y14.5 standard.
5.21. Would a shaft with a left-hand thread engage with a mating part by rotating the shaft clockwise?
5.22. What distance will a triple thread shaft advance in 360-degree revolution.
5.23. How many threads per inch does a 1/4-20 thread have?
5.24. In what units is roughness measured in inches? In metric?
5.25. Why is reading material specifications important?
5.26. Is aluminum a metal? If yes, is it ferrous?
5.27. Can heat treatment affect the shape and/or size of the part? If yes, how?
5.28. Does a CNC programmer have to pay attention to drawing notes describing part coatings and plating? If yes, why?

BIBLIOGRAPHY

Abbreviations for Use on Drawings and Text	ANSI/ASME Y1.1-1989
American National Standard Drafting Practice:	
Metric Drawing Sheet Size and Format	ANSI/ASME Y14.1M-1995
Decimal Inch Drawing Sheet Size and Format	ANSI/ASME Y14.1M-1995
Line Conventions and Lettering	ANSI/ASME Y14.2M-1992
Multi and Sectional View Drawings	ANSI/ASME Y14.3M-1994
Revision of Engineering Drawings	ANSI/ASME Y14.35M-1992
Dimensioning and Tolerancing	ANSI/ASME Y14.5M-1994
Screw Thread Representations	ANSI/ASME Y14.6-1993
Screw Thread Representations (Metric Supplement)	ANSI/ASME Y14.6aM-1993
Types and Applications of Engineering Drawings	ANSI/ASME Y14.24M-1989
Parts Lists, Data Lists, and Index Lists	ANSI/ASME Y14.34M-1989
Surface Texture Symbols	ANSI/ASME Y14.36-1993

Brown W. C. *Blueprint Reading for Industry.* Goodheart-Willcox, South Holland, Illinois, 1989.

Degarmo E. P., Black J. T., Kohser R. A. *Materials and Processes in Manufacturing.* Prentice Hall, Upper Saddle River, NJ, 1997.

Kirkpatrick J. M. *Industrial Blueprint Reading and Sketching.* Merrill Publishing, Columbus, OH, 1989.

Krar S. F., Rapisadra M., Check A. F. *Machine Tool and Manufacturing Technology.* Delmar Publishers, Albany, NY, 1998.

Krulikowski A. *Fundamentals of Geometric Dimensioning and Tolerancing.* Delmar Publishers, Albany, NY, 1998.

Machinery's Handbook, 25th Edition. Industrial Press, New York, 1997.

Machining Data Handbook, Vol. 1 and 2. 3rd Edition. Machinability Data Center, Cincinnati, Ohio, 1980.

Madsen D. A., Shumaker T. M., Turpin J. L., Stark C., *Engineering Drawing and Design.* Delmar Publishers, Albany, NY, 1996.

Maruggi E. A. *Current Practices for Interpreting Engineering Drawings.* West Publishing, Tucson, AZ, 1995.

Shultz R. *Blueprint Reading for the Machine Trades.* Prentice Hall, Upper Saddle River, NJ, 1988.

Stegman G. K. *Blueprint Reading for the Technician.* Macmillan Publishing Co., New York, 1987.

Tool and Manufacturing Engineers Handbook. Volume 1–6. SME, Dearborn, Michigan.

Wilson B. A., *Design Dimensioning and Tolerancing.* Goodheart-Willcox, South Holland, Illinois, 1992.

CHAPTER

6

Review of Basic Geometric Dimensioning and Tolerancing for CNC Programmers

6–1 CHAPTER OBJECTIVES

At the conclusion of this chapter you will be able to

1. Have a fundamental knowledge of the importance of geometric dimensioning and tolerancing and how it controls the manufacturing of a part.
2. Know how to interpret geometric dimensioning and tolerancing notes and symbols in production drawings.
3. Understand how the part features to be controlled will be inspected after production on a CNC machine.

6–2 INTRODUCTION

Geometric dimensioning and tolerancing (GDT) is a system of specifying tolerances based on how a part is to *function.* It is an important tool in manufacturing interchangeable, high quality, low costs parts via CNC technology. GDT is designed to specify the features that are to be manufactured into a part that conventional tolerances of position and size cannot specify. See Figure 6–1.

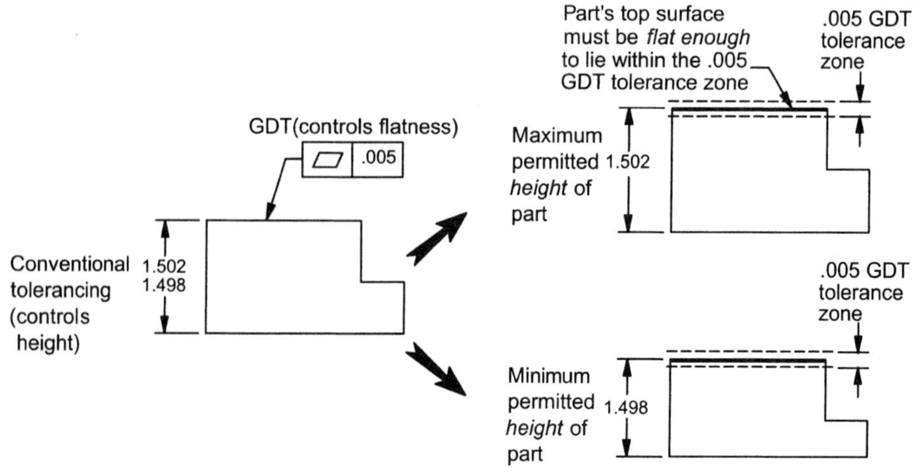

FIGURE 6–1 Conventional and GDT tolerancing.

6–3 GDT TERMINOLOGY

Feature: Any surface, angle, line, hole, etc. that is to be controlled in a part.

GDT Control Symbols: The USA document on GDT, ASME Y14.5M and the European ISO standards use symbols to indicate the type of control being specified. The symbol for straightness, for example, is—while that is for position is ⌖.

GDT Control Frame: A rectangular box containing the GDT control symbol, the tolerance value for the control, and any additional supplementary symbols.

■ EXAMPLE 6–1

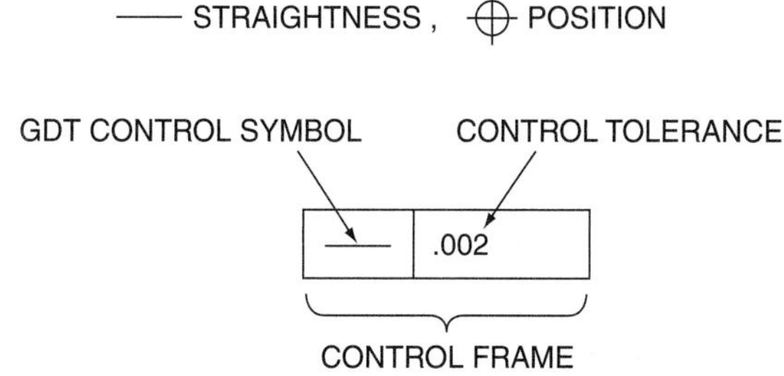

■■

Maximum Material Condition (MMC): The condition at which an external feature is at its largest size and an internal feature is at its smallest size. The important concept is that it is the condition at which the part weighs the most. The GDT symbol for maximum material condition is Ⓜ. In most cases geometric tolerances should be set at Ⓜ.

Least Material Condition (LMC): The condition at which an external feature is at its smallest size and an internal feature is at is largest size. When the part is at LMC it weighs the least. The GDT symbol for least material condition is Ⓛ. See Figure 6–2.

Perfect Form Envelope: An important GDT principle is that the perfect part form envelope exists only at Ⓜ. The part's size may vary between Ⓛ and Ⓜ but must never violate the Ⓜ boundary. Refer to Figure 6–3.

Some parts may have thin or flexible features that cannot be controlled as perfect at Ⓜ. In these cases, a note can be added to the drawing as follows:

PERFECT FORM AT MMC NOT REQUIRED

Regardless of Feature Size (RFS): The specification regardless of feature size means that the same geometric tolerance value for straightness, position, etc., applies to the feature at Ⓜ, Ⓛ, and all part sizes in between. Current changes in ASME Y14.5M-1994 specify that the use of the RFS symbol Ⓢ is to be discontinued. The RFS condition is to be assumed if the symbol Ⓜ or Ⓛ does not appear.

Tolerance Zone: The zone that represents the tolerance and its position relative to the basic size. The diameter symbol Ø is placed before a tolerance value to indicate a cylindrical tolerance zone. See Figure 6–4.

Virtual Condition: The virtual condition of a part is also known as its mating condition. For external features, it is the distortion that results when the part's size at Ⓜ is added to geometric tolerance specified at Ⓜ. For internal features, it is the distortion that results when the part's

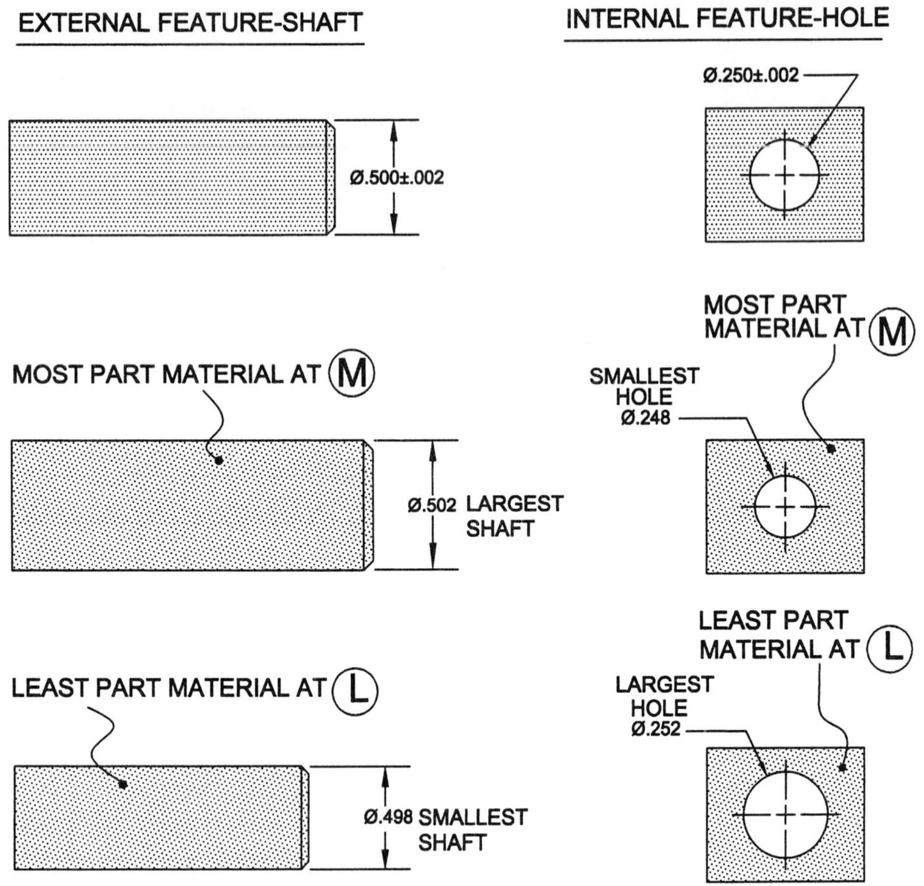

FIGURE 6–2 Maximum Material Condition and Least Material Condition.

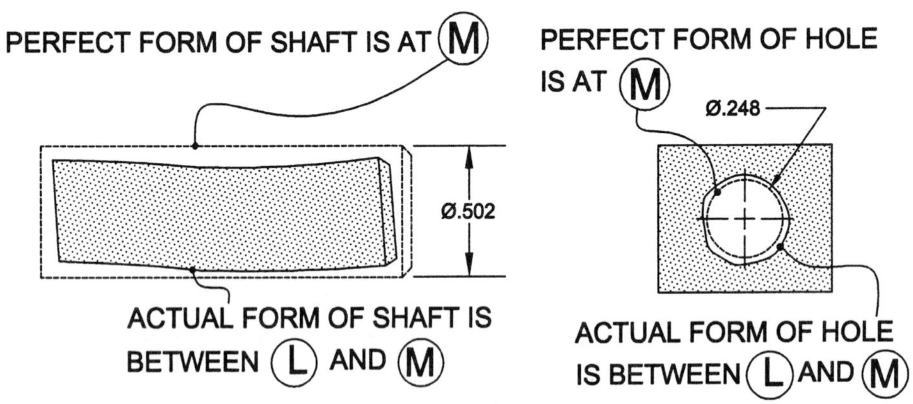

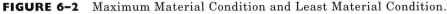

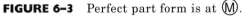

FIGURE 6–3 Perfect part form is at Ⓜ.

size at Ⓜ is subtracted from the geometric tolerance specified at Ⓜ. Refer to Figure 6–5.

Bonus Tolerances: The tolerance value that appears to the left of the symbol Ⓜ applies only at the maximum material condition. As part features depart from Ⓜ, the tolerance can be relaxed. The larger tolerances that are then permitted are called "bonus tolerances." See Figure 6–6.

Functional Gaging: A complete gage fixture that is specially designed to inspect a part for compliance with a GDT specification in a drawing. Functional gages make inspection easy and reduce inspection time and cost. The Ⓜ condition permits the use of functional gaging. Functional

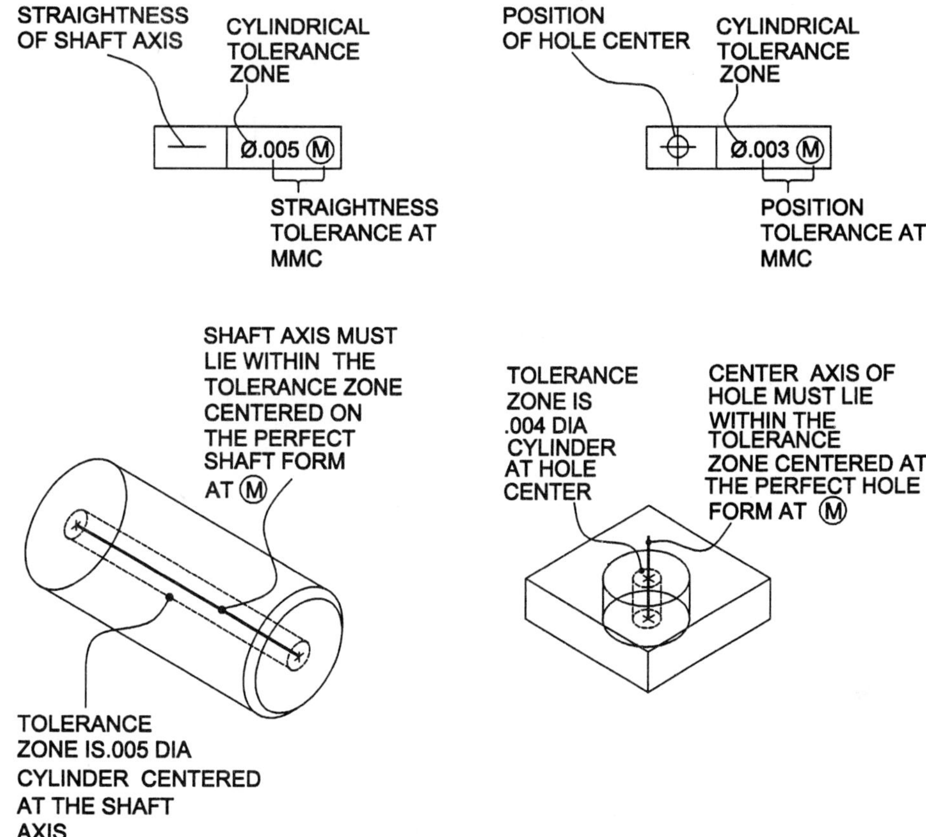

FIGURE 6–4 Examples of a cylindrical tolerance zone used in GDT.

gages are not useful for tolerances smaller than .001 in.

Open Gaging: This method of determining compliance with GDT specifications includes the use of such measuring devices as dial indicators, height gages, surface plates, micrometers, calipers and coordinate measuring machines (CMM's). This type of inspection is more costly and time consuming than is functional gaging. Open gaging must be used when Ⓜ is not permitted and for tolerances smaller than .001 in.

Full Indicator Movement (FIM): This callout in drawings applies to open gaging inspection and complies with the latest ANSI/ASME Y14.5M-1994 standards (formerly referred to as total indicator reading or T.I.R.). It specifies that the difference between the minimum(−) and maximum(+) readings or the *full* movement of the inspection instrument's dial indicator must not exceed the geometric tolerances appearing in the drawing.

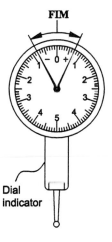

Geometric dimensioning and tolerancing involves the control of five basic types of geometric tolerances: form, profile, location, orientation, and runout.

6–4 PART FORM CONTROLS

Form tolerances control the degree to which a specific part surface or part feature is permitted to vary from its basic exact dimension on a drawing. Datums are not needed to measure form tolerances. Refer to Table 6–1 for a listing of GDT tolerances of form.

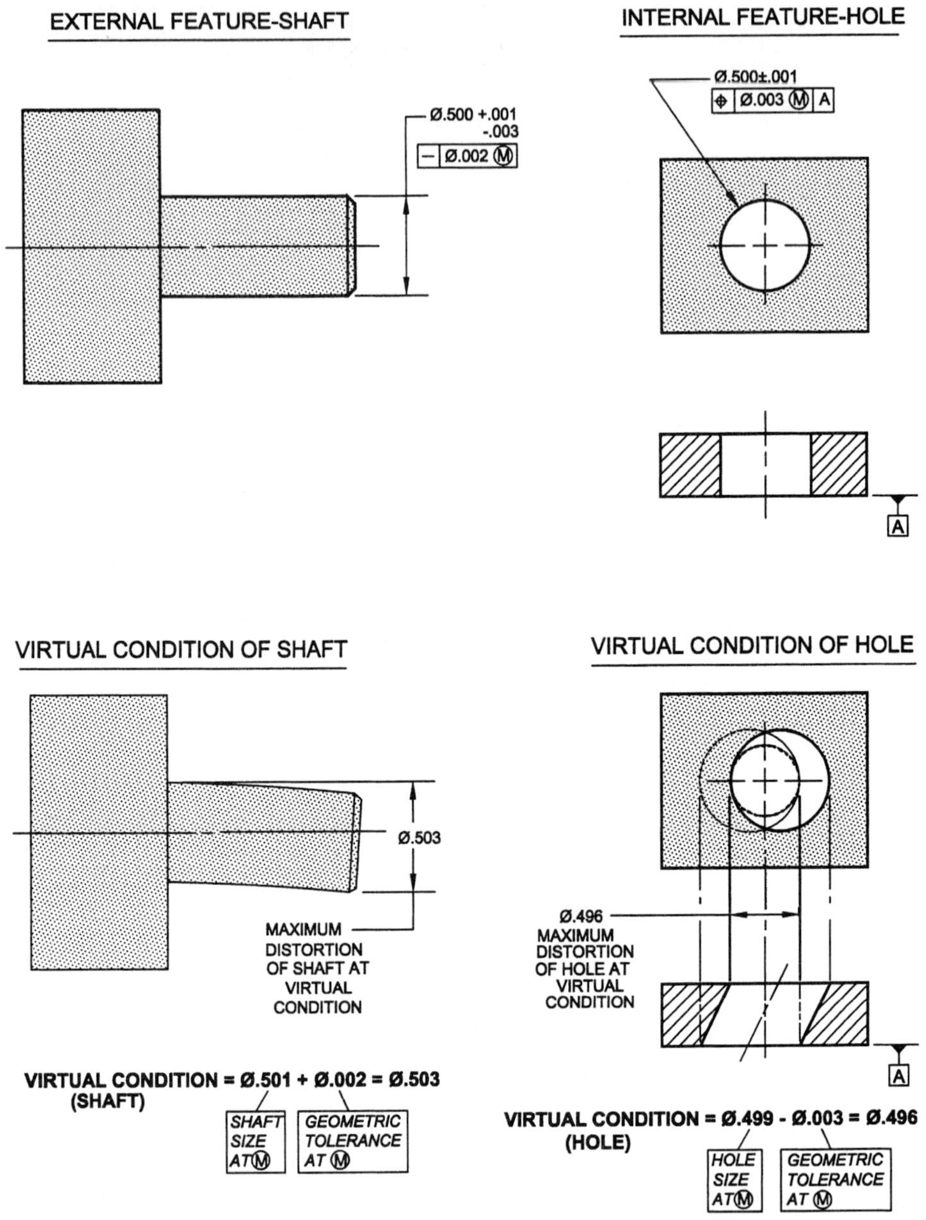

EXTERNAL FEATURE-SHAFT
Ø.500 +.001
-.003
— Ø.002 Ⓜ
INTERNAL FEATURE-HOLE
Ø.500±.001
⌖ Ø.003 Ⓜ A
A
VIRTUAL CONDITION OF SHAFT
Ø.503
MAXIMUM DISTORTION OF SHAFT AT VIRTUAL CONDITION
VIRTUAL CONDITION = Ø.501 + Ø.002 = Ø.503
(SHAFT)
SHAFT SIZE ATⓂ
GEOMETRIC TOLERANCE AT Ⓜ
VIRTUAL CONDITION OF HOLE
Ø.496
MAXIMUM DISTORTION OF HOLE AT VIRTUAL CONDITION
A
VIRTUAL CONDITION = Ø.499 - Ø.003 = Ø.496
(HOLE)
HOLE SIZE ATⓂ
GEOMETRIC TOLERANCE AT Ⓜ

FIGURE 6–5 Virtual condition for external and internal features.

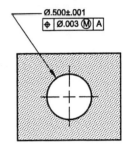

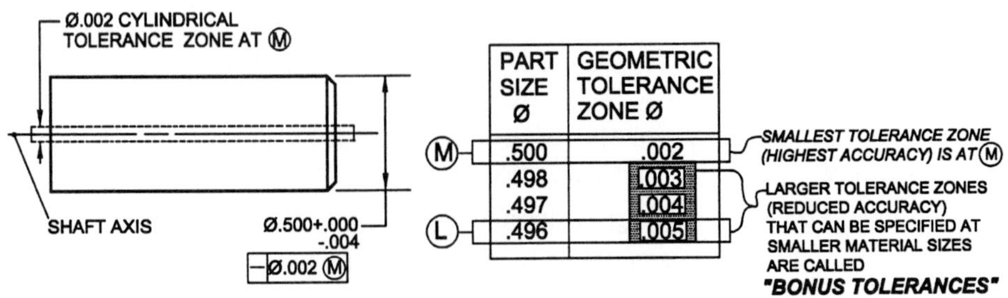

Ø.002 CYLINDRICAL TOLERANCE ZONE AT Ⓜ
SHAFT AXIS
Ø.500+.000
-.004
— Ø.002 Ⓜ
PART SIZE Ø
GEOMETRIC TOLERANCE ZONE Ø
Ⓜ .500 .002
.498 .003
.497 .004
Ⓛ .496 .005
SMALLEST TOLERANCE ZONE (HIGHEST ACCURACY) IS ATⓂ
LARGER TOLERANCE ZONES (REDUCED ACCURACY) THAT CAN BE SPECIFIED AT SMALLER MATERIAL SIZES ARE CALLED
"BONUS TOLERANCES"

FIGURE 6–6 Larger "bonus tolerances" are permitted as part features depart from Ⓜ.

TABLE 6-1

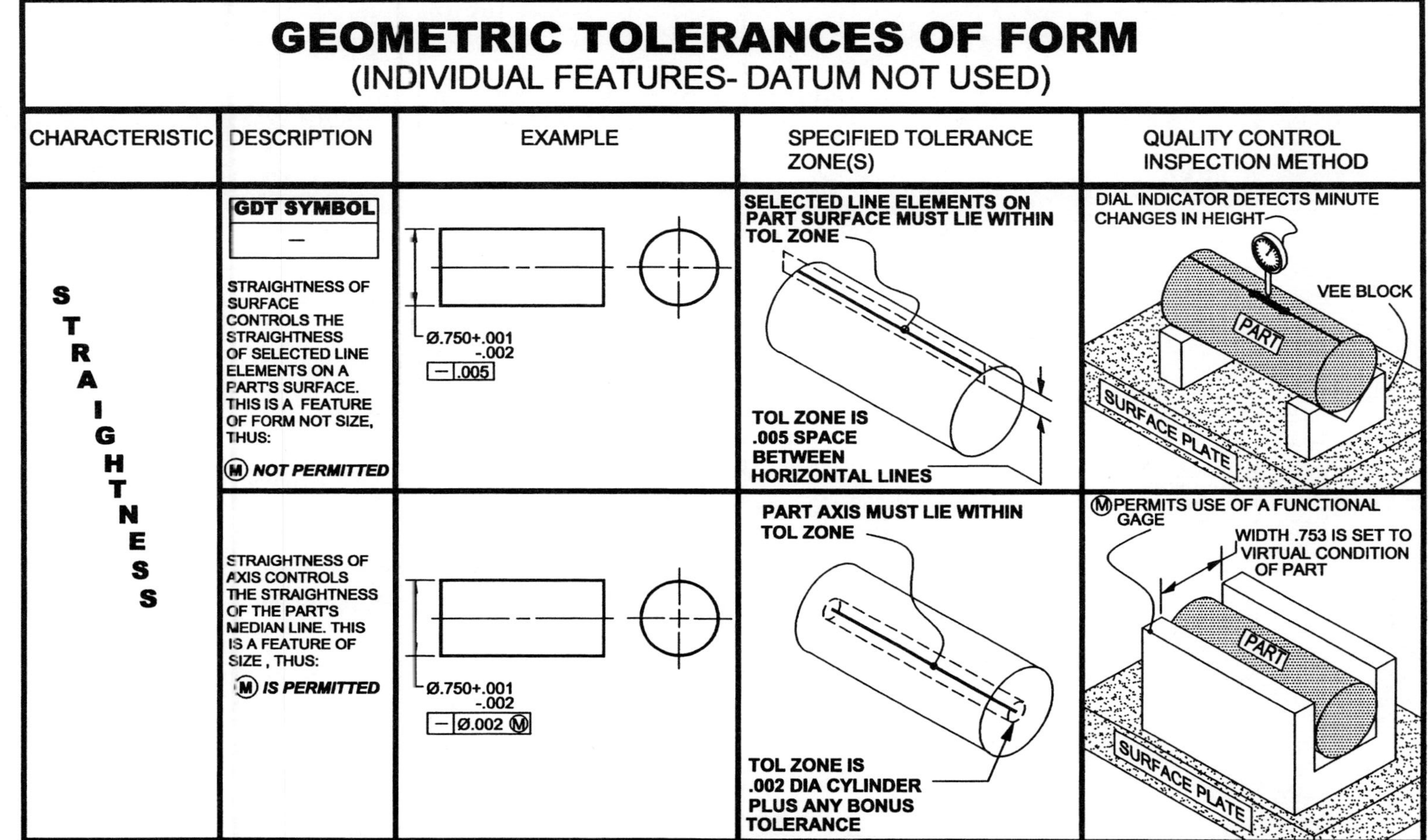

GEOMETRIC TOLERANCES OF FORM
(INDIVIDUAL FEATURES- DATUM NOT USED)

CHARACTERISTIC	DESCRIPTION	EXAMPLE	SPECIFIED TOLERANCE ZONE(S)	QUALITY CONTROL INSPECTION METHOD
STRAIGHTNESS	GDT SYMBOL: — STRAIGHTNESS OF SURFACE CONTROLS THE STRAIGHTNESS OF SELECTED LINE ELEMENTS ON A PART'S SURFACE. THIS IS A FEATURE OF FORM NOT SIZE, THUS: Ⓜ ***NOT PERMITTED***	Ø.750+.001 -.002 | — | .005 |	SELECTED LINE ELEMENTS ON PART SURFACE MUST LIE WITHIN TOL ZONE TOL ZONE IS .005 SPACE BETWEEN HORIZONTAL LINES	DIAL INDICATOR DETECTS MINUTE CHANGES IN HEIGHT VEE BLOCK PART SURFACE PLATE
	STRAIGHTNESS OF AXIS CONTROLS THE STRAIGHTNESS OF THE PART'S MEDIAN LINE. THIS IS A FEATURE OF SIZE , THUS: Ⓜ ***IS PERMITTED***	Ø.750+.001 -.002 | — | Ø.002 Ⓜ |	PART AXIS MUST LIE WITHIN TOL ZONE TOL ZONE IS .002 DIA CYLINDER PLUS ANY BONUS TOLERANCE	Ⓜ PERMITS USE OF A FUNCTIONAL GAGE WIDTH .753 IS SET TO VIRTUAL CONDITION OF PART PART SURFACE PLATE

TABLE 6–1 *(continued)*

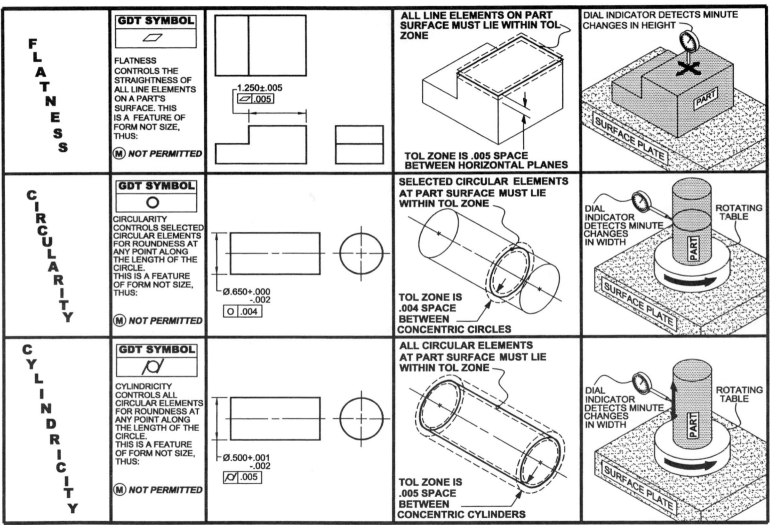

FLATNESS	GDT SYMBOL ⏥ FLATNESS CONTROLS THE STRAIGHTNESS OF ALL LINE ELEMENTS ON A PART'S SURFACE. THIS IS A FEATURE OF FORM NOT SIZE, THUS: Ⓜ ***NOT PERMITTED***	1.250±.005 ⏥ .005	**ALL LINE ELEMENTS ON PART SURFACE MUST LIE WITHIN TOL ZONE** **TOL ZONE IS .005 SPACE BETWEEN HORIZONTAL PLANES**	DIAL INDICATOR DETECTS MINUTE CHANGES IN HEIGHT PART SURFACE PLATE
CIRCULARITY	GDT SYMBOL ○ CIRCULARITY CONTROLS SELECTED CIRCULAR ELEMENTS FOR ROUNDNESS AT ANY POINT ALONG THE LENGTH OF THE CIRCLE. THIS IS A FEATURE OF FORM NOT SIZE, THUS: Ⓜ ***NOT PERMITTED***	Ø.650+.000 -.002 ○ .004	**SELECTED CIRCULAR ELEMENTS AT PART SURFACE MUST LIE WITHIN TOL ZONE** **TOL ZONE IS .004 SPACE BETWEEN CONCENTRIC CIRCLES**	DIAL INDICATOR DETECTS MINUTE CHANGES IN WIDTH ROTATING TABLE PART SURFACE PLATE
CYLINDRICITY	GDT SYMBOL ⌭ CYLINDRICITY CONTROLS ALL CIRCULAR ELEMENTS FOR ROUNDNESS AT ANY POINT ALONG THE LENGTH OF THE CIRCLE. THIS IS A FEATURE OF FORM NOT SIZE, THUS: Ⓜ ***NOT PERMITTED***	Ø.500+.001 -.002 ⌭ .005	**ALL CIRCULAR ELEMENTS AT PART SURFACE MUST LIE WITHIN TOL ZONE** **TOL ZONE IS .005 SPACE BETWEEN CONCENTRIC CYLINDERS**	DIAL INDICATOR DETECTS MINUTE CHANGES IN WIDTH ROTATING TABLE PART SURFACE PLATE

6–5 DATUMS

As was discussed in Section 5–10, a datum is a common reference point from which measurements are made. The same datums are used for CNC manufacturing as well as GDT inspection. The tolerances to be discussed in the sections to follow control how one part feature relates to another, for example, the position of one hole with respect to another or the parallelism between holes. The basic rule here is that measuring one feature relative to another is not permitted. All measurements should be made from common datums.

A part must be fixed in space in order to be manufactured and inspected. A datum reference frame is used to fix the part and locate all its features in 3D space. In practice, exact or theoretical reference frames must be simulated by using actual part features. A datum feature is any feature of a part used to establish a datum. The GDT symbol for a datum feature A, for example, is [A]

A simulated datum plane is created when the datum feature is firmly seated in the manufacturing or inspection fixture as shown in Figure 6–7. Usually, only one datum is needed for controlling orientation features, but positional relationships often require a datum system consisting of two or more datum planes. The planes are mutually perpendicular, as shown in Figure 6–8, and are placed in their order of importance as

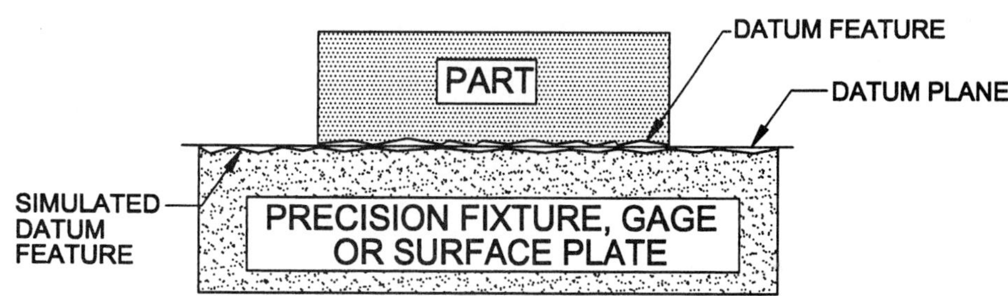

FIGURE 6–7 A datum plane is created from a part datum feature.

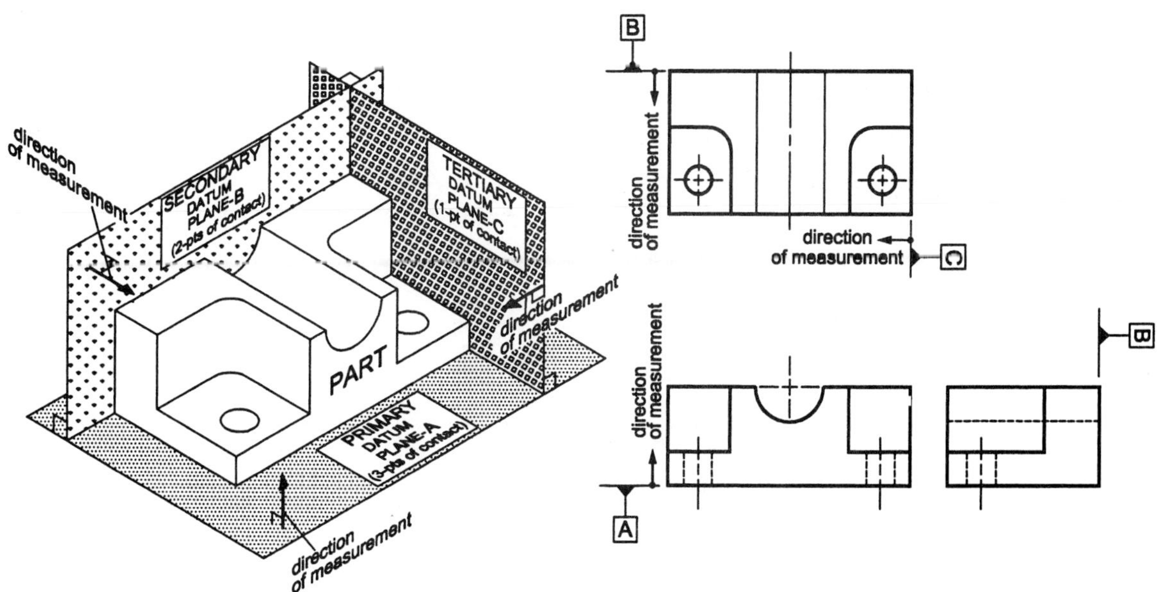

FIGURE 6–8 A datum reference frame for a rectangular part.

Primary(A): This is the datum from which most size and location dimensions are taken. A minimum of three points of contact are needed to establish datum plane A.

Secondary(B): Proper orientation is established via the secondary datum plane. A minimum of two contact points are required to create datum plane B.

Tertiary(C): This datum plane completes the system and fixes the part in 3D space. Only one point of contact is needed to establish datum plane C.

For cylindrical parts, the axis of a cylinder formed by the intersection of two mutually perpendicular center planes can be used as a datum feature. A datum axis feature is shown in Figure 6–9.

6–6 PART PROFILE CONTROLS

Profile tolerances control the degree to which a uniform part boundary is permitted to vary from an exact boundary shape. Individual profile tolerances apply to an individual part feature and do not require the use of datums. Related profile tolerances measure a feature relative to a datum.

Table 6–2 presents a listing of geometric tolerances of profile.

6–7 PART LOCATION, ORIENTATION, AND RUNOUT CONTROLS

Location controls use "true position dimensioning" which requires that the location of a feature be contained within a "cylindrical" tolerance zone.

A comprehensive listing of tolerances of location, orientation and runout are shown in Table 6–3.

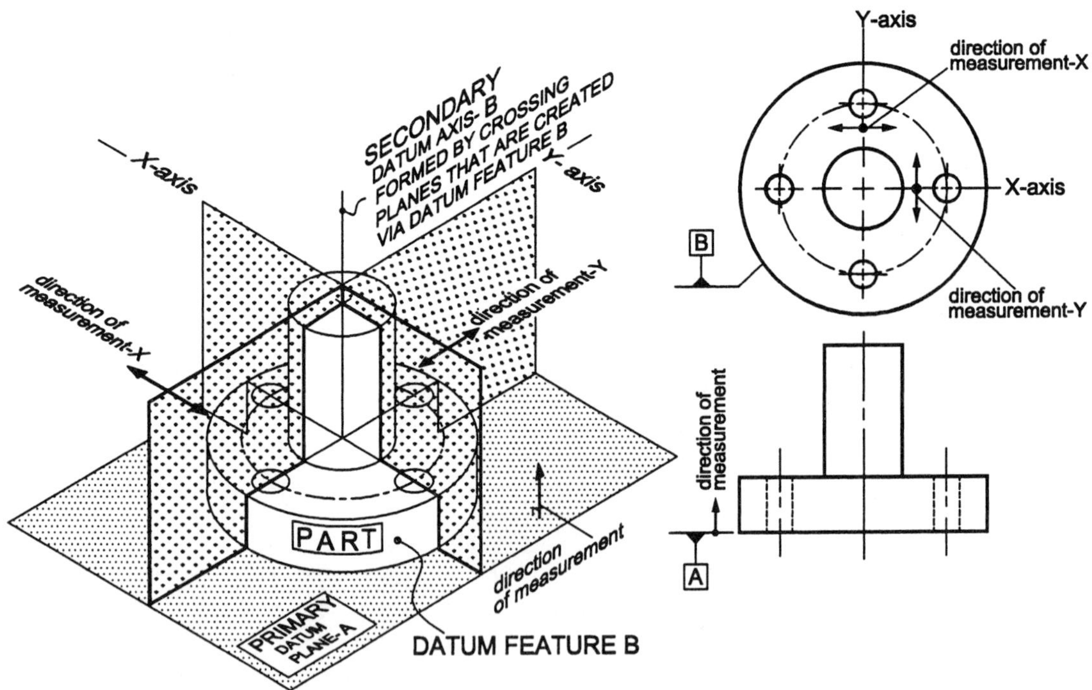

FIGURE 6–9 A datum axis for a cylindrical part.

TABLE 6–2

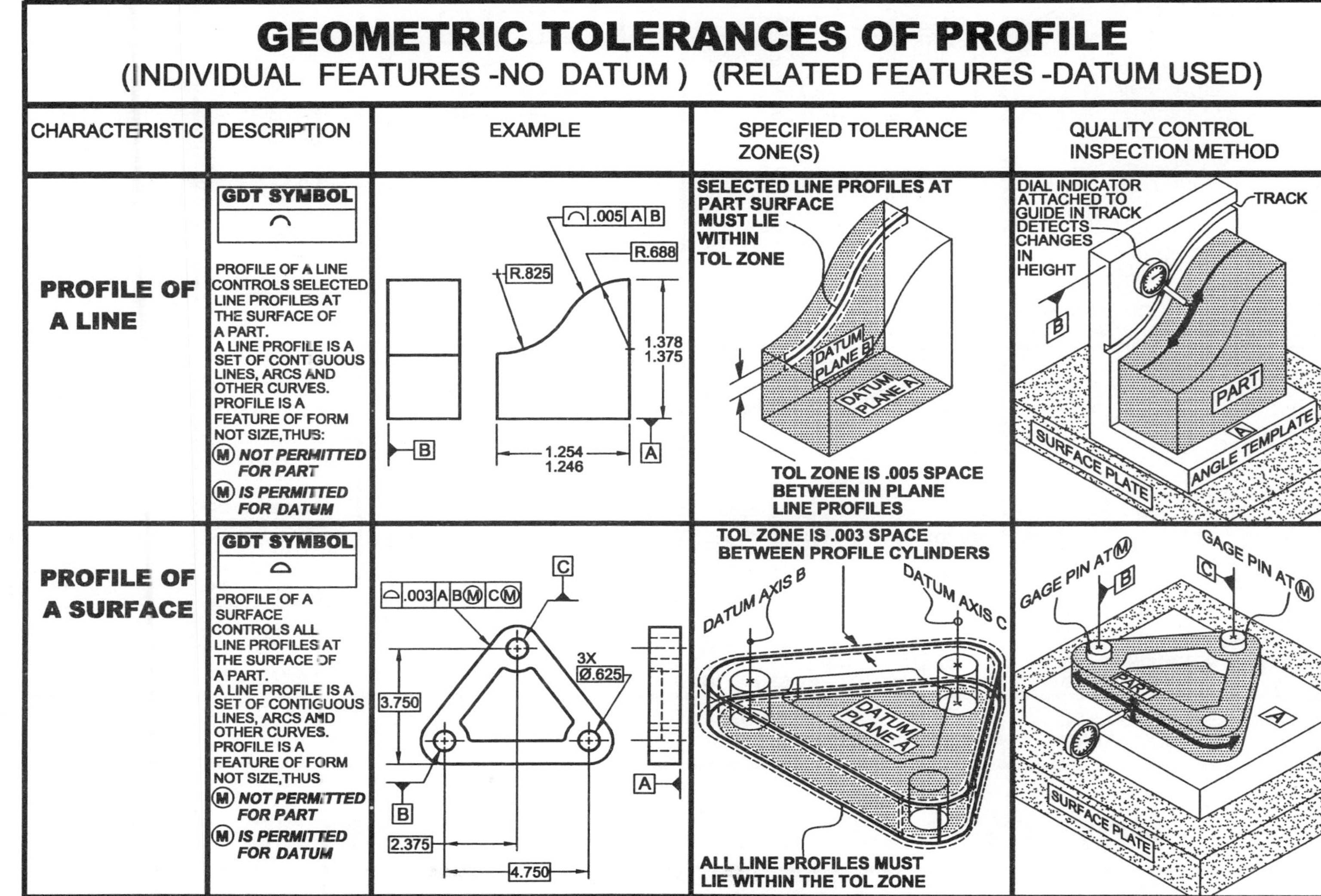
GEOMETRIC TOLERANCES OF PROFILE
(INDIVIDUAL FEATURES -NO DATUM) (RELATED FEATURES -DATUM USED)
CHARACTERISTIC
DESCRIPTION
EXAMPLE
SPECIFIED TOLERANCE ZONE(S)
QUALITY CONTROL INSPECTION METHOD
PROFILE OF A LINE
GDT SYMBOL
PROFILE OF A LINE CONTROLS SELECTED LINE PROFILES AT THE SURFACE OF A PART.
A LINE PROFILE IS A SET OF CONT GUOUS LINES, ARCS AND OTHER CURVES.
PROFILE IS A FEATURE OF FORM NOT SIZE,THUS:
M NOT PERMITTED FOR PART
M IS PERMITTED FOR DATUM
.005 A B
R.688
R.825
1.378
1.375
1.254
1.246
B
A
SELECTED LINE PROFILES AT PART SURFACE MUST LIE WITHIN TOL ZONE
DATUM PLANE B
DATUM PLANE A
TOL ZONE IS .005 SPACE BETWEEN IN PLANE LINE PROFILES
DIAL INDICATOR ATTACHED TO GUIDE IN TRACK DETECTS CHANGES IN HEIGHT
TRACK
PART
SURFACE PLATE
ANGLE TEMPLATE
PROFILE OF A SURFACE
GDT SYMBOL
PROFILE OF A SURFACE CONTROLS ALL LINE PROFILES AT THE SURFACE OF A PART.
A LINE PROFILE IS A SET OF CONTIGUOUS LINES, ARCS AND OTHER CURVES.
PROFILE IS A FEATURE OF FORM NOT SIZE,THUS
M NOT PERMITTED FOR PART
M IS PERMITTED FOR DATUM
.003 A B M C M
C
3X
Ø.625
3.750
B
2.375
4.750
A
TOL ZONE IS .003 SPACE BETWEEN PROFILE CYLINDERS
DATUM AXIS B
DATUM AXIS C
DATUM PLANE A
ALL LINE PROFILES MUST LIE WITHIN THE TOL ZONE
GAGE PIN AT M
GAGE PIN AT M
B
C
PART
A
SURFACE PLATE

TABLE 6–3

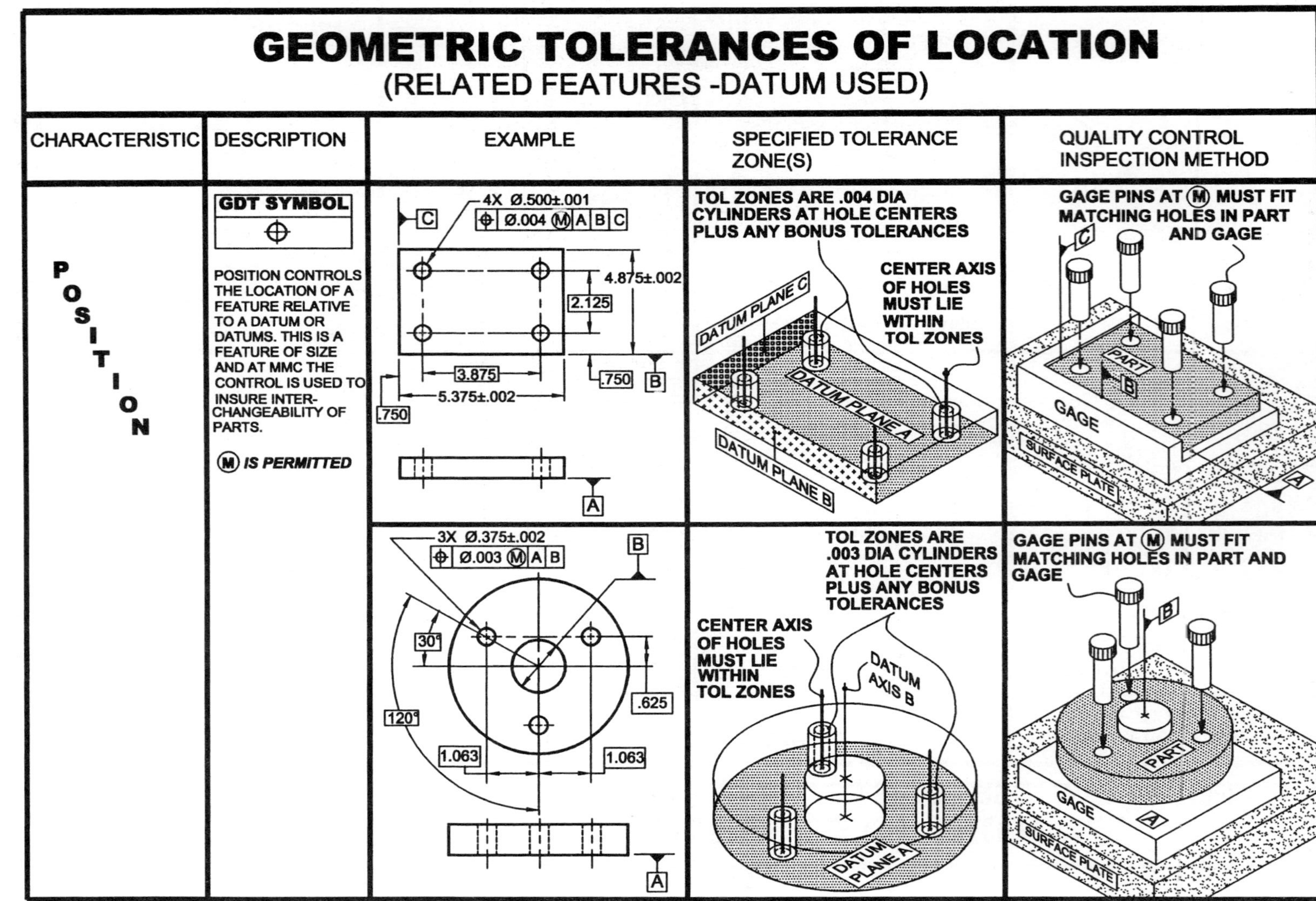

GEOMETRIC TOLERANCES OF LOCATION
(RELATED FEATURES -DATUM USED)

CHARACTERISTIC	DESCRIPTION	EXAMPLE	SPECIFIED TOLERANCE ZONE(S)	QUALITY CONTROL INSPECTION METHOD
POSITION	GDT SYMBOL ⌖ POSITION CONTROLS THE LOCATION OF A FEATURE RELATIVE TO A DATUM OR DATUMS. THIS IS A FEATURE OF SIZE AND AT MMC THE CONTROL IS USED TO INSURE INTER-CHANGEABILITY OF PARTS. Ⓜ IS PERMITTED	4X Ø.500±.001; ⌖ Ø.004 Ⓜ A B C; C; 4.875±.002; 2.125; 3.875; 5.375±.002; .750; .750; B; A	TOL ZONES ARE .004 DIA CYLINDERS AT HOLE CENTERS PLUS ANY BONUS TOLERANCES; CENTER AXIS OF HOLES MUST LIE WITHIN TOL ZONES; DATUM PLANE C; DATUM PLANE A; DATUM PLANE B	GAGE PINS AT Ⓜ MUST FIT MATCHING HOLES IN PART AND GAGE; C; PART; B; GAGE; SURFACE PLATE; A
		3X Ø.375±.002; ⌖ Ø.003 Ⓜ A B; B; 30°; 120°; .625; 1.063; 1.063; A	TOL ZONES ARE .003 DIA CYLINDERS AT HOLE CENTERS PLUS ANY BONUS TOLERANCES; CENTER AXIS OF HOLES MUST LIE WITHIN TOL ZONES; DATUM AXIS B; DATUM PLANE A	GAGE PINS AT Ⓜ MUST FIT MATCHING HOLES IN PART AND GAGE; B; PART; GAGE; A; SURFACE PLATE

TABLE 6–3 *(continued)*

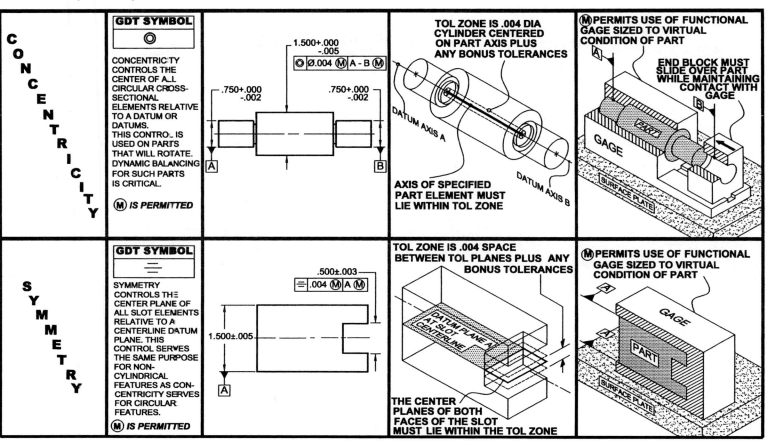

CONCENTRICITY	**GDT SYMBOL** ◎ CONCENTRICITY CONTROLS THE CENTER OF ALL CIRCULAR CROSS-SECTIONAL ELEMENTS RELATIVE TO A DATUM OR DATUMS. THIS CONTROL IS USED ON PARTS THAT WILL ROTATE. DYNAMIC BALANCING FOR SUCH PARTS IS CRITICAL. Ⓜ *IS PERMITTED*	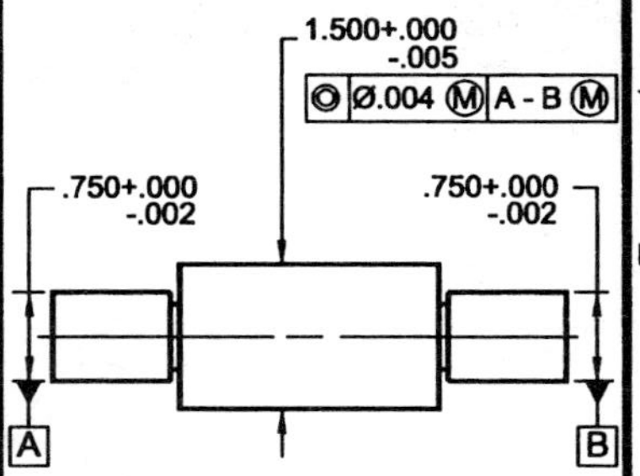 1.500+.000 -.005 ◎ Ø.004 Ⓜ A - B Ⓜ .750+.000 -.002 .750+.000 -.002 A B	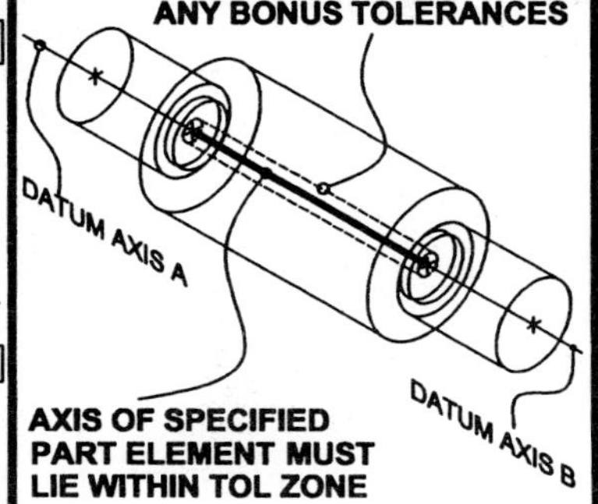 TOL ZONE IS .004 DIA CYLINDER CENTERED ON PART AXIS PLUS ANY BONUS TOLERANCES DATUM AXIS A DATUM AXIS B AXIS OF SPECIFIED PART ELEMENT MUST LIE WITHIN TOL ZONE	Ⓜ PERMITS USE OF FUNCTIONAL GAGE SIZED TO VIRTUAL CONDITION OF PART A END BLOCK MUST SLIDE OVER PART WHILE MAINTAINING CONTACT WITH GAGE B PART GAGE SURFACE PLATE
SYMMETRY	**GDT SYMBOL** ⌯ SYMMETRY CONTROLS THE CENTER PLANE OF ALL SLOT ELEMENTS RELATIVE TO A CENTERLINE DATUM PLANE. THIS CONTROL SERVES THE SAME PURPOSE FOR NON-CYLINDRICAL FEATURES AS CONCENTRICITY SERVES FOR CIRCULAR FEATURES. Ⓜ *IS PERMITTED*	.500±.003 ⌯ .004 Ⓜ A Ⓜ 1.500±.005 A	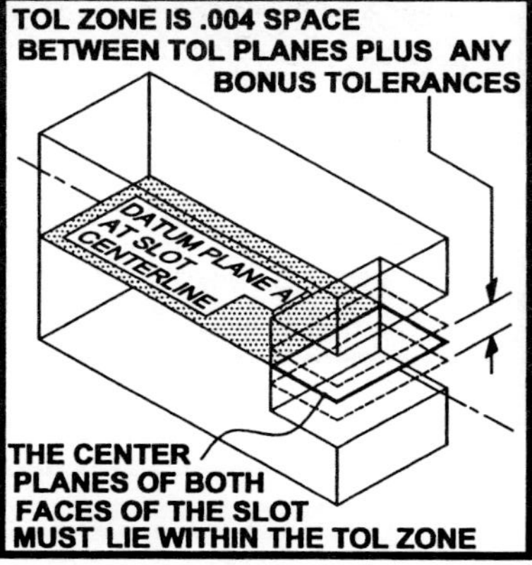 TOL ZONE IS .004 SPACE BETWEEN TOL PLANES PLUS ANY BONUS TOLERANCES DATUM PLANE A AT SLOT CENTERLINE THE CENTER PLANES OF BOTH FACES OF THE SLOT MUST LIE WITHIN THE TOL ZONE	Ⓜ PERMITS USE OF FUNCTIONAL GAGE SIZED TO VIRTUAL CONDITION OF PART A A GAGE PART SURFACE PLATE

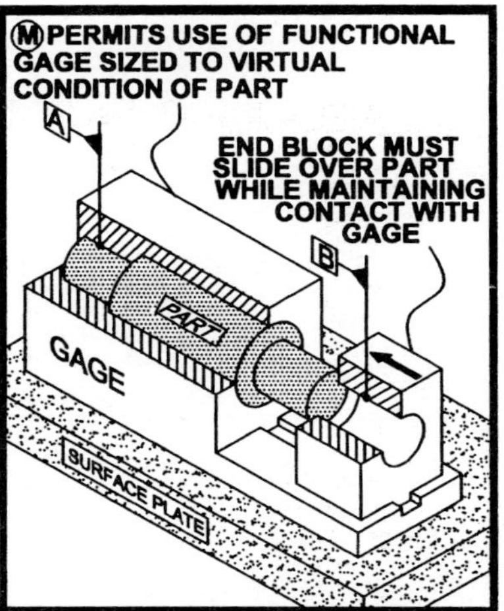

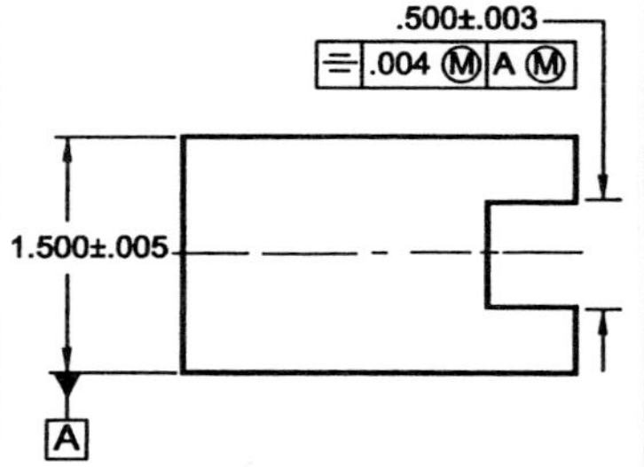

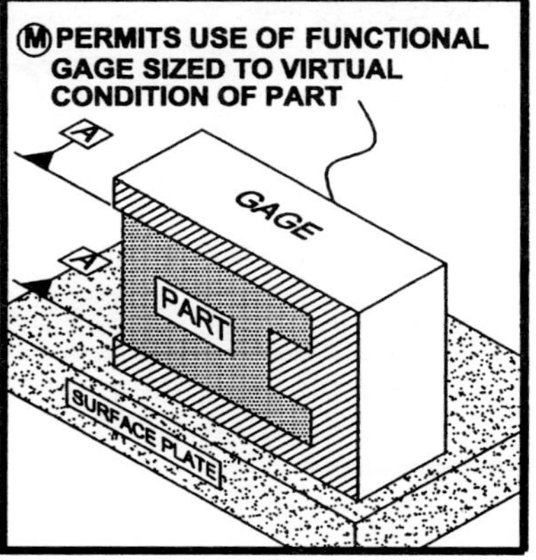

TABLE 6–3 *(continued)*

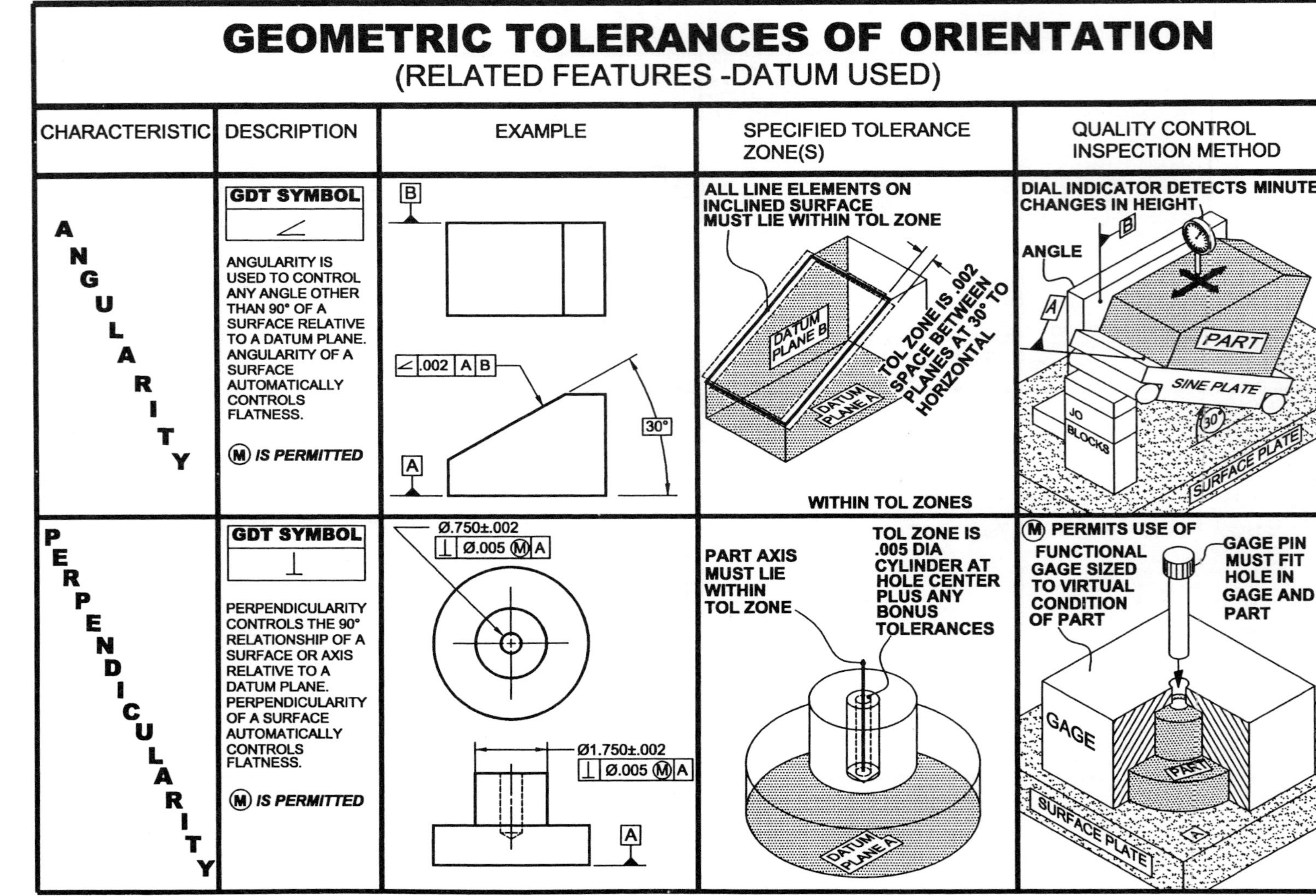

GEOMETRIC TOLERANCES OF ORIENTATION
(RELATED FEATURES -DATUM USED)

CHARACTERISTIC	DESCRIPTION	EXAMPLE	SPECIFIED TOLERANCE ZONE(S)	QUALITY CONTROL INSPECTION METHOD
ANGULARITY	GDT SYMBOL ∠ ANGULARITY IS USED TO CONTROL ANY ANGLE OTHER THAN 90° OF A SURFACE RELATIVE TO A DATUM PLANE. ANGULARITY OF A SURFACE AUTOMATICALLY CONTROLS FLATNESS. Ⓜ *IS PERMITTED*	B ∠ .002 A B 30° A	ALL LINE ELEMENTS ON INCLINED SURFACE MUST LIE WITHIN TOL ZONE DATUM PLANE B DATUM PLANE A TOL ZONE IS .002 SPACE BETWEEN PLANES AT 30° TO HORIZONTAL WITHIN TOL ZONES	DIAL INDICATOR DETECTS MINUTE CHANGES IN HEIGHT B ANGLE A PART SINE PLATE JO BLOCKS 30° SURFACE PLATE
PERPENDICULARITY	GDT SYMBOL ⊥ PERPENDICULARITY CONTROLS THE 90° RELATIONSHIP OF A SURFACE OR AXIS RELATIVE TO A DATUM PLANE. PERPENDICULARITY OF A SURFACE AUTOMATICALLY CONTROLS FLATNESS. Ⓜ *IS PERMITTED*	Ø.750±.002 ⊥ Ø.005 Ⓜ A Ø1.750±.002 ⊥ Ø.005 Ⓜ A A	PART AXIS MUST LIE WITHIN TOL ZONE TOL ZONE IS .005 DIA CYLINDER AT HOLE CENTER PLUS ANY BONUS TOLERANCES DATUM PLANE A	Ⓜ PERMITS USE OF FUNCTIONAL GAGE SIZED TO VIRTUAL CONDITION OF PART GAGE PIN MUST FIT HOLE IN GAGE AND PART GAGE PART SURFACE PLATE A

TABLE 6–3 *(continued)*

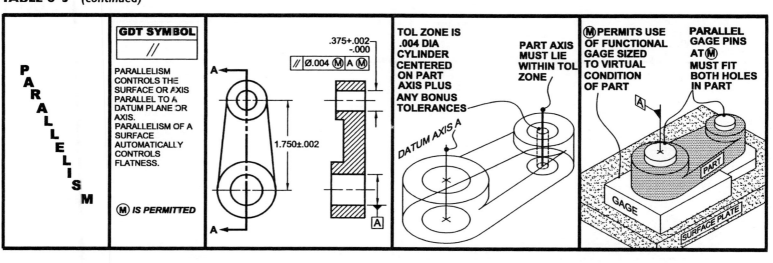

PARALLELISM	GDT SYMBOL // PARALLELISM CONTROLS THE SURFACE OR AXIS PARALLEL TO A DATUM PLANE OR AXIS. PARALLELISM OF A SURFACE AUTOMATICALLY CONTROLS FLATNESS. Ⓜ IS PERMITTED	A A 1.750±.002 .375+.002 -.000 // Ø.004 Ⓜ A Ⓜ A	TOL ZONE IS .004 DIA CYLINDER CENTERED ON PART AXIS PLUS ANY BONUS TOLERANCES PART AXIS MUST LIE WITHIN TOL ZONE DATUM AXIS A	Ⓜ PERMITS USE OF FUNCTIONAL GAGE SIZED TO VIRTUAL CONDITION OF PART PARALLEL GAGE PINS AT Ⓜ MUST FIT BOTH HOLES IN PART A PART GAGE SURFACE PLATE

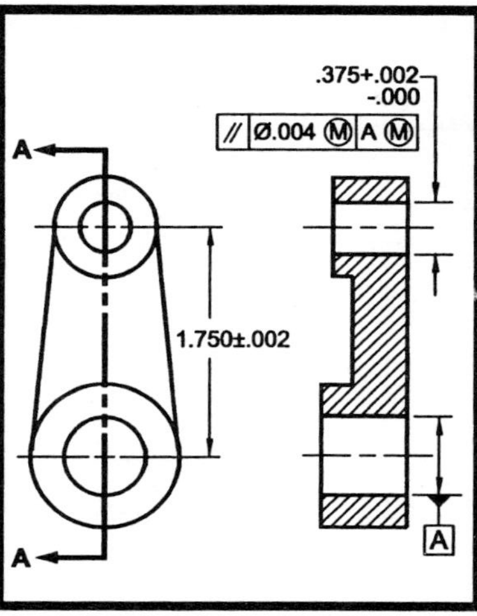

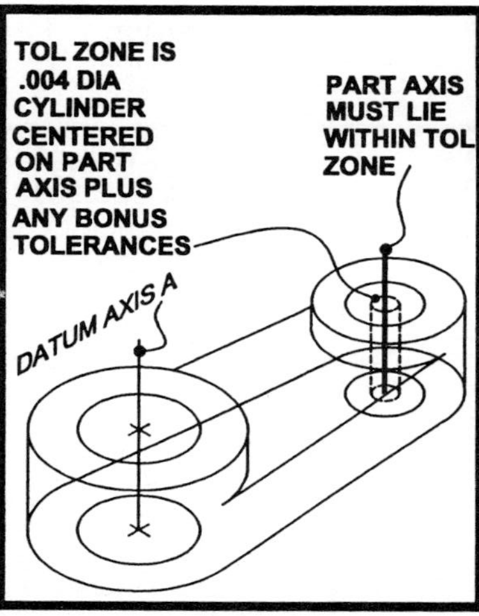

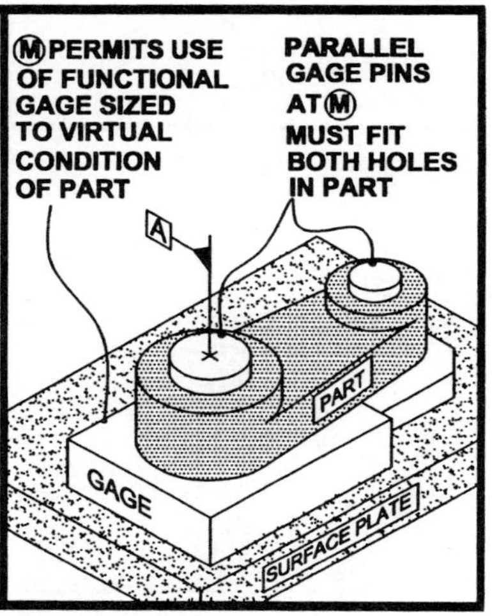

TABLE 6–3 *(continued)*

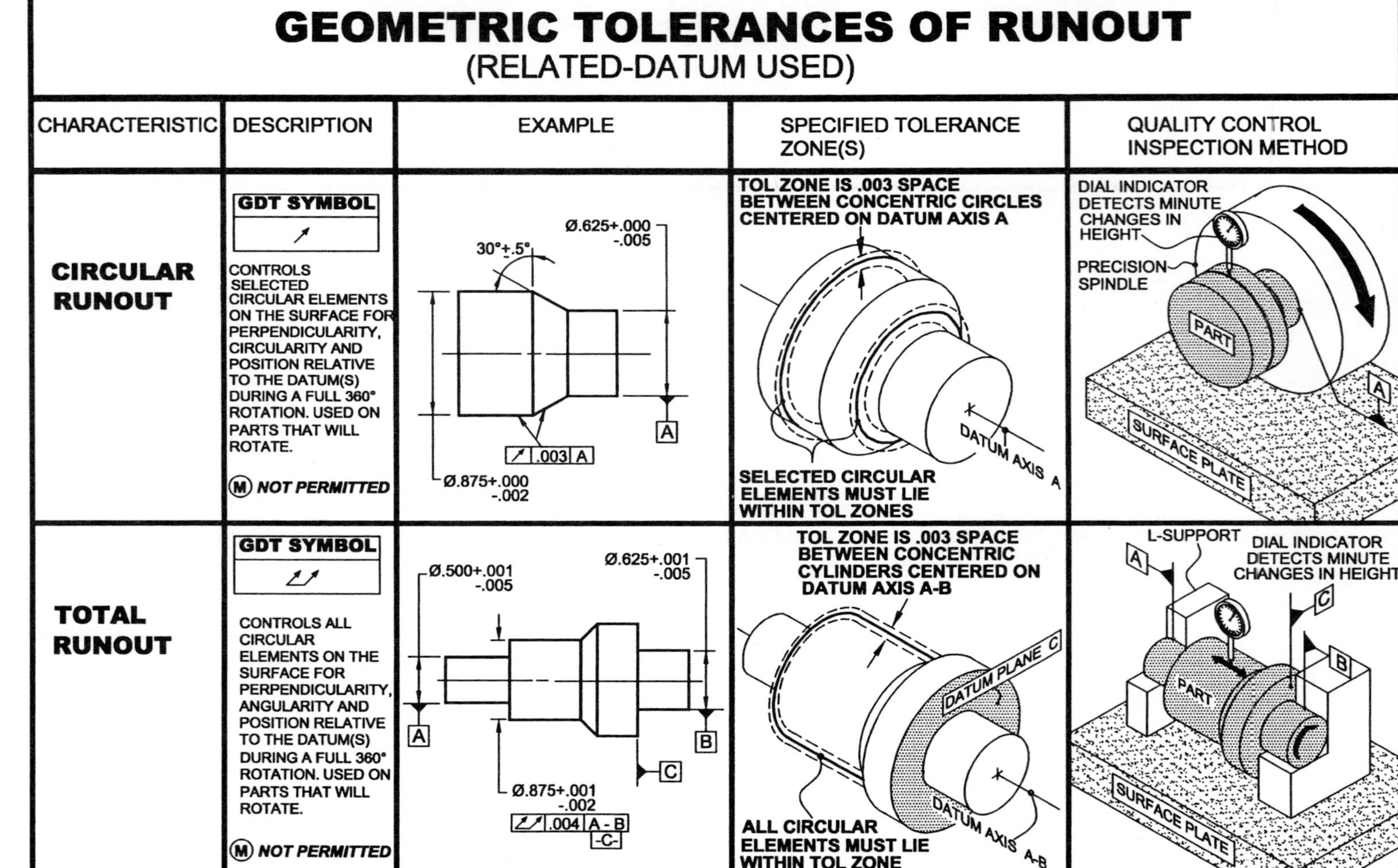

GEOMETRIC TOLERANCES OF RUNOUT (RELATED-DATUM USED)				
CHARACTERISTIC	DESCRIPTION	EXAMPLE	SPECIFIED TOLERANCE ZONE(S)	QUALITY CONTROL INSPECTION METHOD
CIRCULAR RUNOUT	GDT SYMBOL ↗ CONTROLS SELECTED CIRCULAR ELEMENTS ON THE SURFACE FOR PERPENDICULARITY, CIRCULARITY AND POSITION RELATIVE TO THE DATUM(S) DURING A FULL 360° ROTATION. USED ON PARTS THAT WILL ROTATE. Ⓜ ***NOT PERMITTED***	30°±.5° Ø.625+.000 -.005 Ø.875+.000 -.002 ↗ .003 A A	**TOL ZONE IS .003 SPACE BETWEEN CONCENTRIC CIRCLES CENTERED ON DATUM AXIS A** DATUM AXIS A **SELECTED CIRCULAR ELEMENTS MUST LIE WITHIN TOL ZONES**	DIAL INDICATOR DETECTS MINUTE CHANGES IN HEIGHT PRECISION SPINDLE PART A SURFACE PLATE
TOTAL RUNOUT	GDT SYMBOL ⌰ CONTROLS ALL CIRCULAR ELEMENTS ON THE SURFACE FOR PERPENDICULARITY, ANGULARITY AND POSITION RELATIVE TO THE DATUM(S) DURING A FULL 360° ROTATION. USED ON PARTS THAT WILL ROTATE. Ⓜ ***NOT PERMITTED***	Ø.500+.001 -.005 Ø.625+.001 -.005 Ø.875+.001 -.002 ⌰ .004 A - B -C- A B C	**TOL ZONE IS .003 SPACE BETWEEN CONCENTRIC CYLINDERS CENTERED ON DATUM AXIS A-B** DATUM PLANE C DATUM AXIS A-B **ALL CIRCULAR ELEMENTS MUST LIE WITHIN TOL ZONE**	L-SUPPORT DIAL INDICATOR DETECTS MINUTE CHANGES IN HEIGHT A C B PART SURFACE PLATE

6–8 CHAPTER SUMMARY

The following key concepts were discussed in this chapter:

1. Geometric dimensioning and tolerancing (GDT) is a system designed to control part features that cannot be specified by conventional positioning and size tolerancing.
2. GDT involves the control of five basic types of geometric tolerances: form, profile, location, orientation, and runout.
3. Geometric dimensioning and tolerancing insures parts made on CNC machines will be high quality, low cost, and interchangeable.
4. Maximum material condition Ⓜ is the condition at which an external feature (shaft) is largest and an internal feature (hole) is smallest. The greatest GDT accuracy is specified at Ⓜ.
5. Least material condition Ⓛ is the condition at which an external feature (shaft) is smallest and an internal feature (hole) is largest. The least GDT accuracy is specified at Ⓛ.
6. The larger tolerances that are permitted as part features depart from Ⓜ toward Ⓛ are called "bonus tolerances."
7. The same datums are used for CNC manufacturing and GDT inspection.

REVIEW EXERCISES

6.1. What additional part features are controlled by GDT that cannot be specified by conventional tolerancing?

6.2. What is maximum material condition Ⓜ?

6.3. What is least material condition Ⓛ?

6.4. What are the numbers ⟨A⟩ called?

	Shaft diameter	GDT Tolerance Zone	
Ⓜ----	1.500	.004	
	1.498	.005	⟨A⟩
	1.496	.006	⟨A⟩
	1.494	.007	⟨A⟩
Ⓛ----	1.492	.008	⟨A⟩

6.5. The tolerance zone represents the tolerance and its position relative to the basic size. True or False?

6.6. Is the virtual condition of a part the same as mating condition? In determining compliance with GDT, which way of measuring is less costly: functional gaging or open gaging?

6.7. What does part form tolerance control?

6.8. Describe the positional relationship between GDT datum planes.

6.9. Profile tolerances control the degree to which a uniform boundary is permitted to vary from an exact boundary shape. True or False?

6.10. Describe what is specified in the following GDT control frames.

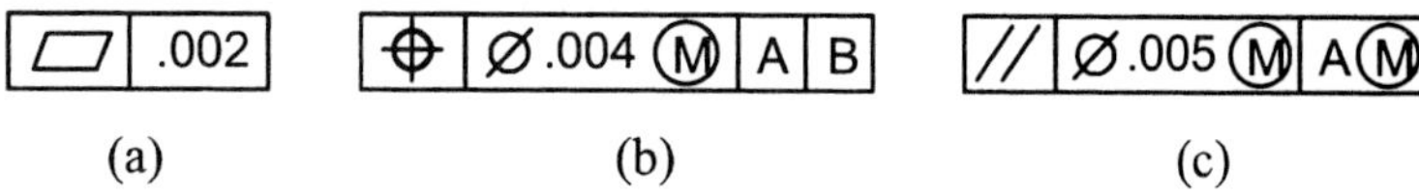

(a) (b) (c)

BIBLIOGRAPHY

American Society of Mechanical Engineers. *Dimensioning and Tolerancing: ASME Y14.5-1994 [Revision of ANSI Y14.5M-1982 (R1988)] N.Y: ASME 1995.*
Foster W. L. *Geo-Metrics.* Addison-Wesley Publishing, Reading, Mass., 1994.
Gooldy G. *Geometric Dimensioning & Tolerancing.* Prentice Hall, Upper Saddle River, N.J. 1995.
Krulikowski A. *Fundamentals of Geometric Dimensioning and Tolerancing.* Delmar Publishers, 1998.

CHAPTER

7

Mathematics for CNC Programming

7–1 CHAPTER OBJECTIVES

At the conclusion of this chapter you will be able to

1. Compute the sides of right triangles.
2. Compute the angles of right triangles.

7–2 INTRODUCTION

Many problems in CNC programming involve finding the X and Y coordinates of tool motion based on a given length and angle. These tasks can usually be accomplished by using basic right-triangle trigonometry. Thus, knowledge of this subject is a must for manual programming. A review of the essential principles is given in this chapter.

7–3 DETERMINING SIDES OF RIGHT TRIANGLES

A right triangle is a three-sided figure; one of whose angles is 90°. See Figure 7–1. The longest side of the triangle is called the hypotenuse.

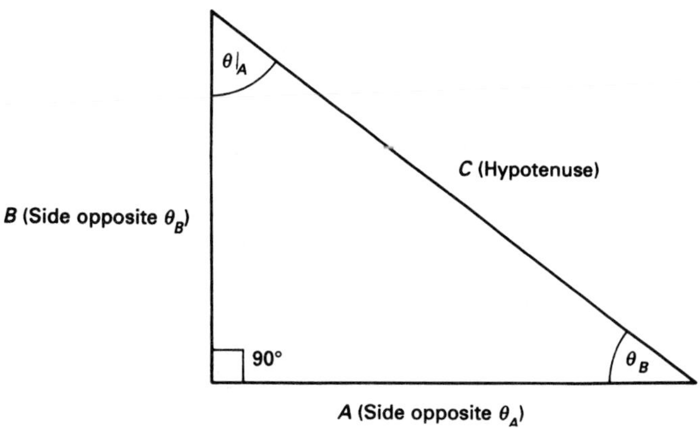

FIGURE 7–1 A right triangle.

The relationships between any two sides and the included angle are given by the following general side-angle formulas:

$$\sin(\theta) = \frac{\text{Side opposite }(\theta)}{\text{Hypotenuse}}$$

$$\cos(\theta) = \frac{\text{Side adjacent }(\theta)}{\text{Hypotenuse}}$$

$$\tan(\theta) = \frac{\text{Side opposite }(\theta)}{\text{Side adjacent }\theta}$$

Specific side-angle formulas are given in Table 7–1.

TABLE 7–1

DETERMINING THE SIDES OF RIGHT TRIANGLES	
Known angle	**Side-angle formulas**
θ_A	$\frac{A}{C} = \sin(\theta_A)$
θ_A	$\frac{B}{C} = \cos(\theta_A)$
θ_A	$\frac{A}{B} = \tan(\theta_A)$
θ_B	$\frac{B}{C} = \sin(\theta_B)$
θ_B	$\frac{A}{C} = \cos(\theta_B)$
θ_B	$\frac{B}{A} = \tan(\theta_B)$
The sides of a right triangle are also related by the Pythagorean formula: $C^2 = A^2 + B^2$	

7–4 USEFUL ANGLE CONCEPTS

The following facts concerning angles are of use when dealing with problems involving triangles.

1. A circle contains 360°. See Figure 7–2.

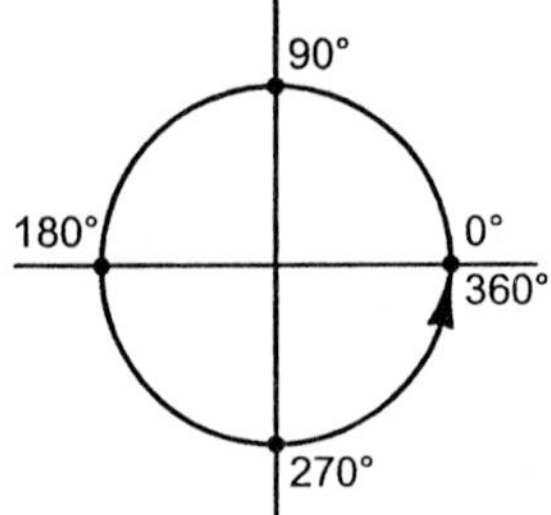

FIGURE 7–2

2. Vertical angles are equal. See Figure 7–3.

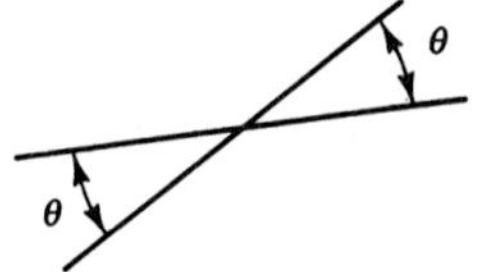

FIGURE 7–3

3. Alternate interior angles are equal. See Figure 7–4.

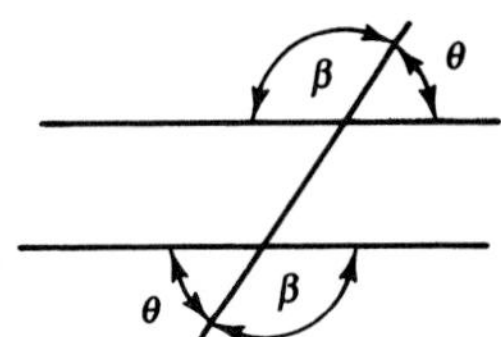

FIGURE 7–4

4. A minute (′) is defined as 1⁄60 of a degree. Convert minutes to fractions of a degree before using the trigonometric functions:

$$\text{Degrees} = \text{Minutes} \times \frac{1}{60'}$$

■ EXAMPLE 7–1

Determine the unknown side in the triangles shown in Figures 7–5 through 7–7.

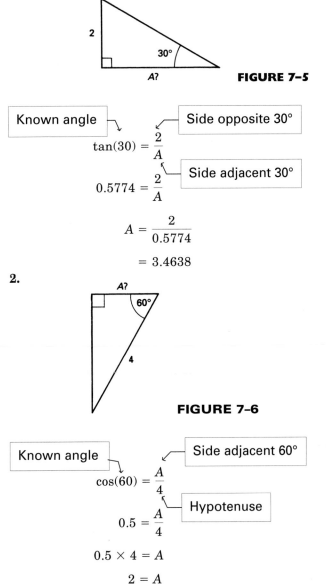

1.

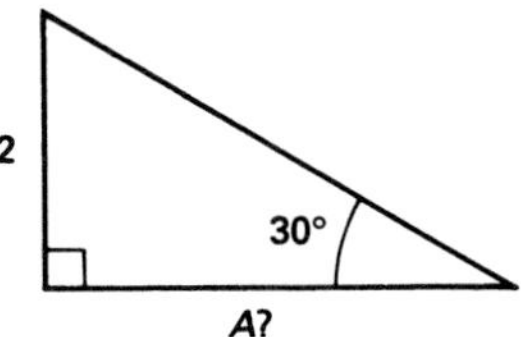

FIGURE 7–5

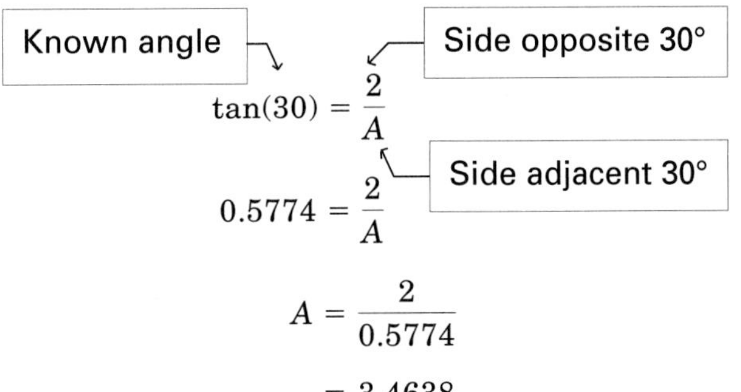

$$\tan(30) = \frac{2}{A}$$

$$0.5774 = \frac{2}{A}$$

$$A = \frac{2}{0.5774}$$

$$= 3.4638$$

2.

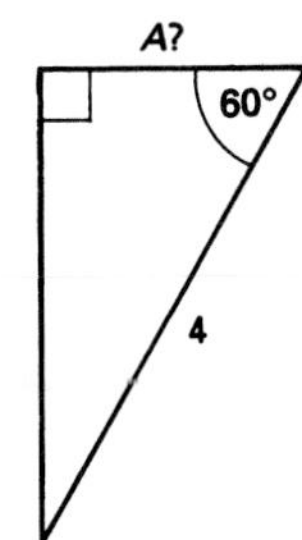

FIGURE 7–6

Known angle | Side adjacent 60° | Hypotenuse

$$\cos(60) = \frac{A}{4}$$

$$0.5 = \frac{A}{4}$$

$$0.5 \times 4 = A$$

$$2 = A$$

3.

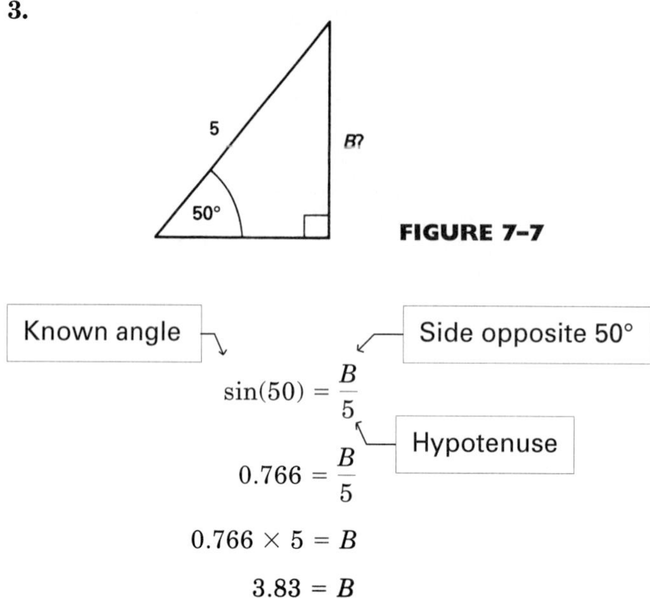

FIGURE 7–7

$$\sin(50) = \frac{B}{5}$$

$$0.766 = \frac{B}{5}$$

$$0.766 \times 5 = B$$

$$3.83 = B$$

■■

■ EXAMPLE 7–2

Assume you are working on a CNC machine with no rotary table. Determine the coordinates that must be inputted for drilling hole 2. Refer to Figure 7–8.

Because the machine has no rotary table, hole 2 cannot be located by directly inputting its angle from 0° and distance from the center of rotation *C*. Instead, its X_2 and Y_2 absolute coordinates must be determined. A circle contains 360° and there are four equally spaced holes in the circular pattern. Thus, each hole makes an angle of 360°/4 = 90° with its neighbor. Refer to Figure 7–9.

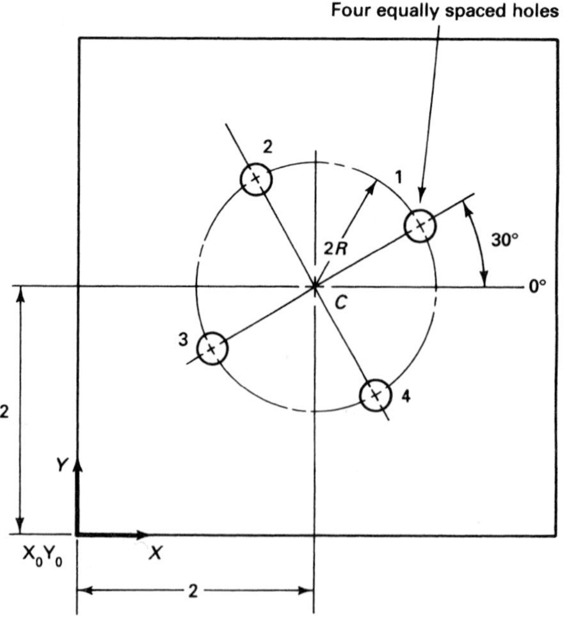

FIGURE 7–8

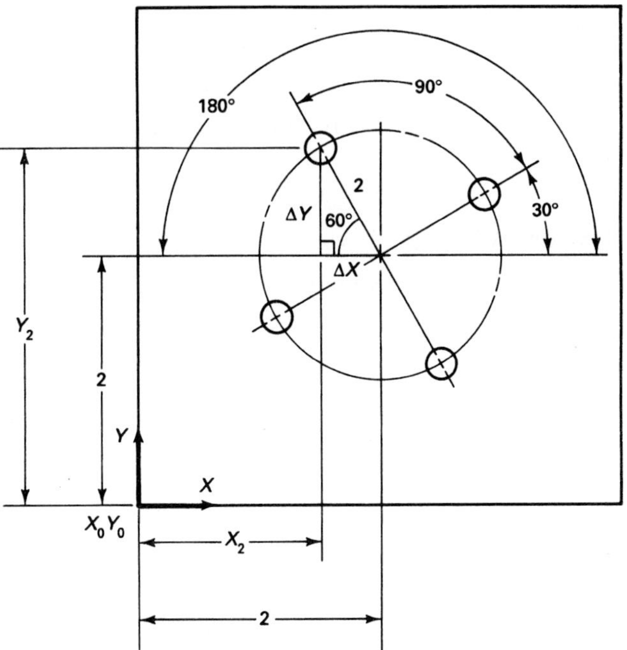

FIGURE 7–9

The required angle θ is given by

$\theta = 180° - 90° - 30°$

$= 60°$

The sides of the right triangle can then be found:

Known angle → $\cos(60) = \dfrac{\Delta X}{2}$ (Side adjacent 60° / Hypotenuse)

$$\cos(60) = \frac{\Delta X}{2}$$

$$0.5 = \frac{\Delta X}{2}$$

$$2 \times 0.5 = \Delta X$$

$$1 = \Delta X$$

Known angle → $\sin(60) = \dfrac{\Delta Y}{2}$ (Side opposite 60° / Hypotenuse)

$$\sin(60) = \frac{\Delta Y}{2}$$

$$0.866 = \frac{\Delta Y}{2}$$

$$0.866 \times 2 = \Delta Y$$

$$1.7321 = \Delta Y$$

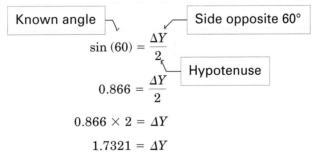

Thus, the absolute coordinates of hole 2 are

$X_2 = 2 - \Delta X$
$X_2 = 2 - 1$
$X_2 = 1$

$Y_2 = 2 + \Delta Y$
$Y_2 = 2 + 1.7321$
$Y_2 = 3.7321$

The same calculations can be made when drilling holes using a milling machine. ■■

■ EXAMPLE 7–3

Determine the unknown side in the triangle shown in Figure 7–10.

Convert minutes to fractions of a degree:

FIGURE 7–10

$$26°40' = 26° + 40' \times \frac{1°}{60'} = 26.6667°$$

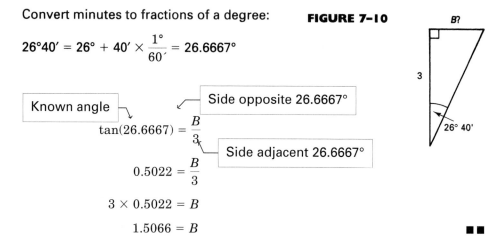

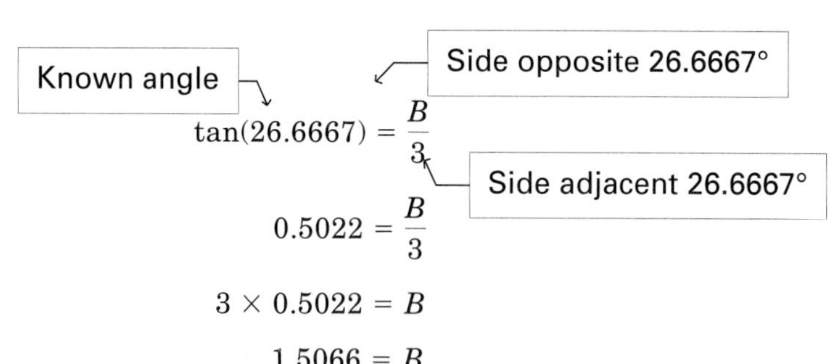

$$\tan(26.6667) = \frac{B}{3}$$

$$0.5022 = \frac{B}{3}$$

$$3 \times 0.5022 = B$$

$$1.5066 = B$$

■■

7–5 DETERMINING ANGLES OF RIGHT TRIANGLES

The side-angle formulas given in Table 7–1 can be inverted to determine the included angle when any two sides of a right triangle are known. It should also be noted that the sum of all the interior angles of a triangle always equals 180°. Refer to Figure 7–1 when using Table 7–2.

In Table 7–2, $\sin^{-1}$, $\cos^{-1}$, and $\tan^{-1}$ are the inverse or arc sine, cosine, and tangent functions found on standard scientific calculators.

TABLE 7–2

DETERMINING THE ANGLES OF RIGHT TRIANGLES	
Known sides	**Inverted side-angle formulas**
A,C	$\theta_A = \sin^{-1}\left(\frac{A}{C}\right)$
B,C	$\theta_A = \cos^{-1}\left(\frac{B}{C}\right)$
A,B	$\theta_A = \tan^{-1}\left(\frac{A}{B}\right)$
B,C	$\theta_B = \sin^{-1}\left(\frac{B}{C}\right)$
A,C	$\theta_B = \cos^{-1}\left(\frac{A}{C}\right)$
B,A	$\theta_B = \tan^{-1}\left(\frac{B}{A}\right)$
Inside angle formula: $\theta_A + \theta_B + 90° = 180°$	

■ EXAMPLE 7–4

Find the required angle in the triangles shown in Figures 7–11 through 7–13.

1.

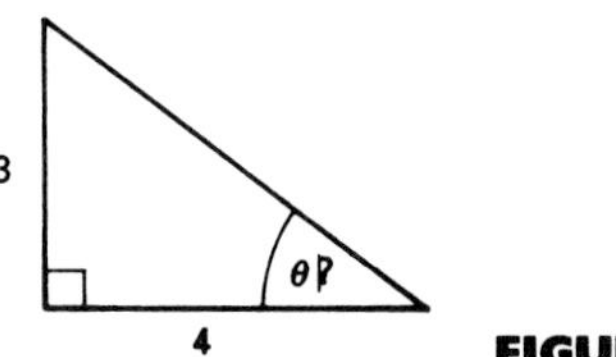

FIGURE 7–11

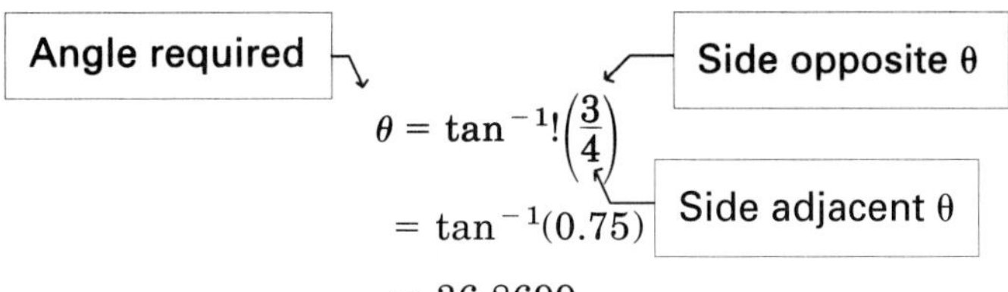

$$\theta = \tan^{-1}\left(\frac{3}{4}\right)$$
$$= \tan^{-1}(0.75)$$
$$= 36.8699$$

2.

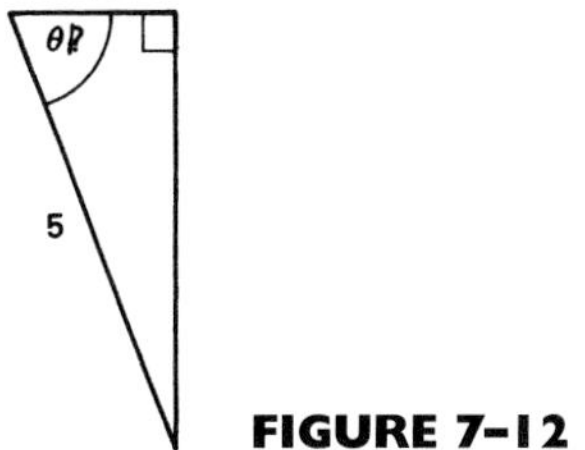

FIGURE 7–12

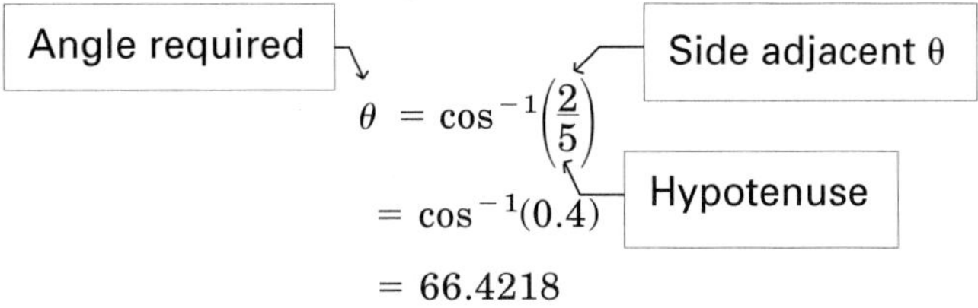

$$\theta = \cos^{-1}\left(\frac{2}{5}\right)$$
$$= \cos^{-1}(0.4)$$
$$= 66.4218$$

3.

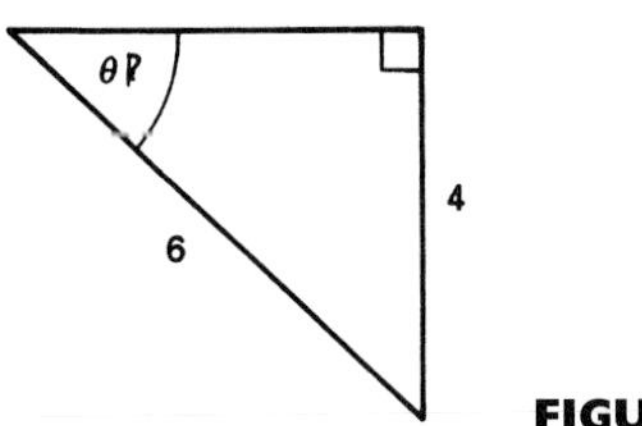

FIGURE 7–13

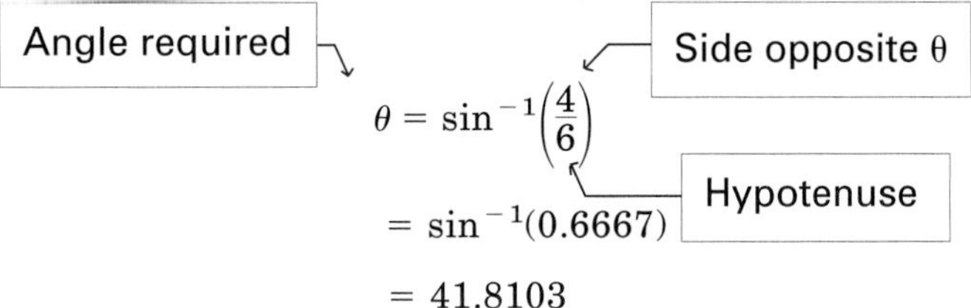

$$\theta = \sin^{-1}\left(\frac{4}{6}\right)$$
$$= \sin^{-1}(0.6667)$$
$$= 41.8103$$

■■

7–6 OBLIQUE TRIANGLES

An oblique triangle is a three-sided figure, none of whose sides is 90°. See Figure 7–14. Formulas for determining the included angle when any two sides are known are given in Table 7–3.

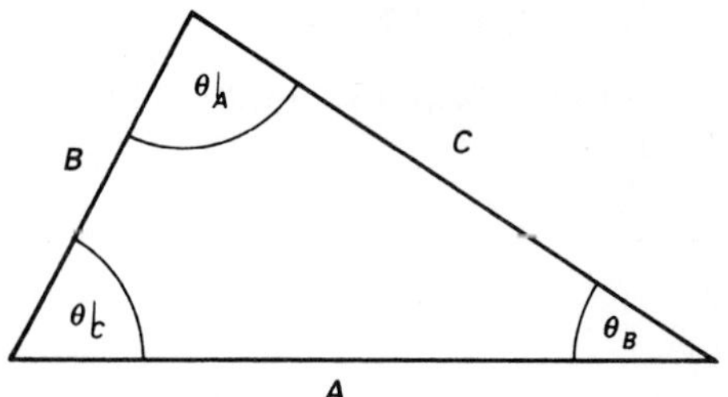

FIGURE 7–14 An oblique triangle.

TABLE 7–3

DETERMINING THE SIDES AND ANGLES OF OBLIQUE TRIANGLES

Law of sines

$$\frac{A}{\sin(\theta_A)} = \frac{B}{\sin(\theta_B)} = \frac{C}{\sin(\theta_C)}$$

Law of cosines

$$a^2 = b^2 + c^2 - 2bc\cos(\theta_A)$$
$$b^2 = a^2 + c^2 - 2ac\cos(\theta_B)$$
$$c^2 = a^2 + b^2 - 2ab\cos(\theta_C)$$

Inside angle formula: $\theta_A + \theta + \theta_C = 180$

7–7 CHAPTER SUMMARY

The following key concepts were discussed in this chapter:

1. Problems involving tool location based on a given angle can be solved by right-triangle trigonometry.
2. If an angle and a side are known, any other side of a triangle can be found. The sin, cos, and tan functions are used to solve for sides.
3. If any two sides are known, any interior angle of a right triangle can be found. The inverse functions $\sin^{-1}$, $\cos^{-1}$, and $\tan^{-1}$ are employed for finding angles.

REVIEW EXERCISES

7.1. Determine the unknown side in each right triangle shown in Figure 7–15.

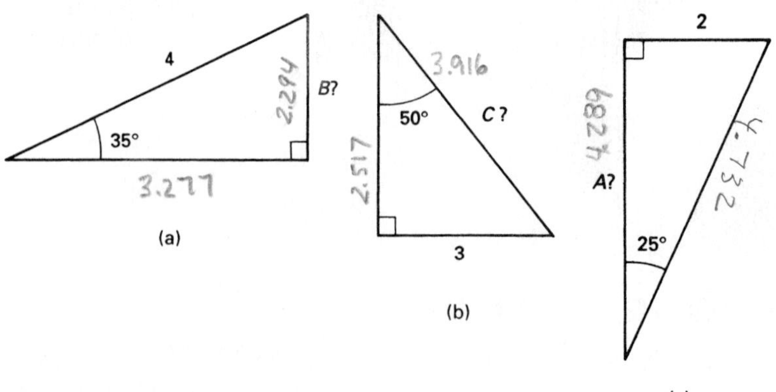

FIGURE 7–15

7.2. Determine the unknown angle for each triangle shown in Figure 7–16.

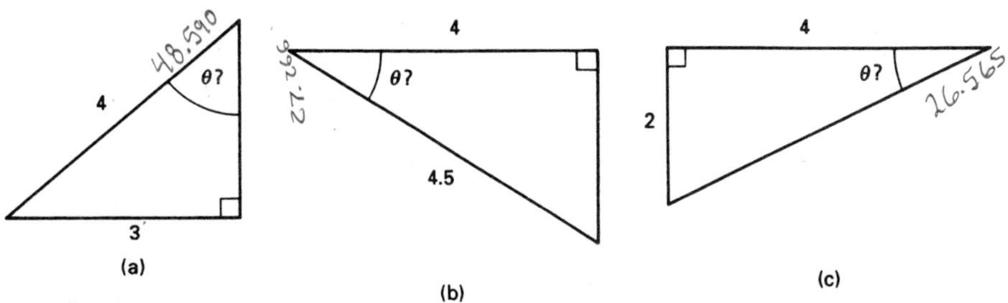

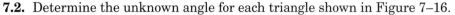

FIGURE 7–16

7.3. Find the distances ΔX and ΔY in the part shown in Figure 7–17.

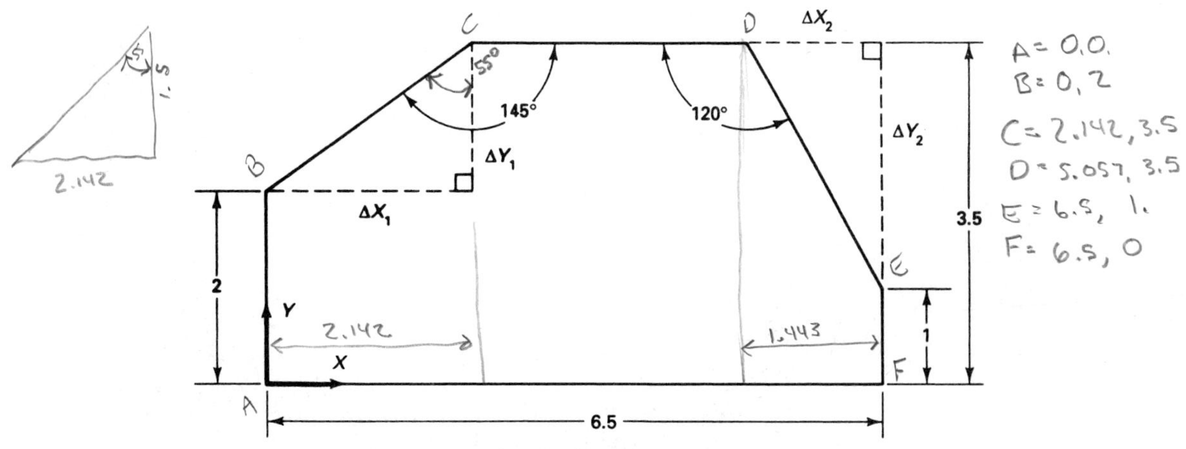

FIGURE 7–17

7.4. Find the absolute coordinates X and Y as indicated for the metric part shown in Figure 7–18.

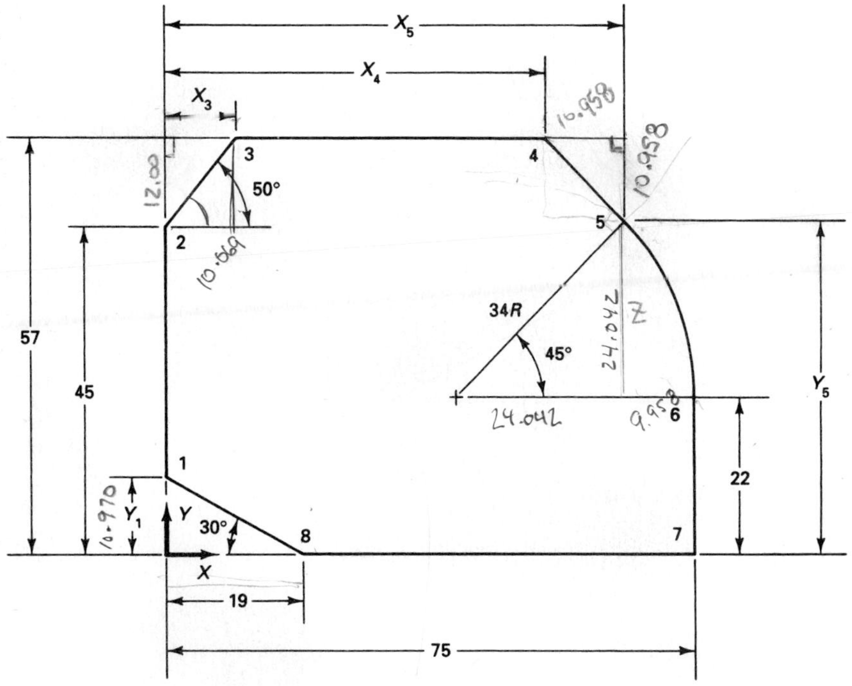

FIGURE 7–18

7.5. Given the bolt circle shown in Figure 7–19, compute the absolute X and Y coordinates of each equally spaced hole. Use the table at the top of page 152 given as a guide.

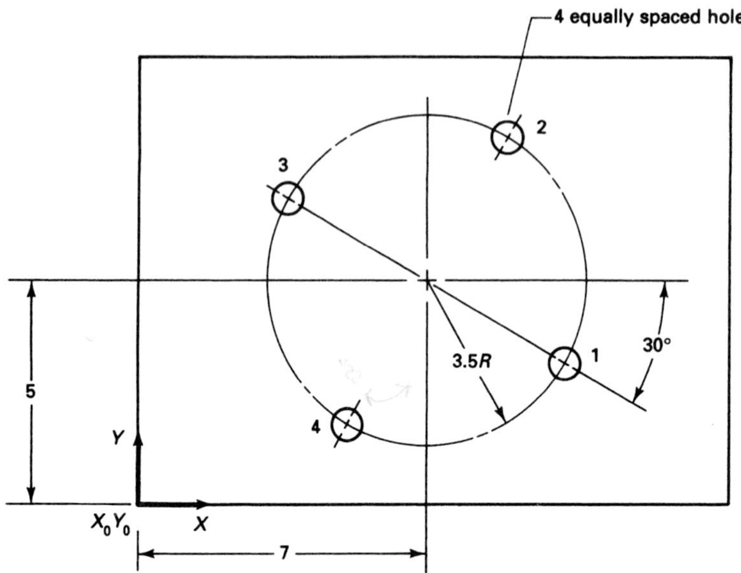

FIGURE 7–19

Hole	Δx	Δy	x	y
1				
2				
3				
4				

7.6. Determine the unknown side in each right triangle in Figure 7–20.

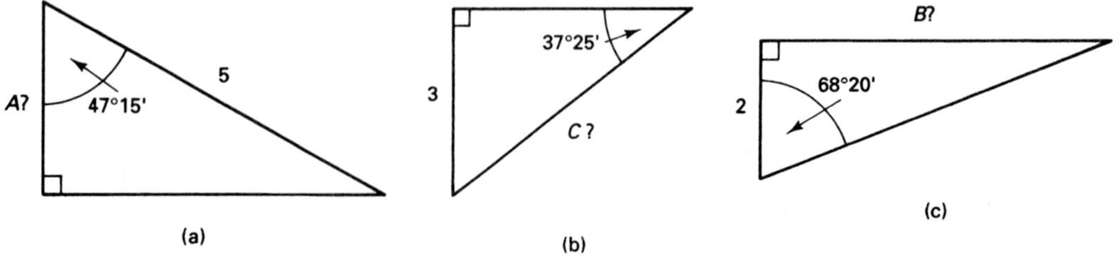

FIGURE 7–20

7.7. For the machining shown in Figure 7–21:

(a) Determine the absolute *X* and *Y* coordinates of the .5 Dia cutter center.

(b) Determine the *I* and *J* distances from the cutter center to the 1R arc center.

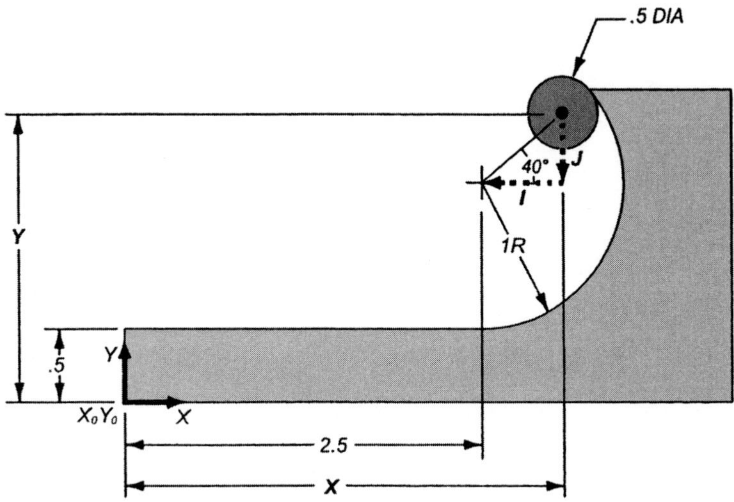

FIGURE 7–21

CHAPTER 8

AN OVERVIEW OF CNC SHOP ACTIVITIES

8-1 CHAPTER OBJECTIVES

At the conclusion of this chapter you will be able to

1. **Identify the three major areas of activities in a CNC machine shop.**
2. **Explain the importance and the order of CNC process planning.**
3. **Describe what factors influence the programmer's selection of a CNC machine for a job run.**
4. Explain the conditions for proper work holding.
5. State what information is important in setup and machining documentation.
6. Understand the terms machine home, part origin, and tool change position.
7. Explain the significance of tool length offset.
8. Identify the functions of the programmer and setup person in getting a job through setup and prove-out.

8-2 INTRODUCTION

This chapter concentrates on the many actions that a CNC shop must perform in order to produce cost-effective, quality parts. It cannot be stressed enough that the programmer must be thoroughly familiar with the part drawing and the manufacturing specifications. The order of machining is decided according to the required accuracy and orientation of the workpiece. In order to get the most out of the CNC machine, the programmer must also select tools that are the most appropriate for the intended cutting operation.

8-3 ESSENTIAL CNC SHOP ACTIVITIES

The most important activities normally carried out by the departments in a CNC shop are presented in the form of a flowchart in Figure 8–1.

A more detailed description of the activities shown in the flowchart is given in Sections 8–4 through 8–15.

8-4 PART DRAWING STUDY

The first step to be taken before preparing any part program is to thoroughly study the part drawing. The drawing indicates what the part looks like, its

Safety Rules for Job Setup on a CNC Machining Center

- ➡ Fixture workpieces securely using adequate blocking and clamping.
- ➡ Use safety goggles and follow personal safety precautions.
- ➡ For cutting operations above OSHA limits wear a face mask.

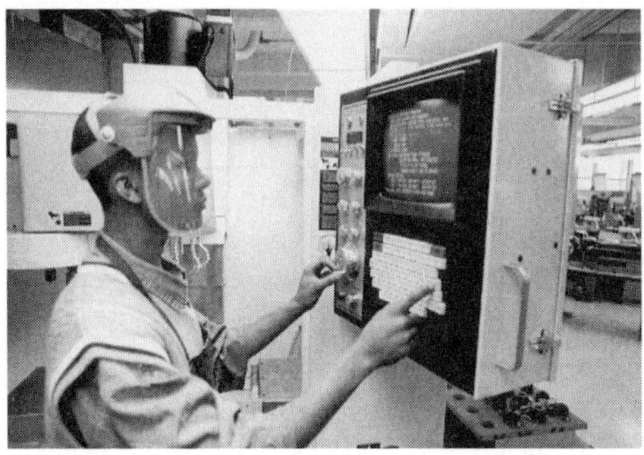

- ➡ Make sure the workpiece is free of burrs and foreign particles.
- ➡ Completely check all machining operations:
  - Read the setup instructions and check for axis positioning accuracy and operation correctness.
  - Make a dry run.
- ➡ Adjust cutting speeds and feeds for each operation within the recommended limits.
- ➡ Be sure there is sufficient clearance between the cutting tool and surrounding objects when moving each axis manually.
- ➡ After operations, be sure to remove all chips and thoroughly clean the working area.

material, tolerances, surface finish, material treatments (if any), and any other requirements. The programmer determines from the part drawing whether or not the part can be machined on a machining center. See Figure 8–2.

Important factors in the decision to utilize CNC operations in the part's manufacture include the quantity of parts to be machined, the part quality requirements, the tooling costs, the fixture manufacturing, and the design and cost of running the CNC machine. An operations sheet is written following a decision as to how the part is to be manufactured.

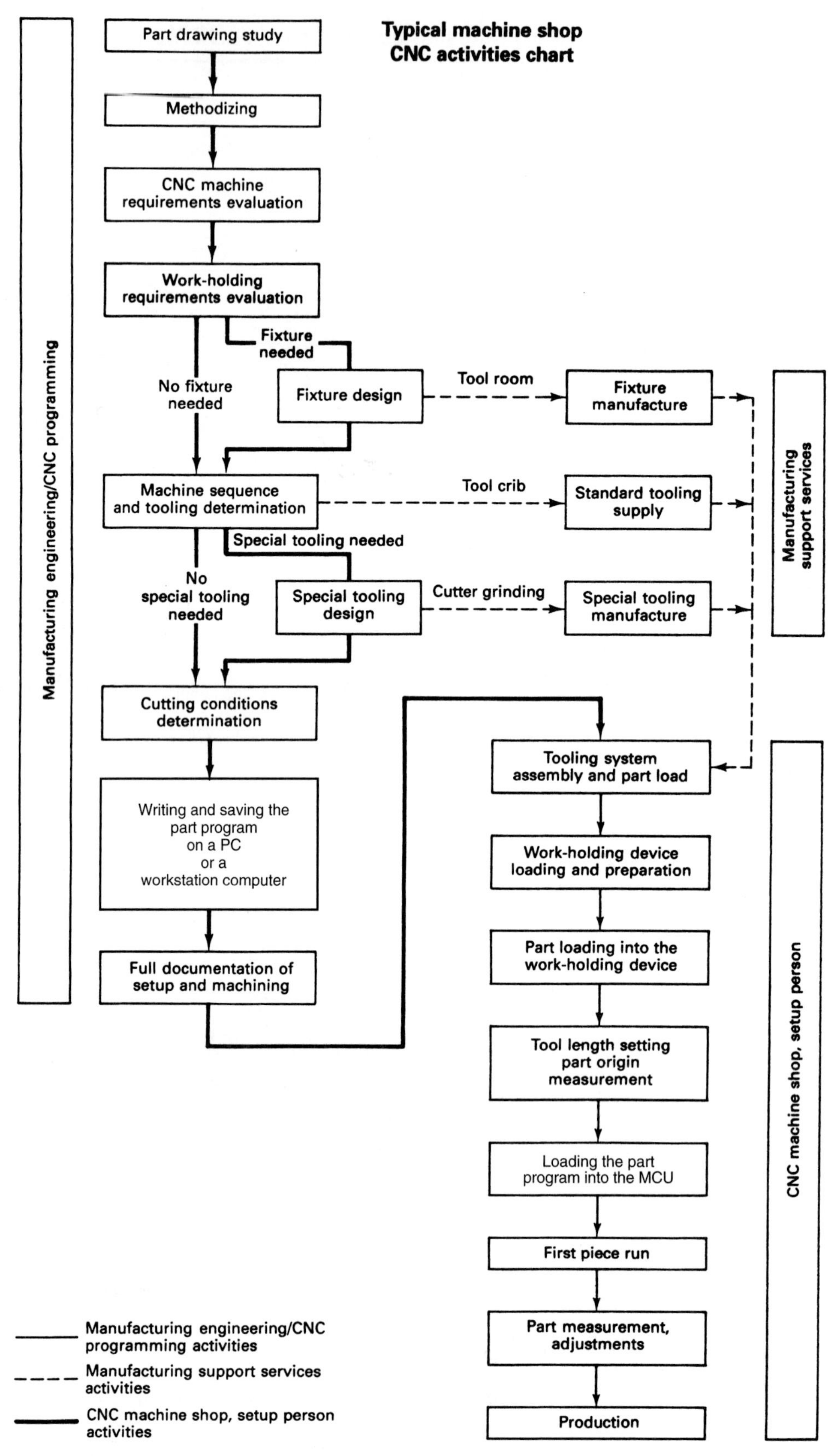

FIGURE 8–1 Typical machine shop CNC activities chart.

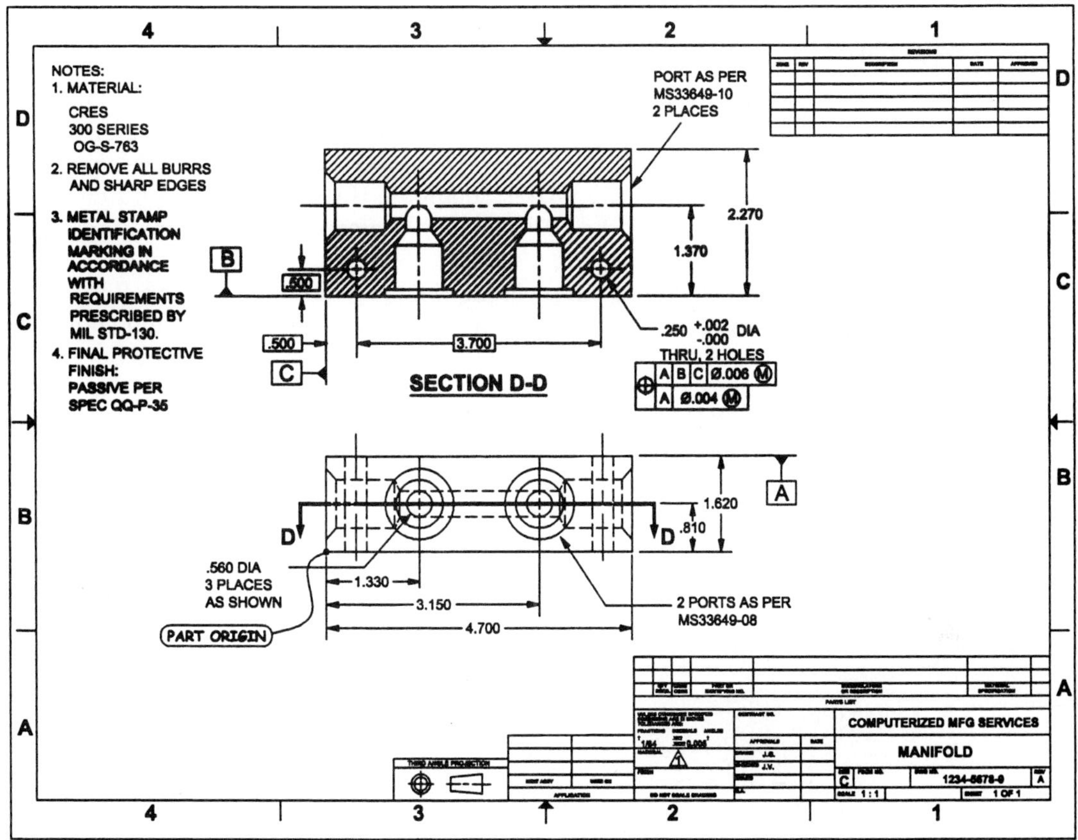

FIGURE 8–2 A sample part production drawing.

8–5 METHODIZING OF OPERATIONS

Methodizing involves the creation of a plan indicating the sequence and methods in which operations are to be carried out in order to produce a part. In a small machine shop where no methods engineer is available, the programmer takes the responsibility for breaking down the manufacturing of the part into operations. In a larger shop specially trained manufacturing engineers handle this important job. The operations sheet contains information the programmer needs concerning the condition of the part material prior to machining. The operations sheet would indicate, for example, if the part material is in the form of a plate, casting, or forging. Whether the plate is to be precut to specific dimensions or if the castings and forgings are to be annealed before machining would also be taken into account. The sheet also specifies if there should be material left for finishing due to shrinkage, expansion, or distortion after heat treating or plating. Operations sheets are used in conjunction with the drawing.

8–6 DECIDING ON A CNC MACHINE

The programmer selects a CNC machine based on its ability to optimize the cutting operations required to manufacture the part. Availability as well as the capacity of the CNC machine must be known. The machine's capacity is measured by its travel horsepower, tooling magazine size, accuracy, and number of machine axes movements.

8–7 METHODS OF HOLDING THE PART DURING MACHINING

It is the programmer's job to decide how a part is to be held during machining operations. Acting on a sketch provided by the programmer, the setup person carries out the task of securing the part in the CNC machine. For simple shape machining a setup vise is adequate. If the part requires some peripheral machining, it could be positioned on a plate with three banking pins and held in place with clamps.

If the part is complicated, the programmer designs a fixture. All necessary arresting and support devices such as stops, pushers, and clamps and their locations must be carefully documented. Later, the programmer must avoid these obstacles when programming tool movements.

Once the fixture is designed, it is sent to the toolroom for manufacturing. The programmer also provides the toolmaker with any additional information needed to complete the fixture. The fixture is then sent to the CNC shop. Some typical methods of holding the part during machining are shown in Figure 8–3.

FIGURE 8–3 Typical methods of holding the part in the CNC machine. [(a) Photo courtesy of American SIP Corp. (b) Photo courtesy of Chick Machine Tool Company Inc. (c) Photo courtesy of Macro Tool and Machine Company Inc.]

8-8 MACHINING DETERMINATION

While the fixture is being completed in the toolroom, the programmer continues with the job of deciding on the machining sequences and the corresponding tooling required.

In general, machining should proceed in the following order:

- Rough machining
- Rough boring
- Drilling and tapping
- Finishing surfaces
- Finish boring
- Finish reaming

The cutting sequence together with the required cutting tools are documented. If a special cutting tool is needed, a sketch is made and submitted to the cutter grinding department for manufacture.

After documenting the entire cutting process, the programmer sends a list to the tool crib where the required tools are compiled.

In the end, the tools as well as the fixture must arrive at the CNC machine when the job is scheduled to run.

8-9 CUTTING CONDITIONS

The spindle speed and feed rate must be determined and entered in the part program. A discussion of how these parameters are calculated for milling and drilling was presented in Chapter 3. Once entered, they are optimized (adjusted up or down) after carrying out a trial machining. The selection of the proper cutting tools for an operation and optimization of the speed and feed during the cutting process will cut down on machining time and subsequently increase productivity.

8-10 WRITING A PART PROGRAMMING MANUSCRIPT

The program is ready to be coded after all the planning is complete. The programmer makes certain programming decisions based on the drawing requirements and fixturing configuration. These include

- Establishing the location of programming zero or part origin (X_0 Y_0)
- Establishing the location of part Z_0
- Determining the mode of programming (absolute or incremental)
- Determining the units of programming (English or metric)

The tool change position is determined by the machine specifications. For X-axis tool removal the stroke from the spindle must be taken into consideration. In addition, the spindle must be fully retracted with part Z set to maximum before executing any X or Y movements to the tool change position.

8-11 INPUTTING PROGRAMS TO THE MCU

Several methods can be used to send the part program to the CNC machine's MCU unit.

Method 1: The program file has been created on a PC or workstation and saved. An RS-232 cable runs from the computer's serial port to the MCU.

For this scenario, the operator can use file transfer programs such as Laplink or PC-talk to download the program file from the computer to the MCU. See Figure 8–4.

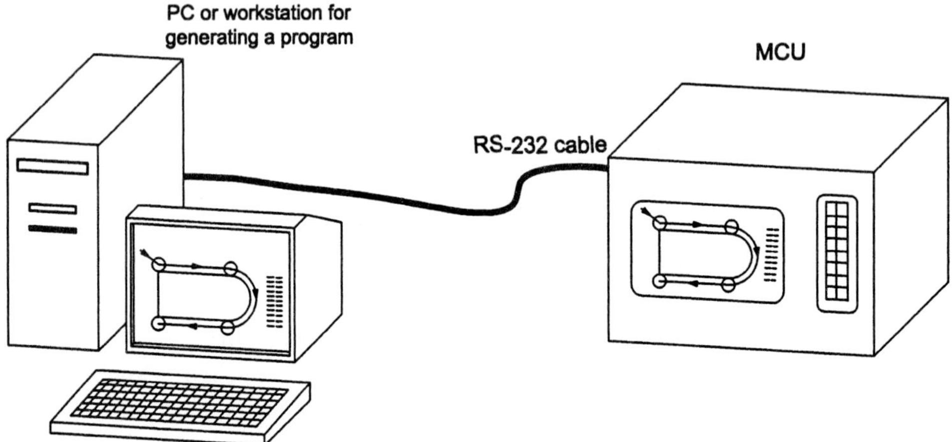

FIGURE 8–4 Downloading a program from PC or workstation to the MCU.

Method 2: A floppy disk, Zip disk, or CD-ROM disk drive unit is attached directly to the CNC machine and connected to its MCU as shown in Figure 8–5. The operator inserts the floppy disk, Zip disk or CD-ROM containing the part program into the drive and downloads the part program.

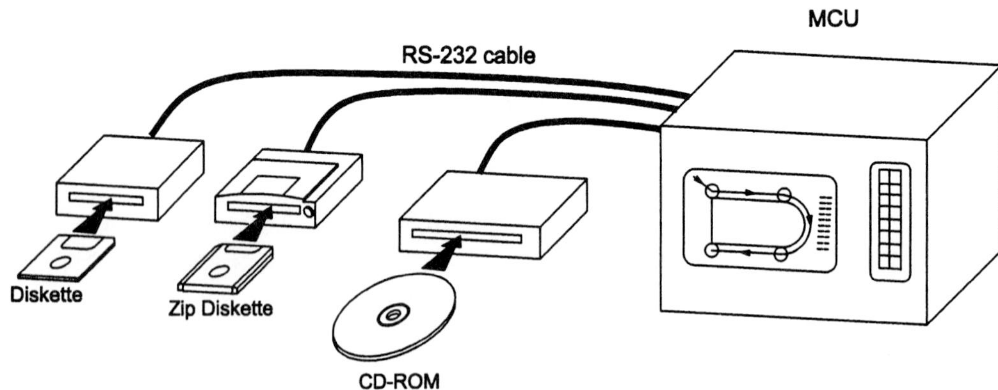

FIGURE 8–5 Downloading a program to the MCU via diskette, Zip diskette or CD-ROM devices.

Method 3: For small machining jobs, the programmer can use the alphanumeric keyboard on the control of the CNC machine tool. EDIT mode is first selected to create a program page. The program is then manually entered into the MCU. This is known as manual data input or MDI. Note: modern MCUs are PC based and contain dedicated online programming software that facilitates the MDI method of program creation and entry. See Figure 8–6.

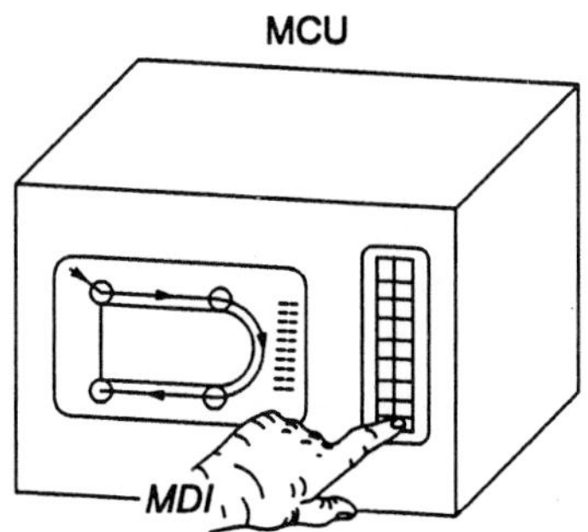

FIGURE 8–6 Manual data entry of a program into the MCU.

8–12 SETUP AND MACHINING DOCUMENTATION

It was previously mentioned that all the necessary information for fixturing, tooling, and setup must be recorded. Good documentation will ensure that the job is set up quickly and with a minimum of correction. A typical CNC setup sheet is shown in Figure 8–7. The circled numbers correspond to the following:

① Part number (drawing number)
② Job number (assigned by CNC shop management)
③ Operation number (operation performed)
④ Programmer's name
⑤ Programming date
⑥ Page number
⑦ Work-holding information (how and where work holding is set down: vise, fixture, etc.)
⑧ Location of part origin
⑨ Setup sketch (illustration of how part is held down in the work-holding device)
⑩ Tool number/station number
⑪ Tool description (diameter, number of flutes, cutting length, overhang, etc.)
⑫ Cutting operation performed by a tool

CNC department

CNC SETUP SHEET

PART # ① JOB # ② OP # ③
BY ④ DATE ⑤ / / PAGE # ⑥

Set-up notes: ⑦ | ⑨

"O"X
"O"Y ⑧
"O"Z

TOOL #	TOOL DESCRIPTION	OPERATION PERFORMED
⑩	⑪	⑫

FIGURE 8–7 Elements of a typical setup sheet for machining a part.

8–13 SETUP PROCEDURE

The setup operation can begin after the fixtures, tooling, program tape, setup sheets, and part blank arrive at the CNC machine. Usually the setup person starts by securing the cutting tools in the tool holders. The tooling assemblies are loaded into the tool magazine according to the order outlined in the setup sheets. Next, the work-holding device is put in place on the machine table. It may require clamping and some minor machining. The part blank is loaded into the work-holding device as requested by the programmer. After loading the program into the MCU, the setup person enters another important piece of information left blank by the programmer: the location of the part origin or part X_0 Y_0 with respect to machine home.

The machine home position or machine zero is a location set once by the machine manufacturer. The CNC machine is homed at the start of the part program setup. When homed, the machine retracts the spindle to its maximum height above the table (machine Z_0) and moves the table to a preset machine zero position (machine X_0 Y_0).

The part origin is the (0, 0, 0) location of the part *XYZ* coordinate system. In absolute coordinate programming, all tool movements are taken with respect to this origin. See Figure 8–8 for an illustration of these positions.

The point from which dimensions are taken on the part print is usually considered the part origin. See Figure 8–2. This could be the center of a hole, the edge of the part or a certain predetermined distance from the edge. The setup person must determine the XYZ locations of the part origin from machine home. This is accomplished by placing an edge finder or end mill with known diameter in the spindle and using the MCU's jog buttons to move or "tram" the edge finder probe from machine home to the part edge or hole center. See Figure 8–9. The X, Y, and Z values displayed at the MCU as a result of tramming indicate the location of the part origin from machine home.

There are two main methods of programming this location into the MCU. The first is called the fixture offset method. This involves using the MCU keypad to enter the XYZ values directly into the MCU unit. The MCU's fixture offset storage area is opened and the values are placed here. Later, when the part program is run, a fixture offset code in the program directs the MCU to use the XYZ values previously placed in its offset storage area.

The second method requires the setup person to open the part program at the MCU and enter the XYZ values into the program prior to running it. Some shops utilize this method while others have a strict policy of not allowing setup personnel to edit part programs. These shops consider the fixture offset method a safer approach.

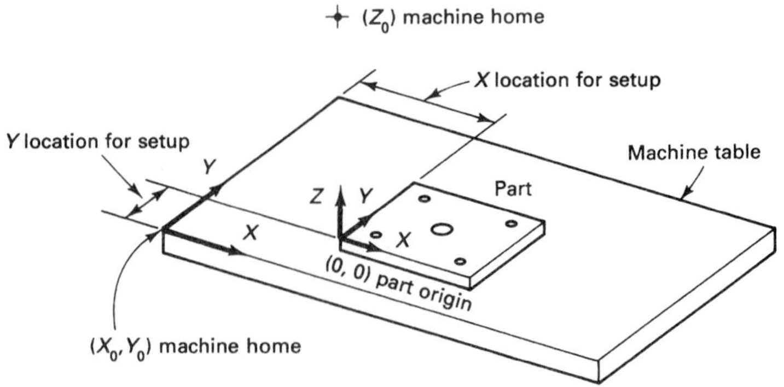

FIGURE 8–8 Machine home and part origin locations.

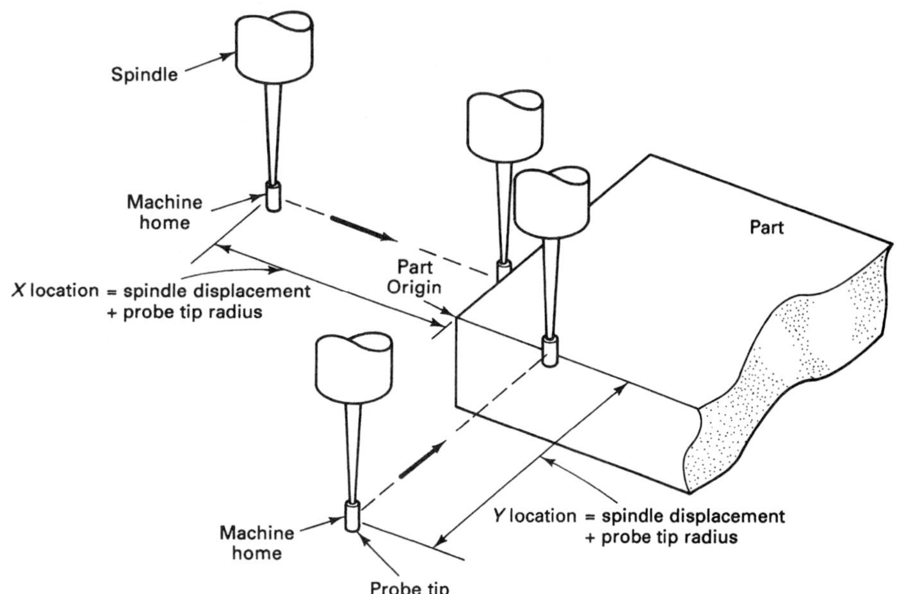

FIGURE 8–9 Edge finding a part origin.

The setup person must next measure and enter the values of the tool length offsets of each tool. The different tools used for the machining operations in a program may vary in length. The system must be direct to compensate for these variations when moving a tool in the Z direction. It can only do so if it knows the initial distance between the bottom of the tool and the part Z_0. The setup person sets the spindle to full retract height or machine Z_0. A tool is loaded and the distance between the bottom of the tool and the part Z_0 is measured and recorded. The next tool is loaded and a similar measurement is recorded. This is done for each tool used in the part program. The values are then entered and stored in the control in a location called the tool length offsets and under the number assigned by the programmer. In some cases a tool can have a few tool length offsets. See Figure 8–10.

Upon running the program, the system knows the tool length offset for each tool it places in the spindle.

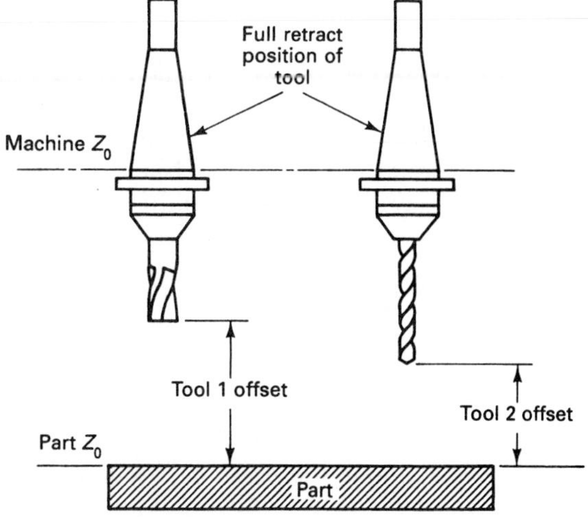

FIGURE 8–10 Tool length offsets.

8-14 DEBUGGING AND VERIFYING THE PROGRAM

The debugging and verification process begins after the part program has been successfully loaded.

Graphical checkout can be made using PC or workstation-based simulation software. Such software is contained on the CD at the back of this text. Similar checkout can also be made at the MCU itself. During program playback, modern MCU units have the capability of displaying the tool path created by a current block of code. These methods allow the operator a quick visual check of the path of tool as it machines the part. This type of checking is referred to as "off line" since it can be done while the CNC machines other parts.

Physical checking is very important for proveout of items that many graphical simulation programs may not check. This includes possible collisions between the tool and fixtures or workholding devices.

Physical checking proceeds as follows:

First, the setup person locks the machine and runs the program using only the output from the MCU. This is done to check if the controller recognizes all the codes in the program. If this test is successful, the program can be run with the machine *Z*-axis locked. This will guard against any possible collisions between the tool and the work holding or part itself. Next is the so-called "dry" run with the part removed. During this test the setup person slows down the rapid feeds and speeds up the actual feeds. These tests will indicate whether the program runs and if there are any extraordinary moves that could cause a collision of the work and/or work holding with the cutting tool. To verify if the program produces a proper part, a blank must be loaded and cut. If the production blanks are made of costly material, some shops may first cut a test part using aluminum, wax, wood, or styrofoam. This saves on material, cutting tools, and prove-out time.

The actual cutting test is run in single-block mode (versus automatic mode). This is done to give the setup person time to see the effects of each command and aids in spotting a wrong move. The program is further optimized by eliminating any unnecessary moves. After the part is completed, it is measured to determine if the drawing, operation sheets, and programmer specifications have been satisfied. Adjustments are made. If necessary, another part is made and checked by the quality control department. If everything is found to be satisfactory, the part goes into production.

8-15 PART PRODUCTION

A completed and checked program is ready for executing production runs only if it is capable of producing quality parts in the time projected. Programs that meet this criterion are a direct reflection of the programmer's professionalism.

The setup person tries to optimize speeds and feeds when parts are manufactured in production. This is done to speed up the cutting cycle, increase tool life, and improve part quality.

8-16 CHAPTER SUMMARY

The following key concepts were discussed in this chapter:

1. The programmer must make a thorough study of the part drawing as a first step in preparing a part program.
2. Important factors entering into the decision to utilize CNC operations include the quantity of parts to be manufactured, the part quality, the tooling costs, the fixture manufacturing, and the cost of running the CNC machine.

3. Methodizing is the process by which a sequence of operations for producing a part is formulated.
4. The operations sheet contains important information concerning the condition of the part material prior to machining.
5. The CNC setup sheet contains information pertaining to fixturing, part setup, tooling, and cutting operations to be performed.
6. The programmer is responsible for writing the part program, selecting the required tooling, and specifying the work-holding and fixturing design.
7. The setup person is responsible for loading the tooling into the tool magazine as well as securing the work-holding device and part blank in the CNC machine. Other setup activities include loading the part program into the MCU, measuring the location of the part origin, entering the tool length offsets, and running the part program.
8. Tool length offset is the distance from the bottom of the tool to the part Z_0 reference plane with the spindle fully retracted.
9. A part is ready for production only after the program has been tested and proven to manufacture quality parts in the time proposed.

REVIEW EXERCISES

8.1. Name five factors that enter into the decision to manufacture a part by CNC operations.

8.2. **(a)** What is methodizing?
(b) In a large shop the _____ person is responsible for this job, whereas in a small shop it is the _____ person.

8.3. **(a)** What is the purpose of the operations sheet?
(b) List three important specifications that can be given in the operations sheet.

8.4. List three factors that must be considered when selecting a CNC machine for an operation.

8.5. Does the programmer or the setup person carry out each of the following CNC shop activities?
(a) Planning operations for part manufacture
(b) Fixture design
(c) Loading fixture and part into the CNC machine
(d) Specifying tooling required
(e) Entering the part program into the MCU
(f) Determining the part origin and tool change position
(g) Entering the location of the part origin and tool length offsets into the MCU
(h) Testing and prove-out of the part program

8.6. **(a)** What is the purpose of the setup sheet?
(b) List four important specifications that are contained in the setup sheet.

8.7. List all the items that must be delivered to the CNC machine in preparation for setting up a job.

8.8. Define the terms machine home, part origin, and tool change position.

8.9. Identify two methods by which the setup person can locate the part origin.

8.10. **(a)** What is the tool length offset?
(b) How is the tool length offset measured?

8.11. Explain the purpose of each of the following tests:
(a) MCU run with the CNC machine locked
(b) Dry run
(c) Test part run

8.12. What two important criteria must a part program satisfy in order to qualify for use in production runs?

CHAPTER 9

Word Address Programming

9-1 CHAPTER OBJECTIVES

At the conclusion of this chapter you will be able to

1. Understand the meaning of important terminology connected with word address programming.
2. Explain what comprises a program.
3. State how addresses should be arranged in a block and identify their use.
4. Describe the importance of G and M codes in a program.

9-2 INTRODUCTION

Fundamental concepts concerning the word address format for programming are considered. The reader is introduced to such terms as programming characters, addresses, words, and blocks. The general arrangement of information in a command block is given and discussed in detail. G and M codes are central to word address programming. A listing of these codes, together with an explanation of their use in programs, is given.

9-3 PROGRAMMING LANGUAGE FORMAT

A program format is a system of arranging information so that it is suitable for input to a CNC controller. Several different types of formats exist. The format to be used in this text is known as the word address. It was originally developed for use with NC tapes and has been retained for CNC programming. Programming of CNC equipment involves the entry of code to a CNC machine for the purpose of precisely controlling all machine movements.

Programming language is similar to the English language and is divided into the following parts:

English language	CNC programming language
English characters	Program characters
English word	Program word
English sentence	Program block
Period	End of block

It is important to mention that there are different programming language "dialects." Each CNC controller is designed to accept a particular dialect. Thus, the format that works on one controller may have to be changed slightly when running it on another controller. It would be impractical and most likely impossible to describe all the different codes for each controller. Therefore, we will give the most general coding pattern that can be followed for any type of programming application. Point-to-point as well as contouring programs will be written for a CNC machine with a Fanuc controller. Many of the same commands can be used on other controllers but some may not. Variations even exist within the same family of Fanuc controllers. The reader is advised to use this text to learn programming technique and as a general reference guide to commands.

In each case, the programming manual of a particular CNC machine must be consulted before attempting to apply the programs presented here to that machine.

9–4 PROGRAMMING LANGUAGE TERMINOLOGY

The following terminology is important when using the word address format.

Programming Character

A programming character is an alphanumeric character or punctuation mark.

■ EXAMPLE 9–1

The following are programming characters:

N G ; ■■

Addresses

An address is a letter that describes the meaning of the numerical value following the address.

■ EXAMPLE 9–2

Identify the address and the number in the codes G00 and X–3.75:

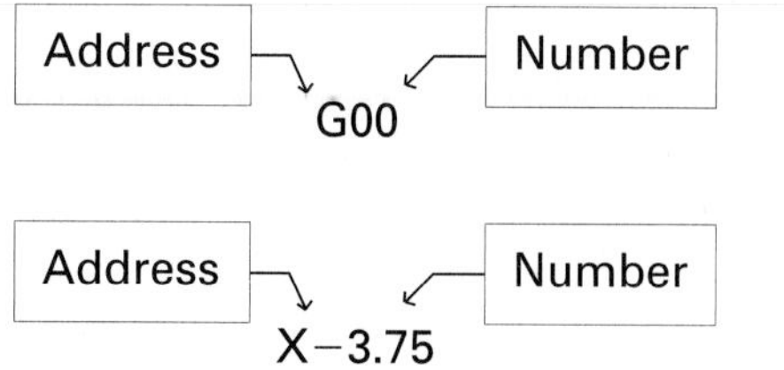

■■

> It is important to note that a minus (−) sign may be inserted between the address and the numeric value. Positive values do not need a plus sign.

Words

Characters are used to form words. Program words are composed of two main parts: an address followed by a number. Words are used to describe such important information as machine motions and dimensions in programs.

Blocks

A block is a complete line of information to the CNC machine. It is composed of one word or an arrangement of words. Blocks may vary in length. Thus, the programmer need only include in a block those words required to execute a particular machine function.

■ EXAMPLE 9–3

Point out the components of the block N0020 G90G00X−2.5Y3.75S1000;.

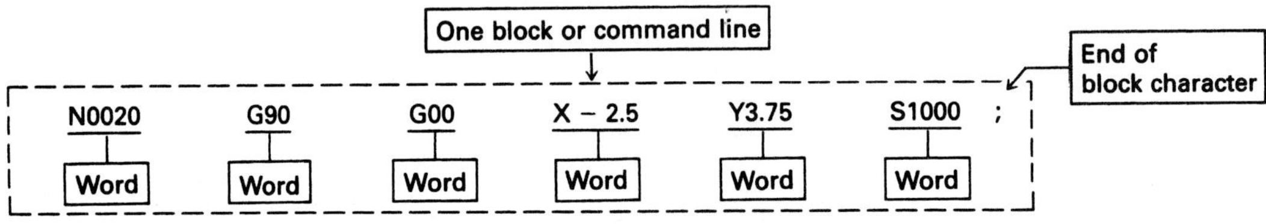
One block or command line
End of block character
N0020
G90
G00
X − 2.5
Y3.75
S1000
;
Word
Word
Word
Word
Word
Word

■■

Each block is separated from the next by an end-of-block (;) code.

> ***Note:*** The End-Of-Block character is automatically generated when the programmer enters a carriage return at the computer or tape preparation machine. The same holds true when the end-of-block key is depressed at the machine control unit during manual data entry. Thus, this character will not appear in the regular program listings in this text.

Programs

A program is a sequence of blocks that describe in detail the motions a CNC machine is to execute in order to manufacture a part. The MCU executes a program block by block. The order in which the blocks appear is the order in which they are processed.

■ EXAMPLE 9–4

Illustrate the order in which the MCU executes the following program.

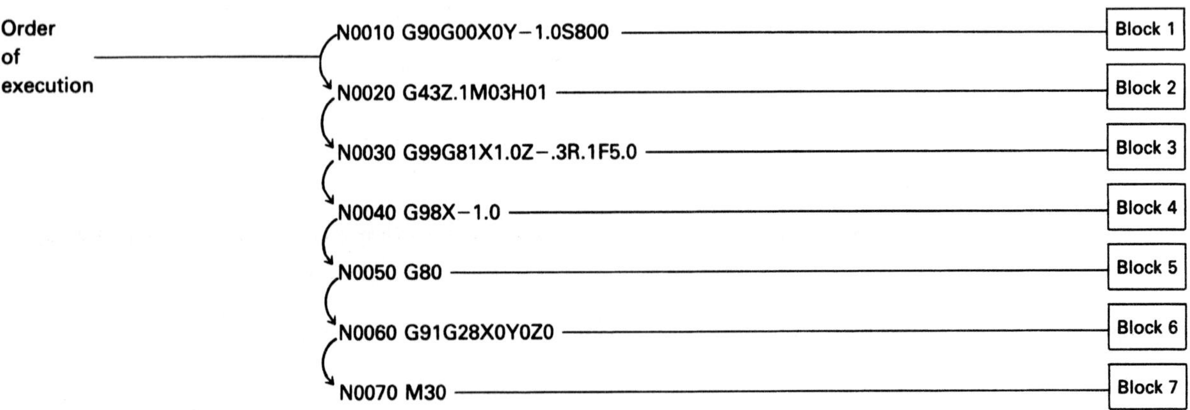
Order of execution
N0010 G90G00X0Y−1.0S800
Block 1
N0020 G43Z.1M03H01
Block 2
N0030 G99G81X1.0Z−.3R.1F5.0
Block 3
N0040 G98X−1.0
Block 4
N0050 G80
Block 5
N0060 G91G28X0Y0Z0
Block 6
N0070 M30
Block 7

9–5 ARRANGEMENT OF ADDRESSES IN A BLOCK

The order in which addresses appear in a block can vary. The following sequence, however, is normally used:

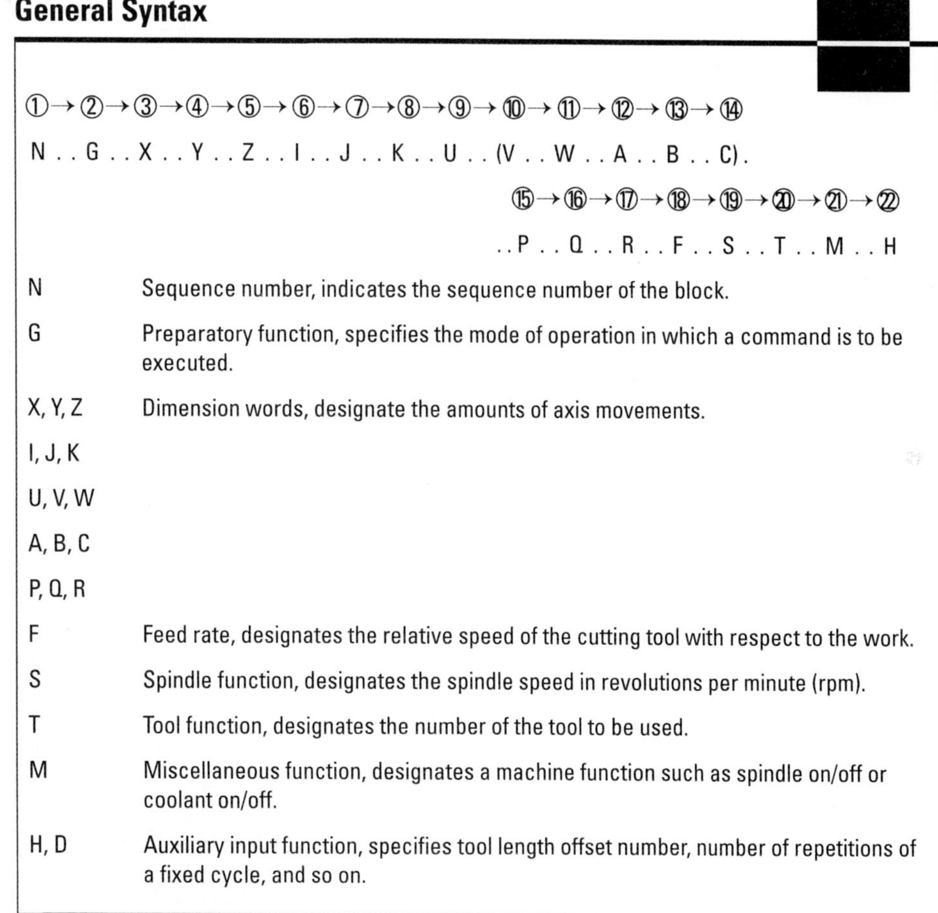

General Syntax

①→②→③→④→⑤→⑥→⑦→⑧→⑨→⑩→⑪→⑫→⑬→⑭

N . . G . . X . . Y . . Z . . I . . J . . K . . U . . (V . . W . . A . . B . . C) .

⑮→⑯→⑰→⑱→⑲→⑳→㉑→㉒

. . P . . Q . . R . . F . . S . . T . . M . . H

N	Sequence number, indicates the sequence number of the block.
G	Preparatory function, specifies the mode of operation in which a command is to be executed.
X, Y, Z	Dimension words, designate the amounts of axis movements.
I, J, K	
U, V, W	
A, B, C	
P, Q, R	
F	Feed rate, designates the relative speed of the cutting tool with respect to the work.
S	Spindle function, designates the spindle speed in revolutions per minute (rpm).
T	Tool function, designates the number of the tool to be used.
M	Miscellaneous function, designates a machine function such as spindle on/off or coolant on/off.
H, D	Auxiliary input function, specifies tool length offset number, number of repetitions of a fixed cycle, and so on.

■ EXAMPLE 9–5

Give an example of addresses arranged in a block.

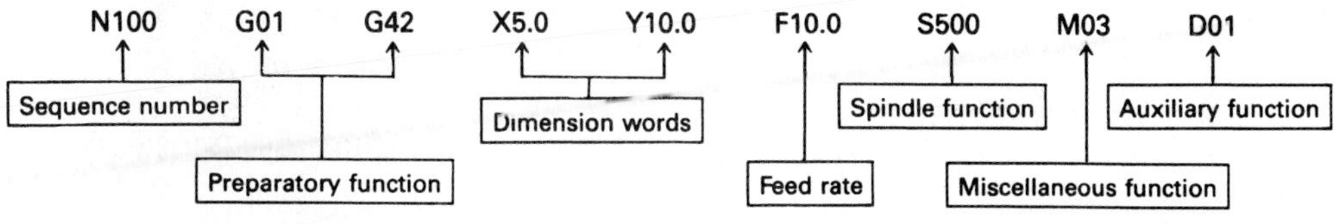

■■

9–6 PROGRAM AND SEQUENCE NUMBERS (O, N CODES)

Program Number (O)

Programs are stored in the MCU memory by program number. The machine recognizes programs according to a numeric code. Most machines can store several different programs at a time. Program numbers range from O1 to O9999. See Figure 9–1.

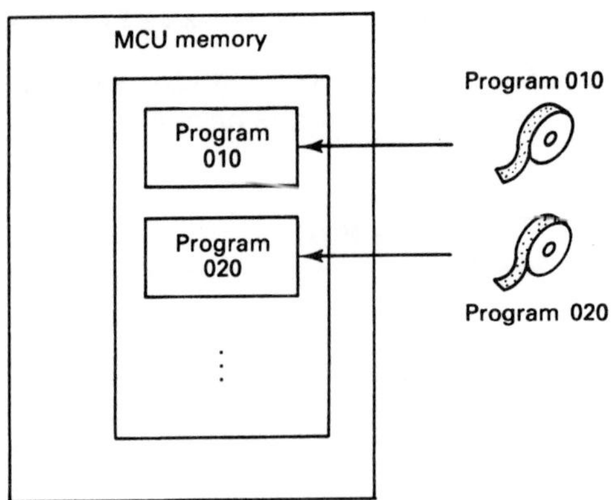

FIGURE 9-1

Sequence Number (N)

A sequence number is an optional tag that can be coded at the beginning of a block if needed. The MCU will execute program blocks in the order in which they appear regardless of the sequence number entered.

Sequence numbers are used so that operators can locate specific lines of a program when entering data or performing checkout operations.

These numbers range from N1 to N9999.

■ EXAMPLE 9-6

Give an example of the use of a program number and sequence numbers.

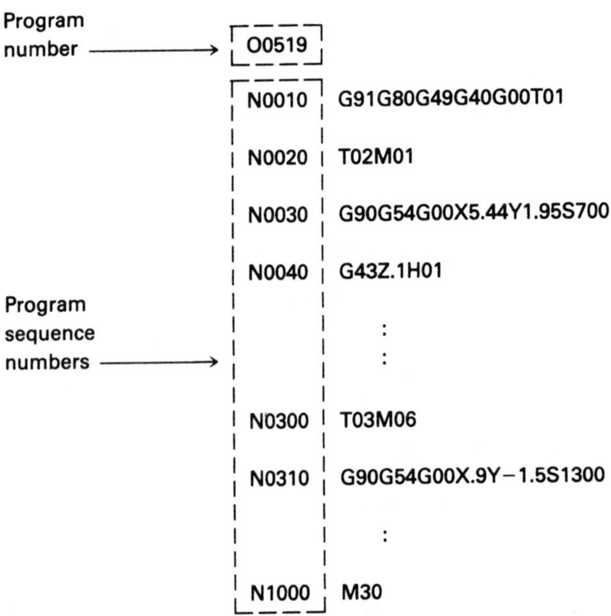

9-7 PREPARATORY FUNCTIONS (G CODES)

A preparatory function is designated by the address G followed by one or two digits to specify the *mode* in which a CNC machine moves along its programmed axes. The term *preparatory* signifies that the word (G address and digit code) prepares the control system for the information that is to follow in the block. Preparatory functions are also referred to as G codes. A G code is usually placed at the beginning of a block so it can set the control for a particular mode when acting on the other words in the block.

G CODES FALL INTO ONE OF THE FOLLOWING TWO MAJOR CATEGORIES:

G code category	Effect
Modal	The G code specification will remain in effect for all subsequent blocks unless replaced by another modal G code.
Nonmodal	The G code specification will only affect the block in which it appears.

Many G codes have been standardized and others are unique to a particular CNC control. It should also be noted that there are differences between the G codes used for CNC machining centers and those used for CNC lathes. The following G codes are especially useful when executing operations involving Fanuc controllers.

G code	Mode	Specification
G00	Modal	Rapid positioning mode. The tool is moved to its programmed *XYZ* location at maximum feed rate. The tool will travel at a 45° angle with both drive motors running at top speed until one or the other runs out of travel.
G01	Modal	Linear interpolation mode. The tool is to be moved along a straight-line path at the programmed feed rate.
G20	Modal	Inch mode for all units. This code is entered at the start of a CNC program to specify units are in inches.
G21	Modal	Specifies metric (mm) mode for all units.
G28	Nonmodal	Returns tool to reference point.
G43	Modal	Specifies tool length offset (positive direction).
G49	Modal	Cancels the tool length offset.

G code category		Effect
G53	Modal	Cancels G54-G59 fixture offsets
G54-G59	Modal	Specifies fixture offset locations. G54 specifies fixture offset location 1 G55 ″ ″ location 2 : : G59 ″ ″ location 6
G80	Modal	Cancels any fixed cycle. This command should appear prior to starting a new program and at the end of a program to cancel any fixed cycles. Refer to Chapter 8.
G90	Modal	Specifies absolute position programming.
G91	Modal	Specifies incremental position programming.
G92	Modal	Directs the system to shift the absolute zero point for programmed moves from the current tool position to

		the part origin. If the current tool position is at machine home, the shift will be from machine home to the part origin. The *X* and *Y* values following G92 are measured from the part origin to the current tool position.
G98	Modal	Specifies a return to the initial point in a machining cycle that had been created by a modal G code.

A more comprehensive listing of G codes is given in Appendix B.

■ EXAMPLE 9–7

Explain the effect of modality of the G codes on the following coding:

N0010	G91G01×1.5F10.0	G91 and G01 remain in effect for these blocks
N0020	X.2	
N0030	Y.2	
N0040	G90G00Y0	G91 and G01 are replaced with G90 and G00

■■

In the following chapters, we will explain in more detail the use and application of G codes in commands.

9–8 DIMENSION WORDS (X, Y, Z . . . CODES)

As was stated previously, dimension words specify the movement of the programming axes. Programming axes were discussed in Chapter 2 but will also be considered here. Recall that programming axes are laid out according to the Cartesian coordinate system. The positive or negative direction of movement along an axis is given by the right-hand rule. For a review, see Figure 9–2.

Address	Information stored
X, Y, Z	Linear axes
A, B, C	Rotary axes
U, V, W	Axes parallel to X, Y, Z axes
I, J, K	Axes used as auxiliary of X, Y, Z axes
R, Q	Axes used as auxiliary of Z axis

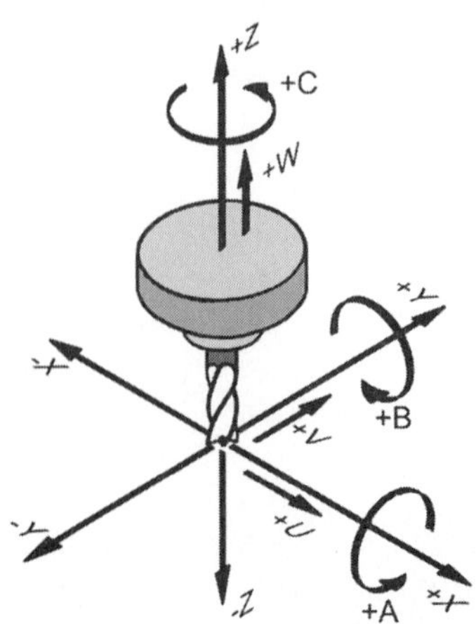

FIGURE 9–2

■ **EXAMPLE 9–8**

Explain the meaning of the following code:

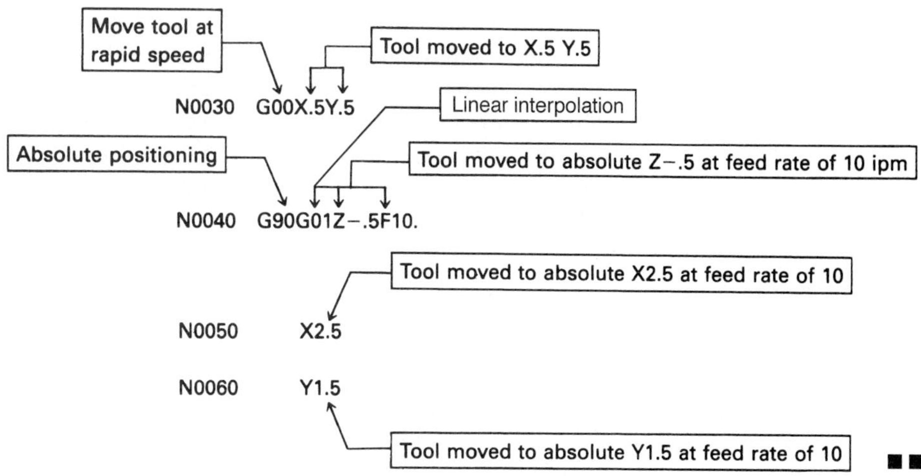

■■

9–9 FEED RATE (F CODE)

The feed rate is the rate at which the cutting tool moves along a programming axis. It is specified by the numerical value following the address F. In the English system, the feed rate is expressed in inches per minute and in the metric system in millimeters per minute.

An F specification is modal and remains in effect in a program for all subsequent tool movements. The feed rate can be changed by entering a new F command. Decimal point input is acceptable with an F command.

■ **EXAMPLE 9–9**

Explain the meaning of the feed rate codes F10 and F10. in the English system.
F10 specifies a feed rate of 0.001 in./min (ipm).
F10. specifies a feed rate of 10 in./min (ipm). ■■

9–10 SPINDLE SPEED (S CODE)

S codes control the speed at which the spindle rotates. A numerical value up to four digits maximum is entered following the address S.

■ **EXAMPLE 9–10**

Explain the meaning of the code S1600.
S1600 directs the control to set the CNC machine spindle at 1600 revolutions per minute or 1600 rpm. ■■

The S command is modal and remains in effect for every subsequent command until replaced by a new S code or canceled by a spindle-off (M05) command.

Spindle rotation should be specified prior to entering blocks containing cutting commands. An S code is usually entered in the same block containing an axis movement instruction. Upon executing the block, the controller will direct the CNC machine to start the spindle turning and move it along the programmed axis.

9-11 MISCELLANEOUS MACHINE FUNCTIONS (M CODES)

M codes specify CNC machine functions not related to dimensional or axes movements. Unlike G codes, they do not prepare the controller to act in a particular mode when processing the other words in a block. Instead, they direct the controller to immediately execute the machine function indicated.

The numbers following the address M call for such miscellaneous machine functions as spindle on/off, coolant on/off, program stop/automatic tool change, program end, and so on. M codes are usually classified into two main groups:

Type A: Those executed with the start of axis movements in a block.
Type B: Those executed after the completion of axis movements in a block.

The following M codes are especially useful in many programming applications.

M code	Type	Specification
M00	B	Causes a program stop.
M02	B	Causes a program end. An M02 code must be the last command in a program. If used, do not use M30.
M03	A	Turns spindle on clockwise.
M04	A	Turns spindle on counterclockwise.
M05	B	Turns spindle off. Usually used prior to a tool change and at the end of a program.
M06	B	Stops the program and calls for an automatic tool change.
M07	A	Turns the coolant tap oil on.
M08	A	Turns the external coolant on.
M09	B	Turns the coolant off.
M30	B	Directs the system to end program processing, rewind the tape, or reset the memory unit. This code must be the last statement in a program. If used, do not use M02.

A complete list of M codes is given in Appendix B.

9-12 AUTOMATIC TOOL CHANGING (M06 CODE)

To execute a tool change on a machining center, the programmer must first rapid the spindle up to a safe *Z* distance. Retracting to a safe height will ensure that the tool does not strike the part or fixturing when moving in the *XY* plane. The tool must then be repositioned from its current location to the tool change position. On many machining centers this is accomplished by returning the tool to the *Z* and *Y* machine home reference points (Z_0Y_0). The code G91 G28 Z0 Y0 will accomplish this task. It should be noted that the spindle does not have to be reset to the X_0 reference point. On some machines a move to the tool change position only requires that the spindle return to the Z_0 reference position.

The number of the new tool to be used is identified by the T word. The word M06 directs the machine to change to the new tool.

■ EXAMPLE 9–11

Explain the meaning of the code:

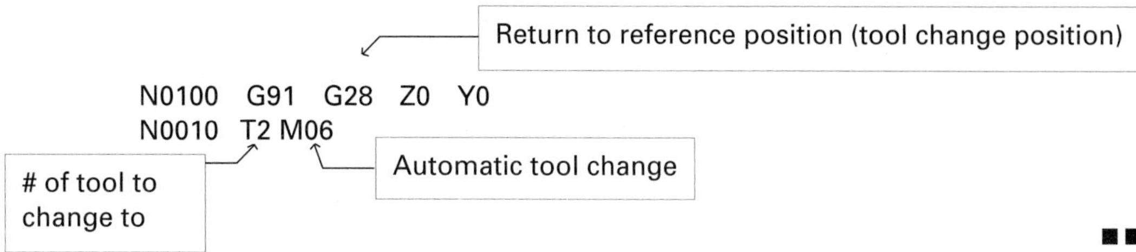
Return to reference position (tool change position)
N0100 G91 G28 Z0 Y0
N0010 T2 M06
# of tool to change to
Automatic tool change

■■

9–13 TOOL LENGTH OFFSET AND CUTTER RADIUS COMPENSATION (H, D CODES)

The H code is used to specify where the values of the tool length offset and the tool position offset are located. The D code indicates where the value of the cutter radius compensation for a tool is to be found if needed.

A two-digit number ranging from 01 to 99 is used with the H or D address. The number used for the H specification should not be used for the D specification. However, it is recommended that the same number be used for both the H and T codes. Thus, T01 should be used with H01 and T10 with H10, and so forth.

■ EXAMPLE 9–12

Explain the meaning of the following code:

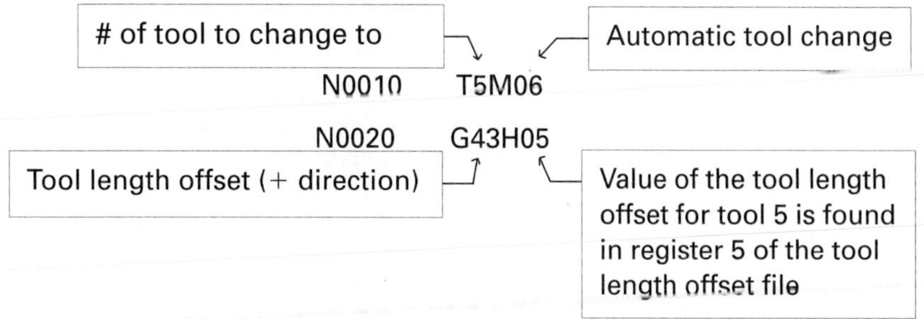
# of tool to change to
Automatic tool change
N0010 T5M06
N0020 G43H05
Tool length offset (+ direction)
Value of the tool length offset for tool 5 is found in register 5 of the tool length offset file

■■

9–14 COMMENTS

A comment is labeling text that is displayed with a program. Information that is written between the left and right parenthesis ( ) is considered to be a comment and is ignored by the controller. The end of tape character % cannot be used in a comment. Long comments placed in the middle of a program will interrupt motion for a long time. It is good practice to insert comments at places where movement can be interrupted or where no movement is specified.

■ EXAMPLE 9–13

The following comments are used for labeling purposes only and are ignored by the controller:

Comment displayed with program but ignored by controller

N0010 (PROGRAM TO MACHINE HOLES)
N0020 (TOOL 1: DRILL 3 .25 DIA HOLES THRU) ■■

9–15 CHAPTER SUMMARY

The following key concepts were discussed in this chapter:

1. The word address is a format for writing programs.
2. A program word is used to describe important machine motions and dimensions in a program.
3. A block is composed of words and represents a complete command to the CNC machine.
4. Program blocks may vary in length depending upon the information needed to command a CNC operation.
5. A preparatory function (G code) is used to specify the mode in which a CNC machine moves along programming axes.
6. A miscellaneous function (M code) is entered to specify machine functions other than dimensional or axis movements.

REVIEW EXERCISES

9.1. What is a program format?
9.2. Briefly describe the word address format.
9.3. Match the terms on the left with the definitions on the right:

Character	A letter describing the meaning of a number following the letter
Address	A sequence of blocks
Word	Alphanumeric or punctuation mark
Block	An address followed by a number
Program	A complete command to the CNC machine

9.4. What is a G code?
9.5. What is the difference between modal and nonmodal G codes?
9.6. Explain the mode and use of the following G codes:
- **(a)** G49
- **(b)** G01
- **(c)** G80
- **(d)** G00
- **(e)** G92

9.7. Describe what an M code is.
9.8. Into what two main groups do M codes fall?
9.9. State the type and use of the following M codes:
- **(a)** M06
- **(b)** M30
- **(c)** M00
- **(d)** M03
- **(e)** M05

9.10. Explain the effect of each block on the CNC machine.

(a) N0010 G01X1.5F5.0

(b) N0100 G00X2Y0S500

(c) N0050 T04M06
N0060 G43H02

(d) N0020 G92X–2.Y5.

(e) N0070 G90G20

CHAPTER

10

PROGRAMMING HOLE OPERATIONS

10-1 CHAPTER OBJECTIVES

At the conclusion of this chapter you will be able to

1. Understand the meaning of a fixed or "canned" cycle.
2. Explain the purpose of the most important fixed cycles for hole operations.
3. Identify the different codes of information required to program a fixed cycle.
4. Describe the general sequence of operations to be followed in programming hole operations.
5. Write simple hole operation programs with the aid of canned cycles.

10-2 INTRODUCTION

The concept of a fixed or "canned" cycle is discussed in this chapter. The most important fixed cycles as related to such hole operations as drilling, counterboring, deep drilling, tapping, and boring are explained in detail. The reader is introduced to the format that can be followed when writing hole operation programs for a vertical machining center. Complete programs are given, together with a comprehensive explanation of the code.

The student is encouraged to use the hole machining simulation software described in Chapter 20 and contained on the CD at the back of the text.

10-3 FIXED OR CANNED CYCLES

The simplest operations to program are those related to producing holes. These include drilling, boring, tapping, spot drilling, and counterboring. The simplicity of programming lies in the fact that the programmer only needs to specify the coordinates of a hole center and the type of machine motions to be performed at the center. A fixed cycle, if used properly, takes over and causes the machine to execute the required movements. The controller stores a number of fixed cycles that can be recalled for use in programs when needed. This reduces the programming time and the length of tape required. See Figure 10–1.

A fixed cycle is programmed by entering in one block of information: the *X* and *Y* coordinates, the *Z*-axis clearance plane (*R*), and the final *Z*-axis depth. A fixed cycle for hole operations will cause the following sequence of operations to occur:

1. Rapid move to the X and Y coordinate of the hole center.
2. Rapid move to the Z-axis clearance plane (R).

Safety Rules for Hole Operations

- Make sure the fixture setup is rigid enough to withstand the high thrusts generated by drilling.
- Take necessary precautions when drilling:
  - Make sure machine safety guard is up.

  - Wear safety goggles.
  - Avoid skin contact with cutting fluids.
- For long boring cuts select the heaviest possible boring bar with the shortest overhang.
- Set cutting speeds and feeds to values recommended by the tool manufacturer.
- Adjust recommended values to allow for accuracy, quality of surface finish, rate of tool wear, chip control, and machine capability.
- Press the EMERGENCY STOP button before entering the work area to remove chips or clean the CNC machine.
- Be mindful of overhead obstructions when leaning into the working area.

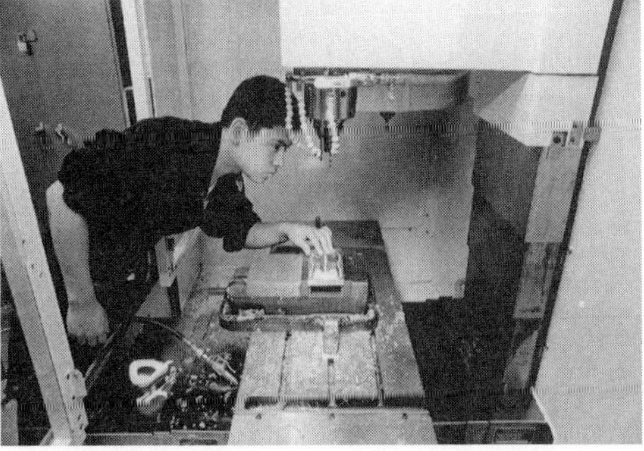

3. Feed to the Z-axis final depth.
4. Rapid back to either the Z-axis initial position or the Z-axis clearance plane (R).

To perform the same operations in the next block, the programmer need only enter the X and Y coordinates of the next hole center.

10–4 HOLE OPERATION COMMANDS

Listed in this section are the word address commands for executing various hole operations on a vertical milling machine equipped with a Fanuc controller. The cycles described are the most important ones used and are not a full listing of

FIGURE 10–1 A boring operation executed using a fixed cycle. (Photo courtesy of American SIP Corp.)

all the cycles that can be obtained. The variety and availability of these codes vary from machine to machine depending upon the machine builder and the options selected.

> ***Note:***
>
> 1. The cycles described are modal. They operate automatically on all subsequent data blocks that include rapid movement (G00) in the *XY* plane. A cycle is canceled by a G80 code or replaced by another autocycle.
> 2. $Z_{initial}$ is the current Z-axis position when a fixed cycle is initiated.
> 3. The fonting in the diagrams carries the following meanings:
>
>   ⇣ represents rapid feed
>
>   ↓ represents programmed feed rate

General Syntax

Drill, Center Drill, or Ream Cycle

A G81 cycle causes the machine to ① rapid the tool from the $Z_{initial}$ position to the R_{plane},

② drill the hole to a depth Zn at feedrate Fn, ③ rapid back to either the R_{plane} or the $Z_{initial}$ position,

④ rapid to the center of the next hole if the Xn Yn coordinates of that hole are programmed in the next block.

G81 Xn Yn Zn Rn Fn

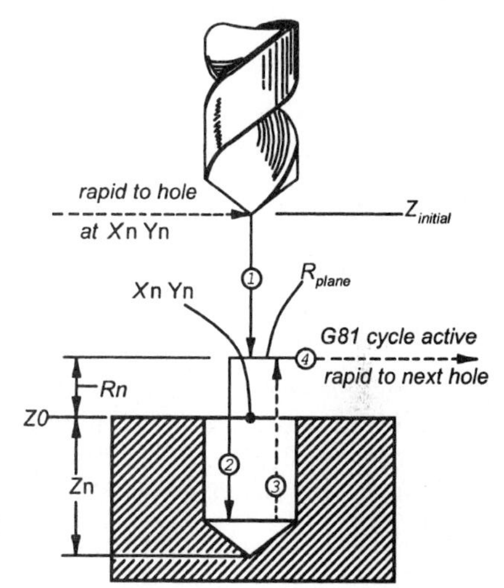

G81 Specifies simple drilling

Xn Yn Numeric values of n specify the absolute X and Y coordinates of the hole center. Coordinates can be absolute (G90) or incremental (G91).

Zn Numeric value of n specifies the depth of the drill in absolute (G90) or the distance below the R_{plane} to the hole bottom in incremental (G91) code.

Rn Numeric value of n specifies the distance to the R_{plane} in absolute (G90) or the distance below the initial tool position to the R_{plane} in incremental (G91) code. If not programmed, the last active R_{plane} is used and, if none is specified, the tool returns to the $Z_{initial}$ position.

Fn Numeric value of n specifies the feed rate of the tool [(in/min) or (mm/min)] into the hole. If not programmed, the system will use the last programmed feed rate.

Counterbore Cycle or Spotface Cycle

A G82 cycle causes the machine to ① rapid the tool from the $Z_{initial}$ position to the R_{plane},

② bore the hole to a depth Zn at feedrate Fn, ③ dwell for Pn seconds at depth Zn,

④ rapid back to either the R_{plane} or the $Z_{initial}$ position,

⑤ rapid to the center of the next hole if the Xn Yn coordinates of that hole are programmed in the next block.

G82 Xn Yn Zn Rn Fn Pn

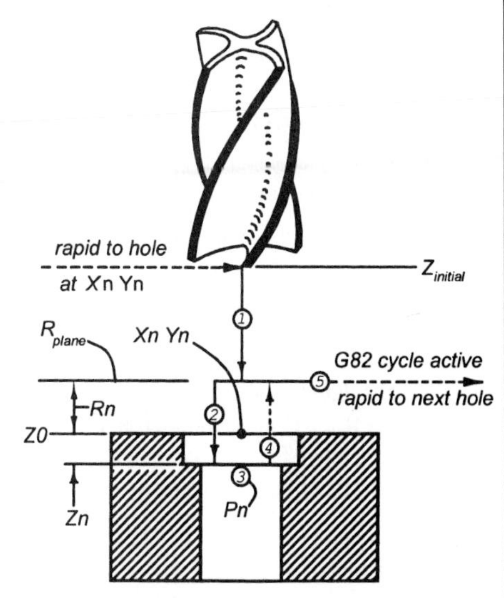

G82 Specifies a counterbore cycle

Xn Yn Numeric values of n specify the absolute X and Y coordinates of the hole center. Coordinates can be absolute (G90) or incremental (G91).

Zn Numeric value of n specifies the depth of the drill in absolute (G90) or the distance below the R_{plane} to the hole bottom in incremental (G91) code.

Rn Numeric value of n specifies the distance to the R_{plane} in absolute (G90) or the distance below the initial tool position to the R_{plane} in incremental (G91) code. If not programmed, the last active R_{plane} is used and, if none is specified, the tool returns to the $Z_{initial}$ position.

Fn Numeric value of n specifies the feed rate of the tool [(in/min) or (mm/min)] into the hole. If not programmed, the system will use the last programmed feed rate.

Pn Numeric value of n specifies the dwell time at the bottom of the hole in seconds (0.01 to 99.99).

Deep Drill or Peck Drill Cycle

A G83 cycle causes the machine to ① rapid the tool from the $Z_{initial}$ position to the R_{plane}, ② drill into the hole a specific peck distance Qn at feedrate Fn, ③ rapid to either the R_{plane} or the $Z_{initial}$ position, ④ drill again at feedrate to a depth of $2Q_n$, ⑤ rapid back to either the R_{plane} or the $Z_{initial}$ position, repeatedly execute this process with uniform increases in the peck depth until the total hole depth ⑥ is reached. Upon reaching the total hole depth rapid back ⑦ to either the R_{plane} or the $Z_{initial}$ position, ⑧ rapid to the center of the next hole if the Xn Yn coordinates of that hole are programmed in the next block.

```
G83 Xn Yn Zn Qn Rn Fn
```

G83 — Specifies a peck drill cycle

Xn Yn — Numeric values of n specify the absolute X and Y coordinates of the hole center. Coordinates can be absolute (G90) or incremental (G91).

Zn — Numeric value of n specifies the depth of the drill in absolute (G90) or the distance below the R_{plane} to the hole bottom in incremental (G91) code.

Qn — Numeric value of n specifies the first peck distance below the R_{plane}. This value is added successively to the last total for each pass until the final hole depth is reached.

Rn — Numeric value of n specifies the distance to the R_{plane} in absolute (G90) or the distance below the initial tool position to the R_{plane} in incremental (G91) code. If not programmed, the last active R_{plane} is used, and if none is specified, the tool returns to the $Z_{initial}$ position.

Fn — Numeric value of n specifies the feed rate of the tool [(in/min) or (mm/min)] into the hole. If not programmed, the system will use the last programmed feed rate.

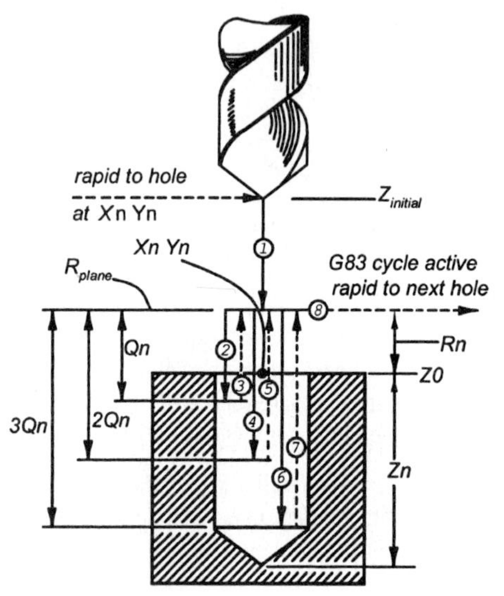

Tap Cycle

A G84 cycle causes the machine to ① rapid the tool from the $Z_{initial}$ position to the R_{plane},

② the threads are cut as the tool advances in at feedrate Fn, ③ at maximum depth the spindle is reversed automatically and the tool is retracted at feedrate to either the $Z_{initial}$ position or the R_{plane},

④ rapid to the center of the next hole if the Xn Yn coordinates of that hole are programmed in the next block. Because the Z-axis motion is instantaneously reversed for up motion when the spindle is reversed, it is advisable to use a floating tap holder.

```
G84 Xn Yn Zn Rn Fn
```

G84 — Specifies a tap cycle

Xn Yn — Numeric values of n specify the absolute X and Y coordinates of the hole center. Coordinates can be absolute (G90) or incremental (G91).

Zn — Numeric value of n specifies the thread depth in absolute (G90) or the distance below the R_{plane} to the bottom of the thread in incremental (G91) code.

Rn — Numeric value of n specifies the distance to the R_{plane} in absolute (G90) or the distance below the initial tool position to the R_{plane} in incremental (G91) code. If not programmed, the last active R_{plane} is used and, if none is specified, the tool returns to the $Z_{initial}$ position.

Fn — Numeric value of n specifies the feed rate [(in/min) or (mm/min)] of the tool into or out of the hole. *The feed rate should be (rpm × lead of tap).* If omitted, the last programmed feedrate is used.

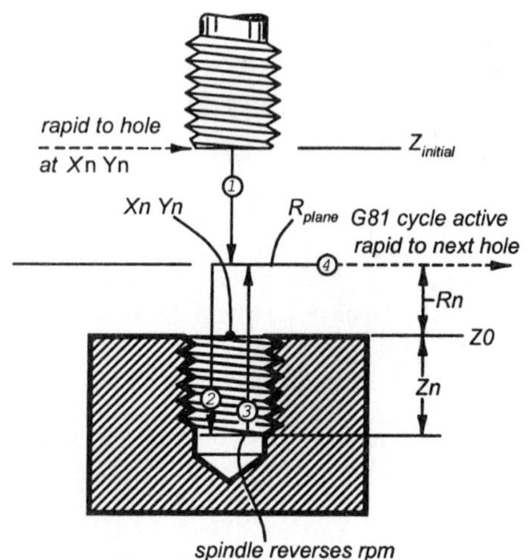

Bore Cycle

A G85 cycle causes the machine to ① rapid the tool from the $Z_{initial}$ position to the R_{plane}, ② advance the tool into the hole at feedrate Fn, ③ retract the tool at feedrate to either the $Z_{initial}$ position or R_{plane}, ④ rapid to the center of the next hole if the Xn Yn coordinates of that hole are programmed in the next block.

G85 Xn Yn Zn Rn Fn

G85	Specifies a bore cycle
Xn Yn	Numeric values of n specify the absolute X and Y coordinates of the hole center. Coordinates can be absolute (G90) or incremental (G91).
Zn	Numeric value of n specifies the depth of the drill in absolute (G90) or the distance below the R_{plane} to the hole bottom in incremental (G91) code.
Rn	Numeric value of n specifies the distance to the R_{plane} in absolute (G90) or the distance below the initial tool position to the R_{plane} in incremental (G91) code. If not programmed, the last active R_{plane} is used and, if none is specified, the tool returns to the $Z_{initial}$ position.
Fn	Numerical value of n specifies the feed rate of the tool [(in/min) or (mm/min)] into or out of the hole. If not programmed, the system will use the last programmed feed rate.

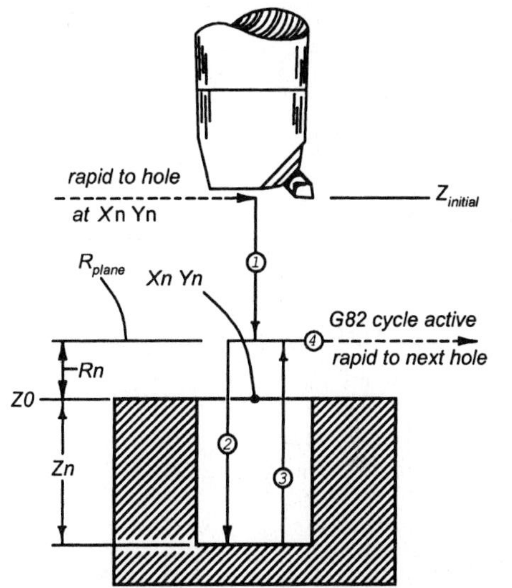

■ EXAMPLE 10–1

For each hole operation shown in Figures 10–2 through 10–6, write the appropriate G code commands in both absolute and incremental mode.

1. Drill autocycle

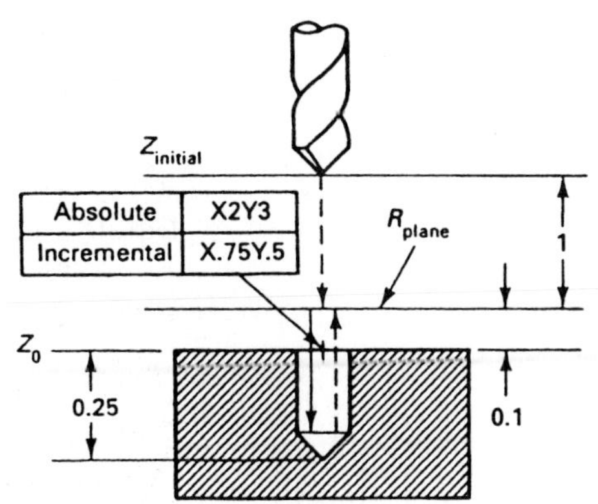

FIGURE 10–2

Absolute
G90
G81 X2. Y3. Z .25 R.1 F7.
Incremental
G91
G81 X.75 Y.5 Z−.35 R−1. F7.

2. Counterbore autocycle

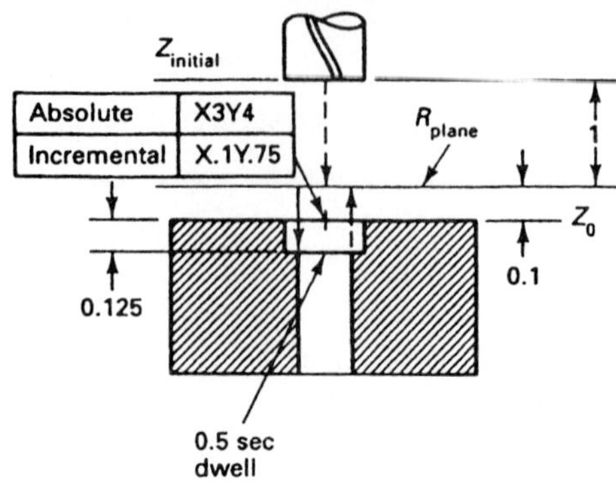

FIGURE 10-3

Absolute
G90
G82 X3. Y4. Z−.125 R.1 P.5 F6.
Incremental
G91
G82 X1. Y.75 Z−.225 R−1 P.5 F6.

3. Deep drill autocycle

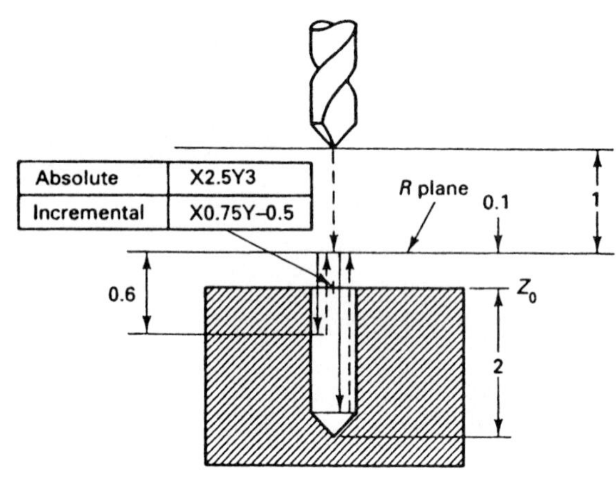

FIGURE 10-4

Absolute
G90
G83 X2.5 Y3. Z−2. R.1 Q.6 F6.
Incremental
G91
G83 X.75 Y−.5 Z−2.1 R−1. Q.6 F6.

4. Tap autocycle

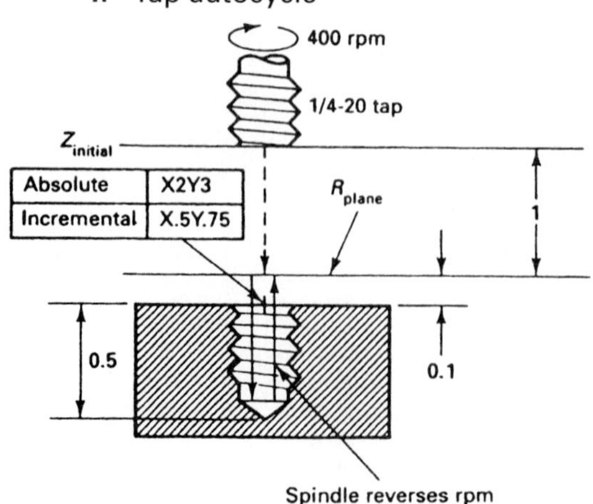

FIGURE 10-5

Determination of feed rate (*F*)
F = rpm × Lead of tap
= 400 rpm × 1/20 in./rev = 20 ipm
Absolute
G90
G84 X2. Y3. Z−.5 R.1 F20.
Incremental
G91
G84 X.5 Y.75 Z−.6 R−1. F20.

5. Bore autocycle

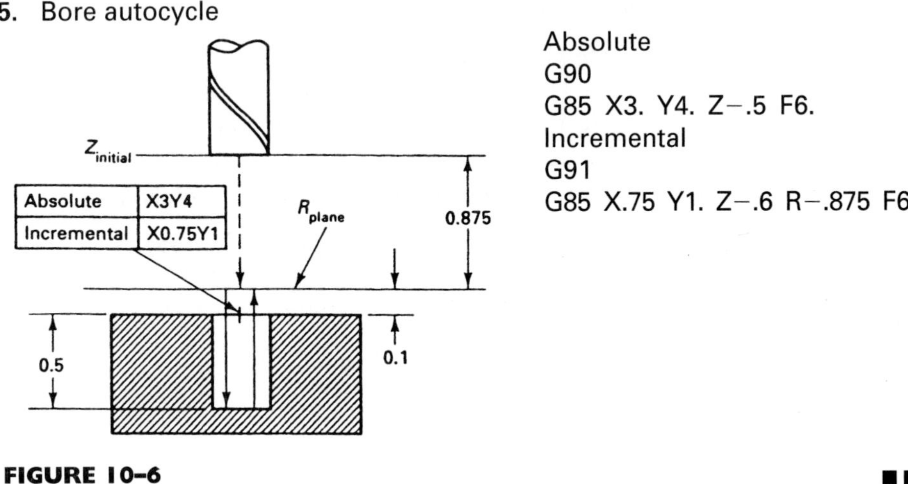

Absolute
G90
G85 X3. Y4. Z–.5 F6.
Incremental
G91
G85 X.75 Y1. Z–.6 R–.875 F6.

FIGURE 10–6

10–5 WRITING A HOLE OPERATION PROGRAM

A suggested pattern or format that can be followed for programming a typical vertical machining center with a Fanuc 6M controller is shown in Figure 10–7.

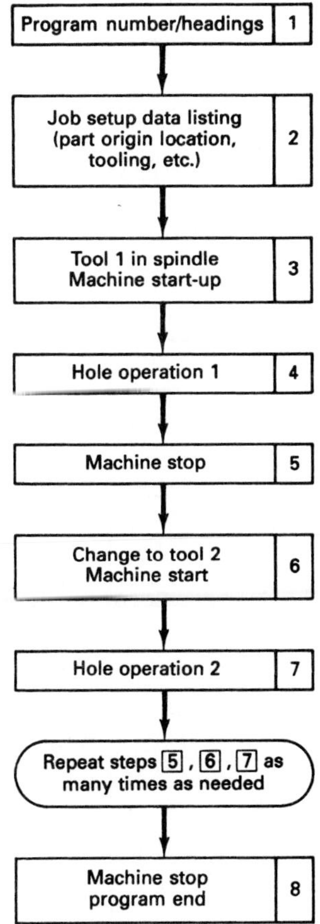

FIGURE 10–7

■ EXAMPLE 10–2

Write a CNC program for executing the hole operations as shown in Figure 10–8. Use the data table information given for specifying operation(s) performed, tooling, tool diameter, tool speed, and feed.

Tool	Operation	Tooling	Speed (rpm)	Feed (ipm)
1	Deep drill 4 holes thru	0.125*D** drill	1600	5

**D* indicates *diameter.*

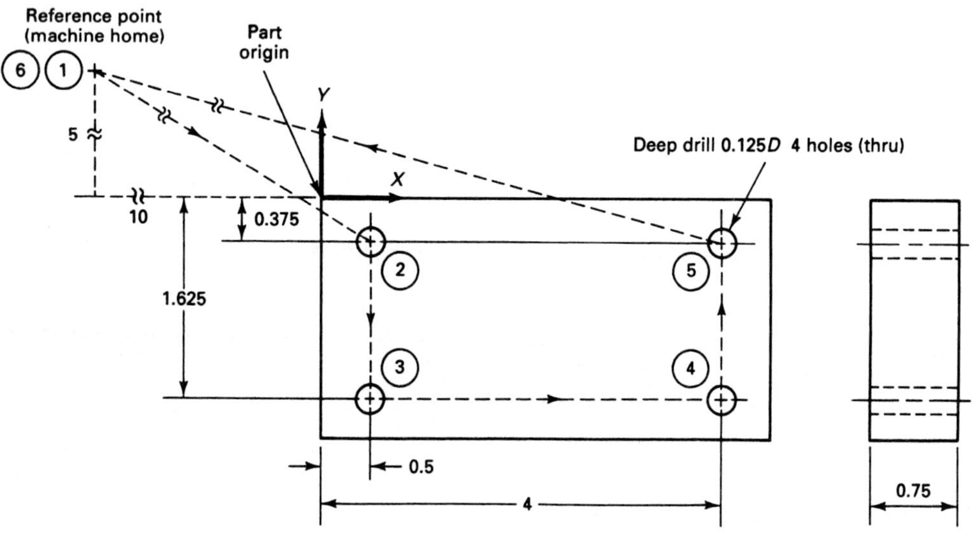

FIGURE 10–8

Programming Pattern

Job setup data listing

Machine start-up sequence

Change to tool 1

Word address command	*Meaning*
%	End of tape. Used to separate programs.
O1001	Program number.
N0010 (X0Y0 IS THE UPPER LEFT HAND CORNER)	Comments.
N0020 (Z0 IS THE TOP OF THE PART)	
N0030 (TOOL 1: 1/8 DRILL)	
N0040 G90 G20 G40 G80	Absolute, inch mode, cancel cutter diameter compensation and fixed cycles.
N0050 G91 G28 X0 Y0 Z0	Return to reference point ①.
N0060 G92 X−10.0 Y5.0 Z0	Preset absolute zero point.
N0070 T1 M06	Change to tool 1.
N0080 (TOOL 1: DRILL 4 .125 DIA HOLES THRU)	
N0090 G00 G90 X.5 Y−.375 Z0. S1600 M03	Rapid to ②. Start spindle (CW) at 1600 rpm.
N0100 G43 Z.1 H01	Rapid tool 1 to 0.1 above part.
N0110 M08	Coolant on.

Hole operation 1 sequence	N0120 G83 X.5 Y−.375 Z−.8876 R.1 Q.1 F5.0	Start deep hole cycle at ②. Final depth is 0.8876, 0.1 depth of peck at feed 5. Return to 0.1 above part.
	N0130 Y−1.625	Deep drill hole at X.5Y−1.625 ③.
	N0140 X4.0	Deep drill hole at X4.Y−1.625 ④.
	N0150 Y −.375	Deep drill hole at X4.Y−.375 ⑤.
	N0160 G80	Cancel any fixed cycles (G83 cycle cancelled).
Machine stop Program end sequence	N0170 G00 G90 Z1.0 M05	Rapid 1.0 above part. Stop spindle.
	N0180 M09	Coolant off.
	N0190 G91 G28 X0 Y0 Z0	Rapid to *XYZ* reference point ⑥.
	N0200 M30	Program end, memory reset.
	%	End of tape.

Machining simulation for this example is contained in the job file EX 10–2 stored on the disk at the back of this text. Refer to Chapter 20 for instructions.

■ EXAMPLE 10–3

Write a program to execute the hole operations as indicated in Figure 10–9 and outlined in the following data table. The program is to be created in metric.

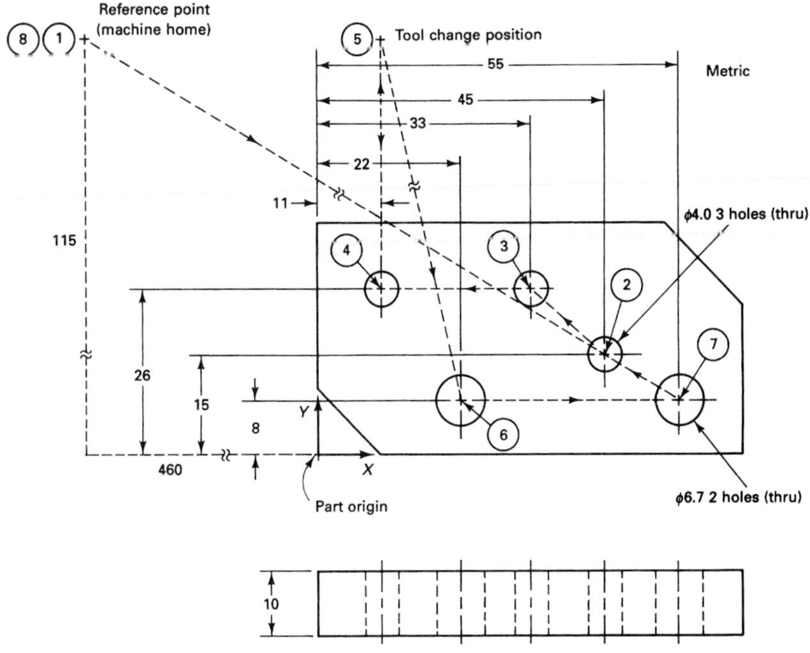

FIGURE 10–9

■■

Tool	Operation	Tooling	Speed (rpm)	Feed (mm/min)
1	Drill (3) ∅4 holes	∅4-mm drill	2200	60
2	Drill (2) ∅6.7 holes	∅6.7-mm drill	1800	80

Programming Pattern

Programming Pattern	*Word address command*	*Meaning*
	%	End of tape.
	O1002	Program number.
Job setup data listing	N0020 (X0Y0 IS THE LOWER LEFT HAND CORNER)	Comments.
	N0030 (TOOL 1: 4 MM CENTER DRILL)	
	N0040 (TOOL 2: 6.7 MM DRILL)	
	N0050 G90 G21 G40 G80	Absolute, metric mode, cancel fixed cycles.
Machine start-up sequence Change to tool 1	N0060 G91 G28 X0 Y0 Z0	Return to reference point ①.
	N0070 G92X−460.0 Y115.0 Z0	Preset absolute zero point.
	N0080 T1 M06	Change to tool 1.
	N0090 (TOOL 1: DRILL 34.0 MM DIA HOLES THRU)	
	N0100 G00 G90 X45. Y15. Z0. S2200 M03	Rapid to ②. Start spindle at 2200 rpm (CW).
	N0110 G43 Z2.5 H01	Rapid tool 1 to 2.5 mm above part.
	N0120 M08	Coolant on.
Hole operation 1 sequence	N0130 G83 X45. Y15. Z−13.5 R2.5 Q2.5 F60.0	Start deep drill cycle at ②. Final depth is 13.5 mm, 2.5-mm peck depth at feed 60 mm/min. Return to 2.5 mm above part.
	N0140 X33.0 Y26.0	Deep drill at X33Y26 mm ③.
	N0150 X11.0	Deep drill at X11Y26 mm ④.
	N0160 G80	Cancel any fixed cycles (G83 cycle cancelled).
Machine stop sequence	N0170 G00 G90 Z2.5 M05	Rapid to 2.5 mm above part.
	N0180 M09	Coolant off.
Change to tool 2 Machine start sequence	N0190 G91 G28 Z0 Y0	Rapid to tool change position ⑤.
	N0200 M06	Change to tool 2.
	N0210 (TOOL 2: DRILL 2 6.7 MM DIA HOLES THRU)	
	N0220 G00 G90 X22.0 Y8.0 Z0 S1800 M03	Rapid to ⑥. Start spindle at 1800 rpm (CW).
	N0230 G43 Z2.5 H02	Rapid tool 2 to 2.5 mm above part.
	N0240 M08	Coolant on.

Hole operation 2 sequence	N0250 G83 X22.0 Y8.0 Z −14.6 R2.5 Q3. F90.0	Start deep drill cycle at ⑥. Final depth is 14.6 mm, 3-mm peck depth at feed 90 mm/min. Return to 2.5 mm above part.
	N0260 X55.0	Deep drill at X55Y8 mm ⑦.
	N0270 G80	Cancel any fixed cycles (G83 cycle cancelled).
Machine stop Program end sequence	N0280 G00 G90 Z2.5 M05	Rapid to 2.5 mm above part. Spindle stop.
	N0290 M09	Coolant off.
	N0300 G91 G28 X0 Y0 Z0	Rapid to *XYZ* reference point ⑧.
	N0310 M30	Program end, memory reset.
	%	End of tape.

Machining simulation for this example is contained in the job file EX 10–3 stored on the disk at the back of this text. Refer to Chapter 20 for instructions.

■ EXAMPLE 10–4

Write a program for directing a vertical CNC machining center to execute the hole operations shown in Figure 10–10.

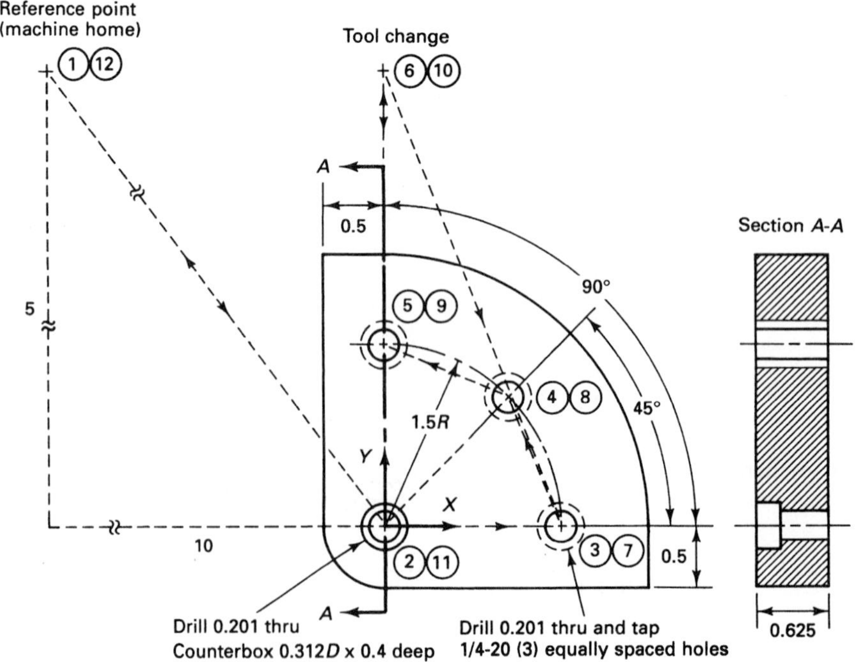

FIGURE 10–10

Tool	Operation	Tooling	Speed (rpm)	Feed (ipm)
1	Drill 0.201*D* (4) places	No. 7 drill	1400	6
2	Tap ¼-20 (3) places	¼-20 tap	400	20
3	Counterbore 0.312*D* × 0.40 deep	5/16-2 FLT end mill	800	4

Notes: 1. Set X_0Y_0 at the center of 0.201 diameter.
2. Z_0 is the top of the part.
3. The part is held by three clamps against three pins.
4. Outside contour is finished.

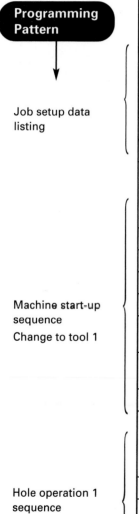

Programming Pattern

	Word address command	Meaning
	%	End of tape.
	O1003	Program number.
Job setup data listing	N0010 (X0Y0 IS AT THE CENTER OF 5/16 HOLE)	Comments.
	N0020 (Z0 IS THE TOP OF THE PART)	
	N0030 (TOOL 1: No. 7 DRILL)	
	N0040 (TOOL 2: 1/4-20 TAP)	
	N0050 (TOOL 3: 5/16-2 FLT END MILL)	
Machine start-up sequence	N0060 G90 G20 G40 G80	Absolute, inch mode, cancel cutter diameter compensation and fixed cycles.
	N0070 G91 G28 X0 Y0 Z0	Return to reference point ①.
	N0080 G92 X−10.0 Y5.0 Z0	Preset absolute zero point.
	N0090 T1 M06	Change to tool 1.
Change to tool 1	N0100 (TOOL 1: DRILL 4 .201 DIA HOLES THRU)	
	N0110 G00 G90 X0 Y0 Z0 S1400 M03	Rapid to ②. Spindle on (CW) at 1400 rpm.
	N0120 G43 Z.1 H01	Rapid tool 1 to 0.1 above part.
	N0130 M08	Coolant on.
Hole operation 1 sequence	N0140 G83 X0 Y0 Z−.7854 R.1 Q.1 F6.0	Deep drill cycle at ②. Final depth is 0.785, 0.1 peck depth at feed rate 6 in./min. Rapid to 0.1 above part.
	N0150 X1.5	Deep drill hole at ③.
	N0160 X1.0607 Y1.0607	Deep drill hole at ④.
	N0170 X0 Y1.5	Deep drill hole at ⑤.
	N0180 G80	Cancel any fixed cycles (G83 cycle cancelled).

Machine stop sequence	N0190 G00 G90 Z1.0 M05	Rapid to 1.0 above part. Stop spindle.
	N0200 M09	Coolant off.
Change to tool 2 Machine start sequence	N0210 G91 G28 Z0 Y0	Rapid to tool change position ⑥.
	N0220 T2 M06	Change to tool 2.
	N0230 (TOOL 2: TAP 3 1/4-20 HOLES THRU)	
	N0240 G00 G90 X1.5 Y0 Z0 S400 M03	Rapid to ⑦. Spindle on (CW) at 400 rpm.
	N0250 G43 Z.1 H02	Rapid tool 2 to 0.1 above part.
	N0260 M08	Coolant on.
Hole operation 2 sequence	N0270 G84 X1.5 Y0 Z−.8001 R.1 F20.	Tap cycle at ⑦. Tap to depth 0.8001 at feed rate 20 in./min. Return to 0.1 above part at feed rate 20 in./min.
	N0280 X1.0607 Y1.0607	Tap hole at ⑧.
	N0290 X0. Y1.5	Tap hole at ⑨.
	N0300 G80	Cancel any fixed cycles.
Machine stop sequence	N0310 G00 G90 Z1.0 M05	Rapid to 1.0 above part. Spindle off.
	N0320 M09	Coolant off.
Change to tool 3 Machine start sequence	N0330 G91 G28 X0 Y0 Z0	Rapid to tool change position ⑩.
	N0340 T3 M06	Change to tool 3.
	N0350 (TOOL 3: COUNTERBORE 1 5/16 DIA .400 DEEP)	
	N0360 G00 G90 X0 Y0 Z0 S800 M03	Rapid to ⑪. Start spindle (CW) at 800 rpm.
	N0370 G43 Z.1 H03	Rapid tool 3 to 0.1 above part.
	N0380 M08	Coolant on.
Hole operation 3 sequence	N0390 G82 X0 Y0 Z−.4 P.5 R.1 F4.0	Counterbore cycle at ⑪. Drill to depth 0.4 at feed rate 4 in./min. for 0.5 s. Rapid to 0.1 above part.
	N0400 G80	Cancel any fixed cycles.
Machine stop Program end sequence	N0410 G00 G90 Z1.0 M05	Rapid to 1. above part. Stop spindle.
	N0420 M09	Coolant off.
	N0430 G91 G28 X0 Y0 Z0	Rapid to *XYZ* reference point ⑫.
	N0440 M30	Program end, memory reset.
	%	End of tape.

■■

Machining simulation for this example is contained in the job file EX 10–4 stored on the disk at the back of this text. Refer to Chapter 20 for instructions.

10-6 CHAPTER SUMMARY

The following key concepts were discussed in this chapter:

1. A fixed or "canned" cycle is a complete set of machine movements that are initiated by issuing a single G code command.
2. The fixed cycles for hole operations are modal. Once programmed they are repeatedly executed at various locations specified in a part. They are canceled by a G80 code or replaced by another fixed-cycle code.
3. The spindle should be retracted to a sufficient height before programming an *XY*-axis move. This will ensure that the tool will clear any clamps or obstructions.
4. The spindle should be retracted to a safe height before executing a tool change and at the end of a program before returning to machine home (machine $X_0Y_0Z_0$) position.

Use the machining editor and simulator contained on the CD at the back of this text to write and verify the word address programs given in these exercises. Refer to Chapter 20 for instructions.

REVIEW EXERCISES

10.1 A program has been written for carrying out the data table operations on the part shown in Figure 10–11. Fill in the meaning of each of the program blocks.

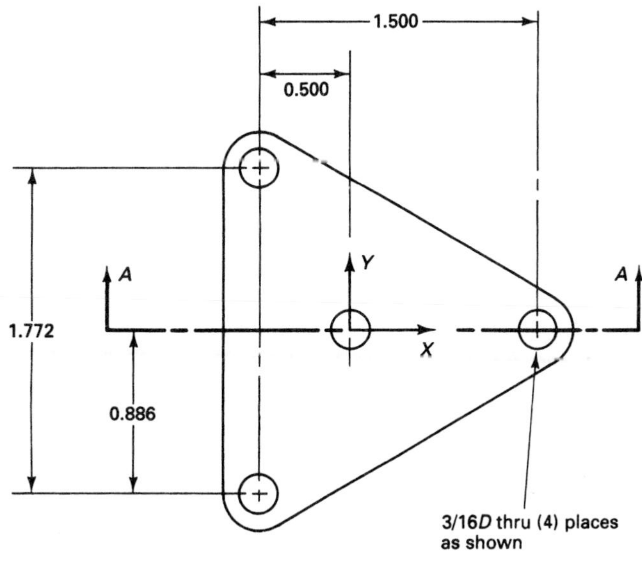

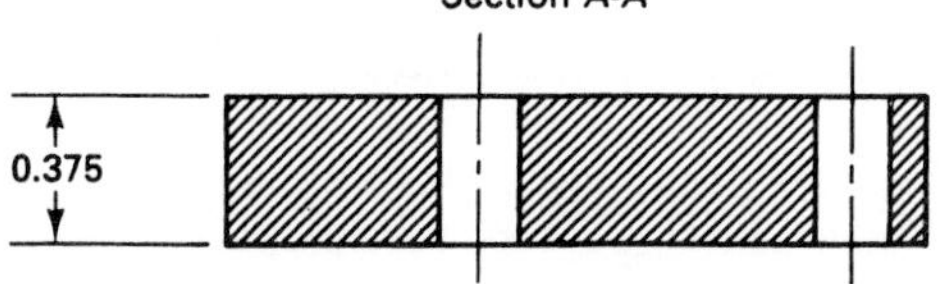

FIGURE 10–11

Tool	Operation	Tooling	Speed (rpm)	Feed (ipm)
1	Drill (4) 3/16*D* thru	3/16 drill	1800	5

Notes: 1. Set X_0 Y_0 at the center of the part.
2. Z_0 is the top of the part.
3. The part is held by three clamps against three pins.
4. Finish holes as per the print.

Word address command	***Meaning***
01004	
N0010 (X0Y0 IS THE CENTER OF THE PART)	
N0020 (Z0 IS THE TOP OF THE PART)	
N0030 (TOOL 1: 3/16*D* DRILL)	
N0040 G90 G20 G40 G80	
N0050 G91 G28 X0 Y0 Z0	
N0060 G92 X10.0 Y5.0 Z0	
N0070 T1 M06	
N0080 (TOOL 1: DRILL 4 3/16 DIA THRU)	
N0090 G00 G90 X0 Y0 Z0 S1800 M03	
N0100 G43 Z.1 H01	
N0110 M08	
N0120 G83 X0 Y0 Z−.5363 R.1 Q.12 F5.0	
N0130 X1.0	
N0140 X−.5 Y .866	
N0150 Y−.866	
N0160 G80	
N0170 G00 G90 Z1.0 M05	
N0180 M09	
N0190 G91 G28 X0 Y0 Z0	
N0200 M30	
%	

For the remaining exercises in this chapter, write a CNC program containing the following general format:

(a) Program number/headings
(b) Job setup data listing
(c) Tool 1 in spindle, machine start-up
(d) Hole operation 1 sequence
(e) Machine stop
(f) Change to tool 2, machine start
(g) Hole operation 2

.
.
.

(h) Machine stop, program end sequence

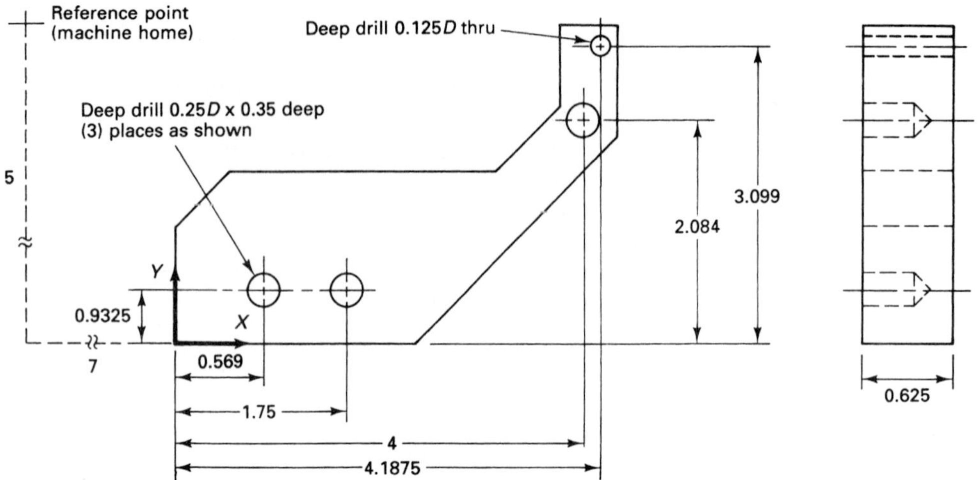

FIGURE 10–12

The machine home reference point in each case is a suggested location. The true location must be determined at the time of setup.

10.2. Program number: O1005 (Figure 10–12)

		Speed	Feed	
Tool	**Operation**	**Tooling**	**(rpm)**	**(ipm)**
1	Deep drill (3) .25*D* hole ×.35 deep	0.25*D* drill	1500	5
2	Deep drill (1) .125*D* hole thru	0.125*D* drill	2000	5

Notes: 1. The part is held by three clamps against three pins.
2. Set $X_0 Y_0$ at the lower left-hand corner.
3. Z_0 is the top of the part.
4. Finish holes as per the print.

10.3 Program number: O1006 (Figure 10–13)

Tool	Operation	Tooling	Speed (rpm)	Feed (ipm)
1	Center drill (6) equally spaced holes × 0.15 deep	No. 2 center drill	1500	6
2	Deep drill (6) holes × 0.5 deep	0.147*D* drill	1500	6
3	Tap (6) holes ×0.4 deep	10–24 UNC tap	70	2

Notes: 1. The part is held in a vise with a stop on the left.
2. Set X_0 Y_0 at upper left-hand corner.
3. Z_0 is the top of the part.
4. Determine the proper feed rate for tapping.

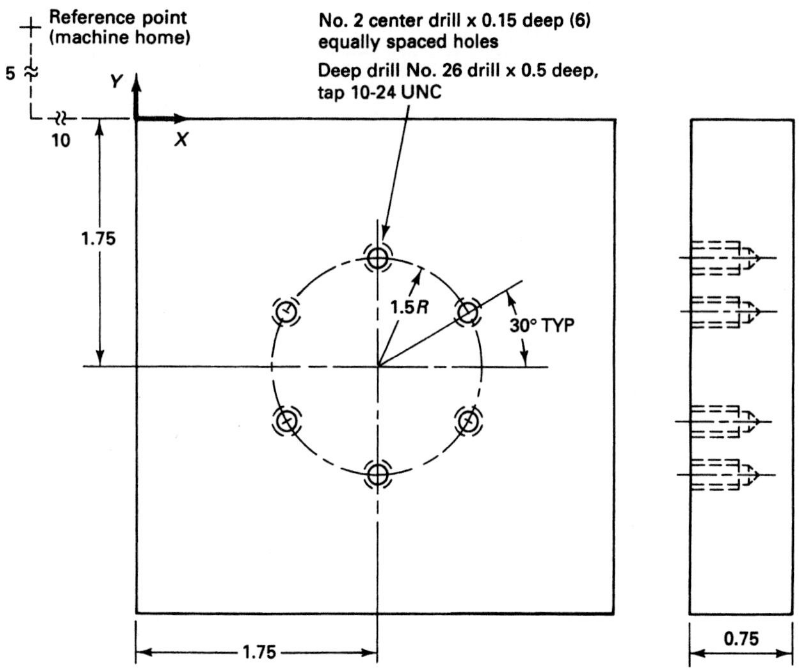

FIGURE 10-13

10.4. Program number: O1007 (Figure 10–14)
Note: Positional tolerance ±0.0005 is only for 0.2500 $^{+0000}_{-0.0002}$. Use ±0.005 elsewhere.

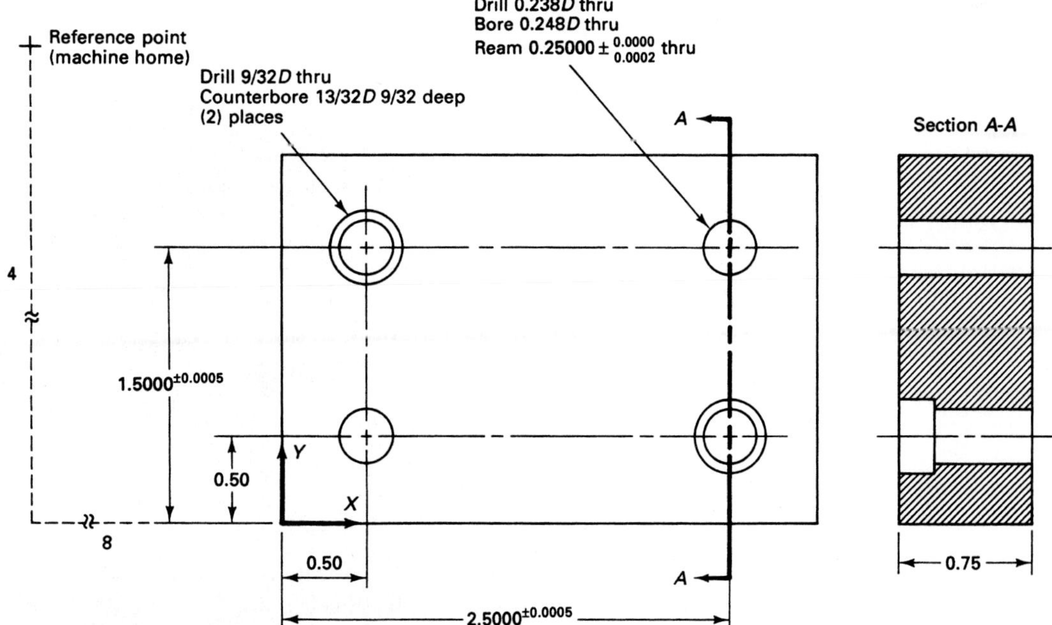

FIGURE 10-14

Tool	Operation	Tooling	Speed (rpm)	Feed (ipm)
1	Drill (2) places	9/32*D* drill	1200	8
2	Counterbore (2) places	13/32−2 FLT end mill	600	6
3	Drill 0.2500*D* to 0.238*D*	B drill	1400	8
4	Bore 0.2500*D* to 0.248*D*	0.248*D* B/bar	1000	2
5	Ream 0.2500 + 0000 −0.0002	0.2499*D* reamer	500	15

Notes: 1. The part is held in a vise with a stop on the left side.
2. Set X_0 Y_0 at the lower left-hand corner.
3. Z_0 is the top of the part.

10.5. Program number: O1008 (Figure 10–15)

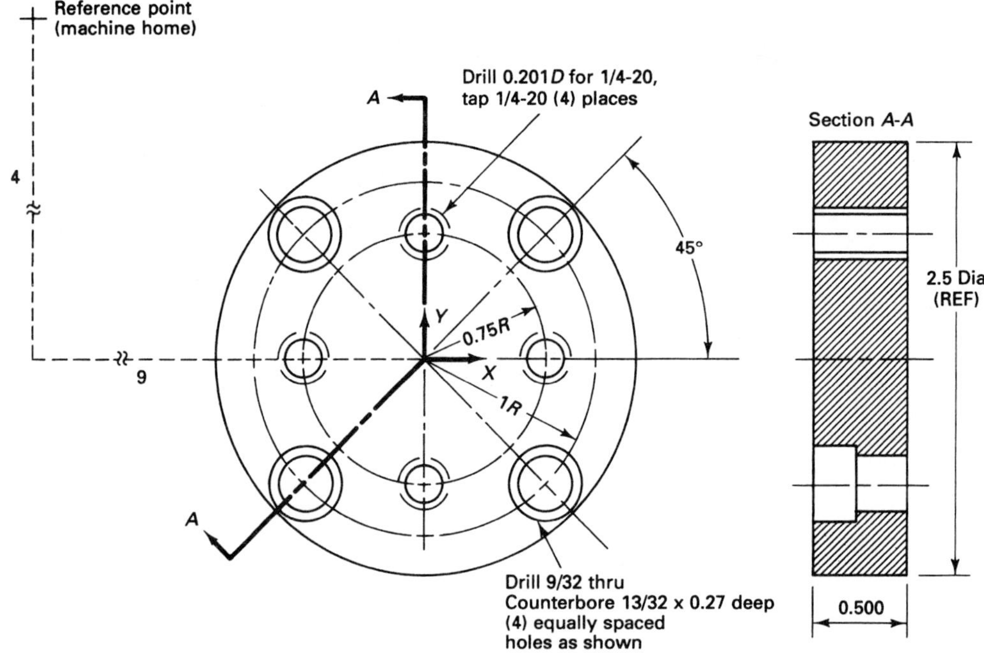

FIGURE 10–15

Tool	Operation	Tooling	Speed (rpm)	Feed (ipm)
1	Drill (4) 9/32 holes thru	9/32*D* drill	1800	7
2	Counterbore (4) 13/32 ×0.27 deep	13/32−2 FLT/end mill	1000	6
3	Drill (4) places for 1/4-20 UNC	No. 7 drill	2000	8
4	Tap 1/4-20 holes	1/4-20 tap	400	20

Notes: 1. The part is held in a three-jaw chuck.
2. Set X_0 Y_0 at the center of the part.
3. Z_0 is the top of the part.
4. Determine the feed rate for tapping.

CHAPTER 11

Programming Linear Profiles

11-1 CHAPTER OBJECTIVES

At the conclusion of this chapter you will be able to

1. Explain what linear profiling is.
2. Understand the word address command for executing linear profiling.
3. Understand the sequence of commands required for executing typical linear profiling operations.
4. Write complete CNC programs for linear profiling.
5. Compute cutter offsets for inclined line profiles.

11-2 INTRODUCTION

The important operation of linear profiling is covered in this chapter. See Figure 11–1. Its general syntax in word address is given and explained in detail. A generalized pattern for writing profiling programs is discussed and applied to a profile formed by horizontal and vertical lines. The reader is introduced to the trigonometry involved in computing cutter offsets for profiles consisting of inclined lines. Detailed examples of computing cutter offsets and writing a corresponding CNC program are also given.

The student is encouraged to use the profile machining simulation software described in Chapter 20 and contained on the CD at the back of the text.

11-3 LINEAR INTERPOLATION COMMANDS

When executing linear interpolation, a continuous-path machine moves the tool along a straight-line path from the initial point to the final point. The machine creates the linear path by simultaneously actuating its drive motors. Linear profiling involves cutting contours composed of straight lines only. The lines may be horizontal, vertical, or at any angle.

The following word address commands are used extensively for cutting linear profiles on a vertical or horizontal machining center that utilizes a Fanuc 6M controller.

> ***Note:*** ↓ represents tool motion at the programmed feed rate.

11-4 WRITING A LINEAR PROFILING PROGRAM

A typical pattern that can be followed for planning many types of basic profiling programs is given in Figure 11–2. This pattern is used in the examples that follow.

FIGURE 11–1 Machining a linear profile. (Photo courtesy of Ingersoll Cutting Tools.)

Safety Rules for Milling Operations

- For climb milling, feed work in the direction of cutter rotation.
- A deep first cut should be made on castings, forgings, and other rough surfaces. This practice will enable the tool to penetrate the hard outer scale.
- Position the side milling cutter such that the cutting edges are not brought against the vise or fixture.
- Observe the following practices when milling:
  - Hold the tool as close as possible to the work fixture and arbor support arm.
  - Direct the cutting forces toward the spindle.
  - Move the tool toward the solid vise jaw or vertical leg of an angle plate.
- Supply a continuous flow of cutting fluid when machining cast iron or steel with carbide end mills.
- Wear hearing protection for noise levels exceeding OSHA standards.

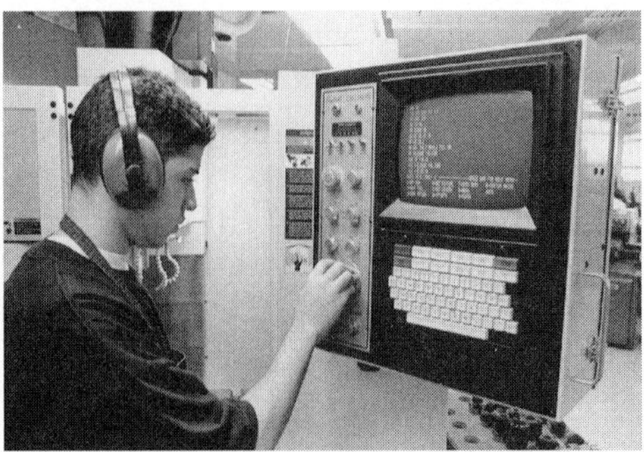

General Syntax

Linear Profile Milling

G01 Zn Fn

X_1Y_1

X_2Y_2

.

.

.

etc.

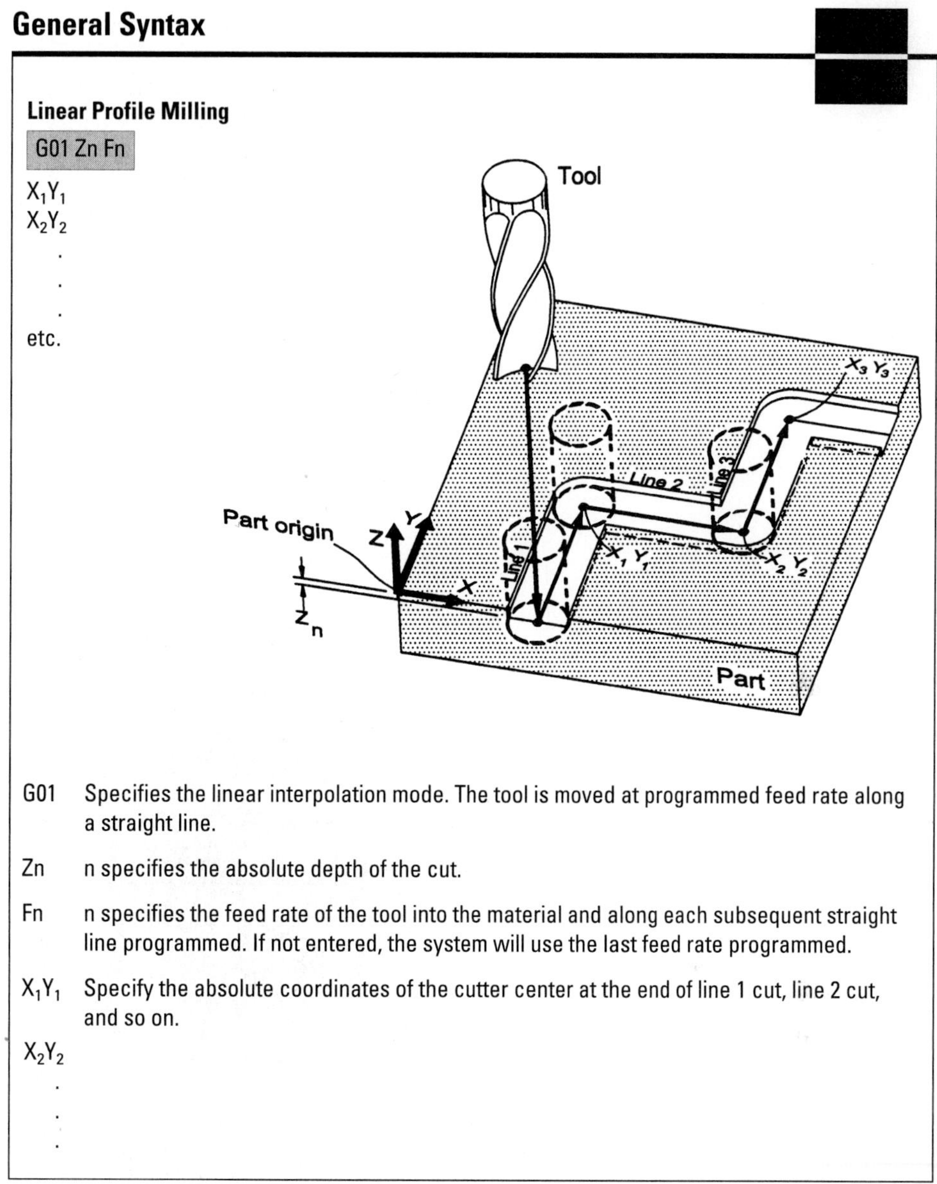

G01 Specifies the linear interpolation mode. The tool is moved at programmed feed rate along a straight line.

Zn n specifies the absolute depth of the cut.

Fn n specifies the feed rate of the tool into the material and along each subsequent straight line programmed. If not entered, the system will use the last feed rate programmed.

X_1Y_1 Specify the absolute coordinates of the cutter center at the end of line 1 cut, line 2 cut, and so on.

X_2Y_2

.

.

.

■ EXAMPLE 11–1

Write a CNC program to profile mill the contour shown in Figure 11–3. Use the data table given.

Tool	Operation	Tooling	Speed (rpm)	Feed (ipm)
1	Profile mill contour × 0.52 deep	0.5 D end mill	1200	8

Notes: 1. Set X_0 Y_0 0.03 in. from the lower left corner.
2. The blank for the part is 5 1/16 in. × 4 1/16 in. × 0.5 in.

.5 x 4 1/16 x 5 1/16

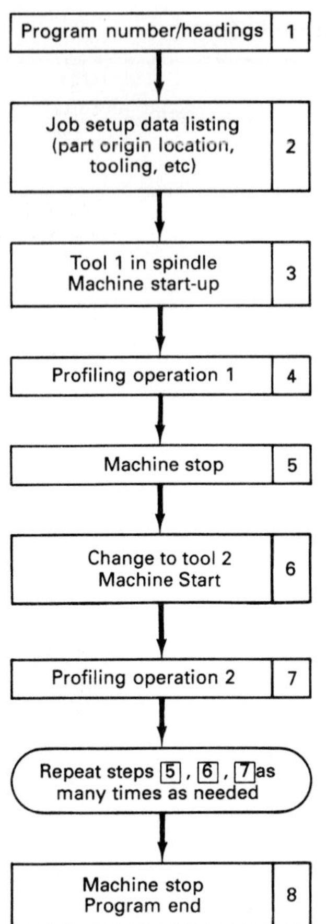

FIGURE 11-2

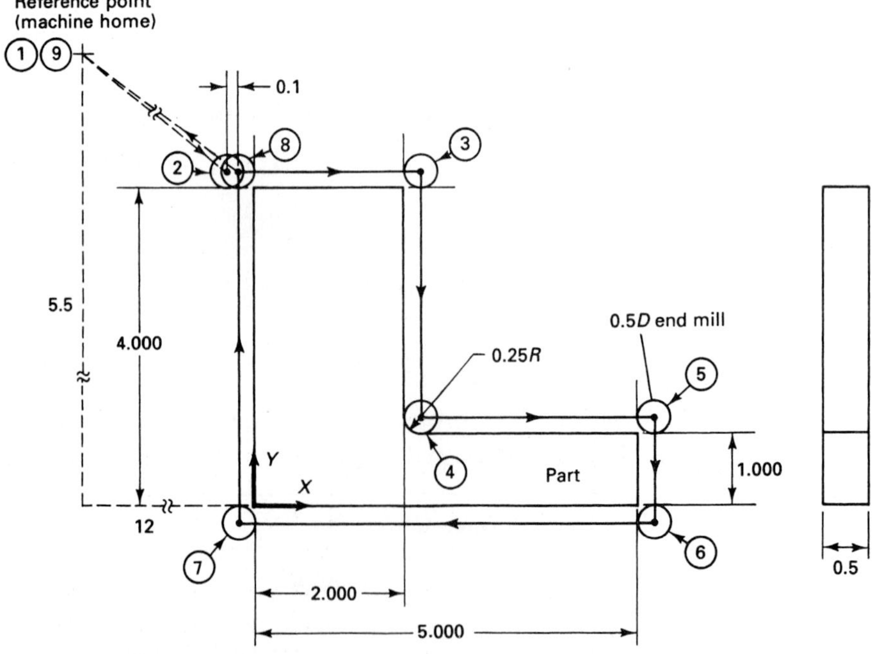

FIGURE 11-3

1/2 ENDMILL

		Absolute coordinates	
Position	Calculation	X	Y
②	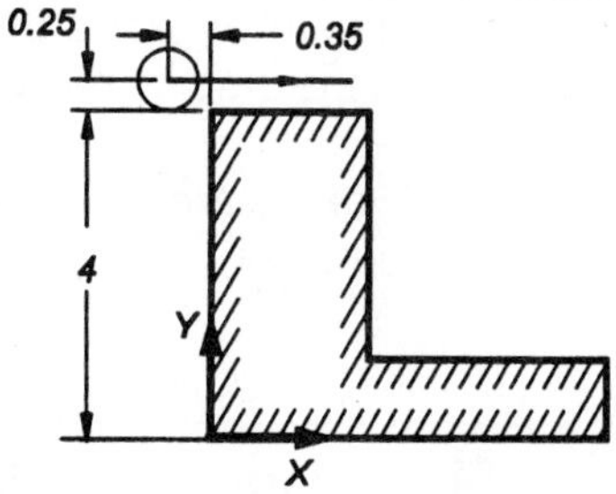 $X = -0.25 - 0.1 = -0.35$ $Y = 4 + 0.25 = 4.25$	−0.35	4.25
③	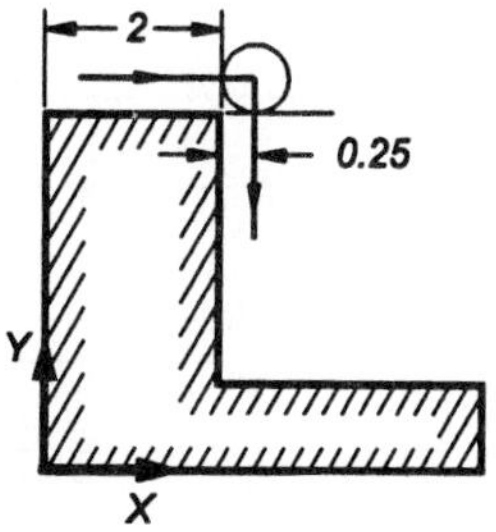 $X = 2 + 0.25 = 2.25$ $Y = 4.25$	2.25	4.25
④	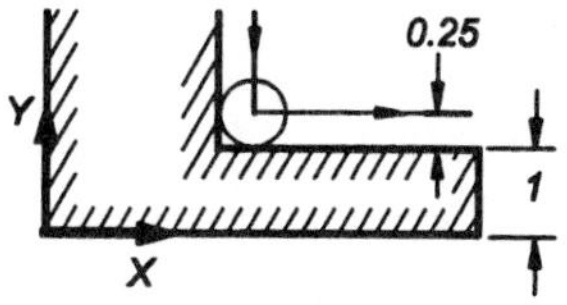 $X = 2.25$ $Y = 1 + 0.25 = 1.25$	2.25	1.25
⑤	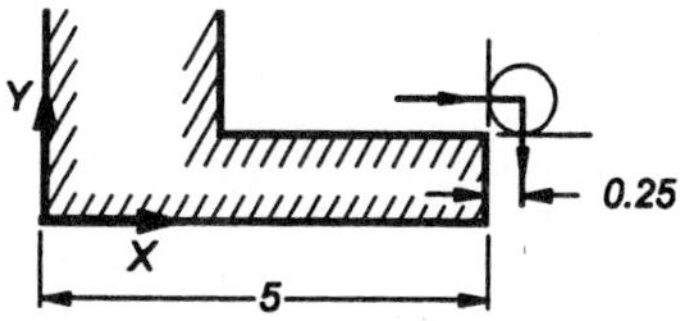 $X = 5 + 0.25 = 5.25$ $Y = 1.25$	5.25	1.25
⑥	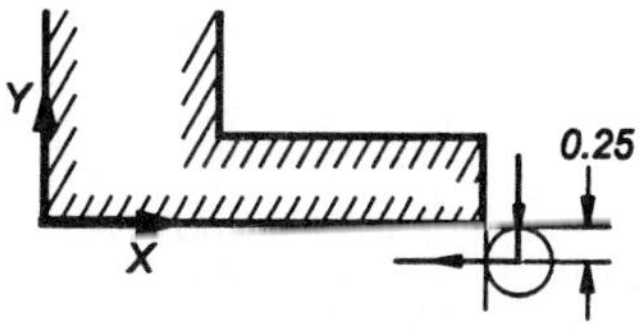 $X = 5.25$ $Y = -0.25$	5.25	−0.25
⑦	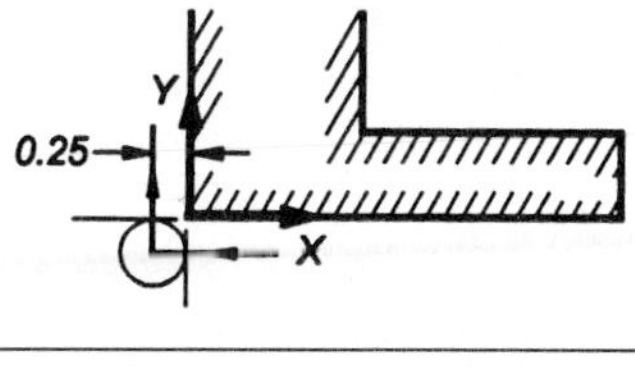 $X = -0.25$ $Y = -0.25$	−0.25	−0.25
⑧	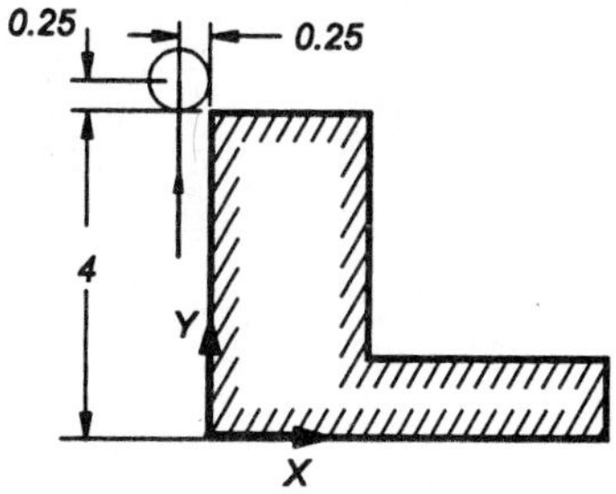 $X = -0.25$ $Y = 4 + 0.25 = 4.25$	−0.25	4.25

The cutter can be considered as a circle whose diameter is the tool diameter. For profiling, the tool must be positioned such that it is always tangent to the line it follows. Specifying the *X, Y* locations of the cutter center is relatively easy when dealing with horizontal and vertical line profiles. The programmer simply adds the cutter radius to the part geometry or subtracts the radius as required. This has been done for the part shown in Figure 11–3:

The computed *X, Y* values are then used in the CNC program. A complete listing is given below. ■■

Programming Pattern

Job setup data listing

Machine start-up sequence
Change to tool 1

Profile operation 1 sequence

Machine stop
Program end sequence

CNC Program—Absolute Coordinates

Word address command	*Meaning*
%	End of tape.
01101	Program number.
N0010 (X0Y0 is 1/32 IN FROM LOWER)	Comments.
N0020 (LEFT HAND CORNER)	
N0030 (TOOL 1: 0.5 DIA END MILL)	
N0040 G90 G20 G40 G80	Absolute, inch mode, cancel cutter diameter compensation and fixed cycles.
N0050 G91 G28 X0 Y0 Z0	Return to reference point ①.
N0060 G92 X−12.0 Y5.5 Z0	Preset absolute zero point.
N0070 T1 M06	Change to tool 1.
N0080 (TOOL 1: FINISH CONTOUR AS PER PRINT)	
N0090 G00 G90 X−.35 Y4.25 Z0. S1200 M03	Rapid to ② spindle on (CW) at 1200 rpm.
N0100 G43 Z. 1 H01	Rapid tool 1 to 0.1 above part.
N0110 M08	Coolant on.
N0120 G01 Z−.52 F8.0	Linear profile mode. Plunge to Z−.52 at feed rate 8 ipm.
N0130 X2.25	Cut to X2.25Y0 ③.
N0140 Y1.25	Cut to X2.25 Y1.25 ④.
N0150 X5.25	Cut to X5.25 Y1.25 ⑤.
N0160 Y−.25	Cut to X5.25 Y−.25 ⑥.
N0170 X−.25	Cut to X−.25 Y−.25 ⑦.
N0180 Y4.25	Cut to X−.25 Y4.25 ⑧.
N0190 G00 Z. 1 M05	Rapid to 0.1 above part.
N200 M09	Coolant off.
N0210 G91 G28 X0 Y0 Z0	Rapid to *XYZ* reference point ⑨.
N0220 M30	Program end, memory reset.
%	End of tape.

Machining simulation for this example is contained in the job file EX 11–1 stored on the disk at the back of this text. Refer to Chapter 20 for instructions.

11–5 DETERMINING CUTTER OFFSETS FOR INCLINED LINE PROFILES

Many parts contain profiles consisting of inclined lines. In these cases the programmer cannot simply add or subtract the cutter radius. Right-triangle trigonometry must be applied to determine the position of the cutter center. The ΔX and ΔY values given in Figures 11–4 through 11–6 are used to compensate for inclined lines. In each case, they are added to or subtracted from the part geometry. The resulting value gives the proper location of the cutter center.

1. Inclined profile case 1

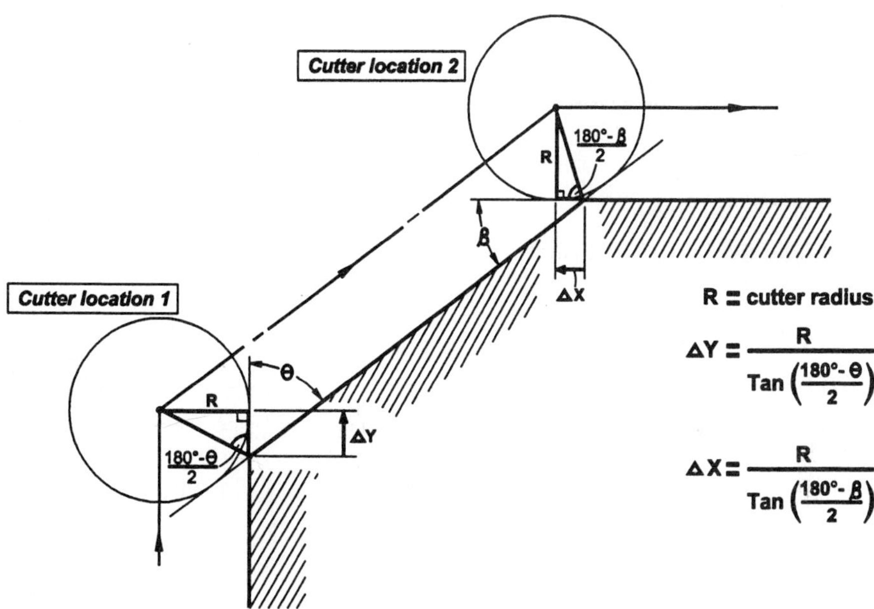

FIGURE 11–4

2. Inclined profile case 2

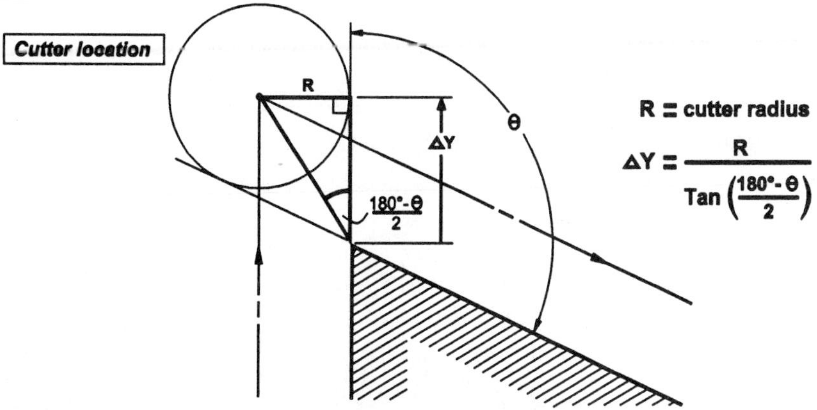

FIGURE 11–5

3. Inclined profile case 3

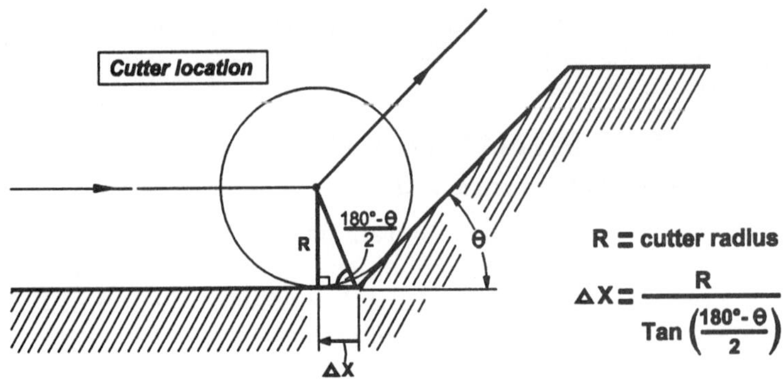

FIGURE 11–6

Example 11–2 illustrates the use of these formulas.

An Approach to Determining Cutter Offsets for *Inclined Line* Profiles

Step	Procedure	Visual
1	***EXTERIOR PROFILES*** EXTEND THE PART PROFILE LINES	PART PROFILE LINES EXTENDED PART
	INTERIOR PROFILES EXTENSIONS ARE ALREADY DEFINED BY THE PART PROFILE LINES	REQUIRED EXTENSIONS PART
2	DRAW A CIRCLE REPRESENTING THE CUTTING TOOL TANGENT TO THE TWO EXTENSION LINES	CUTTING TOOL PART
3	DRAW A LINE FROM THE CENTER OF THE CIRCLE TO THE INTERSECTION OF THE PART PROFILE LINES	PART
4	DROP A PERPENDICULAR TO THE SIDE THAT WAS CUT	R PART

5	**DETERMINE θ, THE ANGLE THE SECOND PROFILE LINE MAKES WITH THE EXTENSION OF THE FIRST PROFILE LINE** **LABEL THE SIDES OF THE RIGHT TRIANGLE** **R - REPRESENTS THE CUTTER RADIUS** **ΔY - REPRESENTS CUTTER OFFSET**	
6	**THE ACUTE ANGLE OF THE RIGHT TRIANGLE AT THE INTERSECTION OF THE TWO PROFILE LINES WILL ALWAYS BE GIVEN AS:** $\frac{180° - \theta}{2}$ **CALCULATE THE TANGENT OF THE ACUTE ANGLE**	$t = \mathrm{Tan}\left(\frac{180° - \theta}{2}\right)$
7	**THE CUTTER OFFSET IS DETERMINED BY DIVIDING THE CUTTER RADIUS, R, BY THE TANGENT VALUE, t, DETERMINED IN STEP 6**	$\Delta Y = \frac{R}{t}$

■ EXAMPLE 11–2

Determine the absolute coordinates of the center of the cutter for milling the profile shown in Figure 11–7.

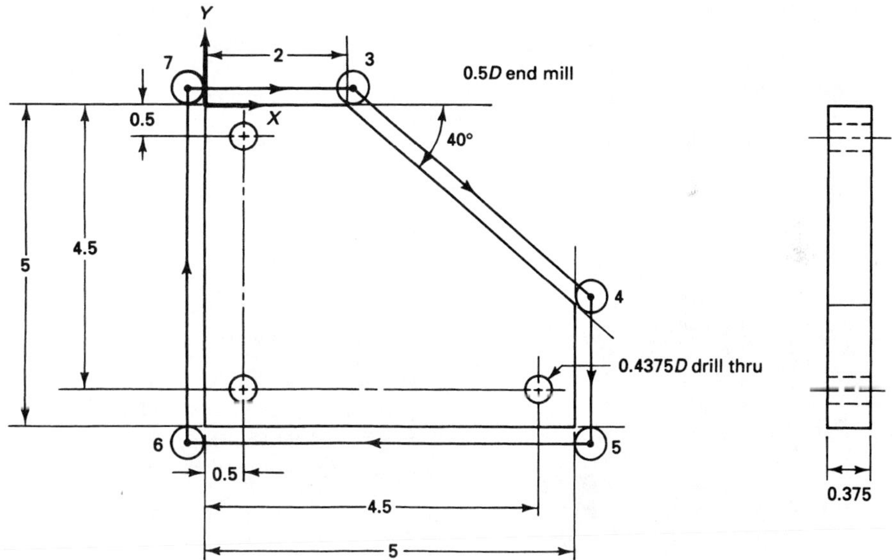

FIGURE 11–7

The required calculations and coordinates are as follows:

		Absolute coordinates	
Position	**Calculation**	**X**	**Y**
⑦		−0.25	0.25
③	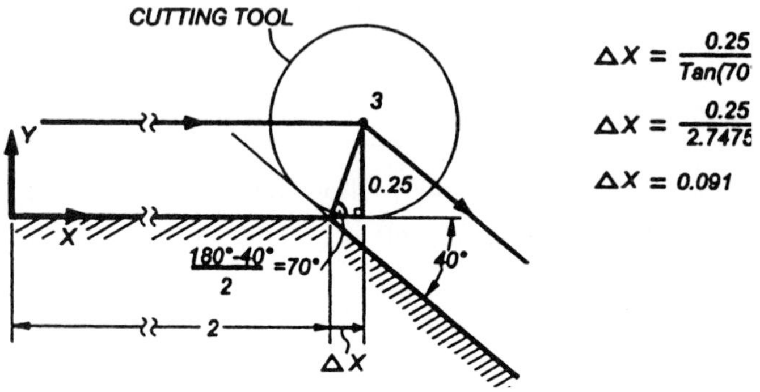 CUTTING TOOL; $\Delta X = \frac{0.25}{Tan(70}$; $\Delta X = \frac{0.25}{2.7475}$; $\Delta X = 0.091$; $\frac{180°-40°}{2}=70°$; 0.25; 40°; 2; ΔX $X = 2 + \overbrace{0.091}^{\Delta X} = 2.091$ $Y = 0.25$	2.091	0.25
④	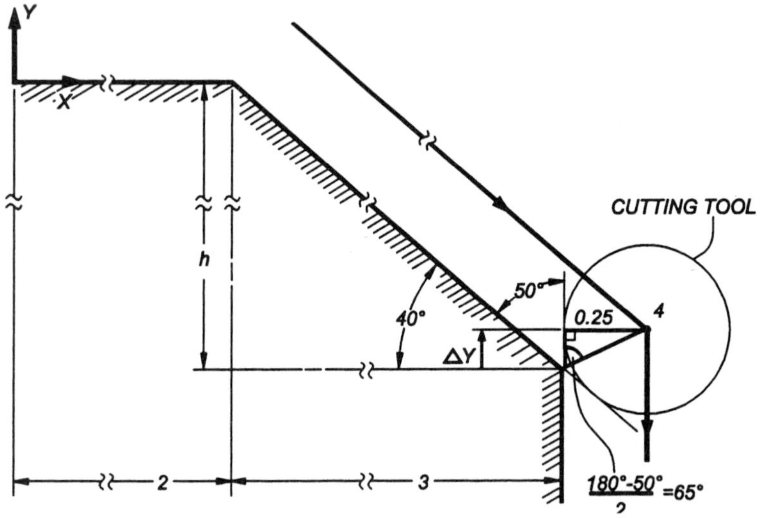 CUTTING TOOL; h; 40°; 50°; 0.25; ΔY; 2; 3; $\frac{180°-50°}{2}=65°$ $\frac{h}{3} = \text{Tan}(40°)$ $\Delta Y = \frac{0.25}{\text{Tan}(65°)}$ $h = 3 \times 0.8391$ $\Delta Y = \frac{0.25}{2.1445}$ $h = 2.5173$ $\Delta Y = 0.1166$ $X = 5.25$ $Y = -2.5173 + \overbrace{0.1166}^{\Delta Y} = -2.4007$	5.25	−2.4007
⑤	$X = 5 + 0.25$ $Y = -5 - 0.25$	5.25	−5.25
⑥	$X = -0.25$ $Y = -5 - 0.25$	−0.25	−5.25
⑦	$X = -0.25$ $Y = 0.25$	−0.25	0.25

■ EXAMPLE 11–3

Write a CNC program to profile mill the contour given in Example 11–2. Additionally, include statements for drilling the holes shown in Figure 11–8. Use the data table as an aid in writing the program.

The coordinates for the tool center for profiling the contour were calculated in Example 11–2 and will be used in the program that follows.

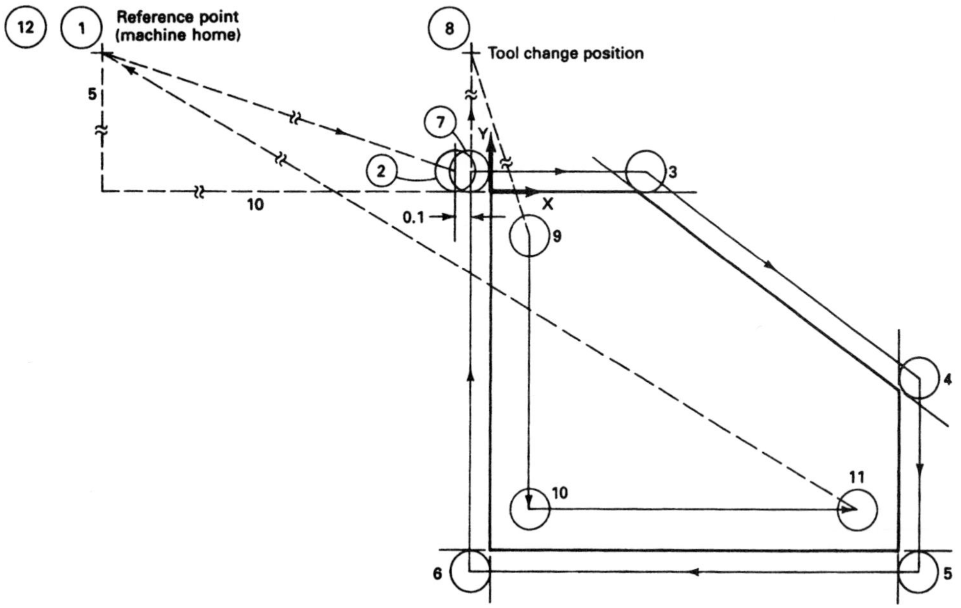

FIGURE 11–8

Tool	Operation	Tooling	Speed (rpm)	Feed (ipm)
1	Profile mill contour × 0.375 deep	0.5*D* end mill	1500	7
2	Deep drill all holes thru	0.4375*D* drill	1500	7

The CNC program written with absolute coordinates is given below.

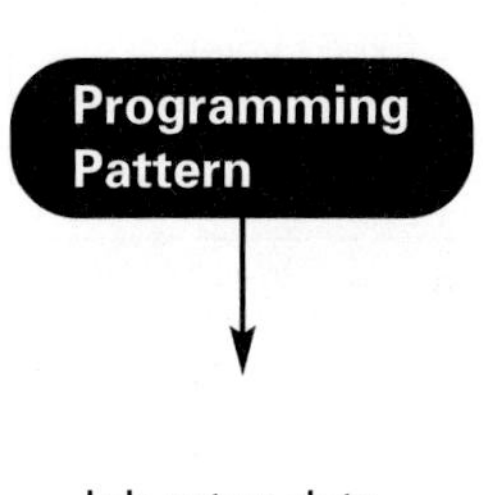

Job setup data listing

CNC Program—Absolute Coordinates

Word address command	*Meaning*
%	End of tape. Used to separate programs.
O1102	Program number.
N0010 (X0Y0 IS THE UPPER LEFT HAND CORNER)	Comments.
N0020 (Z0 IS THE TOP OF THE PART)	
N0030 (TOOL 1: 1/2 DIA END MILL)	
N0040 (TOOL 2: 7/16 STUB DRILL)	

Sequence	Program	Description
Machine start-up sequence Change to tool 1	N0050 G90 G20 G40 G80	Absolute, inch mode, cancel cutter diameter compensation and fixed cycles.
	N0060 G91 G28 X0 Y0 Z0	Return to reference point ①.
	N0070 G92 X−10.0 Y5.0 Z0	Preset absolute zero point.
	N0080 T1 M06	Change to tool 1.
	N0090 (TOOL 1: FINISH CONTOUR AS PER PRINT)	
	N0100 G00 G90 X−.35 Y.25 Z0. S1000 M03	Rapid to ②, spindle on (CW) at 1000 rpm.
	N0110 G43 Z.1 H01	Rapid tool 1 to 0.1 above the part.
	N0120 M08	Coolant on.
Profile operation 1 sequence	N0130 G01 Z−.4 F6.0	Linear profile mode. Plunge to Z−.375 at feed rate of 6 ipm.
	N0140 X2.091	Cut to X2.081 Y0 ③.
	N0150 X5.25 Y−2.4007	Cut to X5.25 Y−2.4007 ④.
	N0160 Y−5.25	Cut to X5.25 Y−5.25 ⑤.
	N0170 X−.25	Cut to X−.25 Y−5.25 ⑥.
	N0180 Y.25	Cut to X−.25 Y.25 ⑦.
Machine stop sequence	N0190 G00 G90 Z1.0 M05	Rapid to 0.1 above part. Stop spindle.
	N0200 M09	Coolant off.
Change to tool 2 Machine start sequence	N0210 G91 G28 Z0 Y0	Rapid to *YZ* reference point ⑧.
	N0220 T2 M06	Change to tool 2.
	N0230 (TOOL 2: DRILL 3 .4375 DIA HOLES)	
	N0240 G00 G90 X.5 Y−.5 Z0. S1500 M03	Rapid to ⑨. Start spindle (CW) at 1500 rpm.
	N0250 G43 Z.1 H02	Rapid tool 2 to 0.1 above part.
	N0260 M08	Coolant on.
Hole operation 2 sequence	N0270 G83 X.5 Y−.5 Z−.515 R.1 Q.1 F8.0	Start deep hole cycle at ⑨. Final depth is 0.515, 0.1 depth of peck, at feed rate 8. Return to 0.1 above part.
	N0280 Y−4.5	Deep drill hole at X.5Y−.45 ⑩.
	N0290 X4.5	Deep drill hole at X4.5Y−.45 ⑪.
	N0300 G80	Cancel any fixed cycles.
Machine stop Program end sequence	N0310 G00 G90 Z1.0 M05	Rapid to 0.1 above part. Stop spindle.
	N0320 M09	Coolant off.
	N0330 G91 G28 X0 Y0 Z0	Rapid to *XYZ* reference point ⑫.
	N0340 M30	Program end, memory reset.
	%	End of tape.

Machining simulation for this example is *created* as job file EX 11–3. This process is described starting at section 20–2 in Chapter 20.

■ EXAMPLE 11–4

Write a program for profile milling the outside and inside of the part shown in Figure 11–9.

Tool	Operation	Tooling	Speed (rpm)	Feed (ipm)
1	Profile inside contour	3/16*D* center cut end mill	2000	4
2	Profile outside contour	1/4*D* end mill	1700	6

Notes: 1. Set X_0 Y_0 at the lower left-hand corner.
2. Z_0 is the top of the part.
3. For tool 1 the part is held by two clamps from outside (see Figure 11–11). After tool 2 clamp the part from inside and remove the outside clamps (see Figure 11–12).

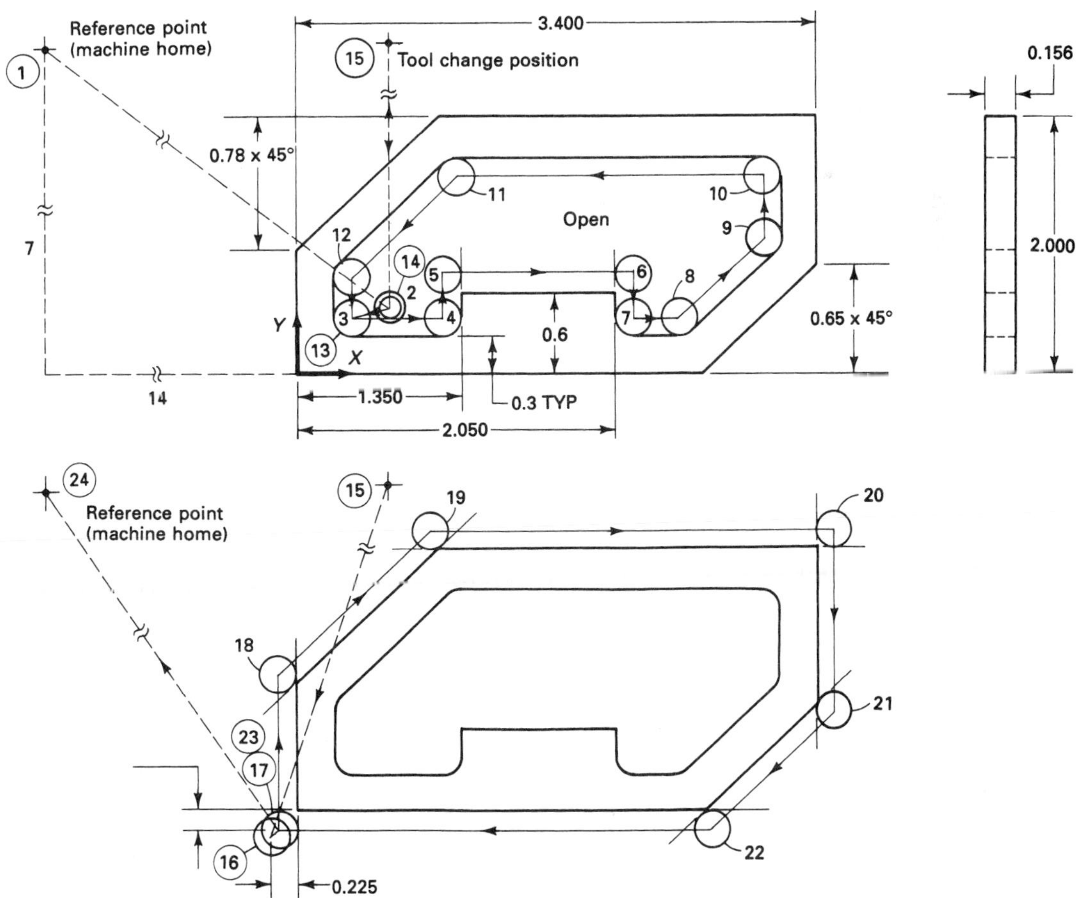

FIGURE 11–9

CNC Program—Absolute Coordinates

Programming Pattern	*Word address command*	*Meaning*
	%	End of tape. Used to separate programs.
	O1103	Program number.
Job setup data listing	N0010 (X0Y0 IS THE LOWER LEFT HAND CORNER)	Comments.
	N0020 (Z0 IS THE TOP OF THE PART)	
	N0030 (TOOL 1: 3/16 DIA END MILL)	
	N0040 (TOOL 2: 1/4 DIA END MILL)	
Machine start-up sequence Change to tool 1	N0050 G90 G20 G40 G80	Absolute, inch mode, cancel cutter diameter compensation and fixed cycles.
	N0060 G91 G28 X0 Y0 Z0	Return to reference point ①.
	N0070 G92 X−14.0 Y7.0 Z0	Preset absolute zero point.
	N0080 T1 M06	Change to tool 1.
	N0090 (TOOL 1: FINISH INSIDE CONTOUR)	
	N0100 G00 G90 X.5 Y.5 Z0. S2000 M03	Rapid to ②. Start spindle (CW) at 2000 rpm.
	N0110 G43 Z.1 H01	Rapid tool 1 to 0.1 above the part.
	N0120 M08	Coolant on.
Profile operation 1 sequence	N0130 G01 Z−.18 F4.0	Linear profile mode. Plunge to Z−.18 at feed rate of 4.
	N0140 X.3938 Y.3938	Cut to position ③.
	N0150 X1.2563	Cut to position ④.
	N0160 Y.6938	Cut to position ⑤.
	N0170 X2.1437	Cut to position ⑥.
	N0180 Y.3938	Cut to position ⑦.
	N0190 X2.5869	Cut to position ⑧.
	N0200 X3.0063 Y.8131	Cut to position ⑨.
	N0210 Y1.6063	Cut to position ⑩.
	N0220 X.9431	Cut to position ⑪.
	N0230 X.3937 Y1.0569	Cut to position ⑫.
	N0240 Y.3938	Cut to position ⑬.
	N0250 X.5 Y.5 Z0	Cut to position ⑭.
Machine stop sequence	N0260 G00 Z1.0 M05	Rapid to 1.0 above part. Stop spindle.
	N0270 M09	Coolant off.

Sequence	Block	Comment
Change to tool 2 Machine start sequence	N0280 G91 G28 Z0 Y0	Rapid to *YZ* reference point ⑮.
	N0290 M00 (CYCLE STOP—RECLAMP—CYCLE START)	
	N0300 T2 M06	Change to tool 2.
	N0310 (TOOL 2: FINISH OUTSIDE CONTOUR)	
	N0320 G00 G90 X−.225 Y−.225 Z0. S1700 M03	Rapid to ⑯. Start spindle (CW) at 1700 rpm.
	N0330 G43 Z1.H02	Rapid tool 2 to 1.0 above part.
	N0340 M08	Coolant on.
Profile operation 2 sequence	N0350 G01 Z−.18 F6.0	Linear profile mode. Plunge to Z−.18 at feed rate 6.
	N0360 X−.125 Y−.125	Cut to position ⑰.
	N0370 Y1.2718	Cut to position ⑱.
	N0380 X.7282 Y2.125	Cut to position ⑲.
	N0390 X3.525	Cut to position ⑳.
	N0400 Y.5982	Cut to position ㉑.
	N0410 X2.8018 Y−.125	Cut to position ㉒.
	N0420 X−.125	Cut to position ㉓.
Machine stop Program end sequence	N0430 G00 G90 Z1.0 M05	Rapid to 1.0 above part. Stop spindle.
	N0440 M09	Coolant off.
	N0450 G91 G28 X0 Y0 Z0	Rapid to *XYZ* reference point ㉔.
	N0460 M30	Program end, memory reset.
	%	End of tape.

Machining simulation for this example is contained in the job file EX 11–4 stored on the disk at the back of this text. Refer to Chapter 20 for instructions.

■ EXAMPLE 11–5

The metric part illustrated in Figure 11–10 is to be machined. Write a CNC program to carry out the operations as outlined in the data table.

Tool	Operation	Tooling	Speed (rpm)	Feed (mm/min)
1	Rough inside contour 0.25 full	∅6-mm roughing end mill	2100	120
2	Finish inside contour as required	∅4-mm end mill	2600	100
1	Rough outside contour 0.25 full	∅6-mm roughing end mill	2100	120
2	Finish outside contour as required	∅4-mm end mill	2600	100

Notes:
1. Set X_0 Y_0 at the lower left-hand corner.
2. Z_0 is the top of the part.
3. The part is roughed and finished in the slot by holding the blank against three pins by two clamps. (See Figure 11–11).
4. For outside machining, stop the machine. Place clamps from inside and remove.

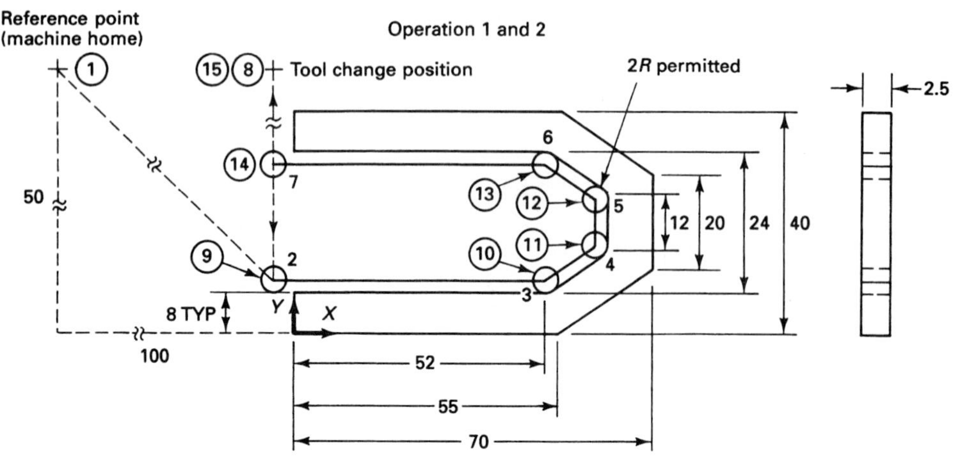

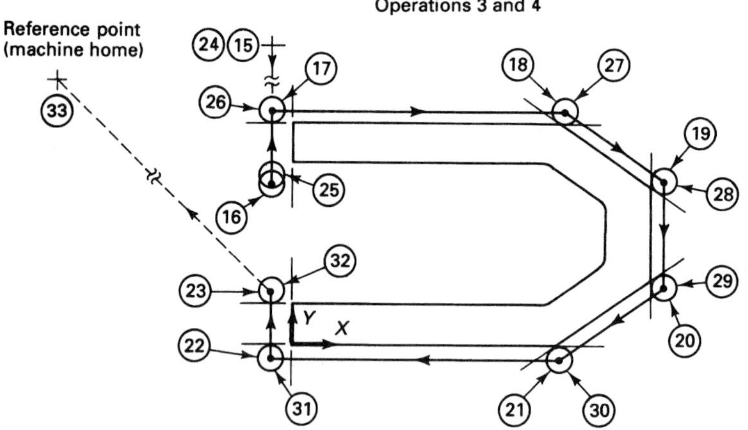

FIGURE 11–10

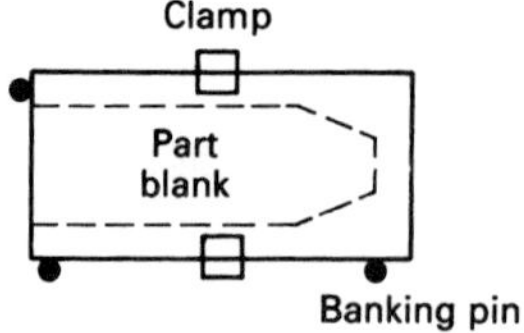

FIGURE 11–11

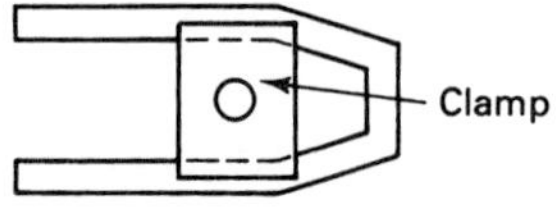

FIGURE 11–12

Programming Pattern

CNC Program—Absolute Coordinates

Programming Pattern	*Word address command*	*Meaning*
	%	End of tape. Used to separate programs.
	O1104	Program number.
Job setup data listing	N0010 (X0Y0 IS THE LOWER LEFT HAND CORNER)	Comments.
	N0020 (Z0 IS THE TOP OF THE PART)	
	N0030 (TOOL 1: 6MM ROUGHING END MILL)	
	N0040 (TOOL 2: 4MM FINISHING END MILL)	
Machine start-up sequence Change to tool 1	N0050 G90 G21 G40 G80	Absolute, metric mode, cancel cutter diameter compensation and fixed cycles.
	N0060 G92 X−100 Y50 Z0	Preset absolute zero point.
	N0070 G91 G28 X0 Y0 Z0	Return to reference point ①.
	N0080 T1 M06	Change to tool 1.
	N0090 (TOOL 1: ROUGH INSIDE .25MM FULL .3MM DEEP)	
	N0100 G00 G90 X−5.25 Y11.25 Z0. S2100 M03	Rapid to ②. Spindle on (CW) at 1500 rpm.
	N0110 G43 Z2.5 H01	Rapid tool 1 to 2.5 above the part.
	N0120 M08	Coolant on.
Profile operation 1 sequence	N0130 G01 Z−3.0 F120.0	Linear profile mode. Plunge cutter to Z−3 at feed rate 120.
	N0140 X51.0998	Cut to position ③.
	N0150 X58.75 Y15.8401	Cut to position ④.
	N0160 Y24.1599	Cut to position ⑤.
	N0170 X51.0998 Y28.75	Cut to position ⑥.
	N0180 X−4.25	Cut to position ⑦.
Machine stop sequence	N0190 G00 G90 Z1.0 M05	Rapid to 1.0 above part. Stop spindle.
	N0200 M09	Coolant off.
Change to tool 2 Machine start sequence	N0210 G91 G28 Z0 Y0	Rapid to *YZ* reference point ⑧.
	N0220 T2 M06	Change to tool 2.
	N0230 (TOOL 2: FINISH INSIDE AS REQUIRED)	
	N0240 G00 G90 X−4. Y10. Z0. S2600 M03	Rapid to ⑨. Start spindle (CW) at 2600 rpm.
	N0250 G43 Z2.5 H02	Rapid tool 2 to 2.5 above part.
	N0260 M08	Coolant on.

CNC PROGRAM ***Continued***

	Word address command	***Meaning***
Profile operation 2 sequence	N0270 G01 Z−3. F100.0	Linear profile mode. Plunge to Z−3 at feed rate 100.
	N0280 X51.440	Cut to position ⑩.
	N0290 X60. Y15.1324	Cut to position ⑪.
	N0300 Y24.8676	Cut to position ⑫.
	N0310 X51.446 Y30.0	Cut to position ⑬.
	N0320 X−3.0	Cut to position ⑭.
Machine stop sequence	N0330 G00 G90 Z1.0 M05	Rapid to 1.0 above part. Stop spindle.
	N0340 M09	Coolant off.
Change to tool 1; Machine start sequence	N0350 G91 G28 Z0 Y0	Rapid to *YZ* reference point ⑮.
	N0360 M00	Program stop.
	N0370 (RECLAMP PART FROM OUTSIDE)	
	N0380 (REMOVE PINS AND OUTSIDE CLAMPS)	
	N0390 T1 M06	Change to tool 1.
	N0400 (TOOL 1: ROUGH OUTSIDE .25MM FULL 3MM DEEP)	
	N0410 G00 G90 X−3.25 Y26.75 Z0. S2100 M03	Rapid to ⑯. Start spindle (CW) at 2100 rpm.
	N0420 G43 Z2.5 H01	Rapid tool 1 to 2.5 above part.
	N0430 M08	Coolant on.
Profile operation 3 sequence	N0440 G01 Z−3. F120.0	Linear profile mode. Plunge to Z−3. at feed rate 120.
	N0450 Y43.25	Cut to position ⑰.
	N0460 X55.1161	Cut to position ⑱.
	N0470 X73.25 Y32.3697	Cut to position ⑲.
	N0480 Y7.6303	Cut to position ⑳.
	N0490 X55.1161 Y−3.25	Cut to position ㉑.
	N0500 X−3.25	Cut to position ㉒.
	N0510 Y12.25	Cut to position ㉓.
Machine stop sequence	N0520 G00 G90 Z1.0 M05	Rapid to 1.0 above part. Stop spindle.
	N0530 M09	Coolant off.

Sequence	Block	Description
Change to tool 2 Machine start sequence	N0540 G91 G28 Z0 Y0	Rapid to *YZ* reference point (24).
	N0550 T2 M06	Change to tool 2.
	N0560 (TOOL 2: FINISH OUTSIDE AS REQUIRED)	
	N0570 G00 G90 X−2. Y28. Z0. S2600 M03	Rapid to (25). Start spindle (CW) at 2600 rpm.
	N0580 G43 Z2.5 H02	Rapid tool 2 to 2.5 above part.
	N0590 M08	Coolant on.
Profile operation 4 sequence	N0600 G01 Z−3. F100.0	Linear profile mode. Plunge to Z−3. at feed rate 100.
	N0610 Y42.0	Cut to position (26).
	N0620 X54.7698	Cut to position (27).
	N0630 X72. Y31.6619	Cut to position (28).
	N0640 Y8.3381	Cut to position (29).
	N0650 X54.7698 Y-2.0	Cut to position (30).
	N0660 X−2.0	Cut to position (31).
	N0670 Y11.0	Cut to position (32).
Machine stop Program end sequence	N0680 G00 G90 Z1.0 M05	Rapid to 1.0 above part. Spindle off.
	N0690 M09	Coolant off.
	N0700 G91 G28 X0 Y0 Z0	Rapid to *XYZ* reference point (33).
	N0710 M30	Program end, memory reset.
	%	End of tape.

Machining simulation for this example is contained in the job file EX 11–5 stored on the disk at the back of this text. Refer to Chapter 2C for instructions.

■ EXAMPLE 11–6

A 0.250-in.-deep pocket is to be machined in the part shown in Figure 11–13. Write a program for executing this operation as per the data table.

Tool	Operation	Tooling	Speed (rpm)	Feed (ipm)
1	Finish pocket depth leave 0.01 on the sides	½*D* center cutting end mill	1500	6
2	Finish sides	⅜*D* end mill	2000	5

Notes:
1. Set X_0 Y_0 at the lower left-hand corner.
2. Z_0 is the top of the part.
3. The part is held in a vise against a stop on the left. The outside profile is assumed to be finished. (See Figure 11–14).

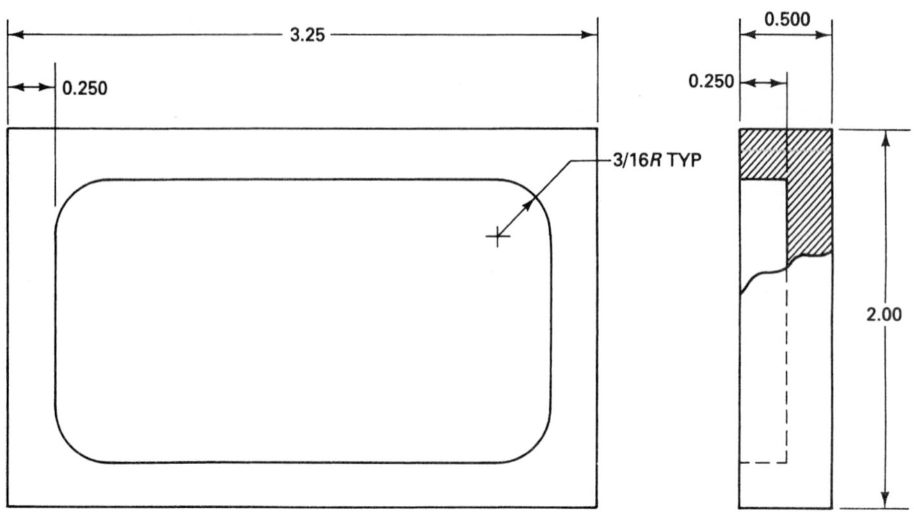

FIGURE 11–13

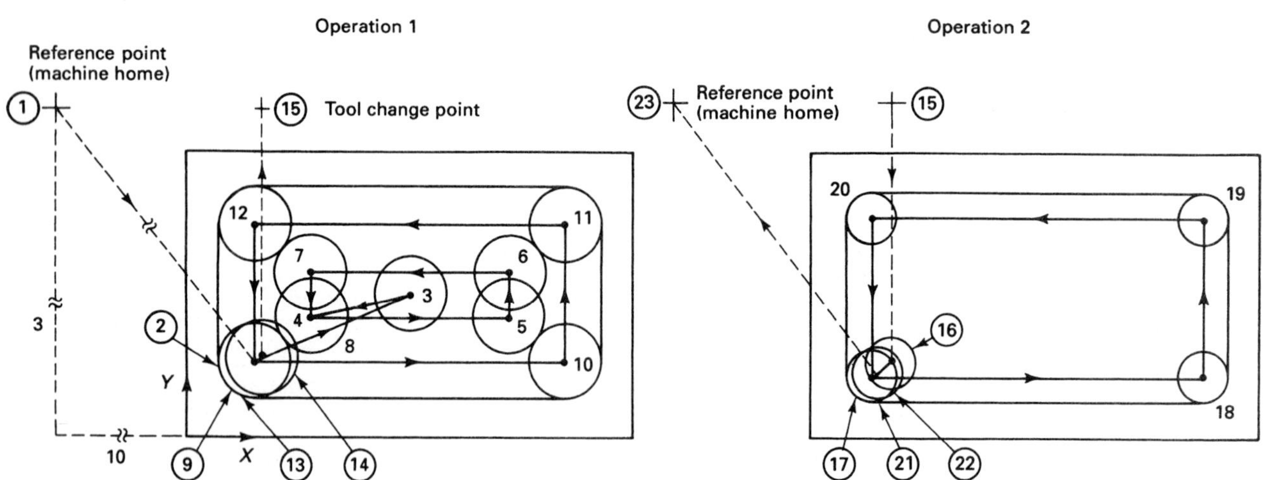

FIGURE 11–14

Programming Pattern

CNC Program—Absolute Coordinates

	Word address command	*Meaning*
	%	End of tape. Used to separate programs.
	O1105	Program number.
Job setup data listing	N0010 (X0Y0 IS THE LOWER LEFT HAND CORNER)	Comments.
	N0020 (Z0 IS THE TOP OF THE PART)	
	N0030 (TOOL 1: .5 DIA CENTER CUTTING END MILL)	
	N0040 (TOOL 2: .375 DIA FINISHING END MILL)	
Machine start-up sequence Change to tool 1	N0050 G90 G20 G40 G80	Absolute, inch mode, cancel fixed cycles.
	N0060 G91 G28 X0 Y0 Z0	Return to reference point ①.
	N0070 G92 X−10.0 Y3.0 Z0	Preset absolute zero point.
	N0080 T1 M06	Change to tool 1.
	N0090 (TOOL 1: FINISH POCKET DEPTH; LEAVE .01 ON THE SIDES)	
	N0100 G00 G90 X.51 Y.51 Z0. S1500 M03	Rapid to ②. Spindle on (CW) at 1500 rpm.
	N0110 G43 Z.1 H01	Rapid tool 1 to 0.1 above the part.
	N0120 M08	Coolant on.
Pocketing operation 1 sequence	N0130 G01 Z0. F6.0	Linear profile mode. Plunge to Z0 at feed rate 6.
	N0140 X1.625 Y1. Z−.25 (RAMP INTO THE PART)	Cut to position ③.
	N0150 X.86 Y.86	Cut to position ④.
	N0160 X2.39	Cut to position ⑤.
	N0170 Y1.14	Cut to position ⑥.
	N0180 X.86	Cut to position ⑦.
	N0190 Y.86	Cut to position ⑧.
	N0200 X.51 Y.51	Cut to position ⑨.
	N0210 X2.74	Cut to position ⑩.
	N0220 Y1.49	Cut to position ⑪.
	N0230 X.51	Cut to position ⑫.
	N0240 Y.51	Cut to position ⑬.
	N0250 X.5375 Y.5375	Cut to position ⑭.
Machine stop sequence	N0260 G00 G90 Z1.0 M05	Rapid to 1.0 above part. Spindle off.
	N0270 M09	Coolant off.

Sequence	Program	Explanation
Change to tool 2 Machine start sequence	N0280 G91 G28 Z0 Y0	Rapid to *YZ* reference point (15).
	N0290 T2 M06	Change to tool 2.
	N0300 (TOOL 2: FINISH SIDES .25 ALL AROUND)	
	N0310 G00 G90 X.5375 Y.5375 Z0. S2000 M03	Rapid to (16) spindle on (CW) at 2000 rpm.
	N0320 G43 Z.1 H02	Rapid tool 2 to 0.1 above part.
	N0330 M08	Coolant on.
Pocketing operation 2 sequence	N0340 G01 Z−.25 F5.0	Linear profile mode. Plunge to Z−.25 at feed rate 5.
	N0350 X.4375 Y.4375	Cut to position (17).
	N0360 X2.8125	Cut to position (18).
	N0370 Y1.5625	Cut to position (19).
	N0380 X.4375	Cut to position (20).
	N0390 Y.4375	Cut to position (21).
	N0400 X.4875 Y.4875	Cut to position (22).
Machine stop Program end sequence	N0410 G00 G90 Z1.0 M05	Rapid to 1.0 above part. Spindle off.
	N0420 M09	
	N0430 G91 G28 X0 Y0 Z0	Rapid to *XYZ* reference point (23).
	N0440 M30	Program end, memory reset.
	%	End of tape.

Machining simulation for this example is contained in the job file EX 11–6 stored on the disk at the back of this text. Refer to Chapter 20 for instructions.

11–6 CHAPTER SUMMARY

The following key concepts were discussed in this chapter:

1. Linear profiling involves cutting a contour composed of straight-line segments.
2. The word address code for linear profiling is G01.
3. The cutter circle must be tangent to the profile line it cuts.
4. Right-triangle trigonometry must be used to compute cutter offsets for the case of inclined line profiles.

REVIEW EXERCISES

For each of the exercises in this chapter, write a CNC program containing the following general format:

1. Program number/headings
2. Job setup data listing

3. Tool 1 in spindle, machine start-up
4. Profiling operation 1
5. Machine stop
6. Change to tool 2, machine start
7. Profiling operation 2

.
.
.

8. Machine stop, program end sequence

The machine home reference point is a suggested location. The true location must be determined at the time of setup.

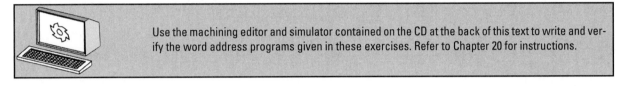

Use the machining editor and simulator contained on the CD at the back of this text to write and verify the word address programs given in these exercises. Refer to Chapter 20 for instructions.

11.1. Program number: O1106 (Figure 11–15)

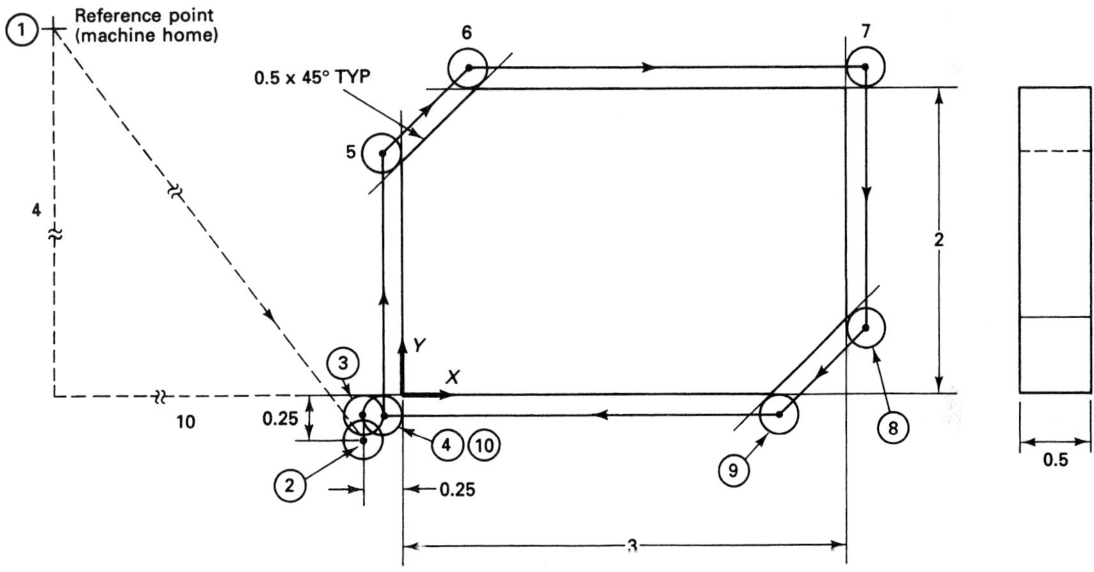

FIGURE 11–15

Tool	Operation	Tooling	Speed (rpm)	Feed (ipm)
1	Profile mill contour × 0.5 deep	0.25*D* end mill	2000	5

Notes: 1. Set X_0 Y_0 at the lower left-hand corner.
2. Z_0 is the top of the part.

11.2. Program number: O1107 (Figure 11–16)

Tool	Operation	Tooling	Speed (rpm)	Feed (ipm)
1	Profile mill contour × 0.25 deep	0.25*D* end mill	1600	8

Notes: 1. Set X_0 Y_0 at the lower left-hand corner.
2. Z_0 is the top of the part.

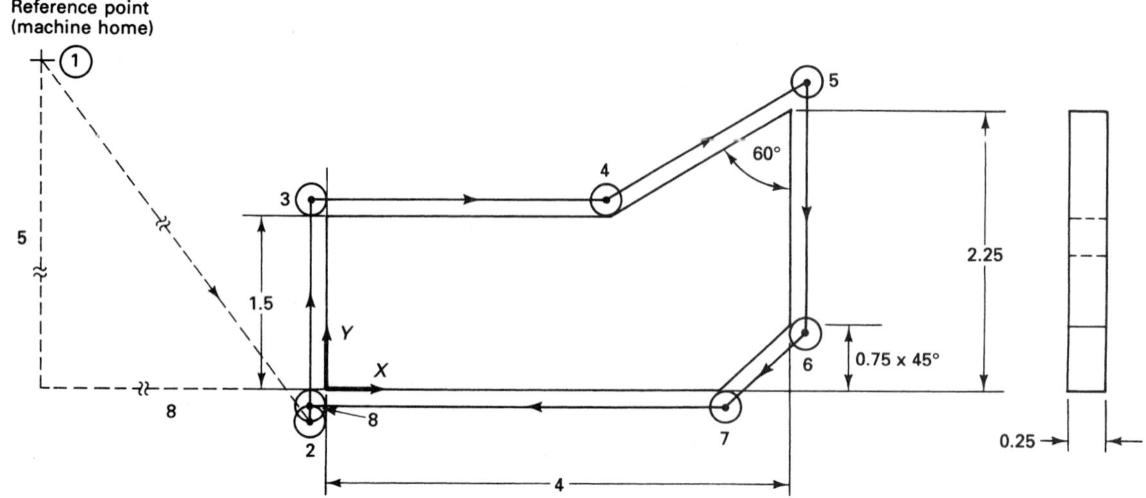

FIGURE 11-16

11.3. Program number: O1108 (Figure 11–17)

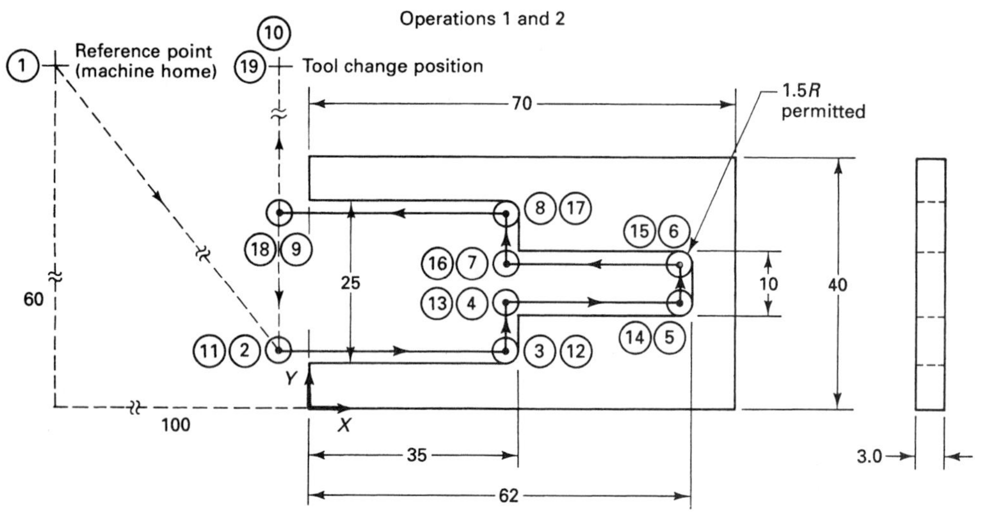

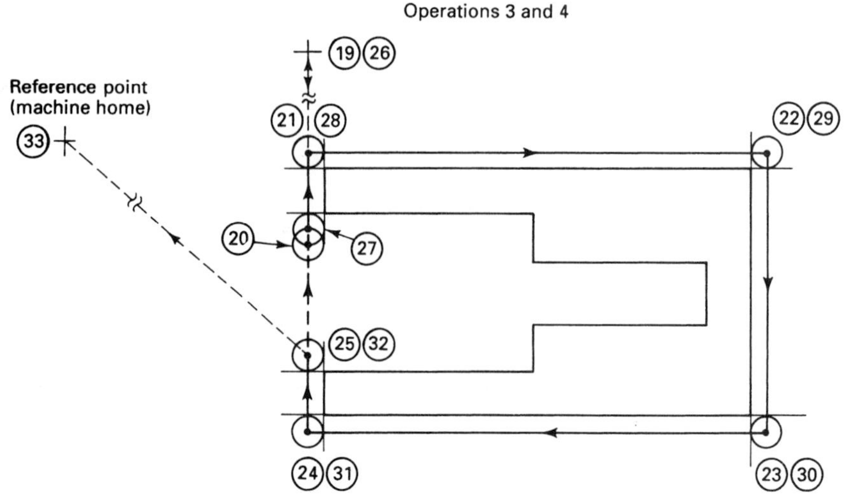

FIGURE 11-17

Tool	Operation	Tooling	Speed (rpm)	Feed (mm/min)
1	Rough inside 0.25 mm full × 0.35 mm deep	5-mm roughing end mill	2000	100
2	Finish inside as required	3-mm end mill	2400	70
3	Rough outside 0.25 mm full × 0.35 mm deep	5-mm roughing end mill	2000	100
4	Finish outside as required	3-mm end mill	2400	70

Notes:
1. Set X_0 Y_0 at the lower left-hand corner.
2. Z_0 is the top of the part.
3. The part is roughed and finished in the slot by holding the blank against three pins by two clamps. (See Figure 11–18).
4. For outside machining, stop the machine. Place clamps from inside and remove outside clamps and pins. Finish as required. (See Figure 11–19).

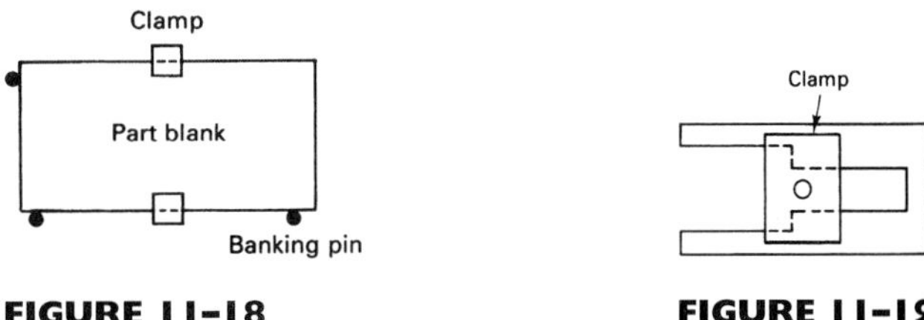

FIGURE 11-18 **FIGURE 11-19**

11.4. Program number: O1109 (Figure 11–20)

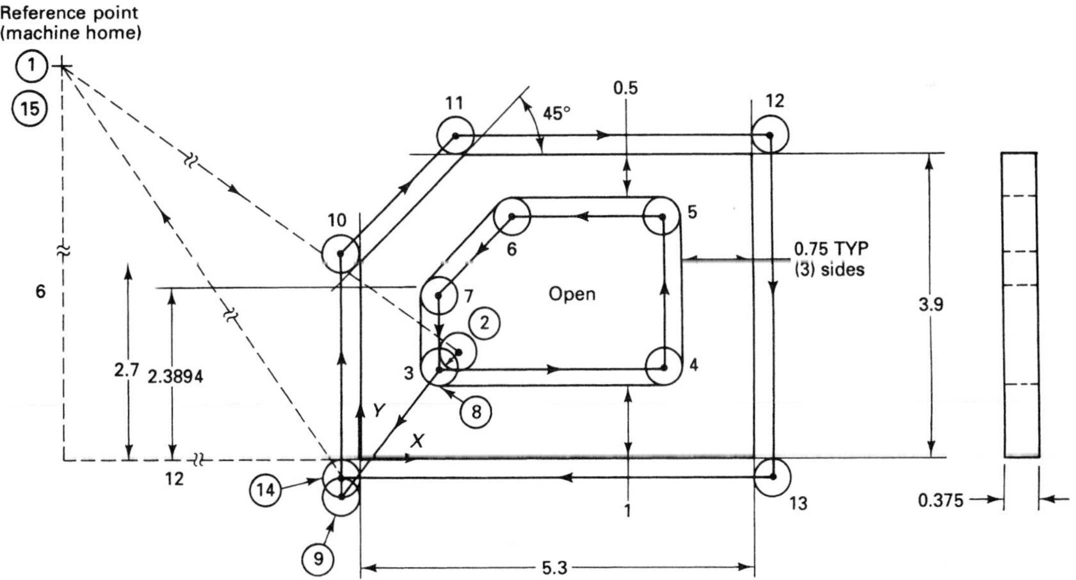

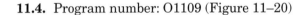

FIGURE 11-20

Tool	Operation	Tooling	Speed(rpm)	Feed (ipm)
1	Profile inside and outside × 0.40 deep	0.5*D* end mill	2000	4

Notes:
1. Set X_0 Y_0 at the lower left-hand corner.
2. Z_0 is the top of the part.

11.5. Program number: O1010 (Figure 11–21)

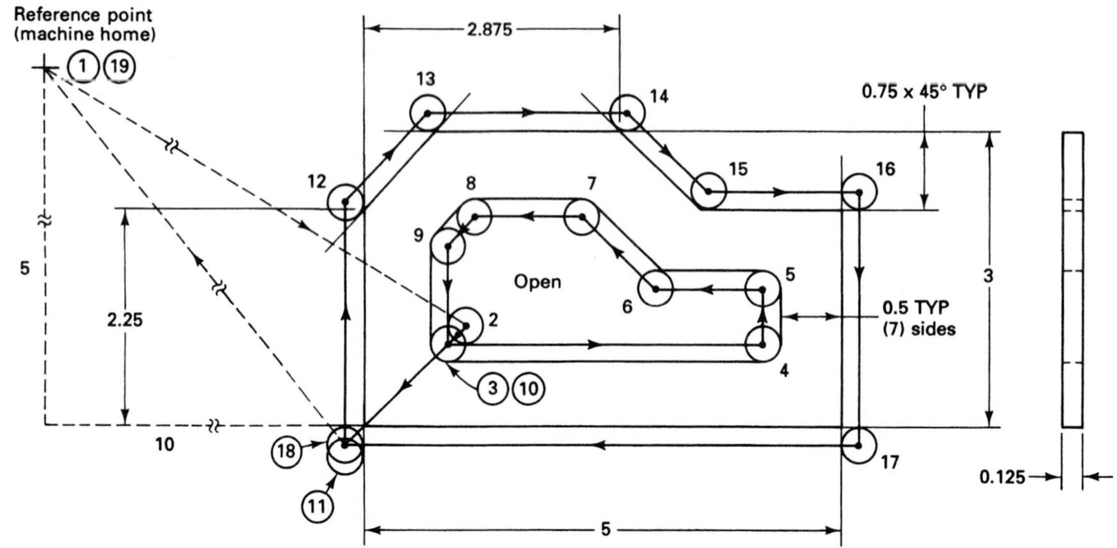

FIGURE 11–21

Tool	Operation	Tooling	Speed (rpm)	Feed (ipm)
1	Profile inside and outside × 0.150 deep	0.25*D* end mill	1800	8

Notes: 1. Set X_0 Y_0 at the lower left-hand corner.
2. Z_0 is the top of the part.

11.6. Program number: O1111 (Figure 11–22)

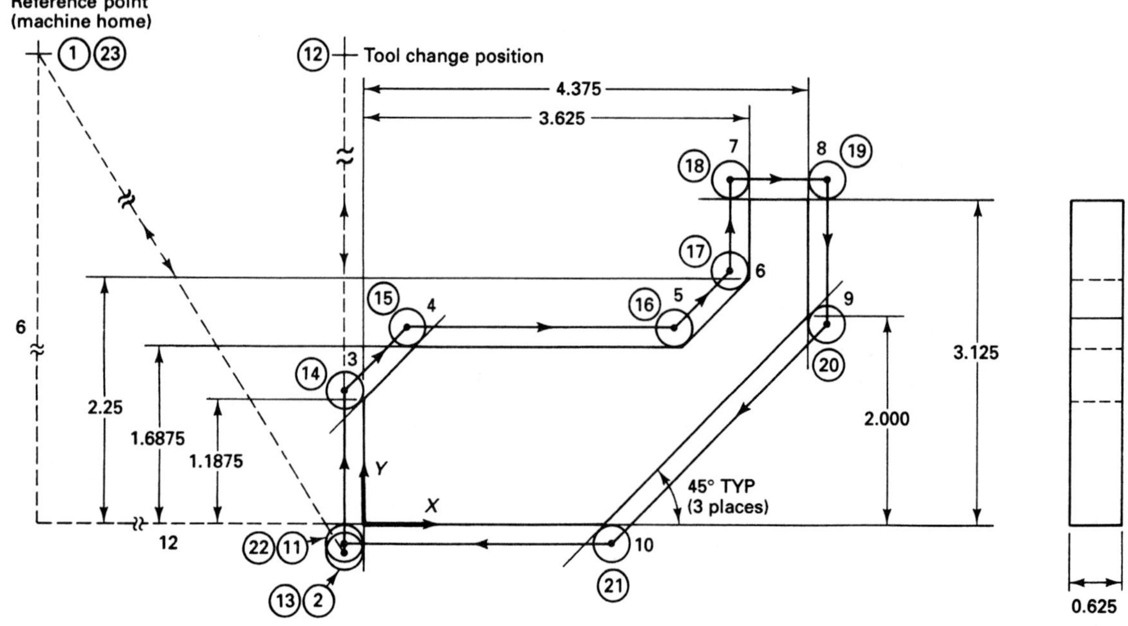

FIGURE 11–22

Tool	Operation	Tooling	Speed (rpm)	Feed (ipm)
1	Rough profile × 0.650 deep	⅜ roughing end mill	2200	12
2	Finish profile contour × 0.650 deep	0.25*D* end mill	2000	8

Notes: 1. Set X_0 Y_0 at the lower left-hand corner.
2. Z_0 is the top of the part.

11.7. Program number: O1112 (Figure 11–23)

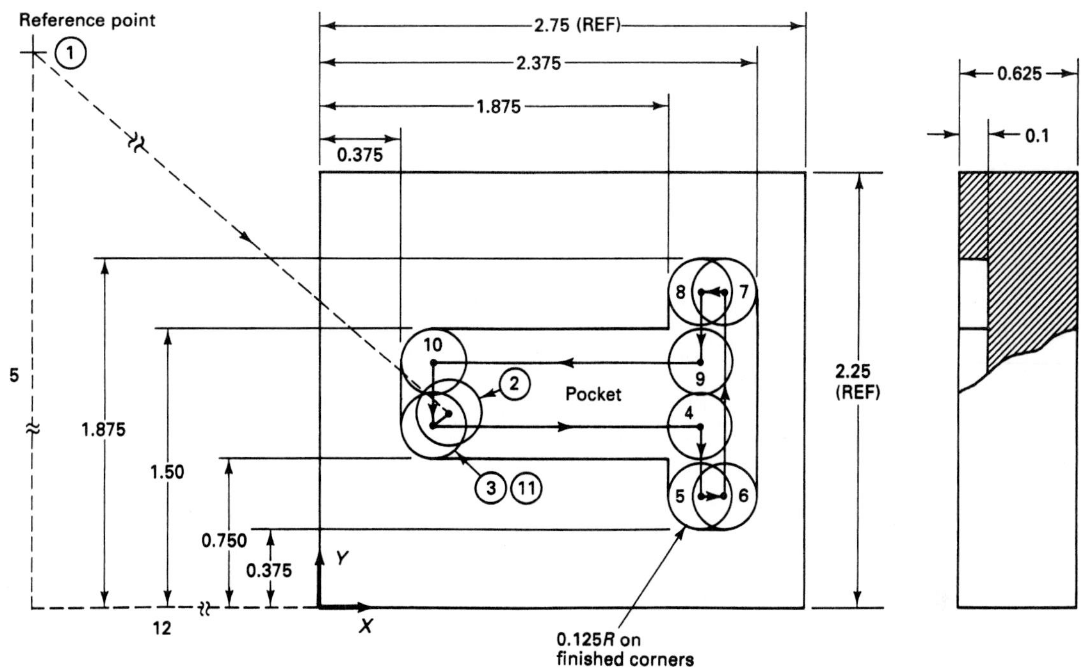

FIGURE 11–23

Tool	Operation	Tooling	Speed (rpm)	Feed (ipm)
1	Finish pocket to depth. Leave 0.01 per side	0.375 center cutting end mill	1600	7
2	Finish sides of the pocket	0.25*D* end mill	1800	5

Notes: 1. Set X_0 Y_0 at the lower left-hand corner.
2. Z_0 is the top of the part.
3. The part is held in a vise against a stop on the left.
4. The outside profile is assumed to be finished.

CHAPTER 12

Programming Circular Profiles

12-1 CHAPTER OBJECTIVES

At the conclusion of this chapter you will be able to

1. Explain what circular interpolation is.
2. Describe how to specify the plane for executing circular interpolation.
3. Explain the commands used for clockwise as well as counterclockwise circular interpolation.
4. Compute cutter offsets for important line-arc profiling cases.
5. Code complete CNC programs for executing line-arc profiling operations.

12-2 INTRODUCTION

Milling of contours with circular arc elements will be discussed in this chapter. See Figure 12–1. These machining operations involve circular interpolation for determining tool movements in the clockwise direction as well as the counterclockwise direction. The specific word address commands for executing such interpolations will also be listed and explained. Useful formulas for determining cutter offsets for important circular arc contours are given. Finally, complete CNC programs for executing typical circular profiling operations are illustrated and discussed.

12-3 SPECIFYING THE PLANE FOR CIRCULAR ARC INTERPOLATION

Circular interpolation is executed on a plane. Thus, the programmer must first identify this plane before issuing any circular interpolation commands. The following word address codes are used for this purpose.

12-4 CIRCULAR INTERPOLATION COMMANDS

The commands for circular interpolation direct the system to move the tool simultaneously in the *X* and *Y* directions such that a programmed circle or portion of a circle is cut. As was stated previously, the programmer first defines the plane in which interpolation is to occur. The tool must then be moved to the start point of the circular arc to be cut. Upon receiving a command to execute circular interpolation, the controller will determine the radius between the start point and the center of the circular arc. It will move the tool from the start

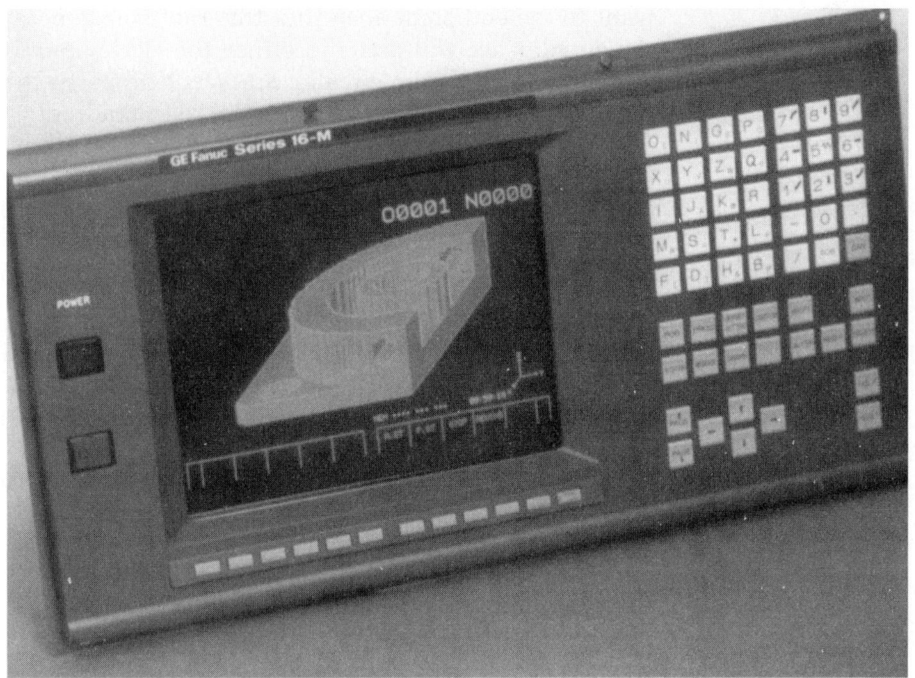

FIGURE 12–1 Machine control display of the part after circular profiling. (Photo courtesy of GE Fanuc Automation North America, Inc.)

General Syntax

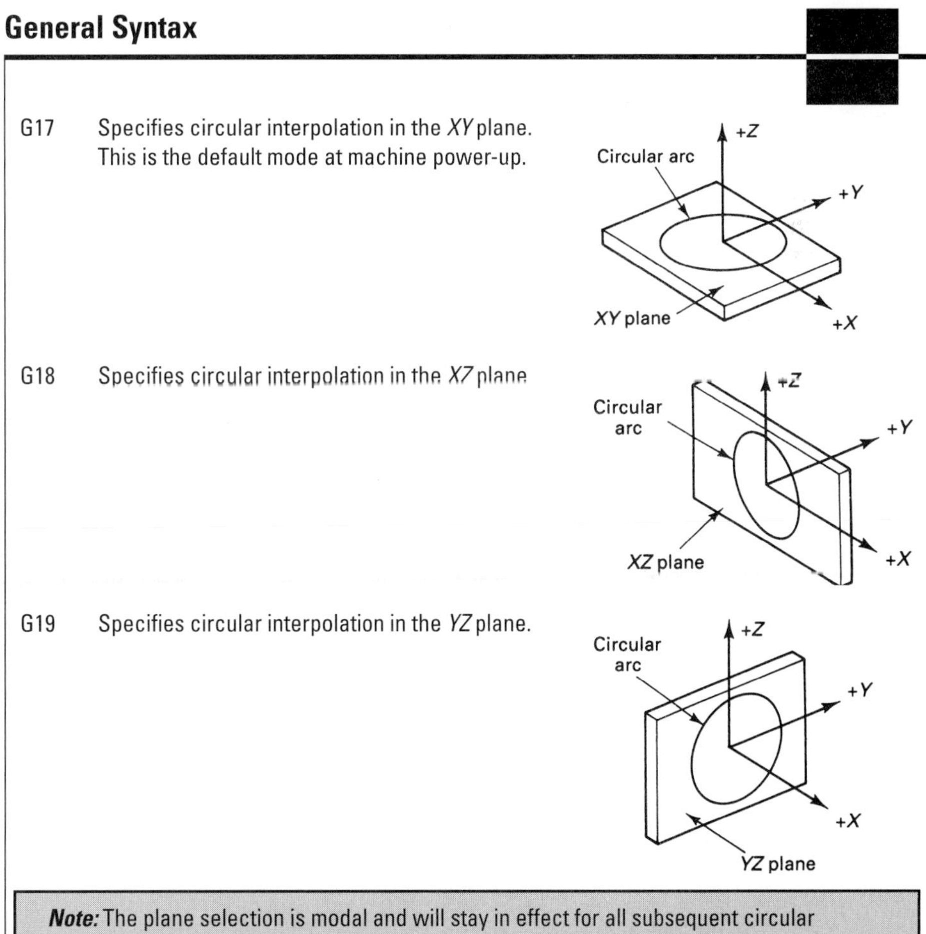

G17 Specifies circular interpolation in the *XY* plane. This is the default mode at machine power-up.

G18 Specifies circular interpolation in the *XZ* plane.

G19 Specifies circular interpolation in the *YZ* plane.

Note: The plane selection is modal and will stay in effect for all subsequent circular interpolation commands until canceled by another plane selection code.

point to the end point such that this radius is generated. Tool motion around the circular arc will occur in either the clockwise (G02) or counterclockwise (G03) direction. If the circular arc is cut properly, the distance from the end point to the center point will be equal to the radius.

The programmer can execute circular interpolation in either the absolute (G90) mode or the incremental (G91) mode.

Below are listed the word address commands for circular interpolation on a vertical milling machine under the control of a Fanuc 6M. The programmer should take care that the following operations have been performed prior to issuing a circular interpolation command:

1. Plane selection (G17, G18, or G19) has been coded if needed.
2. The tool has been positioned at the starting point of the arc.
3. The control is in the linear interpolation mode (G01).
4. A feed rate has been specified.
5. The system is in the proper motion mode (G90, absolute, or G91, incremental).

General Syntax

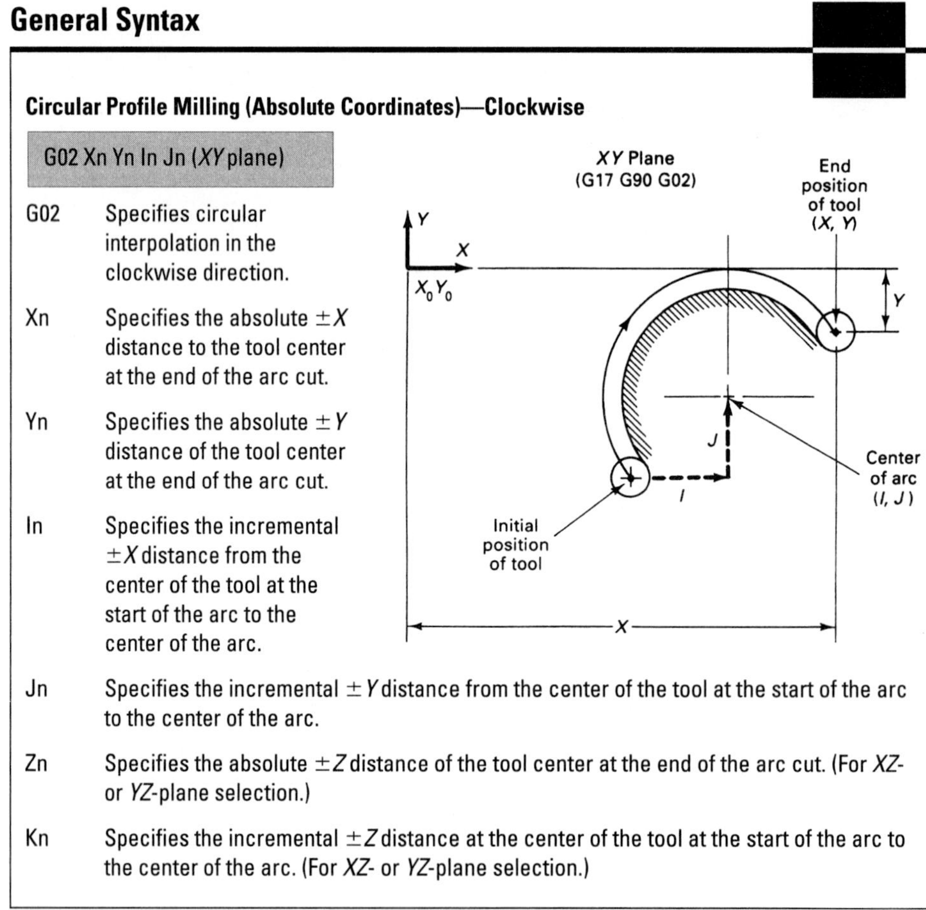

Circular Profile Milling (Absolute Coordinates)—Clockwise

G02 Xn Yn In Jn (*XY* plane)

G02	Specifies circular interpolation in the clockwise direction.
Xn	Specifies the absolute ±*X* distance to the tool center at the end of the arc cut.
Yn	Specifies the absolute ±*Y* distance of the tool center at the end of the arc cut.
In	Specifies the incremental ±*X* distance from the center of the tool at the start of the arc to the center of the arc.
Jn	Specifies the incremental ±*Y* distance from the center of the tool at the start of the arc to the center of the arc.
Zn	Specifies the absolute ±*Z* distance of the tool center at the end of the arc cut. (For *XZ*- or *YZ*-plane selection.)
Kn	Specifies the incremental ±*Z* distance at the center of the tool at the start of the arc to the center of the arc. (For *XZ*- or *YZ*-plane selection.)

Circular Profile Milling (Absolute Coordinates)—Counterclockwise

G03 Xn Yn In Jn (*XY* plane)

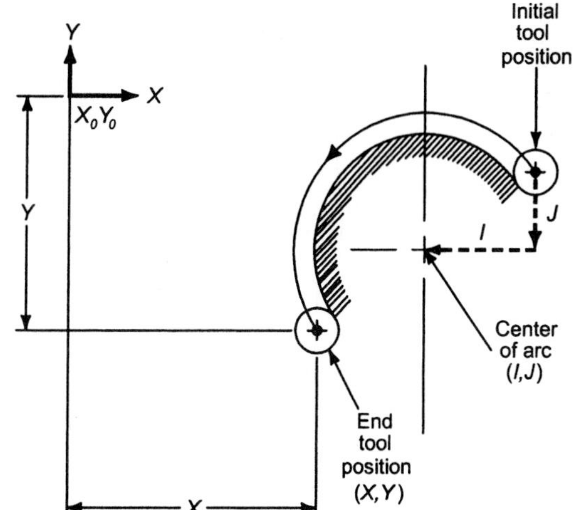

G03 Specifies circular interpolation in the counterclockwise direction.

Xn Specifies the absolute ±*X* distance to the tool center at the end of the arc cut.

Yn Specifies the absolute ±*Y* distance to the tool center at the end of the arc cut.

In Specifies the incremental ±*X* distance from the tool center at the start of the arc to the arc center.

Jn Specifies the incremental ±*Y* distance from the tool center at the start of the arc to the arc center.

Zn Specifies the absolute ±*Z* distance to the tool center at the end of the arc cut. (For *XZ*- or *YZ*-plane selection.)

Kn Specifies the incremental ±*Z* distance from the tool center at the start of the arc to the arc center. (For *XZ*- or *YZ*-plane selection.)

Circular Profile Milling (Incremental Coordinates)—Clockwise

G02 Xn Yn In Jn (*XY* plane)

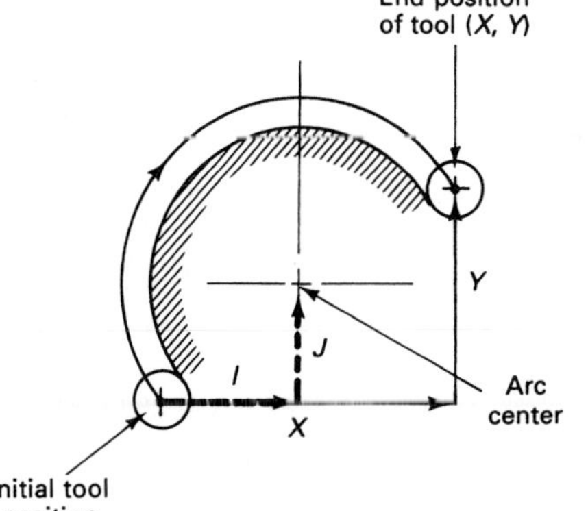

G02 Specifies circular interpolation in the clockwise direction.

Xn Specifies the incremental ±*X* distance from the tool center at the start of the arc to the tool center at the end of the arc.

Yn Specifies the incremental ±*Y* distance from the tool center at the start of the arc to the tool center at the end of the arc.

In Specifies the incremental ±*X* distance from the tool center at the start of the arc to the arc center.

Jn Specifies the incremental ±*Y* distance from the tool center at the start of the arc to the arc center.

Zn Specifies the incremental ±*Z* distance from the tool center at the start of the arc to the tool center at the end of the arc. (For *XZ*- or *YZ*-plane selection.)

Kn Specifies the incremental ±*Z* distance from the tool center at the start of the arc to the arc center. (For *XZ*- or *YZ*-plane selection.)

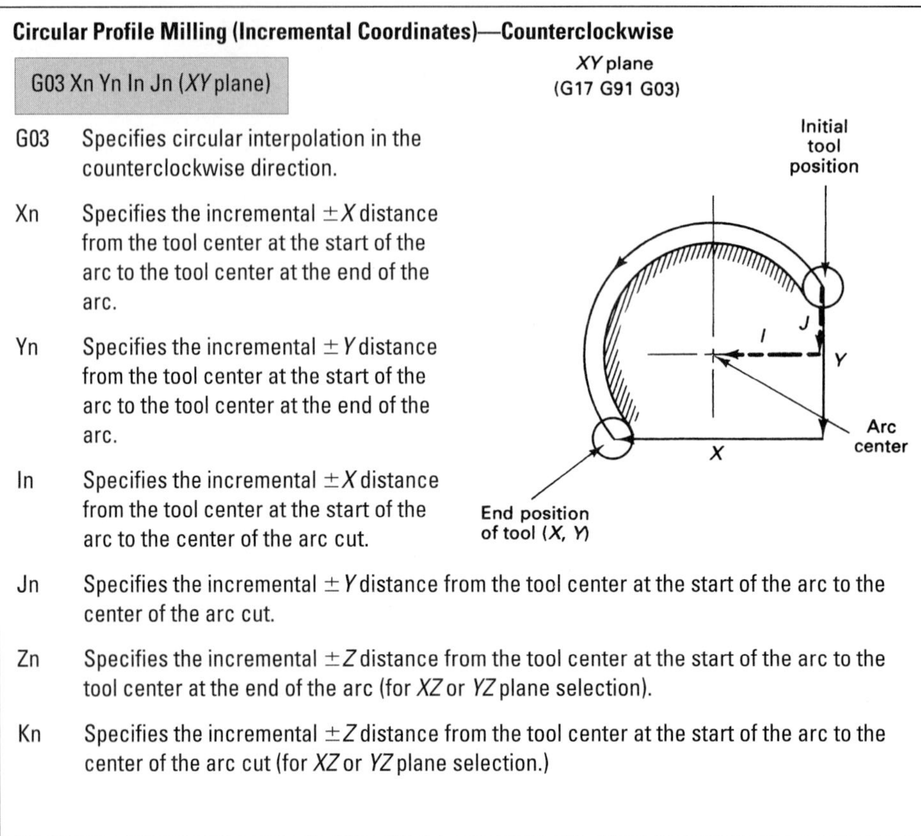

Circular Profile Milling (Incremental Coordinates)—Counterclockwise

G03 Xn Yn In Jn (*XY* plane)

G03 Specifies circular interpolation in the counterclockwise direction.

Xn Specifies the incremental ±*X* distance from the tool center at the start of the arc to the tool center at the end of the arc.

Yn Specifies the incremental ±*Y* distance from the tool center at the start of the arc to the tool center at the end of the arc.

In Specifies the incremental ±*X* distance from the tool center at the start of the arc to the center of the arc cut.

Jn Specifies the incremental ±*Y* distance from the tool center at the start of the arc to the center of the arc cut.

Zn Specifies the incremental ±*Z* distance from the tool center at the start of the arc to the tool center at the end of the arc (for *XZ* or *YZ* plane selection).

Kn Specifies the incremental ±*Z* distance from the tool center at the start of the arc to the center of the arc cut (for *XZ* or *YZ* plane selection.)

■ EXAMPLE 12–1

Write the appropriate plane selection, linear cut to position, and circular interpolation commands for profiling the arc shown in Figure 12–2 in the absolute coordinate mode.

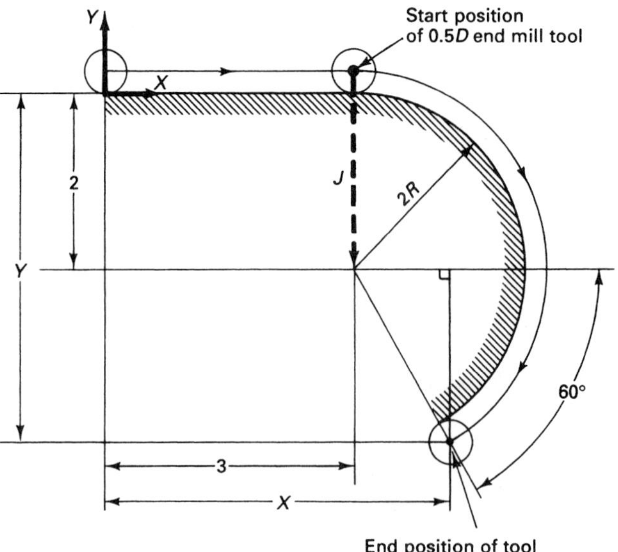

FIGURE 12–2

```
G90
G17
G01 X3.0 Y.25 F10.
G02 X4.125 Y−3.9486 10. J−2.25
```

ΔX	ΔY	X	Y	I	J
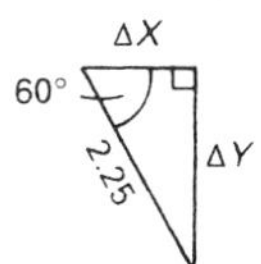					
$\Delta X = 2.25\cos(60) = 1.125$	$\Delta Y = -2.25\sin(60) = -1.9486$	$3 + 1.125 = 4.125$	$-2 - 1.9486 = -3.9486$	0	−2.25

EXAMPLE 12–2

The circular arc shown in Figure 12–3 is to be cut using absolute coordinate programming. Write the appropriate linear cut and circular interpolation commands.

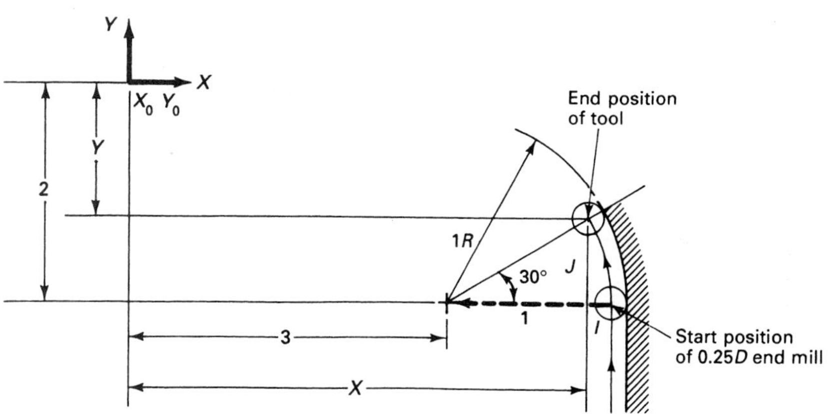

FIGURE 12–3

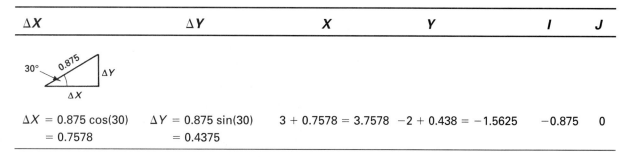

```
G90
G17
G01 X3.875 Y−2. F10.
G03 X3.757 Y−1.5625 I−.875
```

ΔX	ΔY	X	Y	I	J
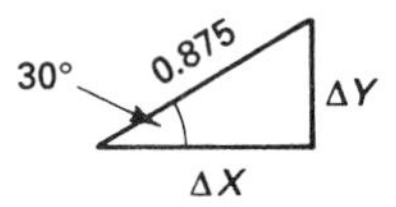					
$\Delta X = 0.875\cos(30) = 0.7578$	$\Delta Y = 0.875\sin(30) = 0.4375$	$3 + 0.7578 = 3.7578$	$-2 + 0.438 = -1.5625$	−0.875	0

■ EXAMPLE 12–3

Incremental coordinates are to be used when cutting the circular arc shown in Figure 12–4. Write the necessary plane selection, linear cut, and circular interpolation commands.

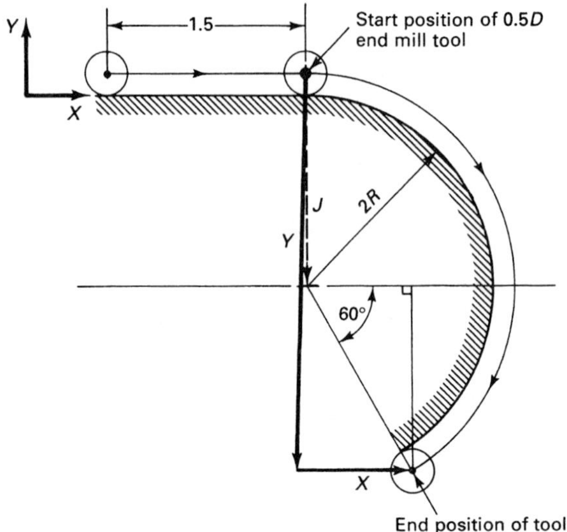

FIGURE 12–4

```
G91
G17
G01 X1.5 F6.0
G02 X1.125 Y−4.1986 J−2.25
```

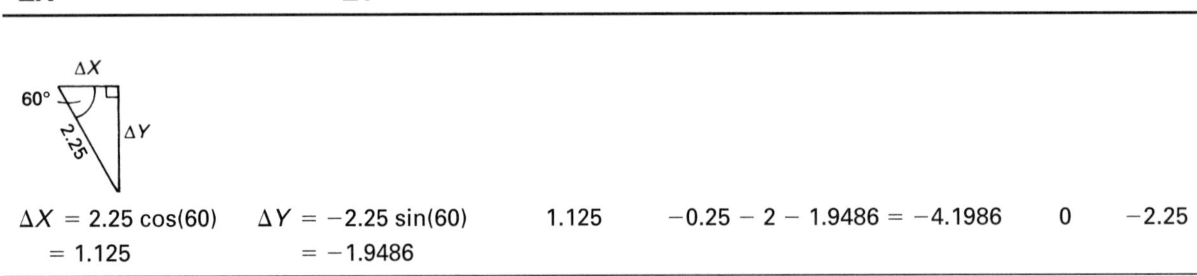

ΔX	ΔY	X	Y	I	J
$\Delta X = 2.25 \cos(60)$ $= 1.125$	$\Delta Y = -2.25 \sin(60)$ $= -1.9486$	1.125	$-0.25 - 2 - 1.9486 = -4.1986$	0	-2.25

■■

12–5 CIRCULAR INTERPOLATION VIA DIRECT RADIUS SPECIFICATION

The programmer also has the option of coding the radius (R) of the circular arc instead of I, J, and K. The arc center can be located at either of two positions: center 1 or center 2. These centers are determined as follows:

1. Connect line 1 between the arc starting point and ending point.
2. Construct line 2 perpendicular to line 1 at its midpoint.
3. Construct arc 1 of radius R from the arc starting point.
4. Centers 1 and 2 are then located at the intersection of arc 1 and line 2 as shown in Figure 12–5.

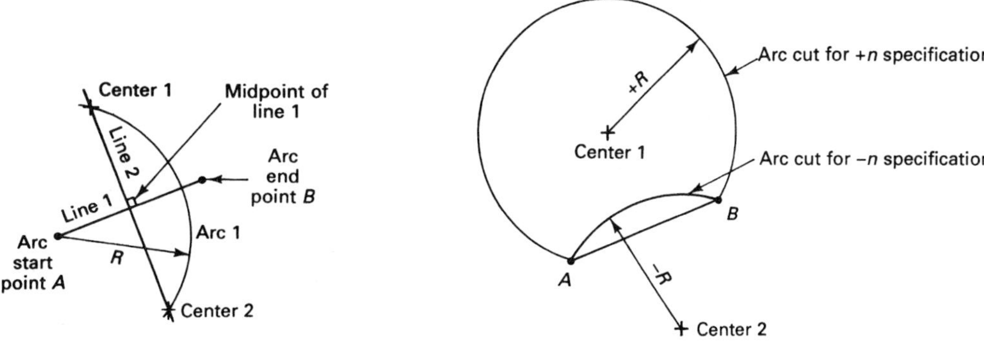

FIGURE 12–5

General Syntax

Circular Profile Milling (Incremental Coordinates)

G02(G03) Xn Yn R_n (XY plane)

G02 Specifies circular interpolation in the clockwise direction.

G03 Specifies circular interpolation in the counterclockwise direction.

Xn Specifies the incremental +*X* distance from the tool center at the start of the arc to the tool center at the end of the arc cut.

Yn Specifies the incremental +*Y* distance from the tool center at the start of the arc to the tool center at the end of the arc cut.

Zn Specifies the incremental +*Z* distance from the tool center at the start of the arc to the tool center at the end of the arc cut. (For *XZ*- or *YZ*-plane selection.)

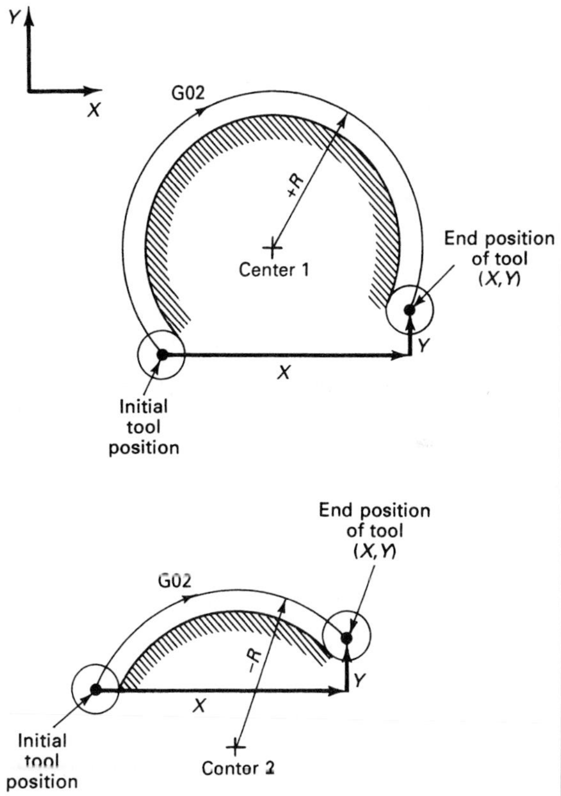

Rn Specifies the radius of the tool path when the arc is cut.
+n is the tool path radius from center 1.
−n is the tool path radius from center 2.

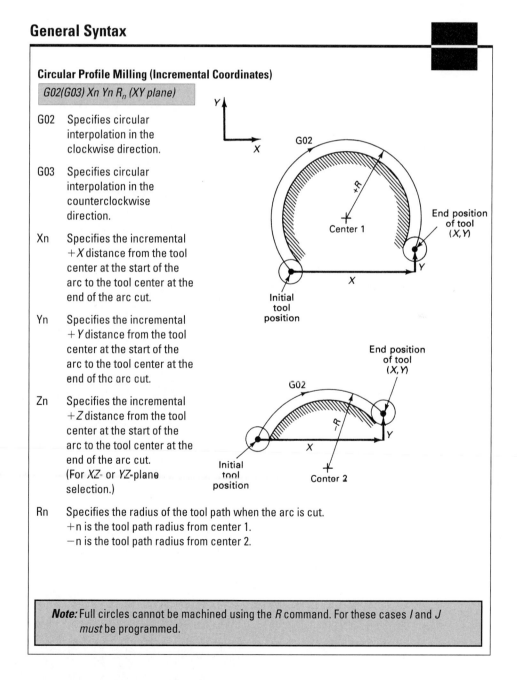

Note: Full circles cannot be machined using the *R* command. For these cases *I* and *J* *must* be programmed.

■ EXAMPLE 12–4

Code the direct radius command to machine the circular arcs shown in Figures 12–6 and 12–7. A feed rate of 2 ipm is to be used.

G17G91
G02 *X*4.125 Y2.125 R−5. F2.

(>180°)

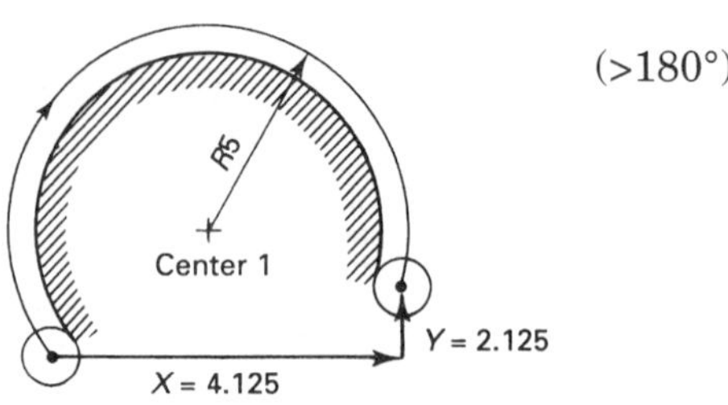

FIGURE 12–6

G17G91
G02 *X*4.125 Y2.125 R−5. F2.

(>180°)

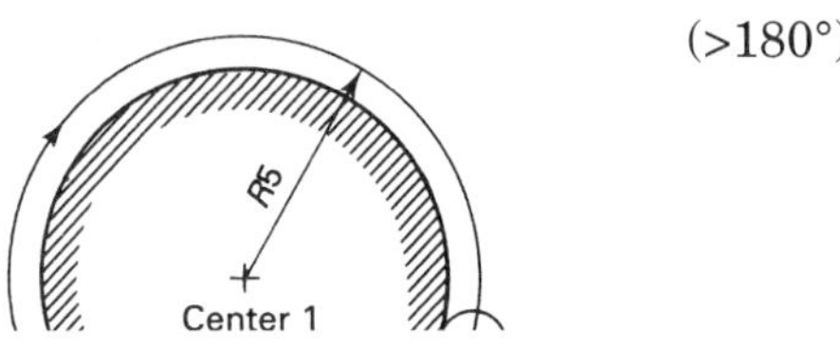

FIGURE 12–7

■■

12–6 PROFILING ARCS AT CONSTANT FEED RATE

For the purposes of achieving a required surface finish and/or machining characteristics for a material, it is often necessary to maintain a constant feed rate while profiling a contour. During linear interpolation operations, the distance traveled at the center of the tool is the same as that traveled at the edge of the tool. For outside circular motion, however, the distance traveled by the tool center is greater than that traveled by the tool's edge. This is because the tool center follows the longer arc based on the tool path radius, just as the tool edge follows the shorter arc based on the part radius. Because the center of the tool must travel a longer distance, the feed rate of the tool center must be adjusted upward. The following formula can be used to adjust the feed rate of the tool center such that a constant feed rate is maintained on the periphery of the tool as it moves from a linear cut to a circular cut:

$$\frac{\textit{Cutter path radius}}{\textit{Part radius}} \times \textit{Linear feed rate} = \textit{Constant feed rate}$$

■ EXAMPLE 12–5

A 0.500-diameter end mill is to be used to machine a radius of 1 in. See Figure 12–8. A constant feed rate of 5 ipm on the tool periphery is to be maintained.

$$\frac{(1 + .5/2)}{1.000} \times 5 = 6.25 \text{ ipm}$$

Therefore, a feed rate of 6.25 ipm for the tool center will ensure that the tool maintains a feed rate of 5 ipm on its periphery when machining a 1.25*R* arc.

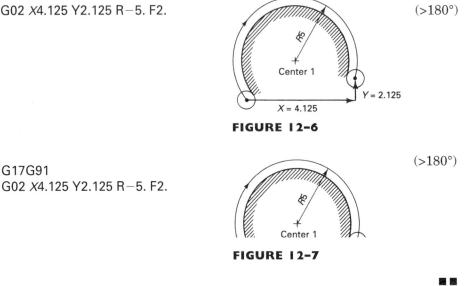

■■

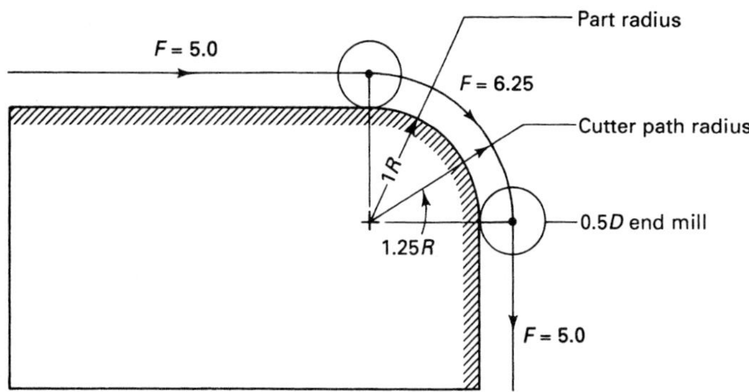

FIGURE 12–8

12–7 DETERMINING CUTTER OFFSETS FOR LINE-ARC PROFILES

Right-triangle trigonometry is again applied in calculating the locations of the tool center as it moves from a line to an arc and vice versa. Some of the most common interface cases are illustrated in Figures 12–9 through 12–11.

1. Line-arc interface case 1

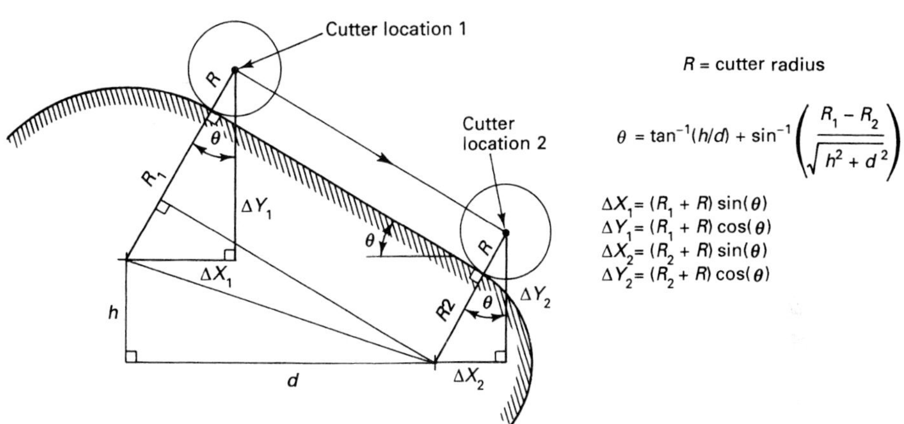

Cutter location 1
Cutter location 2
R = cutter radius
$\theta = \tan^{-1}(h/d) + \sin^{-1}\left(\frac{R_1 - R_2}{\sqrt{h^2 + d^2}}\right)$
$\Delta X_1 = (R_1 + R)\sin(\theta)$
$\Delta Y_1 = (R_1 + R)\cos(\theta)$
$\Delta X_2 = (R_2 + R)\sin(\theta)$
$\Delta Y_2 = (R_2 + R)\cos(\theta)$

FIGURE 12–9

2. Line-arc interface case 2

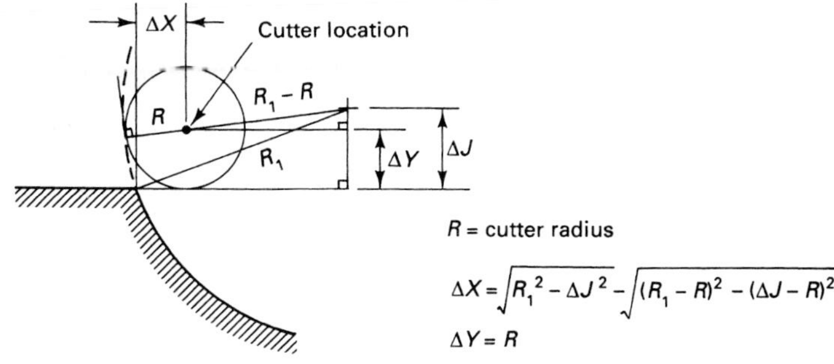

FIGURE 12–10

3. Line-arc interface case 3

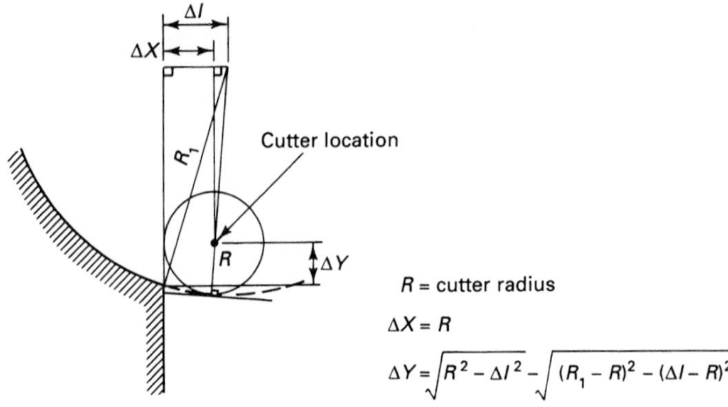

FIGURE 12–11

■ EXAMPLE 12–6

Determine the coordinates for specifying circular interpolation of the profile shown in Figure 12–12. Use the coordinates to write the CNC program for machining the profile.

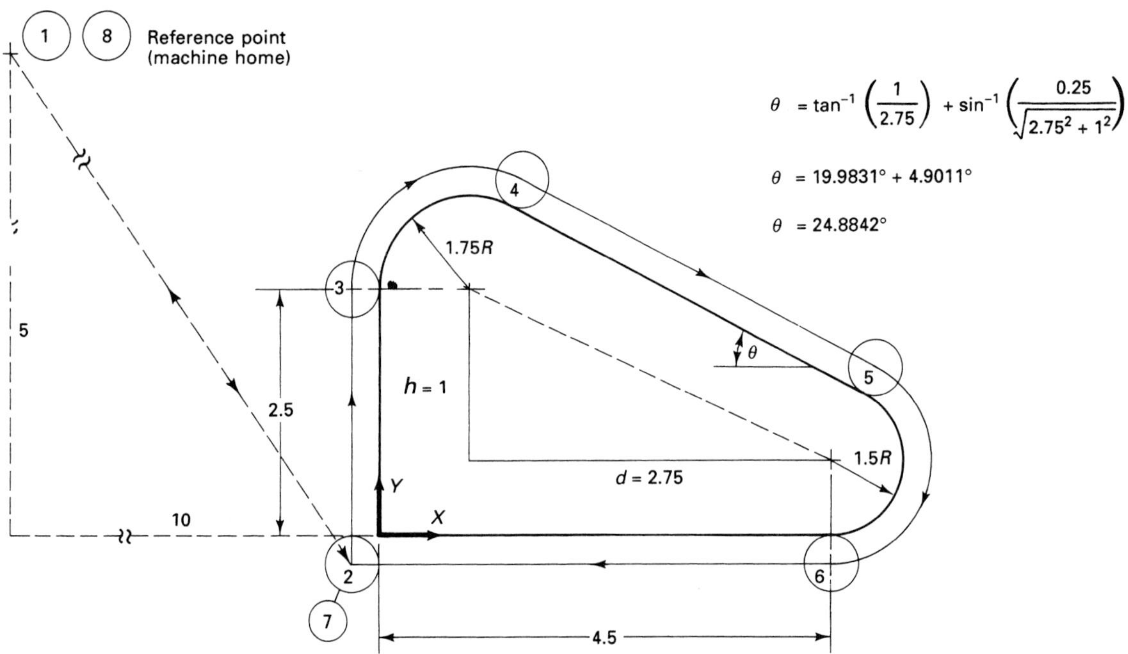

FIGURE 12–12

Tool	Operation	Tooling	Speed (rpm)	Feed (ipm)
1	Profile mill contour × 0.25 deep	0.5*D* end mill	1200	6

Position	ΔX	ΔY	X	Y	I	J	
②			-0.25	-0.25			
③			-0.25	2.5	2.0	0	Arc start
④	(1.75+0.25) × Sin(24.8842°) = 0.8416 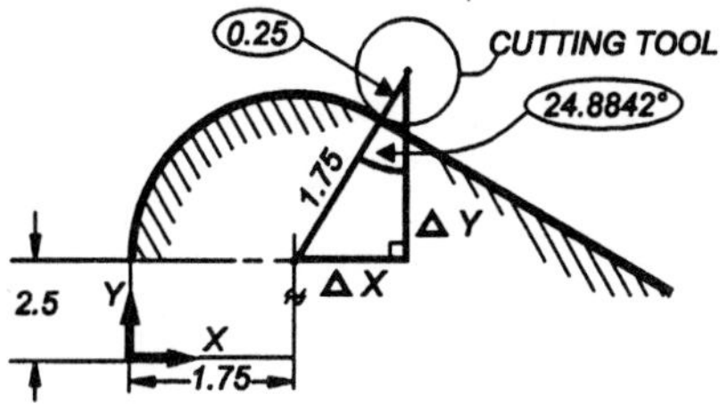	2 × Cos(24.8842°) =1.8143	0.8416+1.75 = 2.5916	2.5 + 1.8143 = 4.3143			Arc end
⑤	(1.5+0.25) × Sin(24.8842°) = 0.7364 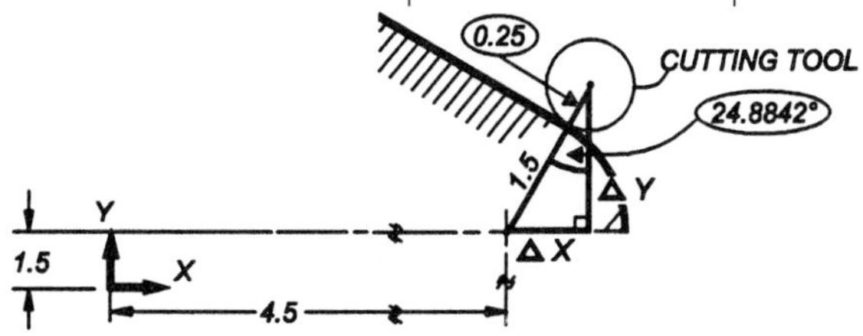	1.75 × Cos(24.8842°) = 1.5875	4.5 + 0.7364 = 5.2364	1.5 + 1.5875 = 3.0875	−0.7364	−0.5875	Arc start
⑥			4.5	−0.25			Arc end
⑦			−0.25	−0.25			

Programming Pattern

CNC Program—Absolute Coordinates

	Word address command	*Meaning*
	%	End of tape. Used to separate programs.
	O1201	Program number.
Job setup data listing	N0010 (X0Y0 IS THE LOWER LEFT HAND CORNER)	Comments.
	N0020 (Z0 IS THE TOP OF THE PART)	
	N0030 (TOOL 1: .5 DIA END MILL)	
	N0040 (TOOL 1 MUST BE IN SPINDLE PRIOR TO START)	
Machine start-up sequence	N0050 G90 G20 G40 G80	Absolute, inch mode, cancel cutter diameter compensation and fixed cycles.
	N0060 G91 G28 X0 Y0 Z0	Return to reference point ①.
	N0070 G92 X−10. Y5. Z0	Preset absolute zero point.
	N0080 T1 M06	Change to tool 1.
Change to tool 1	N0090 (TOOL 1: PROFILE MILL CONTOUR .25 DEEP)	
	N0100 G00 G90 X − .25 *Y* − .25 Z0. S1200 M03	Rapid to ②. Spindle on (CW) at 1200 rpm.
	N0110 G43 Z.1 H01	Rapid tool 1 to 0.1 above part.
	N0120 M08	Coolant on.

Profile operation 1 sequence	N0130 G01 Z − .25 F6.0	Linear profile mode. Plunge cutter to Z−.25 at feed rate 6.
	N0140 Y2.5	Cut to ③.
	N0150 G02 X2.5916 Y4.3143 I2.0 J0.0	Cut 1.75*R* arc to ④.
	N0160 G01 X5.2364 Y3.0875	Cut to ⑤.
	N0170 G02 X4.5 Y − .25 I − .7364 J − 1.5875	Cut 1.5*R* arc to ⑥.
	N0180 G01 X − .25	Cut to ⑦.
Machine stop Program end sequence	N0190 G00 G90 Z1.0 M05	Rapid to 1.0 above part. Stop spindle.
	N0200 M09	Coolant off.
	N0210 G91 G28 X0 Y0 Z0	Return to *XYZ* reference point ⑧.
	N0220 M30	Program end, memory reset.
	%	End of tape.

Machining simulation for this example is contained in the job file EX 12–6 stored on the disk at the back of this text. Refer to Chapter 20 for instructions.

■ EXAMPLE 12–7

The slot shown in Figure 12–13 is to be machined as per the data table given. Write the required CNC program in absolute coordinates.

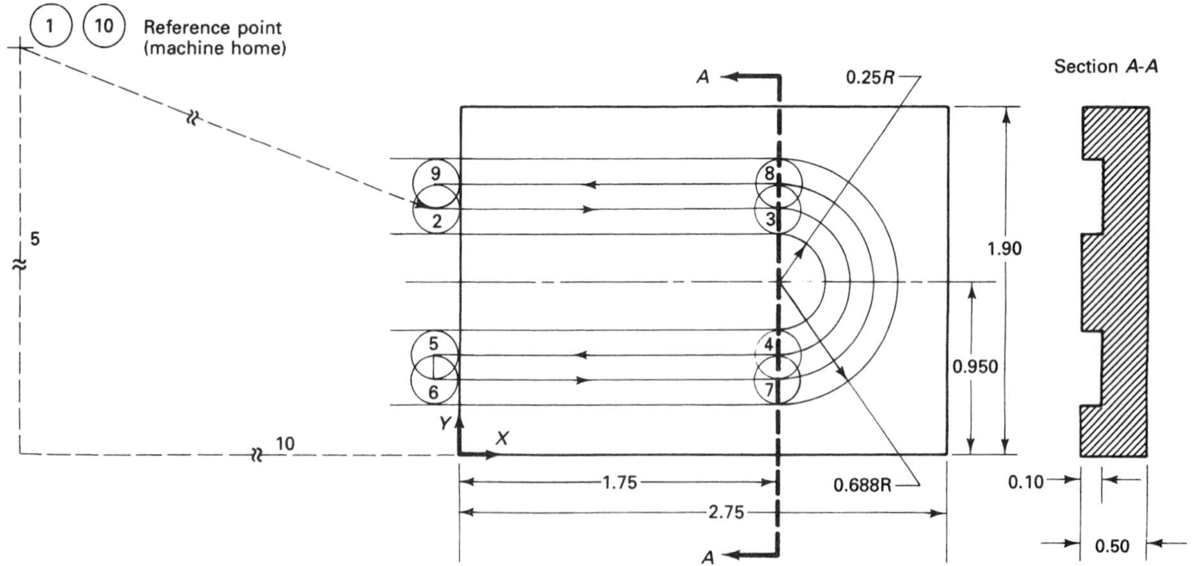

FIGURE 12–13

Tool	Operation	Tooling	Speed (rpm)	Feed (ipm)
1	Finish slot to depth 0.10 in.	0.25*D* stub end mill	2000	6

Notes: 1. Set X_0 Y_0 at the lower left-hand corner.
2. Z_0 is the top of the part.
3. Finish the slot by holding the part in a vise and banking it against a stop on the left.

Programming Pattern

CNC Program—Absolute Coordinates

Programming Pattern	*Word address command*	*Meaning*
	%	End of tape. Used to separate programs.
	O1202	Program number.
Job setup data listing	N0010 (X0Y0 IS THE LOWER LEFT HAND CORNER)	Comments.
	N0020 (Z0 IS THE TOP OF THE PART)	
	N0030 (TOOL 1: .25 DIA STUB END MILL)	
Machine start-up sequence Change to tool 1	N0040 G90 G20 G40 G80	Absolute, inch mode, cancel cutter diameter compensation and fixed cycles.
	N0050 G91 G28 X0 Y0 Z0	Return to reference point ①.
	N0060 G92 X − 10.0 Y5.0 Z0	Preset absolute zero point.
	N0070 T1 M06	Change to tool 1.
	N0080 (TOOL 1: FINISH SLOT TO DEPTH)	
	N0090 G00 G90 X − .225 Y1.325 Z0. S2000 M03	Rapid to ②. Spindle on.
	N0100 G43 Z.1 H01	Rapid tool 1 to 0.1 above part.
	N0110 M08	Coolant on.
Profiling operation 1 sequence	N0120 G01 Z − .1 F6.0	Linear profile mode. Plunge cutter to Z−.1 at feed rate 6.
	N0130 X1.75	Cut to ③.
	N0140 G02 X1.75 Y.575 I0. J − .375	Cut 0.25R arc to ④.
	N0150 G01 X − .125	Cut to ⑤.
	N0160 Y.387	Cut to ⑥.
	N0170 X1.75	Cut to ⑦.
	N0180 G03 X1.75 Y1.513 I0. J.563	Cut 0.688R arc to ⑧.
	N0190 G01 X − .125	Cut to ⑨.
Machine stop Program end sequence	N0200 G00 G90 Z1.0 M05	Rapid to 1.0 above part. Stop spindle.
	N0210 G91 G28 X0 Y0 Z0	Rapid to *XYZ* reference point ⑩.
	N0220 M30	Program end, memory reset.
	%	End of tape.

Machining simulation for this example is contained in the job file EX 12–7 stored on the disk at the back of this text. Refer to Chapter 20 for instructions.

■ EXAMPLE 12–8

Given the part shown in Figure 12–14, code a CNC program in absolute coordinates to execute the hole and profiling operations as per the data table.

Tool	Operation	Tooling	Speed (rpm)	Feed (ipm)
1	Drill 7⁄16*D* thru	7⁄16*D* drill	1200	8
2	Finish 1.375*D* × 0.25 deep Rough 0.625*D* to 0.620 ± .002*D*	0.5*D* end mill	1200	5
3	Finish bore 0.625 ± .0005*D* hole	0.6250*D* boring bar	600	2

Notes: 1. Set X_0Y_0 at the center of the part.
2. Z_0 is the top of the part.
3. Finish the part by holding the blank in a vise minimum 0.3 above the jaws. Bank against a stop on the left.
4. The blank is supplied with 2.000 square dimension completed. See Figures 12–15 and 12–16.

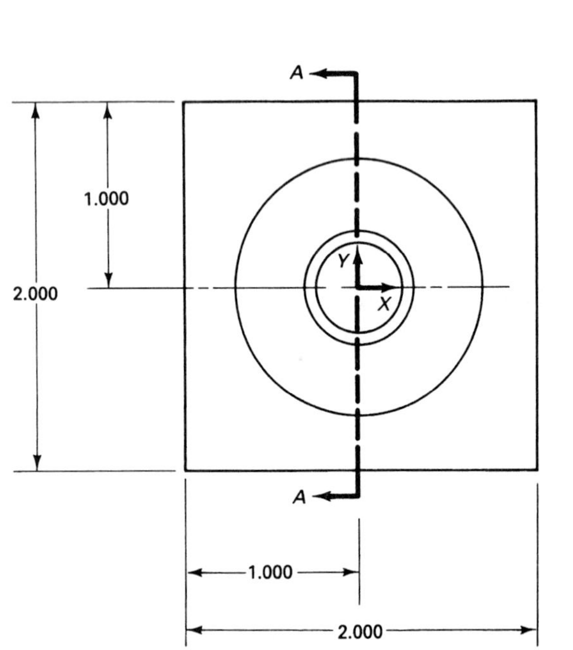

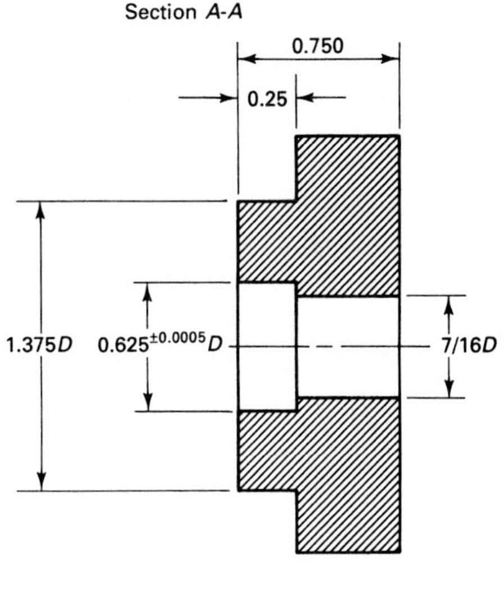

FIGURE 12–14

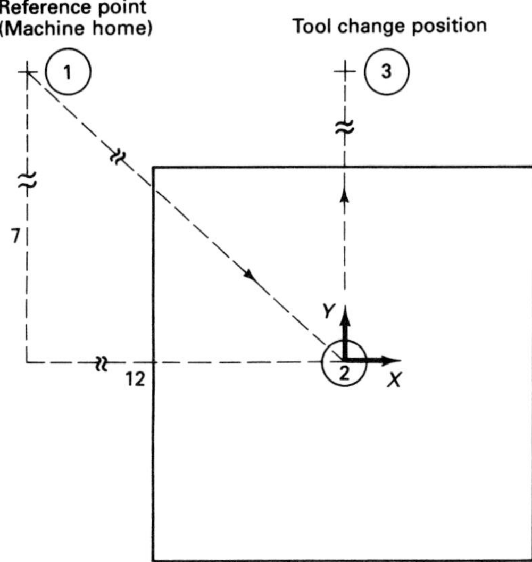

FIGURE 12–15

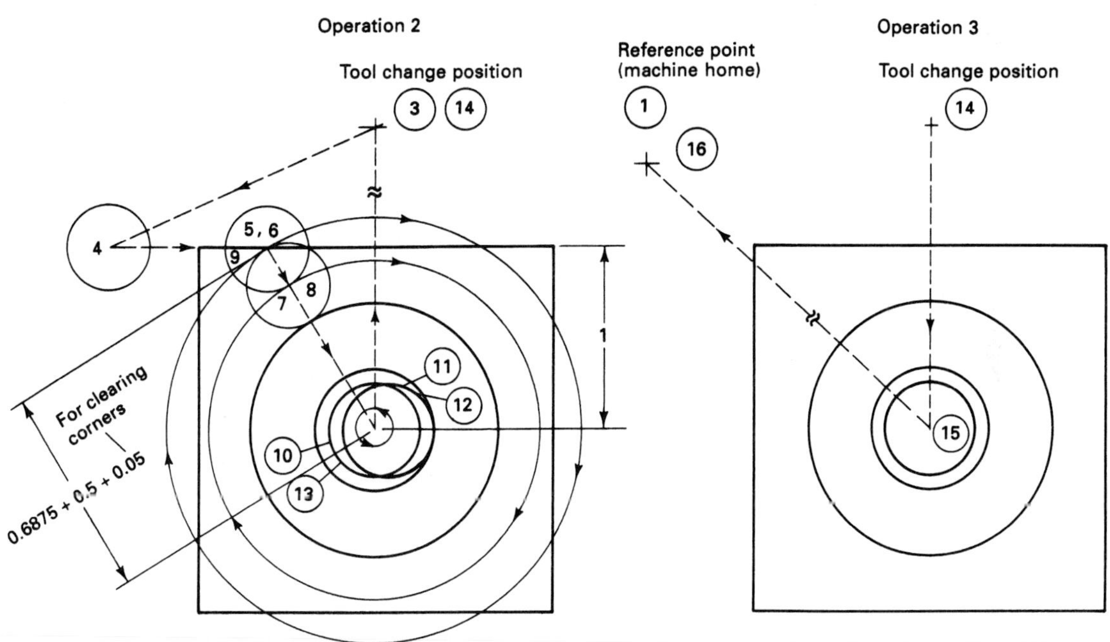

FIGURE 12–16

CNC Program—Absolute Coordinates

Programming Pattern	*Word address command*	*Meaning*
	%	End of tape.
	O1203	Program number.
Job setup data listing	N0010 (X0Y0 IS THE CENTER OF THE PART)	Comments.
	N0020 (Z0 IS THE TOP OF THE PART)	
	N0030 (TOOL 1: 7/16*D* DRILL)	
	N0040 (TOOL 2: .5 DIA END MILL)	
	N0050 (TOOL 3: 6250 DIA BORING BAR)	
	N0060 G90 G20 G40 G80	Absolute, inch mode, cancel fixed cycles.
Machine start-up sequence Change to tool 1	N0070 G91 G28 X0 Y0 Z0	Return to reference point ①.
	N0080 G92 X−12.0 Y7.0 Z0	Preset absolute zero point.
	N0090 TI M06	Change to tool 1.
	N0100 (TOOL 1: DRILL 7/16 DIA HOLE THRU)	
	N0110 G00 G90 X0. Y0. Z0. S1200 M03	Rapid to ②. Spindle on (CW) at 1200 rpm.
	N00120 G43 Z.1 H01	Rapid tool 1 to 0.1 above part.
	N0130 M08	Coolant on.
Hole operation 1 sequence	N0140 G83 X0. Y0. Z−.91 R.1 Q.15 F8.0	Start deep drill cycle at ②. Final depth is 0.91, 0.15 peck depth at feed 8. Return to 0.1 above part.
	N0150 G80	Cancel any fixed cycles.
Machine stop sequence	N0160 G00 G90 Z1.0 M05	Rapid to 1.0 above part. Spindle off.
	N0170 M09	Coolant off.
Change to tool 2 Machine start sequence	N0180 G91 G28 Z0 Y0	Rapid to tool change position ③.
	N0190 T2 M06	Change to tool 2.
	N0200 (TOOL 2: FINISH 1.375 DIA/ROUGH .625 DIA)	
	N0210 G00 G90 X−1.55 Y1.0 Z0 S1200 M03	Rapid to ④. Spindle on (CW) at 1200 rpm.
	N0220 G43 Z1.0 H02	Rapid tool 2 to 1.0 above part.
	N0230 M08	Coolant on.

CNC PROGRAM ***Continued***

Profile operation 2 sequence	N0240 G01 Z−.25 F5.0	Linear profile mode. Plunge to Z−.25 at feed rate 5.
	N0250 X−.729	Cut to position ⑤.
	N0260 G02 X−.729 Y1.0 I.729 J−1.0	Cut 0.9875*R* arc to ⑥.
	N0270 G01 X−.5523 Y.7576	Cut to position ⑦.
	N0280 G02 X−.5523 Y.7576 I.5523 J−.7576	Cut 0.6875*R* arc to ⑧.
	N0290 G01 X−.5125 Y.85	Cut to position ⑨.
	N0300 G00 Z.1	Rapid to 0.1 above part.
	N0310 X0 Y0	Rapid to position ⑩.
	N0320 G01 Z−.25	Plunge to Z−.25.
	N0330 X.06	Cut to position ⑪.
	N0340 G03 X.06 Y0. I−.06 J0	Cut arc to position ⑫.
	N0350 G01 X0	Cut to position ⑬.
Machine stop sequence	N0360 G00 G90 Z1.0 M05	Rapid to 1.0 above part. Spindle off.
	N0370 M09	Coolant off.
Change to tool 3 Machine start sequence	N0380 G91 G28 Z0 Y0	Rapid to tool change position ⑭.
	N0390 T3 M06	Change to tool 3.
	N0400 (TOOL 3: BORE .625 DIA HOLE)	
	N0410 G00 G90 X0 Y0 Z0 S600 M03	Rapid to ⑮. Spindle on (CW) at 600 rpm.
	N0420 G43 Z.1 H03	Rapid tool 3 to 0.1 above part.
	N0430 M08	Coolant on.
Bore operation 3 sequence	N0440 G85 X0 Y0 Z−.25 R.1 F2.0	Start bore cycle at ⑮. Bore to Z−.25 and return to 0.1 above part at feed rate 2.
	N0450 G80	Cancel any fixed cycles.
Machine stop Program end sequence	N0460 G00 G90 Z1.0 M05	Rapid to 1.0 above part. Stop spindle.
	N0470 M09	Coolant off.
	N0480 G91 G28 X0 Y0 Z0	Return to *XYZ* reference point ⑯.
	N0490 M30	Program end, memory reset.
	%	End of tape.

Machining simulation for this example is contained in the job file EX 12–8 stored on the disk at the back of this text. Refer to Chapter 20 for instructions.

■ ■

■ EXAMPLE 12–9

Code a CNC program in absolute metric coordinates for roughing and finishing the contour shown in Figure 12–17.

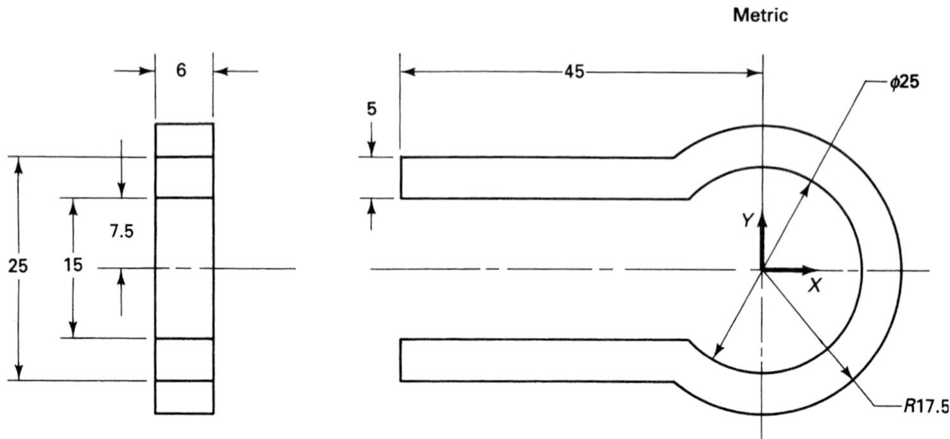

FIGURE 12–17

Tool	Operation	Tooling	Speed (rpm)	Feed (ipm)
1	Rough inside contour 0.2 full	0 6-mm roughing end mill	2100	120
2	Finish inside contour	0 6-mm finishing end mill	2600	100
1	Rough outside contour 0.2 full	0 6-mm roughing end mill	2100	120
2	Finish outside contour	0 6-mm finishing end mill	2600	100

Notes: 1. The blank for the part is 65 × 38 × 6.
2. X_0 is 18.5 mm from the right-hand side.
Y_0 is 18.5 mm from the far side.
Z_0 is the top of the part.
3. The part is roughed and finished from inside by holding the blank against three pins by two clamps. For outside machining, stop the spindle. Place clamps from inside and remove outside clamps and pins. Finish as required. See Figure 12–18.

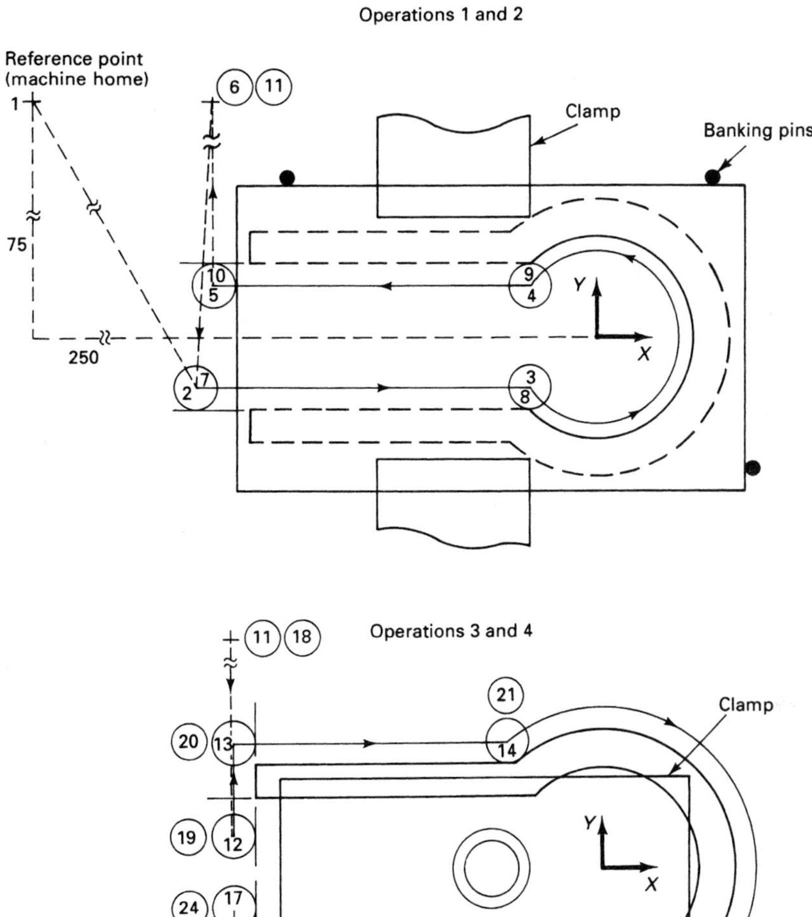

FIGURE 12-18

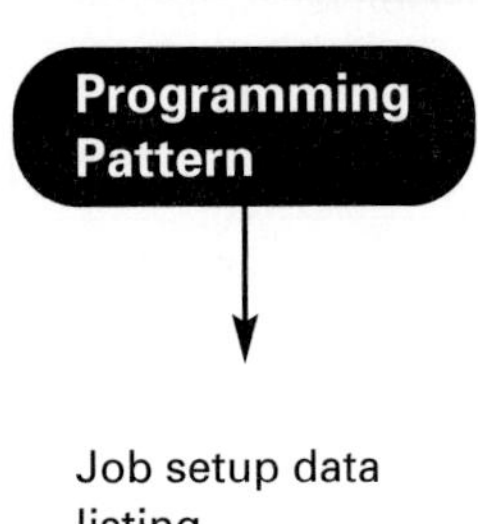

Job setup data listing

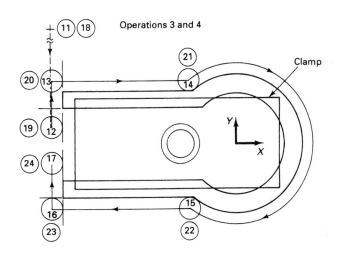

CNC Program—Absolute Coordinates

Word address command	*Meaning*
%	End of tape. Used to separate programs.
O1204	Program number.
N0010 (X0Y0 IS 18.5 MM FROM UPPER RIGHT CORNER)	Comments.
N0020 (Z0 IS THE TOP OF THE PART)	
N0030 (TOOL 1: 6MM ROUGHING END MILL)	
N0040 (TOOL 2: 6MM FINISHING END MILL)	

Sequence	Program	Description
Machine start-up sequence Change to tool 1	N0050 G90 G21 G40 G80	Absolute, metric mode, cancel cutter diameter compensation and fixed cycles.
	N0060 G91 G28 X0 Y0 Z0	Return to reference point ①.
	N0070 G92 X−250.0 Y75.0 Z0	Preset absolute zero point.
	N0080 TI M06	Change to tool 1.
	N0090 (TOOL 1: ROUGH INSIDE CONTOUR 6.5 MM DEEP)	
	N0100 G00 G90 X−50.2 Y−4.3 Z0 S2100 M03 T2	Rapid to ②. Spindle on (CW) at 2100 rpm. Prepare tool 2 in ready position.
	N0110 G43 Z2.0 H01	Rapid tool 1 to 2 mm above part.
	N0120 M08	Coolant on.
Profile operation 1 sequence	N0130 G01 Z−6.5 F120.0	Linear profile mode. Plunge to Z−6.5 mm at feed rate 120 mm/min.
	N0140 X−8.2462	Cut to position ③.
	N0150 G03 X−8.2462 Y4.3 I8.2462 J4.3	Rough 12.5*R* arc to ④.
	N0160 G01 X−48.2	Cut to position ⑤.
Machine stop sequence	N0170 G00 G90 Z1.0 M05	Rapid to 1 mm above part. Spindle off.
	N0180 M09	Coolant off.
Change to tool 2 Machine start sequence	N0190 G91 G28 Z0 Y0	Rapid to tool change position ⑥.
	N0200 T2 M06	Change to tool 2.
	N0210 (TOOL 2: FINISH INSIDE CONTOUR 6 MM DEEP)	
	N0220 G00 G90 X−50.0 Y−4.5 Z0 S2600 M03	Rapid to ⑦. Spindle on (CW) at 2600 rpm.
	N0230 G43 Z2.0 H02	Rapid tool 2 to 2 mm above part.
	N0240 M08	Coolant on.
Profile operation 2 sequence	N0250 G01 Z−6.5 F100.0	Linear profile mode. Plunge to Z−6.5 mm at feed rate 100 mm/min.
	N0260 X−8.3666	Cut to position ⑧.
	N0270 G03 X−8.3666 Y4.5 I8.3666 J4.5	Finish 17.5*R* arc to ⑨.
	N0280 G01 X−48.0	Cut to position ⑩.
Machine stop sequence	N0290 G00 G90 Z1.0 M05	Rapid to 1 mm above part. Stop spindle.
	N0300 M09	Coolant off.

CNC PROGRAM ***Continued***

	Word address command	*Meaning*
Change to tool 1 Machine start sequence	N0310 G91 G28 Z0 Y0	Rapid to tool change position (11).
	N0320 T1 M06	Change to tool 1.
	N0330 M00 (RECLAMP PART FROM INSIDE)	Program stop.
	N0340 (ROUGH OUTSIDE CONTOUR 6.5MM DEEP)	
	N0350 G00 G90 X−48.2 Y2.3 Z0 S2100 M03	Rapid to (12). Spindle on (CW) at 2100 rpm.
	N0360 G43 Z2.0 H01	Rapid tool 1 to 2 mm above part.
	N0370 M08	Coolant on.
Profiling operation 3 sequence	N0380 G01 Z−6.5 F120.0	Linear profile mode. Plunge to Z−6.5 mm at feed rate 120 mm/min.
	N0390 Y15.7	Cut to position (13).
	N0400 X−13.4907	Cut to position (14).
	N0410 G02 X−13.4907 Y−15.7 I13.4907 J−15.7	Rough 17.5*R* arc to (15).
	N0420 G01 X−48.2	Cut to position (16).
	N0430 Y−4.3	Cut to position (17).
Machine stop sequence	N0440 G00 G90 Z1.0 M05	Rapid to 1 mm above part. Spindle off.
	N0450 M09	Coolant off.
Change to tool 2 Machine start sequence	N0460 G91 G28 Z0 Y0	Rapid to tool change position (18).
	N0470 T2 M06	Change to tool 2.
	N0480 (TOOL 2: FINISH OUTSIDE CONTOUR 6.5 MM DEEP)	
	N0490 G00 G90 X−48.0 Y2.5 Z0 S2600 M03	Rapid to (19). Spindle on (CW) at 2600 rpm.
	N0500 G43 Z.1 H02	Rapid tool 2 to 0.1 mm above part.
	N0510 M08	Coolant on.
Profiling operation 4 sequence	N0520 G01 Z−6.5 F100.0	Linear profile mode. Plunge to Z−6.5 mm at feed rate 100 mm/min.
	N0530 Y15.5	Cut to position (20).
	N0540 X−13.4164	Cut to position (21).
	N0550 G02 X−13.4164 Y−15.5 I13.4164 J−15.5	Finish 17.5*R* arc to (22).
	N0560 G01 X−48.0	Cut to position (23).
	N0570 Y−4.5	Cut to position (24).

Machine stop Program end sequence	N0580 G00 G90 Z1.0 M05	Rapid to 1 mm above part. Stop spindle.
	N0590 M09	Coolant off.
	N0600 G91 G28 X0 Y0 Z0	Return to *XYZ* reference point.
	N0610 M30	Program end, memory reset.
	%	End of tape.

Machining simulation for this example is contained in the job file EX 12–9 stored on the disk at the back of this text. Refer to Chapter 20 for instructions.

12-8 CHAPTER SUMMARY

The following key concepts were discussed in this chapter:

1. Circular profiling involves cutting a circular contour in either a clockwise (G02) or counterclockwise (G03) direction.
2. The following operations should be executed prior to coding a circular interpolation command:
   - **a.** The plane selection (G17, G18, G19) is coded if needed.
   - **b.** The tool is positioned at the arc starting point.
   - **c.** The controller is in the linear interpolation (G01) mode.
   - **d.** The feed rate is specified.
   - **e.** The absolute (G90) or incremental (G91) mode is specified.
3. Right-triangle trigonometry is used to compute cutter offsets for many profiles composed of lines and circular arcs.
4. To maintain a constant feed rate at the outside periphery of the tool when cutting arcs, the feed rate specification for the tool center must be increased.

REVIEW EXERCISES

For each of the exercises in this chapter, write a CNC program containing the following general format:

- **a.** Program number/headings
- **b.** Job setup data listing
- **c.** Tool 1 in spindle, machine start-up
- **d.** Operation 1 (drilling or profiling)
- **e.** Machine stop
- **f.** Change to tool 2, machine start
- **g.** Operation 2 (drilling or profiling)
- **h.** Machine stop, program end sequence

Use the machining editor and simulator contained on the CD at the back of this text to verify the word address programs given in these exercises. Refer to Chapter 20 for instructions.

12.1. Program number: O1 205 (Figure 12–19)

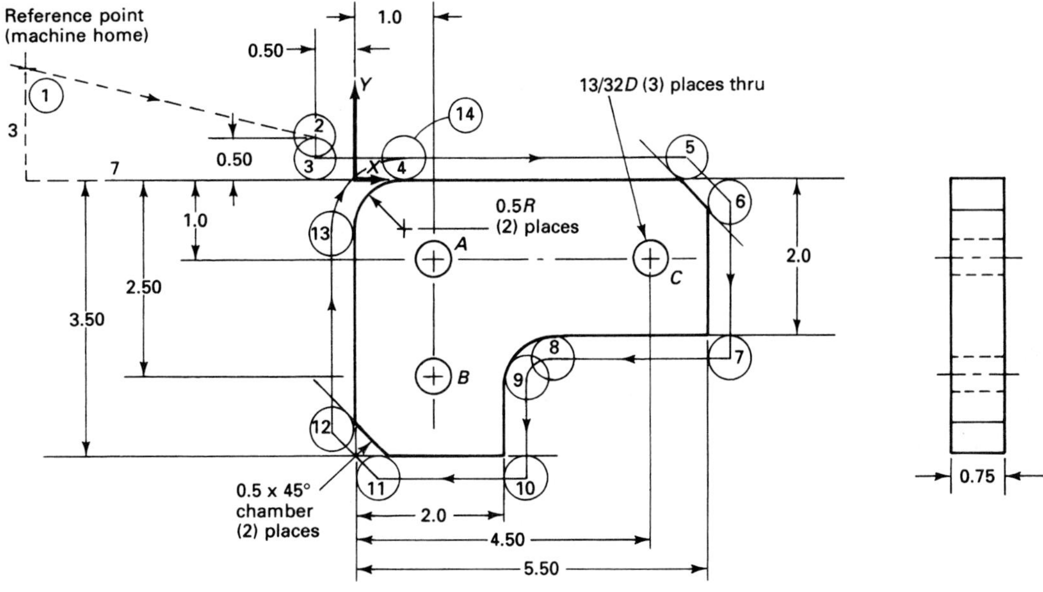

FIGURE 12–19

Tool	Operation	Tooling	Speed (rpm)	Feed (ipm)
1	Peck drill (3) 13/32D holes	13/32D drill	900	6
2	Profile contour as required	0.5D end mill	1000	4

Notes:
1. Set X_0Y_0 at the upper left-hand corner.
2. Z_0 is the top of the part.
3. The part is drilled by holding the blank against three pins by two clamps. See Figure 12–20.
4. Stop the machine after using tool 1. Hold down the part by three 3/8 screws thru holes that have been drilled. Remove three pins and two clamps. Finish contour.

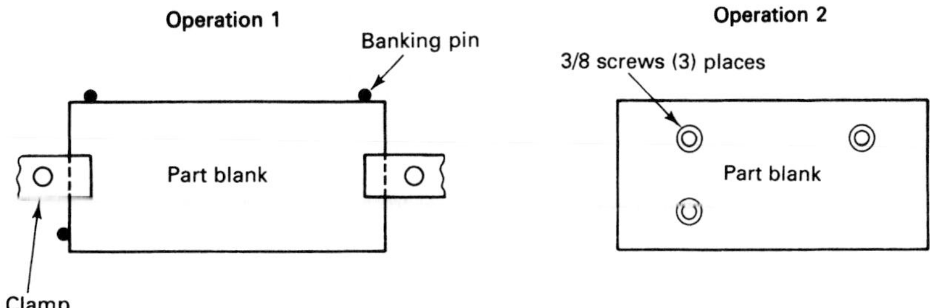

FIGURE 12–20

Use the following coordinate table as an aid in determining the cutter locations.

Position	ΔX	ΔY	*X*	*Y*	*I*	*J*
A						
B						
C						
1						
2						
3						
4						
5						
6						
7						
8						
9						
10						
11						
12						
13						
14						

12.2. Program number: O1206

The CNC program for the part shown in Figure 12–21 has been coded. Using the coordinate chart format given in exercise 12–1, provide the calculations needed for determining the cutter locations.

Tool	Operation	Tooling	Speed (rpm)	Feed (ipm)
1	Drill (5) 9⁄32*D* holes thru	9⁄32*D* drill	1800	6
2	Mill profile as required	0.5*D* end mill	1200	5

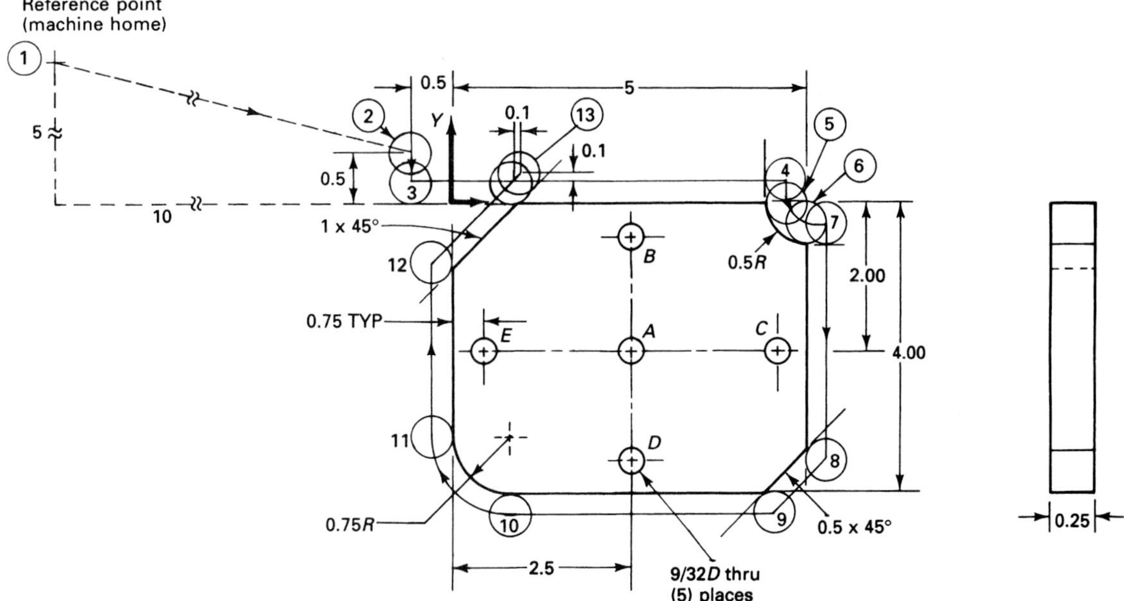

FIGURE 12–21

```
%
O1206
N0010 (X0Y0 IS THE UPPER LEFT HAND CORNER)
N0020 (Z0 IS THE TOP OF THE PART)
N0030 (TOOL 1: 9/32 DIA DRILL)
N0040 (TOOL 2: .5 DIA END MILL)
N0050 G90 G20 G40 G80
N0060 G91 G28 X0 Y0 Z0
N0070 G92 X − 10.0 Y5.0 Z0
N0080 T1 M06
N0090 (TOOL 1: DRILL 5 9/32 DIA HOLES THRU)
N0100 G00 G90 X2.5 Y − 2. Z0. S1800 M03
N0110 G43 Z.1 H01
N0120 M08
N0130 G81 X2.5 Y − 2.0 Z − .48 R.1 F6.0
N0140 Y − .75
N0150 X4.25 Y − 2.0
N0160 X2.5 Y − 3.25
N0170 X.75 Y − 2.0
N0180 G80
N0190 G00 G90 Z1.0 M05
N0200 M09
N0210 G91 G28 Z0 Y0
N0220 T2 M06
N0230 (TOOL 2: MILL PROFILE AS REQUIRED)
N0240 G00 G90 X − .5 Y.5 Z0. S1200 M03
N0250 G43 Z.1 H02
N0260 M08
N0270 G01 Z − .27 F5.0
N0280 Y.25
N0290 X4.75
N0300 Y0
N0310 G03 X5.0 Y − .25 I.25 J0
```

```
N0320 G01 X5.25
N0330 Y – 3.6036
N0340 X4.6036 Y – 4.25
N0350 X.75
N0360 G02 X – .25 Y – 3.25 I0. J1.0
N0370 G01 Y – .8964
N0380 X.9964 Y.35
N0390 G00 G90 Z1. M05
N0400 M09
N0410 G91 G28 X0 Y0 Z0
N0420 M30
%
```

12.3. Program number: O1207 (Figure 12–22)

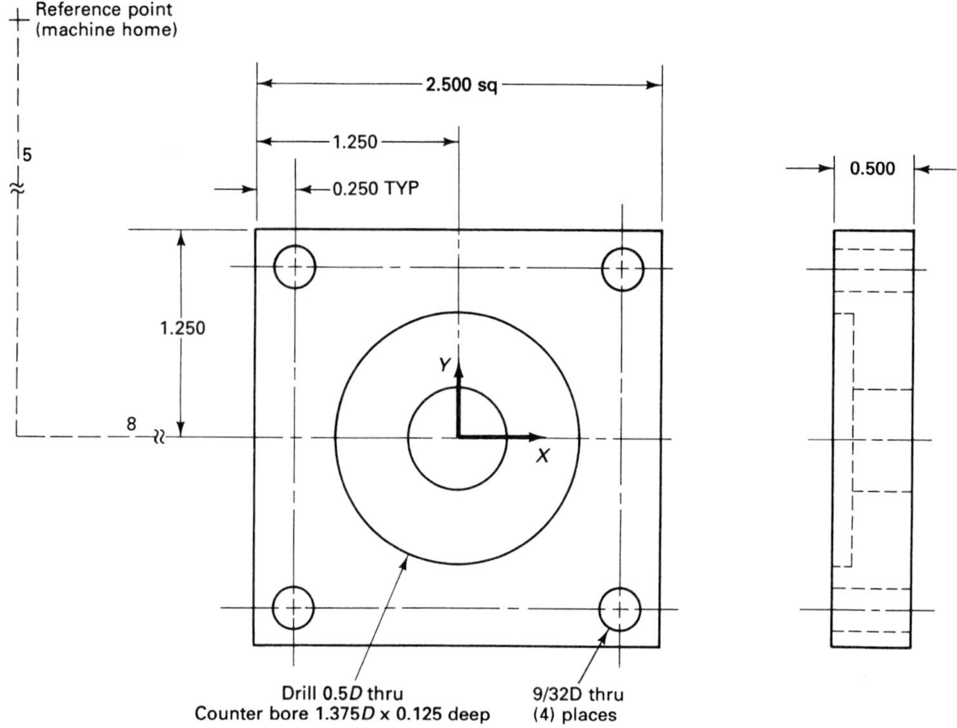

FIGURE 12–22

Tool	Operation	Tooling	Speed (rpm)	Feed (ipm)
1	Drill 0.50*D* hole thru	0.5*D* drill	1200	7
2	Drill (4) 9⁄32*D* holes thru	9⁄32*D* drill	1800	6
3	Finish 1.375 counterbore × 0.125 deep	0.5*D* end mill	1000	7

Notes: 1. Set X_0Y_0 at the center of the part.
2. Z_0 is the top of the part.
3. Finish all hole operations by holding the part in a vise and banking against a stop on the left.
4. The blank is supplied with 2.5 square dimension completed.

12.4. Program number: O1208 (Figure 12–23)

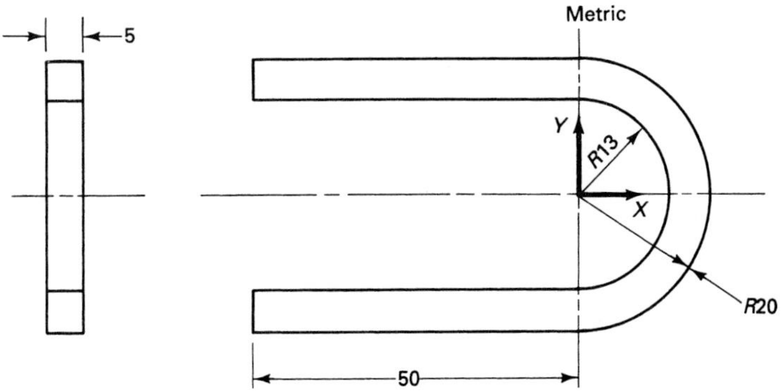

FIGURE 12–23

Tool	Operation	Tooling	Speed (rpm)	Feed (ipm)
1	Rough contour inside and outside 0.4 full	0 10-mm roughing end mill	1800	120
2	Finish contour inside and outside as required	0 8-mm end mill	2200	100

Notes:
1. The blank for the part is 72 mm × 42 mm × 5 mm.
2. X_0 is 21 mm from the right-hand side. Y_0 is 21 mm from the far side. Z_0 is the top of the part.
3. The part is roughed and finished from inside by holding the blank against three pins by two clamps.
4. For outside operations, stop the machine. Place clamps from inside and remove outside clamps and pins. Finish as required. See Figure 12–24.

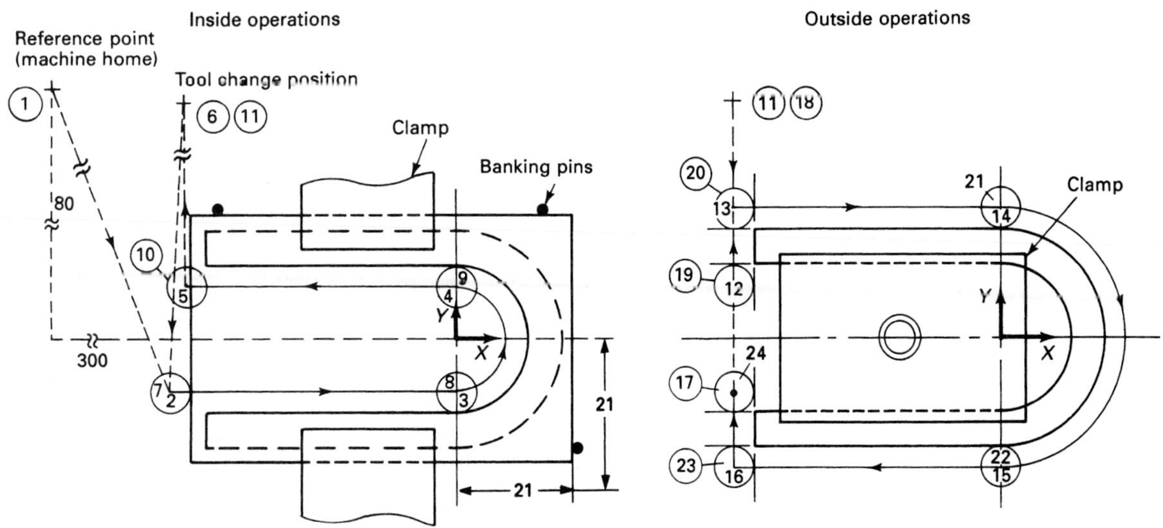

FIGURE 12–24

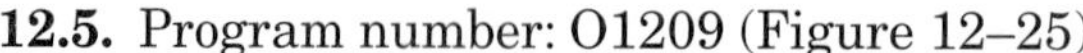

12.5. Program number: O1209 (Figure 12–25)

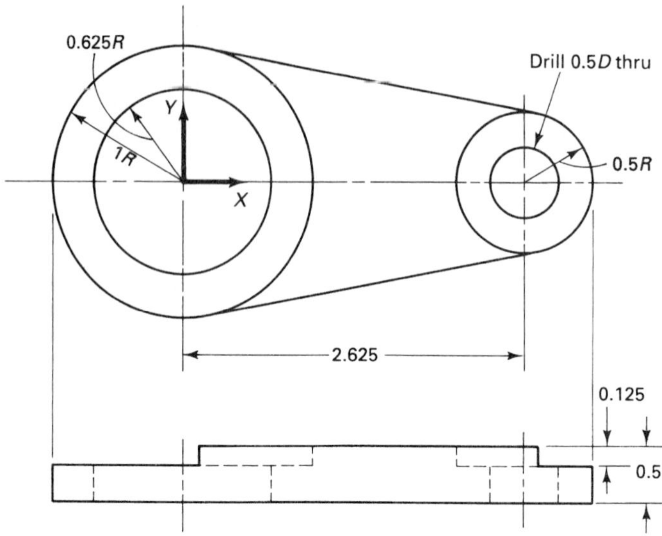

FIGURE 12–25

Tool	Operation	Tooling	Speed (rpm)	Feed (ipm)
1	Drill 0.5*D* hole thru (2) places	0.5*D* drill	1100	7
2	Profile 0.625*R* counterbore	0.5*D* end mill	1200	8
2	1*R* and .5*R* counterbores	0.5*D* end mill	1200	8
2	Reclamp from inside Finish outside profile	0.5*D* end mill	1200	8

Notes:
1. The blank size is 4.25 in. × 2.125 in. × 0.5 in.
2. Hold blank against three pins by two clamps. See Figure 12–26.
3. X_0Y_0 is 1.06 from the upper left-hand corner.
4. Z_0 is the top of the part.
5. After finishing counterbores with tool 3, stop the machine. Reclamp from the inside. Finish contour as required. See Figures 12–27 through 12–30.

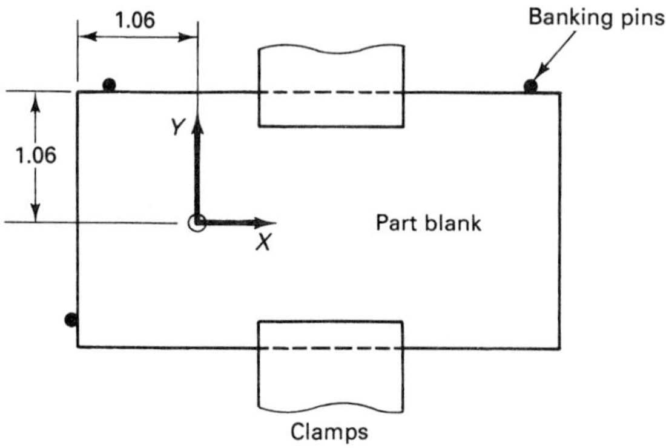

FIGURE 12–26

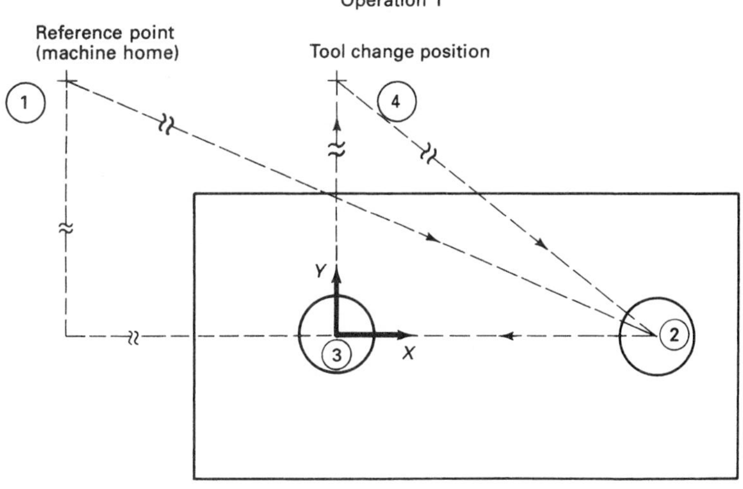

Drill 0.5*D* holes thru

FIGURE 12–27

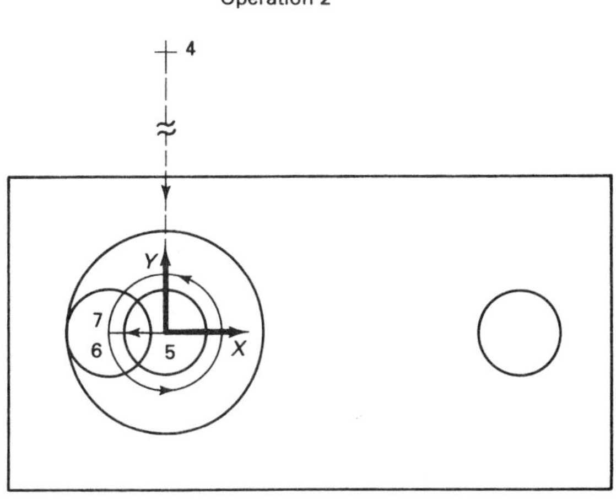

Profile 0.625*R* counterbore

FIGURE 12–28

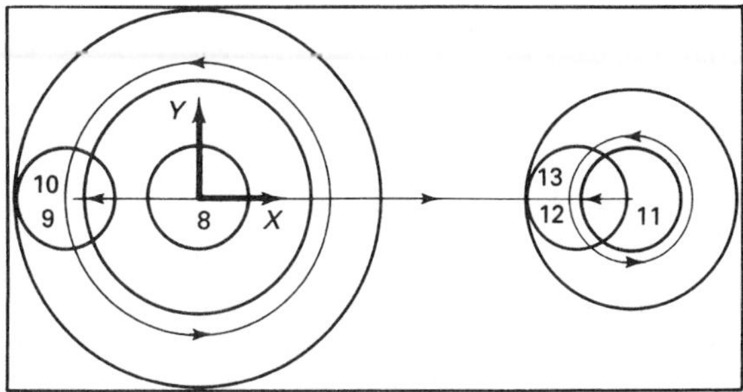

Reclamp from inside/profile 1*R* and 0.5*R* counterbores

FIGURE 12–29

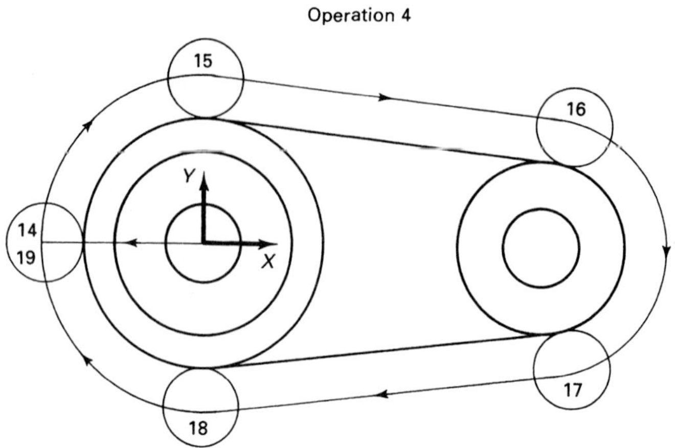

FIGURE 12-30

12.6. Program number: O1210 (Figure 12–31)

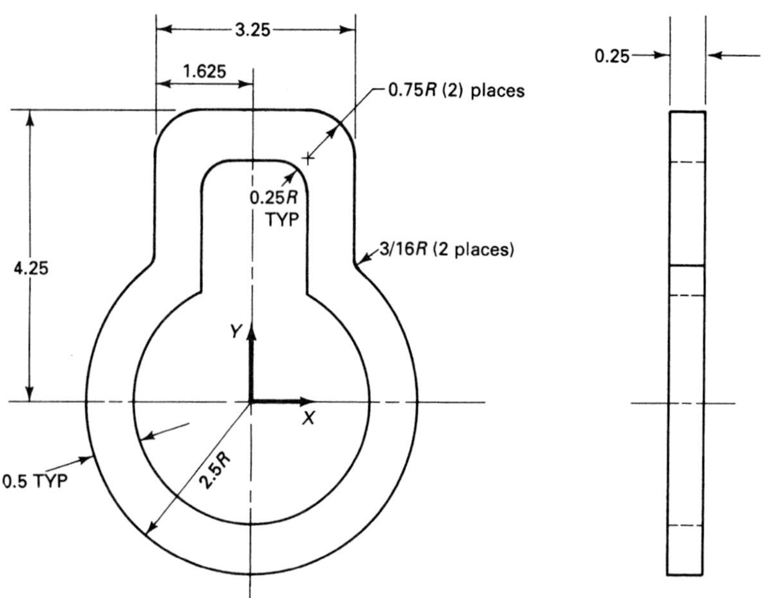

FIGURE 12-31

Tool	Operation	Tooling	Speed (rpm)	Feed (ipm)
1	Finish inside profile Reclamp from inside Finish outside contour	⅜*D* center cutting end mill	1250	6

Notes:
1. The blank has size 6.875 in. × 5.125 in. × 0.25 in.
2. Hold the blank against three pins by four clamps. See Figure 12–32.
3. X_0 is located at 2.56 from the left side.
   Y_0 is located at 2.56 from the near side.
   Z. is the top of the part.
4. After tool 1 finishes the inside contour, stop the machine. Reclamp from the inside and remove outside clamps and (3) pins. Finish part complete. See Figures 12–33 and 12–34.

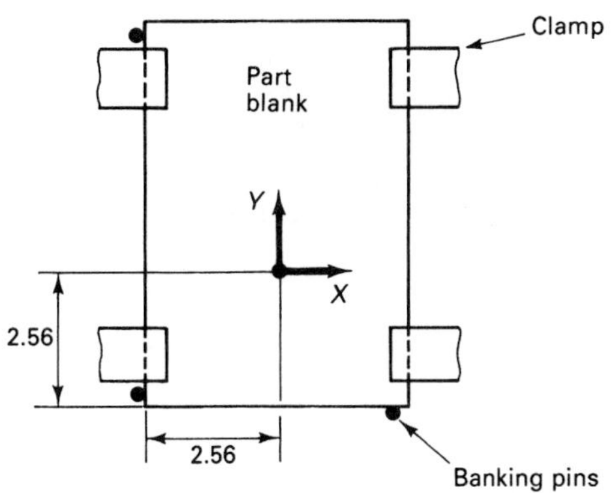

FIGURE 12–32

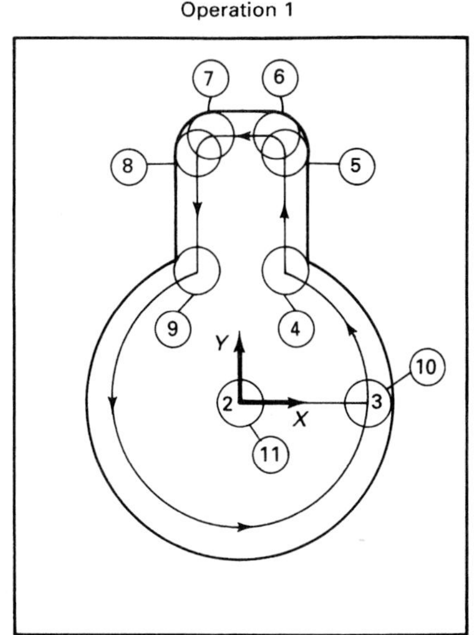

Finish inside profile

FIGURE 12–33

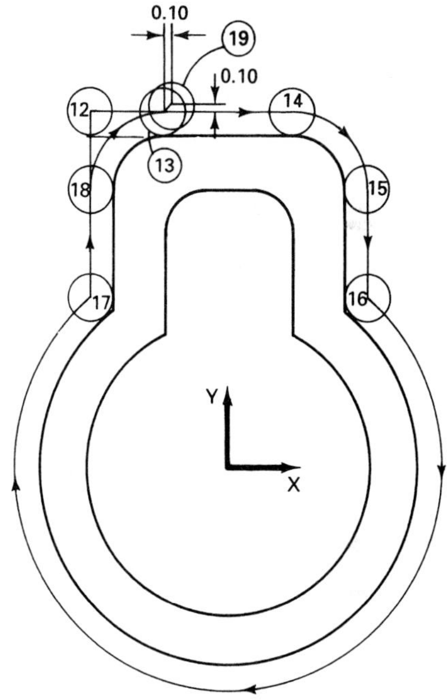

Finish outside contour after reclamp

FIGURE 12–34

CHAPTER 13

Programming with Cutter Diameter Compensation

13-1 CHAPTER OBJECTIVES

At the conclusion of this chapter you will be able to

1. Explain what cutter diameter compensation is.
2. Recognize the advantages of invoking cutter diameter compensation in programs.
3. State the restrictions involved in applying cutter diameter compensation.
4. Apply cutter diameter compensation in programming milling operations.

13-2 INTRODUCTION

The important technique of cutter diameter compensation is presented in this chapter. A description of tool motion with cutter diameter compensation is considered first. The main advantages of using compensation are listed and discussed. The reader is introduced to the various restrictions and conditions that apply in using compensation with most MCU controllers. The word address commands for initiating diameter compensation as well as canceling this mode are given and explained. Many examples are given to illustrate how this technique can be applied to manufacturing parts.

The student is encouraged to use the profile machining simulation software described in Chapter 20 and contained on the CD at the back of the text.

13-3 CUTTER DIAMETER COMPENSATION

Up to now the center of the cutter has been programmed to move around the part geometry. It has been found that more complex part geometries having inclined lines, lines tangent to arcs, and lines intersecting arcs involve substantial trigonometric computations to determine the center of the cutter. Cutter diameter compensation involves programming the part geometry directly instead of the tool center. In effect, the programmer considers the cutter radius as zero when programming the tool path. The actual cutter radius is entered separately in the MCU tool register prior to running the program. Upon executing the program, the machine will automatically compensate by offsetting the programmed tool path by the radius of the cutter read. In many cases this approach will allow for highly accurate side milling and true circle cutting. Furthermore, it will enable the programmer to use the same tool path for different cutter types and radii. Refer to Figure 13–1 and 13–2 for an illustration of this type of programming.

FIGURE 13-1 Machine control display of tool paths for a part program. (Photo courtesy of GE Fanuc Automation North America, Inc.)

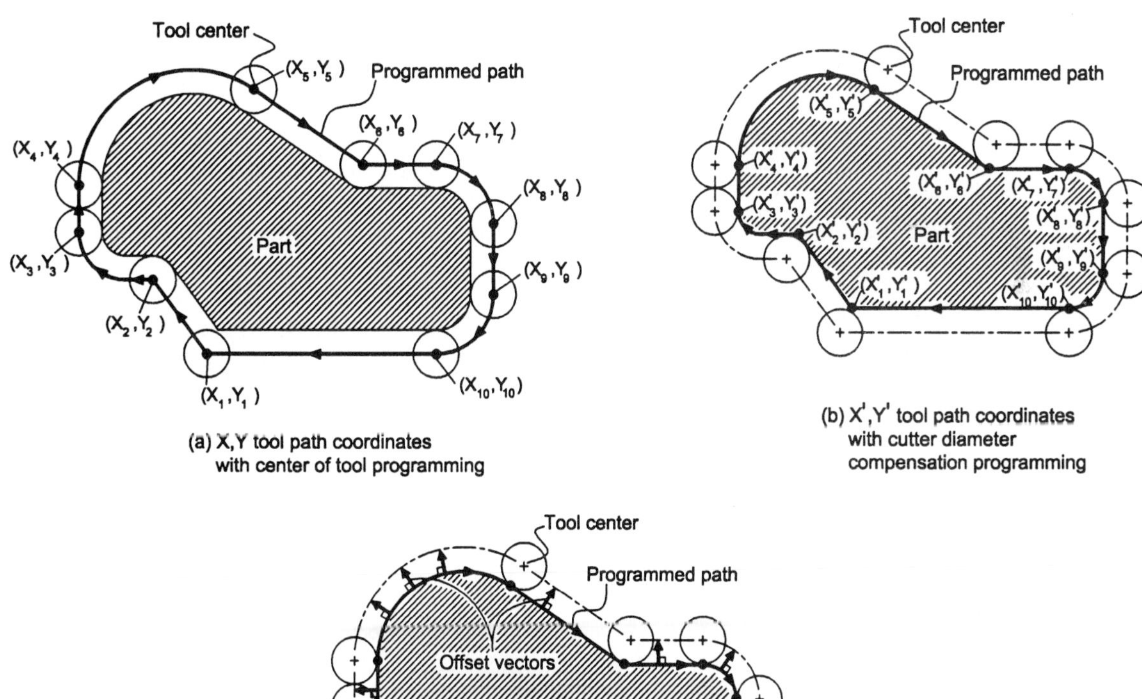

FIGURE 13-2 Tool center verses cutter diameter compensation programming.

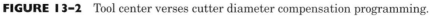

13-4 ADVANTAGES OF USING CUTTER DIAMETER COMPENSATION

Cutter diameter compensation offers the following main advantages:

1. The mathematical computations for determining a tool path are greatly simplified.
2. Because the part's geometry and not the tool center is programmed, the same program can be used for a variety of different cutter diameters.
3. The same program can be used for roughing as well as finishing cuts.
4. Inside as well as outside cuts can be programmed.

13-5 SOME RESTRICTIONS WITH CUTTER DIAMETER COMPENSATION

The restrictions listed in this section apply to cutter diameter compensation moves within the Fanuc family of controllers. Similar restrictions will also apply when other types of controllers are used.

The plane in which compensation is to occur is specified by the codes G17, G18, or G19. In the discussion to follow, it is assumed that a G17 has been entered; thus interpolation occurs in the *XY* plane. The conditions given will apply for most controllers.

1. A cutter diameter compensation G code entered in a separate block in a program must be followed by an *X, Y* linear tool motion code. The linear motion code must be the first or second command directly in sequence. The motion command will signal the controller to initiate (ramp on) or cancel (ramp off) cutter diameter compensation. This is required because most MCUs read program commands two blocks ahead of the current block to determine the next cutter move. Circular interpolation (G02 and G03) blocks may be programmed after the initial linear motion block.
2. The first *X,Y* linear tool movement following a cutter diameter compensation block must be equal to or greater than the radius of the cutter being used. See Figures 13–3 and 13–4.
3. The MCU will apply compensation by offsetting the cutter in a direction perpendicular to the next *X, Y* axis tool movement. The offset will be equal to the cutter radius previously entered at setup. See Figure 13–5.

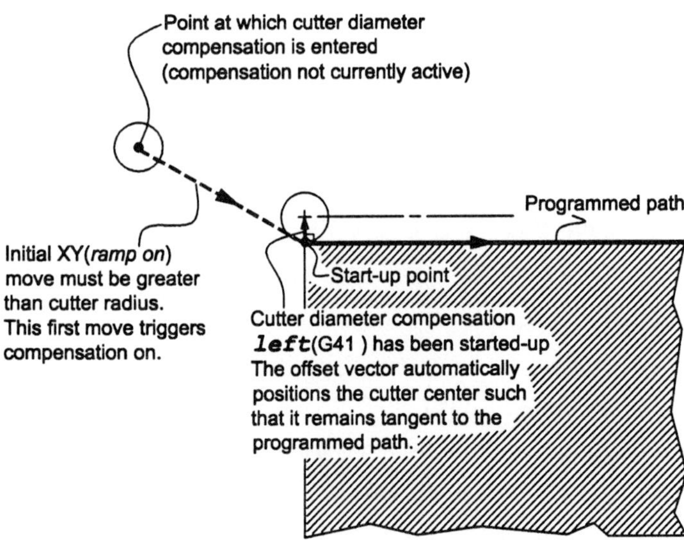
Point at which cutter diameter compensation is entered (compensation not currently active)
Programmed path
Initial XY(*ramp on*) move must be greater than cutter radius. This first move triggers compensation on.
Start-up point
Cutter diameter compensation left(G41) has been started-up The offset vector automatically positions the cutter center such that it remains tangent to the programmed path.

FIGURE 13–3 Ramp on move to initiate cutter diameter compensation left.

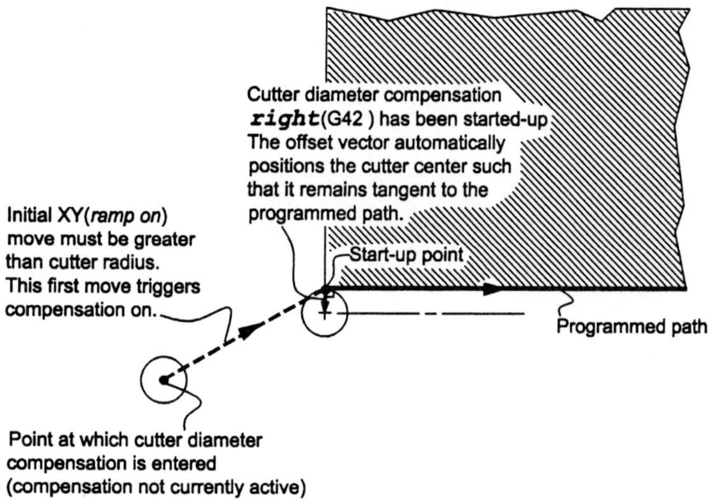
Cutter diameter compensation
right(G42) has been started-up
The offset vector automatically
positions the cutter center such
that it remains tangent to the
programmed path.
Initial XY(ramp on)
move must be greater
than cutter radius.
This first move triggers
compensation on.
Start-up point
Programmed path
Point at which cutter diameter
compensation is entered
(compensation not currently active)

FIGURE 13–4 Ramp on move to initiate cutter diameter compensation right.

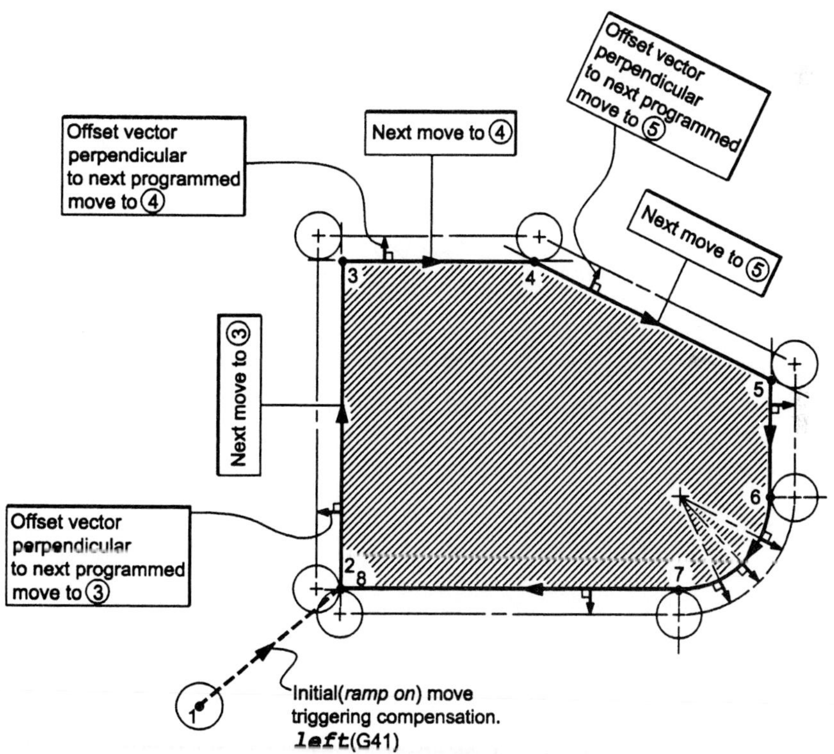
Offset vector
perpendicular
to next programmed
move to ⑤
Offset vector
perpendicular
to next programmed
move to ④
Next move to ④
Next move to ⑤
Next move to ③
Offset vector
perpendicular
to next programmed
move to ③
Initial(ramp on) move
triggering compensation.
left(G41)

FIGURE 13–5 Offset vectors generated by MCU when cutter diameter compensation is active.

4. The first move for an inside cut should be to a location away from an inside corner. This will prevent the cutter from notching the part. See Figure 13–6.
5. A step-down cut smaller than the radius of the cutter will result in overcutting or notching the part. See Figure 13–7.
6. When cutter diameter compensation is in effect, it is not possible to program two blocks in succession that contain tool motion of the *XY*

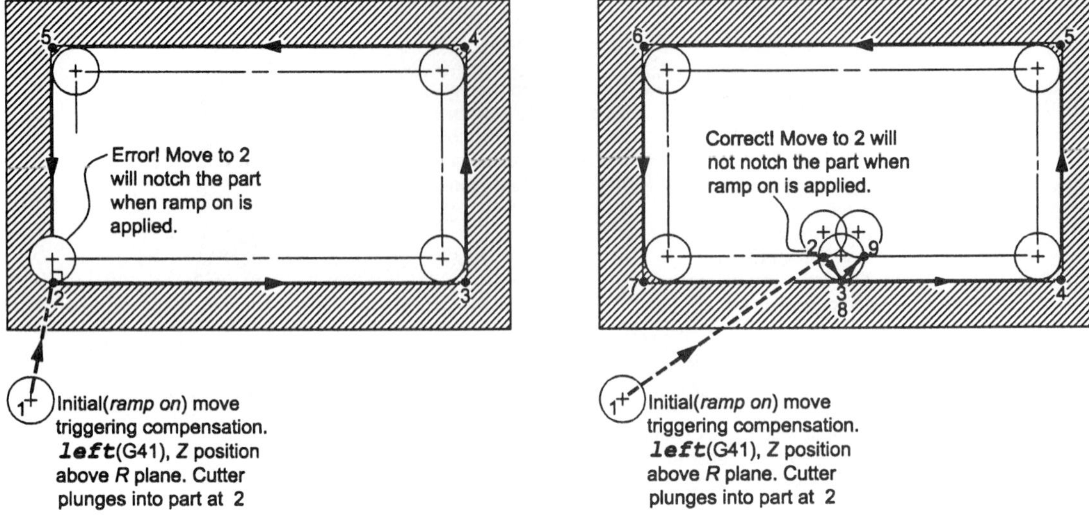

FIGURE 13-6 Ramp on errors.

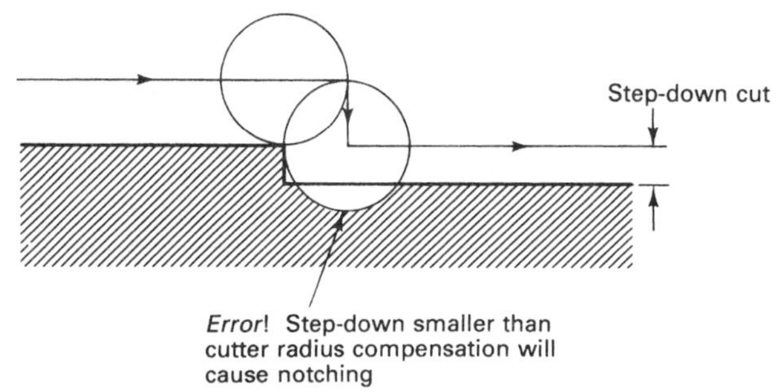

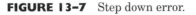

FIGURE 13-7 Step down error.

plane. For example, with a G17, *XY*-plane selection, it is not possible to utilize cutter diameter compensation for two successive blocks containing *Z*-axis moves. Additionally, some controls will not compensate rapid moves, or inside line intersections of 90° or less.

13-6 CUTTER DIAMETER COMPENSATION COMMANDS

Cutter diameter compensation is invoked by programming a G41 or G42 word. The system offsets the tool (ramp on) in the direction of the offset vector by the value previously inputted in the D register. The direction of the vector will be automatically adjusted such that the tool remains tangent to the part boundary as it moves along. The first two motion commands after a G41 or G42 code contain information for properly executing ramp on. The first move read must be a G00 or G01 linear motion code, and it creates the offset vector. The next move is used by the controller to indicate the direction of the vector.

General Syntax

Cutter Diameter Compensation Left

G41 Xn Yn Dn (XY plane)

G41 Directs the controller to offset the tool (ramp on) to the *left side of upward tool motion.* The offset will occur on the *next linear XY-axis move.*

Dn Specifies the address in memory where the cutter radius offset value is stored. n indicates the number of the register containing the offset value.

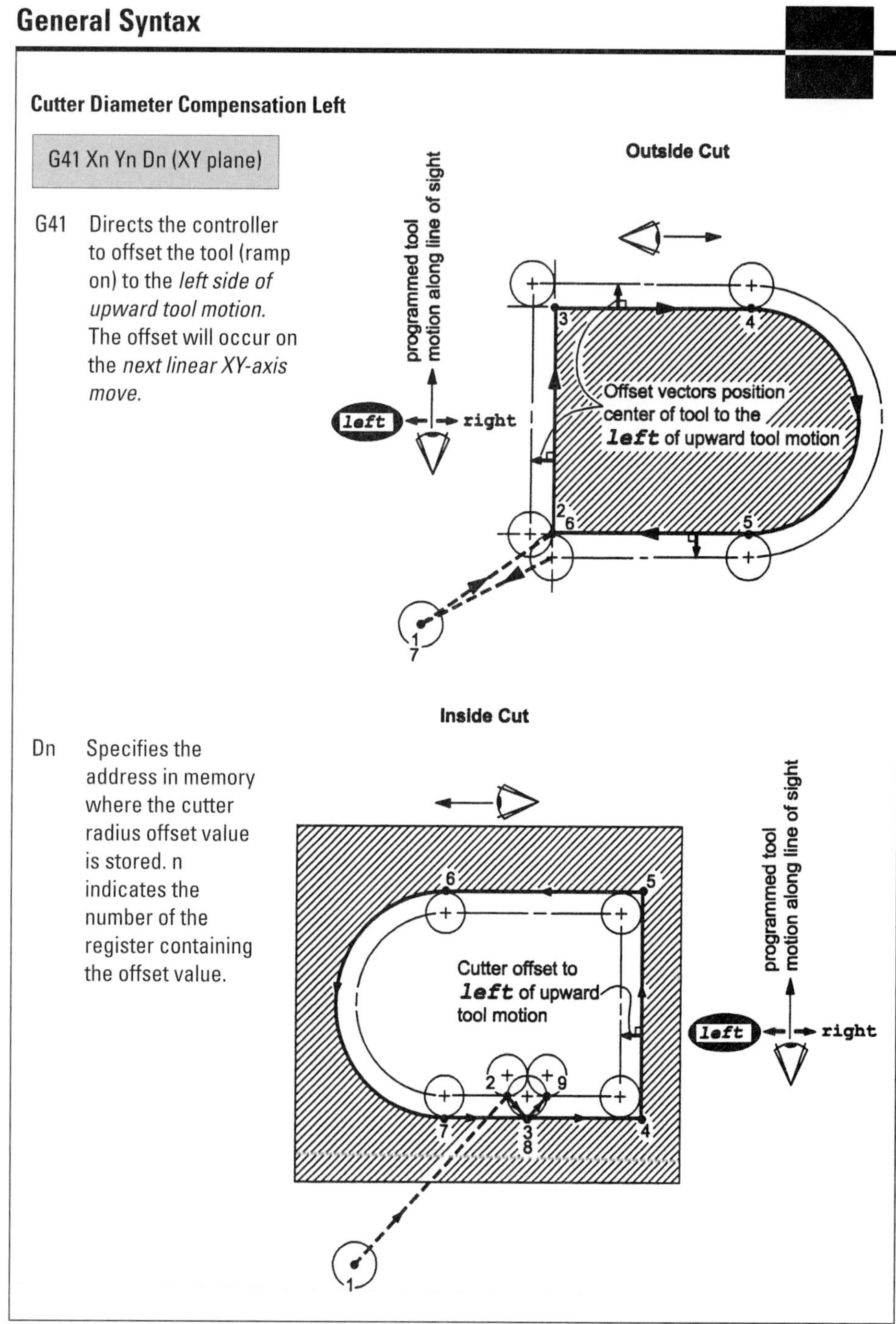

General Syntax

Cutter Diameter Compensation Right

G42 Xn Yn Dn (XY plane)

G42 Directs the controller to offset the tool (ramp on) to the *right side of upward tool motion.* The offset will occur on the *next linear XY-axis move.*

Dn Specifies the address in memory where the cutter radius offset value is stored. n indicates the number of the register containing the offset value.

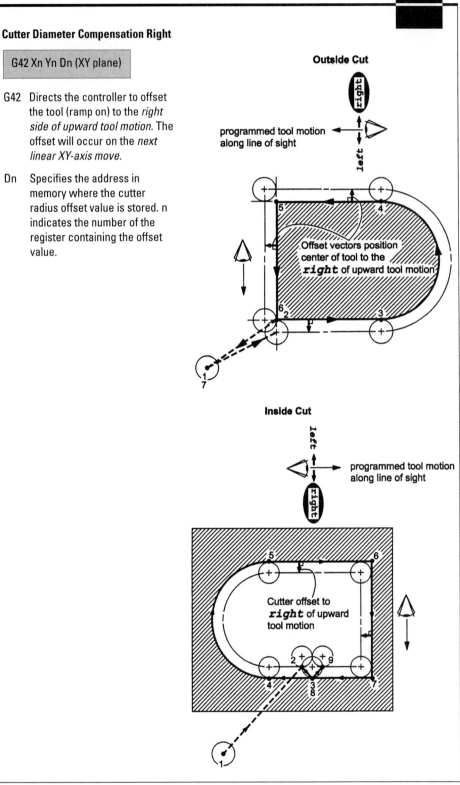

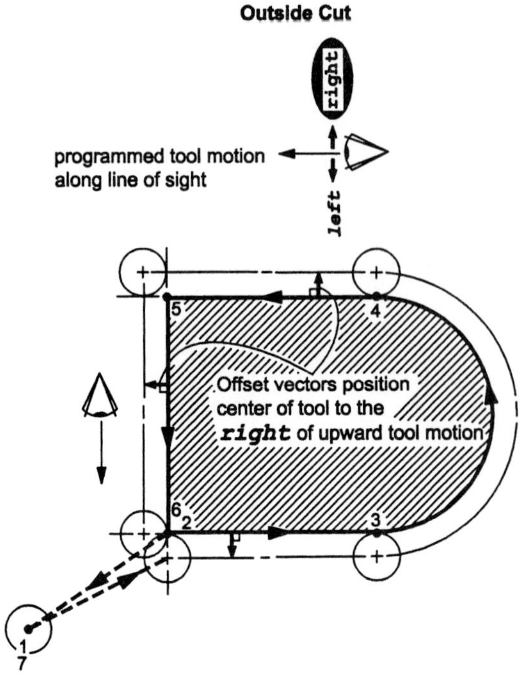

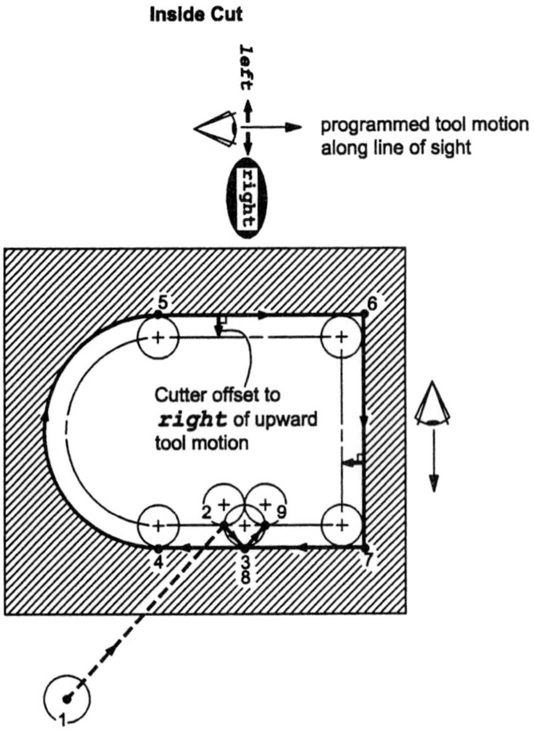

Outside Cut

right

programmed tool motion along line of sight

left

Offset vectors position center of tool to the **right** of upward tool motion

Inside Cut

left

programmed tool motion along line of sight

right

Cutter offset to **right** of upward tool motion

General Syntax

Cutter Diameter Compensation Cancel

G40 Xn Zn

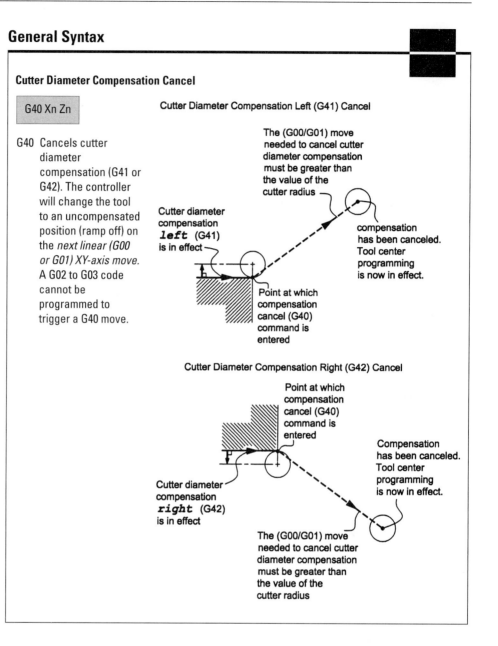

G40 Cancels cutter diameter compensation (G41 or G42). The controller will change the tool to an uncompensated position (ramp off) on the *next linear (G00 or G01) XY-axis move.* A G02 to G03 code cannot be programmed to trigger a G40 move.

Cutter Diameter Compensation Left (G41) Cancel

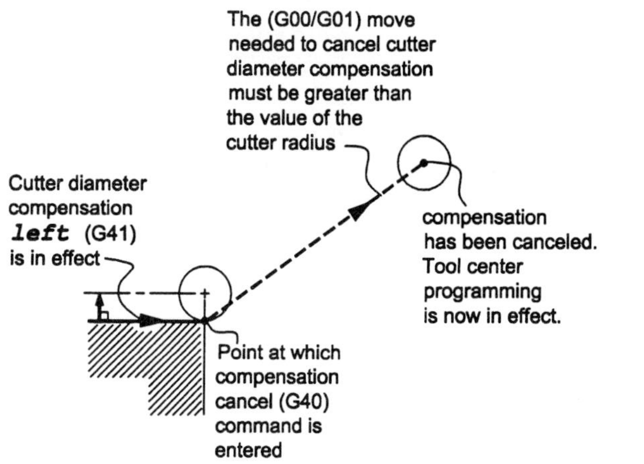

The (G00/G01) move needed to cancel cutter diameter compensation must be greater than the value of the cutter radius
Cutter diameter compensation **left** (G41) is in effect
compensation has been canceled. Tool center programming is now in effect.
Point at which compensation cancel (G40) command is entered

Cutter Diameter Compensation Right (G42) Cancel

Point at which compensation cancel (G40) command is entered
Compensation has been canceled. Tool center programming is now in effect.
Cutter diameter compensation **right** (G42) is in effect
The (G00/G01) move needed to cancel cutter diameter compensation must be greater than the value of the cutter radius

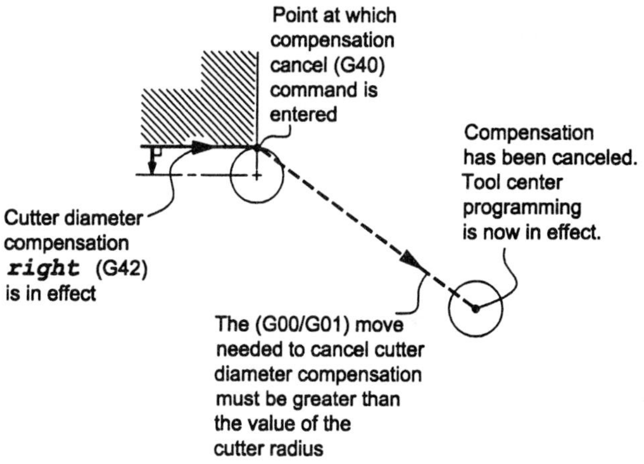

Note:

1. More than one D code can be programmed with any tool. This allows for the use of the same tool for executing rough as well as finish cuts. Furthermore this feature can be used to compensate for change in tool diameter due to tool wear.
2. G02/G03 motion cannot be used to initiate a G41, G42, or G40 word in a program.
3. G41/G42 are modal. This means they remain in effect for all subsequent tool motions until canceled by a G40 word. G40 is also modal and remains in effect until canceled by a G41/G42 word.
4. The initial state of the control at machine start-up is G40.

■ EXAMPLE 13–1

Utilize cutter diameter compensation in writing code to profile the rectangle shown in Figure 13–8.

Word address command	*Meaning*
G90 G00 X −11.0 Y6.0 S800	Rapid to position ②.
G01 Z −.5 M03 F10.0	Plunge to −0.5. Spindle on (CW).
G41 X −10.5 D21	Ramp on to left of upward tool motion on next move to ③. Offset tool by value in address D21.
G01 X10.0	Cut to ④ at feed rate 10.
Y −6.0	Cut to ⑤.
X −10.0	Cut to ⑥.
Y6.5	Cut to ⑦.
G00 G40 Y7.0	Ramp off on the next move to ⑧.
Z.1	Rapid to 0.1 above the part.

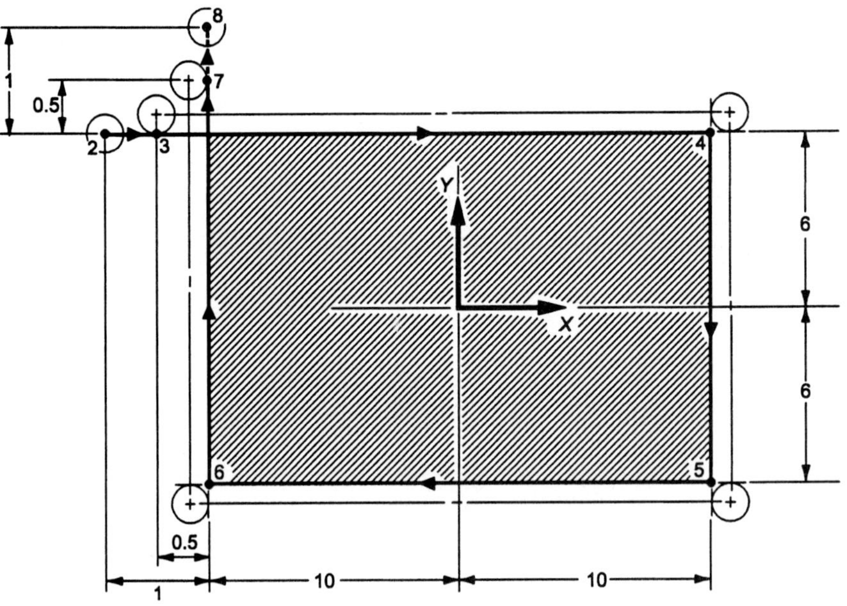

FIGURE 13–8

■■

■ EXAMPLE 13-2

Use cutter diameter compensation to machine the outside of the circle in Figure 13-9.

Word address command	*Meaning*
G90 G00 X−1.0 Y6.0 S400	Rapid to position ②.
G01 Z−.5 M03 F10.0	Plunge to −0.5. Spindle on (CW).
G41 X−.5 D20	Ramp on to the left of upward tool motion on the next move to X−.5. ③. Offset the tool by the value in the address D20.
G01 X0	Cut to ④.
G02 X0 Y0 J−6.0	Cut arc to ⑤.
G01 X.5	Cut to ⑥.
G00 G40 X1.0	Ramp off on the next move to X1.0 ⑦.

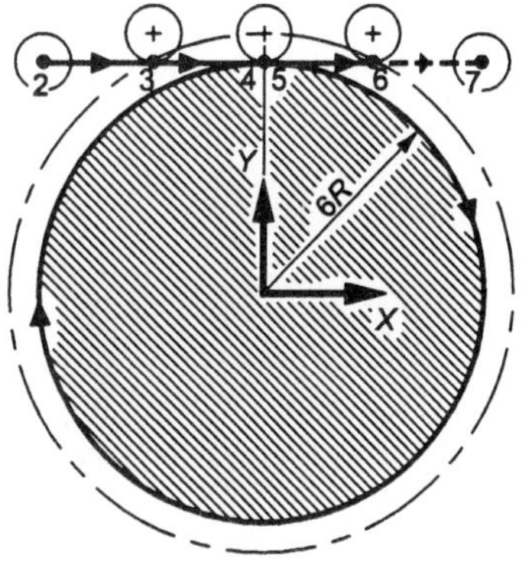

FIGURE 13-9

■■

■ EXAMPLE 13-3

The inside circle shown in Figure 13-10 is to be machined via cutter diameter compensation. Use this technique for executing operation 1 (rough profiling) and operation 2 (finish profiling). To complete operation 2, see Figure 13-11.

Word address command	*Meaning*
G90 G00 X5.0 Y10.0 S600	Rapid to ②.
G01 Z −.5 M03	Plunge to −0.5. Spindle on (CW).
G01 G41 Y15.0 F6.0 D22	Ramp on to left of upward tool motion on the next move to Y15. Offset cutter by the value in the address D22 ③.
G03 X5 Y15.0 J−5.0	Cut arc to ④.
G40 G00 Y10.0	Ramp off on the next move to Y10.0 ⑤.
Z.1	Rapid to 0.1 above the part.

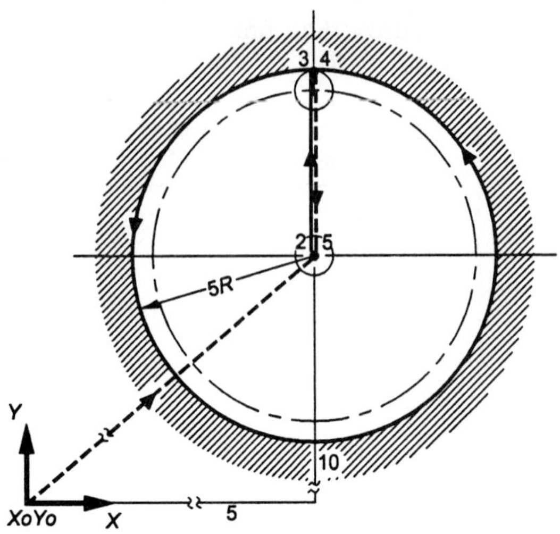

FIGURE 13–10

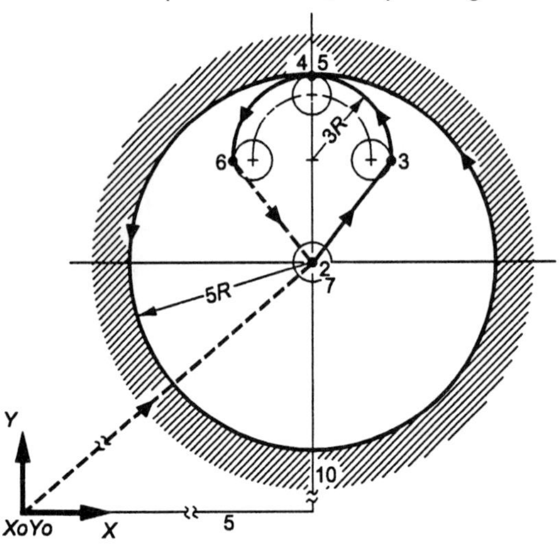

FIGURE 13–11

Word address command	*Meaning*
G90 G00 X5.0 Y10.0 S500	Rapid to ②.
G01 Z −.5 M03 F5.0	Plunge to −0.5. at Feed rate 5 rpm. Spindle on (CW).
G41 X8.0 Y12.0 D23	Ramp on to left of upward tool motion on next move to Y12. Offset cutter by the value in the address D23 ③.
G03 X5.0 Y15.0 R3.0	Cut to 3*R* arc to ④.
J-5.0	Cut 5*R* arc to ⑤.
X2.0 Y12.0 R3.0	Cut 3*R* arc to ⑥.
G40 G00 X5.0 Y10.0	Ramp off on the next move to X5. Y10.0 ⑦.
Z.1	Rapid to 0.1 above the part.

■■

13-7 CUTTER DIAMETER COMPENSATION WITH Z-AXIS MOVEMENT

As was discussed previously, the controller will read two blocks ahead in processing programs. Additionally, for *XY* motion, it is permitted to code only one *Z*-axis move between the block calling for a G41 or G42 move and the next G00 or G01 axis move. If two *Z*-axis moves are programmed, proper compensation will not be applied.

■ EXAMPLE 13–4

The following is correct:

Word address command	*Meaning*
G00 G41 X2.0 Y3.0 D05	Ramp on at X2. Y3.
G01 Z −.5 F2.0	Plunge to −0.5 at feed rate.
X4.0	First cut in new compensation.

The following is incorrect:

Word address command	*Meaning*
G00 G41 X2.0 Y3.0 D05	Ramp on at X2. Y3.
Z.1	Rapid to 0.1 above part.
G01 Z −.5 F2.0	Plunge to −0.5 at feed rate.
X4.0	First *XY* cut will *not* be made with new compensation applied.

■■

13-8 CUTTER DIAMETER COMPENSATION INTERRUPTIONS

Cutter diameter compensation will be interrupted by the following conditions:

Condition	Result
Two or more blocks in a G41 or G42 cutting sequence do not contain motion in the proper plane.	An error in the last cutting block in the proper plane will occur.

■ EXAMPLE 13–5

Compensation is interrupted by two or more blocks containing only the codes:
Dwell: G04
M Code: M05
G Code: G90
S Code: S1500

Condition	Result
A G28, G29, or G92 is programmed in a G41 or G42 cutting sequence.	Compensation is temporarily canceled. It can be reestablished with any subsequent blocks containing motions in the proper plane.

■■

■ EXAMPLE 13–6

Code a CNC program for profile milling the part shown in Figure 13–12. Apply cutter diameter compensation and use two different register numbers in the cutter offset address D.

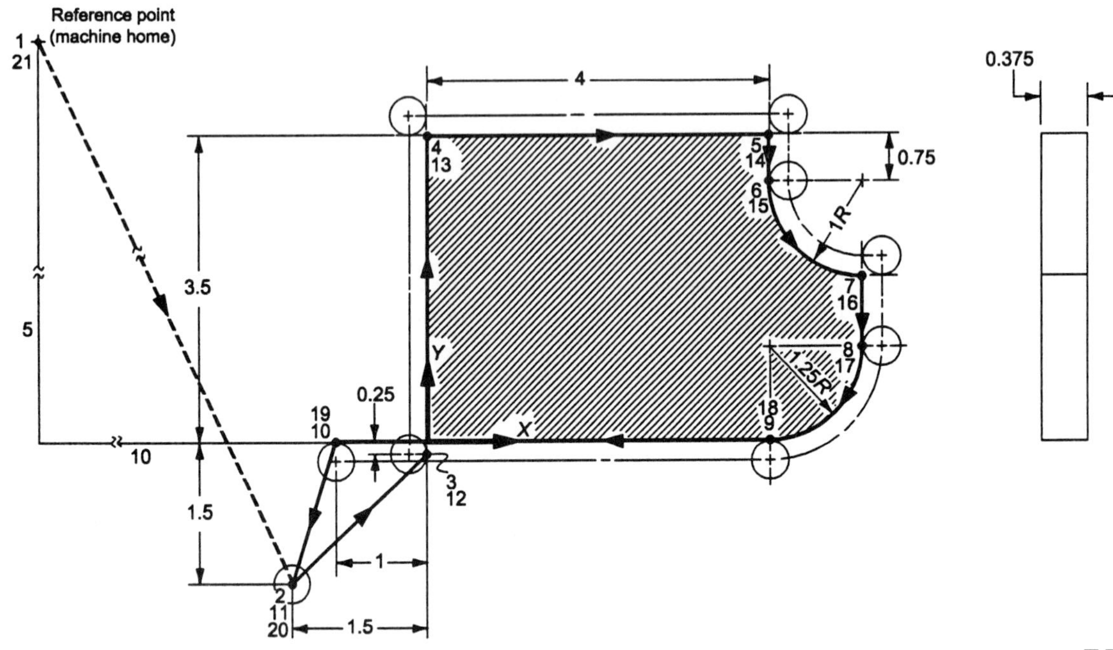

■ ■

FIGURE 13–12

Tool	Operation	Tooling	Speed (rpm)	Feed (ipm)	D*	Effect
1	Rough profile leave .01 for finish	0.5D end mill	600	8	D21=.51	*cutter is offset by .51* (CUTTER, D21=.51, .01, ROUGH PROFILE, FINISH PROFILE)
1	Finish profile	0.5D end mill	800	6	D22=.5	*cutter is offset by .5* (CUTTER, D22=.5, FINISH PROFILE)

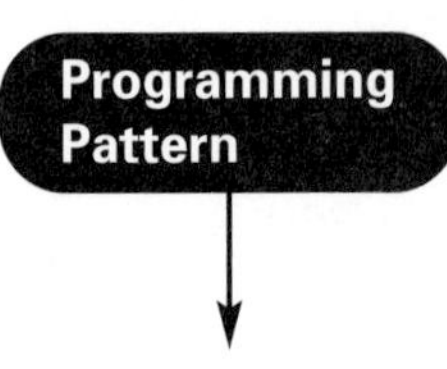

Job setup data listing

CNC Program—Absolute Coordinates

Word address command	*Meaning*
%	End of tape.
O1311	Program number.
N0010 (X0Y0 IS THE LOWER LEFT HAND CORNER)	
N0020 (Z0 IS THE TOP OF THE PART)	
N0030 (TOOL 1: 1.0 DIA END MILL)	Comments.
N0040 (USE D21 FOR ROUGHING AND D22 FOR FINISHING)	
N0050 G90 G20 G40 G80	Absolute, inch mode, cancel cutter diameter compensation and fixed cycles.
N0060 G91 G28 X0 Y0 Z0	Return to reference point ①.

CNC PROGRAM ***Continued***

	Word address command	*Meaning*
	N0070 G92 X−10.0 Y5.0 Z0	Present absolute zero point.
	N0080 (TOOL 1: ROUGH AND FINISH PART CONTOUR)	
Machine start-up sequence Change to tool 1	N0090 T1 M06	Change to tool 1.
	N0100 G00 G90 X−1.5 Y−1.5 Z0 S600 M03	Rapid to ② spindle on (CW) at 600 rpm.
	N0110 G43 Z.1 H01	Rapid tool 1 to 0.1 above part.
	N0120 M08	Coolant on.
	N0130 G01 Z−.385 F8.0	Plunge to −0.385 at feed rate.
	N0140 G41 X0 Y−.25 D21	Ramp on at ③ X0.Y−.25. Offset cutter by the value in D21.
	N0150 Y3.5	Cut to ④.
	N0160 X4.0	Cut to ⑤.
	N0170 Y2.75	Cut to ⑥.
	N0180 G03 X5.0 Y1.75 I1.0 J0	Cut arc to ⑦.
	N0190 G01 Y1.25	Cut to ⑧.
	N0200 G02 X3.75 Y0 I−1.25 J0	Cut arc to ⑨.
Profiling operation 1 sequence	N0210 G01 X−1.0	Cut arc to ⑩.
	N0220 G40 X−1.5 Y−1.5	Ramp off at X−1.5Y−1.5⑪.
	N0230 S800	Set spindle speed to 800.
	N0240 G41 X0 Y−.25 D22 F6.0	Ramp on at⑫ X0.Y−.25. Offset cutter by the value in D22.
	N0250 Y3.5	Cut to⑬.
	N0260 X4.0	Cut to⑭.
	N0270 Y2.75	Cut to⑮.
	N0280 G03 X5.0 Y1.75 I1.0 J0	Cut arc to⑯.
	N0290 G01 Y1.25	Cut to⑰.
	N0300 G02 X3.75 Y0 I−1.25 J0	Cut arc to⑱.
	N0310 G01 X−1.0	Cut to⑲.
	N0320 G40 X−1.5 Y−1.5	Ramp off at⑳ X−1.5Y−1.5.
Machine stop program end sequence	N0330 G00 G90 Z1.0 M05	Rapid to 1.0 above part. Spindle off.
	N0340 M09	Coolant off.
	N0350 G91 G28 X0 Y0 Z0	Rapid to *XYZ* reference point ㉑.
	N0360 M30	Program end, memory reset.
	%	End of tape.

Machining simulation for this example is contained in the job file EX 13–6 stored on the disk at the back of this text. Refer to Chapter 20 for instructions.

■ EXAMPLE 13–7

Program blocks to finish the inside and outside boundaries of the part are shown in Figure 13–13. Use the cutter diameter compensation when specifying tool paths.

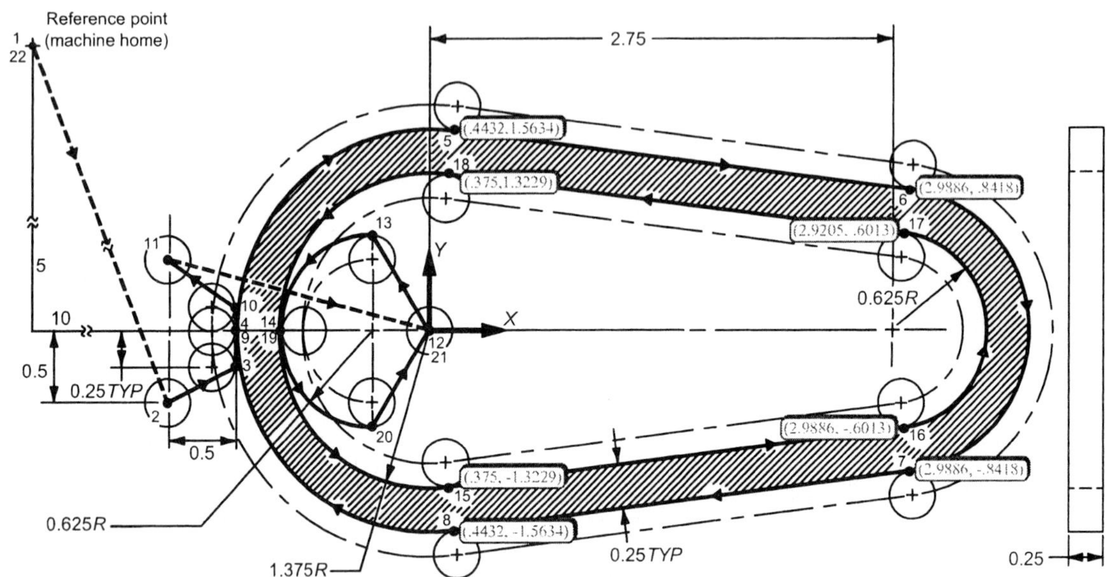

FIGURE 13–13

Tool	Operation	Tooling	Speed (rpm)	Feed (ipm)
1	Finish inside and outside contour, ×0.27 deep	⅜D end mill	1200	6

■■

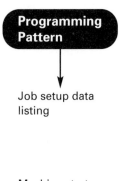

Programming Pattern

Job setup data listing

Machine start-up sequence
Change to tool 1

CNC Program—Absolute Coordinates

Word address command	*Meaning*
%	End of tape.
O1312	Program number.
N0010 (X0Y0 IS THE CENTER OF THE LARGE DIAMETER)	
N0020 (Z0 IS THE TOP OF THE PART)	Comments.
N0030 (TOOL 1: 3/8 DIA END MILL)	
N0040 G90 G20 G40 G80	Absolute, inch mode, cancel fixed cycles.
N0050 G91 G28 X0 Y0 Z0	Return to reference point ①.
N0060 G92 X–10.0 Y5.0 Z0	Present absolute zero point.
N0070 T1 M06	Change to tool 1.
N0080 (TOOL 1: FINISH OUTSIDE AND INSIDE PROFILE)	

CNC PROGRAM ***Continued***

Outside and inside profiling Operations sequence	N0090 G00 G90 X–2.125 Y–.5 Z0 S1200 M03	Rapid to ②. Spindle on (CW) at 1200 rpm.
	N0100 G43 Z.1 H01	Rapid tool 1 to 0.1 above part.
	N0110 M08	Coolant on.
	N0120 G01 Z–.26 F6.0	Plunge to −0.26 at feed rate.
	N0130 G41 X–1.625 Y–.25 D25	Ramp on at ③. Offset cutter by value in D25.
	N0140 Y0	Cut to ④.
	N0150 G02 X.4432 Y1.5634 I1.625 J0	Cut arc to ⑤.
	N0160 G01 X2.9886 Y.8418	Cut to ⑥.
	N0170 G02 X2.9886 Y–.8418 I–.2386 J–.8418	Cut arc to ⑦.
	N0180 G01 X.4432 Y–1.5634 I1.625 J0	Cut to ⑧.
	N0190 G02 X–1.625 Y0 I–.4432 J1.5634	Cut to ⑨.
	N0200 G01 X–1.625 Y.25	Cut to ⑩.
	N0210 G40 X–2.125 Y.5	Ramp off at ⑪.
	N0220 G00 Z.1	Rapid to 0.1 above part.
	N0230 X0 Y0	Rapid to ⑫.
	N0240 G01 Z–.26	Plunge to −0.26 at feed rate.
	N0250 G41 X–.75 Y.625 D25	Ramp on at ⑬. Offset cutter by value in D25.
	N0260 G03 X–1.375 Y0 I0 J–.625	Cut arc to ⑭.
	N0270 X.375 Y–1.3229 I1.375 J0	Cut arc to ⑮.
	N0280 G01 X2.9205 Y–.6013	Cut to ⑯.
	N0290 G03 X2.9205 Y.6013 I–.1705 J.6013	Cut arc to ⑰.
	N0300 G01 X.375 Y1.3229	Cut to ⑱.
	N0310 G03 X–1.375 Y0 I–.375 J–1.3229	Cut arc to ⑲.
	N0320 X–.75 Y–.625 I.625 J0	Cut arc to ⑳.
	N0330 G01 G40 X0 Y0	Cut to ㉑.
Machine stop Program end sequence	N0340 G00 G90 Z1.0 M05	Rapid to 1.0 above part. Spindle off.
	N0350 M09	Coolant off.
	N0360 G91 G28 X0 Y0 Z0	Return to *XYZ* reference point ㉒.
	M0380 M30	Program end, memory reset.
	%	End of tape.

Machining simulation for this example is contained in the job file EX 13–7 stored on the disk at the back of this text. Refer to Chapter 20 for instructions.

■ EXAMPLE 13–8

Utilize cutter diameter compensation in writing a program to profile mill the outside and inside contours of the part shown in Figure 13–14.

Tool	Operation	Tooling	Speed (rpm)	Feed (ipm)
1	Profile mill contour inside and outside ×0.25 deep	0.25*D* end mill	1200	8

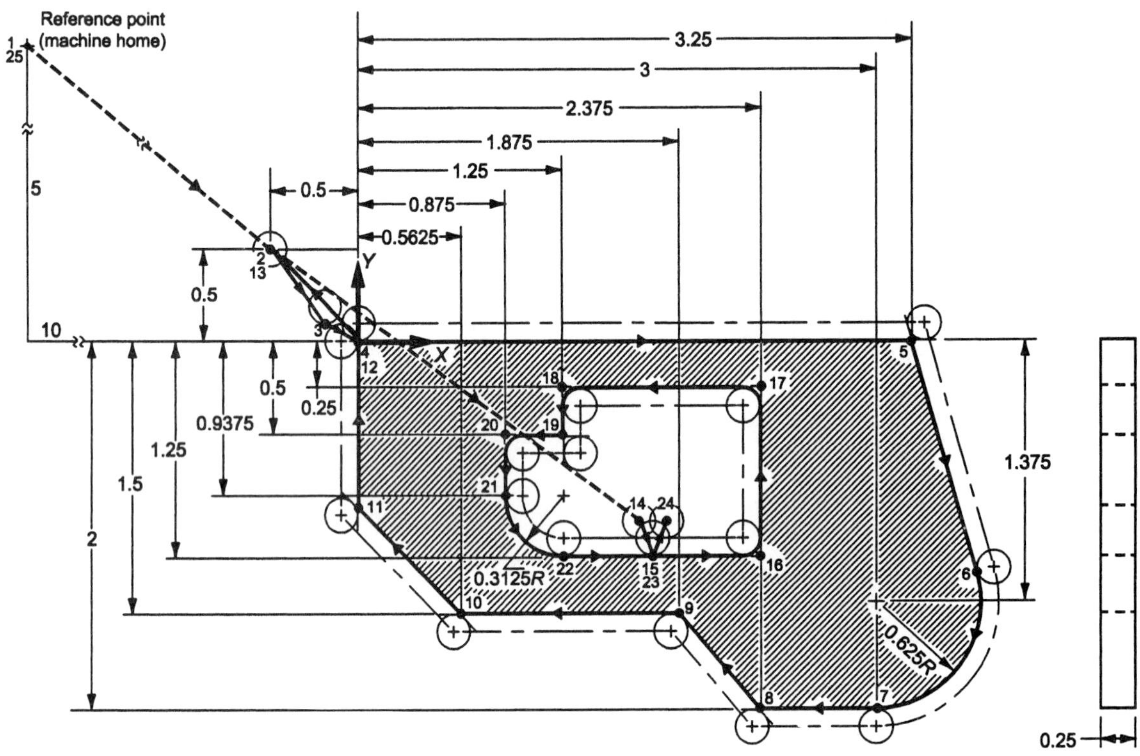

FIGURE 13–14

■■

CNC Program—Absolute Coordinates

Programming Pattern

Job setup data listing

Tool 1 in spindle
Machine start-up sequence

Word address command	*Meaning*
%	End of tape.
O1313	Program number.
N0010 (X0Y0 IS THE UPPER LEFT HAND CORNER)	Comments.
N0020 (Z0 IS THE TOP OF THE PART)	
N0030 (TOOL 1: USE REGISTER 21 TO SET CUTTER DIA)	
N0040 G90 G20 G40 G80	Absolute, inch mode, cancel cutter diameter compensation and fixed cycles.
N0050 G91 G28 X0 Y0 Z0	Return to reference point ①.

CNC PROGRAM ***Continued***

Outside and inside profiling operations sequence	N0060 G92 X–10.0 Y5.0 Z0	Preset absolute zero point.
	N0070 T1 M06	Change to tool 1.
	N0080 (TOOL 1: FINISH INSIDE AND OUTSIDE PROFILES)	
	N0090 G00 G90 X–.5 Y.5 Z0 S1200 M03	Rapid to ②. Spindle on (CW) at 1200 rpm.
	N0100 G43 Z.1 H01	Rapid tool 1 to 0.1 above part.
	N0110 M08	Coolant on.
	N0120 G01 Z–.25 F8.0	Plunge to –0.25 at feed rate.
	N0130 G41 X–.25 Y.25 D21	Ramp on at ③. Offset tool by the value in D21.
	N0140 X0 Y0	Cut to ④.
	N0150 X3.25	Cut to ⑤.
	N0160 X3.6 Y–1.2	Cut to ⑥.
	N0170 G02 X3.0 Y–2.0 I–.625 J–.175	Cut arc to ⑦.
	N0180 G01 X2.375	Cut to ⑧.
	N0190 X1.875 Y–1.5	Cut to ⑨.
	N0200 X.5625	Cut to ⑩.
	N0210 X0 Y–.9375	Cut to ⑪.
	N0220 Y0	Cut to ⑫.
	N0230 G40 X–.5 Y.5	Ramp off at ⑬.
	N0240 G00 Z.1	Rapid to 0.1 above part.
	N0250 X1.625 Y–1.05	Rapid to ⑭.
	N0260 G01 Z–.25	Plunge to –0.25 at feed rate.
	N0270 G41 X1.825 Y–1.25 D21	Ramp on at ⑮. Offset cutter by value in D21.
	N0280 X2.375	Cut to ⑯.
	N0290 Y–.25	Cut to ⑰.
	N0300 X1.25	Cut to ⑱.
	N0310 Y–.5	Cut to ⑲.
	N0320 X.875	Cut to ⑳.
	N0330 Y –.9375	Cut to ㉑.
	N0340 G03 X1.1875 Y–1.25 I.3125 J0	Cut arc to ㉒.
	N0350 G01 X1.875	Cut to ㉓.
	N0360 G40 X2.075 Y–1.05	Ramp off at ㉔.
Machine stop Program end sequence	N0370 G00 G90 Z1.0 M05	Rapid to 1.0 above part.
	N0380 M09	Coolant off.
	N0390 G91 G28 X0 Y0 Z0	Return to *XYZ* reference point ㉕.
	N0400 M30	Program end, memory reset.
	%	End of tape.

Machining simulation for this example is contained in the job file EX 13–8 stored on the disk at the back of this text. Refer to Chapter 20 for instructions.

13-9 CHAPTER SUMMARY

The following key concepts were discussed in this chapter:

1. Cutter diameter compensation involves programming the part geometry directly instead of the cutter center.
2. Because the controller automatically determines the location of the cutter center at each point along a part boundary, mathematical computations are greatly simplified.
3. A cutter diameter compensation program can be used for a variety of cutter sizes.
4. The word address command for cutter diameter compensation left of upward tool motion is G41. Compensation right of upward tool motion is a G42 code and compensation cancel is a G40 code.
5. Compensation is initiated or canceled by the next linear tool motion command directly following a G41, G42, or G40 code. The motion must be in the selected G17, G18, or G19 plane.
6. The first in-plane tool motion command following a G41 or G42 code must be equal to or greater than the cutter radius.
7. Compensation is always applied by the controller in a direction perpendicular to the next in-plane axis move.
8. Care must be used to avoid notching the part when cutting inside corners or step-downs.

REVIEW EXERCISES

For each of the exercises in this chapter, write a CNC program by using cutter diameter compensation. Programs should contain the following general format:

1. Program number headings
2. Job setup data listing
3. Tool 1 in spindle, machine start-up
4. Profiling operation 1
5. Machine stop
6. Change to tool 2
7. Profiling operation 2

.
.
.

8. Machine stop, program end sequence

Use the machining editor and simulator contained on the CD at the back of this text to write and verify the word address programs given in these exercises. Refer to Chapter 20 for instructions.

13.1. Program number: O1314 (Figure 13–15)

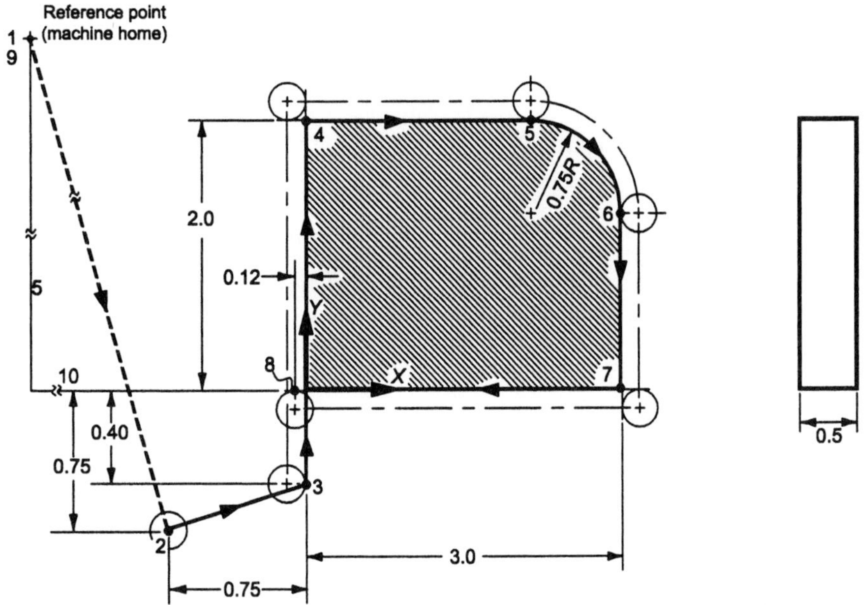

FIGURE 13–15

Tool	Operation	Tooling	Speed (rpm)	Feed (ipm)
1	Finish profile × 0.52 deep	0.75*D* end mill	800	7

13.2. Program number: O1315 (Figure 13–16)

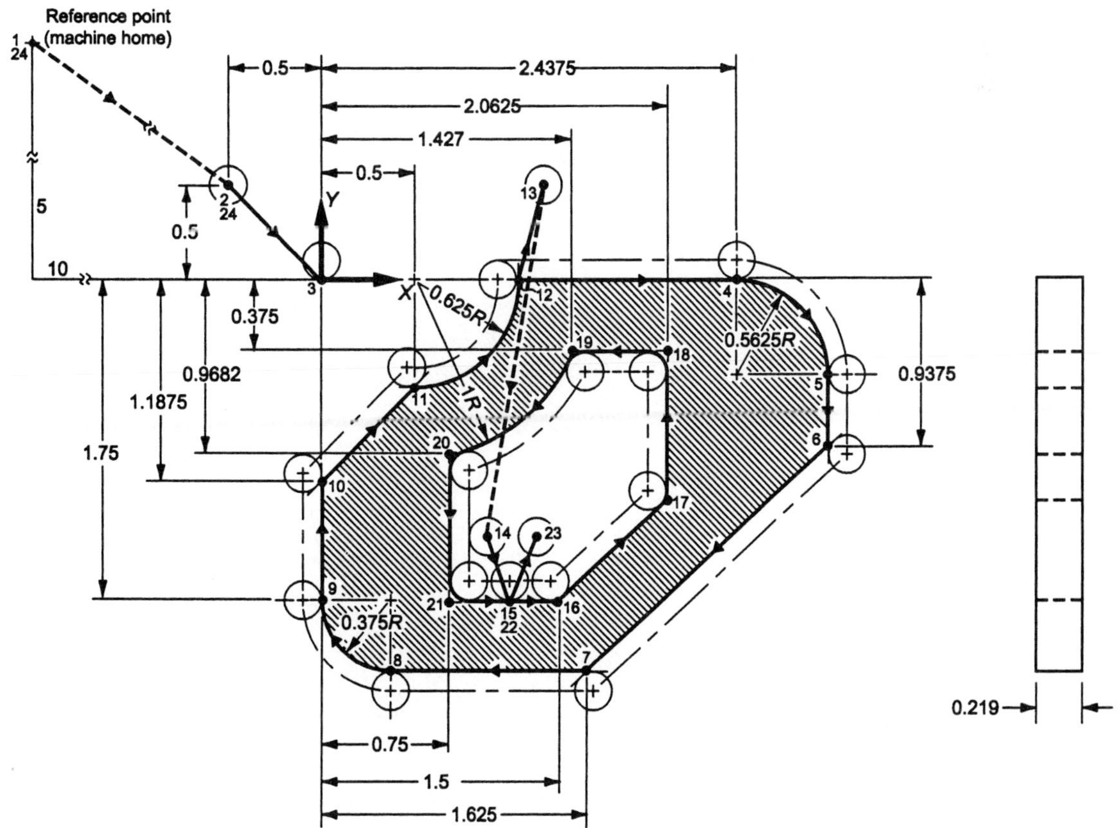

FIGURE 13–16

Tool	Operation	Tooling	Speed (rpm)	Feed (ipm)
1	Finish outside and inside profile as shown	0.5*D* end mill	1200	8

13.3. Program number: O1316 (Figures 13–17 and 13–18)

Tool	Operation	Tooling	Speed (rpm)	Feed (ipm)
1	Rough part inside as shown. Leave 0.005.	1.0*D* end mill	600	10
2	Finish inside as shown	0.75*D* end mill	800	6

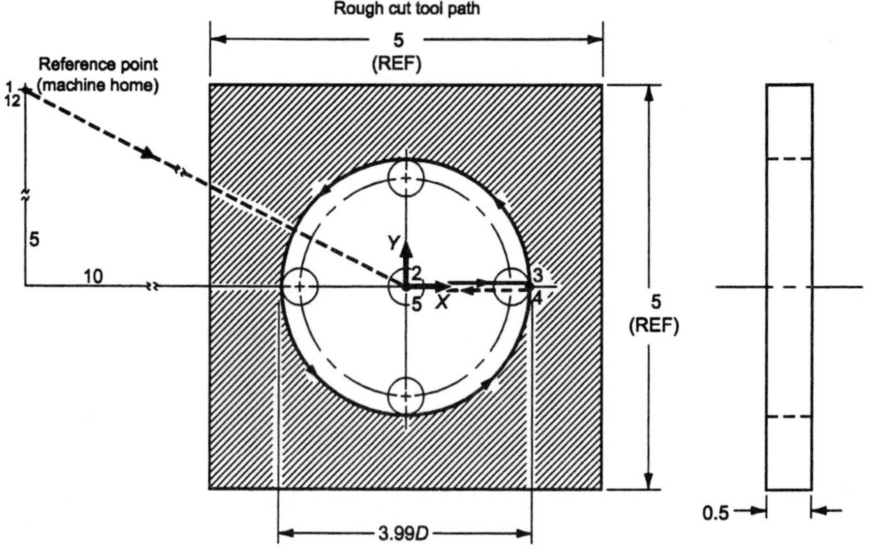

FIGURE 13–17

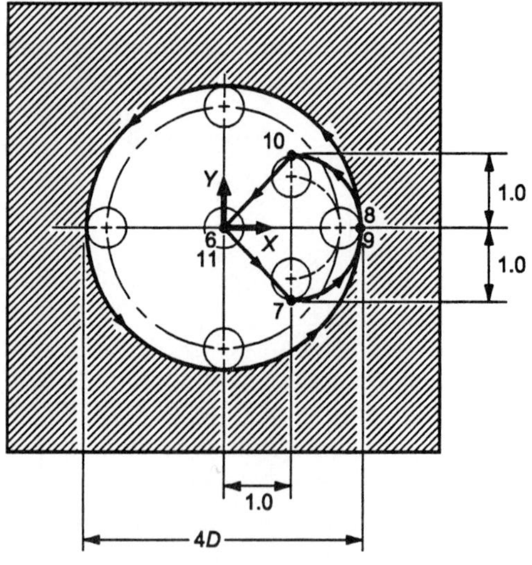

FIGURE 13–18

13.4. Program number: O1317 (Figure 13–19)

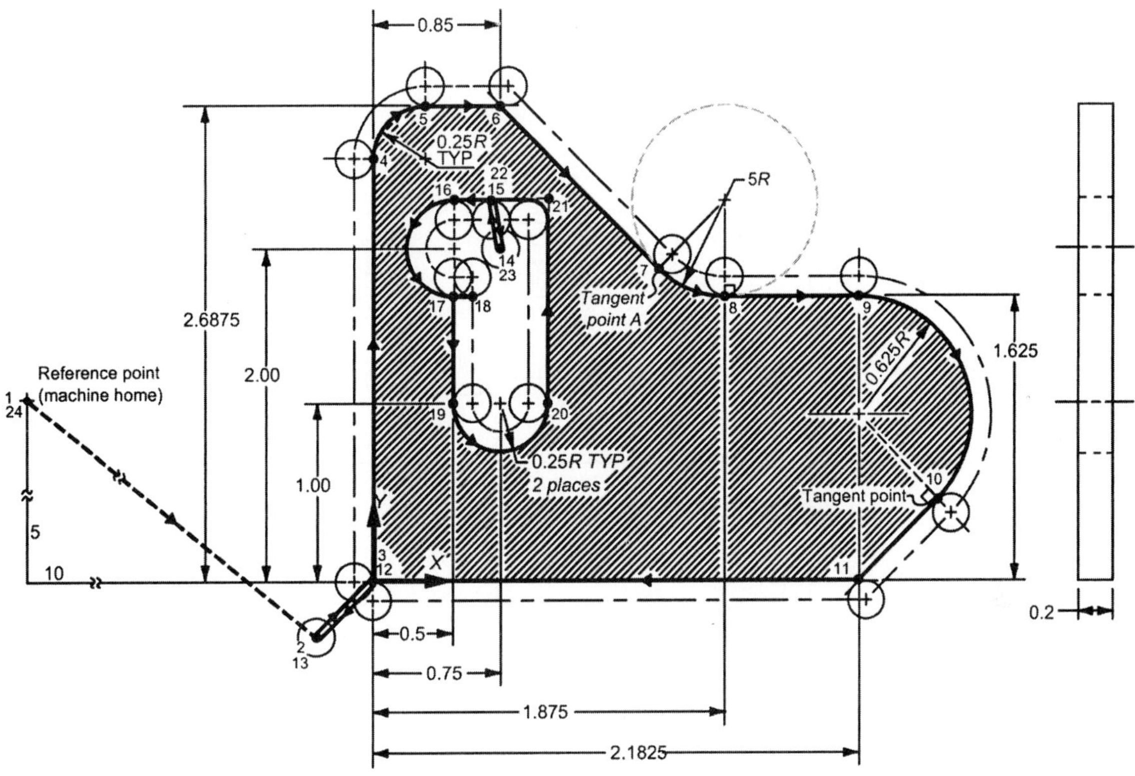

FIGURE 13–19

Tool	Operation	Tooling	Speed (rpm)	Feed (ipm)
1	Finish outside and inside profile as shown	0.1875*D* end mill	1800	4

Note: The following formulas will give the location of the tangent point A:

$$\theta = \tan^{-1}\left(\frac{h}{d}\right) + \sin^{-1}\left(\frac{R}{\sqrt{h^2 + d^2}}\right)$$

$$\Delta X = R\sin(\theta) \qquad \Delta J = R\cos(\theta)$$

$$\Delta Y = R(1 - \cos(\theta))$$

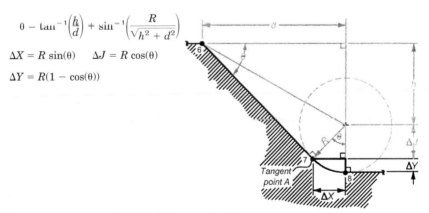

13.5. Program number: O1318 (Figure 13–20)

Tool	Operation	Tooling	Speed (rpm)	Feed (ipm)
1	Finish cut outside and inside profile as shown	.125*D* end mill	1200	8

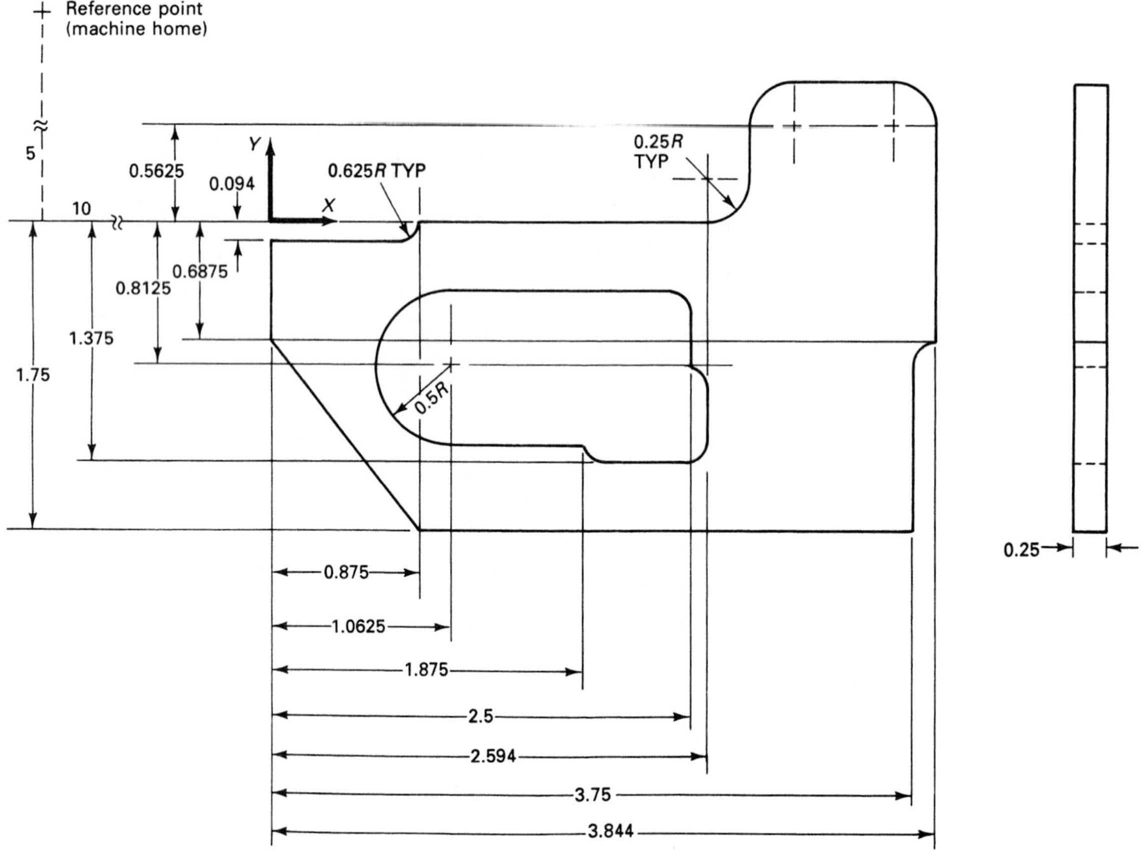
Reference point
(machine home)
5
10
0.5625
0.094
Y
X
0.625R TYP
0.25R
TYP
0.6875
0.8125
1.375
1.75
0.5R
0.875
1.0625
1.875
2.5
2.594
3.75
3.844
0.25

FIGURE 13–20

CHAPTER 14

Programming with Subprograms

14–1 Chapter Objectives

At the conclusion of this chapter you will be able to

1. Explain what a subprogram is.
2. Understand the commands and rules for creating and processing subprograms.
3. Describe the advantages of subprogramming.
4. Write complete programs by utilizing subprogramming.

14–2 Introduction

In this chapter the reader is introduced to the important efficiency device known as subprogramming. The need for using a subprogram is discussed first. The advantages of subprogramming are given. This is followed by a detailed explanation of the commands required to call and execute subprograms for the Fanuc family of CNC controllers. The important restrictions and conditions that must be observed when writing subprograms are given. Several programming examples with explanations are presented throughout the chapter.

14–3 Subprogram Concept

The programmer will often encounter cases where a sequence of machining instructions is to be repeated. An example of this is a part having an identical geometric pattern appearing at different locations within its body. The programmer could code the instructions to machine each pattern individually for the part shown in Figure 14–1.

This would result in a program that has the same machining sequences appearing three times. Thus, the program would be lengthy, the chances for error would be increased, and the program would occupy more memory in the MCU. A better approach is to create a separate program called a subprogram that contains the instructions for machining one pattern. A main program is also created that calls the subprogram as many times as the machining sequence is repeated. In this case, the subprogram would be called three times to machine pattern A. The subprogram can also be used by other main programs any time pattern A is required. In fact, a library of subprograms can be created for machining specific patterns. See Figure 14–2.

It should be noted that both the main program and each subprogram are separate programs that must be stored in the MCU's memory prior to execution.

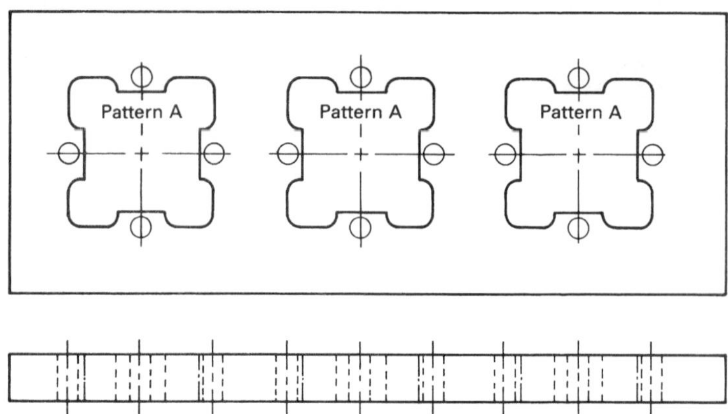

FIGURE 14–1 Subprograms are used to machine repeat patterns in a part.

FIGURE 14–2 Setups for executing multiple machining via subprograms. [(a) Photo courtesy of Chick Machine Tool, Inc. (b) Photo courtesy of Macro Tool and Machine Co., Inc.]

14–4 FANUC COMMANDS FOR CALLING A SUBPROGRAM AND RETURNING TO THE MAIN PROGRAM

The following commands apply to the Fanuc family of CNC controllers.

General Syntax

Calling a Subprogram from a Main Program

G00 Xn Yn

M98 Pn Ln

G00 Xn Yn	Rapids tool to the X_nY_n location where the subprogram is to be executed.
M98	Calls for a jump to the subprogram.
Pn	n is a one-to-four-digit number of the subprogram in memory. The number, n, is usually preceded by the letter, O (EIA), format or : (ISO) format when stored in memory.
Ln	n is the number of times to repeat the call of the subprogram. If omitted, the subprogram is repeated only once.

Format of the Subprogram

Subprograms are prepared according to the following format:

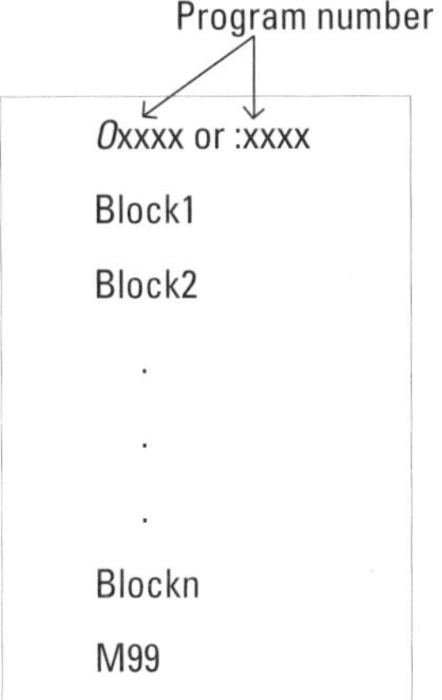

Returning to the Main Program from the Subprogram

M99

M99 Is the last statement in the subprogram and signals a return to the main program. The system will return to the very next statement in the main program following the M98 Pn Ln command.

> ***Note:*** It is also possible to program a motion command in the same block as an M98 to M99 code. Transfer will occur after the motion command has been completed.

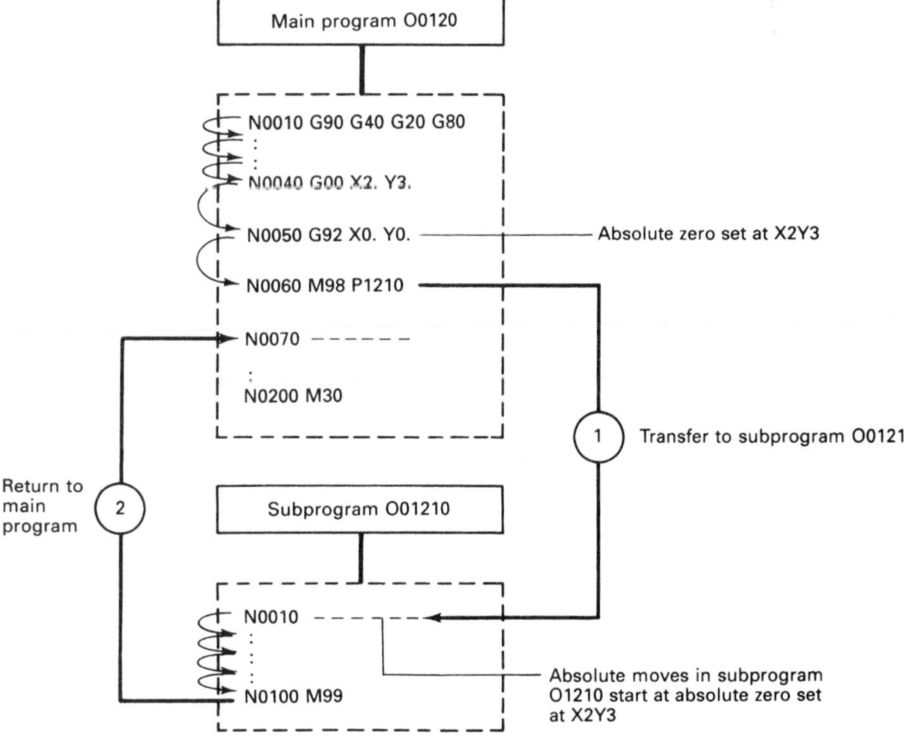

FIGURE 14–3 A main program accessing a single subprogram.

■ EXAMPLE 14–1

The process by which a main program transfers to a subprogram and vice versa is illustrated in Figure 14–3.

A subprogram may call another subprogram. ■■

■ EXAMPLE 14–2

The general programming shown in Figure 14–4 illustrates the action that results when a main program accesses two levels of subprograms.

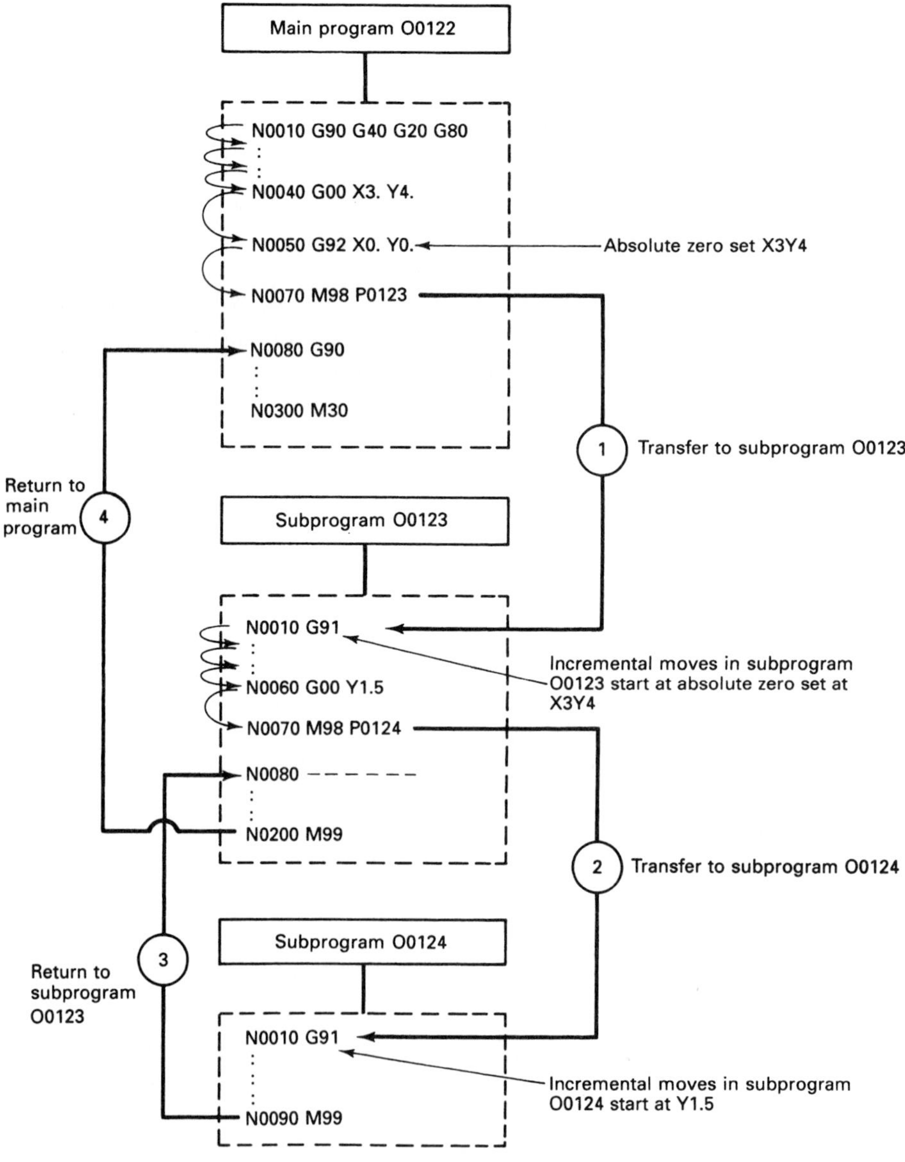

FIGURE 14–4 A main program accessing two levels of subprograms.

■■

14–5 ADDITIONAL SUBPROGRAM CONTROL FEATURES

Programmers can also utilize the following important features when writing subprograms for Fanuc controllers.

1. M99 Pn may be coded in the main program with n as a four-digit block number. This will result in the controller returning to the block whose number has been entered. Thus, a continuous loop can be created.

■ EXAMPLE 14–3

Coding a M99 P(block#) in the main program. See Figure 14–5.

If Pn is omitted, control will return to the top of the main program and execution will continue. ■■

2. M99 Pn may also be coded in the subprogram with n as a one-to-four-digit number of a block in the main program. Upon encountering this code in the subprogram, the controller will transfer to the block in the main program whose number has been coded.

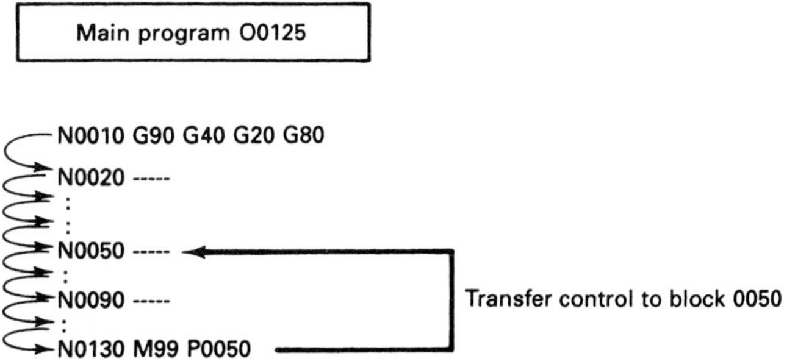
Main program O0125
N0010 G90 G40 G20 G80
N0020 -----
N0050 -----
N0090 -----
N0130 M99 P0050
Transfer control to block 0050

FIGURE 14–5 Coding an M99 P (block#) in the main program.

■ EXAMPLE 14–4

Coding a M99 P (block#) in the subprogram. See Figure 14–6. ■■

3. The option of allowing the operator to skip execution of a block or blocks in a program is controlled by entering skip switch (/) codes before the blocks. The operator decides which blocks are to be skipped. Appropriate skip switches on the MDI control panel are then switched on. The controller will then skip those blocks tagged with the code (/) when processing the program.

This feature is useful in cases where a trial or gauge run is to be made and checked out. After prove-out the skip switch may be turned off and the program continuously run for cutting multiple parts.

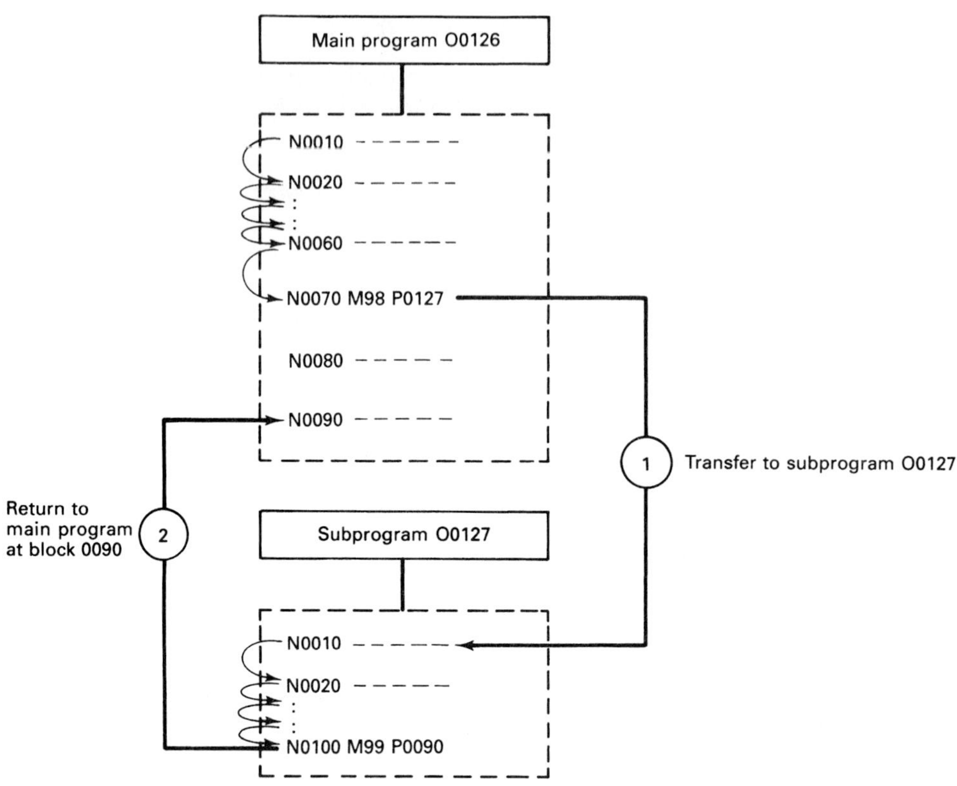

FIGURE 14–6 Coding an M99 P (block#) in the subprogram.

■ **EXAMPLE 14–5**

Coding a skip switch signal (/) in the main program. See Figure 14–7.

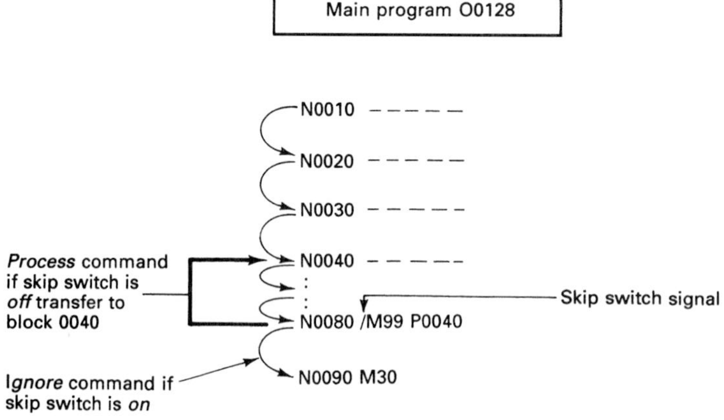

FIGURE 14–7 Coding 2 skip switch signal (1) in the main program. ■■

14–6 GENERAL RULES FOR SUBPROGRAMMING

The following general conditions will apply when writing subprograms with Fanuc controllers.

1. One subprogram may call another subprogram (nesting). The nesting level is limited to four levels deep.
2. Oxxxx (EIA code) is used to indicate subprogram numbers. Words beginning with the address O specify subprogram numbers. Subprogram numbers can range from 1 to 9999.

3. If the ISO coding system is used, a colon (:) is entered as the address for specifying program numbers.
4. The mode of motion (G90 or G91) must be reset to the main program mode just prior to or just after transfer to the main program. This can occur, for example, if the main program is set to G90 (absolute) mode and the subprogram is set to G91 (incremental) mode. In this case a G90 block must be inserted at the end of the subprogram. It could also appear in the main program as the first block to be processed after transfer from the subprogram.
5. An active canned cycle in the main program not required in the subprogram should be canceled with a G80 prior to or just after transfer to the subprogram. Also, an active canned cycle in the subprogram not required in the main program should be canceled with a G80 prior to or just after transfer back to the main program.
6. If cutter diameter compensation is required in the subprogram, it should be set up with a G41 or G42 in the subprogram. This holds true regardless of whether diameter compensation is active in the main program.

■ EXAMPLE 14–6

Write a conventional CNC program for drilling the holes shown in Figure 14–8. Contrast this approach with a more efficient method that utilizes a main program and a subprogram.

Tool	Operation	Tooling	Speed (rpm)	Feed (ipm)
1	Drill 0.25*D* holes ×0.5 deep	0.25*D* drill	2000	10

Method 1: Conventional CNC program

```
%
O0149
N0010 (..........)
N0020 (X0Y0 IS THE LOWER LEFT HAND CORNER)
N0030 (Z0 IS THE TOP OF THE PART)
N0040 (..........)
N0050 (TOOLING LIST:)
N0060 (TOOL 1: .25 DIA DRILL)
N0070 G90 G20 G40 G80
N0080 G91 G28 X0 Y0 Z0
N0090 G92 X10.0 Y5.0 Z0
N0100 G00 G90 X.5 Y.5 Z0 S2000 M03
N0110 T1 M06
N0120 G43 Z.1 H01
N0130 M08
N0140 G83 X.5 Y.5 Z−.5 R.1 Q.1 F10.0
N0150 X1.0
N0160 X1.5
N0170 X2.0
N0180 X2.5
N0190 X3.0
N0200 X3.5
```

```
N0210 X4.0
N0220 X4.5
N0230 X5.0
N0240 Y1.0
N0250 X4.5
N0260 X4.0
    .
    .
    .
    .
    .
    .
    .
    .
    .
N1120 X.5
N1130 G80
N1140 G00 G90 Z1.0 M05
N1150 M09
N1160 G91 G28 X0 Y0 Z0
N1170 M06
N1180 M30
%
```

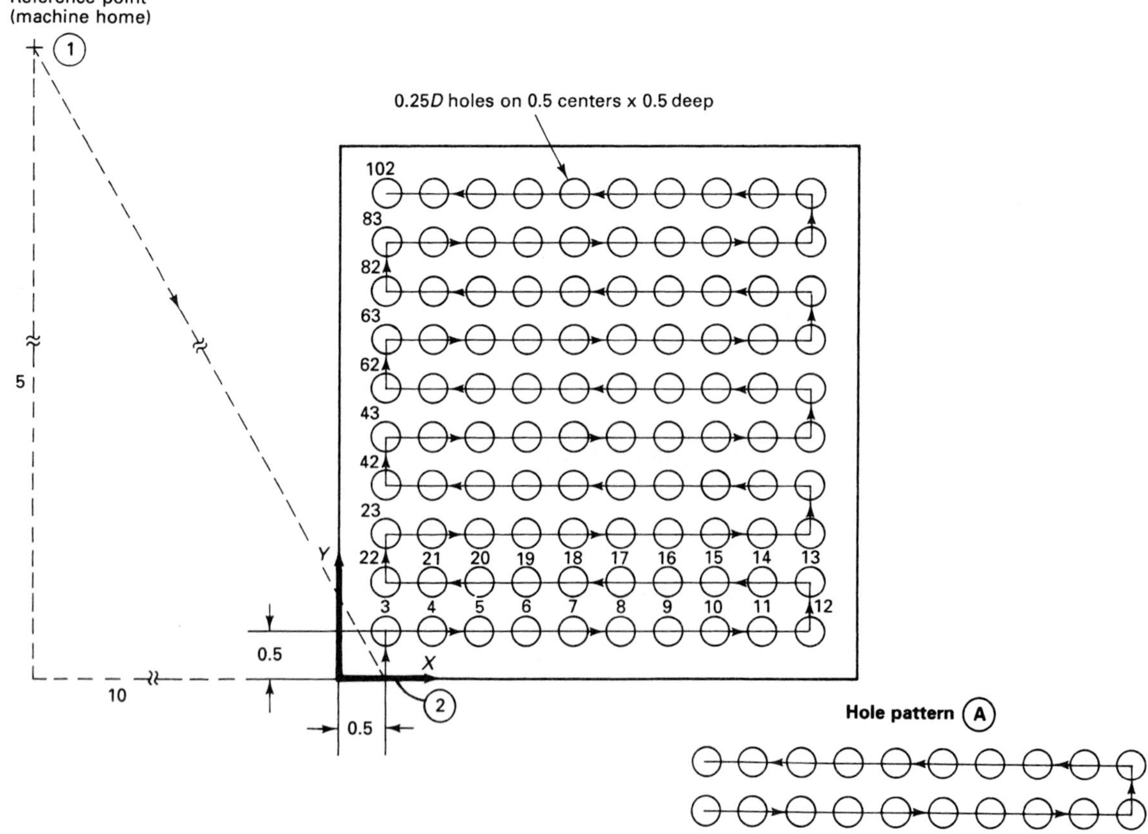

FIGURE 14–8

Programming Pattern

Job setup data listing

Tool 1 in Spindle
Machine start-up sequence
Change to tool 1

Transfers to subprogram O0001 for executing drilling operation sequence

Machine stop
Program end sequence

Method 2: Programming with a Subprogram

Word address command	***Meaning***
%	End of tape.
O0149	Main program number.
N0010 (. . . . .)	Optional setup data.
N0020 (X0Y0 IS THE LOWER LEFT HAND CORNER)	
N0030 (Z0 IS THE TOP OF THE PART)	
N0040 (. . . . .)	
N0050 (TOOLING LIST:)	
N0060 (TOOL 1: .25 DIA DRILL)	
N0070 G90 G20 G40 G80	Absolute, inch mode, cancel cutter diameter compensation and fixed cycles.
N0080 G91 G28 X0 Y0 Z0	Return to reference point.
N0090 G92 X−10.0 Y5.0 Z0	Preset absolute zero point.
N0100 T1 M06	Change to tool 1.
N0110 G00 G90 X−.5 Y0 Z0 S2000 M03	Rapid to ②. Spindle on (CW) at 2000 rpm.
N0120 G43 Z.1 H01	Rapid tool 1 to 0.1 above part.
N0130 M08	Coolant on.
N0140 M98 P0001 L5	Transfer to subprogram 0001 five times. Repeat deep drill of hole pattern A as follows:

Transfer	*Holes drilled*
1	(3) (4) ... (22)
2	(23).... (42)
3	(43) .... (62)
4	(63) .... (82)
5	(83) .... (102)

Word address command	***Meaning***
N0150 G80	Cancel any fixed cycles.
N0160 G00 G90 Z1.0 M05	Reset to absolute mode. Rapid tool to 1.0 above part. Spindle off.
N0170 M09	Coolant off.
N0180 G91 G20 X0 Y0 Z0	Return to *XYZ* reference point.
N0190 M30	Program end, memory reset.
%	

Subprogram 0001 Drill pattern A holes and return sequence

O0001		**Subprogram number**
N0200 G91 G83 Y.5 Z–.6 R.1 Q.1 F10.0	Transfer 1	Set to incremental mode tool movement. Deep drill cycle at ③. Final depth is –0.5, 0.1 peck/depth at feed rate 10 in./min. Rapid to 0.1 above part.
N0210 X.5 L9		Deep drill holes at ④ ⑤ ⑥ . . . ⑫.
N0220 Y.5		Deep drill at ⑬.
N0230 X–.5 L9		Deep drill holes at ⑭ ⑮. . . ㉒.
N0240 M99	Return to the main program at block N0130.	
%		

■ EXAMPLE 14–7

Use a main program and subprograms to drill the hole patterns shown in Figure 14–9.

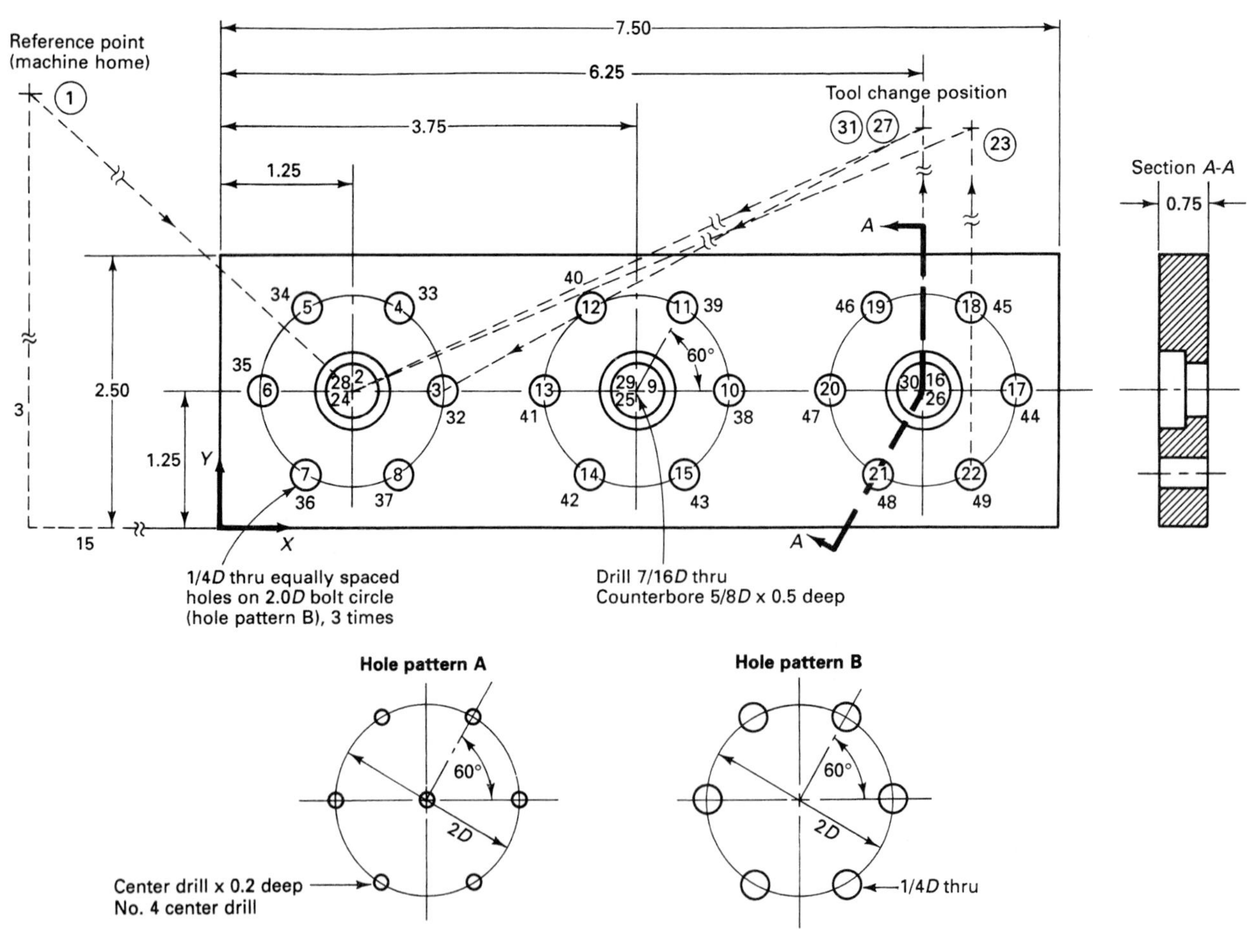

FIGURE 14–9

Tool	Operation	Tooling	Speed (rpm)	Feed (ipm)
1	Center drill all holes × 0.2 deep	No. 4 center drill	1200	8
2	Drill (3) 7⁄16*D* holes	7⁄16*D* drill	1000	10
3	Counterbore (3) 5⁄8*D* × 0.5 deep	5⁄8*D* end mill	600	5
4	Drill 1⁄4*D* thru	1⁄4*D* drill	1600	6

Notes: 1. Set X_0Y_0 at the lower left-hand corner.
2. Z_0 is the top of the part.
3. The part is held in a vise with a stop on the left.
4. The blank supplied is 7.5 in. × 2.5 in. × 0.75 in.

Programming Pattern

Job setup data listing (N0010–N0110)

Machine start-up sequence / Change to tool 1 (N0120–N0190)

CNC Program Utilizing Main and Subprograms

Word address command	*Meaning*
%	End of tape.
O1410	Main program number.
N0010 (..........)	Optional program setup data.
N0020 (X0Y0 IS THE LOWER LEFT HAND CORNER)	
N0030 (Z0 IS THE TOP OF THE PART)	
N0040 (..........)	
N0050 (TOOLING LIST:)	
N0060 (TOOL 1: NO. 4 CENTER DRILL)	
N0070 (TOOL 2: 7⁄16 DIA DRILL)	
N0080 (TOOL 3: 5⁄8 DIA END MILL)	
N0090 (TOOL 4: 1⁄4 DIA DRILL)	
N0100 (..........)	
N0110 (..........)	
N0120 G90 G20 G40 G80	Absolute, inch mode, cancel cutter diameter compensation and fixed cycles.
N0130 G91 G28 X0 Y0 Z0	Return to reference point.
N0140 G92 X−15.0 Y3.0 Z0	Preset absolute zero point.
N0150 T1 M06	Change to tool 1.
N0160 (TOOL 1: CENTER DRILL ALL HOLES)	
N0170 G00 X1.25 Y1.25 Z0 S1200 M03	Rapid to ②. Spindle on (CW) at 1200 rpm.
N0180 G43 Z.1 H01	Rapid tool 1 to 0.1 above part.
N0190 M08	Coolant on.

Sequence	Block	Explanation
Transfers to subprogram 0010 for executing hole operation 1 sequence	N0200 M98 P0010	Transfer 1 to subprogram 0010. Center drill hole pattern A holes ② ③③ ④ ... ⑧.
	N0210 G90 X3.75 Y1.25	Switch back to absolute mode. Rapid tool to ⑨.
	N0220 M98 P0010	Transfer 2 to subprogram 0010. Center drill hole pattern A holes ⑨ ... ⑮.
	N0230 G90 X6.25 Y1.25	Switch back to absolute mode. Rapid tool to ⑯.
	N0240 M98 P0010	Transfer 3 to subprogram 0010. Center drill hole pattern A holes ⑯ ... ㉒.
Machine stop sequence	N0250 G00 G90 Z1.0 M05	Switch back to absolute mode. Rapid to 1.0 above part. Spindle off.
	N0260 M09	Coolant off.
Change to tool 2 Machine start sequence	N0270 G91 G28 Z0 Y0	Rapid to tool change position ㉓.
	N0280 T2 M06	Change to tool 2.
	N0290 (TOOL 2: DRILL (3) 7/16 DIA HOLES THRU)	
	N0300 G00 G90 X1.25 Y1.25 Z0 S1000 M03	Rapid to ㉔. Spindle on (CW) at 1000 rpm.
	N0310 G43 Z.1 H02	Rapid tool 2 to 0.1 above part.
	N0320 M08	Coolant on.
Hole operation 2 sequence	N0330 G83 X1.25 Y1.25 Z−.9 R.1 Q.1 F10.0	Deep drill cycle at ㉔. Final depth is 0.9, 0.1 peck depth at feed rate 10 in./min. Rapid to 0.1 above part.
	N0340 X3.75	Deep drill at ㉕.
	N0350 X6.25	Deep drill at ㉖.
	N0360 G80	Cancel any fixed cycles.
Machine stop sequence	N0370 G00 G90 Z1.0 M05	Rapid to 1.0 above part. Spindle off.
	N0380 M09	Coolant off.
Change to tool 3 Machine start sequence	N0390 G91 G28 Z0 Y0	Rapid to tool change position ㉗.
	N0400 T3 M06	Change to tool 3.
	N0410 (TOOL 3: COUNTERBORE (3) .625 DIA HOLES)	
	N0420 G00 G90 X1.25 Y1.25 Z0 S600 M03 T4	Rapid to ㉘. Start spindle (CW) at 600 rpm. Prepare tool 4 in ready position.
	N0430 G43 Z.1 H03	Rapid tool 3 to 0.1 above part.
	N0440 M08	Coolant on.

CNC PROGRAM ***Continued***

Sequence	Block	Comment
Hole operation 3 sequence	N0450 G82 X1.25 Y1.25 Z−.2 P50 R.1 F5.0	Counterbore cycle at (28). Final depth is 0.2 at feed rate 5. Dwell at depth for 50 s. Rapid to 0.1 above part.
	N0460 X3.75	Counterbore at (29).
	N0470 X6.25	Counterbore at (30).
	N0480 G80	Cancel any fixed cycles.
Machine stop sequence	N0490 G00 G90 Z1.0 M05	Rapid to 1.0 above part.
	N0500 M09	Coolant off.
Change to tool 4 Machine start sequence	N0510 G91 G28 Z0 Y0	Rapid to tool change position (31).
	N0520 T4 M06	Change to tool 4.
	N0530 (TOOL 4: DRILL ¼ DIA HOLES THRU)	
	N0540 G00 G90 X2.25 Y1.25 Z0 S1600 M03	Rapid to (32). Spindle on (CW) at 1600 rpm. Prepare tool 1 in ready position.
	N0550 G43 Z.1 H04	Rapid tool 4 to 0.1 above part.
	N0560 M08	Coolant on.
Transfers to subprogram 0020 for executing hole operation 4 sequence	N0570 M98 P0020	Transfer 1 to subprogram 0020. Drill hole pattern B holes (32) ... (37).
	N0580 G90 X3.75 Y1.25	Change back to absolute mode. Rapid tool to (38).
	N0590 M98 P0020	Transfer 2 to subprogram 0020. Drill hole pattern B holes (38)... (43).
	N0600 G90 X6.25 Y1.25	Change back to absolute mode. Rapid tool to (44).
	N0610 M98 P0020	Transfer 3 to subprogram 0020. Drill hole pattern B holes (44)... (49).
Machine stop Program end sequence	N0620 G00 Z1.0 M05	Rapid tool to 1.0 above part. Spindle off.
	N0630 M09	Coolant off.
	N0640 G91 G28 X0 Y0 Z0	Return to *XYZ* reference point.
	N0650 M30	Program end, memory reset.

Subprogram 0010 for executing hole operation 1 and return

O0010	Subprogram number.	
N0020 G91	Switch to incremental mode.	
N0030 G81 Z−.2 R.1 F5.0	Transfer 1	Drill cycle at ②. Final depth is −0.2 at feed rate 5 in./min. Rapid to 0.1 above part.
N0040 X1.0		Drill hole at ③.
N0050 X−.5 Y.866		Drill hole at ④.
N0060 X−1.0		Drill hole at ⑤.
N0070 X−.5 Y−.866		Drill hole at ⑥.
N0080 X.5 Y−.866		Drill hole at ⑦.
N0090 X1.0		Drill hole at ⑧.
N0100 G80	Cancel any fixed cycles.	
N0110 M99	Return to main program block N0250.	

Subprogram 0020 for executing hole operation 4 and return

O0020	Subprogram number.	
N0010 G91	Change to incremental mode.	
N0020 G81 X1.0 Z−.85 R.1 F6.0	Transfer 1	Drill cycle at ㉜. Final depth is −0.85 at feed rate 6 in./min. Rapid to 0.1 above part.
N0030 X−.5 Y.866		Drill hole at ㉝.
N0040 X−1.0		Drill hole at ㉞.
N0050 X−.5 Y2.866		Drill hole at ㉟.
N0060 X.5 Y−.866		Drill hole at ㊱.
N0070 X1.0		Drill hole at ㊲.
N0080 G80	Cancel any fixed cycles.	
N0090 M99	Return to main program block N0620.	

■ EXAMPLE 14–8

Write a program for milling the part shown in Figure 14–10. Use the technique of subprogramming.

Tool	Operation	Tooling	Speed (rpm)	Feed (ipm)
1	Drill (3) 0.5*D* entrance holes	½*D* drill	800	10
2	Profile mill inside and stop	0.375*D* end mill	1200	8
3	Reclamp and finish outside	0.375*D* end mill	1200	8

Notes:
1. Set X_0Y_0 at the upper left-hand corner.
2. Z_0 is the top of the part.
3. The blank is supplied in a 6 in. × 3.75 in. × 0.375 in.
4. Hold the blank on a plate against three pins by four clamps. See Figure 14–11.
5. After the inside patterns are finished, stop the machine and reclamp from the inside. Remove the outside clamps and finish the outside complete.

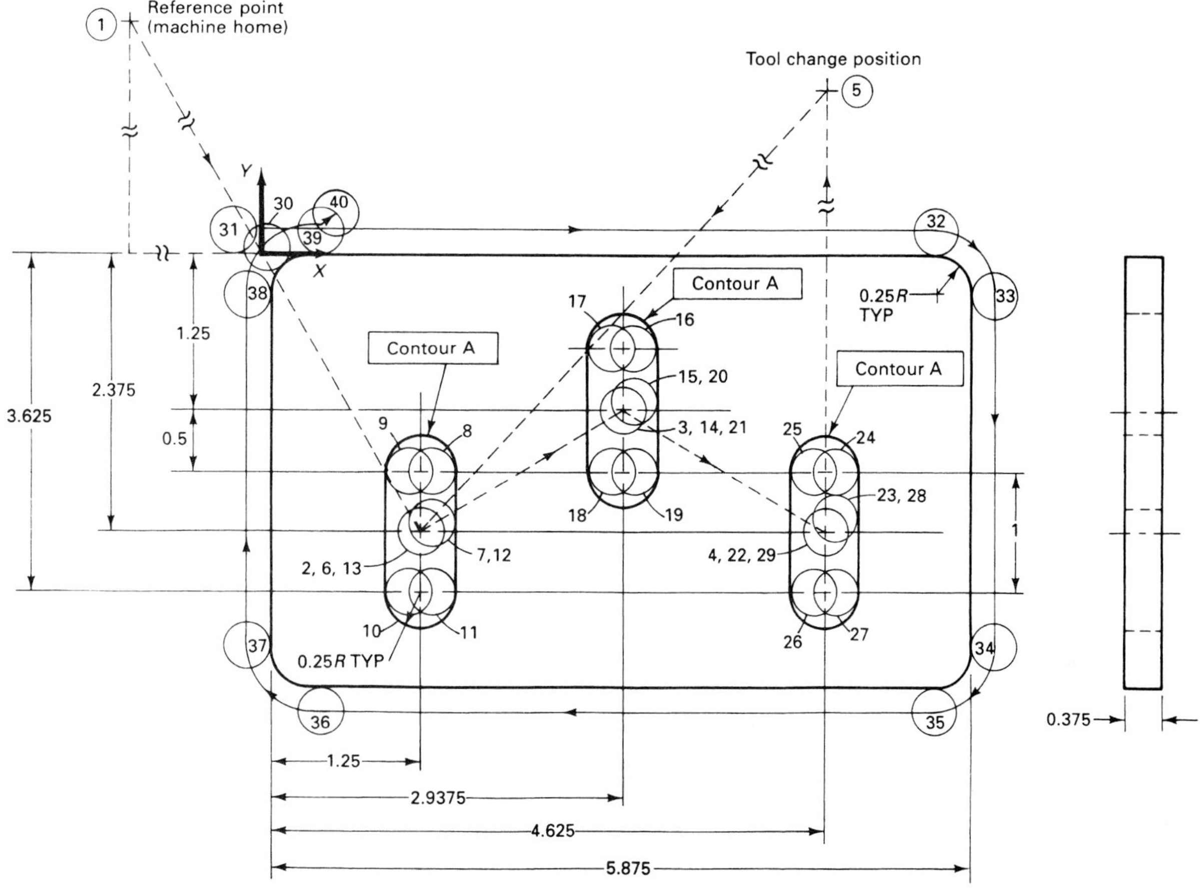

FIGURE 14-10

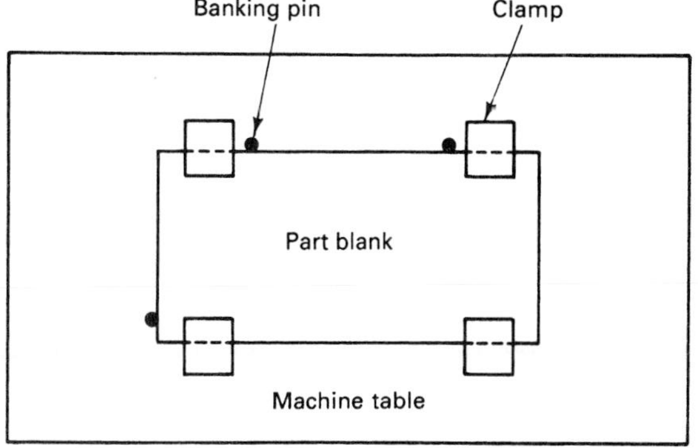

FIGURE 14-11

Programming Pattern

CNC Program Using Subprogramming

Programming Pattern	*Word address command*	*Meaning*
	%	End of tape.
	01411	Main program number.
	N0010 (..........)	Optional setup data.
Job setup data listing	N0020 (X0Y0 IS THE UPPER RIGHT HAND CORNER)	
	N0030 (Z0 IS THE TOP OF THE PART)	
	N0040 (..........)	
	N0050 (TOOLING LIST:)	
	N0060 (Tool 1: .5 DIA DRILL)	
	N0070 (TOOL 2: .375 DIA END MILL)	
	N0080 (..........)	
	N0090 (..........)	
Machine start-up sequence Change to Tool 1	N0100 G90 G20 G40 G80	Absolute, inch mode, cancel cutter diameter compensation and fixed cycles.
	N0110 G91 G28 X0 Y0 Z0	Return to reference point.
	N0120 G92 X−10.0 Y5.0 Z0.0	Preset absolute zero point.
	N0130 T1 M06	Change to tool 1.
	N0140 (TOOL 1: DRILL (3) ENTRANCE HOLES)	
	N0150 G00 G90 X1.25 Y−2.375 Z0 S800 M03 T2	Rapid to ②. Start spindle (CW) at 800 rpm.
	N0160 G43 Z.1 H01	Rapid tool 1 to 0.1 above part.
	N0170 M08	Coolant on.
Hole operation 1 sequence	N0180 G83 X1.25 Y−2.375 Z−.6 R.1 Q.25 F10.0	Deep drill cycle at ②. Final depth is 0.6, 0.25 peck depth at feed rate 10 in./min. Rapid to 0.1 above part.
	N0190 X2.9375 Y−1.25	Deep drill at ③.
	N0200 X4.625 Y−2.375	Deep drill at ④.
	N0210 G80	Cancel any fixed cycle.
Machine stop sequence	N0220 G00 G90 Z1.0 M05	Rapid tool to 1.0 above part. Spindle off.
	N0230 M09	Coolant off.
Change to tool 2 Machine start sequence	N0240 G91 G28 Z0 Y0	Rapid to tool change position ⑤.
	N0250 T2 M06	Change to tool 2.
	N0260 (TOOL 2: PROFILE (3) INSIDE PATTERNS)	
	N0270 G00 G90 X1.25 Y−2.375 Z0 S1200 M03	Rapid to ⑥. Start spindle (CW) at 1200 rpm.
	N0280 G43 Z.1 H02	Rapid tool 2 to 0.1 above part.
	N0290 M08	Coolant on.
	N0300 G92 X0 Y0	Preset absolute zero at ⑥.

CNC PROGRAM ***Continued***

Sequence	Program	Comment
Transfers to subprogram 0030 for executing profiling operation 2 sequence	N0310 G98 P0030	Transfer 1 to subprogram 0030. Cut contour A tool positions ⑥ ... ⑬.
	N0320 G00 X1.6875 Y1.125	Rapid to ⑭.
	N0330 G92 X0 Y0	Preset absolute zero at ⑭.
	N0340 M98 P0030	Transfer 2 to subprogram 0030. Cut contour A tool positions ⑭... ㉑.
	N0350 G00 X1.6875 Y−1.125	Rapid to ㉒.
	N0360 G92 X0 Y0	Preset absolute zero at ㉒.
	N0370 M98 P0030	Transfer 3 to subprogram 0030. Cut contour A tool positions ㉒ ... ㉙.
Machine stop for fixture change sequence	N0380 G00 Z10.0	Rapid tool to 10 in. above part.
	N0390 G00 X−4.625 Y2.375	Rapid to ㉚.
	N0400 G92 X0 Y0	Preset absolute zero at ㉚.
	N0410 M00 (CHANGE CLAMPS)	
	N0420 (FINISH PART COMPLETE)	
Machine start sequence	N0430 M03	Spindle on (CW).
	N0440 M08	Coolant on.
Profile operation 3 sequence	N0450 G00 X−.2875 Y.1875	Rapid to ㉛.
	N0460 Z.1	Rapid to 0.1 above part.
	N0470 G01 Z−.4	Plunge to (−0.4) at feed rate.
	N0480 X5.625	Cut to ㉜.
	N0490 G02 X6.0625 Y−.25 I0.J−.4375	Cut 0.25*R* arc to ㉝.
	N0500 G01 Y−3.375	Cut to ㉞.
	N0510 G02 X5.625 Y−3.8125 I−.4375 J0	Cut 0.25*R* arc to ㉟.
	N0520 G01 X.−5	Cut to ㊱.
	N0530 G02 X.−1875 Y−3.375 I0 J.4375	Cut 0.25*R* arc to ㊲.
	N0540 G01 Y−.25	Cut to ㊳.
	N0550 G02 X.25 Y.1875 I.4375 J0	Cut 0.25*R* arc to ㊴.
	N0560 G01 X.35 Y.2375	Cut to ㊵.
Machine stop Program end sequence	N0570 G00 G90 Z1.M05	Rapid to 1.0 above part. Spindle off.
	N0580 M09	Coolant off.
	N0590 G91 G28 X0 Y0 Z0	Return to *XYZ* reference point.
	N0600 M30	Program end, memory reset.

Subprogram 0030 for executing profiling operation 2 and return

O0030	Subprogram number.	
N0010 G01 Z−.4 F8.0	Plunge to −0.4 at feed rate.	
N0020 X.0625 Y.0625	Transfer 1	Cut to ⑦.
N0030 Y.5		Cut to ⑧.
N0040 G03 X−.0625 Y.5 I−.0625 J0		Cut .25*R* arc to ⑨.
N0050 G01 Y−.5		Cut to ⑩.
N0060 G03 X.0625 Y−.5 I.0625 J0		Cut 0.25*R* arc to ⑪.
N0070 G01 Y.0625		Cut to ⑫
N0080 G01 X0 Y0		Cut to ⑬.
N0090 G00 Z.1	Rapid to 0.1 above part.	
N0100 M99	Return to main program block.	

■ EXAMPLE 14–9

The multiple-part shapes shown in Figure 14–12 are to be machined. Write a CNC program for executing the operations described in the data table. Use main and subprograms.

Tool	Operation	Tooling	Speed (rpm)	Feed (ipm)
1	Drill (2) 0.257*D* holes (8) times	F Ltr drill	1600	8
2	Rough profile 0.01 full (4) times	⅜ rough stub end mill	2000	10
3	Finish profile (4) times	¼ stub end mill	2200	6

Notes: 1. X_0Y_0 is given on the fixture illustration.
2. Z_0 is the top of the part.
3. After operation 1, stop the machine and add ¼-20 screws thru holes. Finish setup complete.

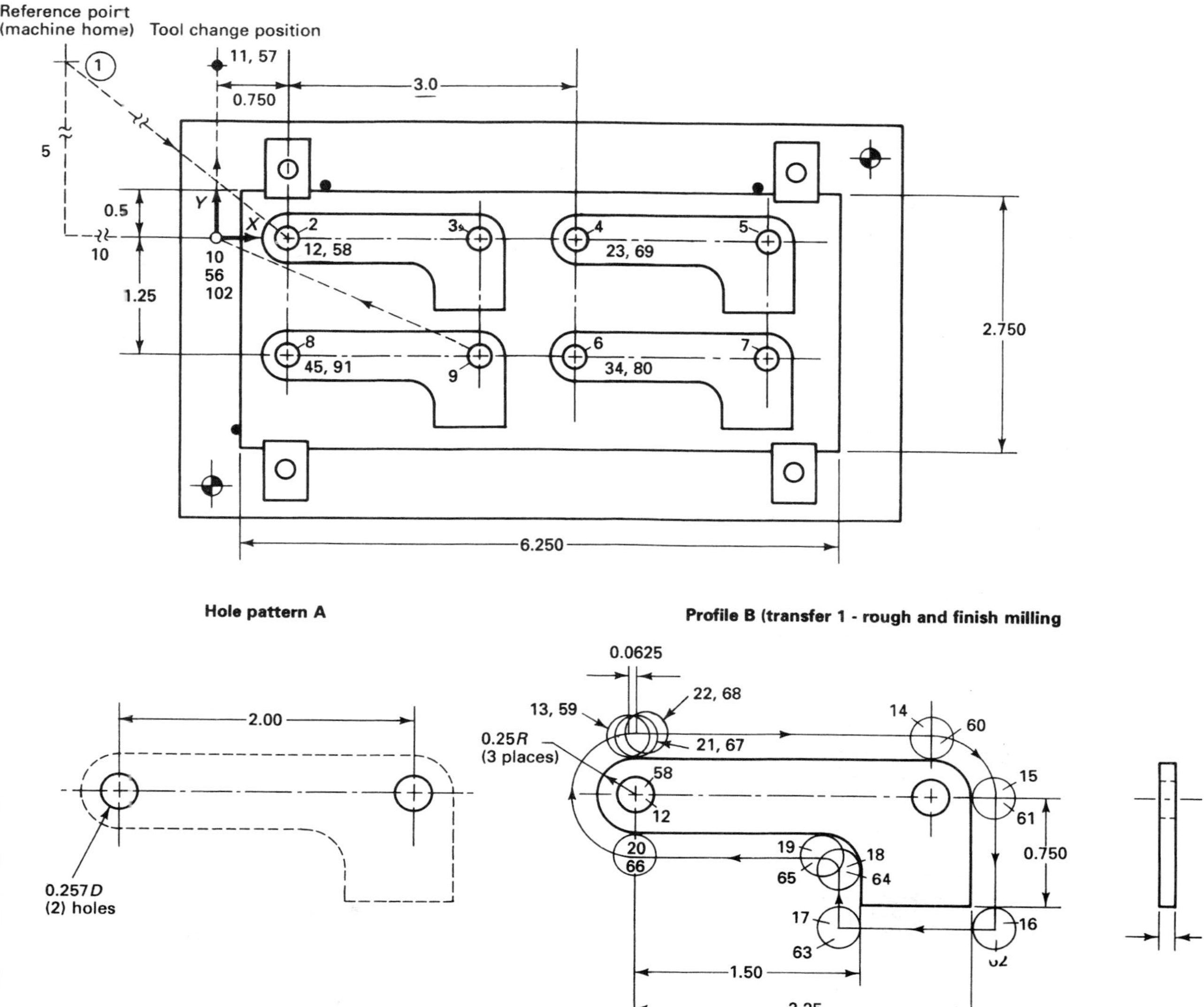
Reference point
(machine home)
Tool change position
1
11, 57
0.750
3.0
5
0.5
10
1.25
10
56
102
Y
X
2
12, 58
3
4
23, 69
5
8
45, 91
9
6
34, 80
7
2.750
6.250
Hole pattern A
Profile B (transfer 1 - rough and finish milling
0.0625
13, 59
22, 68
21, 67
0.25R
(3 places)
58
12
14
60
15
61
20
66
19
65
18
64
17
63
16
0.750
1.50
2.25
2.00
0.257D
(2) holes
0.125

FIGURE 14–12

Programming Pattern

CNC Program Utilizing Main Subprograms

	Word address command	*Meaning*
	%	End of tape.
	01312	Main program number.
	N0010 (..........)	Optional setup data.
Machine setup data listing	N0020 (X0Y0 IS A .375 HOLE IN THE FIXTURE)	
	N0030 (Z0 IS THE TOP OF THE PART)	
	N0040 (..........)	
	N0050 (TOOLING LIST:)	
	N0060 (Tool 1: 'F' LTR DRILL)	
	N0070 (TOOL 2: 3/8D ROUGH STUB END MILL)	
	N0080 (TOOL 3: .250 DIA STUB END MILL)	
	N0090 (..........)	
	N0100 (..........)	
Tool 1 in spindle Machine start sequence	N0110 G90 G20 G40 G80	Absolute, inch mode, cancel cutter diameter compensation and fixed cycles.
	N0120 G91 G28 X0 Y0 Z0	Return to reference point.
	N0130 G92 X−10.0 Y 5.0 Z0	Preset absolute zero point.
	N0140 T1 M06	Change to tool 1.
	N0150 (TOOL 1: DRILL .275 DIA HOLES)	
	N0160 G00 G90 X.75 Y0 Z0 S1600 M03	Rapid to ②. Spindle on (CW) at 1600 rpm.
	N0170 G43 Z.1 H01	Rapid tool 1 to 0.1 above part.
	N0108 M08	Coolant on.
	N0190 G92 X0 Y0	Preset absolute zero at ②.
Transfers to subprogram 0001 for executing hole operation 1 sequence	N0200 M98 P0001	Transfer 1 to subprogram 0001. Drill hole pattern A, holes ② ③.
	N0210 G00 X3.0 Y0	Rapid to ④.
	N0220 G92 X0 Y0	Preset absolute zero at ④.
	N0230 M98 P0001	Transfer 2 to subprogram 0001. Drill hole pattern A, holes ④ ⑤.
	N0240 G00 X0 Y−1.25	Rapid to ⑥.
	N0250 G92 X0 Y0	Preset absolute zero at ⑥.
	N0260 M98 P0001	Transfer 3 to subprogram 0001. Drill hole pattern A, holes ⑥ ⑦.
	N0270 G00 X−3.0 Y0	Rapid to ⑧.
	N0280 G92 X0 Y0	Preset absolute zero at ⑧.
	N0290 M98 P0001	Transfer 4 to subprogram 0001. Drill hole pattern A, holes ⑧ ⑨.

CNC PROGRAM ***Continued***

Sequence	Program	Comment
Preset absolute zero sequence	N0300 G00 X−.75 Y1.25	Rapid to ⑩.
	N0310 G92 X0 Y0	Preset absolute zero at ⑩.
Machine stop sequence	N0320 G00 G90 Z1.0 M05	Rapid to 1.0 above part. Spindle off.
	N0330 M09	Coolant off.
Change to tool 2 Machine start sequence	N0340 G91 G28 Z0 Y0	Rapid to tool change position ⑪.
	N0350 T2 M06	Change to tool 2.
	N0360 (TOOL 2: ROUGH PROFILE .01 FULL)	
	N0370 G00 G90 X.75 Y0 Z0 S2000 M03	Rapid to ⑫. Spindle on (CW) at 2000 rpm. Prepare tool 3 in ready position.
	N0380 G43 Z.1 H02	Rapid tool 2 to 0.1 above part.
	N0390 M08	Coolant on.
	N0400 G92 X0 Y0	Preset absolute zero at ⑫.
Transfers to subprogram 0002 for executing profiling operation 2 sequence	N0410 M98 P0002	Transfer 1 to subprogram 0002. Rough cut profile B, tool positions ⑫ ... ㉒.
	N0420 G00 X3.0 Y0	Rapid to ㉓.
	N0430 G92 X0 Y0	Preset absolute zero at ㉓.
	N0440 M98 P0002	Transfer 2 to subprogram 0002. Rough cut profile B, tool positions ㉒ ... ㉜.
	N0450 G00 X0 Y−1.25	Rapid to ㉞.
	N0460 G92 X0 Y0	Preset absolute zero at ㉞.
	N0470 M98 P0002	Transfer 3 to subprogram 0002. Rough cut profile B, tool positions ㉞ ... ㊹.
	N0480 G00 X−3.0 Y0	Rapid to ㊺.
	N0490 G92 X0 Y0	Preset absolute zero at ㊺.
	N0500 M98 P0002	Transfer 4 to subprogram 0002. Rough cut profile B, tool positions ㊺ ... (55).
Preset absolute zero sequence	N0510 G00 X−.75 Y1.25	Rapid to (56).
	N0520 G92 X0 Y0	Preset absolute zero at (56).
Machine stop sequence	N0530 G00 G90 Z1.0 M05	Rapid to 1.0 above part. Spindle off.
	N0540 M09	Coolant off.
Change to tool 3 Machine start sequence	N0550 G91 G28 Z0 Y0	Rapid to tool change position (57).
	N0560 T3 M06	Change to tool 3.
	N0570 (TOOL 3: FINISH PROFILE FOUR TIMES)	
	N0580 G00 G90 X.75 Y0 Z0 S2200 T1	Rapid to (58) Spindle on (CW) at 2200 rpm. Prepare tool 1 in ready position.
	N0590 G43 Z.1 H03	Rapid tool 3 to 0.1 above part.
	N0600 M08	Coolant on.
	N0610 G92 X0 Y0	Preset absolute zero at (58).

Transfers to subprogram 0003 for executing profiling operation 3 sequence	N0620 M98 P0003	Transfer 1 to subprogram 0003. Finish cut profile B, tool positions (58) ... (68).
	N0630 G00 X3.5 Y0	Rapid to (69).
	N0640 G92 X0 Y0	Preset absolute zero at (69).
	N0650 M98 P0003	Transfer 2 to subprogram 0003. Finish cut profile B, tool positions (69) ... (79).
	N0660 G00 X0.5 Y−1.25	Rapid to (80).
	N0670 G92 X0 Y0	Preset absolute zero at (80).
	N0680 M98 P0003	Transfer 3 to subprogram 0003. Finish cut profile B, tool positions (80) ... (90).
	N0690 G00 X−3.0 Y0	Rapid to (91).
	N0700 G9 X0 Y0	Preset absolute zero at (91).
	N0710 M98 P0003	Transfer 4 to subprogram 0003. Finish cut profile B, tool positions (91) ... (101).
Preset absolute zero sequence	N0720 G00 X−.75 Y1.25	Rapid to (102)
	N0730 G92 X0 Y0	Preset absolute zero at (102).
Machine stop Program end sequence	N0740 G90 Z1.0 M05	Rapid to 1.0 above part. Spindle off.
	N0750 M09	Coolant off.
	N0760 G91 G28 X0 Y0 Z0	Return to *XYZ* reference point.
	N0770 M30	Program end, memory reset.
	%	

Subprogram 0001 for executing hole operation 1 and return sequence	O0001	Subprogram number	
	N0010 G81 X0 Y0 Z−.35 R.1 F8.0	Transfer 1	Drill cycle at ②. Final depth is −0.35 at feed rate 8 in./min. Rapid to 0.1 above part.
	N0020 X2.5 Y.5	Drill hole at ③.	
	N0030 G80	Cancel any fixed cycles.	
	N0040 M99	Return to main program block.	

Subprogram 0002 for executing profiling operation 2 and return sequence

O0002	Subprogram number	
N0010 G00 G90 X−.0625 Y.4475	Transfer 1	Rapid to (13).
N0020 G01 Z−.14 F10.0		Plunge to −0.14 at feed rate.
N0030 X2.0		Cut to (14).
N0040 G02 X2.4475 Y0 I0 J2−.4475		Cut 0.25*R* arc to (15).
N0050 G01 Y−1.1975		Cut to (16).
N0060 X1.3025		Cut to (17).
N0070 Y−.5		Cut to (18).
N0080 G03 X2.25 Y−.4475 I−.0525 J0		Cut 0.25*R* arc to (19).
N0090 G01 X0		Cut to (20).
N0100 G02 X0 Y.4475 I0 J.4475		Cut to 0.25*R* arc to (22).
N0110 G01 X.0625 Y.51		Cut to (23).
N0120 G00 Z.1	Rapid to 0.1 above part.	
N0130 M99	Return to main program block.	

Subprogram 0003 for executing profiling operation 3 and return sequence

O0003	Subprogram number	
N0010 G00 G90 X−.0625 Y.375	Transfer 1	Rapid to (59).
N0020 G01 Z−.14 F6.0		Plunge to −0.14 at feed rate.
N0030 X2.0		Cut to (60).
N0040 G02 X2.375 Y0 I0 J−.375		Cut 0.25*R* arc to (61).
N0050 G01 Y−1.125		Cut to (62).
N0060 X1.375		Cut to (63).
N0070 Y−.5		Cut to (64).
N0080 G03 X2.25 Y−.375 I−.125 J0		Cut 0.25*R* arc to (65).
N0090 G01 X0		Cut to (66).
N0100 G02 X0 Y.375 I0 J.375		Cut 0.25*R* arc to (67).
N0110 G01 X.0625 Y.4375		Cut to (68).
N0120 G00 Z.1	Rapid to 0.1 above part.	
N0130 M99	Return to main program block.	

14-7 CHAPTER SUMMARY

The following key concepts were discussed in this chapter:

1. Subprogramming is an efficiency device used to make the task of programming complex parts easier and less error prone.
2. The subprogramming technique can significantly reduce the number of program blocks for a job. A program written with subprograms requires less MCU memory space and executes faster.
3. Subprogramming should be used when a machining pattern is repeated at several locations in a part.
4. The main program and subprogram are separate programs. Each must be assigned a unique program number and title in the MCU.

REVIEW EXERCISES

For each of the parts given in this section, write a main program and the appropriate subprograms required for executing the machining operations. Use the data table information in each case for specifying operations, tooling, tool speeds, and tool feeds.

14.1. Main program number: 1413 (Figure 14–13)
Subprogram number: 1414

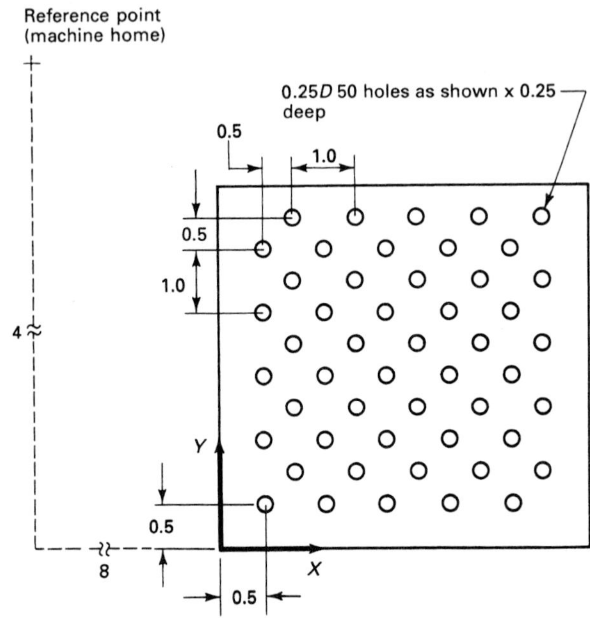

FIGURE 14–13

Tool	Operation	Tooling	Speed (rpm)	Feed (ipm)
1	Drill (50) 0.25*D* holes ×.25 deep	0.25*D* drill	1800	8

Notes: 1. X_0 Y_0 is at the lower left-hand corner.
2. Z_0 is the top of the part.
3. The part is held in a vise with a stop on the left.
4. The blank supplies is 6 in. × 6 in. square.

14.2. Main program number: 1415 (Figure 14–14)
Subprogram number: 1416

Tool	Operation	Tooling	Speed (rpm)	Feed (ipm)
1	Drill (6) entrance holes	3/8*D* drill	1800	8
2	Profile mill (6) contour A stop, reclamp inside, finish outside profile	0.25*D* end mill	1300	7

Notes: 1. X_0 Y_0 is at the upper left-hand corner.
2. Z_0 is the top of the part.
3. The part blank is supplied in size 5.125 in. × 3.375 in. × 0.25 in.
4. Hold the blank on a plate against three pins by four clamps. See Figure 14–15.
5. After the inside contours are finished, stop the machine. Reclamp from the inside. Remove the outside clamps and finish the part complete.

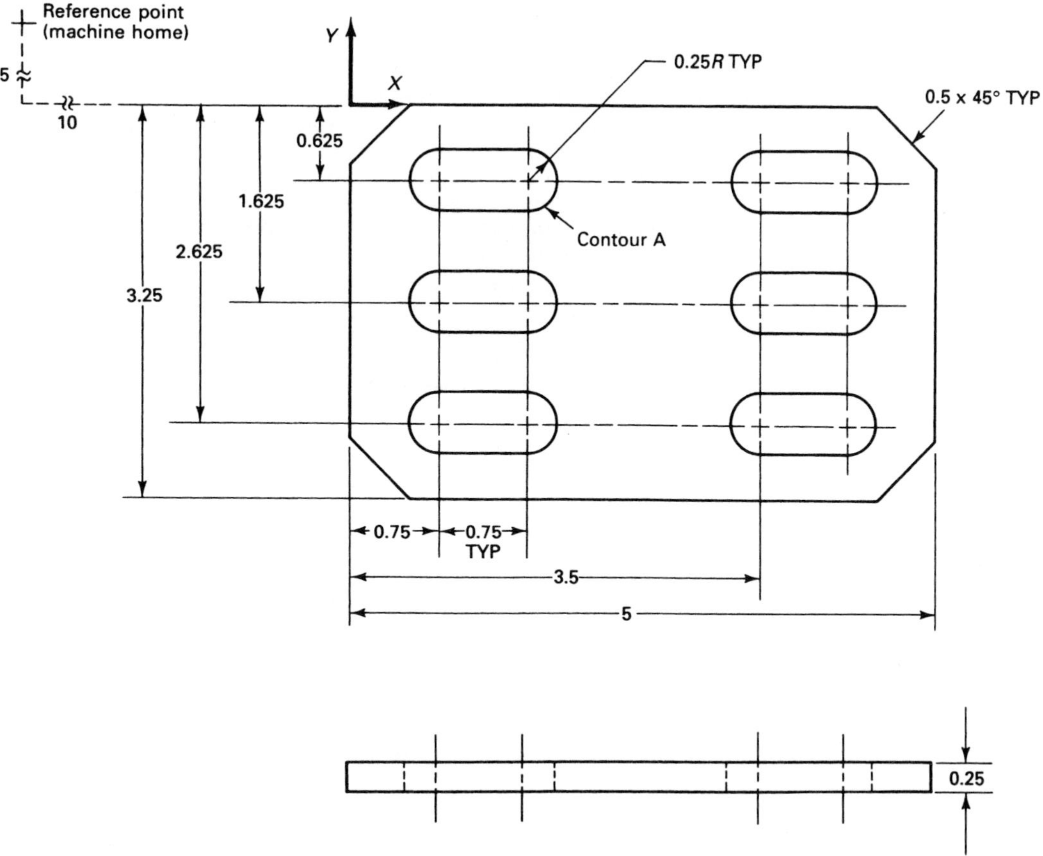

FIGURE 14–14

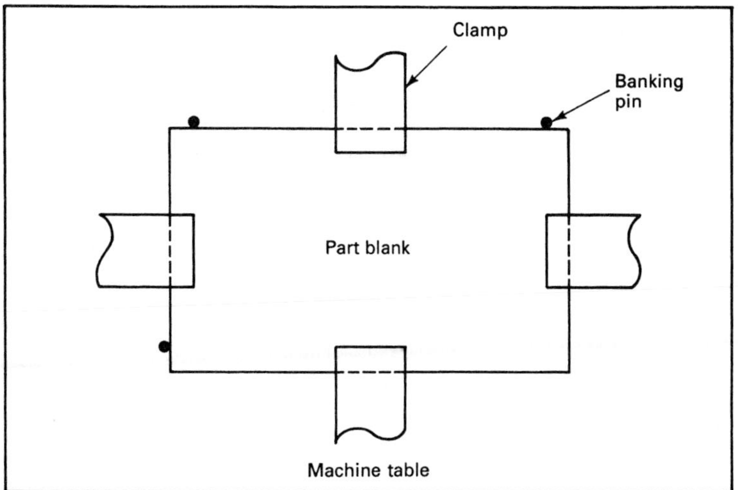

FIGURE 14–15

14.3. Main program number: 1417 (Figure 14–16)
Subprogram number: 1418

Tool	Operation	Tooling	Speed (rpm)	Feed (ipm)
1	Drill (3) entrance holes	½*D* drill	1000	10
2	Finish (3) pattern A stop, reclamp from inside and finish outside	0.5*D* end mill	1200	8

Notes: 1. X_0 Y_0 is at the lower left-hand corner.
2. Z_0 is the top of the part.
3. The part blank is supplied as 6.125 in. × 4.625 in. × 0.5 in.
4. Hold the blank on a plate against three pins by four clamps. See Figure 14–17.
5. After the inside patterns are finished, stop the machine. Reclamp from the inside. Remove two clamps from the left-hand side and finish complete.

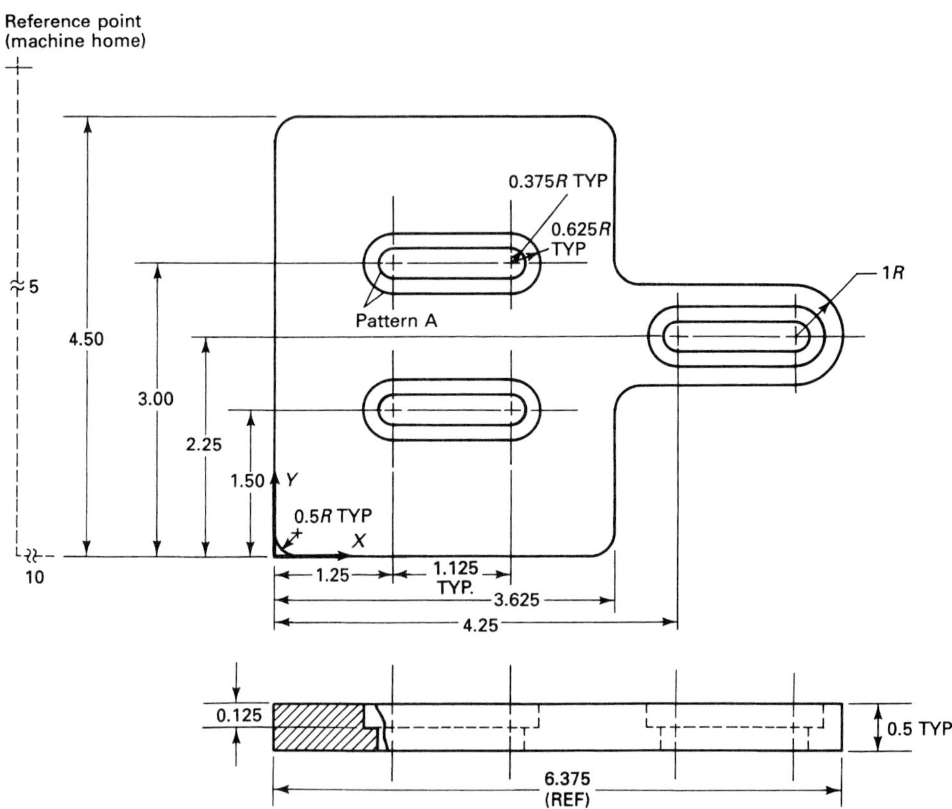

FIGURE 14–16

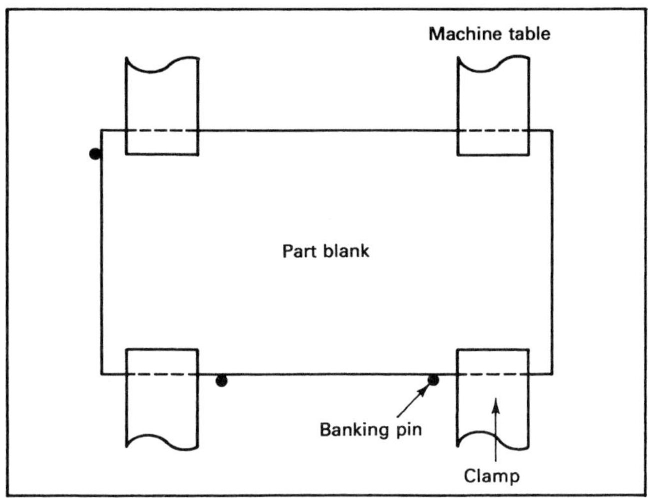

FIGURE 14–17

14.4. Main program number: 1419 (Figure 14–18)
Subprogram 1 number: 1420
Subprogram 2 number: 1421
Subprogram 3 number: 1422
Subprogram 4 number: 1423
Subprogram 5 number: 1424

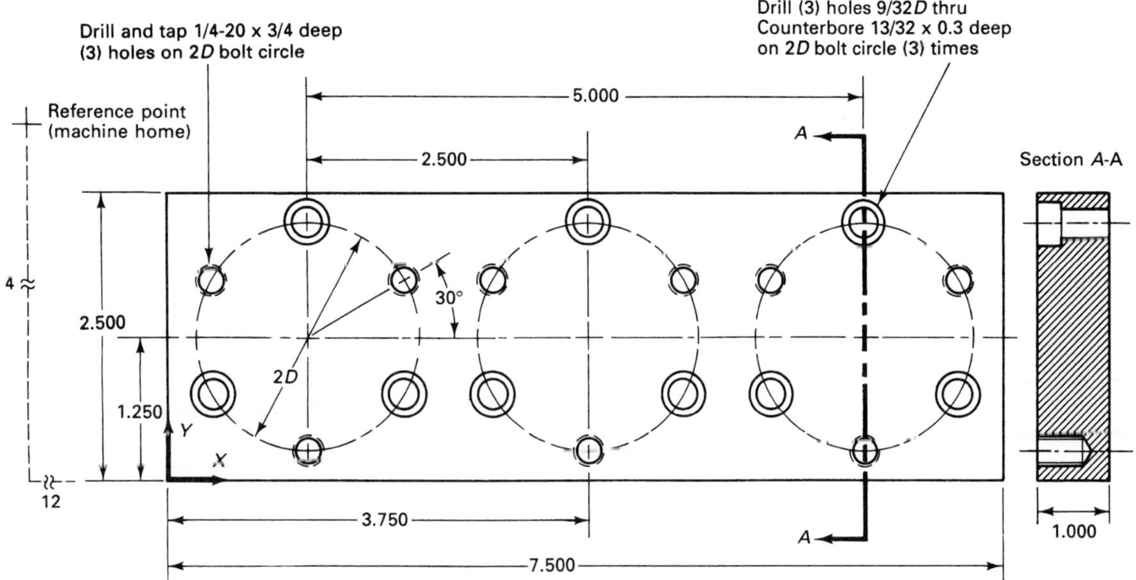

FIGURE 14–18

Tool	Operation	Tooling	Speed (rpm)	Feed (ipm)
1	Center drill all holes	No. 2 center drill	1800	8
2	Drill 9/32*D* holes thru	9/32*D* drill	1500	7
3	Counterbore 13/32*D* × 0.3 deep	13/32*D* flat bottom drill	600	4
4	Drill for 1/4-20 tap	No. 7 (0.201*D*) drill	2000	6
5	Tap 1/4-20 holes	1/4-20 tap in floating holder	400	20

Notes: 1. X_0 Y_0 is at the lower left-hand corner.
2. Z_0 is the top of the part.
3. The part is held in a vise with a stop on the left.
4. The part blank supplied is 7.5 in. × 2.5 in. × 1 in.

14.5. Main program number: 1425 (Figure 14–19)
Subprogram 1 number: 1426
Subprogram 2 number: 1427
Subprogram 3 number: 1428

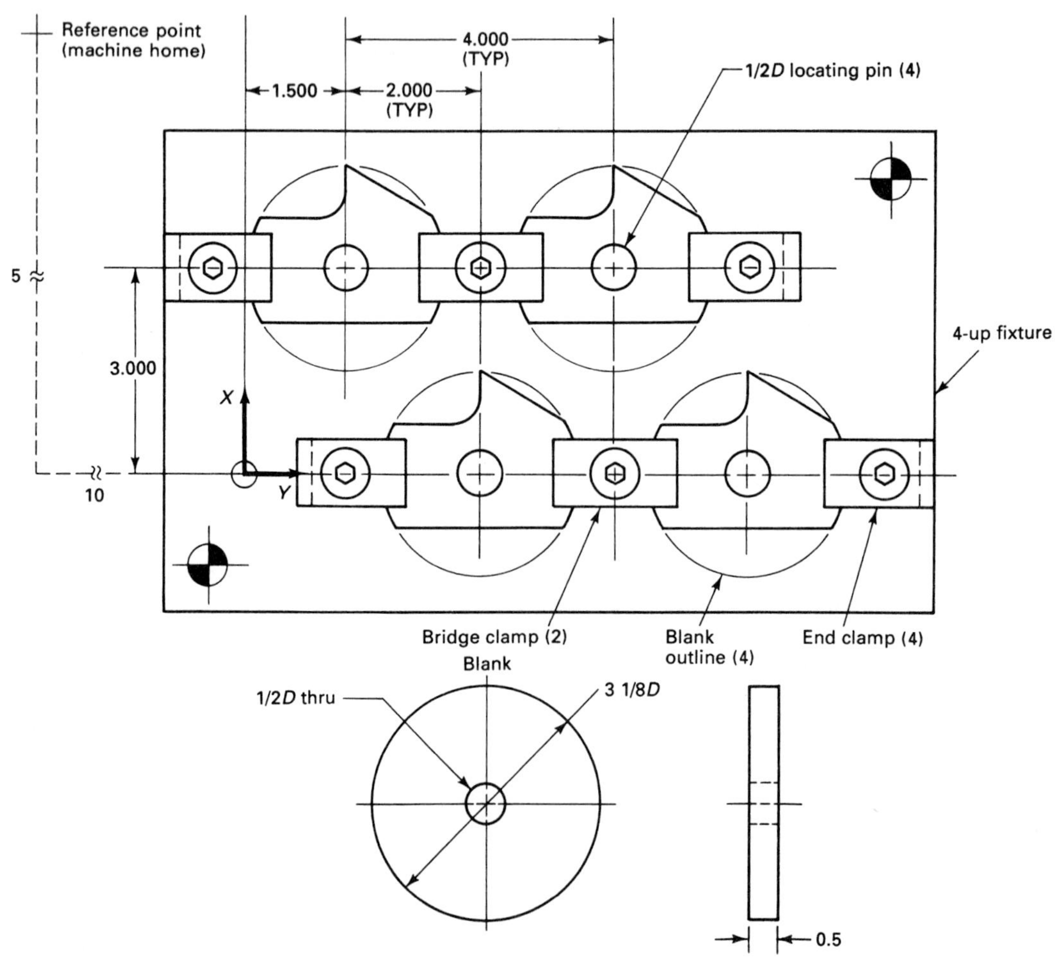

FIGURE 14–19

Tool	Operation	Tooling	Speed (rpm)	Feed (ipm)
1	Rough 0.625*D* and counterbore and profile	1/2*D* stub rough end mill	800	8
2	Finish 0.625*D* to 0.627 and counterbore and profile	1/2*D* stub finish end mill	1000	6
3	Finish .625 ± .0005 diameter hole	0.6250*D* boring bar	500	2

Notes: 1. X_0 Y_0 is given on the fixture.
2. Z_0 is the top of the part.
3. The blank for the part is turned on the lathe to the dimensions shown for the blank.
4. The part is machined on a 4-up fixture in the following order:
   - Locate the blanks off the ½ diameter holes by pins.
   - Hold the blanks down by clamps.
   - Remove the pins and machine the blanks as required. See Figure 14–20.

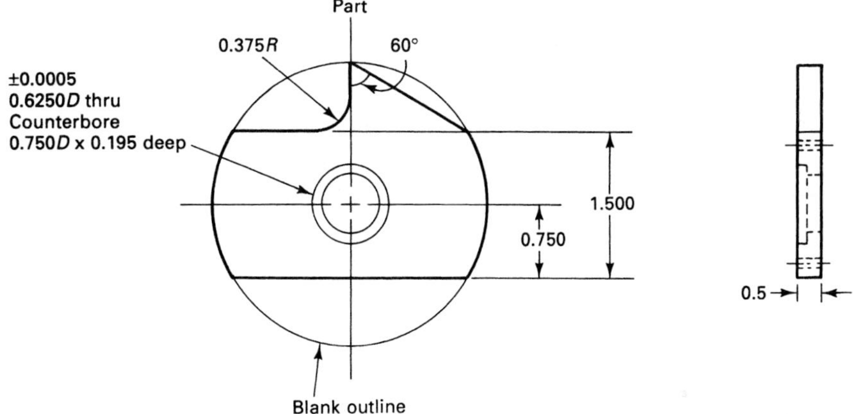

FIGURE 14–20

CHAPTER 15

Introduction to the CNC Lathe

15-1 CHAPTER OBJECTIVES

At the conclusion of this chapter you will be able to

1. Understand the basic elements comprising the CNC lathe.
2. Identify the axis of motion for CNC lathes.
3. Describe the most important cutting operations performed on the CNC lathe.
4. State the different types of tooling used for CNC lathe cutting operations.
5. Name the important components of CNC lathe tooling.
6. Explain lathe feeds and speeds.
7. Understand the importance of tool feed direction and rake angles for cutting materials.

15-2 INTRODUCTION

This chapter considers the basic elements of modern CNC lathes and turning centers. The lathe axes are discussed and illustrated. The most important lathe operations are examined, together with the type of tooling required for each operation. Speeds and feeds for lathe work are very important for prolonging tool life and producing a desired quality of cut. Thus, a concise presentation of these parameters and a brief table of values for various materials are given. The concepts of tool clearance angles and back rake angles are fundamental and a brief discussion of their significance is presented. The chapter ends with a discussion of tool feed directions and rake angle recommendations for cutting different materials.

15-3 COMPONENTS OF MODERN CNC LATHES

CNC Lathe Description

The CNC lathe is a machine tool designed to remove material from stock that is clamped and rotated around an axis. Most metal cutting is done with a sharp single-point cutting tool. Modern CNC lathes use turrets to rigidly hold and move cutting tools. The turret is also used to quickly index an old tool out and a new tool into its cutting position. A front turret is built to move tools from below the spindle centerline up to the work. A rear turret, on the other hand, moves tools from above the spindle centerline down to the work. Machines equipped with front and rear turrets can execute cutting operations simultaneously from above and below the work. Refer to Figure 15–1.

General Safety Rules for Lathe Operations

- Prior to operation, confirm that there are no other personnel in the immediate vicinity of the machine.
- Be sure the machine has completely stopped before entering work area.

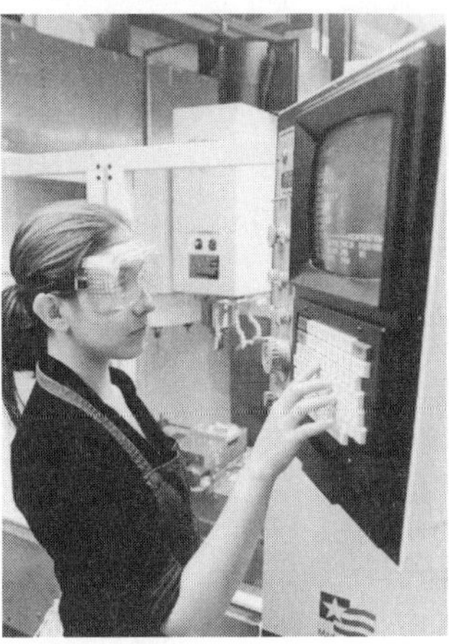

- Wear ANSI approved safety goggles. Secure all loose clothing. Put long hair up.
- Observe the following practices when cleaning the machine:
  - Clean the spindle bore and collets.
  - Use a wire brush to remove foreign particles from the threads.
  - Do not use an air hose.
- Check all seating areas on the tool holder to be sure each tool rests solidly.
- Revolve the spindle by hand before starting any operation. Check to see that the chuck jaws work and clear the carriage.
- Avoid excessive overhang of the cutting tool or boring bar.
- Position splash guards to contain the coolant and chips.
- Keep area around machine clean and dry and free of obstructions.

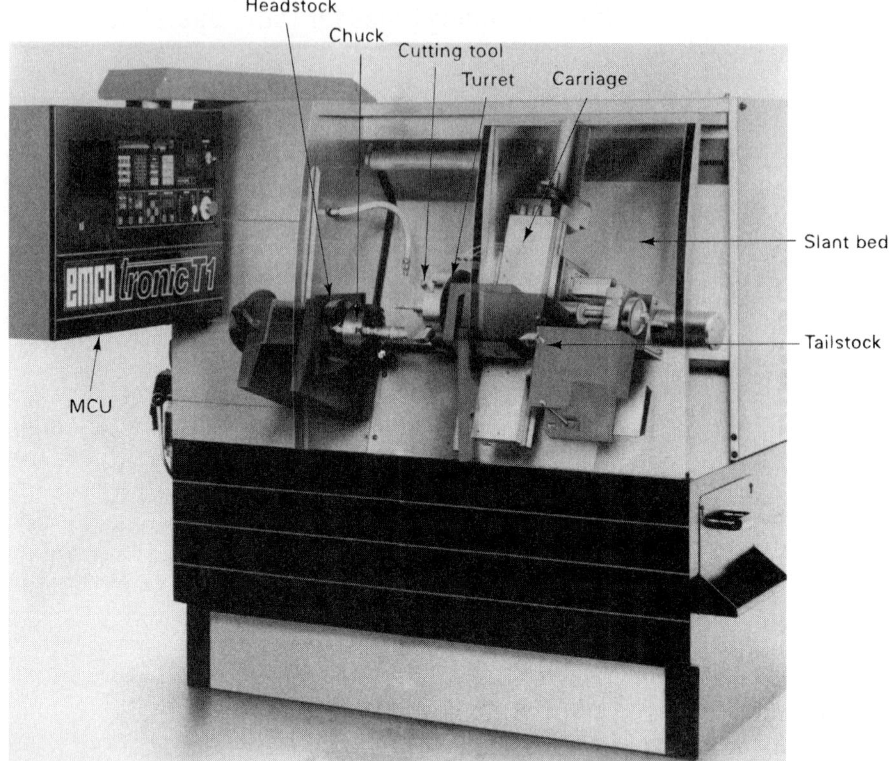

FIGURE 15–1 Components of the modern CNC lathe. (Photo courtesy of Emco Maier Corp.)

The bed of a CNC lathe is usually slanted to allow chips to fall away easily. A typical CNC turret lathe or turning center is shown in Figure 15–2. The components of the lathe are as follows:

Headstock: Contains the spindle shaft and transmission gearing used to rotate the spindle.

Chuck: Connects to the spindle and clamps the work.

FIGURE 15–2 A lathe equipped with front and rear turrets. (Photo courtesy of Monarch Machine Tool Company.)

Turret: Holds the cutting tool and replaces an old tool with a new tool during a tool change.

Carriage: Moves the cutting tool into the revolving work. It contains the saddle, cross-slide, and apron.

Tailstock: Supports the right end of the work.

Slant bed: Supports all the components just listed and provided a path for chips as they fall.

MCU: A computer used to store and process the CNC programs entered.

Tool Changing Mechanism

As was stated previously, a turret on a CNC lathe is used to quickly change tools when required. Each tool position on the turret is numbered for identifying the tool it holds. Tools can be mounted on the face of the turret as well as on its sides. Upon receiving a tool change command from the MCU, the turret moves to a safe tool change location and indexes the old tool out and the new tool in. It then proceeds to the proper coordinates programmed for cutting the part with the new tool. For turning centers with programmable tailstocks, the tailstock may have to be moved back before a tool change is executed. The motion of the turret during a typical tool change operation is illustrated in Figure 15–3.

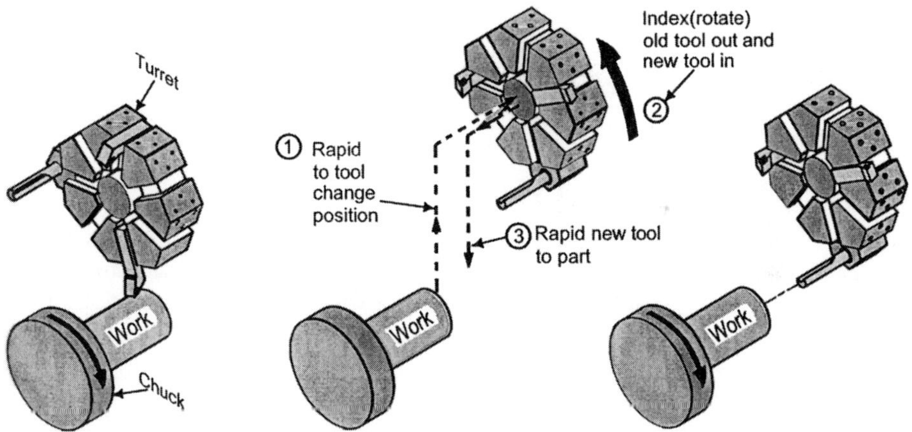

FIGURE 15–3 Turret motion during a tool change.

15–4 CNC LATHE AXES OF MOTION

In this text we will only consider programming basic two-axis machine motions when dealing with CNC lathes. The axes of importance are the Z and X axes. The Z axis is in the direction of the spindle. Positive $+Z$ is motion away from the spindle and work and negative $-Z$ is motion toward the spindle and into the work. The X axis controls the cross-slide movements. Positive $+X$ is in the direction away from the spindle centerline and negative $-X$ is motion toward the spindle centerline and into the work. Refer to Figure 15–4. In some machines with programmable tailstocks, the W axis is used to designate the movement of the tailstock. This accounts for a third axis on these machines. More complex turning centers may also have a fourth axis.

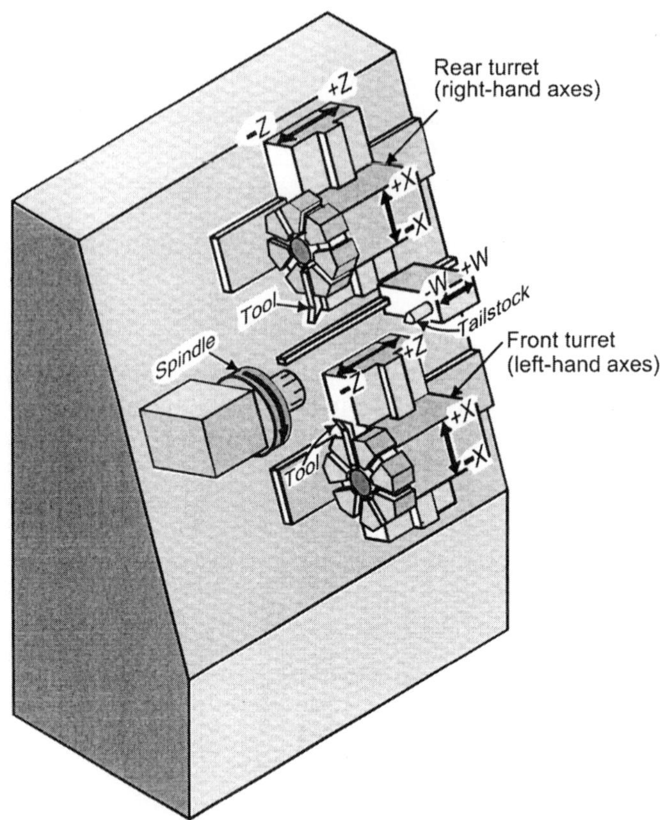

FIGURE 15–4 Typical machine axes for CNC lathes.

15–5 BASIC LATHE OPERATIONS

This text will consider programming the most basic lathe operations: facing, turning, grooving, parting, drilling, boring, and threading. Some of these cuts are performed on both the outside surface of the part (OD operations) as well as the inside surface (ID operations). These are illustrated in Figures 15–5 through 15–11.

Facing

This operation involves cutting the end of the stock such that the resulting end is perpendicular or square with respect to the stock centerline. A smooth flat end surface should be produced. The tool is fed into the work in a direction perpendicular to the stock centerline.

Turning

Turning involves the removal of material from the outside diameter of rotating stock. Different profile shapes can be created including tapers, contours, and shoulders. A rough cut pass is usually made first. This is followed by one or more finishing passes.

Grooving

OD and ID grooving requires that the tool be fed into the work in a direction perpendicular to the work's centerline. The cutting edge of the tool is on its end.

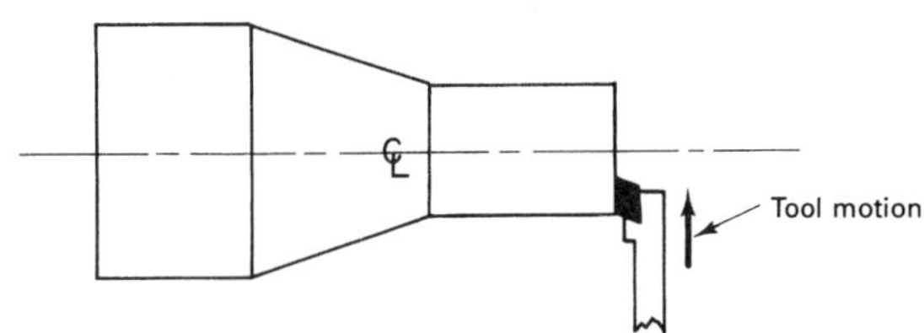

FIGURE 15–5 OD facing.

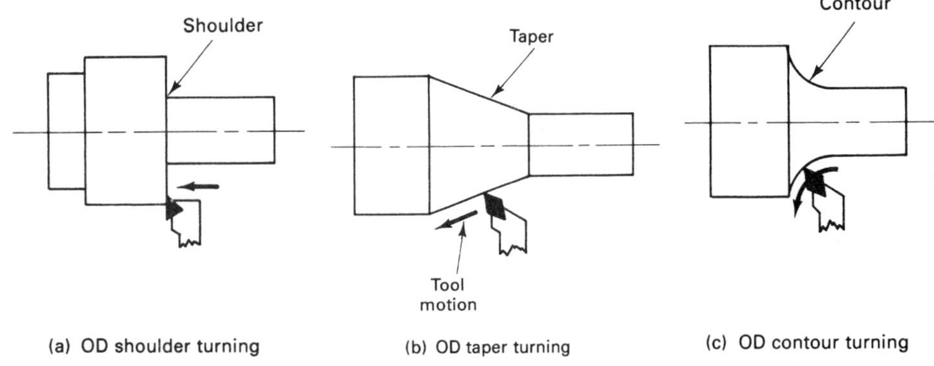

FIGURE 15–6 Turning.

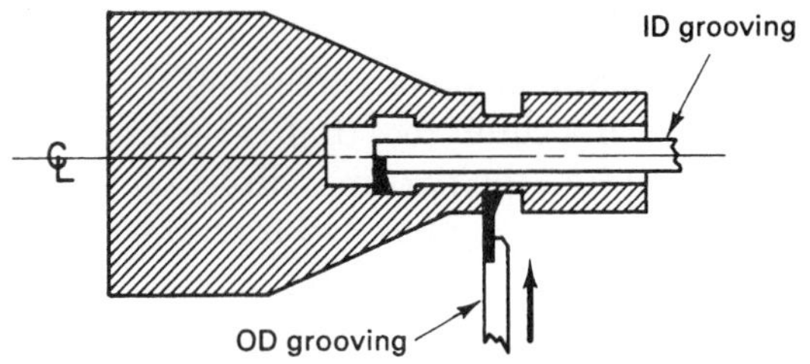

FIGURE 15–7 Grooving.

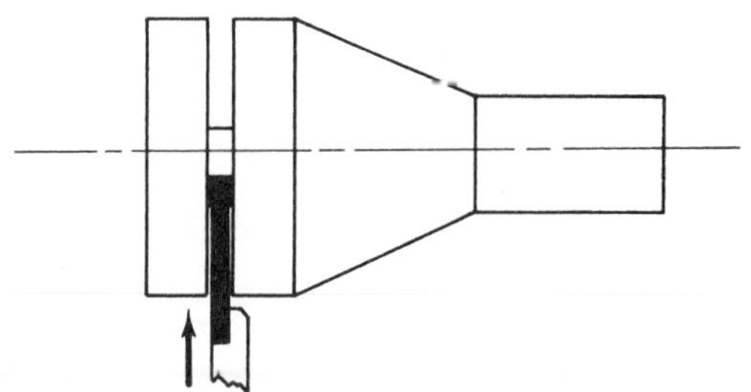

FIGURE 15–8 Parting.

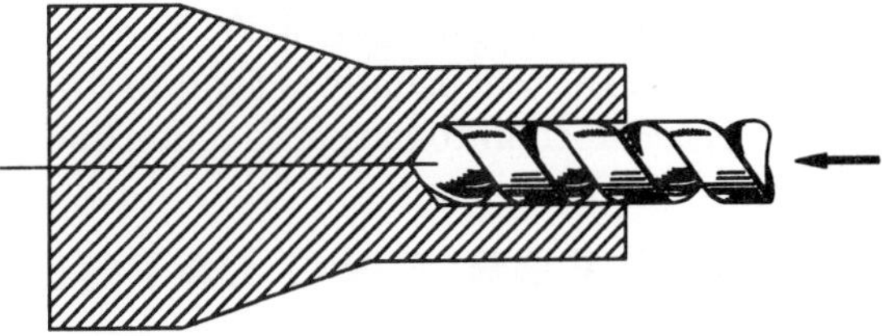

FIGURE 15–9 Drilling.

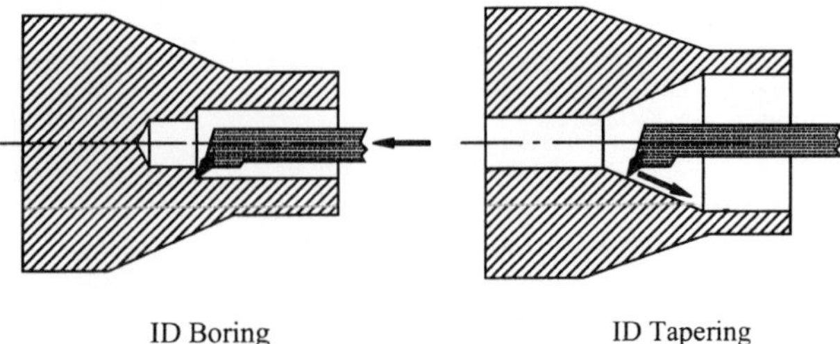

FIGURE 15–10 Boring.

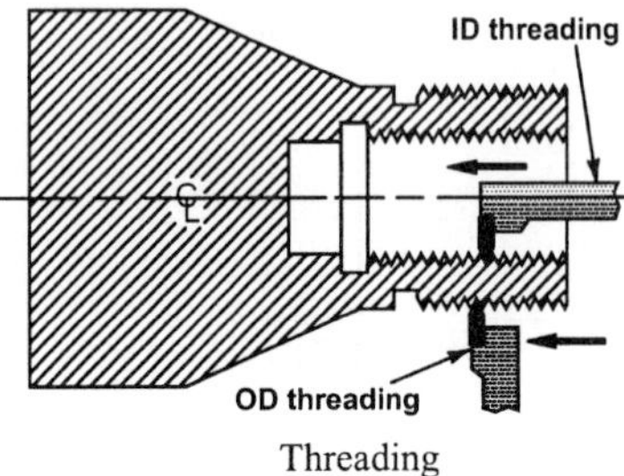

FIGURE 15–11 Threading.

Grooving for thread relief is usually done prior to threading to ensure that the resulting threads will be fully engaged up to a shoulder.

Parting

Parting involves cutting off the part from the main bar stock. Parting is done with a cutoff tool that is tapered and has a cutting edge at its end. The tool is fed into the part in a direction perpendicular to its centerline until the part is completely separated from the main bar stock.

Drilling

The drill is usually mounted in a drill chuck or held in a bushing and fed into the rotating work. A center drill should be applied first before using a high-speed-steel twist drill. This is not necessary if a spade drill or carbide insert drill is to be used.

Boring

Boring is an internal turning operation. It involves the enlargement of a previously drilled hole. Boring can be used to more accurately size and true a hole as well as to create internal tapers and contours.

Threading

This operation involves the cutting of helical grooves on the outside or inside surface of a cylinder or cone. The grooves or threads have a specific angle. Unified threads have an angle of 60°. The distance between the teeth is called the thread pitch. The tool usually is fed into the material at a cut angle of 29° in a direction perpendicular to the part's centerline and at a feed rate equal to the thread pitch.

15–6 TOOLING FOR CNC LATHE OPERATIONS

Modern CNC turning centers utilize tool holders with replaceable/indexable inserts. Tool holders come in a variety of styles. Each style is suited for a particular type of cutting operation as illustrated in Figure 15–12.

Kennanmetal Standard Tool Style

Kennanmetal tools conform to industry standards and are made in styles and sizes for all general types of machining operations:

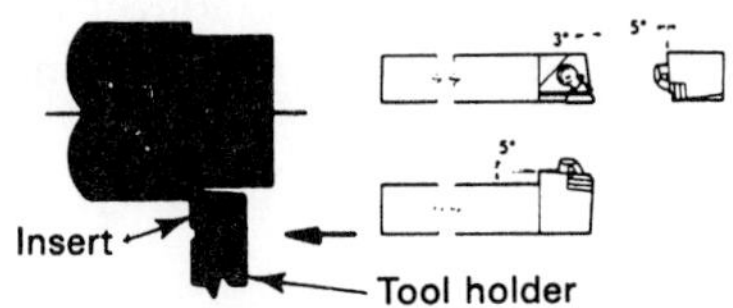

Style A for turning, facing or to a square shoulder.

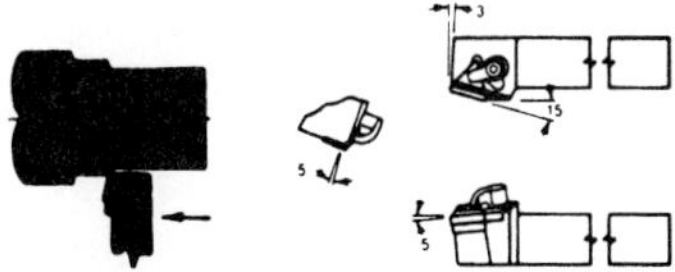

Style B for rough turning, facing or boring where a square shoulder is not required.

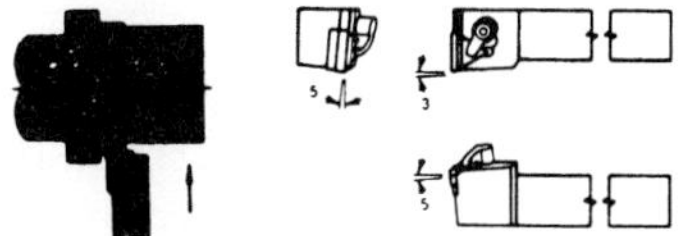

Style F for facing, straddle facing or turning with shank parallel to work axis.

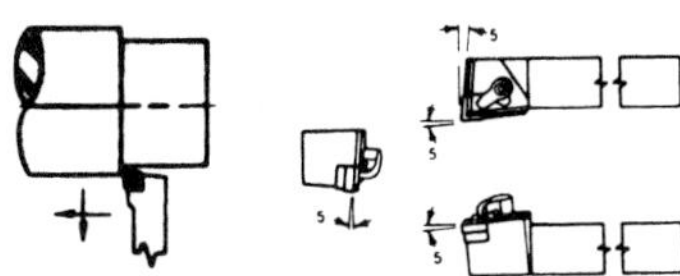

Style L for both turning and facing with same tool. 80° diamond insert.

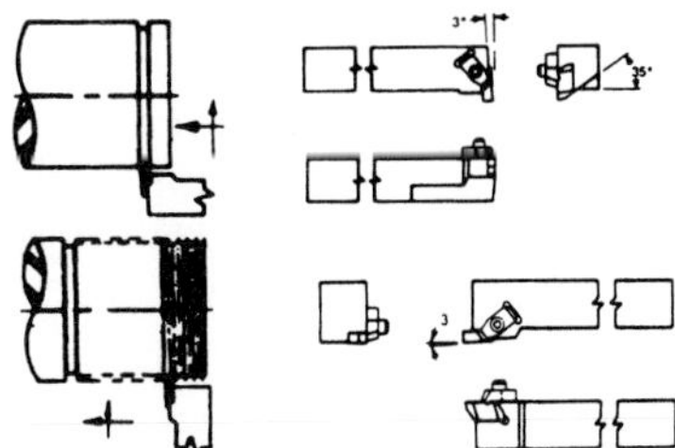

Style NE and style NS for treading and grooving operations. Inserts available for 60° V, Acme, Buttress and API threads.

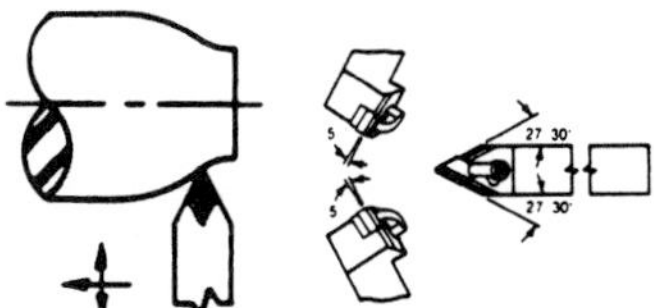

Style P for Profile machining. Insert centrally located.

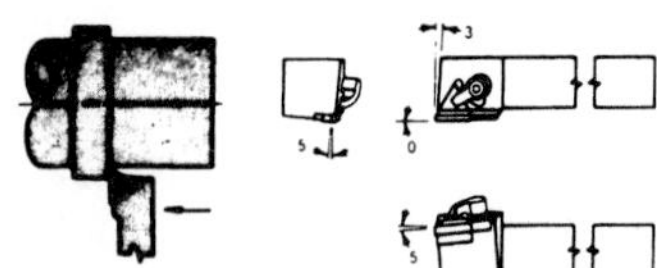

Style G for turning close to chuck or shoulder, or facing to a corner.

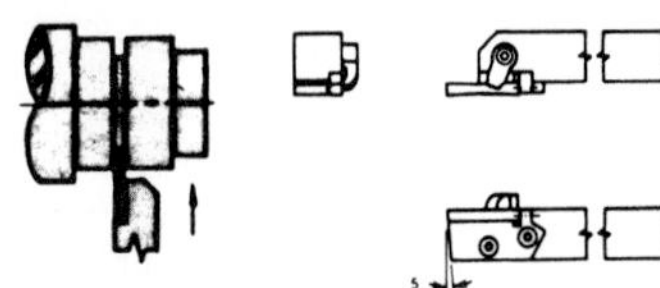

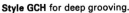

Style GCH for deep grooving.

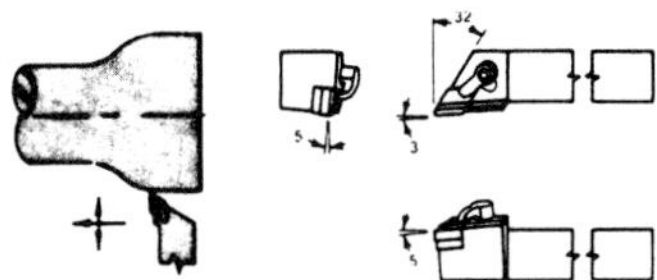

Style J for profiling and finish turning.

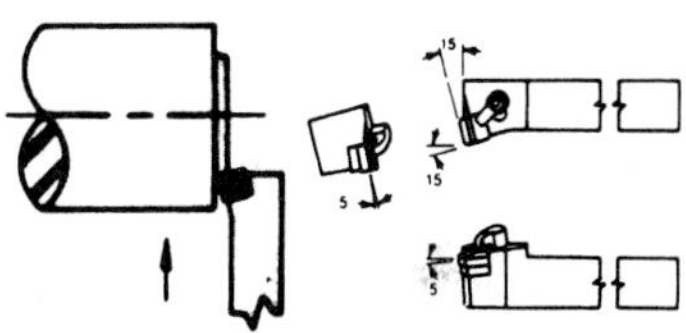

Style K for lead angle facing or turning with shank parallel to work axis.

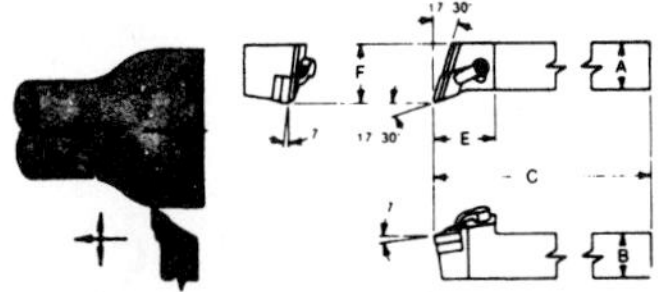

Style Q for profiling machining 55° insert.

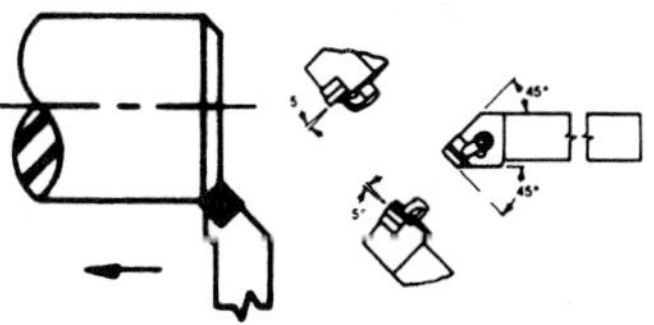

Style S for chamfering and facing 45° lead angle.

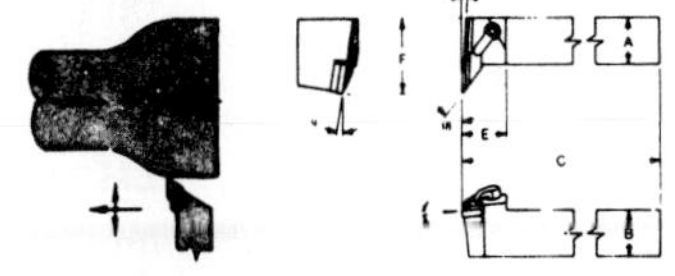

Style U for profile machining 35° insert.

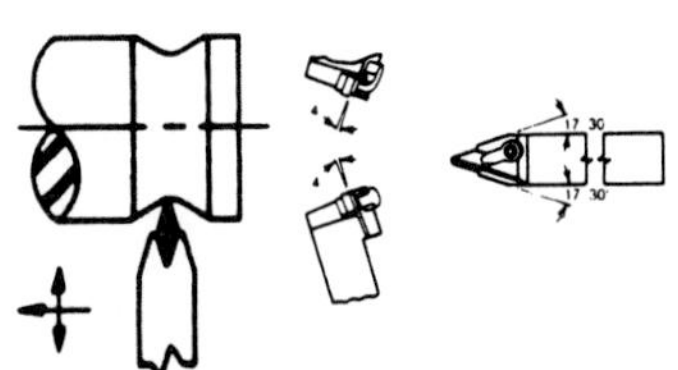

Style V for profile machining. 35° insert.

FIGURE 15–12 Kennanmetal tool holder styles. (Courtesy of Kennanmetal, Inc.)

The advantages of using carbide insert tools were discussed in Chapter 3. These included cutting capabilities at higher speeds (approximately two to three times faster than for high-speed-steel cutters) as well as reductions in tool inventory and elimination of regrinding time and cost. Another advantage of single-point tools is the fact that inserts are made with a precise tool nose radius for cutting. Thus, the location of the tool nose center for any tool can be accurately determined. This makes tool offset specifications and CNC programming a much easier job. See Figure 15–13.

Insert Material

With regard to lathe operations, there are four important types of materials used for inserts.

Cemented Carbides

Cemented carbides are formed by using tungsten carbide sintered in a cobalt matrix. Some grades contain titanium carbide, tantalum carbide, or other materials as additives. Carbides were discussed previously in Chapter 3.

Coated Carbides

The wear resistance of cemented carbides can be improved by 200% to 500% by coating them with wear-resistant materials. Coating materials include titanium carbide and aluminum oxide (a ceramic). The best resistance to abrasion wear for speeds below 500 sfm is achieved when titanium carbide is used. For higher speeds resistance to the chemical reaction between the tool and workpiece is afforded by ceramic-coated inserts. Both coatings offer excellent performance on steels, cast iron, and nonferrous materials.

Ceramics

A ceramic is a very hard material formed without metallic bonding. It displays exceptional resistance to wear and heat load. The most popular material for forming ceramics is aluminum oxide. Often an additive such as titanium oxide or titanium carbide is used. Hard materials can be machined at extremely high cutting speeds with relatively little loss in tool life. In addition, the surface finish is better than with other cutting materials. The main disadvantage with ceramics is that they have low resistance to impact and shock. Thus, they can only be used in operations where impact loading is low.

Diamonds

There are two main types of diamond cutting materials. The first is single-crystal natural diamonds having outstanding wear resistance but low shock

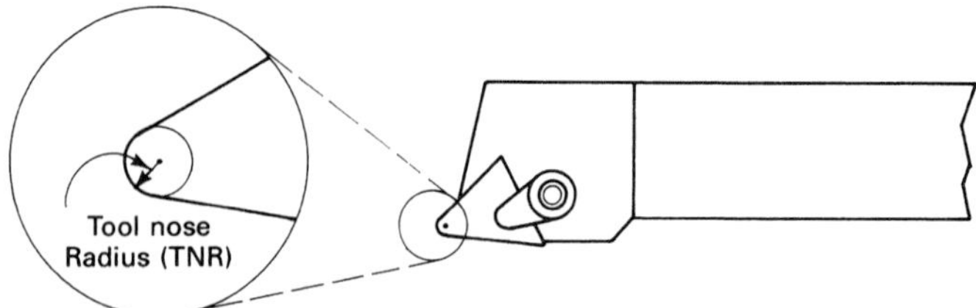

FIGURE 15–13 Tool nose radius.

resistance. The other consists of smaller synthetic diamond crystals fused together at high temperature and pressure into a carbide substrate. This material was developed by GE under the name of Compax. This type of material displays good resistance to shock loading. Diamond tools offer substantial improvements over carbides. Better surface finish and higher cutting speeds with substantial improvements in tool life can be achieved.

Insert Shape

Insert shapes were discussed in Chapter 3. Six basic shapes for turning operations, in order of increasing strength, are as follows: 35° diamond, 55° diamond, triangular, square, 80°/100° diamond, and round. Refer to Figure 15–14 for an illustration of insert shapes and tool holders used in lathe tooling.

Some important rules to be considered when selecting tooling for lathe operations are

1. Select the strongest insert shape possible.
2. Select the smallest practical insert size.
3. Select the largest tool nose radius possible.
4. Select the largest boring bar diameter having the smallest possible overhang.

(a) Ceramic insert tooling

(b) Carbide insert burning tool

(c) Carbide insert boring bars

FIGURE 15–14 [(a) Photo courtesy of Greenleaf Corp. (b, c) Photos courtesy of Kennametal, Inc.]

15-7 TOOL SPEEDS AND FEEDS FOR LATHE OPERATIONS

Tool Speed

For lathe operations, tool speed is defined as the rate at which a point on the circumference of the work passes the cutting tool. It is expressed in surface feet per minute. See Figure 15–15.

$$\text{Spindle rpm} = \frac{4 \times \text{Tool Cutting Speed (sfpm)}}{\text{Work Diameter (in)}}$$

The approximate recommended cutting speeds for high-speed-steel cutting tools are listed in Table 15–1.

The values for tool speed listed in Table 15–1 can be doubled for carbide-tipped cutting tools.

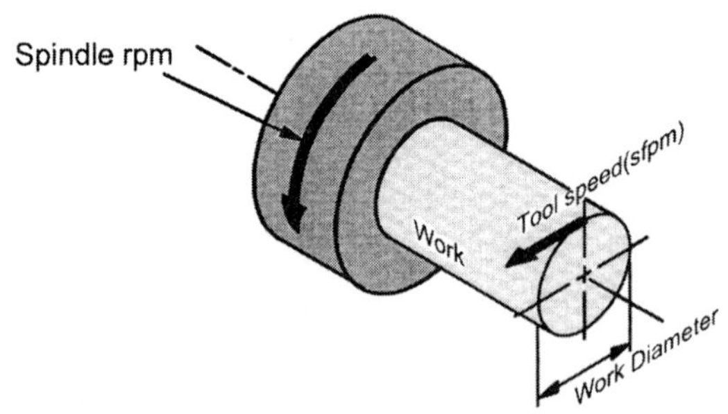

FIGURE 15-15

TABLE 15-1

APPROXIMATE TURNING SPEEDS FOR HIGH-SPEED-STEEL CUTTING TOOLS. DEPTH OF ROUGH CUT IS BETWEEN 0.005 AND 0.01 IN.

	Tool speed (sfpm)	
Material	**Rough cuts**	**Finish cuts**
Aluminum	400	800
Brass	250	500
Cast iron	75	200
Mild steel	100	250
Tool steel	50	100

Tool Feed

Tool feed is the rate at which the tool advances into the work per revolution of the work. Table 15–2 lists approximate recommended feed rates for cutting various materials. A more comprehensive listing of turning speeds and feeds is given in Appendix C.

TABLE 15–2

TURNING FEEDS FOR HIGH-SPEED-STEEL CUTTING TOOLS

Material	Tool feed (ipr)	
	Rough cuts	Finish cuts
Aluminum	0.015–0.025	0.005–0.010
Brass	0.015–0.025	0.003–0.010
Cast iron	0.015–0.025	0.005–0.010
Mild steel	0.010–0.020	0.003–0.010
Tool steel	0.010–0.020	0.003–0.010

15–8 FEED DIRECTIONS AND RAKE ANGLES FOR LATHE OPERATIONS

Feed Directions

Right-handed cutting tools are usually fed from *right* to *left.* The reverse is practiced for left-handed tools. See Figure 15–16.

Tooling Nomenclature

Single-point turning and facing tools must have specific angles at their cutting ends to ensure longer tool life and cutting efficiency. These are shown in Figure 15–17. The components of Figure 15–17 are as follows:

TNR: Tool nose radius for producing acceptable surface finish and longer tool life.

BR: Back rake angle for directing the chips away from the work and toward the tool holder. This is the angle made between the top face of the carbide insert and the tool shank in the length direction.

SR: Side rake angle for directing chips away from the work and toward the side. This is the angle made between the top face of the carbide insert and the tool shank in the width direction.

SC: Side clearance angle for permitting the side of the tool to enter the work.

EC: End clearance angle for permitting the end of the tool to enter the work.

SCEA: Side cutting edge angle for improving the shear cut and producing thinner chips when turning.

ECEA: End cutting edge angle for maintaining clearance between the tool and the work during boring or facing operations.

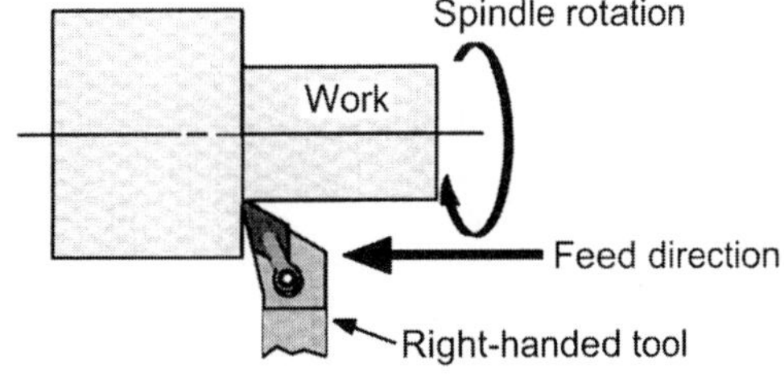

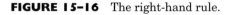

FIGURE 15–16 The right-hand rule.

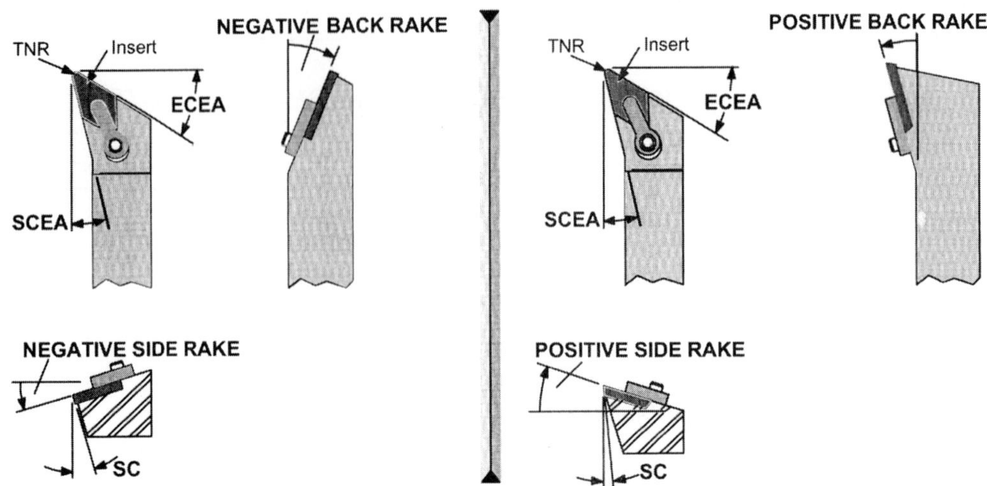

FIGURE 15–17 Single point tool angles (angles shown exaggerated for emphasis).

Rake Angles and Cutting Force

As was stated previously, a tool's rake angles help carry away chips. These angles also help to protect the cutting tool from excessive heating and abrasive action. In general, positive rake angles tend to decrease the force with which the tool cuts the material and negative rake angles increase the force. For most cases tool holders should be used that create negative back rake and side rake. For softer materials and slender parts, positive back rake and side rake can be used. Refer to Figure 15–18.

It is also best to select the tool holder style that creates the largest possible side cutting edge angle with respect to the work. This will produce thinner chips at lower cutting edge temperatures and thus protect the tool nose from excessive wear.

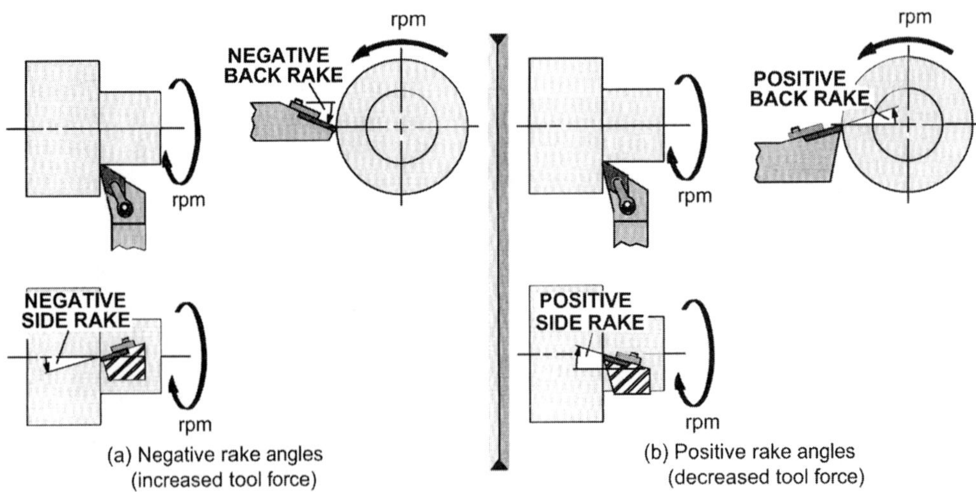

FIGURE 15–18 Rake angles for turning (angles shown exaggerated for emphasis).

15–9 CHAPTER SUMMARY

The following key concepts were discussed in this chapter:

1. A CNC lathe is a machine designed to remove material from stock that is clamped and rotated.
2. Most CNC lathes use turrets for holding and moving tools.
3. The CNC lathe machine axes of motion are: $+Z$ along the spindle axis and toward the spindle, $-Z$ along the spindle axis and away from the spindle, $+X$ transverse to the spindle axis and away from the spindle, $-X$ transverse to the spindle axis and toward the spindle.
4. The seven most important lathe operations are facing, turning, grooving, parting, drilling, boring, and threading.
5. Modern CNC lathes and turning centers utilize indexable insert tooling.
6. Single-point turning and facing tools have specific angles at their cutting ends to ensure longer tool life and cutting efficiency.
7. Tool speed for lathes is the rate at which a point on the revolving circumference of the work passes the cutting tool. It is expressed in surface feet per minute.
8. Tool feed is the rate at which the tool advances into the work per revolution of the spindle. It is expressed in inches per revolution.
9. Right-handed cutting tools are fed into the work from right to left. The opposite is true for left-handed cutting tools.
10. Negative rake angles are used for most turning operations involving carbide insert tools. Positive rake angles tend to decrease tool force and are used with high-speed-steel cutters on softer materials.

REVIEW EXERCISES

15.1. Match the terms on the left with the definitions on the right:

Chuck	Moves tool into work
Carriage	Supports right end of work
Turret	Clamps the work
Tailstock	Stores and executes CNC programs
MCU	Provides a path for falling chips
Headstock	Machinery to rotate spindle
Slant bed	Holds cutting tool

15.2. **(a)** What type of tool changer mechanisms are used on CNC lathes and turning centers?
(b) Describe a typical tool changing operation.

15.3. What is the difference between a front turret and a rear turret?
15.4. Describe the CNC lathe machine axis of motion.
15.5. What are OD versus ID operations?
15.6. Describe the following machining operations and state whether each is OD, ID, or both:

(a) Facing
(b) Turning
(c) Grooving
(d) Parting
(e) Boring
(f) Threading

15.7. List four advantages of using indexable/insert tooling for lathe operations.

15.8. State the advantages and disadvantages of using the following insert materials:

(a) Cemented carbides	**(c)** Ceramics
(b) Coated carbides	**(d)** Diamonds

15.9. List four important rules that should be followed when selecting insert tooling.

15.10. Explain the difference between tool speed and tool feed as regards lathe operations.

15.11. Explain the right-hand rule.

15.12. Match the terms on the left with the definitions on the right:

Side rake angle	Permits side of tool to enter work
Tool nose radius	Clearance for boring/facing
Side clearance angle	Directs chips toward the side
Side cutting edge angle	Produces surface finish
Back rake angle	Controls chip thickness
End cutting edge angle	Directs chips away from work

15.13. **(a)** What is the effect of using negative rake angle tooling?
(b) What is the effect of using positive rake angle tooling?

15.14. What recommendations can be made regarding rake angles and the side cutting edge angle when selecting a tool holder for lathe operations?

CHAPTER 16

FUNDAMENTAL CONCEPTS OF CNC LATHE PROGRAMMING

16–1 CHAPTER OBJECTIVES

At the conclusion of this chapter you will be able to

1. Explain the different types of positioning modes for CNC lathe operations.
2. Know important locations for programming and setup of CNC lathes. These include the reference point, machining origin, and program origin.
3. State the basic setup operations performed on the CNC lathe prior to running a job.
4. Understand the important preparatory (G) codes and miscellaneous (M) codes used in programming lathe operations.
5. Know the codes for specifying the spindle speed and tool feed values.
6. Explain the code for changing tools.

16–2 INTRODUCTION

This chapter deals with the basic parameters and concepts involved in programming CNC lathe operations. See Figure 16–1. Modes for positioning the tool are considered first. Important setup information concerning the reference

Safety Rules for Job Setup on a CNC Turning Center

- Hold the workpiece securely. For bar work, make sure the bar does not extend beyond safe limits.
- Use safety goggles, wear protective clothing, and put long hair up.
- Set the jaw pressure to conform to proper holding conditions.
- Set cutting speeds and feeds for each tool within the limits recommended for the process. Adjust speeds and feeds during an operation to obtain optimum machining conditions.
- When moving an axis in manual mode, make sure there is sufficient clearance between the cutting tool and all surrounding objects (part, fixture, etc.).
- Test the program by making a dry run.
- When loading, make sure the workpiece is free of burrs and foreign particles.
- Maintain a backup copy of setup and the tape record.

(a) Boring

(b) Turning

(c) Parting

FIGURE 16–1 CNC lathe operations. (Photos courtesy of Kennametal, Inc.)

point, part origin, tool change point, and tool offsets are explored in detail. Important preparatory (G) codes and miscellaneous (M) codes for executing basic lathe operations are listed and explained. Additionally, feed rate (F) codes, speed (S) codes, and tool change (T) codes are also studied. Examples of linear, circular, and grooving operations are provided and explained.

16–3 ESTABLISHING LOCATIONS VIA CARTESIAN COORDINATES (CNC LATHES)

The machine axes of motion for CNC lathes were introduced in Section 15–4. It was pointed out that for turning centers, the X axis (cross-slide movement) and the Z axis (longitudinal travel) are used to specify tool locations. See Figure 16–2.

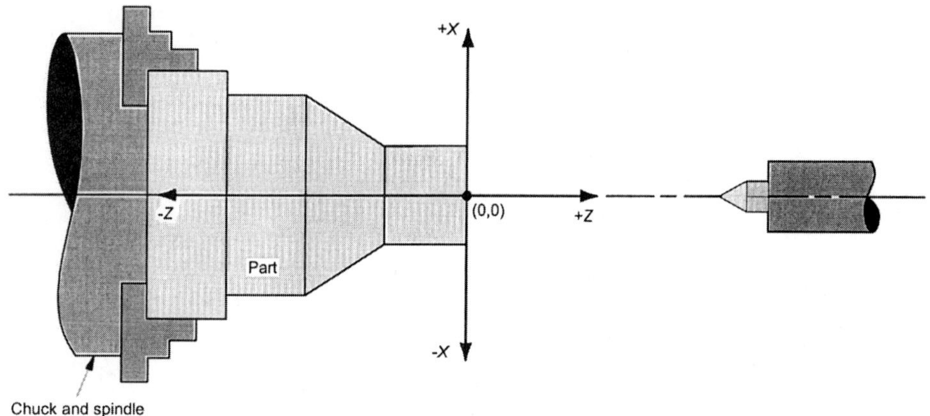

FIGURE 16–2 The Cartesian coordinate system for CNC lathes.

16–4 TYPES OF TOOL POSITIONING MODES (CNC LATHES)

CNC lathe programs can be written to move the tool in the following modes: absolute, incremental, or mixed (incremental and absolute). Furthermore, the *X* axis can be programmed in terms of the diameter or the radius.

Absolute Positioning

When operating in this mode the *new* position of the tool is given by its *X* and *Z* distances from a *fixed home or origin (0, 0).*

Diameter programming: With absolute diameter programming, the *X* position of the tool is specified as *twice* the distance from the spindle centerline. Refer to Figure 16–3.

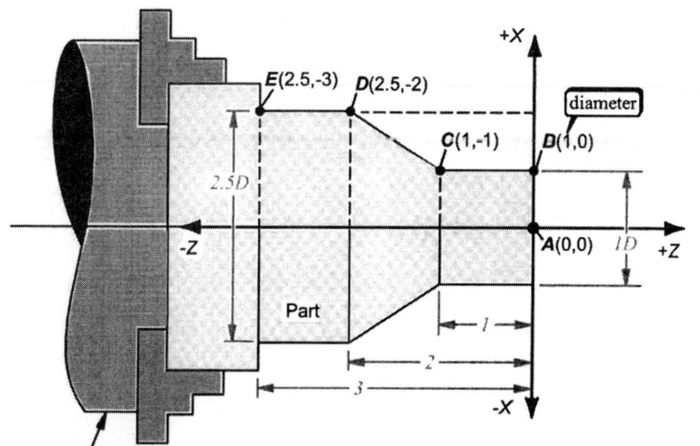

Tool position	Location(absolute)	
	X [diameter]	Z
A	0	0
B	1	0
C	1	-1
D	2.5	-2
E	2.5	-3

FIGURE 16–3 Programming by diameter (absolute positioning).

Radius programming: With absolute radius programming, the *X* position of the tool is specified as the distance from the spindle centerline. See Figure 16–4.

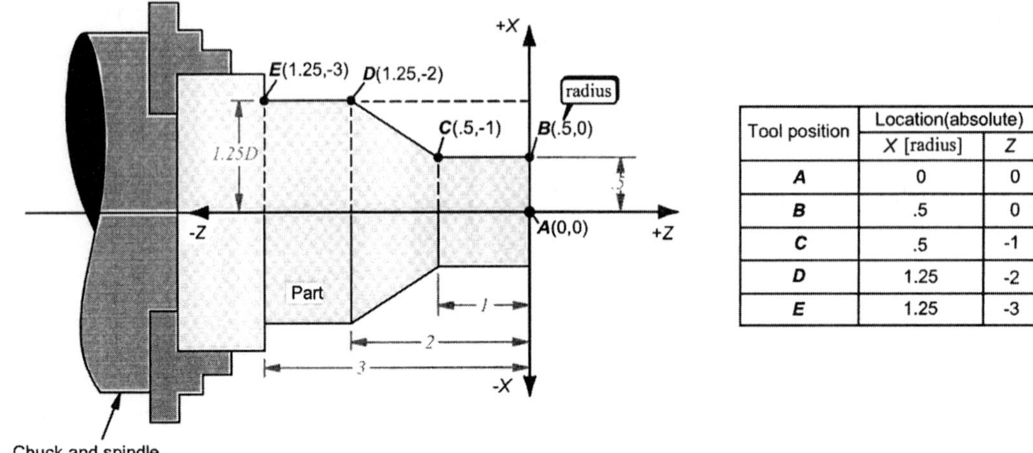

Tool position	Location(absolute)	
	X [radius]	*Z*
A	0	0
B	.5	0
C	.5	-1
D	1.25	-2
E	1.25	-3

FIGURE 16–4 Programming by radius (absolute positioning).

Incremental Positioning

When operating in this mode of programming, the *new* position of the tool is specified by inputting its *direction and distance from the last position* achieved. The address *U* is used to indicate incremental *X*-axis motion and the address *W* is used to indicate incremental *Z*-axis movement.
Motion toward the spindle center on the X *axis is indicated by* –U *and motion away by* +U. *Motion toward the spindle center on the* Z *axis is indicated by* –W *and motion away by* +W.

Diameter programming: With incremental diameter programming, U is entered as *twice* the directed distance from the last position achieved. Refer to Figure 16–5.

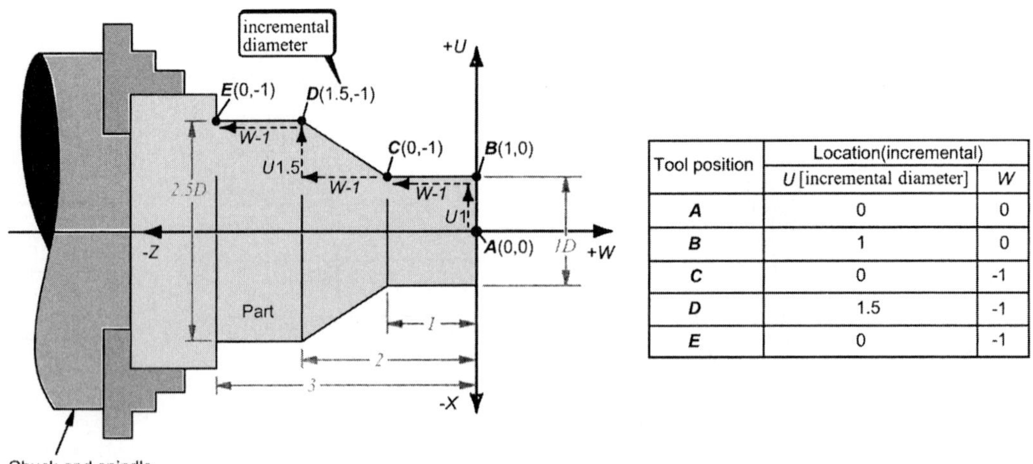

Tool position	Location(incremental)	
	U [incremental diameter]	*W*
A	0	0
B	1	0
C	0	-1
D	1.5	-1
E	0	-1

FIGURE 16–5 Programming by diameter (incremental positioning).

Radius programming: Incremental radius programming mode involves inputting U as the directed distance from the last position achieved. See Figure 16–6.

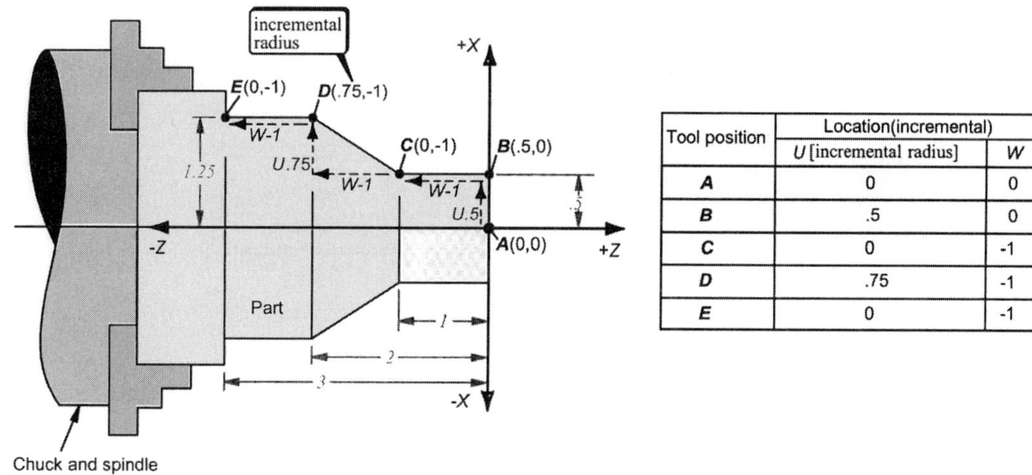

Tool position	Location(incremental)	
	U [incremental radius]	W
A	0	0
B	.5	0
C	0	-1
D	.75	-1
E	0	-1

FIGURE 16–6 Programming by radins (incremental positioning).

16–5 REFERENCE POINT, MACHINING ORIGIN, AND PROGRAM ORIGIN

There are three important origins or zero points for CNC lathes. See Figure 16–7. These are described as follows.

Reference Point (Machine Zero)

This is the position of the turret when the machine's axes are zeroed out. It is set once by the manufacturer.

Machining Origin (Tool Change Position)

The machining origin is determined at setup. This location is inputted at the beginning of the program by means of the "zero offset" command (see Section 16–9). The machine executes all programmed X and Z movements relative to this origin.

Program Origin

The program origin is a zero point from which all dimensions are defined in the part program. The setup person uses tool offsets as a means of locating the program origin with respect to the machining origin. Refer to Section 16–6 for a discussion of tool offsets.

Upon receiving a programmed X and Z move regarding the program origin, the controller will compute the corresponding X and Z move relative to the machining origin. It will then execute the move relative to the machining origin.

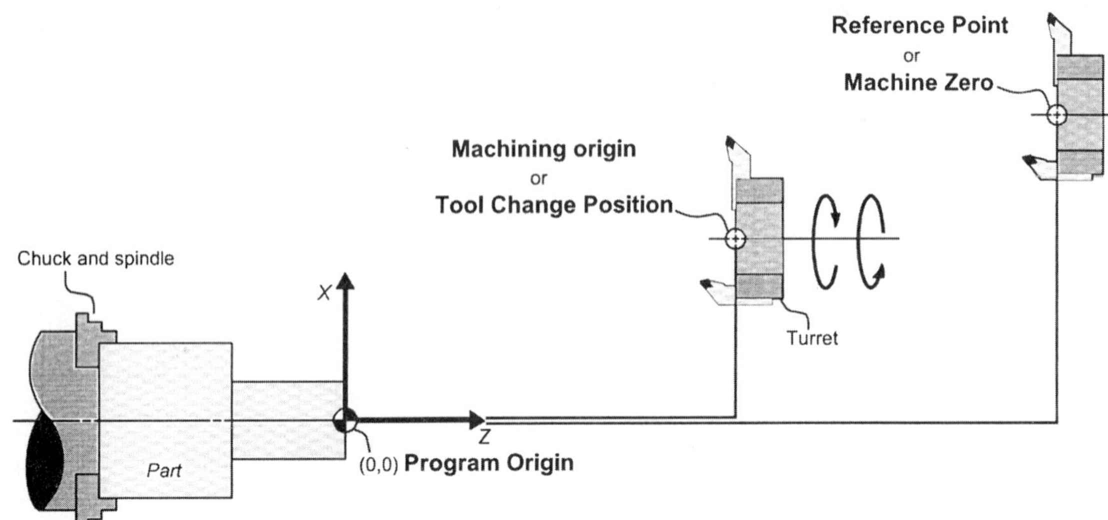

FIGURE 16-7 Important locations for CNC lathes.

The machining origin is determined such that when the turret is at this location the reference (longest) tool is at least 1 in. from any face. The control is zeroed out at this location. The turret is then jogged to the reference point. The *X* and *Z* locations of the machine origin from the reference point are recorded. These numbers are then used in the "zero offset" command.

The tool change position is the safe location the machine returns to when indexing an old tool with a new tool. It is usually set at the machining origin. The operator can manually home the turret by pressing the "return to reference" button on the console. This can be done, for example, if the turret is not homed when the CNC lathe is turned on.

16-6 SETUP PROCEDURES FOR CNC LATHES

The setup operation normally begins with the setup person securing the required OD and ID tools in the turret. The part blank is loaded into the chuck. It is very important that the proper length of stock extends beyond the chuck. If the overhang length is too long, excessive wobbling will occur when the part is machined. If the length is too short, the tool may collide with the chuck or insufficient room may be left for the cutoff tool to operate. The length of stock permitted to extend beyond the chuck is specified by the programmer.

Tool offset values must be determined and entered into the controller's memory. Each offset value is assigned a memory address number. Later, the controller will know the proper offset values for a tool when its memory address number is read in a program.

During setup, each tool is manually moved to the part blank. The tool's *X* and *Z* offset values are entered into memory. The tool offset measurements determine the point on the tool to be programmed as follows:

Tool offset measurement(s)	***Point on tool programmed***
From part blank to tool edge(s)	Tool edge(imaginary tool nose radius) point located at the intersection of the horizontal and vertical tangents to the Tool nose radius(TNR) — Tool, TNR, Tangent, Tangent
From part blank to center of tool nose	Center of tool nose point

The controller uses the tool offsets to convert programmed *X* and *Z* moves relative to the program origin to corresponding moves relative to the machining origin.

It should be noted throughout this text that the *X* offset values are given in terms of diameter. Tool offsets are shown in Figure 16–8.

Upon changing a tool the system is instructed to cancel the current tool offset and rapid the turret to the tool change position.

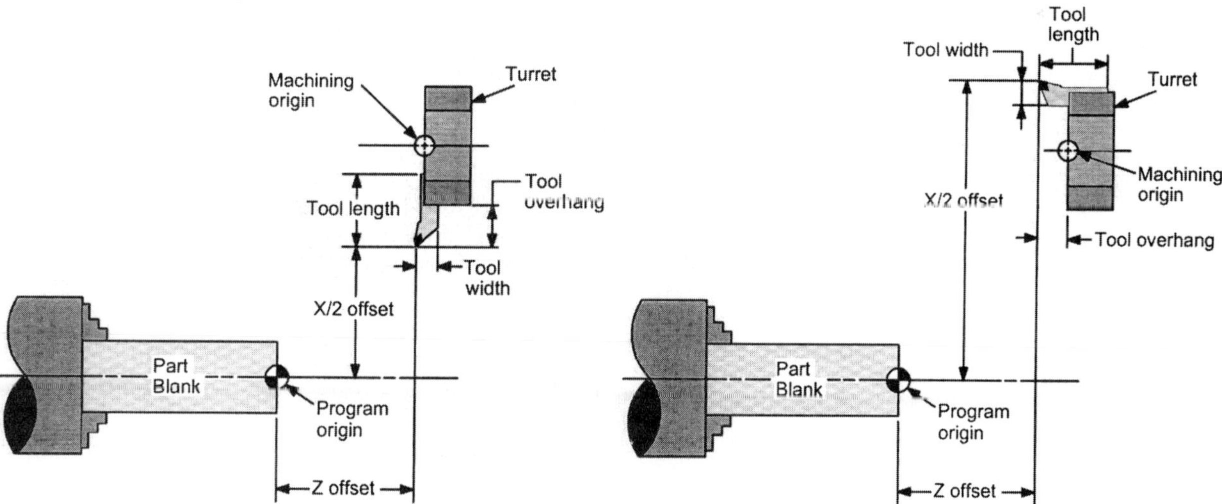

FIGURE 16–8 Tool offsets.

16-7 IMPORTANT PREPARATORY FUNCTIONS (G CODES) FOR LATHES

The following modal and nonmodal G codes are important when programming lathe operations.

G code	Mode	Specification
G00	Modal	Rapid positioning mode. The tool is moved to the programmed *XZ* position at maximum feed rate.
G01	Modal	Linear interpolation mode. The tool is moved along a straight-line path at programmed feed rate.
G02	Modal	Circular interpolation mode (CW).
G03	Modal	Circular interpolation mode (CCW).
G04	Nonmodal	Programmed dwell.
G20	Modal	Inch mode for all units. This code is entered at the start of the CNC program in a separate block. It must appear before a G50 block.
G21	Modal	Metric mode for all units. This code is entered at the start of the CNC program in a separate block. It must appear before a G50 block.
G28	Nonmodal	Return tool to reference point.
G50	Modal	Maximum spindle speed setting for G96 mode. or Set machining origin at the current tool position.
G75	Nonmodal	Grooving in the *X* axis.
G96	Modal	Constant surface speed control.
G97	Modal	Cancel constant surface speed control. (Direct rpm.)
G98	Modal	Feed per minute specification.
G99	Modal	Feed per revolution specification.

16-8 IMPORTANT MISCELLANEOUS FUNCTIONS (M CODES) FOR LATHES

The following miscellaneous functions are often used to initiate machine functions not related to dimensional or axis movements.

M code	Type	Specification
M00	B	Causes a program stop.
M03	A	Turns spindle on (CCW).
M04	A	Turns spindle on (CW).
M05	B	Turns spindle off.
M08	A	Turns external coolant on.
M09	B	Turns coolant off.
M30	B	Directs the system to end program processing, rewind the tape, or reset the memory unit. This code must be the last statement in a program.
M41	A	Shifts spindle into low-gear range.
M42	A	Shifts spindle into intermediate-gear range.
M43	A	Shifts spindle into high-gear range.
M68	A	Clamps the chuck.
M69	B	Opens the chuck.

16–9 "ZERO OFFSET" COMMAND

The machining origin is established, once, at the beginning of the program. Assume the turret has been jogged to the machining origin. Assume further, the setup person knows the *X* and *Z* distances from the program origin to this point.

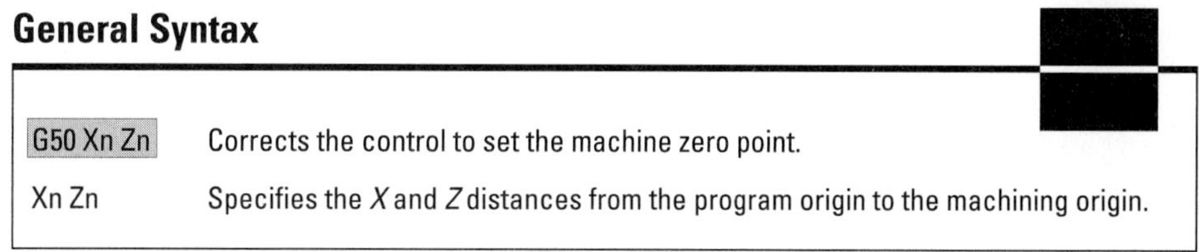

General Syntax

G50 Xn Zn	Corrects the control to set the machine zero point.
Xn Zn	Specifies the *X* and *Z* distances from the program origin to the machining origin.

■ EXAMPLE 16–1

Enter a program block for creating a machining origin at the location shown in Figure 16–9.

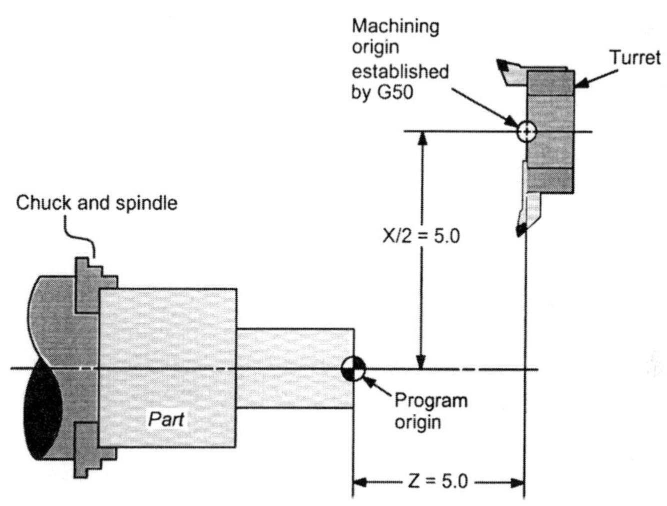

N0020 G50 X10.0 Z5.0

FIGURE 16–9

All subsequent tool movements will then be computed by the controller relative to the machining origin established. ■■

16–10 FEED RATE (F CODE)

The numerical value following the address F specifies the feed rate.

■ EXAMPLE 16–2

Set the feed rate at 10.0 ipm. Use Figure 16–10.

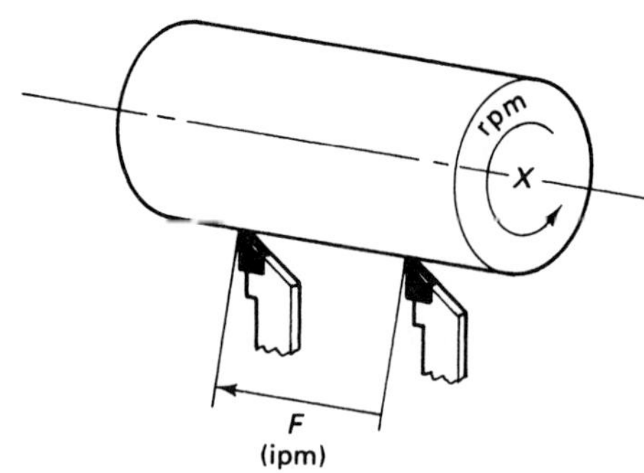

N0030 G98 F10.0

FIGURE 16–10

■ EXAMPLE 16–3

Set the feed rate at 0.020 ipr. Use Figure 16–11.

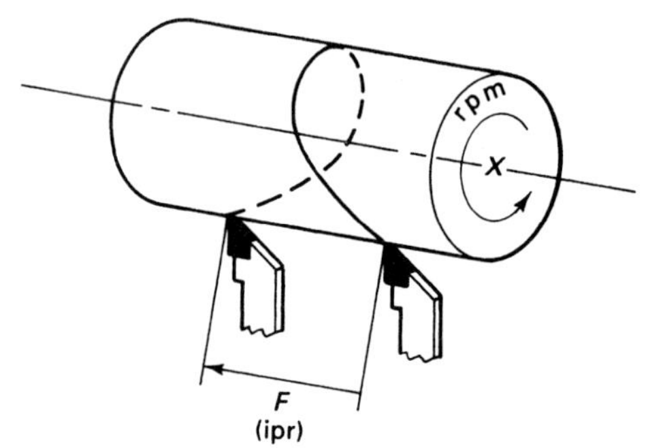

N0030 G99 F.020

FIGURE 16–11

16–11 SPINDLE SPEED (S CODE)

The spindle speed is specified by the S code. When programmed with the preparatory function G97, it indicates the spindle rpm.

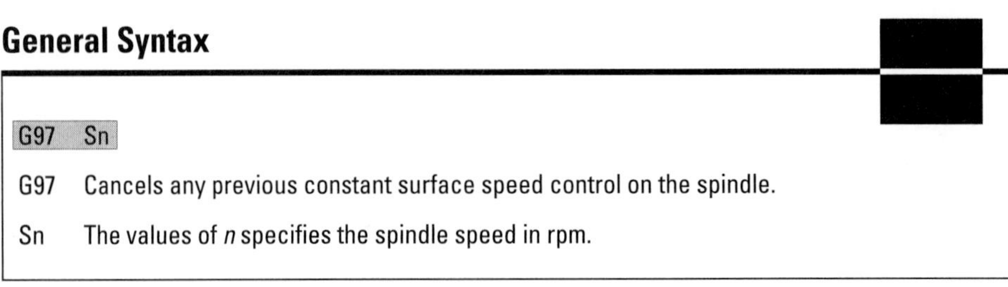

General Syntax

G97 Sn

G97 Cancels any previous constant surface speed control on the spindle.

Sn The values of *n* specifies the spindle speed in rpm.

■ EXAMPLE 16–4

Program a spindle speed of 1600 rpm.
N0050 G97 S1600

16–12 SPINDLE SPEED WITH CONSTANT SURFACE SPEED CONTROL

Recall that the cutting speed is given by the formula:

$$\text{Cutting speed} = \text{rpm} \times D/4$$

where rpm is the spindle speed and D is the work diameter.

As the tool removes material from the outside, the work diameter will decrease. Thus, the cutting speed will also decrease. The opposite will be true for inside operations with the cutting speed increasing as the machining progresses. Tool suppliers recommend specific cutting speeds to be maintained in order for tools to operate at optimum performance and produce a required surface finish. The control can be directed to adjust the rpm of the spindle such that a constant surface speed results as the part diameter changes. A G96 code can be programmed to ensure constant surface speed control. An S word entered with this code no longer indicates spindle rpm but tool cutting speed.

General Syntax

G96 Sn	Specifies constant surface speed control to be in effect.
Sn	The value of *n* indicates the tool cutting speed in surface feet per minute (sfm). This value is to be maintained as the part diameter changes.

■ EXAMPLE 16–5

Write a program block directing the controller to adjust the spindle rpm so that a constant surface speed of 600 sfm is held constant during machining.

N0060 G96 S600

The controller is now instructed to adjust the spindle rpm upward if the tool moves to a smaller diameter and downward if the tool moves to a larger diameter. ■■

16–13 SPINDLE SPEED WITH CLAMP SPEED AND CONSTANT SURFACE SPEED CONTROLS

The G96 command directs the controller to increase or decrease the spindle rpm in increments of 1 rpm. There may be cases where a specific rpm level is not to be reached. This can occur if there are chucking requirements or the work becomes unstable at certain rpms. For these situations it is possible to assign an upper limit on the rpm value the controller is not to exceed. A G50 code will accomplish this task. When S is coded with G50 it indicates spindle rpm.

General Syntax

G50 Sn

G50	Specifies that the controller is not to exceed a programmed rpm value.
Sn	The value of *n* indicates the maximum spindle rpm value not to be exceeded.

■ EXAMPLE 16–6

For the facing to center operation shown in Figure 16–12, the spindle is to start at 700 rpm and is not to exceed 1500 rpm. Write blocks to assign constant surface speed control under these conditions.

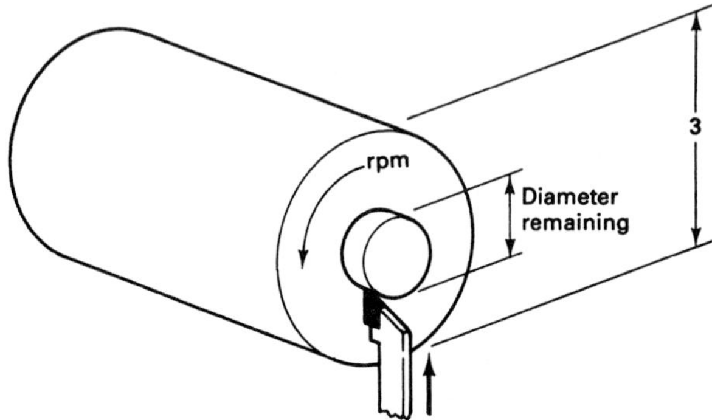

FIGURE 16–12

$$\text{Cutting speed} = \text{rpm} \times D/4$$
$$= 700 \times 3/4$$
$$= 525 \text{ sfm}$$

```
N0020 G50 S1500
   .
   .
   .
   .
N0100 G96 S525
```

The following table lists the spindle rpm values that the controller will assign as the work diameter decreases.

Diameter	rpm
3	700
2	1050
1.4	1500
1	1500
0.5	1500

The controller will continue to increase the spindle rpm as the diameter decreases to 1.4. At diameters less than 1.4, the maximum spindle rpm will be reached and the controller will assign this value as the tool cuts to the spindle centerline. Thus, constant surface speed will be maintained as the tool cuts from 3*D* to 1.4*D* and will decrease thereafter. ■■

16–14 AUTOMATIC TOOL CHANGING

The following general word form is used to program tool changes.

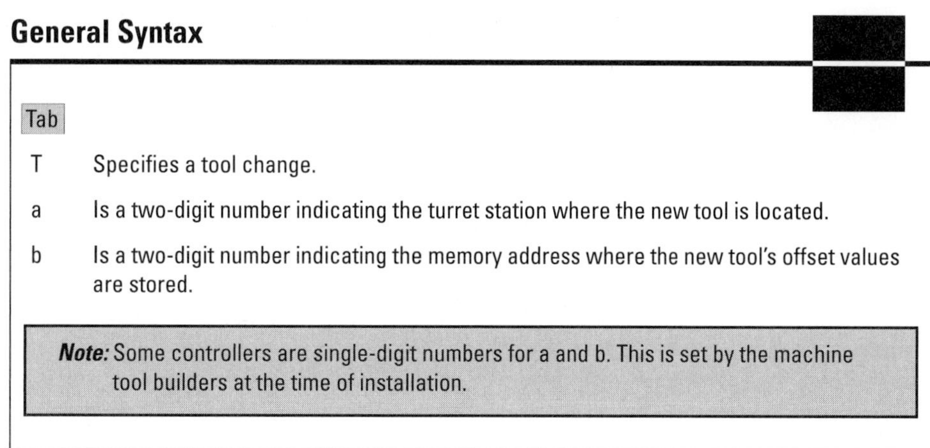

General Syntax

Tab

T	Specifies a tool change.
a	Is a two-digit number indicating the turret station where the new tool is located.
b	Is a two-digit number indicating the memory address where the new tool's offset values are stored.

Note: Some controllers are single-digit numbers for a and b. This is set by the machine tool builders at the time of installation.

Prior to executing a tool change, the programmer should enter a command to return to the tool change position and cancel the tool offsets of the old tool used.

■ EXAMPLE 16–7

Code blocks directing the controller to change from tool 1 to tool 4. Use Figure 16–13.

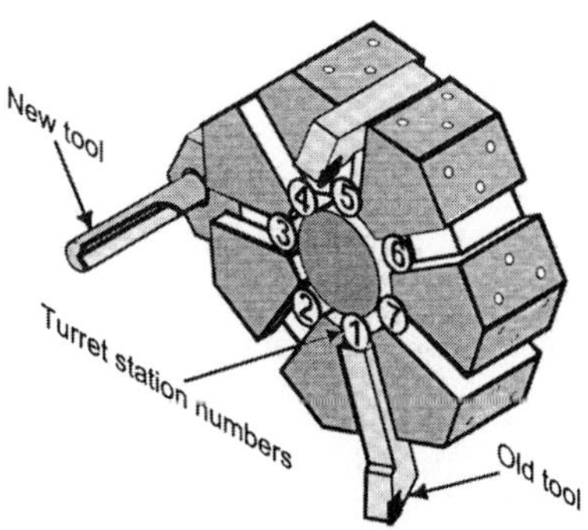

FIGURE 16–13

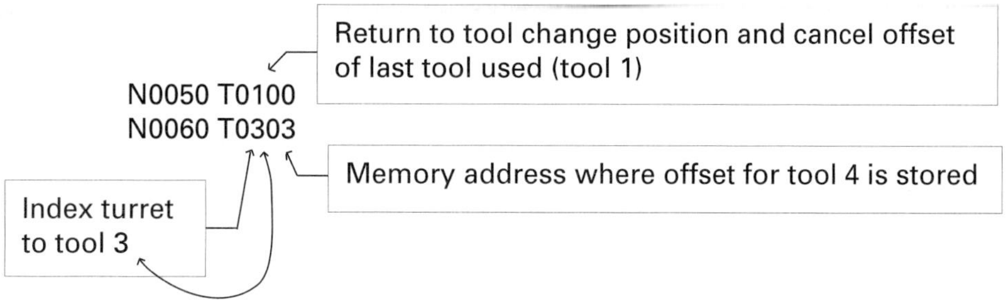

The control will properly move tool 4 after the tool change because it will know the corresponding tool offsets. ■■

16–15 LINEAR INTERPOLATION COMMANDS (CNC LATHES)

For CNC lathes, linear interpolation involves moving the tool along a straight line programmed at the specified feed rate. Linear interpolation is used to execute such operations as turning, facing, taper turning, and taper boring.

In the discussion to follow it is assumed that the center of the tool nose radius is programmed. Tool edge programming with linear interpolation is considered in Chapter 17.

General Syntax

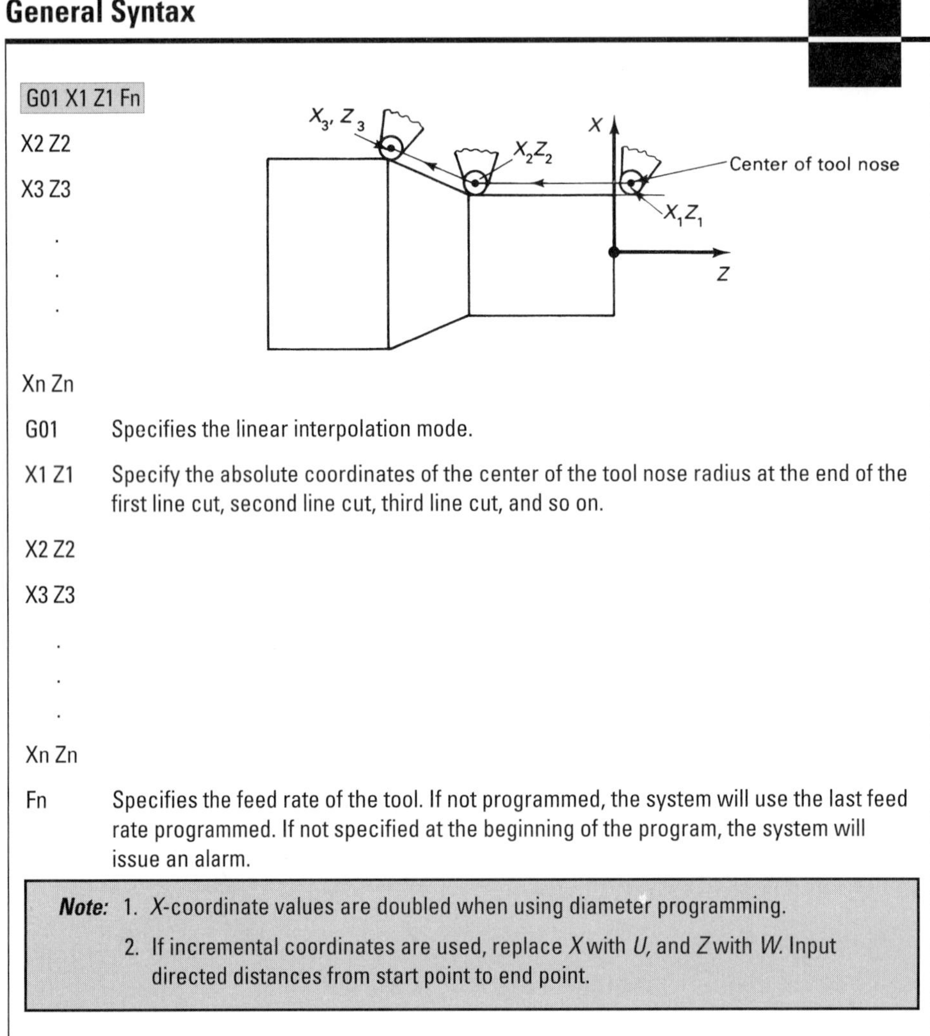

G01 X1 Z1 Fn
X2 Z2
X3 Z3
.
.
.
Xn Zn

G01 Specifies the linear interpolation mode.

X1 Z1 Specify the absolute coordinates of the center of the tool nose radius at the end of the first line cut, second line cut, third line cut, and so on.

X2 Z2

X3 Z3

.
.
.

Xn Zn

Fn Specifies the feed rate of the tool. If not programmed, the system will use the last feed rate programmed. If not specified at the beginning of the program, the system will issue an alarm.

Note: 1. *X*-coordinate values are doubled when using diameter programming.
2. If incremental coordinates are used, replace *X* with *U*, and *Z* with *W*. Input directed distances from start point to end point.

16–16 DETERMINING CUTTER OFFSETS FOR CNC LATHE OPERATIONS

The technique of programming the center of the tool nose radius is identical to programming the cutter center in milling. The only differences are as follows:

- The *Y* axis (mills) is replaced by the *X* axis (lathes).
- The *X* axis (mills) is replaced by the *Z* axis (lathes).
- For diameter programming, the *X*-axis (lathes) values are doubled.

Thus, the formulas and concepts presented in Sections 11–4 and 11–5 can also be used for lathe operations.

■ EXAMPLE 16–8

The profile shown in Figure 16–14 is to be turned. Determine the required *X* and *Z* absolute coordinates of the center of a 0.0625*R* tool nose. Assume diameter programming is used.

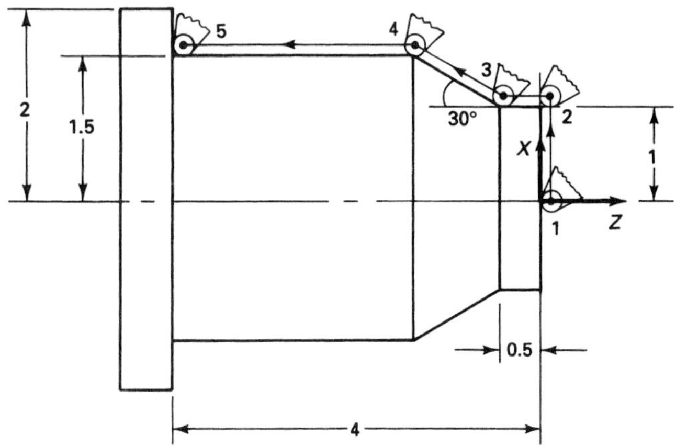

FIGURE 16–14

		Absolute coordinates	
Position	**Calculation**	***X***	***Z***
①	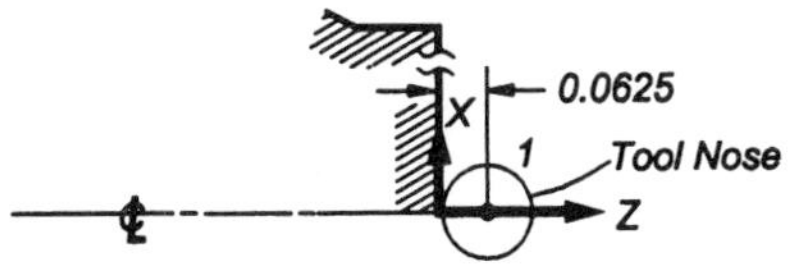 0.0625; Tool Nose	0	0.0625
②	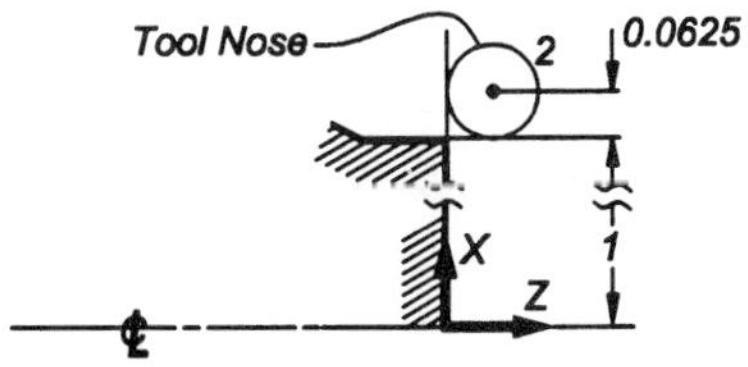 Tool Nose; 0.0625; $X = 2(1 + 0.0625) = 2.125$; $Z = 0.0625$	2.125	0.0625
③	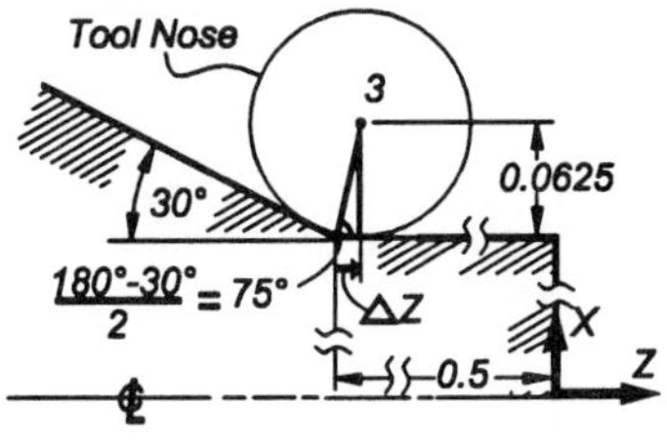 Tool Nose; 30°; $\frac{180°-30°}{2} = 75°$; 0.0625; 0.5; $\Delta Z = \frac{0.0625}{\text{Tan}(75°)}$; $\Delta Z = \frac{0.0625}{3.7321}$; $\Delta Z = 0.0167$; $Z = -0.5 + \overbrace{0.0167}^{\Delta Z} = -0.4833$; $X = 2.125$	2.125	−0.4833

④	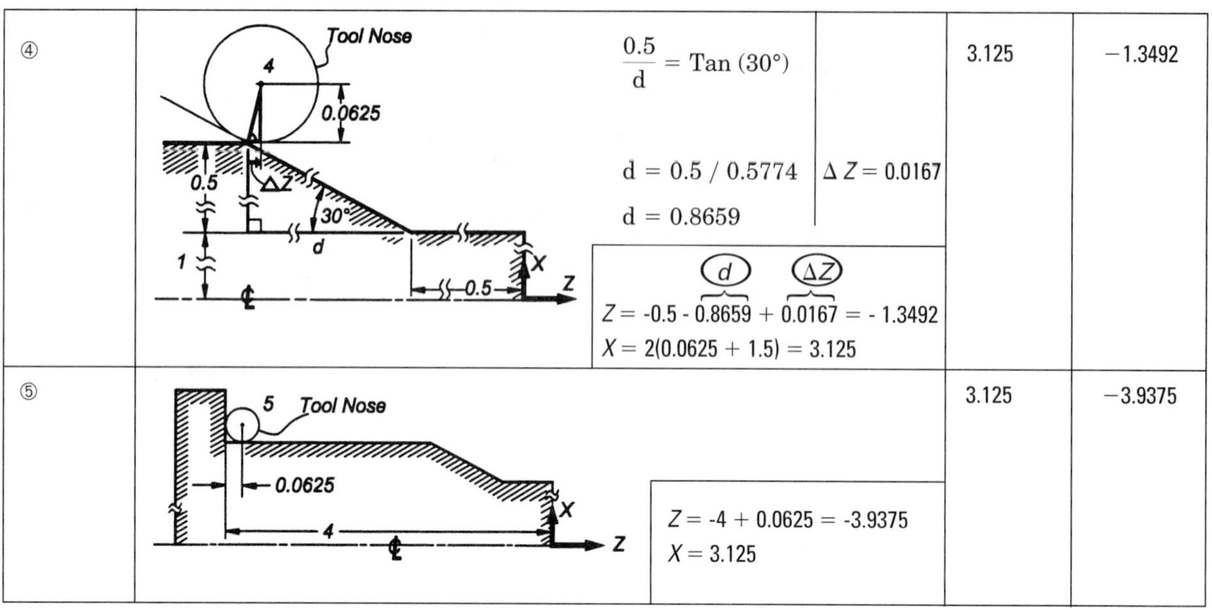	$\frac{0.5}{d} = \text{Tan}(30°)$ d = 0.5 / 0.5774 | Δ Z = 0.0167 d = 0.8659 Z = -0.5 - 0.8659 (d) + 0.0167 (ΔZ) = - 1.3492 X = 2(0.0625 + 1.5) = 3.125	3.125	−1.3492
⑤		Z = -4 + 0.0625 = -3.9375 X = 3.125	3.125	−3.9375

EXAMPLE 16–9

Write linear interpolation blocks to move the tool from position ① to position ② in each machining case shown in Figure 16–15.

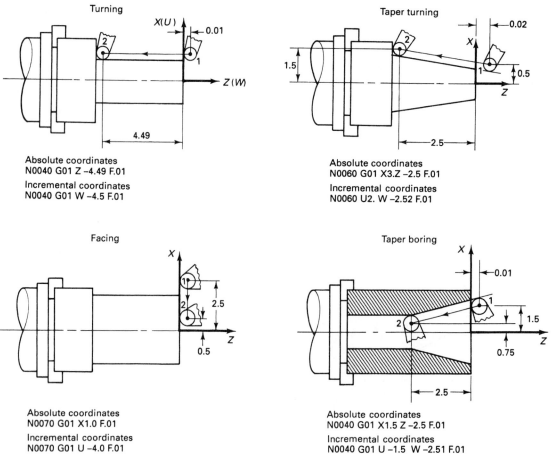

Turning

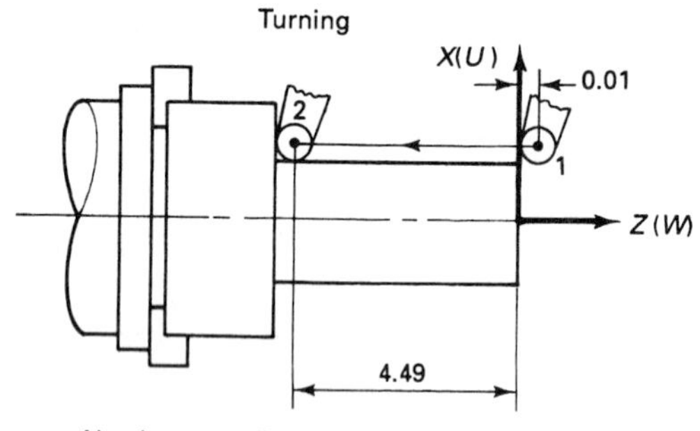

Absolute coordinates
N0040 G01 Z -4.49 F.01

Incremental coordinates
N0040 G01 W -4.5 F.01

Taper turning

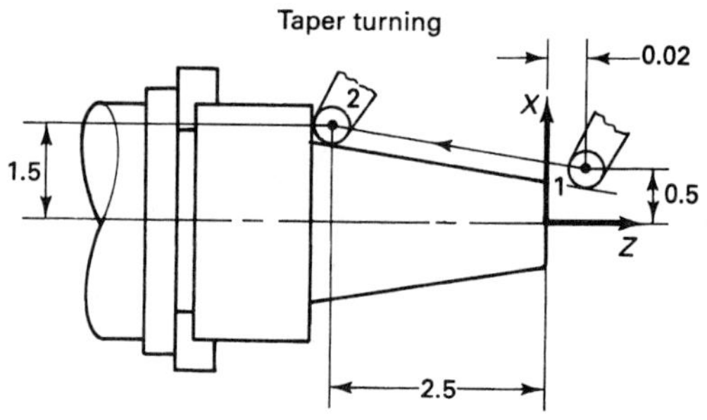

Absolute coordinates
N0060 G01 X3.Z -2.5 F.01

Incremental coordinates
N0060 U2. W -2.52 F.01

Facing

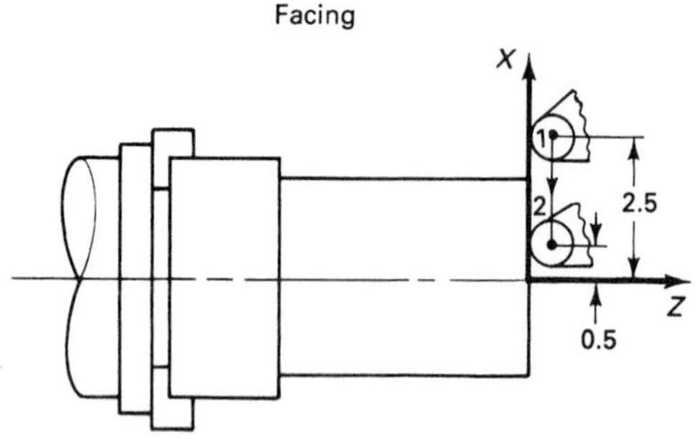

Absolute coordinates
N0070 G01 X1.0 F.01

Incremental coordinates
N0070 G01 U -4.0 F.01

Taper boring

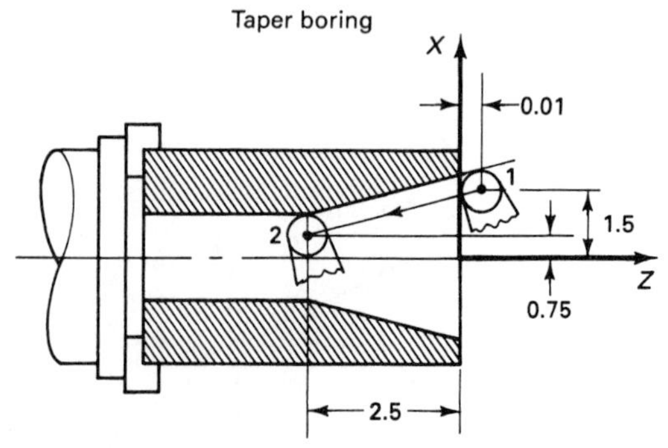

Absolute coordinates
N0040 G01 X1.5 Z -2.5 F.01

Incremental coordinates
N0040 G01 U -1.5 W -2.51 F.01

FIGURE 16–15

16–17 CIRCULAR INTERPOLATION COMMANDS (CNC LATHES)

Circular interpolation for lathe operations involves cutting a circular arc in either a clockwise or a counterclockwise direction. The commands are similar to those used in milling.

Assume, again, that the center of the tool nose is programmed. The technique of programming the edge of the tool with circular interpolation is presented in Chapter 17.

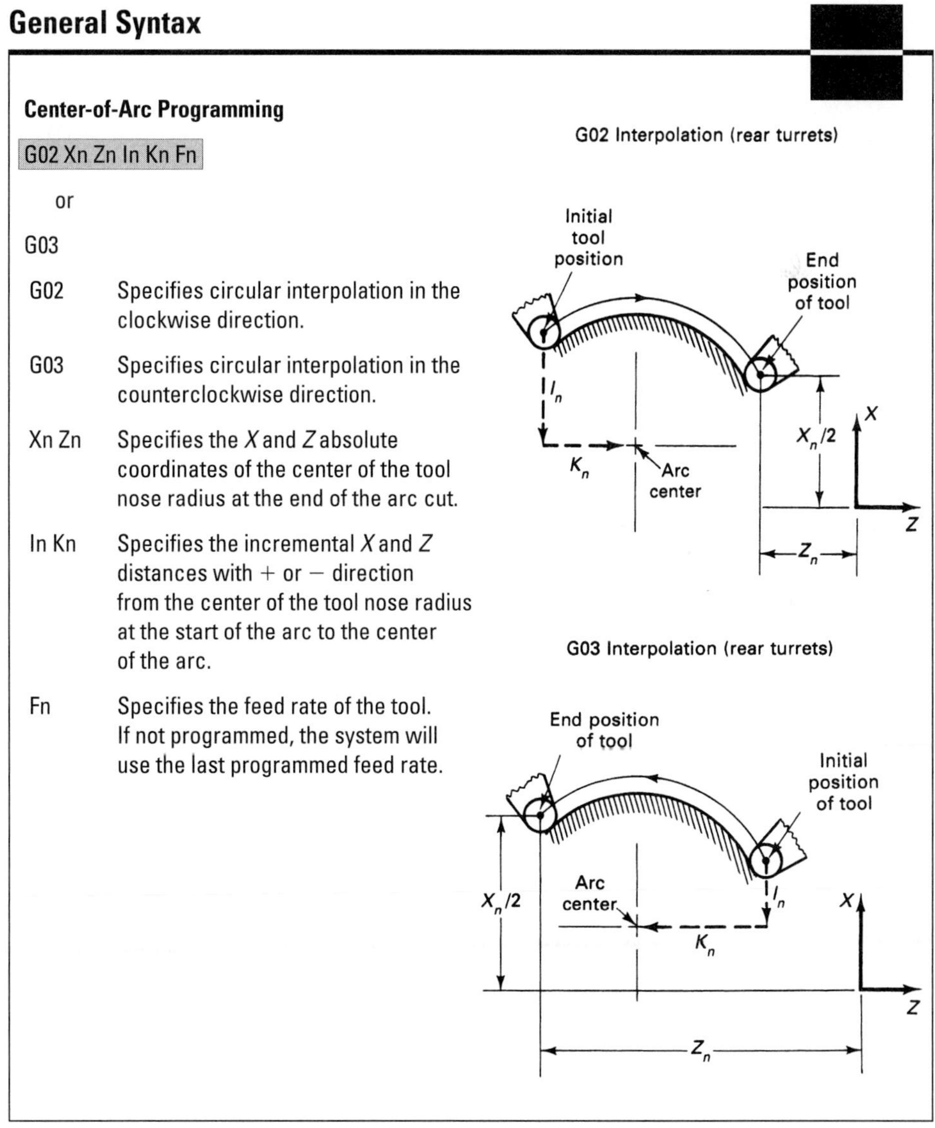

General Syntax

Center-of-Arc Programming

G02 Xn Zn In Kn Fn

or

G03

G02	Specifies circular interpolation in the clockwise direction.
G03	Specifies circular interpolation in the counterclockwise direction.
Xn Zn	Specifies the *X* and *Z* absolute coordinates of the center of the tool nose radius at the end of the arc cut.
In Kn	Specifies the incremental *X* and *Z* distances with + or − direction from the center of the tool nose radius at the start of the arc to the center of the arc.
Fn	Specifies the feed rate of the tool. If not programmed, the system will use the last programmed feed rate.

Radius-of-Arc Programming

G02 Xn Zn Rn Fn

or

G03

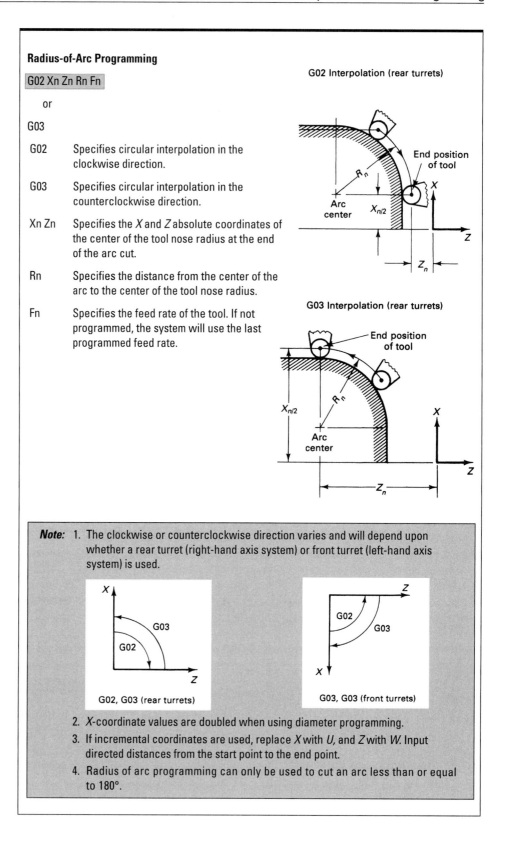

G02 Specifies circular interpolation in the clockwise direction.

G03 Specifies circular interpolation in the counterclockwise direction.

Xn Zn Specifies the *X* and *Z* absolute coordinates of the center of the tool nose radius at the end of the arc cut.

Rn Specifies the distance from the center of the arc to the center of the tool nose radius.

Fn Specifies the feed rate of the tool. If not programmed, the system will use the last programmed feed rate.

G02 Interpolation (rear turrets)

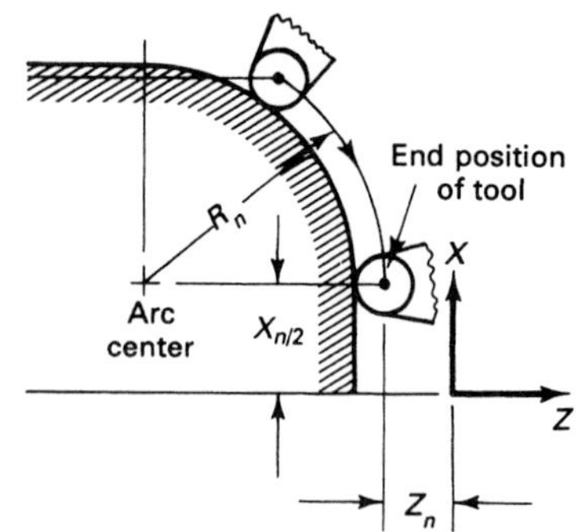

G03 Interpolation (rear turrets)

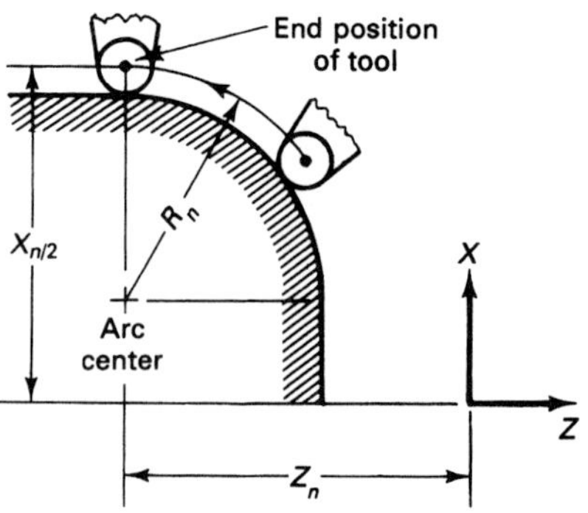

Note: 1. The clockwise or counterclockwise direction varies and will depend upon whether a rear turret (right-hand axis system) or front turret (left-hand axis system) is used.

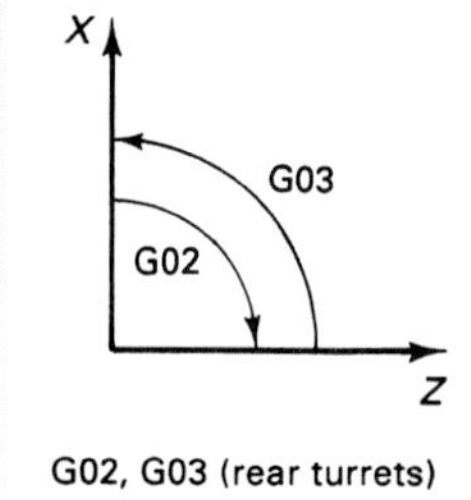

G02, G03 (rear turrets)

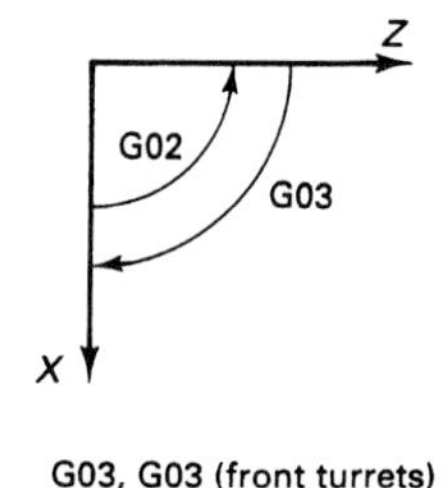

G03, G03 (front turrets)

2. *X*-coordinate values are doubled when using diameter programming.
3. If incremental coordinates are used, replace *X* with *U*, and *Z* with *W*. Input directed distances from the start point to the end point.
4. Radius of arc programming can only be used to cut an arc less than or equal to 180°.

■ EXAMPLE 16–10

Write blocks to execute the machining operation shown in Figure 16–16.

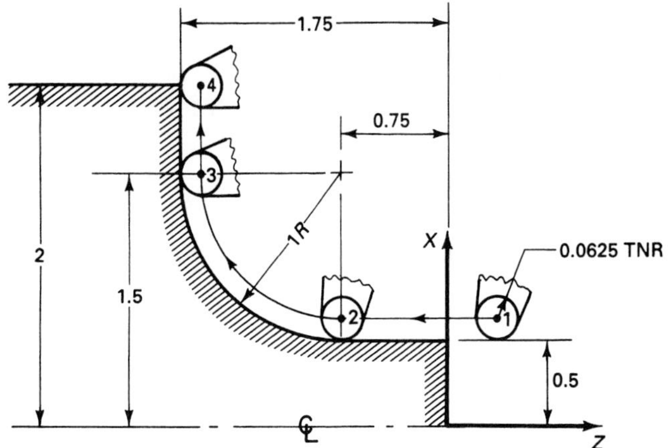

FIGURE 16–16

Word address command	*Meaning*
N0050 G01 Z−.75 F.005	Cut to ②.
N0060 G02 X3. Z−1.6875 I.9375 K0. or N0060 G02 X3. Z−1.6875 R.9375	Cut 1*R* arc to ③.
N0070 G01 X4.	Cut to ④.

■■

■ EXAMPLE 16–11

Code blocks to machine the profile shown in Figure 16–17.

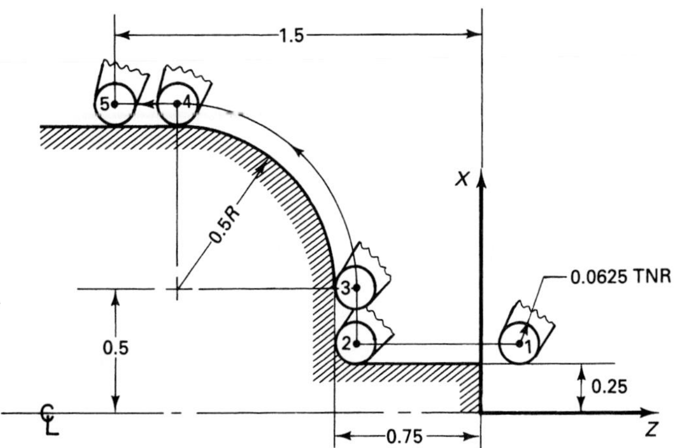

FIGURE 16–17

Word address command	*Meaning*
N0060 G01 Z−.6875 F.005	Cut to position ②.
N0070 X1.0	Cut to position ③.
N0080 G03 X2.125 Z−1.25 I0. K−.5625 or N0080 G03 X2.125 Z−1.25 R.5625	Cut 0.5*R* arc to ④.
N0090 G01 Z−1.5	Cut to position ⑤.

16-18 GROOVING COMMANDS

Grooving is executed by programming a linear cut with a specified dwell. The dwell is necessary to make the diameter of the groove uniform. The tool should be stopped at the bottom of the groove for at least one revolution of the spindle.

General Syntax

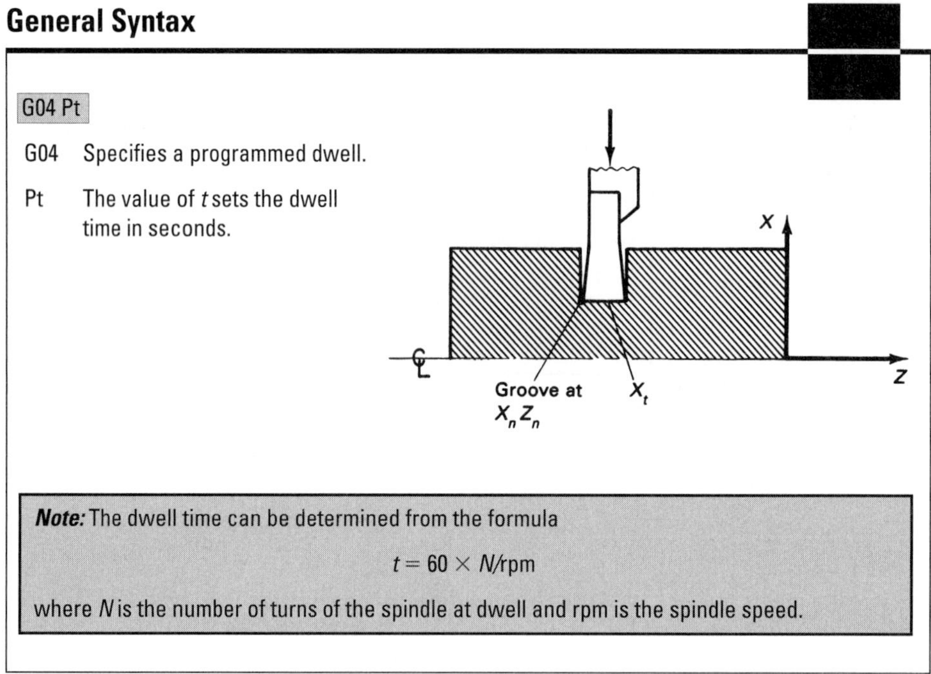

G04 Pt
G04 Specifies a programmed dwell.
Pt The value of *t* sets the dwell time in seconds.
X
Z
Groove at $X_n Z_n$
X_t
Note: The dwell time can be determined from the formula
$t = 60 \times N/\text{rpm}$
where *N* is the number of turns of the spindle at dwell and rpm is the spindle speed.

■ EXAMPLE 16-12

Code program blocks to machine the groove shown in Figure 16-18.

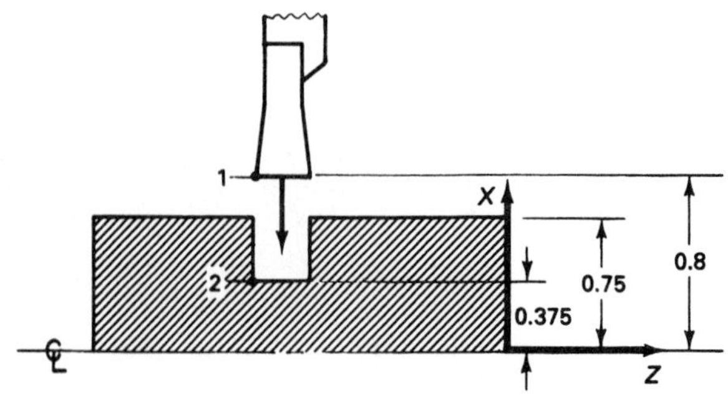

FIGURE 16-18

Spindle speed = 500 rpm

$N = 2$ turns

$t = 60 \times 2/500$

$= 0.24$ s

Word address command	Meaning
N0060 G00 X1.6	Rapid to ①.
N0070 G01 X.75 F.05	Cut to ② at feed rate.
N0080 G04 P.24	Dwell for 0.24 s.
N0090 G00 X1.6	Rapid to ①.

■■

16–19 RETURN TO REFERENCE POINT COMMAND

The CNC lathe can be programmed to automatically move the tool first to an intermediate point inputted and then to the reference point. These moves are made in rapid traverse.

General Syntax

G28 Xn Zn

G28 Specifies a rapid move to the intermediate point and from there to the reference point.

Xn Zn Specifies absolute coordinates of the intermediate point selected.

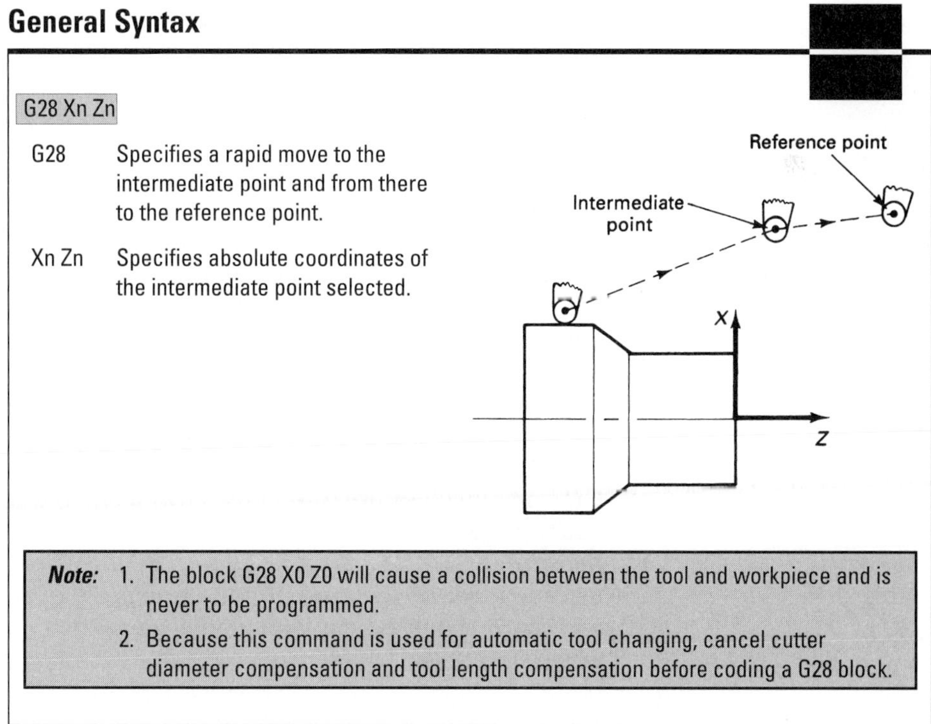

Note: 1. The block G28 X0 Z0 will cause a collision between the tool and workpiece and is never to be programmed.
2. Because this command is used for automatic tool changing, cancel cutter diameter compensation and tool length compensation before coding a G28 block.

■ EXAMPLE 16–13

Program a block to automatically rapid the tool to the reference point by way of the intermediate point shown in Figure 16–19.

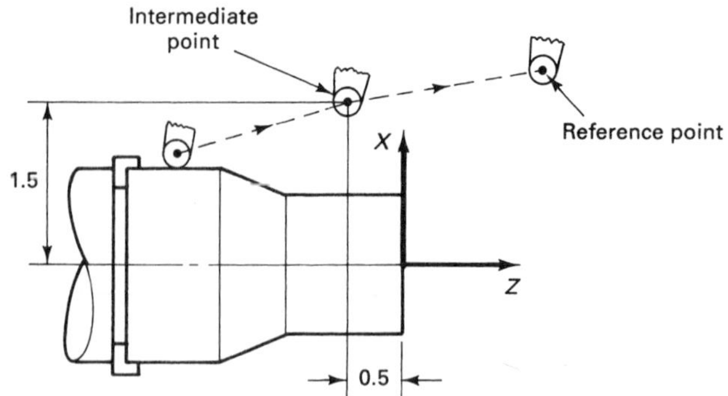

FIGURE 16–19

N0090 G28 X3. Z − 0.5

16–20 CHAPTER SUMMARY

The following key concepts were discussed in this chapter:

1. Lathe tools can be programmed to move in either the incremental mode or the absolute mode.
2. Radius programming involves specifying the X-axis coordinate as the distance from the spindle centerline to the center of the tool nose radius.
3. With diameter programming, twice the distance from the spindle centerline to the center of the tool nose radius is used.
4. The reference point (turret home position) is a point set once by the manufacturer.
5. All tool movements are programmed relative to the part origin.
6. The tool change point is located during setup and entered with a G50 code in the program.
7. Tool offsets for each tool are determined at setup and entered into the controller's memory.
8. Lathe feed rates can be programmed in terms of inches per minute (G98) or inches per revolution (G99).
9. Automatic tool change is programmed as a Tab block, where a is the number of the turret station holding the tool, and b is the memory address number containing the tool offset value.
10. For lathes, linear interpolation at feed rate is programmed with a G01 code.
11. For lathes, circular interpolation is programmed with a G02 (CW) or G03 (CCW) code. If I and K are used, the arc center is specified. If R is used, the arc radius is indicated.
12. Grooving involves programming a dwell (G04) after a linear interpolation (G01) move.
13. Machining at constant surface speed is important in maintaining optimum tool performance. Upon receiving a G96 Sn code, the controller will continuously adjust the spindle rpm to achieve constant surface speed (Sn).

REVIEW EXERCISES

16.1. Write absolute coordinates for the points shown in Figure 16–20 (X is expressed in terms of the diameter).

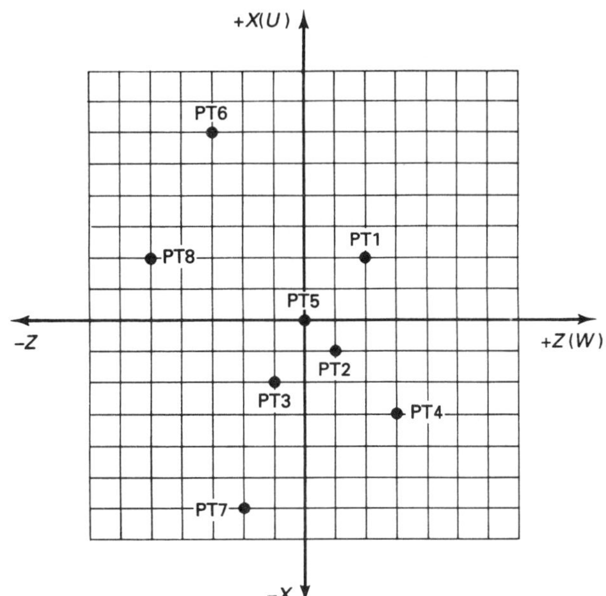

PT	X	Z
1		
2		
3		
4		
5		
6		
7		
8		

FIGURE 16–20

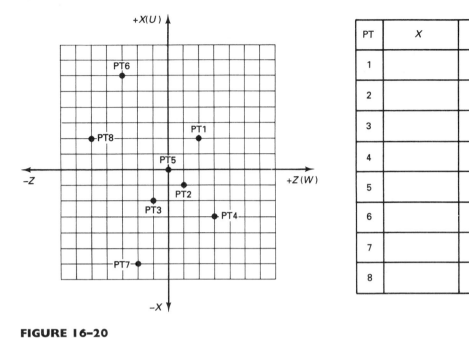

16.2. Write incremental coordinates for the points shown in Figure 16–20. Use the following order: origin to PT1, PT1 to PT2, PT2 to PT3 . . . finish with PT8 (*U* is expressed in terms of the incremental diameter).

PT	U	W
1		
2		
3		
4		
5		
6		
7		
8		

16.3. For the rapid positioning movement shown in Figure 16–21, write a G00 block in absolute (*X, Z*) coordinates and incremental (*U, W*) coordinates.

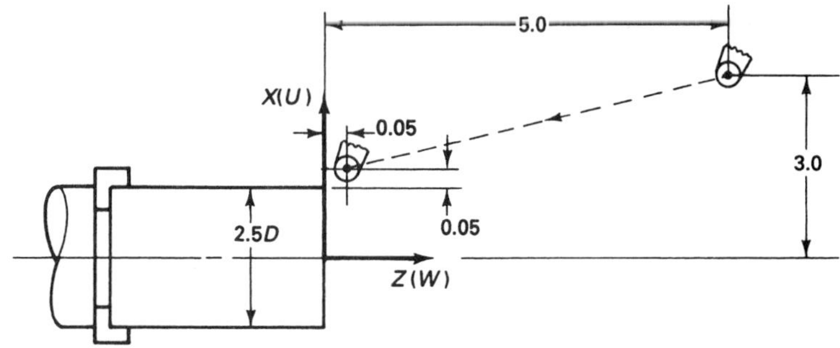

FIGURE 16–21

Absolute coordinates
G00 X _____ Z _____

Incremental coordinates
G00 U _____ W _____

16.4. Code blocks for turning the profile shown in Figure 16–22. Use a feed rate of 0.01 ipr.

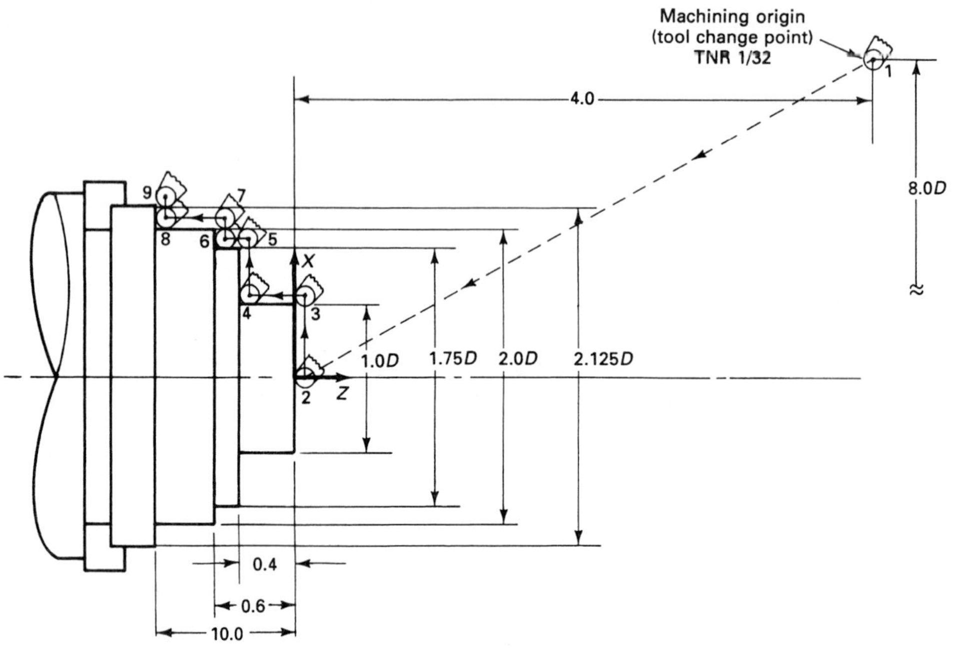

FIGURE 16–22

Absolute coordinates	Incremental coordinates	Position
G00 X_____ Z_____	G00 U_____ W_____	1
G01 X_____ Z_____ F_____	G01 U_____ W_____ F_____	2
X_____ Z_____	U_____ W_____	3
X_____ Z_____	U_____ W_____	4
X_____ Z_____	U_____ W_____	5
X_____ Z_____	U_____ W_____	6
X_____ Z_____	U_____ W_____	7
X_____ Z_____	U_____ W_____	8
X_____ Z_____	U_____ W_____	9

16.5. The profile shown in Figure 16–23 is to be turned at a feed rate of 0.01 ipm. Write blocks for executing the tool movements shown in absolute and incremental coordinates.

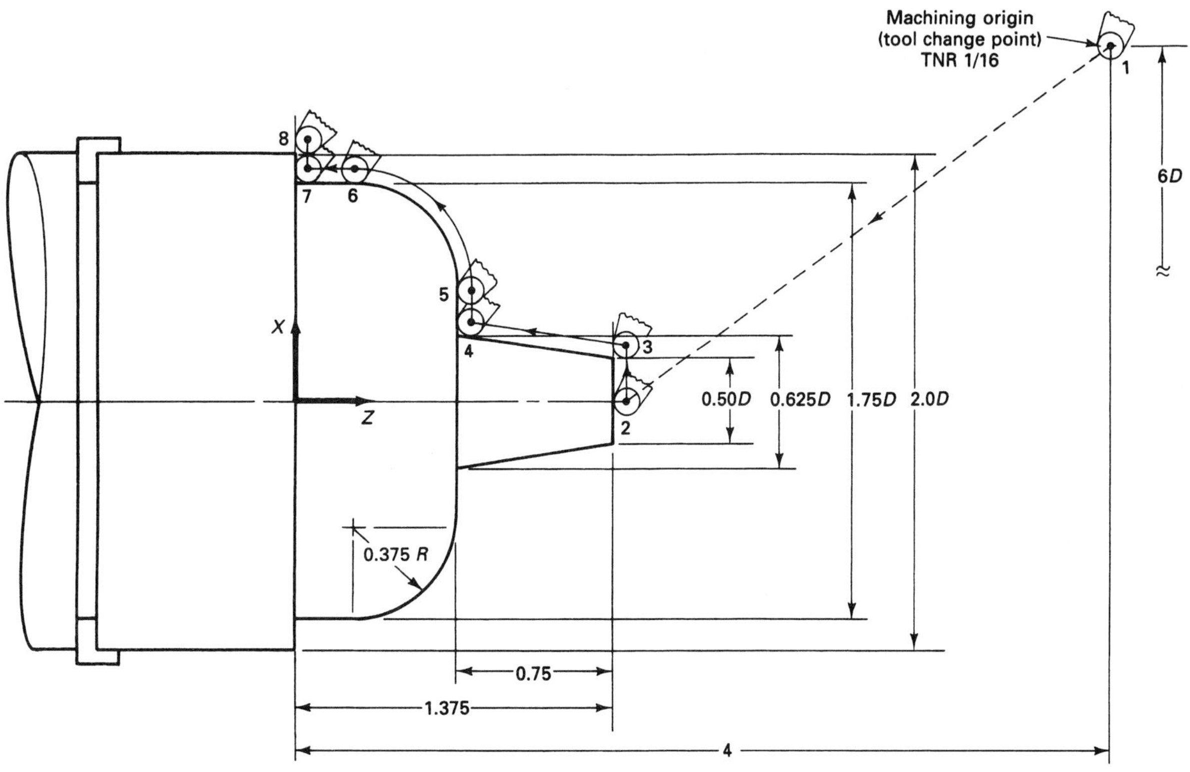

FIGURE 16–23

Absolute coordinates	Incremental coordinates	Position
G00 X ____ Z ____	G00 U ____ W ____	1
G01 X ____ Z ____ F ____	G01 U ____ W ____ F ____	2
X ____ Z ____	U ____ W ____	3
X ____ Z ____	U ____ W ____	4
X ____ Z ____	U ____ W ____	5
G03 X ____ Z ____ I ____ K ____	G03 U ____ W ____ I ____ K ____	6
G01 X ____ Z ____	G01 U ____ W ____	7
X ____ Z ____	U ____ W ____	8

16.6. Explain the significance of the following points:

(a) Reference point
(b) Machining origin
(c) Programming origin
(d) Tool change position

16.7. **(a)** What are tool offsets with regard to CNC lathes?
(b) How are tool offsets used to execute a programmed tool move?

16.8. The tool manufacturer recommends a surface speed of 600 sfm when using the facing tool as shown in Figure 16–24. Machining constraints limit the maximum spindle rpm to 2000.

(a) Write appropriate G50 and G96 blocks directing the controller not to exceed the maximum rpm while maintaining constant surface speed.
(b) What beginning rpm value will the controller compute?
(c) At what minimum work diameter can constant surface speed be maintained?
(d) What will happen to surface speed as the tool begins to decrease the work diameter from that determined in part c?

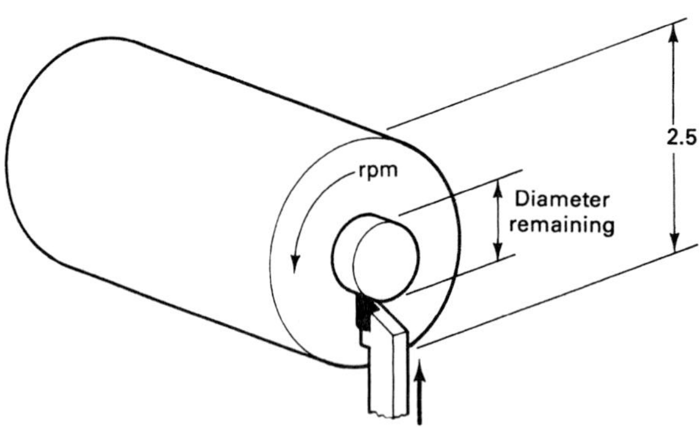

FIGURE 16–24

CHAPTER

17

TECHNIQUES AND FIXED CYCLES FOR CNC LATHE PROGRAMMING

17-1 CHAPTER OBJECTIVES

At the conclusion of this chapter you will be able to

1. Explain the difference between tool center programming and tool edge programming.
2. Understand the advantages and conditions involved in utilizing tool nose radius (TNR) compensation.
3. Use TNR compensation in programming turning and boring operations.
4. Explain and program the rough boring cycle G90 and rough facing cycle G94.
5. Understand and program the multiple repetitive cycle G71 and finishing cycle G70.
6. Describe and program the threading cycles G32, G92, and G76.

17-2 INTRODUCTION

Specific programming techniques are studied in this chapter. Canned cycles for executing important lathe operations are also considered. The chapter begins with a discussion of tool edge programming. This approach is used in conjunction with tool nose radius (TNR) programming. Word address codes for invoking and canceling TNR are discussed. The reader is introduced to the advantages, restrictions, and setup conditions involved with TNR programming.

CNC lathe machining cycles make programming operations easier and more efficient. See Figure 17–1. Cycles for rough turning, boring, and facing linear profiles are presented first. These are followed by a discussion of a cycle for rough machining profiles consisting of arcs and tapered lines. The chapter ends with detailed descriptions of the most important CNC lathe threading cycles.

17-3 TOOL EDGE PROGRAMMING

Recall from Chapter 16 that tool center programming is used when tool offsets are measured to the center of the tool nose radius. Tool edge programming is in effect when offsets are measured to the edge of the tool. With tool edge programming the part geometry may be inputted directly. Because tool edge programming ignores the tool nose radius of cutting tools, it can only be applied to part geometries consisting of horizontal or vertical lines. Errors will result if

Safety Rules for Programming CNC Lathes

- Always check for any interference between the workpiece, the tooling setups, and the indexing of each turret face.
- Always select cutters, holders, and turret accessories that provide maximum rigidity for holding during machining.
- Select speeds and feeds that permit machining at optimum efficiency and safety.
- Reduce the speed and feed:
  - When drilling large-diameter holes to a depth greater than twice the drill size.
  - When cutting thread.
- When finished cutting an internal taper, always feed the tool in the direction of the large diameter.
- During rough turning, pay attention to the following:
  - The effect of longitudinal, tangential, and radial forces.
  - The heat generated between the workpiece and tool.

arcs or tapered lines are programmed without applying additional compensation. See Figures 17–2 and 17–3.

The inherent problems with tool edge programming can be corrected if tool nose radius (TNR) compensation is applied.

17–4 TOOL NOSE RADIUS COMPENSATION PROGRAMMING

Cutter diameter compensation for milling applications was discussed in Chapter 12. Its key advantages were noted as follows:

- Ability to program the part geometry directly
- Ability to use the same program for a variety of cutter types and tool radii

For CNC lathes and turning centers, this feature is called tool nose radius (TNR) compensation. See Figure 17–3.

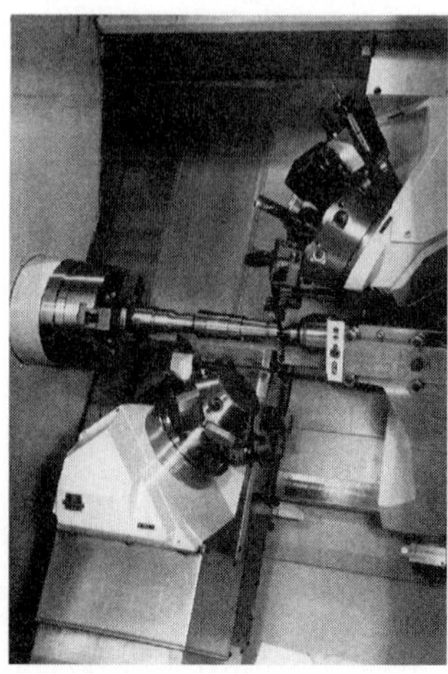

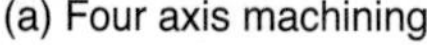

(a) Four axis machining

(b) Finish autocycle

FIGURE 17–1 CNC lathe fixed cycle operations. [(a) Photo courtesy of Index Corp. (b) Photo courtesy of Monarch Machine Tool Company.]

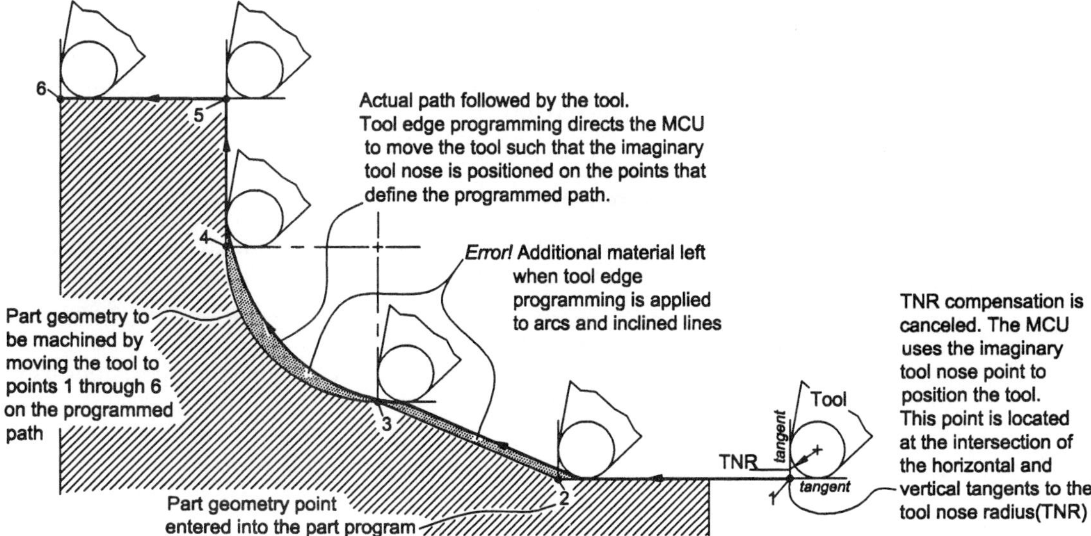

FIGURE 17–2 Programming the part geometry directly with tool edge programming.

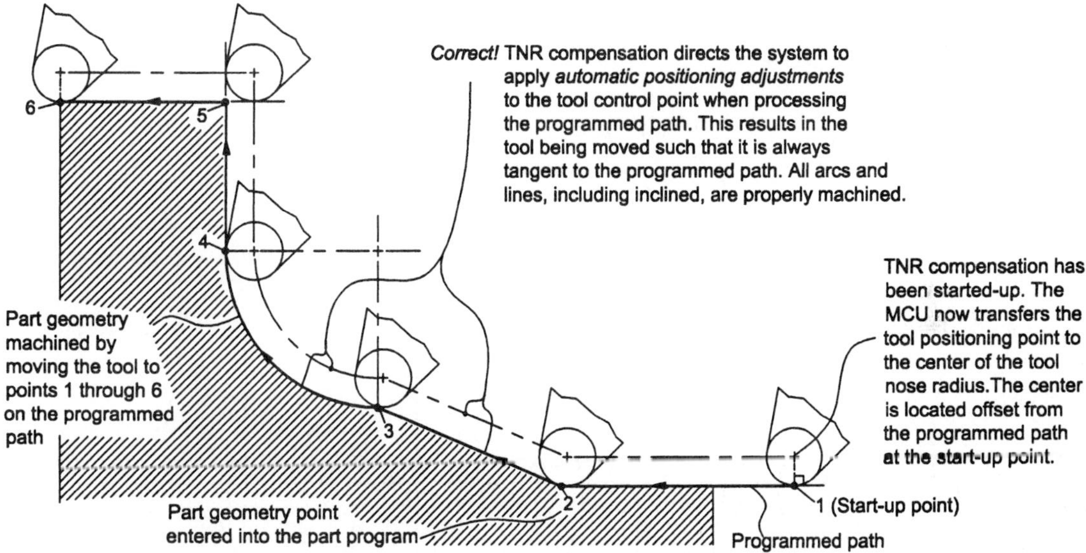

FIGURE 17–3 Programming the part geometry directly with TNR compensation.

17–5 SETTING UP TOOL NOSE RADIUS COMPENSATION

Important information must first be entered into the machine control unit prior to using TNR compensation. During setup the control is instructed to open an offset file in memory. The setup person keys in the following information for each tool:

- X and Z tool offsets to the tool edge
- Size of the tool nose radius
- The tool nose vector

Tool Nose Radius

The size of the tool nose radius is readily available from tool supplier catalogs or the insert package. Standard radii in inches are 1/64 or 0.0156, 1/32 or 0.0312, and 3/64 or 0.0469.

Tool Nose Vector

The tool tip (imaginary tool point) of single-point tools has a specific location from the center of the tool nose radius. The tool nose vector indicates this location to the controller. Standard tool nose vector numbers and the corresponding tool tip locations are shown in Figure 17–4.

Later the controller will use this information to properly determine the tool movement in response to a compensation command.

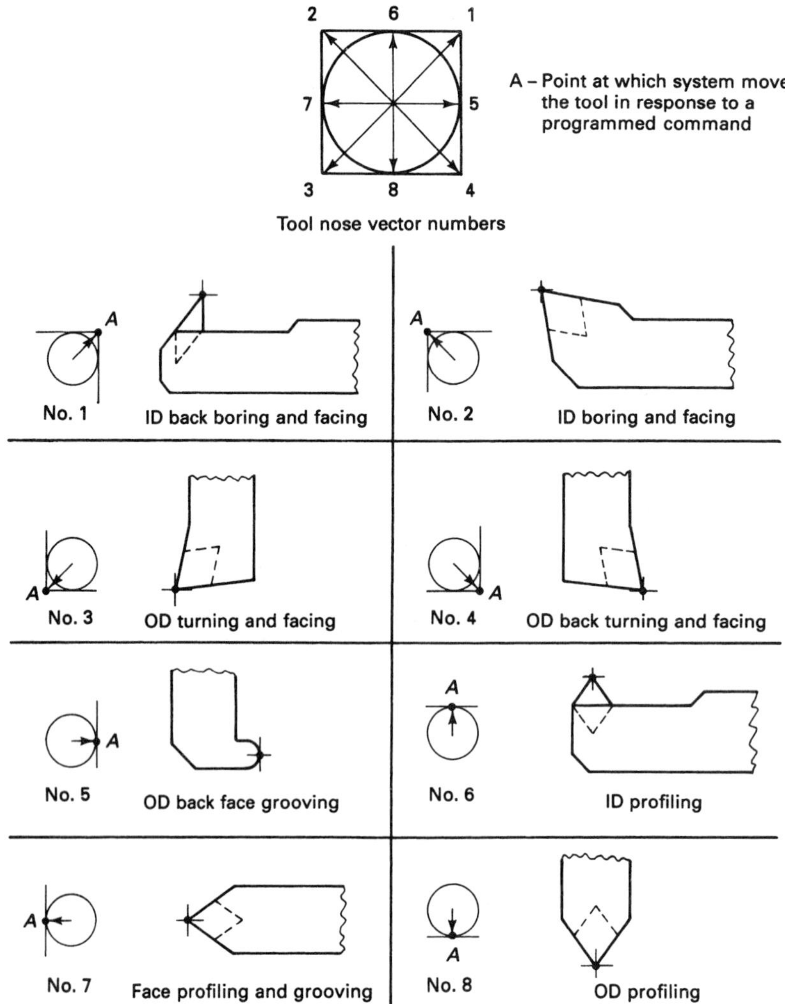

FIGURE 17–4 Tool nose vectors and tool tip locations.

17–6 SOME RESTRICTIONS WITH TOOL NOSE RADIUS COMPENSATION

The following restrictions directly apply to the Fanuc family of controllers. They are also valid for other types of controllers.

1. A G code activating TNR compensation is entered as a separate block in the program. It must be commanded before the tool starts to cut.
2. The first or second block following a TNR compensation code must contain an *XZ* linear motion command. The motion command signals the controller to initiate TNR compensation or cancel compensation. Failure to do this will result in over or under cutting.
3. Motion in the G40, G41, or G42 block must be greater than twice the value of the tool nose radius. Following this rule will ensure that overcutting or undercutting will not result.

17–7 TOOL NOSE RADIUS COMPENSATION COMMANDS

A G41 or G42 word indicates TNR compensation is to start up on the next linear move. When G41 or G42 is active, the MCU *transfers the tool positioning point from the imaginary point A as shown in Figures 17–5, 17–6 to the center of the tool nose radius.* The controller will signal for the tool to be offset at the start-up point by the radius value stored in the MCU's offset file. The tool nose vector number for the particular tool used is also stored in the offset file. This

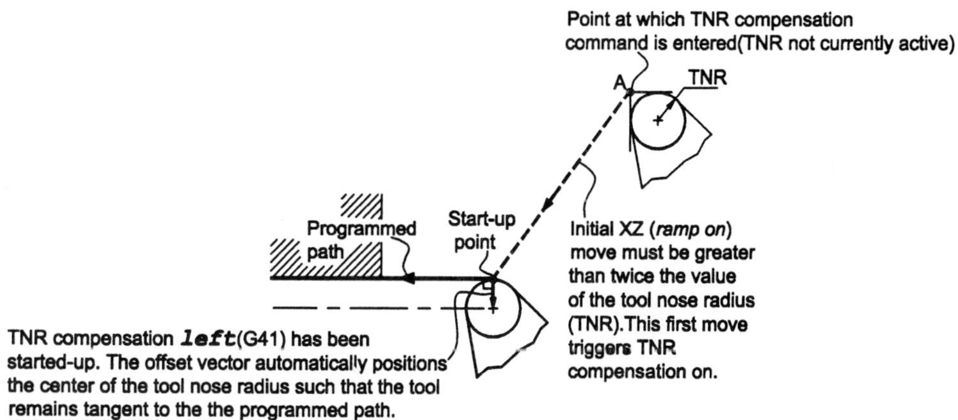
Point at which TNR compensation command is entered(TNR not currently active)
A
TNR
Programmed path
Start-up point
Initial XZ (ramp on) move must be greater than twice the value of the tool nose radius (TNR).This first move triggers TNR compensation on.
TNR compensation left(G41) has been started-up. The offset vector automatically positions the center of the tool nose radius such that the tool remains tangent to the the programmed path.

FIGURE 17–5 Ramp on move to initiate tool nose radius compensation left.

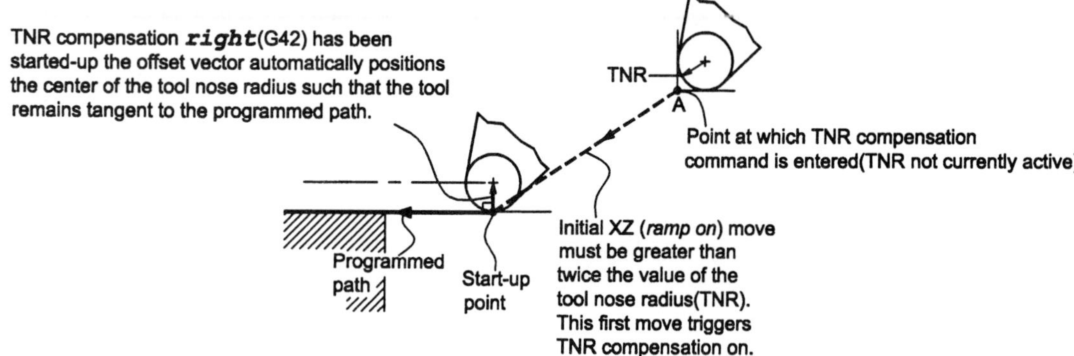
TNR compensation right(G42) has been started-up the offset vector automatically positions the center of the tool nose radius such that the tool remains tangent to the programmed path.
TNR
A
Point at which TNR compensation command is entered(TNR not currently active)
Initial XZ (ramp on) move must be greater than twice the value of the tool nose radius(TNR). This first move triggers TNR compensation on.
Programmed path
Start-up point

FIGURE 17–6 Ramp on move to initiate tool nose radius compensation right.

number enables the MCU to determine the location of the tool edge. Automatic adjustments will then be made such that tool nose is always positioned tangent to the programmed path. When TNR compensation is canceled with a G40 word, the MCU shifts the tool positioning point back to the imaginary point A.

General Syntax

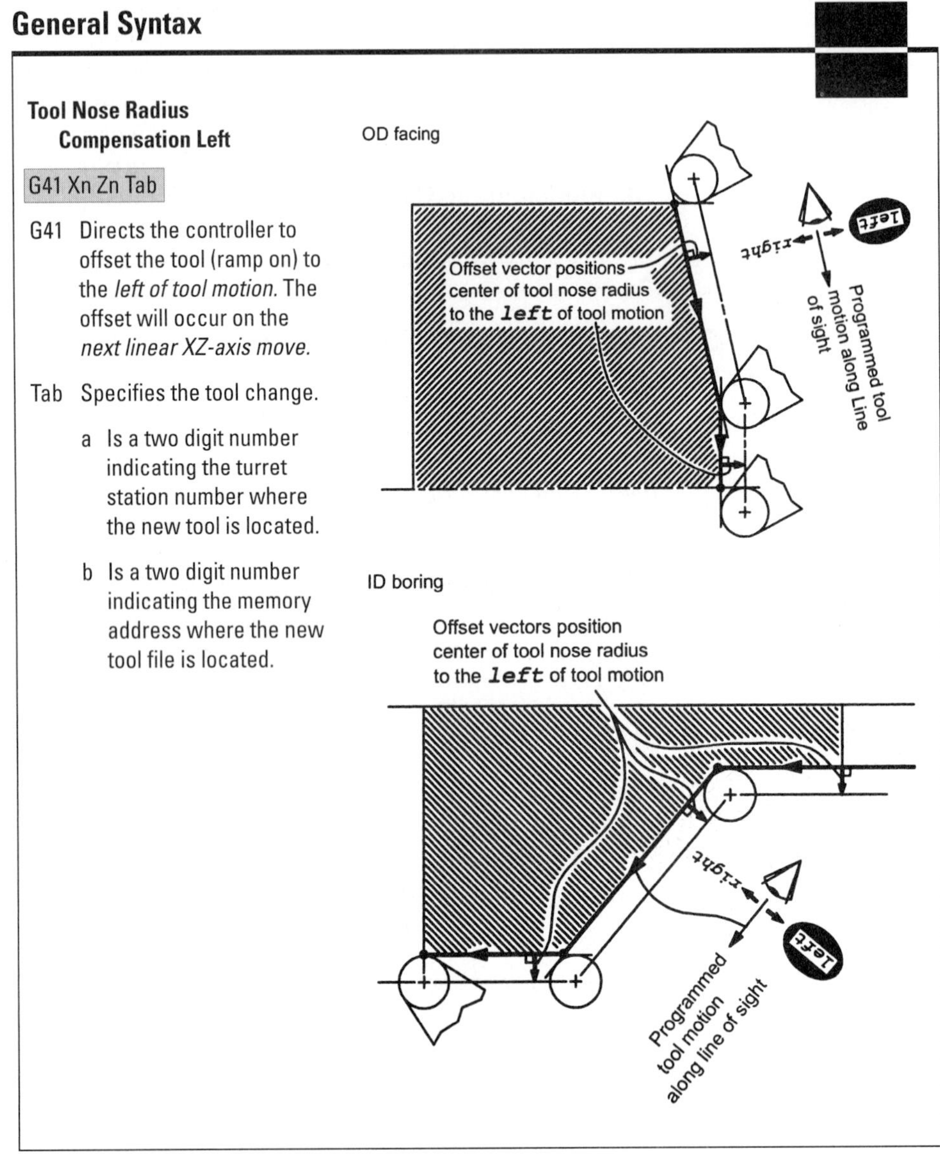
OD facing
Offset vector positions center of tool nose radius to the left of tool motion
right
left
Programmed tool motion along Line of sight
ID boring
Offset vectors position center of tool nose radius to the left of tool motion
right
left
Programmed tool motion along line of sight

Tool Nose Radius Compensation Left

G41 Xn Zn Tab

G41 Directs the controller to offset the tool (ramp on) to the *left of tool motion.* The offset will occur on the *next linear XZ-axis move.*

Tab Specifies the tool change.

a Is a two digit number indicating the turret station number where the new tool is located.

b Is a two digit number indicating the memory address where the new tool file is located.

General Syntax

Tool Nose Radius Compensation Right

G42 Xn Zn Tab

G42 Directs the controller to offset the tool (ramp on) to the *right of tool motion.* The offset will occur on the *next linear XZ-axis move.*

Tab Specifies the tool change.

- a Is a two digit number indicating the turret station number where the new tool is located.
- b Is a two digit number indicating the memory address where the new tool file is located.

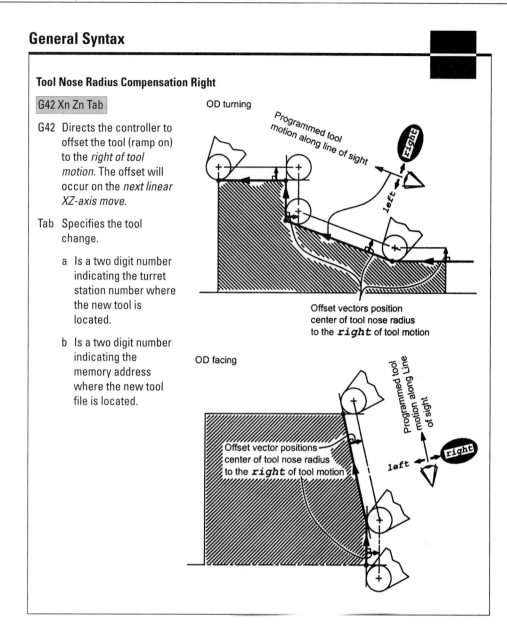

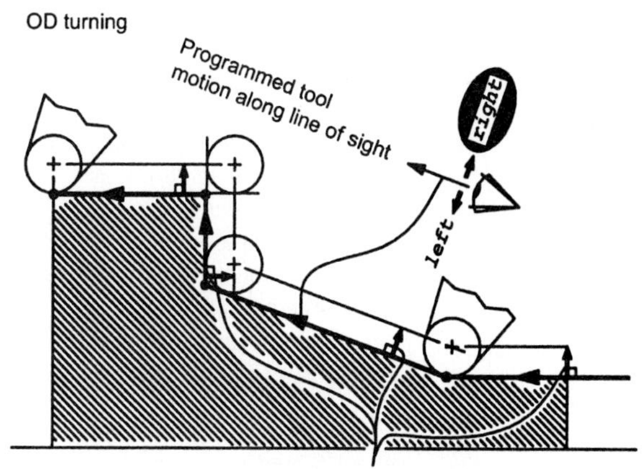

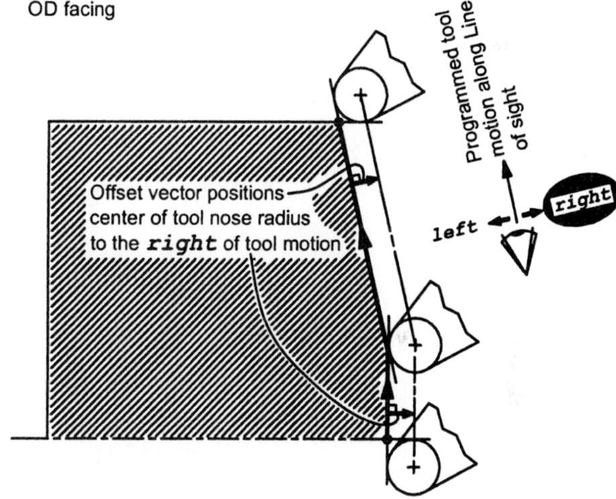

General Syntax

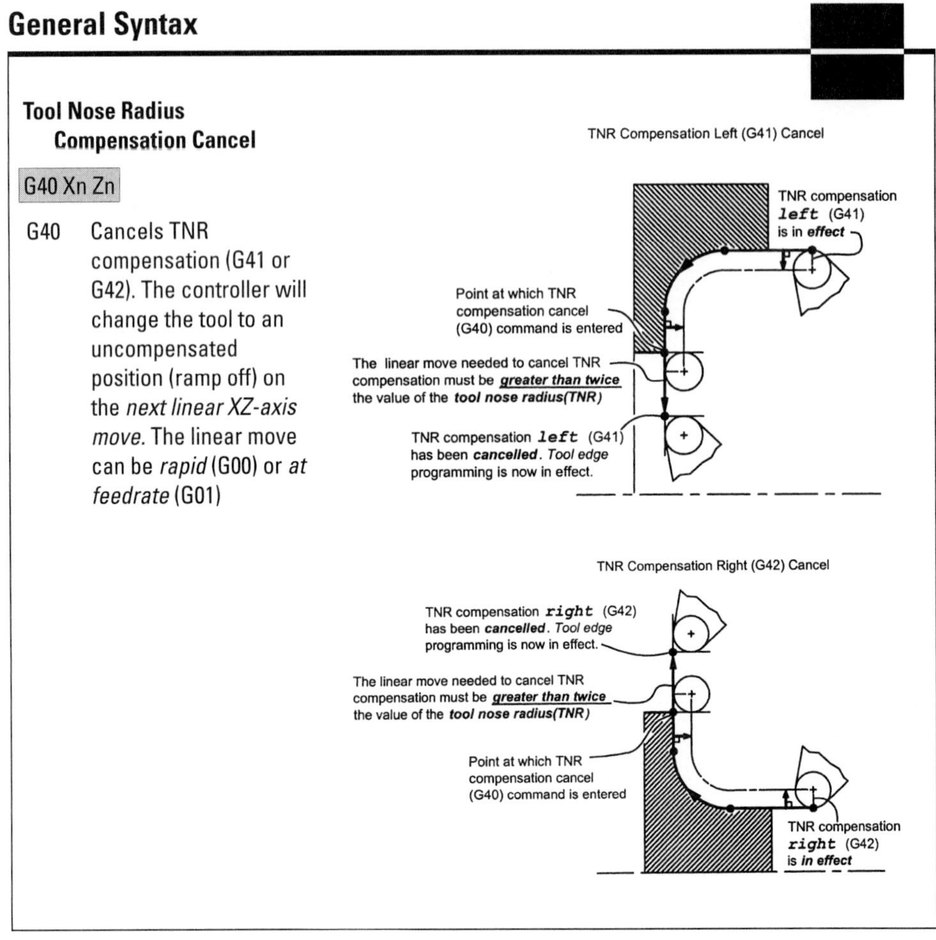
TNR Compensation Left (G41) Cancel
TNR compensation left (G41) is in effect
Point at which TNR compensation cancel (G40) command is entered
The linear move needed to cancel TNR compensation must be greater than twice the value of the tool nose radius(TNR)
TNR compensation left (G41) has been cancelled. Tool edge programming is now in effect.
TNR Compensation Right (G42) Cancel
TNR compensation right (G42) has been cancelled. Tool edge programming is now in effect.
The linear move needed to cancel TNR compensation must be greater than twice the value of the tool nose radius(TNR)
Point at which TNR compensation cancel (G40) command is entered
TNR compensation right (G42) is in effect

Tool Nose Radius Compensation Cancel

G40 Xn Zn

G40 Cancels TNR compensation (G41 or G42). The controller will change the tool to an uncompensated position (ramp off) on the *next linear XZ-axis move.* The linear move can be *rapid* (G00) or *at feedrate* (G01)

Note:

1. Some controllers use single-digit numbers for a and b.
2. More than one b code can be programmed with a tool. This allows for the use of the same tool for executing rough as well as finish cuts or to compensate for tool wear.
3. G02/G03 motion cannot be used to initiate a G41, G42, or G40 word in a program.
4. G41/G42 are modal. This means they remain in effect for all subsequent tool motions until canceled by a G40 word. G40 is also modal and remains in effect until canceled by a G41/G42 word.
5. The initial state of the control at machine start up is G40.

■ EXAMPLE 17–1

The part shown in Figure 17–7 has been rough turned. Use TNR compensation to execute the facing and turning finish pass. Assume the following information has been entered into the offset file during setup:

- Tool offset values for tool edge programming
- Tool nose radius value
- Tool nose vector 3

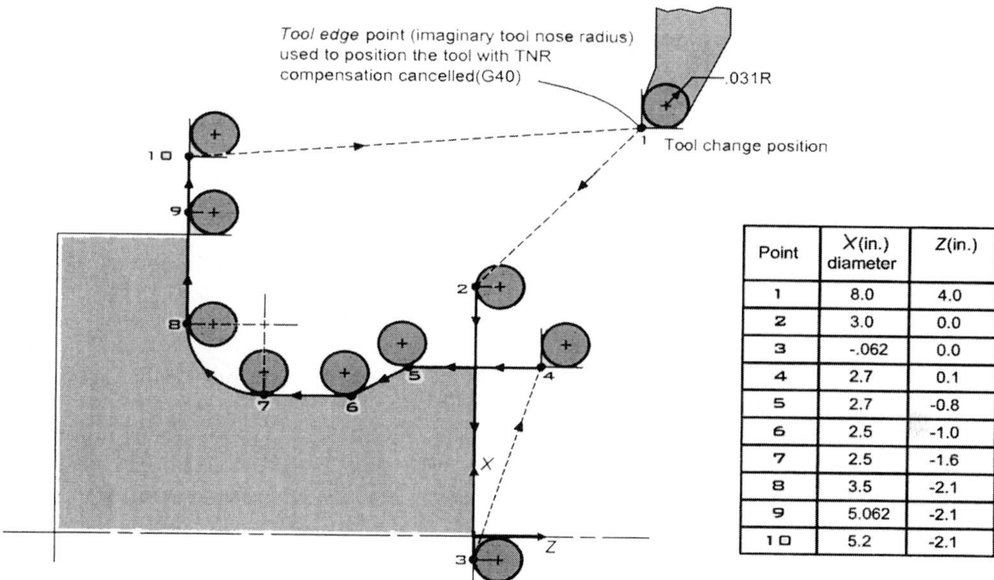

Point	X(in.) diameter	Z(in.)
1	8.0	4.0
2	3.0	0.0
3	-.062	0.0
4	2.7	0.1
5	2.7	-0.8
6	2.5	-1.0
7	2.5	-1.6
8	3.5	-2.1
9	5.062	-2.1
10	5.2	-2.1

FIGURE 17–7

■■

Word address command	*Meaning*
N0040 (TOOL 1: FINISH TURN CONTOUR) N0050 T0100	Return to tool change position ①. Index to tool 1. Cancel values in offset file 1. Coolant on.
N0060 G00 M42	Shift to intermediate-gear range. Rapid on.
N0070 G96 S500 M03	Spindle on (CW) at 500 sfpm constant surface speed.
N0080 G41 X3.0 Z0 T0101 M08	G41 (comp on left) is coded at ①. code move to ② Tool will be offset to the left *of its movement to* ②. Use values in offset file 1. Coolant on.
N0090 G01 X−.031 F.006	Cut to ③ at feed rate 0.006 ipr.
N0100 G00 G40 X2.7 Z.1	G40 (comp off) is coded at ③.Code move to ④. Use tool edge as control point when tool moves to④.
N0110 G01 G42 Z−.8 F.010	G42 (comp on right) is coded at ④. Code move to ⑤ Tool will be offset to the right of its movement to ⑤. Uses value in offset file 1. Coolant on.

N0120 X2.5 Z−1.0	Cut to ⑥.
N0130 Z−1.6	Cut to ⑦.
N0140 G02 X3.5 Z−2.1 R.5	Cut 0.5*R* arc to ⑧.
N0150 G01 X5.064	Cut to ⑨.
N0160 G40 X5.2	G40(comp) is coded at ⑨. code moves to ⑩.Use tool edge as control point when tool moves to ⑩.
N0170 G00 X8.0 Z4.0 T0100 M09	Rapid to tool change position ①. Cancel values in offset file 1. Coolant off.
N0180 M05	Spindle off.

■ EXAMPLE 17–2

Assume the part shown in Figure 17–8 has been previously roughed. Write a CNC program segment using TNR to finish bore the profile. The following information was entered into the offset file at setup:

- Tool offset values for tool edge programming
- Tool nose radius value
- Tool nose vector 2

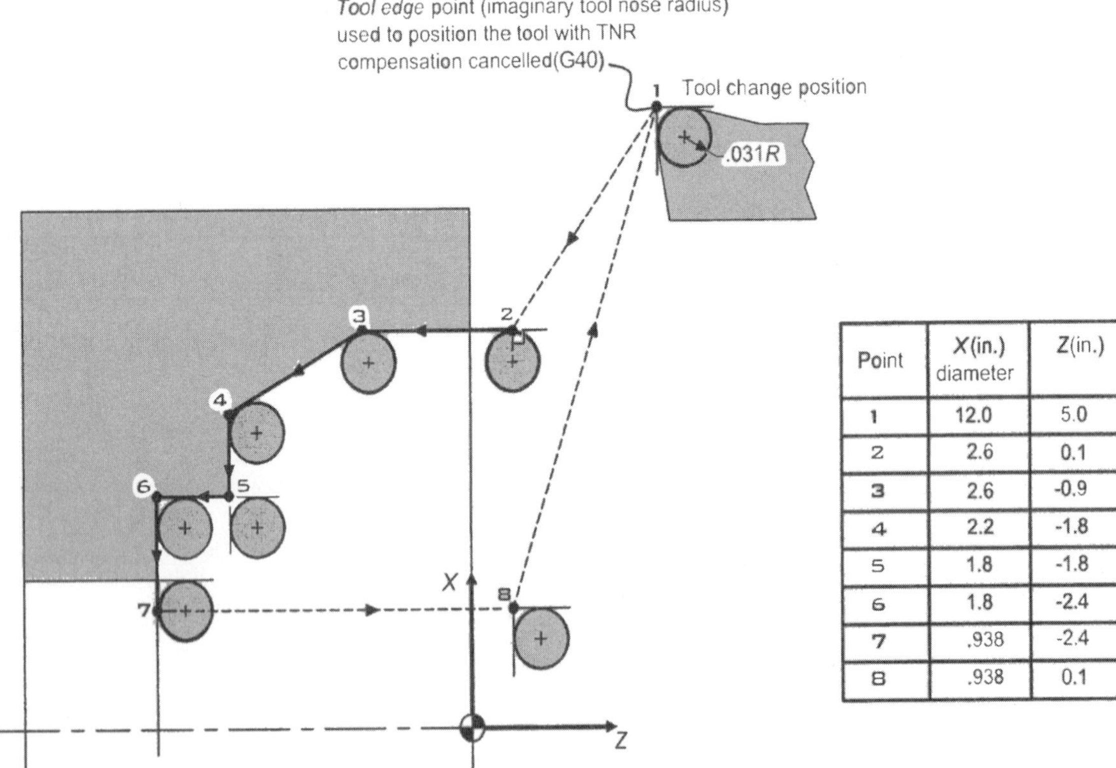

Point	X(in.) diameter	Z(in.)
1	12.0	5.0
2	2.6	0.1
3	2.6	-0.9
4	2.2	-1.8
5	1.8	-1.8
6	1.8	-2.4
7	.938	-2.4
8	.938	0.1

FIGURE 17–8 ■■

Word address command	*Meaning*
N0050 (TOOL 7: FINISH BORE CONTOUR)	
N0060 T0700	Return to tool change position ①. Index to tool 7. Cancel values in offset file 7.
N0070 G00 M42	Shift to intermediate-gear range. Rapid on.
N0080 G96 S500 M03	Spindle on (CW) at 500 sfm constant surface speed.
N0990 G41 X2.6 Z.1 T0707 M08	G41(comp on) left is coded at ①.Code moves to ②. Tool will be offset to the left of its movement to ②. Use values in offset file 1. Coolant on.
N0100 G01 Z−.9 F.012	Cut to ③ at 0.012 ipr feed rate.
N0110 X2.2 Z−1.8	Cut to ④.
N0120 X1.8	Cut to ⑤.
N0130 Z−2.4	Cut to ⑥.
N0140 X1.0	Cut to ⑦.
N0150 G00 G40 X1.7 Z.1	G40(comp off) is coded at ⑦. Code moves to ⑧. Uses tool edge as control point when tool moves to 8.
N0160 X12.0 Z5.0 T0700 M09	Rapid to tool change position ①. Cancel values in offset file 7. Coolant off.
N0170 M05	Spindle off.

17–8 TURNING AND BORING CYCLE: G90

A single block containing a G90 word executes repetitive straight-cut machining passes required to turn or bore a part from stock.

General Syntax

```
G90 Xa Za Fn
    Xb Za
     .
     .
     .
    Xn Za
```

G90 Signals the controller to begin the straight-cut turning or boring cycle.

Xa Za Are the absolute coordinates of the tool at the end of each machining pass (order point).

Xb Za

.

.

.

Xn Za

Fn Value of *n* specifies the feed rate.

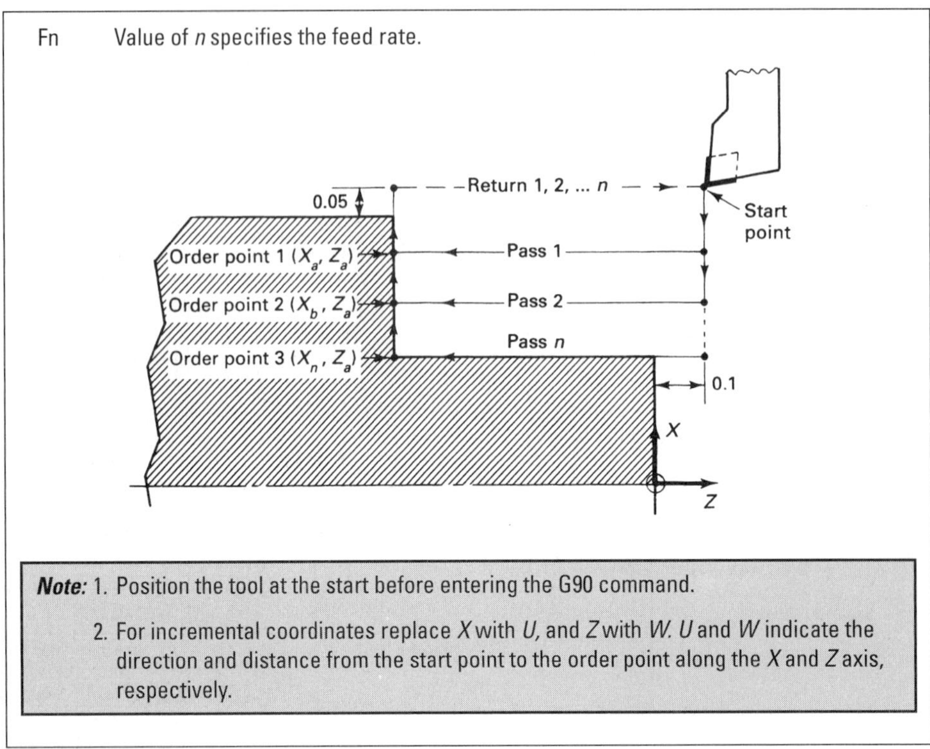

Note: 1. Position the tool at the start before entering the G90 command.

2. For incremental coordinates replace *X* with *U*, and *Z* with *W*. *U* and *W* indicate the direction and distance from the start point to the order point along the *X* and *Z* axis, respectively.

■ EXAMPLE 17–3

Write a program segment for rough turning the part as illustrated in Figure 17–9. Use the G90 cycle and assume all information needed for programming the tool edge has been entered into the controller's offset file.

Because the profile consists only of horizontal and vertical lines, tool edge programming may be directly applied.

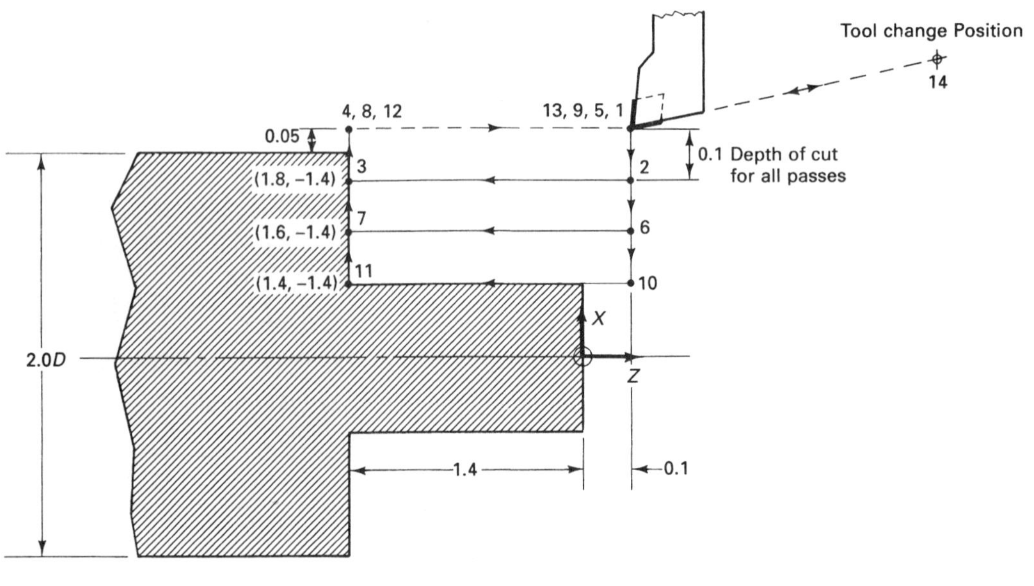

FIGURE 17–9

Word address command	*Meaning*
N0030 (TOOL 1: ROUGH TURN THE PART)	
N0040 T0100	Return to tool change position ⑭. Index to tool 1. Cancel values in offset file 1.
N0050 G00 M42	Shift to intermediate-gear range.
N0060 G96 S500 M03	Spindle on (CW) at 500 sfm constant surface speed.
N0070 X2.1 Z.1 T0101 M08	Rapid to position ①. Use values in offset file 1. Coolant on.
N0080 G90 X1.8 Z−1.4 F.01	Execute multiple-pass straight cutting cycle: rapid to ②; cut to ③ and ④ at feed rate; rapid to ⑤.
N0090 X1.6	Rapid to ⑥; cut to ⑦ and ⑧ at feed rate; rapid to ⑨.
N0100 X1.4	Rapid to ⑩; cut to ⑪ and ⑫ at feed rate; rapid to ⑬.
N0110 G00 X8.0 Z5.0 T0100 M09	Rapid to tool change position ⑭. Cancel values in offset file 1. Coolant off.
N0120 M05	Spindle off.

■ EXAMPLE 17–4

Use the G90 cycle to write a program segment for rough boring the part shown in Figure 17–10. Assume all the information for tool edge programming has been entered into the controller's offset file.

Tool edge programming may be directly applied for executing the horizontal and vertical line cuts.

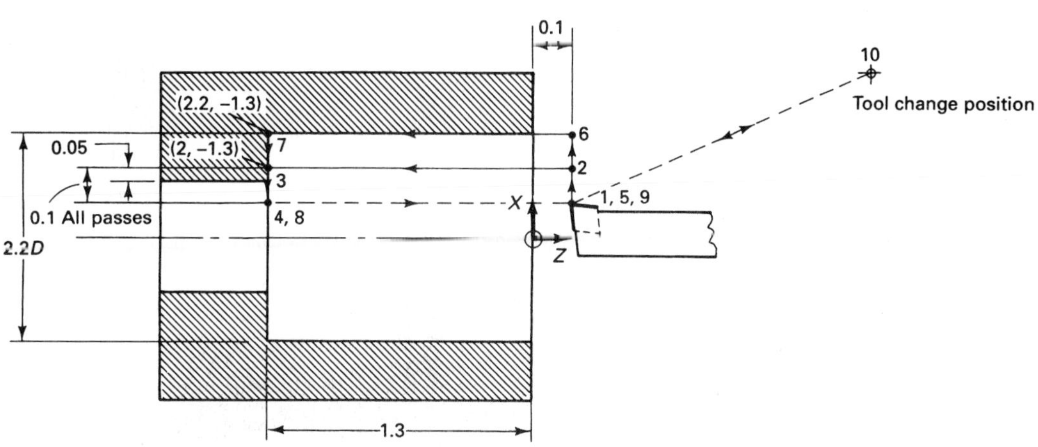

FIGURE 17–10

Word address command	***Meaning***
N0050 (TOOL 3: ROUGH BORE THE PART)	
N0060 T0300	Return to tool change position ⑩. Index to tool 3. Cancel values in offset file 3.
N0070 G00 M42	Shift to intermediate-gear range.
N0080 G96 S600 M03	Spindle on (CW) at 600 sfm constant surface speed.
N0090 G00 X1.8 Z.1 M08	Rapid to position ①. Coolant on.
N0100 G90 X2.0 Z−1.3 F.012	Execute multiple-pass straight cutting cycle: rapid to ②; cut to ③ and ④ at feed rate; rapid to ⑤.
N0110 X2.2	Rapid to ⑥; cut to ⑦ and ⑧ at feed rate; rapid to ⑨.
N0120 G00 X10.0 Z5.0 T0300 M09	Rapid to tool change position ⑩. Cancel values in offset file 3. Coolant off.
N0130 M05	Spindle off.

17-9 FACING CYCLE: G94

The G94 word enables the programmer to execute the multiple straight-cut machining passes needed to face a part from stock.

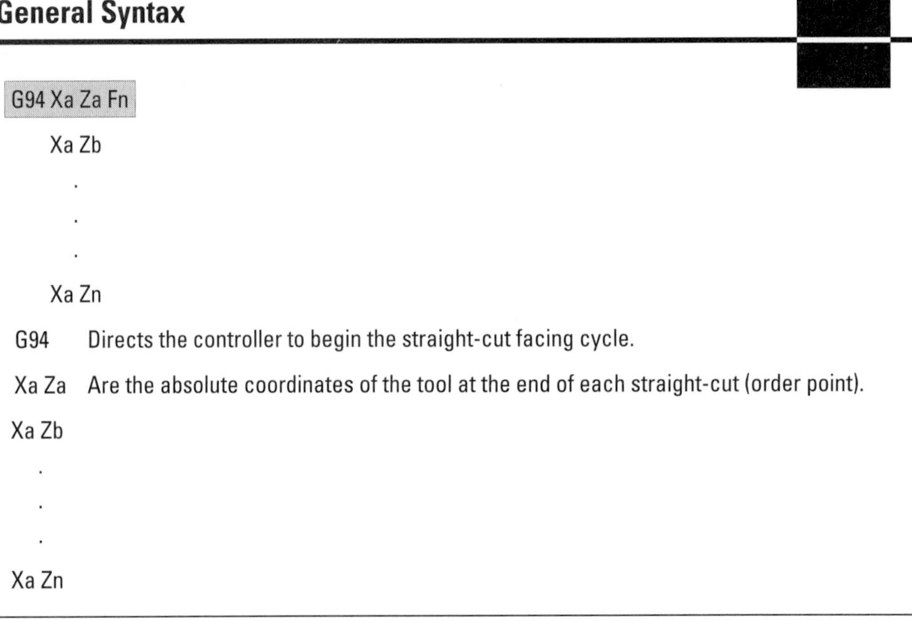

General Syntax

G94 Xa Za Fn
Xa Zb
.
.
.
Xa Zn

G94 Directs the controller to begin the straight-cut facing cycle.

Xa Za Are the absolute coordinates of the tool at the end of each straight-cut (order point).
Xa Zb
.
.
.
Xa Zn

Fn Value of *n* specifies the feed rate.

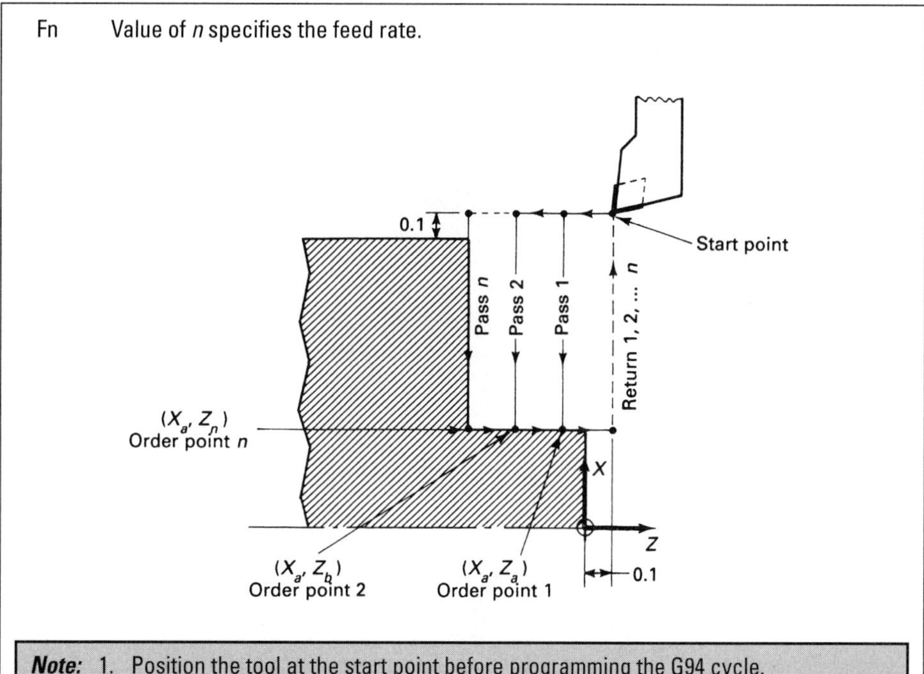

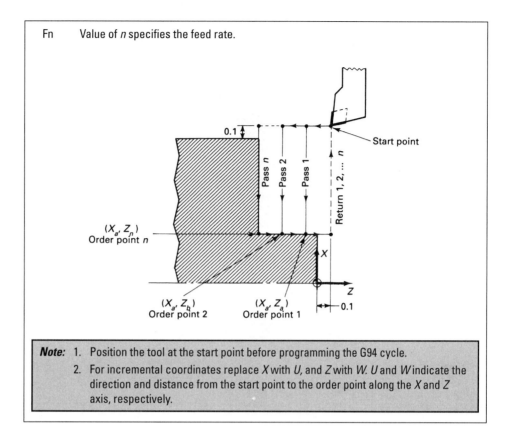

Note: 1. Position the tool at the start point before programming the G94 cycle.
2. For incremental coordinates replace *X* with *U*, and *Z* with *W*. *U* and *W* indicate the direction and distance from the start point to the order point along the *X* and *Z* axis, respectively.

■ EXAMPLE 17–5

The part shown in Figure 17–11 is to be rough faced from stock. Use the G94 cycle to write the required program segment. Assume the tool offset file contains all the information for tool edge programming.

Tool edge programming will, again, be used directly to execute the horizontal and vertical line cuts.

Word address command	***Meaning***
N0040 (TOOL 7: ROUGH FACE PART)	
N0050 T0700	Return to tool change position ⑩. Index to tool 7. Cancel values in offset file 7.
N0060 G00 M42	Shift to intermediate-gear range.
N0070 G96 S600 M03	Spindle on (CW) at 600 sfm constant surface speed.
N0080 G00 X3.2 Z.1 M08	Rapid to position ①. Coolant on.
N0090 G94 X.8 Z−.1 F.012	Execute multiple-pass straight-cut facing cycle: rapid to ②; cut to ③ and ④ at feed rate; rapid to ⑤.
N0100 Z−.2	Rapid to ⑥; cut to ⑦ and ⑧ at feed rate; rapid to ⑨.
N0110 G00 X10.0 Z5.0 T0700 M09	Rapid to tool change position ⑩. Cancel values in offset file 7. Coolant off.
N0120 M05	Spindle off.

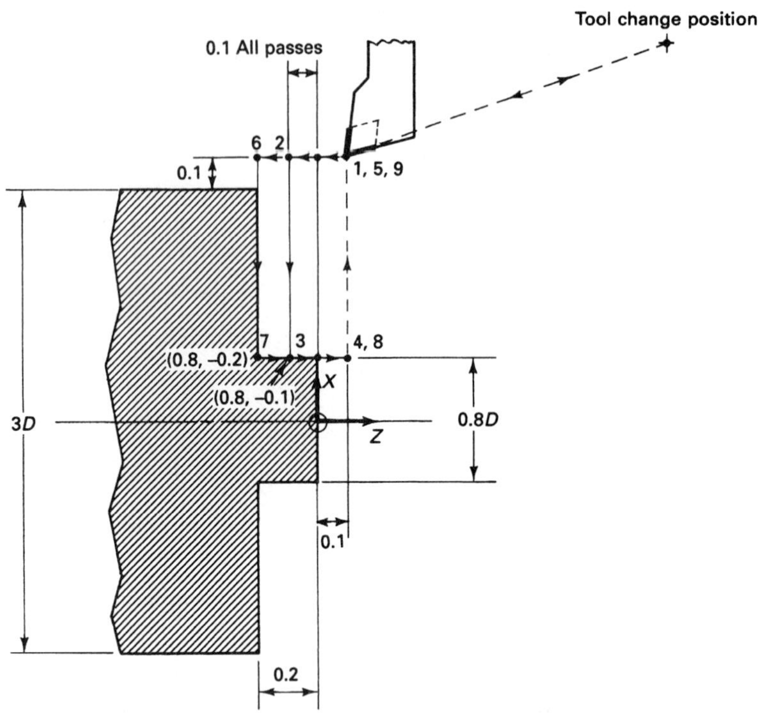

FIGURE 17-11

It should be noted that the G90 and G94 cycles can also be used to rough cut tapered profiles from stock. The reader is advised to consult the CNC programming manual for this information.

17-10 MULTIPLE REPETITIVE CYCLES: G70 TO G75

The repetitive cycles (G70 to G75) reduce the effort involved in programming rough machining operations. These cycles simply require the programmer to specify the dimensions of the finished shape and cutting parameters. The controller then responds by directing the tool to execute the necessary repetitive cutting. The tool will automatically remove material from stock so that the finish shape is achieved.

17-11 STOCK REMOVAL IN TURNING AND BORING CYCLE: G71

The G71 cycle is used to rough cut a general contour from stock. The contour may consist of lines (horizontal, vertical, or tapered) and circular arcs. OD (turning) or ID (boring) operations may be programmed. The format given is for a Fanuc OT controller. Other controllers may have slightly different coding.

General Syntax

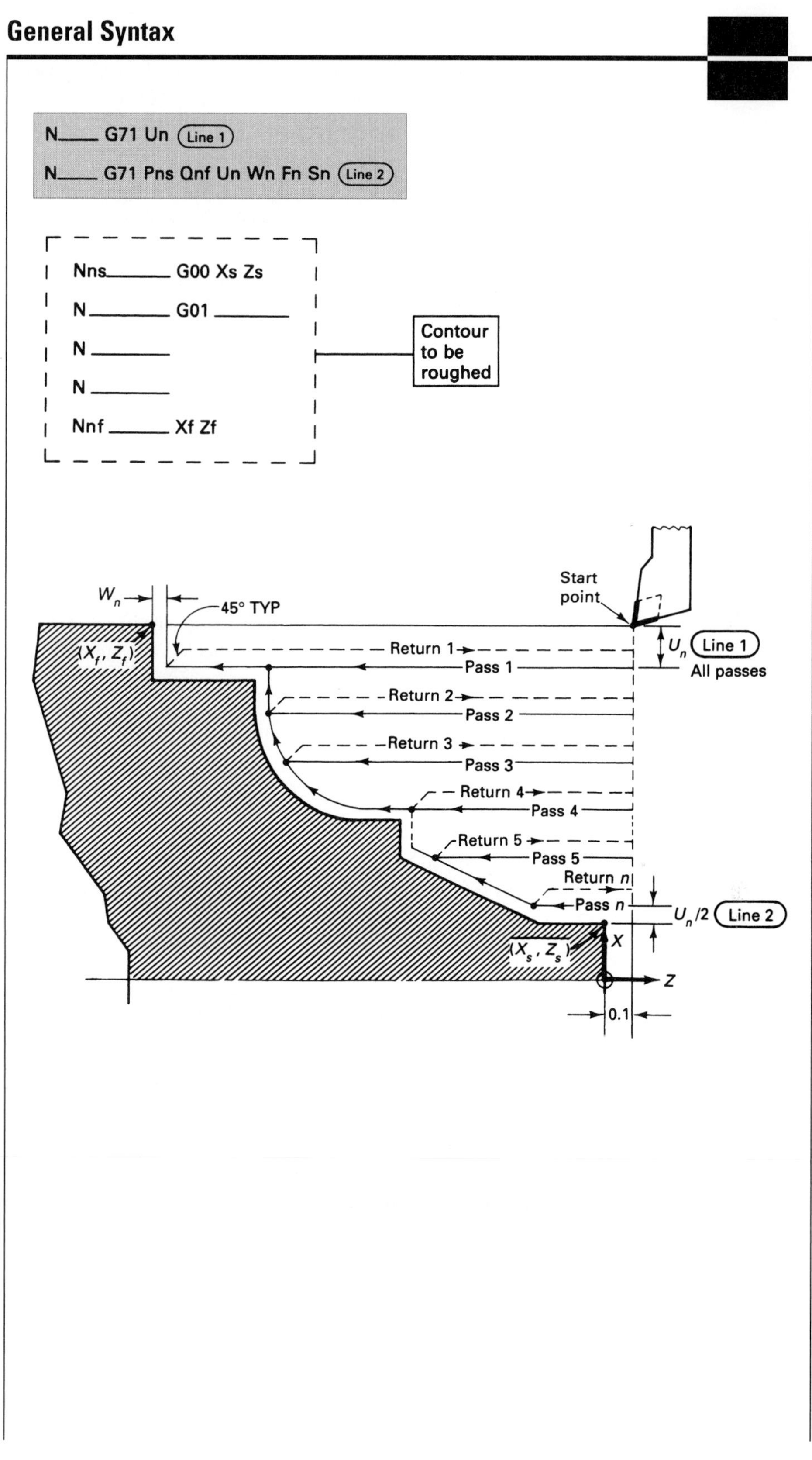

G71	Initiates rough cutting of the contour described in blocks Nns through Nnf.
Un (Line 1)	n, sets the depth of cut per side for each roughing pass. Decimal point not accepted. Value equals the number of 0.0001's. Example: 0.160 per side = 1600 × 0.0001. Thus, U1600 is specified.
Pns	ns, specifies the sequence number of the block, which starts the contour description.
Qnf	nf, specifies the sequence number of the block, which ends the contour description.
Un (Line 2)	n, sets the direction and amount of finishing allowance along the *X* axis (diameter values).
Wn	n, sets the direction and amount of finishing allowance along the *Z* axis.
Fn	n, specifies the roughing feed rate.
Sn	n, specifies the roughing speed. Use G96 prior to G71 to set S to sfm.

Note: 1. Linear interpolation (G01), corner rounding, chamfering, and circular interpolation (G02/G03) may be used to describe the contour in blocks P through Q.

2. Any F and S functions contained in blocks P through Q are ignored. The functions programmed in the G71 block are effective throughout the cycle.

3. For the finishing allowances specification Un (Line 2), the sign of n will vary as follows: n is + for OD turning; n is − for ID boring.

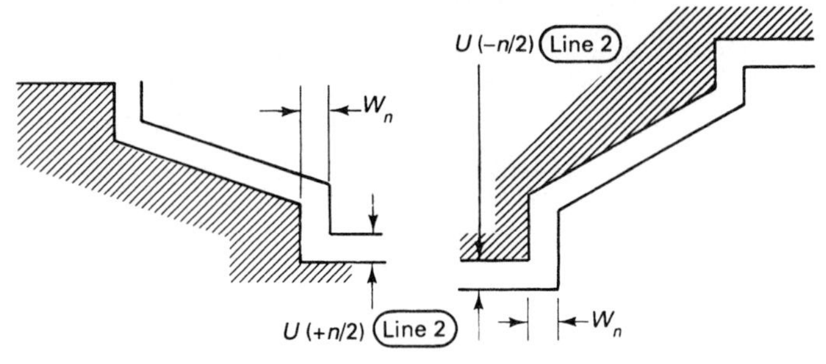

17–12 FINISH TURNING AND BORING CYCLE: G70

The G70 cycle can be used to execute finish passes over a contour that has been roughed by a G71 cycle.

■ EXAMPLE 17–6

The part shown in Figure 17–12 is to be machined from stock. Write a program to execute the necessary roughing passes and the finish pass. The tool offsets for tool edge programming, tool nose radius, and tool nose vector 3 have been entered into the offset file.

General Syntax

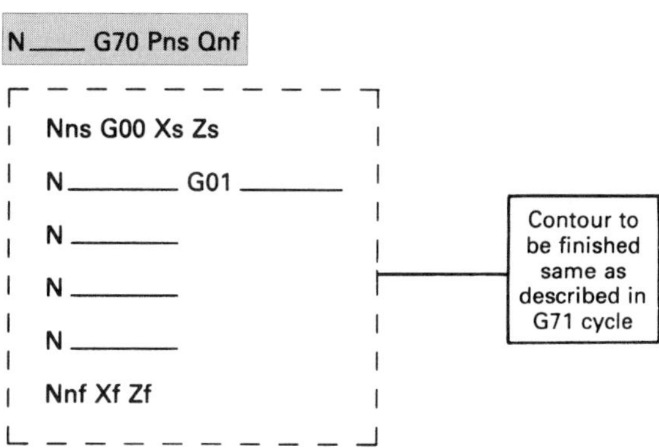

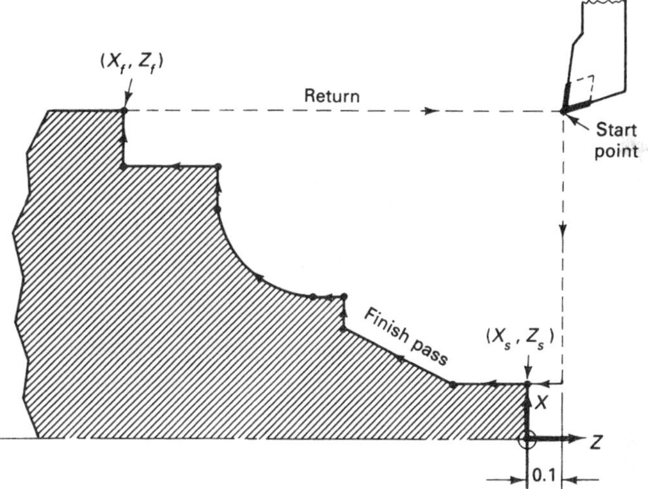

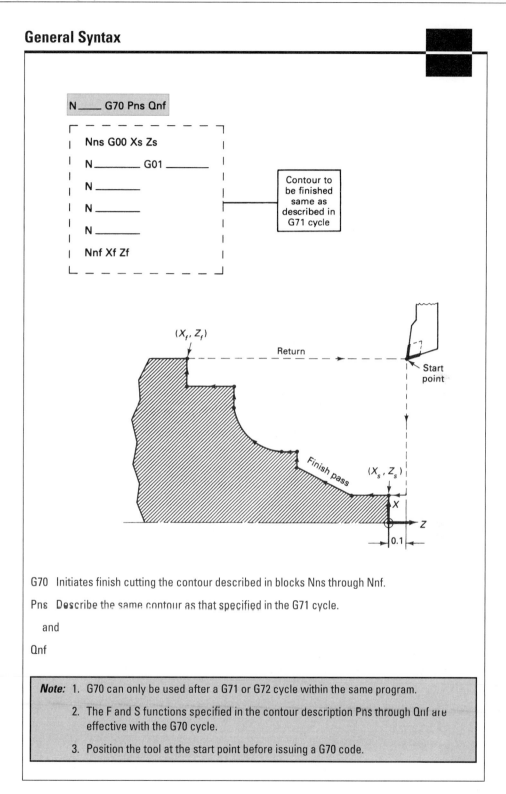

G70 Initiates finish cutting the contour described in blocks Nns through Nnf.

Pns Describe the same contour as that specified in the G71 cycle.

and

Qnf

Note: 1. G70 can only be used after a G71 or G72 cycle within the same program.

2. The F and S functions specified in the contour description Pns through Qnf are effective with the G70 cycle.

3. Position the tool at the start point before issuing a G70 code.

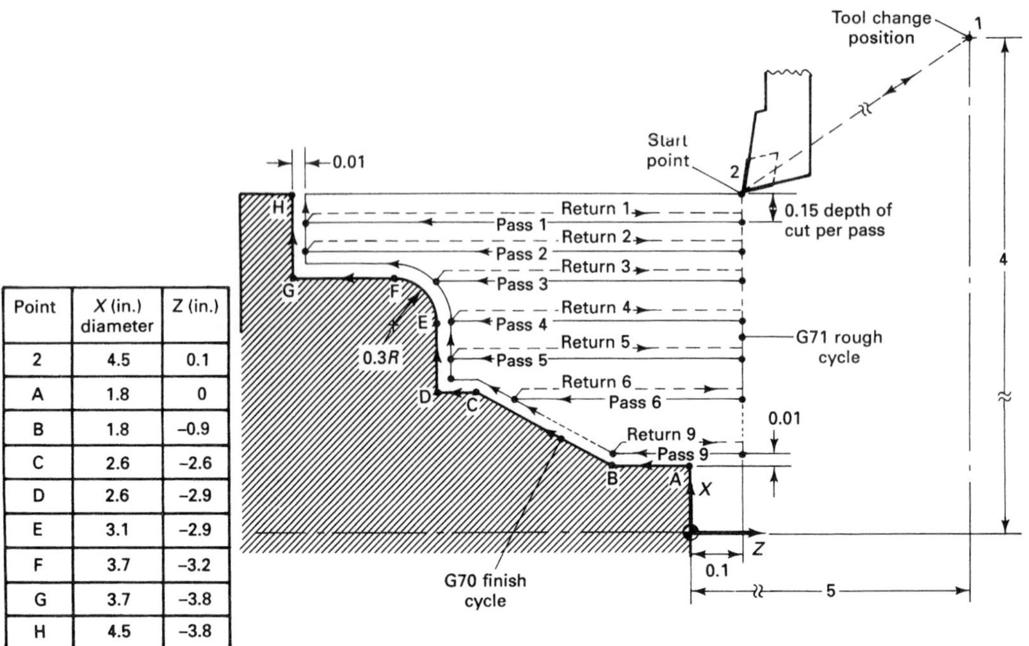

Point	X (in.) diameter	Z (in.)
2	4.5	0.1
A	1.8	0
B	1.8	-0.9
C	2.6	-2.6
D	2.6	-2.9
E	3.1	-2.9
F	3.7	-3.2
G	3.7	-3.8
H	4.5	-3.8

FIGURE 17-12

Number of rough passes

$$= \frac{\text{Stock diameter} - \text{Minor diameter of part} + \text{Material for finish}}{\text{Depth of cut} \times 2}$$

$$= \frac{4.5 - 1.8 + 0.02}{0.15 \times 2} = 9$$

Thus, the system will execute eight full-depth passes and one remaining stock pass. ■■

Programming Pattern	*Word address command*	*Meaning*
	%	End of tape.
	O0171	Program number.
Job setup data listing	N0010 (. . . . .)	
	N0020 (X0 IS ON THE SPINDLE CENTERLINE)	
	N0030 (Z0 IS ON THE BLANK FACE)	
	N0040 (. . . . .)	
	N0050 (TOOLING LIST:)	
	N0060 (TOOL 1: .03125 TNR ROUGHING TOOL)	
	N0070 (TOOL 2: .0625 TNR FINISHING TOOL)	Optional setup statements.
Tool 1 in turret Machine start-up sequence	N0080 G90	Inch mode.
	N0090 G50 X8.0 Z5.0	Set machining origin at ①.
	N0100 T0100	Return to tool change position ①. Index to tool 1. Cancel values in offset file 1.
	N0110 G00 M42	Shift to intermediate-gear range.
	N0120 G96 S300 M03	Spindle on (CW) at 300 sfm constant surface speed.
	N0130 (TOOL 1: ROUGH TURN CONTOUR)	
	N0140 G42 X4.5 Z.1 T0101 M08	Ramp on to right of upward tool motion on next rapid move to ②. Use values in offset file 1. Coolant on.
Operation 1: rough turn the contour	N0150 G71 U1500	Start roughing cycle: cut to 0.15 in. per side for each pass.
	N0160 G71 P0170 Q0240U.02 W.01 F.01	Rough contour defined by blocks 0170–0240. Leave 0.01 in. per side along the *X* and *Z* axis. Machine at feed 0.01 ipr.
	N0170 G00 X1.8 S700	Point Ⓐ on contour.
	N0180 G01 Z–.9 F.006	Point Ⓑ on contour.
	N0190 X2.6 Z–2.6	Point Ⓒ on contour.
	N0200 Z–2.9	Point Ⓓ on contour.
	N0210 X3.1	Point Ⓔ on contour.
	N0220 G03 X3.7 Z–3.2 R.3	Around 0.5*R* arc to point Ⓕ on contour.
	N0230 G01 Z–3.8	Point Ⓖ on contour.
	N0240 X4.5	Point Ⓗ on contour.
Machine stop sequence	N0250 G00 G40 X8.0 Z5.0 T0100 M09	Ramp off on next rapid move to tool change position ①. Cancel values in offset file 1. Coolant off.
	N0260 M05	Spindle off.
	N0270 (TOOL 8: FINISH CONTOUR)	

CNC PROGRAM ***Continued***

Change to tool 8 Machine start sequence	N0280 T0800	Return to tool change position ①. Index to tool 8. Cancel values in offset file 8. Coolant on.
	N0290 G00 M43	Shift to high-gear range.
	N0300 G96 S800 M03	Spindle on (CW) at 800 sfm constant surface speed.
	N0310 G42 X4.5 Z.1 T0808 M08	Ramp on to right of upward tool motion on next rapid move to ②. Use values in offset file 8. Coolant on.
Operation 2: finish turn the contour	N0320 G70 P0170 Q0240	Finish contour defined by blocks 0170–0240. Machine at feed 0.006 ipr and speed 700 rpm.
Machine stop Program end sequence	N0330 G00 G40 X8.0 Z5.0 T0800 M09	Ramp off on next rapid move to the tool change position ①. Cancel offsets for tool 8. Coolant off.
	N0340 M05	Spindle off.
	N0350 M30	Program end, memory reset.
	%	

■ EXAMPLE 17–7

The finished part shown in Figure 17–13 is to be created by boring from stock. Write the CNC program that includes the roughing and finishing cycles. Assume the tool offsets, tool nose radius, and tool nose vector 2 have been entered into the offset file.

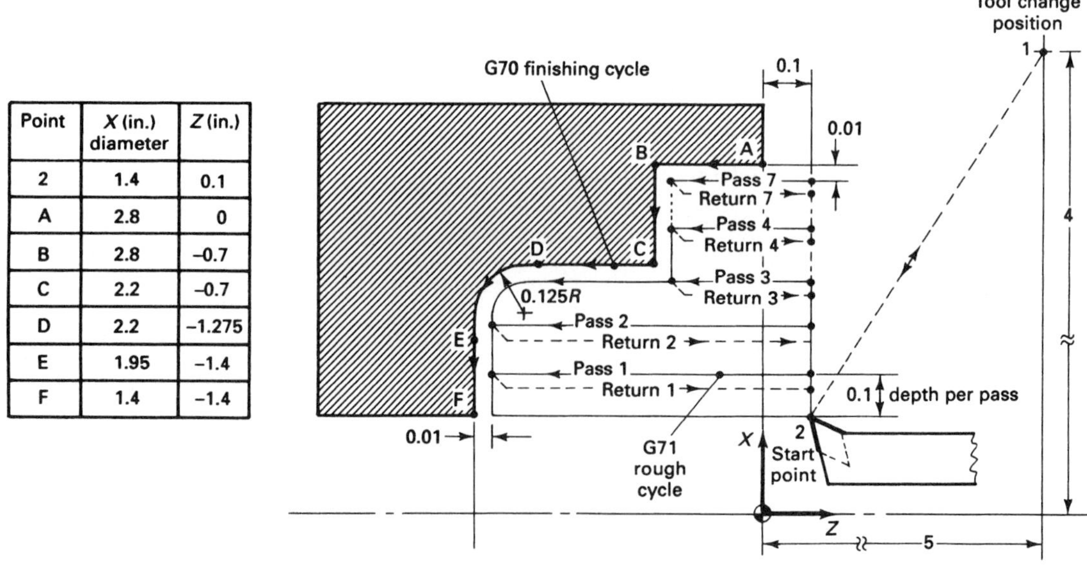

Point	X (in.) diameter	Z (in.)
2	1.4	0.1
A	2.8	0
B	2.8	-0.7
C	2.2	-0.7
D	2.2	-1.275
E	1.95	-1.4
F	1.4	-1.4

FIGURE 17–13

$$\text{Number of rough passes} = \frac{2.8 - 0.02 - 1.4}{0.1 \times 2} = 7$$

■■

Programming Pattern

	Word address command	Meaning
	%	End of tape.
	O0172	Program number.
Job setup data listing	N0010 (. . . . .) N0020 (X0 IS ON THE SPINDLE CENTERLINE) N0030 (Z0 IS ON THE BLANK FACE) N0040 (. . . . .)	Optional setup statements.
	N0050 (TOOLING LIST:)	
	N0060 (TOOL 1: .03125 TNR ROUGHING TOOL)	
	N0070 (TOOL 2: .0625 TNR FINISHING TOOL)	
	N0080 (. . . . .)	
Tool 3 in turret Machine start-up sequence	N0090 G90	Inch mode.
	N0100 G50 X8.0 Z5.0	Set machining origin at ①.
	N0110 T0300	Return to tool change position ①. Index to tool 3. Cancel values in offset file 3.
	N0120 G00 M42	Shift to intermediate-gear range.
	N0130 G96 S200 M03	Spindle on (CW) at 200 sfm constant surface speed.
	N0140 (TOOL 1: ROUGH BORE CONTOUR)	
	N0150 G41 X1.4 Z.1 T0303 M08	Ramp on to left of upward tool motion on next rapid move to ②. Use offset values in register 3. Coolant on.
Operation 1: Rough bore contour	N0160 G71 U1000	Start rough boring cycle. Cut to 0.1 in. per side for each roughing pass.
	N0170 G71 P0180 Q0230 U−.02 W.01 F.01	Rough bore contour defined by blocks 0180–0230. Leave 0.01 in. per side along the *X* and *Z* axes. Machine at feed 0.01 ipr.
	N0180 G00 X2.8 S650	Point Ⓐ on contour.
	N0190 G01 Z−.7 F.006	Point Ⓑ on contour.
	N0200 X2.2	Point Ⓒ on contour.
	N0210 Z−1.275	Point Ⓓ on contour.
	N0220 G03 X1.95 Z−1.4 R.125	Around to 0.125*R* arc to point Ⓔ.
	N0230 G01 X1.4	Point Ⓕ on contour.
	N0240 (TOOL 2: FINISH BORE CONTOUR)	
Machine stop sequence	N0250 G00 G40 X8.0 Z5.0 T0300 M09	Ramp off on next rapid move to tool change position ①. Cancel values in offset file 3. Coolant off.
	N0260 M05	Spindle off.

Change to tool 5 Machine start sequence	N0270 T0500	Index to tool 5. Cancel values in offset file 5.
	N0280 G00 M43	Shift to high-gear range.
	N0290 G96 S800 M03	Spindle on (CW) at 800 sfm constant surface speed.
	N0300 G41 X1.4 Z.1 T0505 M08	Change to tool 5. Ramp on left of upward tool motion on next rapid move to 2. Use values offset file 5. Coolant on.
Operation 2: Finish bore contour	N0310 G70 P0180 Q0230	Finish bore contour defined by blocks 0180–0230. Machine at feed 0.006 ipr and speed 650 rpm.
Machine stop Program end sequence	N0320 G00 G40 X8.0 Z5.0 T0500 M09	Ramp off on next rapid move to tool change position 1. Cancel values in offset file 5. Coolant off.
	N0330 M05	Spindle off.
	N0340 M30	Program end, memory reset.
	%	

17-13 PECK DRILLING AND FACE GROOVING CYCLE: G74

The G74 word initiates a multi-purpose cycle that can be used for peck drilling the stock center as well as peck machining wide or multiple face grooves. The feed interruptions that occur in the cycle cause chips to be broken up and cleared away. This cycle is especially useful for drilling a hole in solid stock prior to executing ID boring or threading operations. The format given is for a Fanuc OT controller. Other controllers may have slightly different coding.

General Syntax

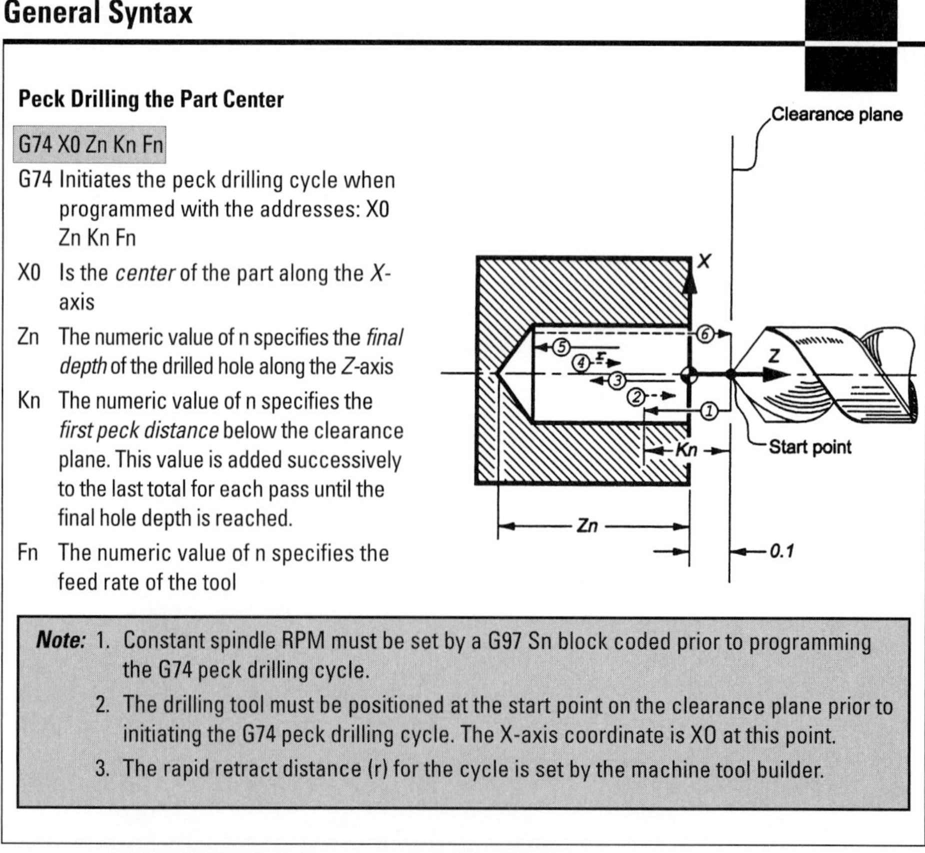

Peck Drilling the Part Center

G74 X0 Zn Kn Fn

G74 Initiates the peck drilling cycle when programmed with the addresses: X0 Zn Kn Fn

X0 Is the *center* of the part along the *X*-axis

Zn The numeric value of n specifies the *final depth* of the drilled hole along the *Z*-axis

Kn The numeric value of n specifies the *first peck distance* below the clearance plane. This value is added successively to the last total for each pass until the final hole depth is reached.

Fn The numeric value of n specifies the feed rate of the tool

Note: 1. Constant spindle RPM must be set by a G97 Sn block coded prior to programming the G74 peck drilling cycle.
2. The drilling tool must be positioned at the start point on the clearance plane prior to initiating the G74 peck drilling cycle. The X-axis coordinate is X0 at this point.
3. The rapid retract distance (r) for the cycle is set by the machine tool builder.

Wide and Multiple Face Grooving

G74 Xn Zn In Kn Fn

G74 Initiates the grooving cycle when programmed with the addresses: Xn Zn In Kn Fn

Zn The numeric value of n specifies the *final depth* of the groove along the *Z*-axis

Kn The numeric value of n specifies the *first peck distance* below the clearance plane. This value is added successively to the last total for each pass until the final groove depth is reached.

Fn The numeric value of n specifies the feed rate of the tool

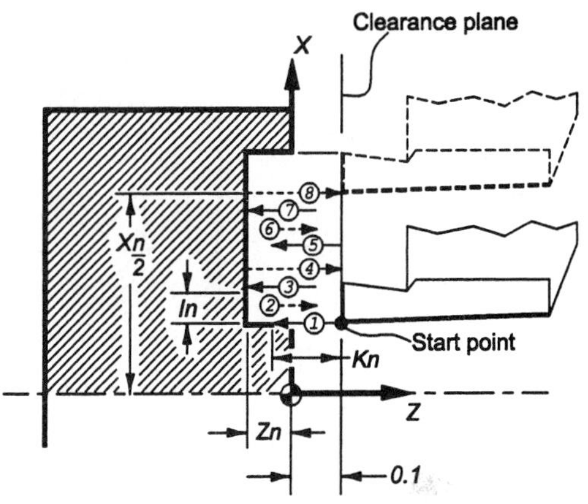

Wide Grooves

Xn The numeric value of n specifies the *final X*-position of the tool for the cycle where n is in terms of *diameter.*

In The numeric value of n specifies the *stepover distance* on the *X*-axis. The tool is successively offset by this distance each time it reaches the groove bottom and is rapided to the clearance plane.

Multiple Grooves

Xn The numeric value of n specifies the location of the *last groove* on the *X*-axis where n is in terms of *diameter.*

In The numeric value of n specifies the *distance between grooves* on the *X*-axis.

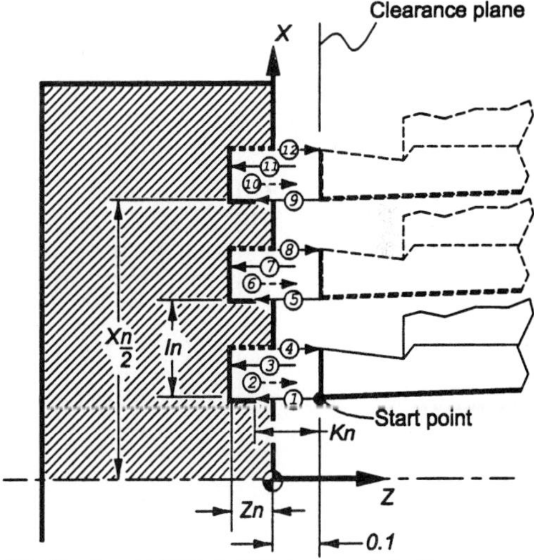

Note: 1. The tool control point influences the value programmed for Xn. In these illustrations it is assumed to be at the lower left corner. The programmer must determine which corner of the tool is the leading edge control point.

2. The controller uses the tool positioning block, which appears just before the G74 block, to establish the proper start point for machining the groove(s).

■ EXAMPLE 17–8

Write a word address program for peck drilling the center of the part mounted on a turning center. See Figure 17–14.

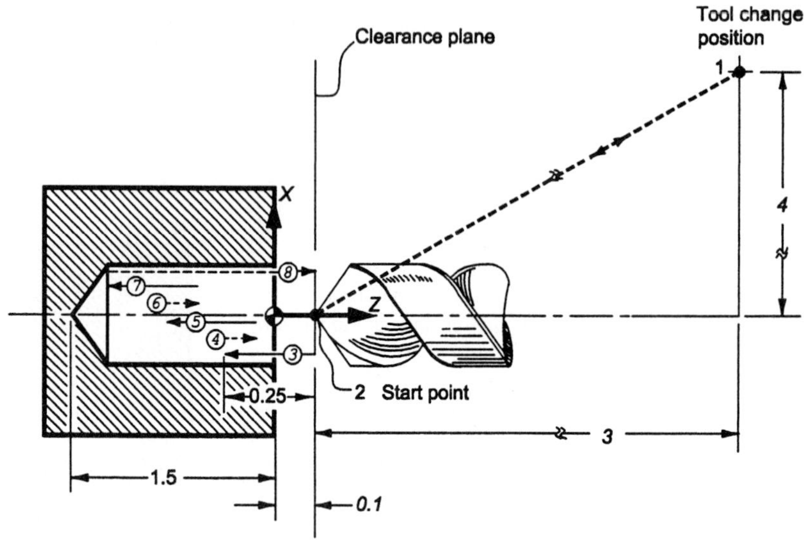

FIGURE 17–14

Programming Pattern

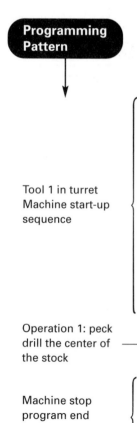

Word address command	Meaning
%	End of tape.
O0173	Program number.
N0010 (. . . . .)	Optional setup statement.
N0020 G90	Inch mode.
N0030 G50 X8.0 Z3.0	Set machining origin at ①.
N0040 (TOOL 1: FINISH DRILL AS PER PRINT)	Optional comment statement.
N0050 T0100	Return to tool change position ① index to tool 1. Cancel values in offset file 1.
N0060 G00 M41	Shift to low gear range.
N0070 G97 S400 M03	Spindle on (CW) at 400 rpm. Constant surface speed cancelled.
N0080 G00 X0 Z.1 T0101 M08	Rapid tool to the start point ②. Use offset values in register 1. Coolant on.
N0090 G74 X0 Z-1.5 K.25 F.012	Peck drill to a depth of 1.5 in. at feed rate .012 ipr. Set peck depth at .25 in.
N0100 G00 X4.0 Z3.0 T0100 M09	Cancel cycle. Rapid to tool change position ①. Cancel values in offset file 1.
N0110 M05	Spindle off.
N0120 M30	Program end, memory reset.
%	

Tool 1 in turret Machine start-up sequence (N0020–N0080)

Operation 1: peck drill the center of the stock (N0090)

Machine stop program end sequence (N0100–N0120)

■ EXAMPLE 17–9

Write a word address program to machine a wide groove into the face of the part mounted on a turning center as shown in Figure 17–15.

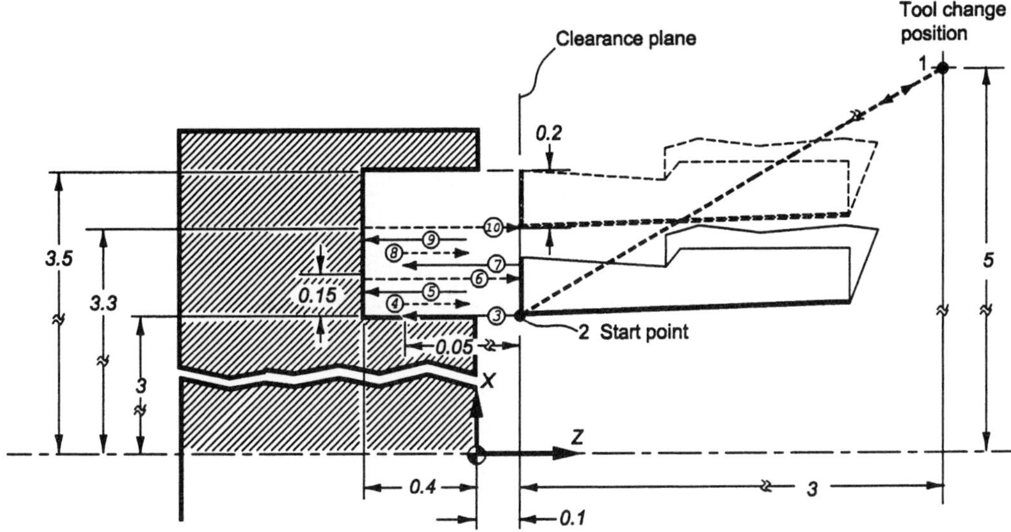

FIGURE 17–15

Note: X is expressed in terms of diameter.

Thus: Xn = X2(3.3) = X6.6 ■■

Programming Pattern

Tool 1 in turret Machine start-up sequence

Operation 3: peck wide face groove into the stock

Word address command	*Meaning*
%	End of tape.
O0174	Program number.
N0010 (. . . . .)	Optional setup statement.
N0020 G20	Inch mode.
N0030 G50 X10.0 Z3.0	Set machining origin at ①.
N0040 (TOOL 3: WIDE GROOVE AS PER PRINT)	Optional comment statement.
N0050 T0300	Return to tool change position ① index to tool 3. Cancel values in offset file 3.
N0060 G00 M42	Shift to intermediate gear range.
N0070 G96 S215 M03	Spindle on (CW) at 215 sfm, constant surface speed.
N0080 G00 X6.0 Z.1 T0303 M08	Rapid tool to the start point ②. Use offset values in register 3. Coolant on.
N0090 G74 X6.6 Z-0.4 I0.15 K0.05 F0.01	Peck wide face groove starting at the smaller diameter of 6 in. Final X position of the tool is at 6.6 in. Depth of the groove is 0.4 in. Stepover distance is set at 0.15 in. The peck depth is 0.05 in. at feed rate 0.01 ipr.

Machine stop program end sequence

N0100 G00 X8.0 Z3.0 T0300 M09	Cancel cycle. Rapid to tool change position ①. Cancel values in offset file 3.
N0110 M05	Spindle off.
N0120 M30	Program end, memory reset.
%	

The machining can also be started at the larger diameter of 6.6 in. In this case the startup blocks would be

N0080 G00 X6.0 Z.1 T0303 M08
N0090 G74 X6.6 Z-0.4 I0.15 K0.05 F0.01

■ EXAMPLE 17–10

For the part given in Example 17–9, instead of a single wide groove, machine a series of grooves into the end face as shown in Figure 17–16.

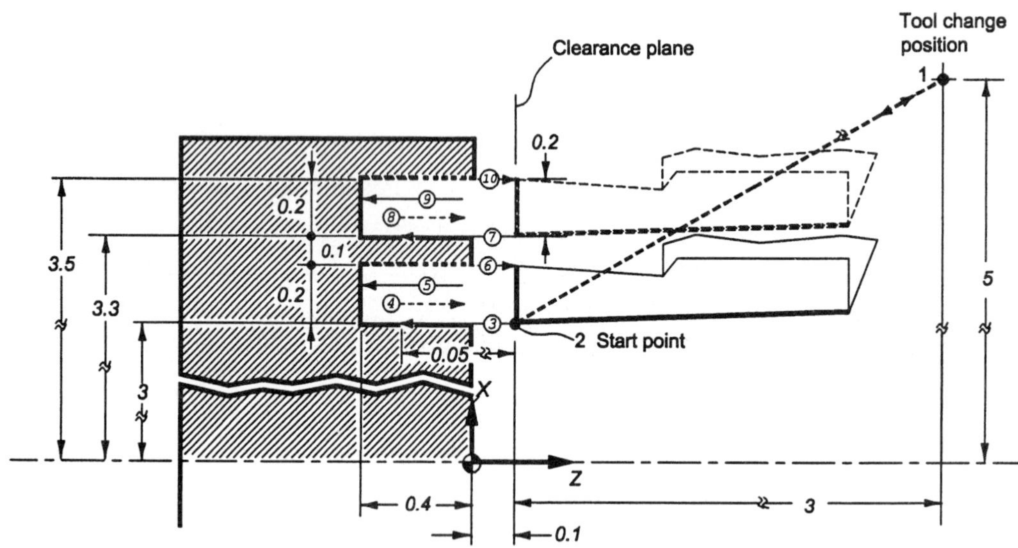

FIGURE 17–16

The same word address program applies as given in Example 17–9 except that the G74 block would be:

Word address command	*Meaning*
N0090 G74 X6.0 Z-0.4 I0.3 K0.05 F0.01	Peck a series of face grooves starting at the smaller diameter of 6.6 in. Final X position of the tool is at 7.2 in. Depth of each groove is 0.4 in. Distance between each groove is set to 0.3 in. The peck depth is 0.05 in. at feed rate 0.01 ipr.

If the tool control point were located at the upper left corner of the tool, as shown in Figure 17–17, and the machining started at the larger diameter of 7 in., the startup blocks would be:

FIGURE 17–17

```
N0080 G00 X7 Z.1 T0303 M08
N0090 G74 X6.4 Z-0.4 10.15 K0.05 F0.01
```

■■

17–14 PECK CUTOFF AND GROOVING CYCLE: G75

The G75 cycle is used to execute cutoff operations as well as wide and multiple grooving. It is identical to the G74 cycle if the *X* and *Z* axis were reversed as well as the I and K addresses.

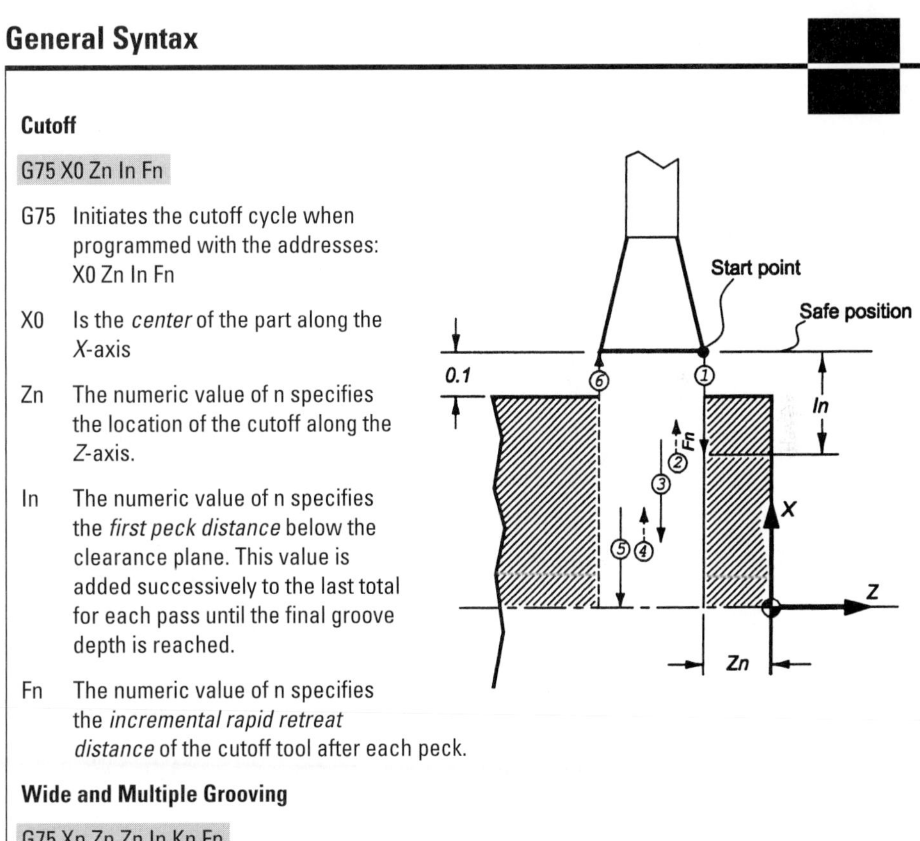

General Syntax

Cutoff

G75 X0 Zn In Fn

G75 Initiates the cutoff cycle when programmed with the addresses: X0 Zn In Fn

X0 Is the *center* of the part along the *X*-axis

Zn The numeric value of n specifies the location of the cutoff along the *Z*-axis.

In The numeric value of n specifies the *first peck distance* below the clearance plane. This value is added successively to the last total for each pass until the final groove depth is reached.

Fn The numeric value of n specifies the *incremental rapid retreat distance* of the cutoff tool after each peck.

Wide and Multiple Grooving

G75 Xn Zn Zn In Kn Fn

G75 Initiates the grooving cycle when programmed with the addresses: Xn Zn In Kn Fn.

Xn The numeric value of n specifies the *final depth* of the groove along the *X*-axis, where n is in terms of diameter.

In The numeric value of n specifies the *first peck distance* below the clearance plane. This value is added successively to the last total for each pass until the final groove depth is reached.

Fn The numeric value of n specifies the *incremental rapid retreat distance* after each peck.

Wide grooves

Zn The numeric value of n specifies the *final Z*-position of the tool for the cycle.

Kn The numeric value of n specifies the *stepover distance* on the *Z*-axis. The tool is successively offset by this distance each time it reaches the groove bottom and is rapided to the safe position.

Multiple Grooves

Zn The numeric value of n specifies the location of the *last groove* on the *Z*-axis.

Kn The numeric value of n specifies the *distance between grooves* on the *Z*-axis.

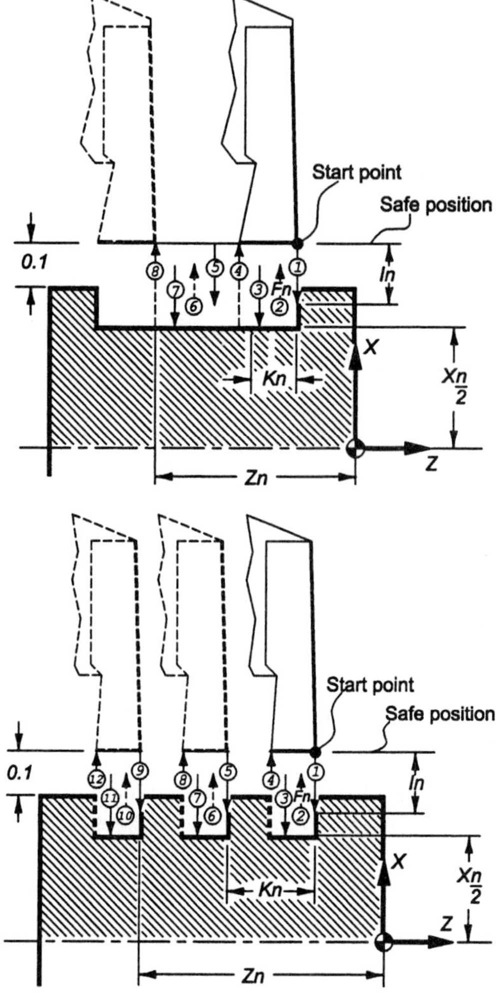

Note: 1. The tool control point influences the value programmed for Zn. In this illustration it is assumed to be at the lower left corner. The programmer must determine which corner of the tool is the leading edge control point.

2. The controller uses the tool positioning block, which appears just before the G75 block, to establish the proper start point for machining the cutoff or groove(s).

3. A feed rate *cannot* be coded with a G75 block. Therefore, the controller will use the *last active feed rate* programmed.

■ EXAMPLE 17–11

Write a word address program to execute peck cutoff of the part shown in Figure 17–18.

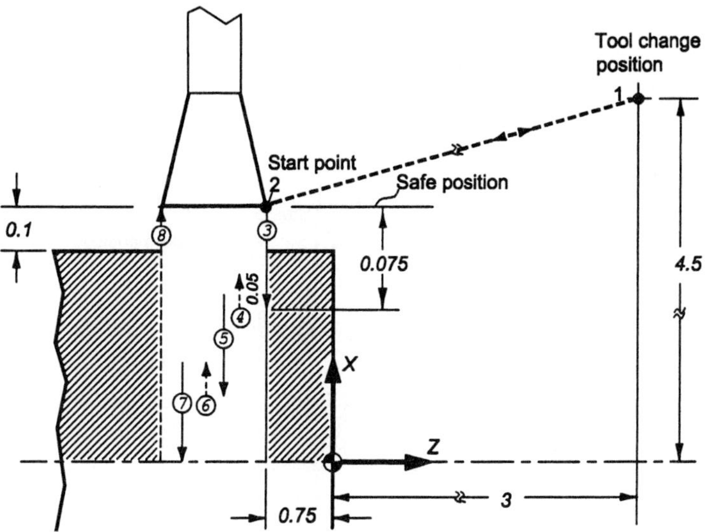

FIGURE 17–18 ■■

Programming Pattern

Tool 1 in turret Machine start-up sequence

Operation 6: peck cutoff the stock

Machine stop program end sequence

Word address command	*Meaning*
%	End of tape.
O0175	Program number.
N000 (. . . . .)	Optional setup statement.
N0020 G90	Inch mode.
N0030 G50 X9.0 Z3.0	Set machining origin at ①.
N0040 (TOOL 6: CUTOFF AS PER PRINT)	Optional comment statement.
N0050 T0600	Return to tool change position ① index to tool 6. Cancel values in offset file 6.
N0060 G00 M42	Shift to intermediate gear range.
N0070 G96 S100 M03	Spindle on (CW) at 100 sfm, constant surface speed.
N0080 G00 X2.2 Z-0.75 T0606 M08 F0.01	Rapid tool to the start point ②. Use offset values in register 6. Coolant on. Set feed rate of 0.01 ipr for blocks that follow.
N0090 G75 X0 I0.075 F0.05	Peck cutoff at location Z-0.75 in. at peck feed rate 0.010 ipr. Set the peck depth to 0.075 in. Set rapid retract distance at 0.05 in.
N0100 G00 X4.0 Z3.0 T0100 M09	Cancel cycle. Rapid to tool change position ①. Cancel values in offset file 1.
N0110 M05	Spindle off.
N0120 M30	Program end, memory reset.
%	

■ EXAMPLE 17–12

A wide groove is to be peck machined into the part shown in Figure 17–19. Write the required word address program.

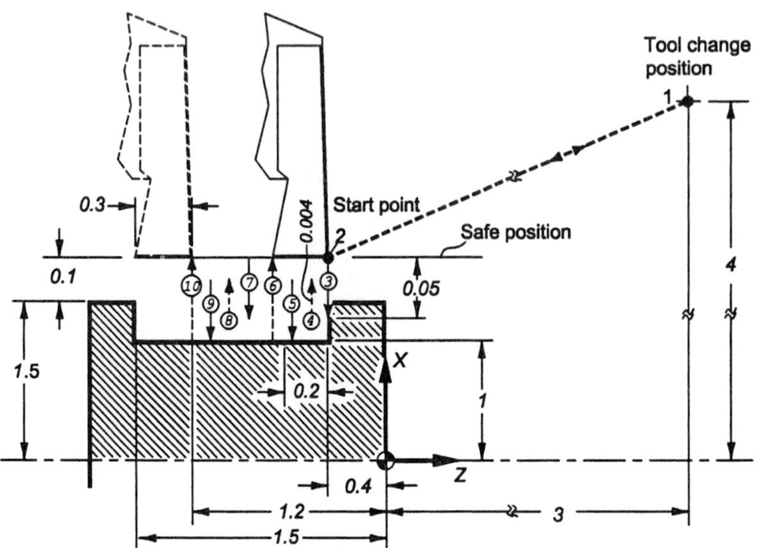

FIGURE 17–19

Note: X is expressed in terms of diameter.

Thus: Xn = X2(1) = X2 ■■

Programming Pattern

Word address command	*Meaning*
%	End of tape.
O0176	Program number.
N0010 (. . . . .)	Optional setup statement.
N0020 G90	Inch mode.
N0030 G50 X8.0 Z3.0	Set machining origin at ①.
N0040 (TOOL 5: WIDE GROOVE AS PER PRINT)	Optional comment statement.
N0050 T0500	Return to tool change position ① index to tool 5. Cancel values in offset file 5.
N0060 G00 M42	Shift to intermediate gear range.
N0070 G96 S100 M03	Spindle on (CW) at 100 sfm, constant surface speed.
N0080 G00 X3.2 Z-.4 T0505 F0.01	Rapid tool to the start point ②. Use offset values in register 5. Coolant on. Set feed rate to 0.01 ipr for blocks that follow.
N0090 G75 X2.0 Z-1.2 I0.05 K0.2 F0.004	Peck wide groove to 2 in. diameter at feed rate 0.010 ipr. Final Z position of the tool is at 1.2 in. The peck depth is 0.05 in. Stepover distance is set to 0.2 in. Set rapid retract distance at 0.004 in.

Tool 1 in turret Machine start-up sequence

Operation 5: peck wide groove into the stock

Machine stop program end sequence	N0100 G00 X8.0 Z3.0 T0300 M09	Cancel cycle. Rapid to tool change position ①. Cancel values in offset file 5.
	N0110 M05	Spindle off.
	N0120 M30	Program end, memory reset.
	%	

■ EXAMPLE 17–13

Instead of machining a single wide groove into the part given in Example 17–12, produce a series of grooves as shown in Figure 17–20.

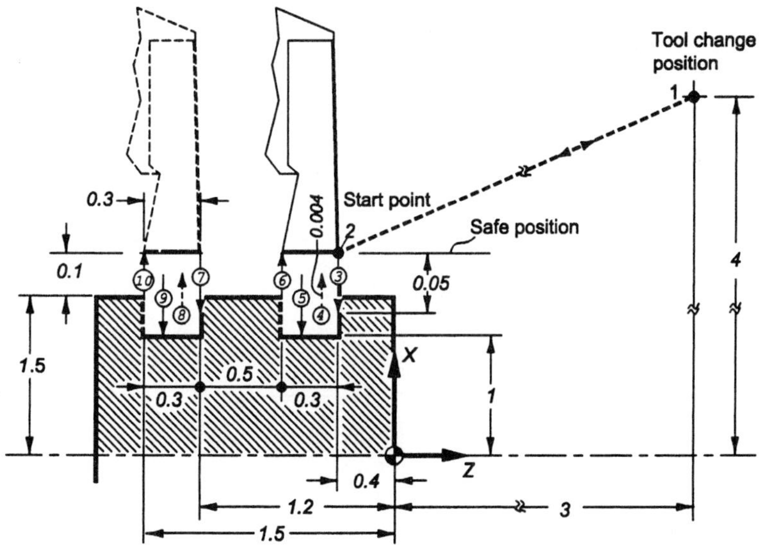

FIGURE 17–20

The word address program is identical to that given in Example 17–12 except that the G75 block would be:

Word address command	*Meaning*
N0090 G75 X2.0 Z-1.2 I0.05 K0.8 F0.004	Peck a series of grooves to 2 in. diameter at feed rate 0.010 ipr. Final *Z* position of the tool is at 1.2 in. The peck depth is 0.05 in. Distance between grooves is set to 0.8 in. Set rapid retract distance at 0.004 in.

■■

17–15 THREAD CUTTING ON CNC LATHES AND TURNING CENTERS

Turning centers can be programmed to machine straight, tapered, or scroll threads. Refer to Figure 17–21.

Threads are machined by a special tool having the thread shape. The tool is positioned at a specific *Z* starting distance from the end of the work. This distance will vary from machine to machine. Its value can be found in the machine's programming manual. A formula giving the approximate starting distance is also given in Section 17–12. Beginning at the start point, the tool accelerates to the feed rate required to cut the threads. The tool creates the thread shape by repeatedly following the same path as axial infeed is applied. For standard V threads the infeed can be applied along a 0°, 29°, or 30° angle. The depth of cut for the first pass is largest. The cutting depth is then decreased for each successive pass until the required thread depth is achieved. A final or spring pass is then made with the tool set at the thread depth.

17–16 SINGLE-PASS THREADING CYCLE: G32

The most basic threading cycle is initiated by a G32 word. This cycle requires four blocks of data to perform one threading pass. The tool must be positioned at the appropriate start point prior to executing the cycle.

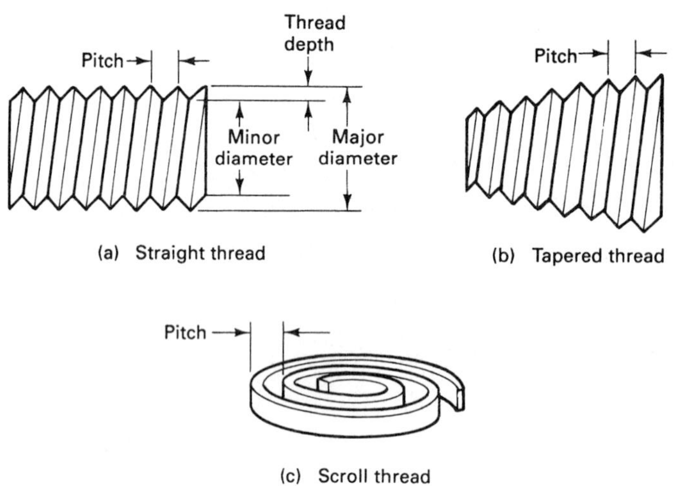

(a) Straight thread (b) Tapered thread (c) Scroll thread

FIGURE 17–21

General Syntax

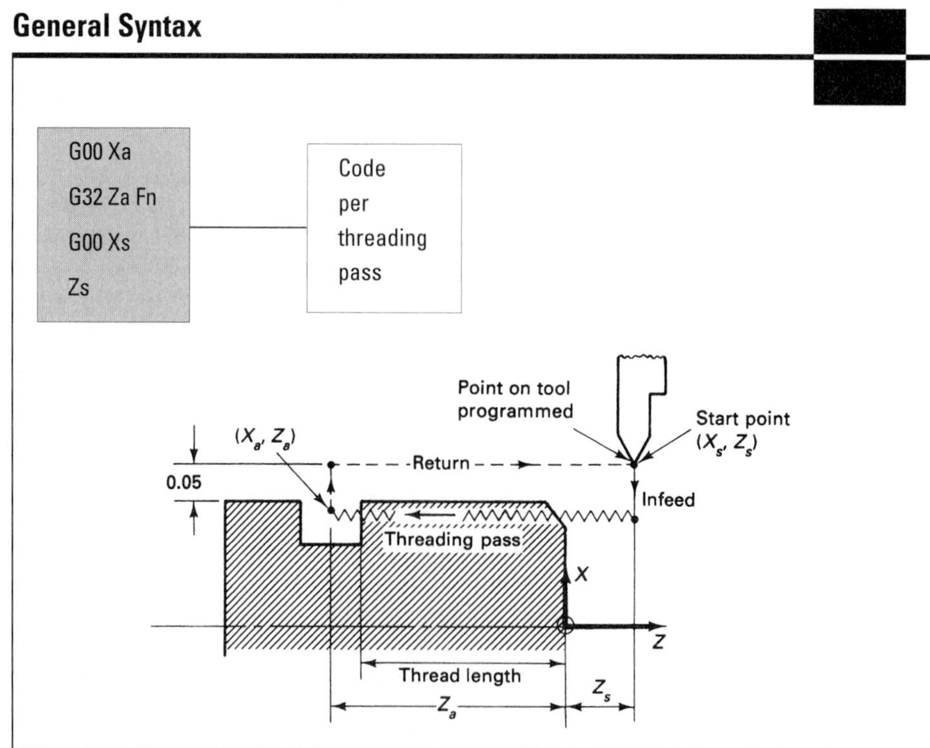

Xa	Specifies the absolute *X* coordinate of the tool after axial infeed.
G32	Initiates the single-pass threading cycle.
Za	Specifies the absolute *Z* coordinate of the tool after the threading pass.
Fn	n specifies the feed rate.
Xs Zs	Specifies the absolute *X* and *Z* coordinates of the start point.

Note: 1. The starting distance Z_s can be determined from the manufacturer's programming manual or estimated by the formula:

$$Z_s = \frac{\text{rpm} \times \text{Pitch}}{1800}$$

2. The stopping distance Z_a can be estimated by the formula:

$$Z_a = Z_s \times 3.605 + \text{thread length}$$

3. Fn must be set to the thread pitch:

$$\text{n} = \text{Pitch} = 1/\text{threads per inch}$$

The value of *n* can have an accuracy to six decimal places.

4. Constant surface speed G96 should not be used during threading.
5. Manual adjustment of the feed rate is not possible when a G32 cycle is in effect.
6. Straight, tapered, and scroll threads may be cut. The cycle can be used for outside as well as inside threading.
7. The infeed direction is controlled by the programmer.

■ EXAMPLE 17–14

Threads are to be cut into the part blank as indicated in Figure 17–22. Use the G32 cycle for writing the program segment.

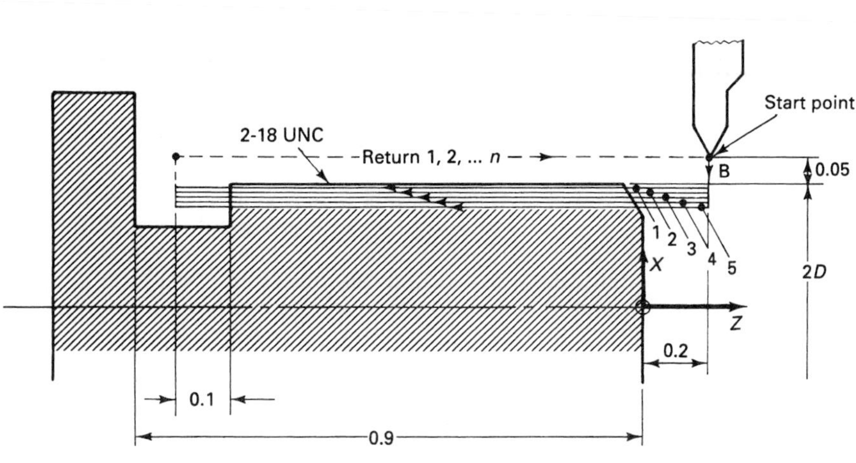

FIGURE 17–22

The following information has been given to the programmer:

Thread: 2–18 threads per inch (tpi)
Major diameter: 2 in.
Minor diameter: 1.9303 in.
Thread depth: $\frac{2 - 1.9303}{2} = 0.0349$ in.
Thread pitch: $\frac{1}{18}$ tpi = 0.055555 in.

The programmer has elected to use five passes for machining the threads to depth. The infeed per pass can be found from the formula:

$$\text{Infeed}_p = d\sqrt{p}$$

where *d* is the infeed for the first pass and *p* is the number of passes. The value of *d* is determined as follows:

$$0.0349 = d\sqrt{5}$$

$$\frac{0.0349}{\sqrt{5}} = d$$

$$0.016 = d$$

The infeed for each pass can then be computed and tabulated.

Pass	Infeed (in.)
①	$0.016\sqrt{1} = 0.016$
②	$0.016\sqrt{2} = 0.023$
③	$0.016\sqrt{3} = 0.0277$
④	$0.016\sqrt{4} = 0.032$
⑤	0.0349

■■

Word address command	*Meaning*
N0040 (TOOL 1: THREAD TO DEPTH)	
N0050 T0101	Return to tool change position Ⓐ. Index to tool 1. Use values in offset file 1.
N0060 G00 M41	Shift to low-gear range.
N0070 G97 S200 M03	Cancel constant surface speed control. Spindle on (CW) at 200 sfm.
N0080 X2.1 Z.2 M08	Rapid tool to start point position Ⓑ. Coolant on.
N0090 X1.968	Rapid tool to first *X* depth position $X = 2-2(0.016) = 1.968$. Execute threading pass 1 and return to start point.
N0100 G32 Z–.8 F.055555	
N0110 G00 X2.1	
N0120 Z.2	
N0130 X1.954	Rapid tool to second *X* depth position $X = 2-2(0.023) = 1.954$. Execute threading pass 2 and return to start point.
N0140 G32 Z–.8 F.055555	
N0150 G00 X2.1	
N0160 Z.2	

Word address command	*Meaning*
N0170 X1.9446	Rapid tool to third X depth position $X = 2-2(0.0277) = 1.9446$. Execute threading pass 3 and return to start point.
N0180 G32 Z–.8 F.055555	
N0190 G00 X2.1	
N0200 Z.2	
N0210 X1.936	Rapid tool to fourth X depth position $X = 2-2(0.32) = 1.936$ Execute threading pass 4 and return to start point.
N0220 G32 Z–.8 F.055555	
N0230 G00 X2.1	
N0240 Z.2	
N0250 X1.9302	Rapid tool to fifth X depth position $X = 2-2(0.0349) = 1.9302$. Execute threading pass 5 and return to start point.
N0260 G32 Z–.8 F.055555	
N0270 G00 X2.1	
N0280 Z.2	
N0290 G00 X8.0 Z5.0 T0100 M09	Rapid to tool change position Ⓐ. Cancel values in offset file 1. Coolant off.
N0300 M00	Program stop.

17–17 MULTIPLE-PASS THREADING CYCLE: G92

The G92 cycle reduces the number of programming blocks required for threading operations. With G92 threading, only one block of data is needed to execute one threading pass. As with G32, the tool must first be positioned at the start point prior to programming the cycle.

General Syntax

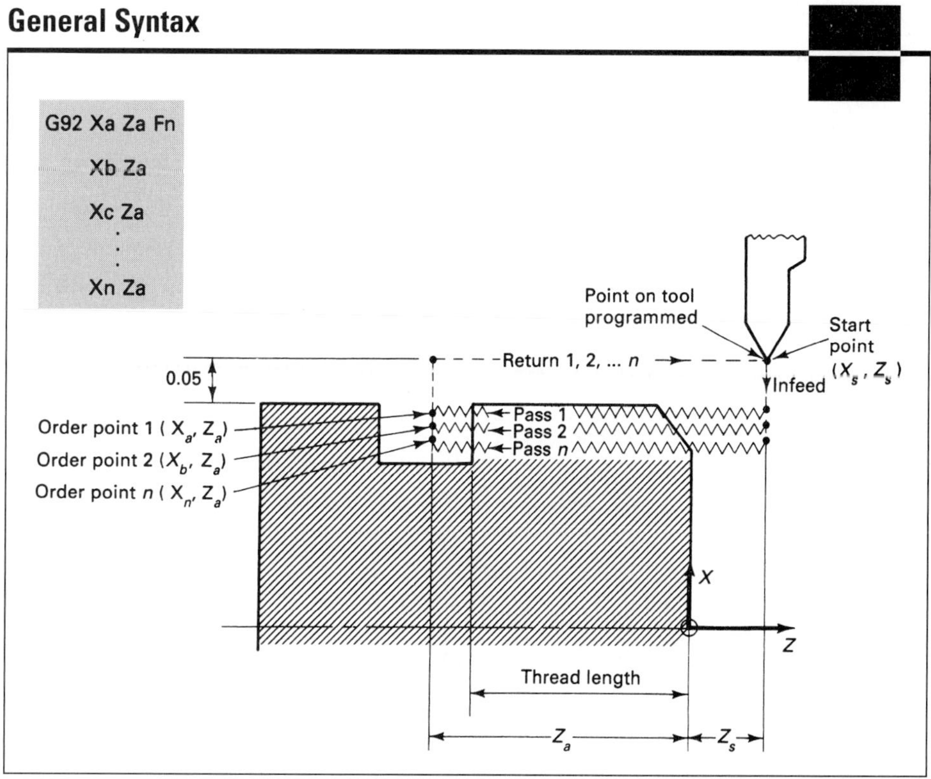

G92 Initiates a multiple-pass threading cycle.
Xa Za Specify the *X* and *Z* absolute.
Xb Za Coordinates of the tool point.
Xc Za After each threading pass (order point).
.
.
.
Xn Za
Fn n specifies the feed rate.

Note: 1. For incremental coordinates replace *X* with *U*, and *Z* with *W*. *W* and *U* indicate the direction and distance from the start point to the order point along the *X* and *Z* axis, respectively.
2. The feed rate (pitch) can have up to six digits accuracy to the right of the decimal point.
3. Only straight and tapered threads can be cut. Outside as well as inside threading is permitted.
4. The Z_s and Z_a formulas used with G32 can also be used with this cycle.
5. The infeed direction is straight in.

■ EXAMPLE 17–15

Using the G92 cycle, rewrite the program segment for threading as outlined in Example 17–14. ■■

Word address command	*Meaning*
N0040 (TOOL 1: THREAD TO DEPTH)	
N0050 T0101	Return to tool change position Ⓐ. Index to tool 1. Use values in offset file 1.
N0060 G00 M41	Shift to low-gear range.
N0070 G97 S200 M03	Cancel constant surface speed control. Spindle on (CW) at 200 sfm.
N0080 X2.1 Z.2 M08	Rapid to start position Ⓑ. Coolant on.
N0090 G92 X1.968 Z−.8 F.055555	Begin multiple pass threading cycle: • Rapid tool to first *X* depth. • Thread at feed rate to order point ① on *Z* axis. • Return at rapid to start point. pass 1
N0100 X1.954	Repeat G92 cycle for pass 2.
N0110 X1.9446	Repeat G92 cycle for pass 3.
N0120 X1.936	Repeat G92 cycle for pass 4.
N0130 X1.9302	Repeat G92 cycle for pass 5.
N0140 X1.9302	Repeat G92 cycle for a spring pass.
N0150 G00 X10.0 Z6.0 T0100 M09	Rapid to tool change position Ⓐ. Cancel values in offset file 1. Coolant off.
N0160 M00	Program stop.

17–18 MULTIPLE REPETITIVE THREADING CYCLE: G76

The G76 cycle further reduces the number of programming blocks involved in executing a threading operation. Using only one block of data as input, this cycle is capable of executing multiple threading passes. The programmer must first position the tool at an appropriate start point before entering the G76 command.

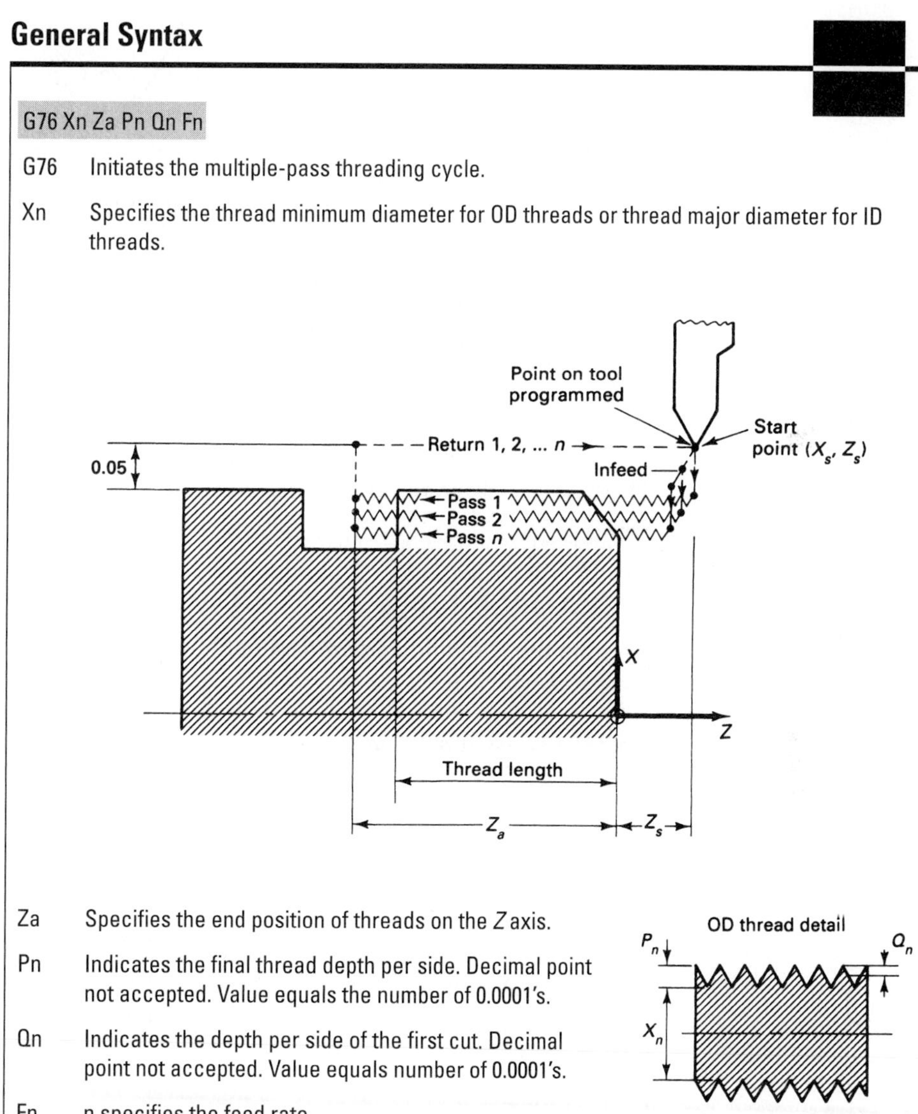

General Syntax

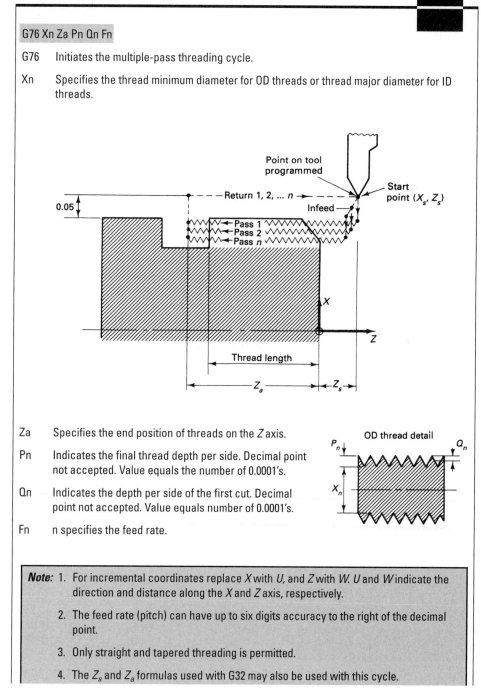

G76 Xn Za Pn Qn Fn

G76 Initiates the multiple-pass threading cycle.

Xn Specifies the thread minimum diameter for OD threads or thread major diameter for ID threads.

Za Specifies the end position of threads on the *Z* axis.

Pn Indicates the final thread depth per side. Decimal point not accepted. Value equals the number of 0.0001's.

Qn Indicates the depth per side of the first cut. Decimal point not accepted. Value equals number of 0.0001's.

Fn n specifies the feed rate.

Note: 1. For incremental coordinates replace *X* with *U*, and *Z* with *W*. *U* and *W* indicate the direction and distance along the *X* and *Z* axis, respectively.

2. The feed rate (pitch) can have up to six digits accuracy to the right of the decimal point.

3. Only straight and tapered threading is permitted.

4. The Z_s and Z_a formulas used with G32 may also be used with this cycle.

The angle at which infeed occurs will be half the tool tip angle specified. The default tool tip value is set at 60. This setting, as well as the other parameters used with G76, can be changed. Consult the CNC machine programming manual for details.

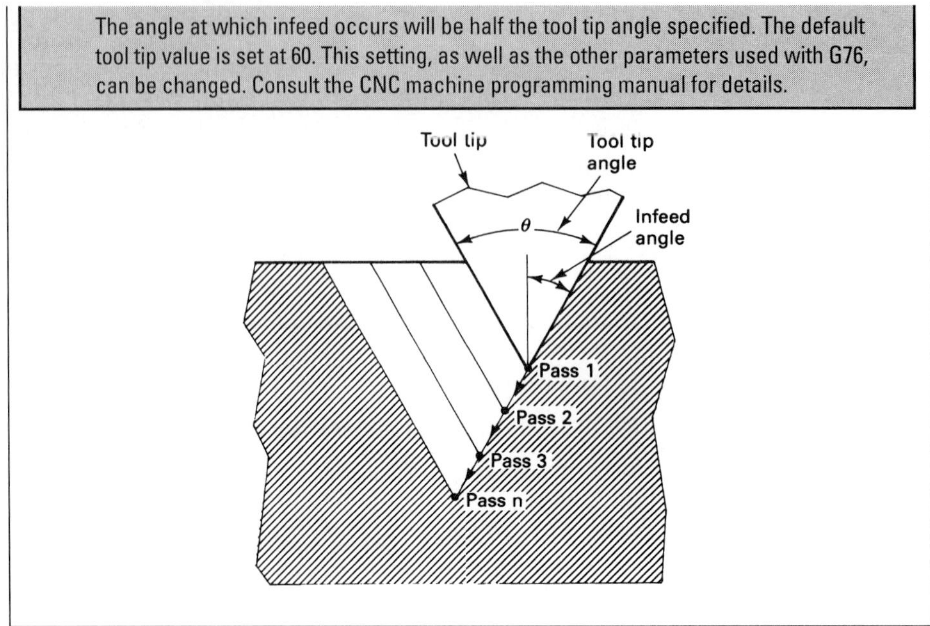

It is possible to machine tapered threads using the G32, G92, or G76 cycles. The CNC machine programming manual will supply the information for these operations.

■ EXAMPLE 17–16

Execute the threading operation of Example 17–15 by using a G76 cycle. ■■

Word address command	***Meaning***
N0040 (TOOL 1: THREAD TO DEPTH)	
N0050 T0101	Return to tool change position Ⓐ. Index to tool 1. Use values in offset file 1.
N0060 G00 M41	Shift to low-gear range.
N0070 G97 S200 M03	Cancel constant surface speed control. Spindle on (CW) at 200 sfm.
N0080 X2.1 Z.2 M08	Rapid tool to start position Ⓑ. Coolant on.
N0090 G76 X1.9303 Z−.8 P349 Q160 F.055555	Initiate multiple repetitive threading:
	Pass 1 • Rapid tool to X1.968. • Thread at feed rate to order point 1. • Rapid return to start point.
	Pass 2 Repeat G76 cycle for $X = 1.954$.
	Pass 3 Repeat G76 cycle for $X = 1.9446$.
	Pass 4 Repeat G76 cycle for $X = 1.936$.

	Pass 5 Repeat G76 cycle for $X = 1.302$.
	Spring Pass Repeat G76 cycle for $X = 1.302$.
N0100 G00 X10.0 Z6.0 T0100 M09	Rapid to tool change position Ⓐ. Cancel values in offset file 1. Coolant off.
N0110 M00	Program stop.

17-19 CHAPTER SUMMARY

The following key concepts were discussed in this chapter:

1. Tool edge programming involves programming the part geometry directly. This technique can only be used to machine horizontal or vertical lines.
2. Tool nose radius (TNR) compensation involves programming the part geometry directly instead of the center of the tool nose radius. The part geometry can include arcs and tapered lines.
3. The word address command for TNR compensation left of upward motion is G41. The G42 word affects compensation right of upward motion.
4. TNR compensation is initiated or canceled by the next linear (G00 or G01) tool motion command following a G41, G42, or G40 word.
5. The setup person prepares the CNC lathe for TNR programming by keying in the following information:
   - *X* and *Z* tool offsets to the tool edge
   - Tool nose radius value
   - Tool nose vector number
6. The tool nose vector indicates to the controller the direction of the tool edge from the TNR center.
7. The G90 cycle is used to rough turn or bore a part from stock. Straight or tapered passes may be programmed.
8. Rough facing from stock can be executed via the G94 cycle. Straight or tapered passes may be programmed.
9. The multiple repetitive cycle G71 is used to rough cut general profiles consisting of arcs and tapered lines. This can be followed by a G70 finishing pass cycle.
10. The following cycles can be used for threading:
    G32—single-pass cycle (four blocks per pass)
    G92—multiple-pass cycle (one block per pass)
    G76—multiple-pass cycle (one block for all passes)

REVIEW EXERCISES

17.1. The part shown in Figure 17–23 has been rough turned. Write a CNC program segment using TNR compensation to perform the finish pass (facing and turning).

Assume the setup person has entered the following information into the offset file:

- Tool offset values for tool edge programming
- Tool nose radius values
- Tool nose vector 3

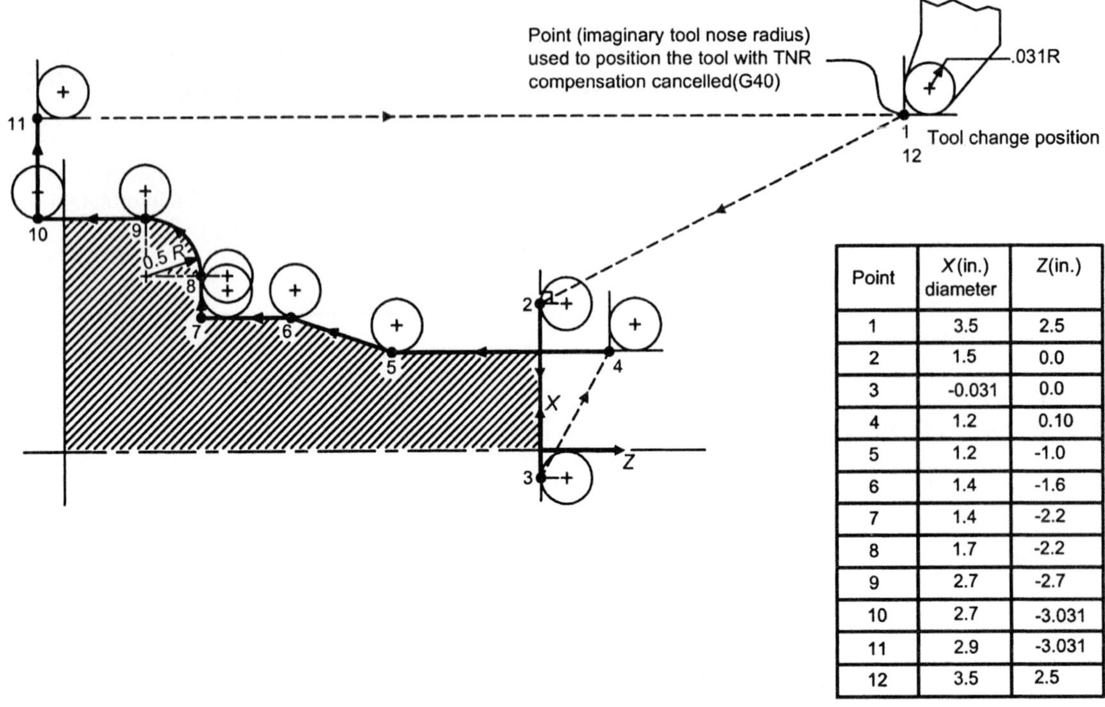

Point	X(in.) diameter	Z(in.)
1	3.5	2.5
2	1.5	0.0
3	-0.031	0.0
4	1.2	0.10
5	1.2	-1.0
6	1.4	-1.6
7	1.4	-2.2
8	1.7	-2.2
9	2.7	-2.7
10	2.7	-3.031
11	2.9	-3.031
12	3.5	2.5

FIGURE 17–23

Notes: **1.** Machine at constant surface speed of 200 sfm.
2. Set the feed rate for facing at 0.005 ipr.
3. Set the feed rate for turning at 0.007 ipr.
4. Use tool 11 with offset 11.

17.2. The part shown in Figure 17–24 has been previously roughed. Write a CNC program segment utilizing TNR compensation to finish bore the profile. Assume the following information has been previously entered into the offset file:

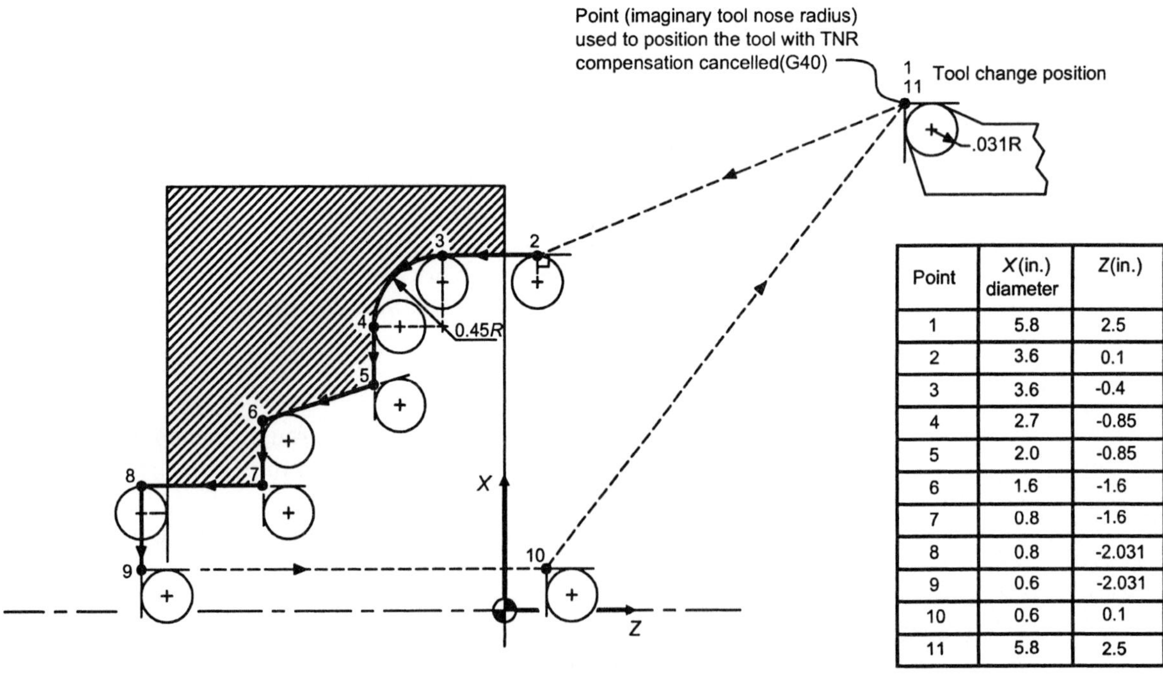

Point	X(in.) diameter	Z(in.)
1	5.8	2.5
2	3.6	0.1
3	3.6	-0.4
4	2.7	-0.85
5	2.0	-0.85
6	1.6	-1.6
7	0.8	-1.6
8	0.8	-2.031
9	0.6	-2.031
10	0.6	0.1
11	5.8	2.5

FIGURE 17–24

- Tool offset values for tool edge programming
- Tool nose radius
- Tool nose vector 2

Notes: **1.** Machine at 350 sfm constant surface speed.
2. Set the boring feed rate at 0.008 ipr.
3. Use tool 10 with offset 10.

17.3. Write a program segment for rough turning the part shown in Figure 17–25. Use the G90 cycle to rough turn the part from stock. The profile consists only of horizontal and vertical lines. Thus, tool edge programming can be applied.

Notes: **1.** Machine at 300 sfm constant surface speed.
2. Use a roughing feed rate of 0.015 ipr.
3. Use tool 9 with offset 9.
4. Remove 0.15 in. of material per turning pass.

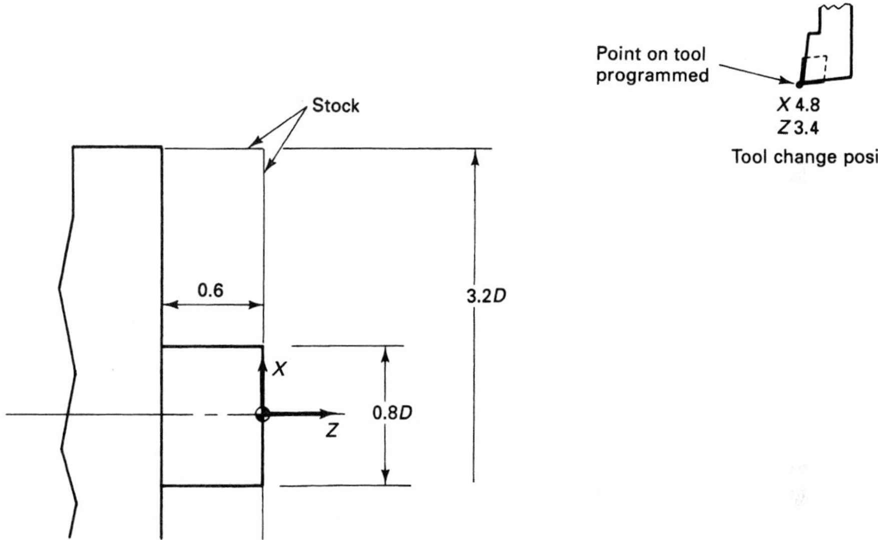

FIGURE 17–25

17.4. The part shown in Figure 17–26 is to be rough faced from stock. Use the rough facing cycle G94 to write the program segment. Tool edge programming may be applied since the profile consists only of horizontal or vertical lines.

Notes: **1.** Rough face the part at 150 sfm constant surface speed.
2. Assign a roughing feed rate of 0.010 ipr.
3. Use tool 8 with offset 8.
4. Remove 0.10 in. of material per facing pass.

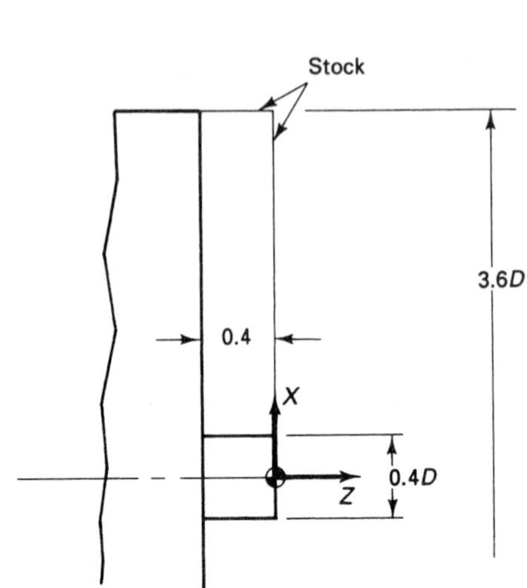

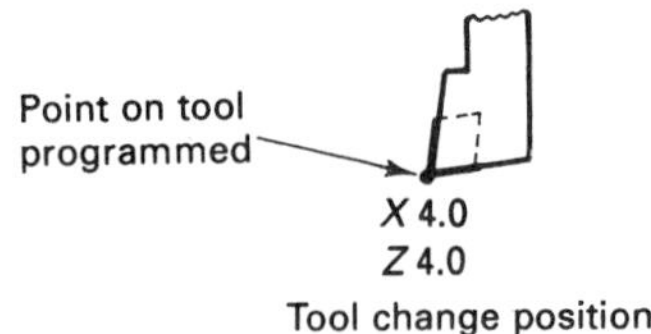

FIGURE 17–26

17.5. The profile shown in Figure 17–27 is to be machined from stock. Write a complete program to execute the required roughing passes and the finish pass. The setup person has entered the tool offsets for tool edge programming, tool nose radius values, and tool nose vector 3 into the offset file.

Notes: For *roughing:*

1. Use a G71 roughing cycle.
2. Rough at constant surface speed of 180 sfm.
3. Use tool 7 with offset 7.

For *finishing:*

4. Use a G70 finishing cycle.
5. Finish at constant surface speed of 250 sfm.
6. Use tool 6 with offset 6.

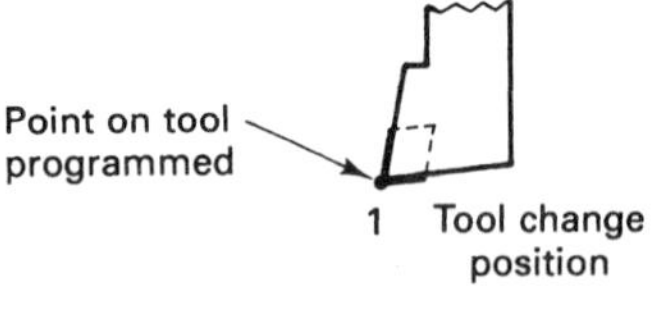

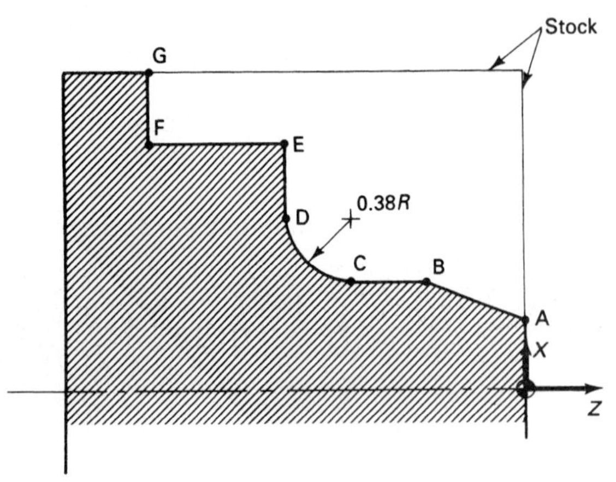

Point	X (in.) diameter	Z (in.)
1	4.4	2.6
A	0.8	0.0
B	1.2	-0.6
C	1.2	-0.92
D	1.96	-1.3
E	2.8	-1.3
F	2.8	-2.1
G	3.6	-2.1

FIGURE 17–27

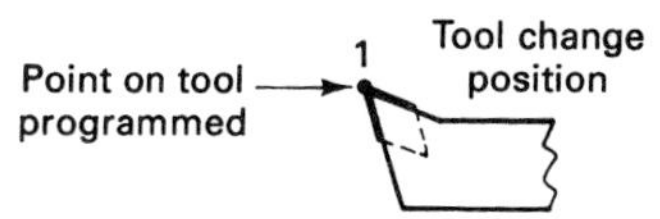

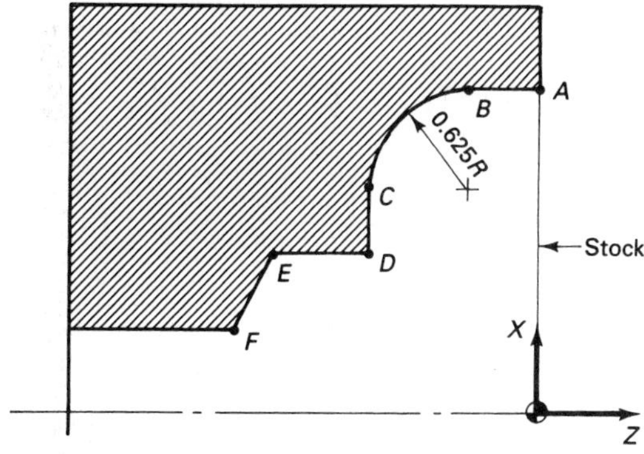

Point	X (in.) diameter	Z (in.)
1	6.0	2.6
A	3.25	0.0
B	3.25	-0.375
C	2.0	−1.0
D	1.6	−1.0
E	1.6	−1.4
F	0.8	−1.6

FIGURE 17–28

17.6. Write a complete program for machining, from stock, the profile shown in Figure 17–28. Apply appropriate roughing and finishing passes. Assume the tool offsets, tool nose radius, and tool nose vector 2 have been entered into the offset file.

Notes: For *roughing:*

1. Employ the G71 roughing cycle.
2. Use a constant surface speed of 120 sfm.
3. Rough bore at a feed rate of 0.015 ipr.
4. Use tool 5 with offset 5.
5. Remove 0.10 in. material per roughing pass.

For *finishing:*

6. Employ the G70 finishing cycle.
7. Use a constant surface speed of 220 sfm.
8. Finish bore at a feed rate of 0.006 ipr.
9. Use tool 4 with offset 4.

17.7. Use G71, G70 to rough and finish machine the part shown in Figure 17–29. Employ G74 peck drilling and G75 grooving cycles to complete the part.

Notes: For *roughing:*

1. Employ the G71 roughing cycle.
2. Rough at constant surface speed of 120 sfm.
3. Use a roughing feed rate of .004 ipr.
4. Use tool 3 with offset 3.
5. Remove 0.01 in. material per roughing pass.

For *finishing:*

6. Employ the G70 finishing cycle.
7. Finish at constant surface speed of 150 sfm.
8. Use a finishing feed rate of .002 ipr.
9. Use tool 4 with offset 4.

For *peck drilling:*

10. Employ the G74 peck drilling cycle.
11. Drill at 600 rpm.
12. Assign a feed rate of 0.012 ipr.
13. Set the peck depth at 0.1 in.

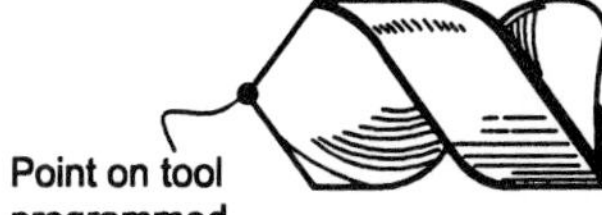

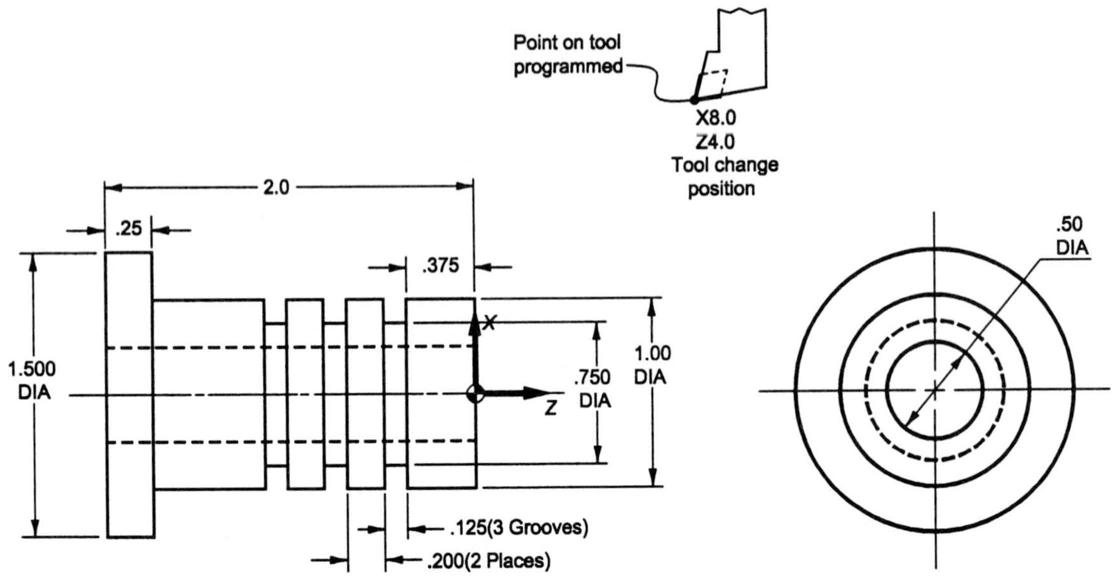

FIGURE 17–29

14. Use tool 5 with offset 5.
    For *grooving:*
15. Program the G75 grooving cycle.
16. Assign a constant surface speed of 100 sfm.
17. Set the grooving feed rate at 0.002 ipr.
18. Set the peck depth at 0.05 in.
19. Use a .125 in. tool assigned as tool 6 with offset 6.

17.8. The part shown in Figure 17–30 is to be rough and finished machined using the G71, G70 cycles. The part is to be completed by machining face grooves via the G74 face grooving cycle.

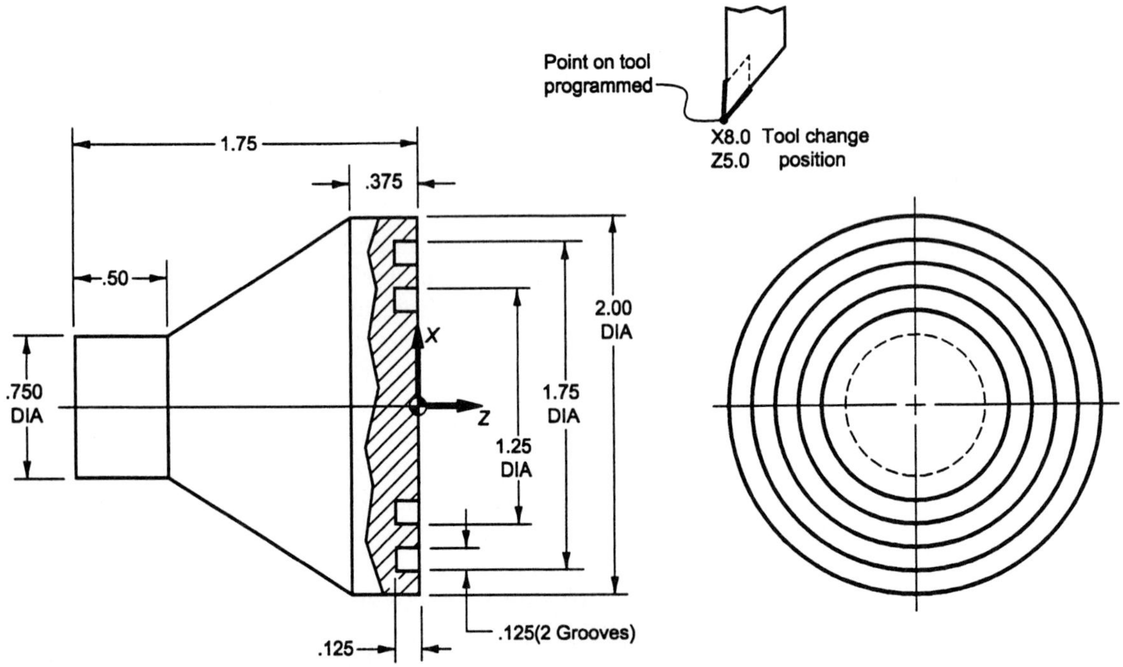

FIGURE 17–30

Notes: For *roughing:*

1. Employ the G71 roughing cycle.
2. Rough at constant surface speed of 130 sfm.
3. Use a roughing feed rate of .004 ipr.
4. Use tool 5 with offset 5.
5. Remove 0.01 in material per roughing pass.

For *finishing:*

6. Employ the G70 finishing cycle.
7. Finish at constant surface speed of 170 sfm.
8. Use a finishing feed rate of .002 ipr.
9. Use tool 6 with offset 6.

For *face grooving:*

10. Program the G74 face grooving cycle.
11. Assign a constant surface speed of 120 sfm.
12. Set the grooving feed rate at 0.002 ipr.
13. Set the peck depth at 0.05 in.
14. Use a .125 in. tool assigned as tool 7 with offset 7.

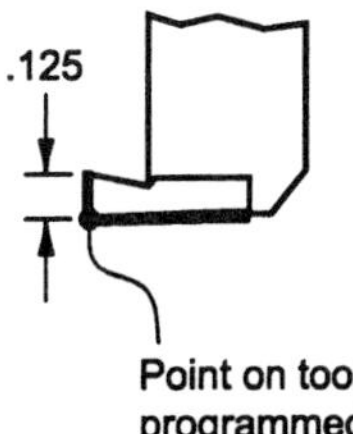

17.9. The part shown in Figure 17–31 is to be produced by first peck drilling from solid stock using the G74 pecking cycle. This is to be followed by rough and finish chamfering and boring via the G71 and G70 cycles.

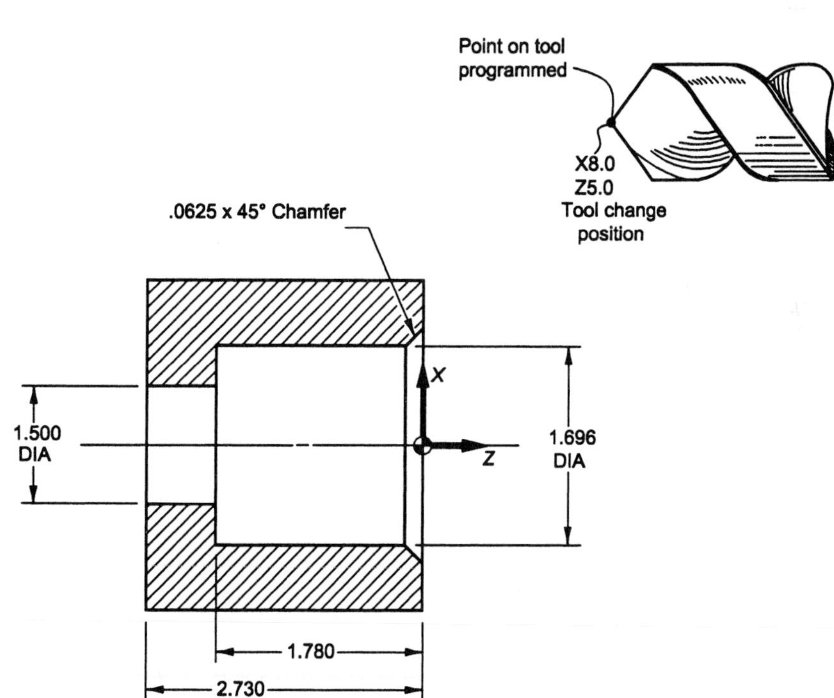

FIGURE 17–31

Notes: For *peck drilling:*

1. Employ the G74 peck drilling cycle.
2. Drill at 400 rpm.
3. Assign a feed rate of 0.01 ipr.
4. Set the peck depth at 0.2 in.
5. Use tool 5 with offset 5.

For *roughing:*

6. Employ the G71 roughing cycle.

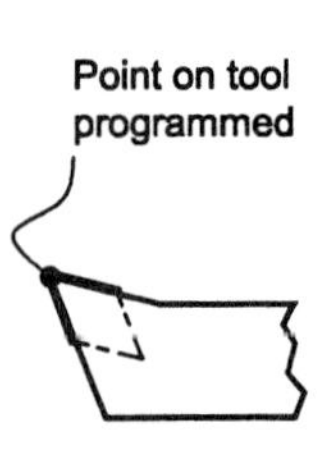

7. Rough at constant surface speed of 120 sfm.
8. Use a roughing feed rate of .004 ipr.
9. Use tool 3 with offset 3.
10. Remove 0.01 in material per roughing pass.

For *finishing:*

11. Employ the G70 finishing cycle.
12. Finish at constant surface speed of 180 sfm.
13. Use a finishing feed rate of .002 ipr.
14. Use tool 4 with offset 4.

17.10. Write the program segment for threading the blank as indicated in Figure 17–32. Use the G32 single-pass cycle.

Notes: **1.** Thread at 300 sfm cutting speed.

2. Use the formula:

$$\text{Feed rate} = 1/\text{threads/in.}$$

to define the required feed rate in the program.

3. Use the formula:

$$Z_s = \frac{\text{rpm} \times \text{Pitch}}{1800}$$

to define the starting distance in the program.

4. Use the formula:

$$Z_a = Z_s \times 3.605 + \text{thread length}$$

to define the stopping distance in the program.

5. Use tool 3 and offset 3.

6. Determine the number of passes to program from the formula:

$$\text{Infeed} = d\sqrt{p}$$

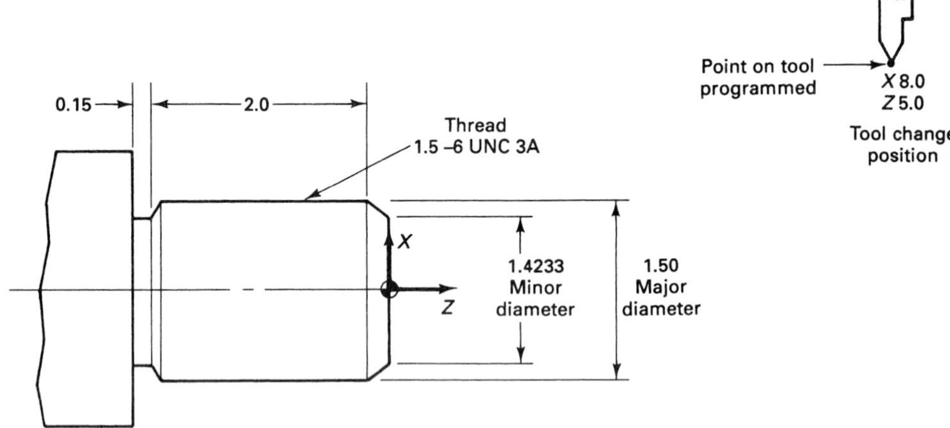

FIGURE 17–32

17.11. A thread relief and internal threads are to be machined into the part produced in exercise 17.9. Use the G75 grooving cycle to machine the thread relief and the G92 cycle to cut the internal thread as indicated in Figure 17–33.

Notes: For *thread relief groove:*

1. Program the G75 grooving cycle.
2. Assign a constant surface speed of 120 sfm.
3. Set the grooving feed rate at 0.002 ipr.
4. Set the peck depth at 0.02 in.
5. Use a .125 in. tool assigned as tool 8 with offset 8.

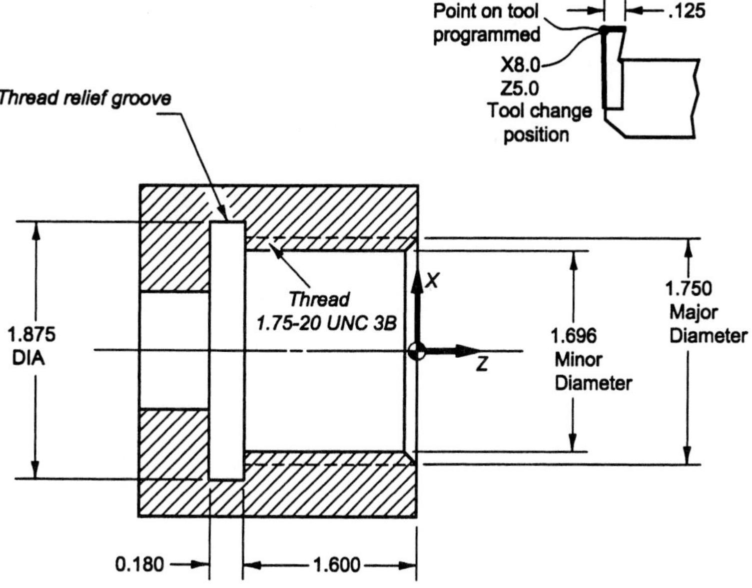

FIGURE 17–33

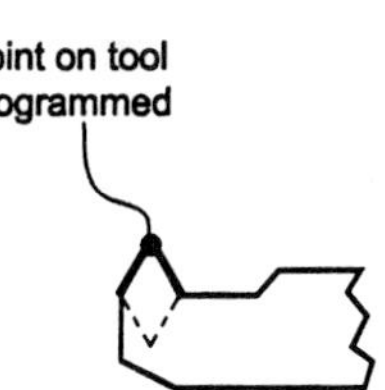

For *threading:*

6. Program the G92 threading cycle.
7. Define the program feed rate as follows:

$$\text{Feed rate} = \frac{1}{\text{threads/in}}$$

8. Define the program starting distance as:

$$Z_s = \frac{\text{rpm} \times \text{Pitch}}{1800}$$

9. Define the program stopping distance as:

$$Z_a = Z_s \times 3.605 + \text{thread length}$$

10. Use tool 2 offset 2.
11. Determine the number of passes to program from the formula:

$$\text{Infeed} = d\sqrt{p}$$

17.12. Use the G76 multiple repetitive threading cycle to write the program segment as outlined in exercise 17.10.

17.13. Use the G76 multiple repetitive threading cycle in writing the program segment described in exercise 17.11.

CHAPTER

18

Modern Computer-Aided Part Programming

18-1 CHAPTER OBJECTIVES

At the conclusion of this chapter you will be able to

1. Describe the advantages of using off-line programming.
2. Understand the advantages of using a computer-aided programming language.
3. Write simple programs in the computer-aided programming language APT.
4. State the key elements comprising a CAD/CAM system.
5. Explain how part programs are developed with the aid of CAD/CAM.
6. Explain the advantages of using CAD/CAM technology.
7. Understand how knowledge-based machining software simplifies the job of creating part programs.

18-2 INTRODUCTION

This chapter introduces the reader to modern developments that are transforming the job of creating part programs. At the center of this revolution is the computer and the modern application software it can process.

Off-line programming is discussed first. This technique allows the programmer to write and store programs at a remote computer terminal. As a result, the CNC machine can spend more time executing machining operations.

The computer-aided programming language APT is discussed next. The reader is introduced to the advantages of using this approach to write part programs.

The chapter continues with a discussion of CAD/CAM systems. The elements comprising CAD/CAM systems are identified. The benefits of using a CAD/CAM system for creating part programs are also explained. The process of creating a part and generating a part program with CAD/CAM is described in detail.

The chapter ends with a presentation of the latest innovation, which is simplifying and making more efficient the task of creating part programs—"knowledge-based" software.

18-3 MODERN DEVELOPMENTS IN THE PROCESS OF CNC PROGRAMMING

In the past, programming was performed manually. The program was written in word address and was tailored to run on a particular controller. Trigonometry was a must for determining tool paths. The CNC machine needed to be halted while the operator keyed the program into the machine control unit. The CNC machine was further tied up for program verification, which involved machining a test part and inspecting the results.

The digital computer, however, has changed the way programs are written, checked, and run. Some of the most dramatic improvements provided by the computer include off-line programming, the use of computer-aided programming languages, and the application of CAD/CAM to programming.

18–4 USING AN OFF-LINE COMPUTER TO WRITE AND STORE PART PROGRAMS

CNC machines operate at optimum when running part programs and machining parts. This goal is achieved when new programs are written and tested at an off-line computer terminal. The program can be composed directly in word address (G and M codes) with the aid of text editing software. It can be tested and then stored on punched tape, disk, Zip disk, or CD-ROM media. If the remote computer has a communication line to the CNC machine, the program can be downloaded directly to the machine's controller. This technique is known as off-line programming. It is illustrated in Figure 18–1.

18–5 AN INTRODUCTION TO COMPUTER-AIDED PROGRAMMING LANGUAGES

The controller of a CNC machine can only operate on word address (G and M) codes. Furthermore, these codes must be entered according to a specific format acceptable to the particular controller. By now the reader has also discovered that word address programming can require trigonometric computations.

A computer-aided programming language is meant to address some of these problems. A computer-aided programming language allows the programmer to write the program using English-like commands, not G and M codes. Because the tool path is described in terms of these commands, trigonometric computations are not needed. All tooling parameters such as tool diameter, tool speeds, and feeds are easily specified. Furthermore, the programmer need not be concerned with tailoring the program to a particular controller. A program known as a postprocessor can automatically translate the program written in a computer-aided

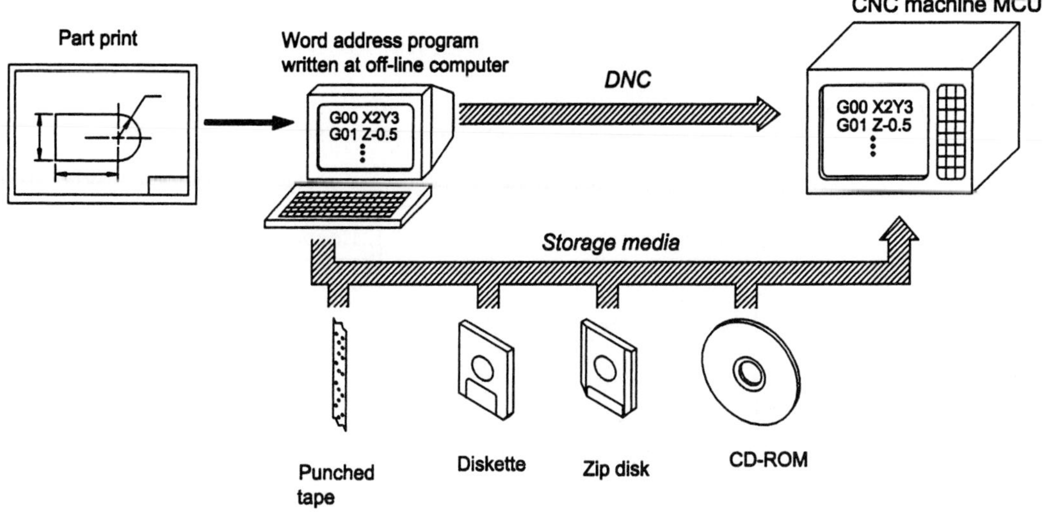

FIGURE 18–1 Off-line programming.

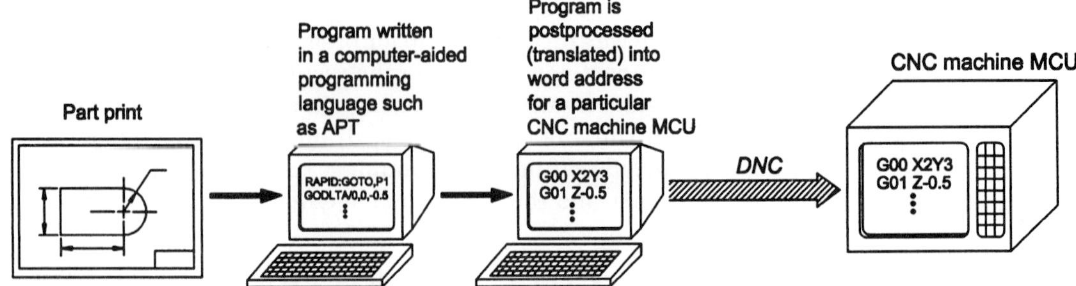

FIGURE 18–2 Programming via a computer-aided programming language.

programming language into a corresponding word address program. Special postprocessors have been written for each machine controller. This allows a single program written in a computer-aided programming language to be applied to many different types of machine controllers. This technique is illustrated in Figure 18–2.

The most popular computer-aided programming language still in use today is APT.

APT

APT is short for Automatically Programmed Tools. It was the first and still is the most powerful computer programming language. APT can be used to program the machining of complex surfaces on four- and five-axis machining centers. It was originally restricted to run on large mainframe computers. Advances in microprocessor technology have allowed it to be available on workstations and personal computers.

18–6 ELEMENTS OF THE APT PROGRAMMING LANGUAGE

The APT language is very large and is composed of many different types of statements. This section is by no means comprehensive. Only the most basic APT commands will be presented. The reader should consult an APT programming manual for further information.

Commands to Specify Geometry

■ EXAMPLE 18–1

The following APT commands are used in part programs to define the points shown in Figure 18–3.

APT command	*Meaning*
P1 = POINT/4,2,0	Create P1 at *X*4, *Y*2, *Z*0.
P2 = POINT/INTOF,L1,L2	Create P2 at intersection of L1 and L2.
P3 = POINT/XSMALL,INTOF,L3,C1	Create P3 at intersection of L3 and C1 in direction of X_{small} (←).
P4 = POINT/YLARGE,INTOF,L3,C1	Create P4 at intersection of L3 and C1 in direction of Y_{large} (↑).
P5 = POINT/C2,ATANGL,30	Create P5 on C2 and at an angle of 30° with the horizontal.

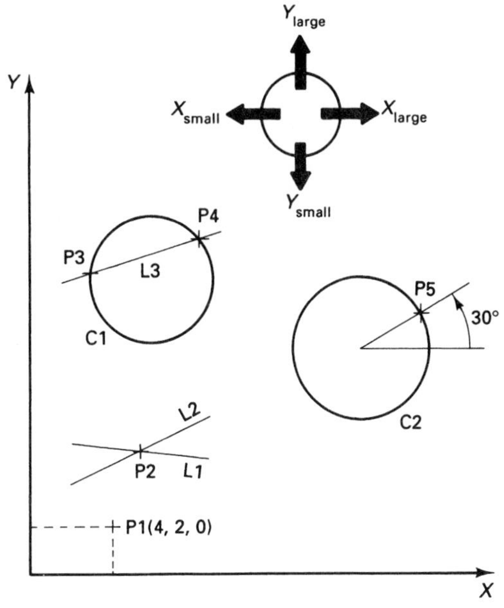

FIGURE 18–3

■ EXAMPLE 18–2

The following APT commands are used in part programs to define the lines shown in Figure 18–4.

APT command	*Meaning*
L1 = LINE/4,2,0,8,5,0 or L1 = LINE/P1,P2	L1 starts at P1(4,2,0) and runs to P2(8,5,0).
L2 = LINE/PARLEL,L1,YLARGE,1.5	L2 is parallel to L1 and offset in the direction of Y_{large} (↑).
L3 = LINE/P3,PERPTO,L2	L3 starts at P3 and runs in a direction perpendicular to L2.
L4 = LINE/P3,ATANGL,30,L2	L4 starts at P3 and runs at an angle of 30° with respect to L2.
L5 = LINE/P4,LEFT,TANTO,C1	L5 starts at P4 and runs tangent to C1. Tangency is left of circle center when viewed along the line of sight.
L6 = LINE/P4,RIGHT,TANTO,C1	L6 starts at P4 and runs tangent to C1. Tangency is right of circle center when viewed along the line of sight.
L7 = LINE/RIGHT,TANTO,C3,RIGHT,TANTO,C2	L7 starts at right tangency with C3 and runs to right tangency with C2.
L8 = LINE/RIGHT,TANTO,C3,LEFT,TANTO,C2	L8 starts at right tangency with C3 and runs to left tangency with C2.

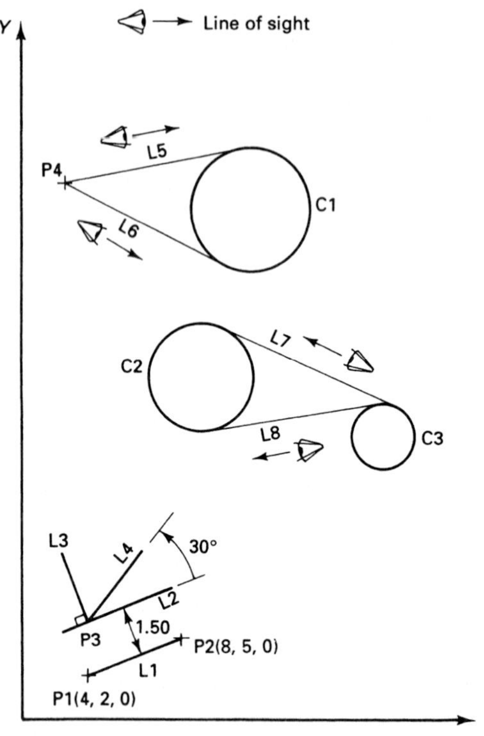

FIGURE 18–4

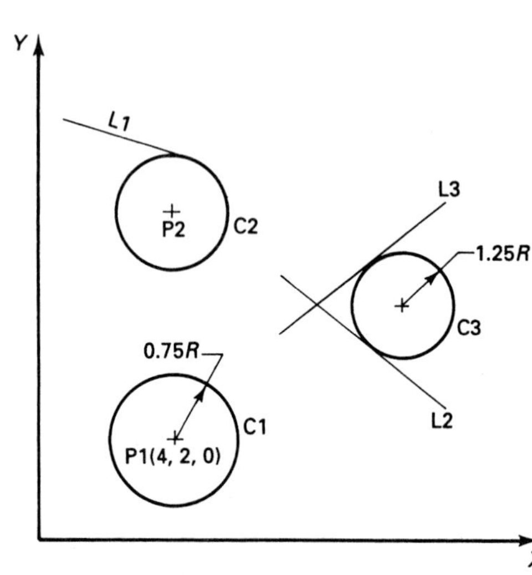

FIGURE 18–5

■ EXAMPLE 18–3

The following APT commands are used in part programs to define the circles shown in Figure 18–5.

APT command	*Meaning*
C1 = CIRCLE/4,2,0,.75 or C1 = CIRCLE/CENTER,P1, RADIUS,.75	Create C1. Center at *X*4, *Y*2, *Z*0. Radius is 0.75.
C2 = CIRCLE/CENTER,P2,TANTO,L1	Create C2. Center at P2 and tangent to L1.
C3 = CIRCLE/XLARGE,L3,YLARGE,L2,RADIUS,1.25	Create C3 located on X_{large} (→) side of L3 and Y_{large} (↑) side of L2. Radius is 1.25.

Commands to Specify Setup Parameters

The commands in this section are used to specify important identification and machining information. This information must be inputted prior to entering any tool motion statements.

■ EXAMPLE 18–4

The APT commands given below specify setup parameters in part programs.

APT command	*Meaning*
PARTNO MILLING EXAMPLE 16.1	Specifies the program ID is MILLING EXAMPLE 16.1. PARTNO must be the first statement in an APT program.
MACHIN/FANUC	Specifies the APT program is to be translated (postprocessed) into a word address program for machine FANUC.
SETPT = POINT/−4,−3,.1	Specifies the location of the tool change position with respect to the part origin.
LOADTL/1,SETOOL,0,0,3.25	Directs the system to change to tool 1. Tool offset value is (0,0,3.25).
CUTTER/.25	Specifies the cutter diameter is 0.25 in.
SPINDL/1500,CLW	Specifies the spindle speed is 1500 rpm clockwise (S1500 M03).
FEDRAT/8,IPM	Specifies the feed rate is 8 in./min (F8.0).
COOLANT/FLOOD	Calls for coolant on in flood mode (M08).

■■

Commands to Specify Tool Motion

The programmer directs the tool along the geometry defined previously by using the tool motion commands. The computer always uses three surfaces to help guide the cutter along its programmed path. These surfaces are identified as follows:

Part surface: Guides the bottom of the tool.
Drive surface: Guides the side of the tool.
Check surface: Stops tool motion along the part and drive surfaces.

See Figure 18–6.

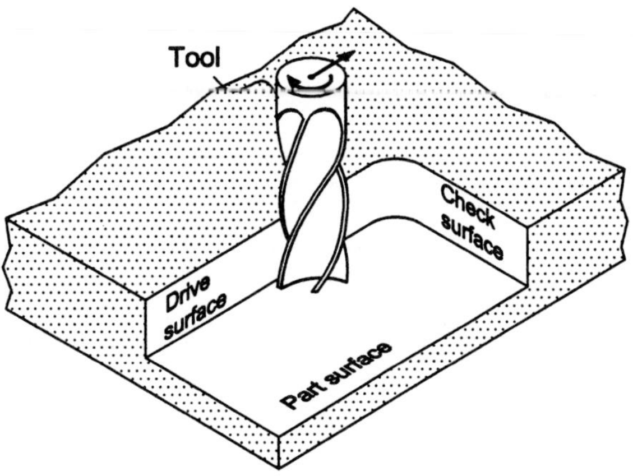

FIGURE 18–6 Part, drive, and check surfaces for guiding tool motion.

■ EXAMPLE 18–5

The APT commands listed below define tool motion in part programs.

APT command	***Meaning***
FROM/SETPT	Start tool motion from reference point (set at tool change position in this program). Position ①. See Figure 18–7.
RAPID; GOTO/L1,NOZ	Move cutter at rapid to check surface at L1. Execute no *Z* motion. Position ②.
TLLFT,GOLFT/L1,TANTO,C1	With cutter on the left of drive surface, execute motion. Move to left of line of sight along L1. Stop at check surface located at tangency with C1. Position ③.
GOFWD/C1,TANTO,L2	Move cutter around C1. Stop cutter at check surface located at tangency with L2. Position ④.
GOFWD/L2,PAST,L3	Move cutter along L2. Stop cutter past check surface at L3. Position ⑤.
GORGT/L3,PAST,L4	Move cutter to right of line of sight along L3. Stop cutter past check surface at L4. Position ⑥.
GORGT/L4,PAST,L1	Move cutter to right of line of sight along L4. Stop cutter past check surface at L1. Position ⑦.

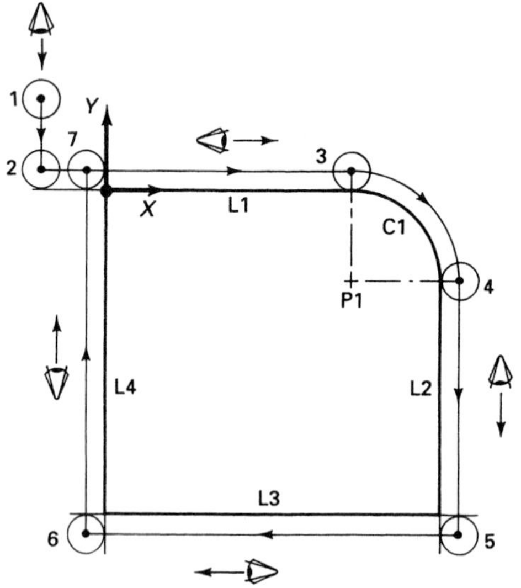

FIGURE 18–7

■■

18–7 WRITING A COMPLETE PROGRAM IN APT

Working with the part drawing, the programmer proceeds to write the required APT program. The program usually consists of four main parts:

- Setup parameter statements describing computer operations and machining specifications
- Geometric definition statements
- Tool motion statements
- End statements for returning the tool to a safe position, homing the machine, and so on

■ EXAMPLE 18–6

Write an APT program for milling and drilling the part shown in Figure 18–8.

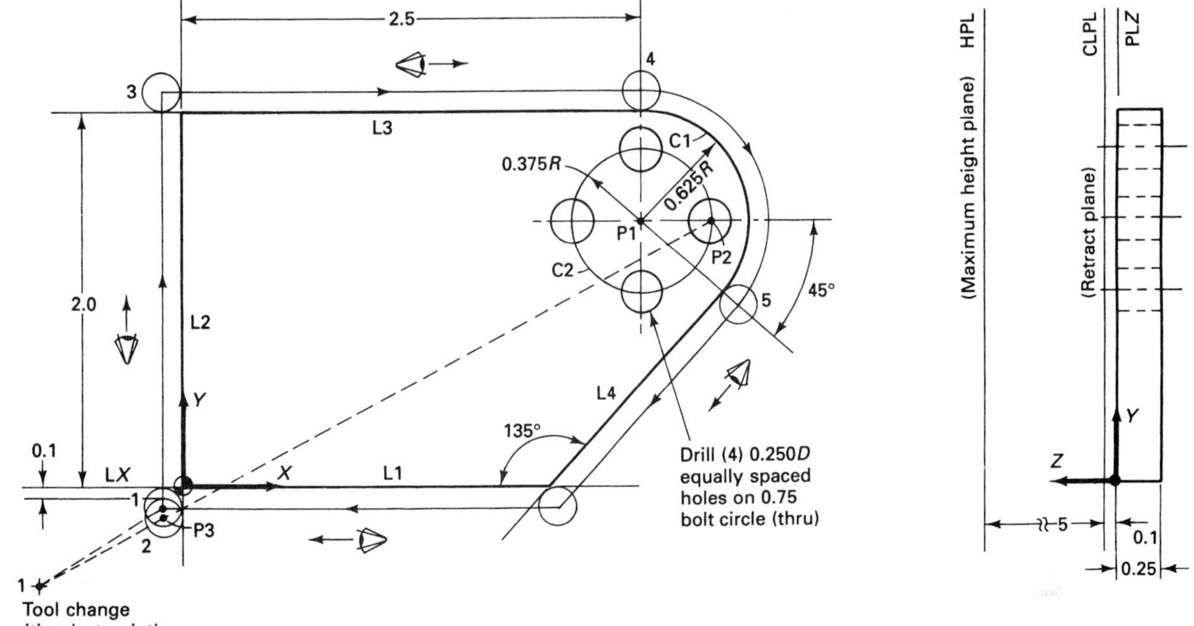

FIGURE 18–8

Tool	Operation	Tooling	Speed (rpm)	Feed (in./min)
1	Profile contour × 0.25 deep	0.25*D* end mill	800	6
2	Drill (4) 0.25*D* equally spaced holes on 0.75 bolt circle (thru)	0.25*D* stub drill	1200	4

APT command	*Meaning*
PARTNO APT DEMO PROGRAM	Program ID statement.
MACHIN/FANUC	Specifies the APT program is to be translated (postprocessed) into a word address program for CNC machine FANUC.
TOLER/.001	Tolerance set at 0.001 in.
CLPRNT	Print the cutter location file resulting from the APT program.
$$***PART GEOMETRY***	Optional comment statement.
LX = LINE/XAXIS	Define line *LX* along the *X* axis.
LY = LINE/YAXIS	Define line L*Y* along the *Y* axis.
L3 = LINE/PARLEL,LX,YLARGE,2.0	Define line L3 offset from *LX* by 2 in the Y_{large} direction (↑).

APT command	***Meaning***
P1 = POINT/2.5,1.375,.1	Define point P1 at (2.5, 1.375, 0.1).
C1 = CIRCLE/CENTER,P1,RADIUS,.625	Define C1. Center at P1, radius 0.625.
L4 = LINE/ATANGL,45,LX,TANTO,C1,XLARGE	Define L4 starting at a 45° angle to *LX* and running tangent to C1 in the X_{large} direction (→).
C2 = CIRCLE/CENTER,P1,RADIUS,.375	Define C2. Center at P1, radius 0.375.
P2 = POINT/C2,ATANGL,0	Define P2 on C2 and at an angle of 0° with the horizontal.
SETPT = POINT/−2.0,−2.0,4.0	Set tool change position from part origin. Position ①.
PLZ = PLANE/0,0,1.0	Define a zero reference plane on the top part surface.
HPL = PLANE/PARLEL,PLZ,ZLARGE,5.0	Define a maximum height plane offset from PL*Z* by 5 in the Z_{large} direction (safe retract height prior to tool change).
CLPL = PLANE/PARLEL,PLZ,ZLARGE,.1	Define a retract plane offset from PL*Z* by 0.1 in the Z_{large} direction.
P3 = POINT/INTOF,(LINE/PARLEL,LX,YSMALL,.35),(LINE/PARLEL,LY,XSMALL,.25)	
	Define P3 at the intersection of two lines. One is offset 0.35 in the Y_{small} direction from L*X*. The other is offset 0.25 in the X_{small} direction (←) from L*Y*.
$$***TOOL MOTION***	Optional comment statement.
PPRINT MILL PERIPHERY PPRINT LOAD .5 DIA END MILL	Print comment into the tape.
CUTTER/.5	Define cutter diameter as 0.5.
FEDRAT/6,IPM	Define feed rate as 6 in./min.
LOADTL/1,SETOOL,0,0,2.5	Change to tool 1. Tool offset is (0,0,2.5).
SPINDL/800,CLW	Spindle on at 800 rpm, CLW.
COOLANT/FLOOD	Coolant on (flood).
FROM/SETPT	Start cutter motion from set point (tool change position in this program).
RAPID; GOTO/PT3,NOZ	Rapid to PT3 with no *Z* motion. Position ②.
RAPID; GO/ON,CLPL	Rapid to the retract plane.
GODLTA/0,0,−.37	Execute −0.37-in. incremental motion in *Z* (plunge 0.37 in. into material).
TLLFT, GOLFT/LY,PAST,L3	With tool on the left side of the drive surface, move left along L*Y*. Stop past L3. Position ③.

GORGT/L3,TANTO,C1	Turn right and move along L3. Stop at the tangency of L3 and C1. Position ④.
GOFWD/C1,TANTO,L4	Move around C1. Stop at the intersection of C1, L4. Position ⑤.
GOFWD/L4,PAST,LX	Move along L4. Stop past the intersection of L4, *LX*. Position ⑥.
GORGT/LX,PAST,LY	Turn right and move along L*X*. Stop past the intersection of L*X*, L*Y*. Position ⑦.
RAPID; GO/ON,HPL	Rapid to the maximum height plane.
SPINDL/OFF	Spindle off.
COOLNT/OFF	Coolant off.
RAPID; GOTO/SETPT	Rapid to the tool change position.
PPRINT DRILL 4 HOLES PPRINT LOAD .25 DIA DRILL	Print comment into the tape.
CUTTER/.25	Set cutter diameter at 0.25.
FEDRAT/4,IPM	Set feed rate at 4 in./min.
LOADTL/2,SETOOL,0,0,1.75	Change to tool 2. Tool offset is (0, 0, 1.75).
SPINDL/1200,CLW	Spindle on at 1200 rpm, CLW.
COOLANT/FLOOD	Coolant on (flood).
INDEX/10	Start of index marker, 10.
RAPID; GOTO/P2,NOZ	Rapid to P2 with no *Z* motion.
RAPID; GO/ON, CLPL	Rapid to the reference plane.
GODLTA/0,0,−.48	Execute −0.48-in. incremental motion in *Z* (plunge 0.48 in. into material).
RAPID; GODLTA/0,0,1.	Rapid to the reference plane.
INDEX/10,NOMORE	End of index marker, 10.
COPY/10,XYROT,90,4	Repeat commands between start of index 10 and end of index 10 as follows: repeat four times, rotate tool 90° from last position each time.
RAPID; GO/ON,HPL	Rapid to the maximum height plane.
RAPID; GOTO/SETPT	Rapid to the tool change position.
SPINDL/OFF	Spindle off.
COOLNT/OFF	Coolant off.
END	End of APT program.
FINI	Serves as a marker for signaling the system to stop reading the tape.

■■

18-8 AN INTRODUCTION TO CAD/CAM TECHNOLOGY

CAD/CAM stands for Computer-Aided Design and Computer-Aided Manufacturing. Experienced programmers use this powerful tool to simplify the task of writing and testing part programs. In the right hands, CAD/CAM can dramatically boost the productivity of CNC shops. The CAD/CAM approach is superior to the technique of using a computer-aided programming language to prepare part programs. Some of the advantages of using CAD/CAM are:

- The system can be used for off-line checking of the program. Resulting tool paths can be graphically displayed on the computer screen. The user can zoom in and view the tool paths at various orientations. The system can generate real time material removal displays and flag tool breakthrough or part violation problems.
- Time and cost of machining the part can be quickly determined.
- The system can determine optimum tooling, speeds, and feeds for the material selected.

18-9 THE ELEMENTS COMPRISING CAD

CAD is a technique of using a computer to create, modify, and refine a design. The user creates the design (model) by applying graphics commands stored in the computer. The design is stored in a file in the computer's memory. The computer sends a graphical picture of the file to the graphics display monitor. The user has complete freedom to zoom and view the model at different orientations. Graphics editing commands allow for easily making modifications as required. Today's CAD systems come in many sizes: large (mainframe based), medium (minicomputer or workstation based), and small (microcomputer based). In general, CAD systems are composed of the following physical components or hardware:

- A system unit that houses a central processing unit (CPU) for executing all inputted commands. Various storage devices such as disk drives, Zip disk drives, and CD-ROM drives are also contained in the system unit.
- A graphics display monitor also known as a cathode ray tube (CRT). This unit is used for displaying model graphics and all information inputted into and received from the system unit.
- A keyboard for entering commands and data.
- A mouse for picking commands from on-screen menu displays, for pointing to model graphic features, and for entering the location of objects.
- Hard copy output units. A printer produces small to medium prints. A plotter is capable of generating large production prints.

Figure 18–9 illustrates the elements that comprise typical CAD systems.

Many CAD software packages are readily available. These include AutoCAD, Solidworks, Solidedge, Ironcad, Cadkey, Microstation, CADAM, Pro-Engineer, etc.

Modeling Techniques

Using basic CAD software, the user can represent the design as a set of points, lines, arcs, and so on. This type of modeling is known as wire frame. A wire frame model is shown in Figure 18–10. More sophisticated CAD software allows the user to fill the void between the wire frame elements with a surface shape. Later a tool path can be applied over the surface. An example of a surfaced model is shown in Figure 18–11. The latest advances in CAD software and hardware enable the model to be created and stored as a solid object. Such models have no voids and all points are filled. Solids eliminate any ambiguities that may arise with wire frame or surface models. A solid model is shown in Figure 18–12.

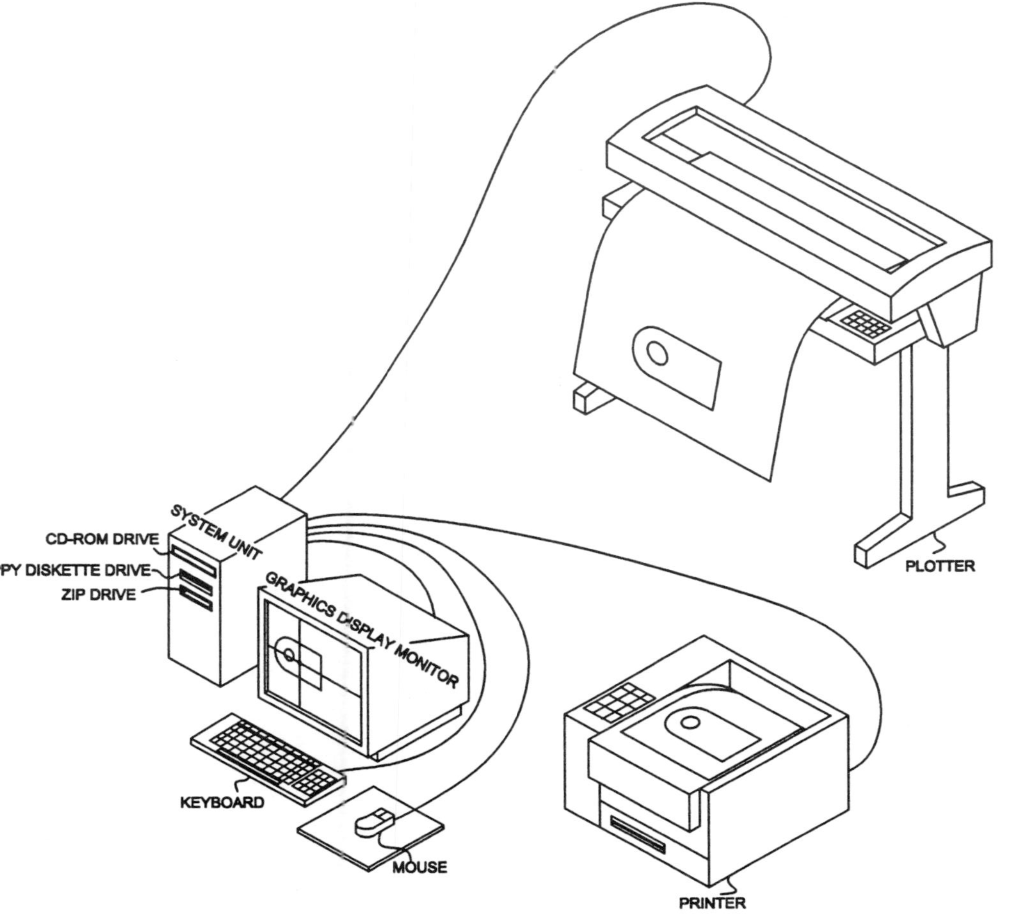

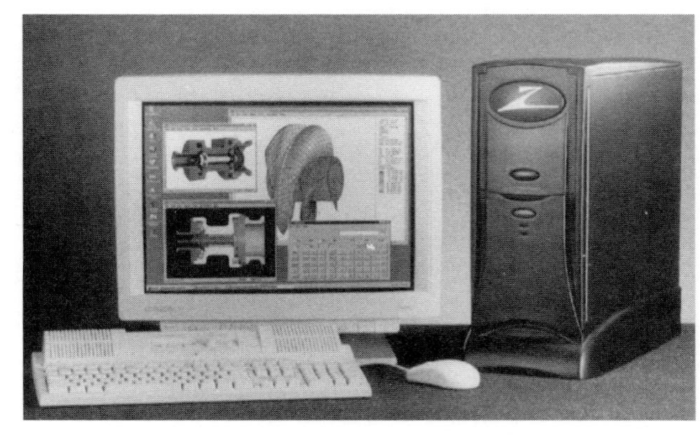

FIGURE 18–9 Elements comprising typical CAD systems.

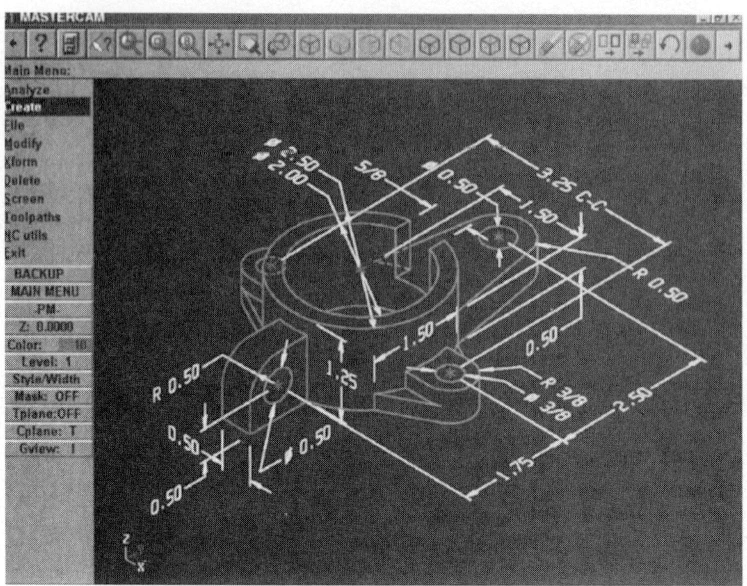

FIGURE 18–10 A wire frame model. (Photo courtesy of CNC Software, Inc.)

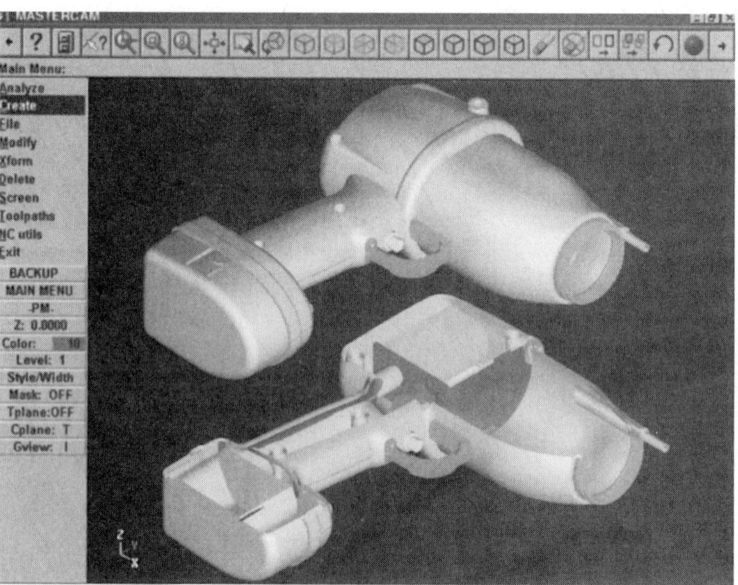

FIGURE 18–11 A wire frame model with surfacing applied. (Photo courtesy of CNC Software, Inc.)

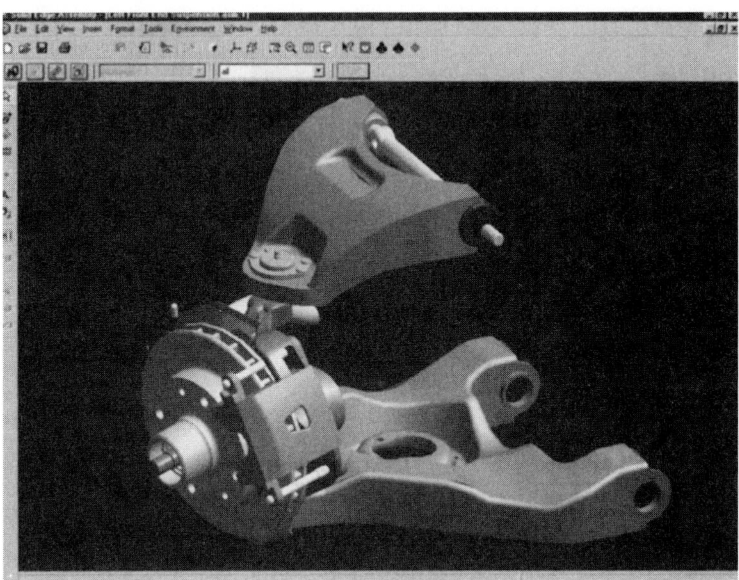

FIGURE 18–12 A solid model. (Photo courtesy of Unigraphics Solutions, Inc.)

Creating a Model by Digitizing

A digitizing system is used in cases where portions of an existing full-scale print are to be entered into the CAD database. The system consists of a sensing tablet and an electronic pointing device or digitizer. The drawing is placed on the tablet. The digitizer is then moved over the drawing, causing the corresponding locations to be entered into the CAD database. In the end, a duplicate drawing is stored in the computer. The accuracy of the final CAD model will depend upon the accuracy of the print. Three-dimensional digitizing systems consist of an arm or wand. These types of systems can be used to convert a full-scale object into a matching three-dimensional CAD model. This approach is useful for modeling such objects as car bodies or irregularly shaped parts.

Creating a Model by Scanning

Scanning is also useful in cases where a complete print is to be converted into a CAD database. The print is placed into the scanner. The system then automatically digitizes the entire drawing and sends this information to the CAD computer. A corresponding CAD print is then created and stored in the computer.

18–10 THE ELEMENTS COMPRISING CAM

CAM system software is designed to generate a complete word address program for machining a part on a particular CNC machine.

Specifically, the following operations are executed offline using CAM software:

1. A CAM file is created.
2. The CAD model is sent to the CAM file for machining.
3. The portion of the model to be machined is identified by the operator. One method of achieving this step is to use the mouse cursor to pick the appropriate model geometry from the display on the CRT. The start point for the machining is also inputted.

4. The tool to be used is identified. Tool speeds and feeds are inputted.
5. A postprocessor file for a particular CNC machine controller is recalled from the computer's memory.
6. The CAM file is retrieved and postprocessed into a word address program file acceptable to the machine controller.
7. The word address program can then be played back for editing. Upon playback the computer can display all resulting tool paths. The tool paths can be zoomed or viewed at any angle.

These steps are illustrated in Figure 18–13.

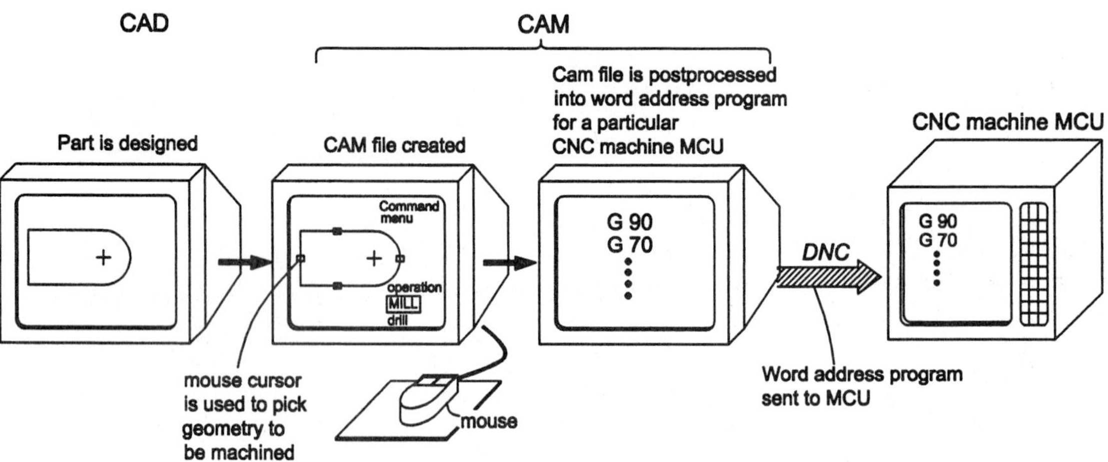

FIGURE 18–13 The CAD/CAM approach to part programming.

CAM system software has undergone a tremendous revolution in the last few years. Today's advances include microcomputer-based systems for two- and three-axis machining. Larger systems consisting of workstations or mainframes can handle more complex four- and five-axis machining operations. A library of postprocessors can be stored in the computer. These include mills, lathes, punch presses, flame, laser, or water jet cutters and wire EDM machines. Surface and solid modeling enables the user to develop tool paths for the most complex part geometries. Some CAM software is designed to automatically avoid clamps and fixtures when developing tool paths. Other software is capable of creating a seamless tool path where multiple surfaces meet or solids intersect. See Figure 18–14. Animation has also reached new heights. Modern CAM software has the capability of showing solid material being removed from a model displaying tool motion in real time. Many different types of CAM software packages are available today. The price of this software has dropped to a point where small shops are finding it cost effective.

The increasing use of CAM software and the proliferation of the microcomputer is gradually rendering obsolete the need for programming in a computer-aided programming language.

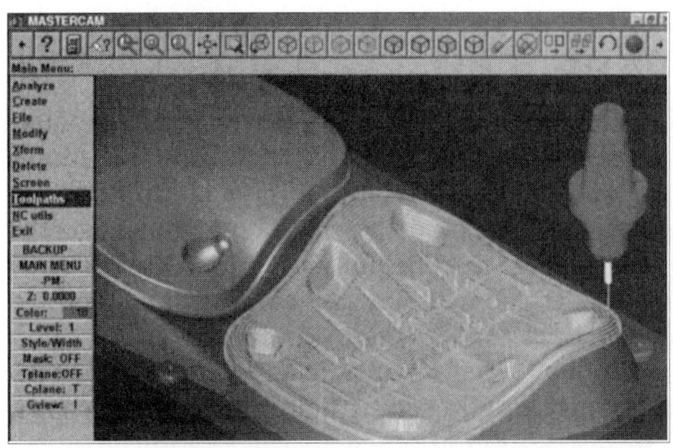

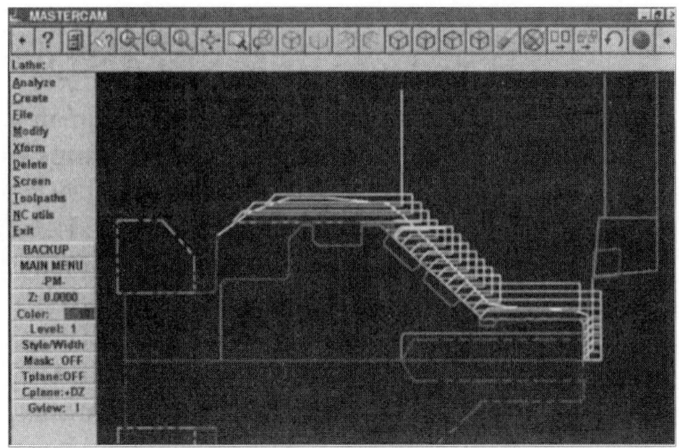

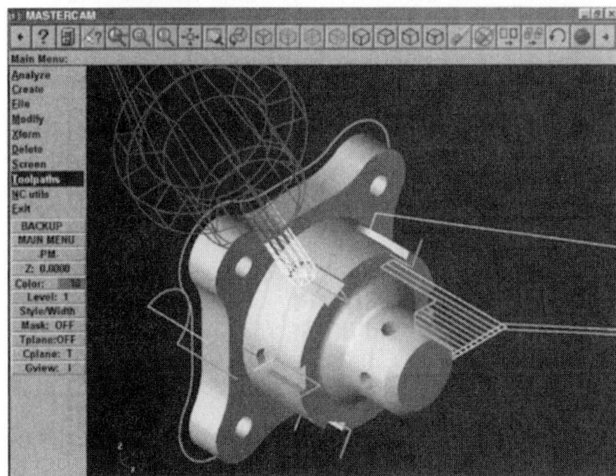

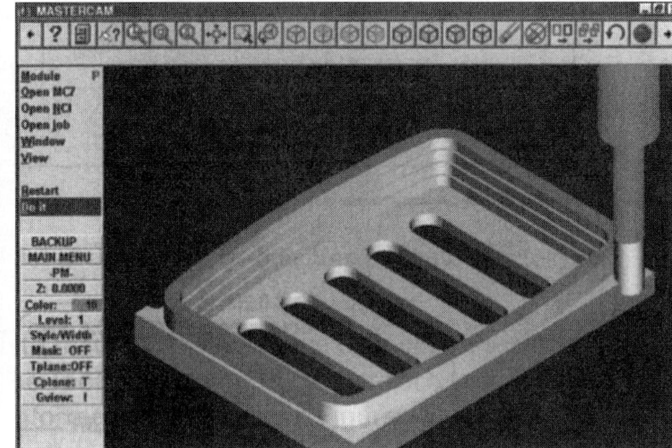

FIGURE 18–14 Computer graphics displays via CAM software. (Photos courtesy of CNC Software, Inc.)

18–11 CREATING A COMPLETE PART PROGRAM USING *MASTERCAM* CNC SOFTWARE

One of the best CNC software packages available today is *Mastercam* from CNC Software, Inc. located at 671 Old Post Road, Tolland, CT 06084. CNC software can also be reached at 860-875-5006 or at their web site *www.mastercam.com*. This software has a short learning time. It presents the user with an easy to follow menu system that works fully with the Windows 95/98/00 operating system. Part geometry can be easily created with *Mastercam's* CAD package. The CAM package enables the operator to quickly select machining operations and cutting tools. It allows the operator to identify the CAD geometry to be selected for a machining operation, then quickly generates the required tool path. The operator simply selects the appropriate postprocessor from the system's library and directs *Mastercam* to generate the corresponding word address part program. A very powerful feature of the software is its ability to verify the part program by animating the entire machining process.

This section will demonstrate how *Mastercam* can be used to create a word address part program.

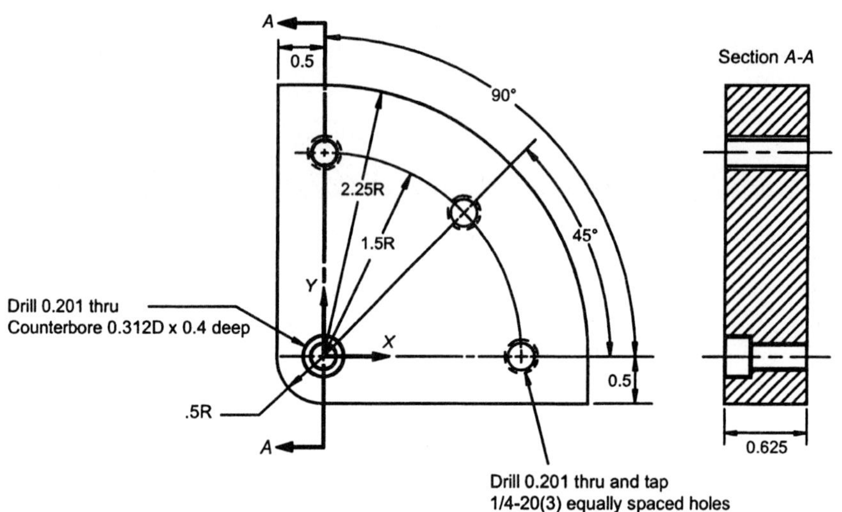

FIGURE 18–15

■ EXAMPLE 18–7

Use *Mastercam* CNC software to create a word address program for directing a vertical CNC machining center to execute the machining operations shown in Figure 18–15.

Tool	Operation	Tooling	Speed (rpm)	Feed (ipm)
1	Drill 0.201*D* (4) places	No. 7 drill	1400	6
2	Tap ¼-20 (3) places	¼-20 tap	400	20
3	Counterbore 0.312*D* × 0.40 deep	5⁄16-2 FLT end mill	800	4
4	Profile mill contour × 0.625 deep	½-2 FLT end mill	1500	7

Notes: 1. Set X_0 Y_0 at the center of 0.201 diameter
2. Z_0 is the top of the part

Step 1 A CAD model of the part is created using *Mastercam* software or it is imported from a compatible CAD software package such as AutoCAD. See Figure 18–16a.

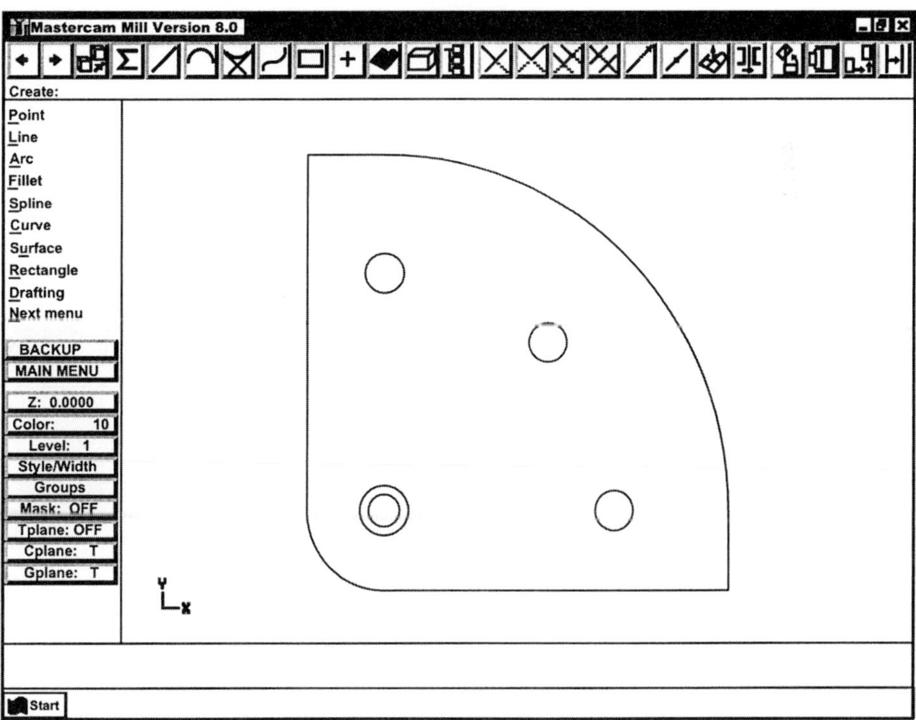

FIGURE 18–16a A CAD model of the part to be machined.

Step 2 The stock to be machined is set up as shown in Figure 18–16b.

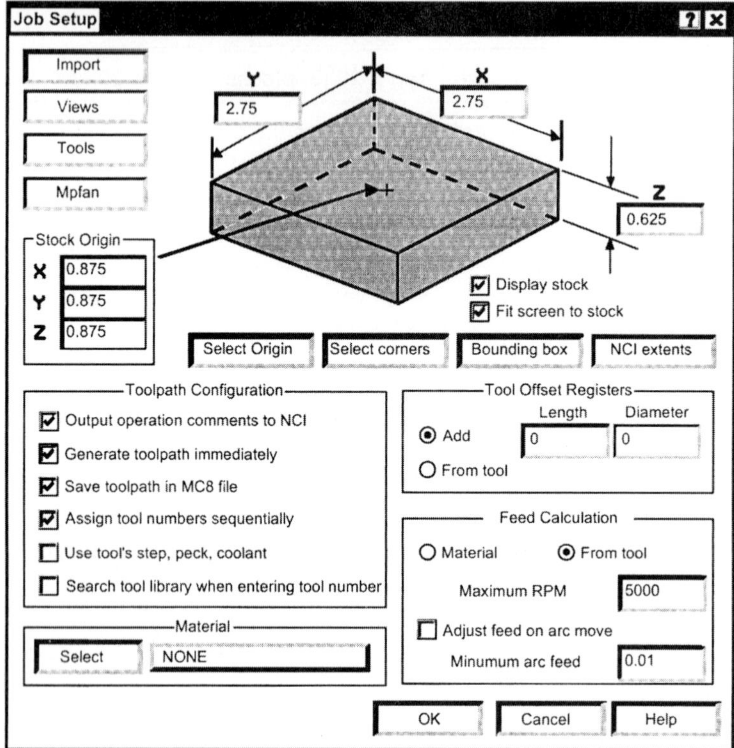

FIGURE 18–16b The stock is set up.

Step 3 The material to be machined is selected. See Figure 18–17.

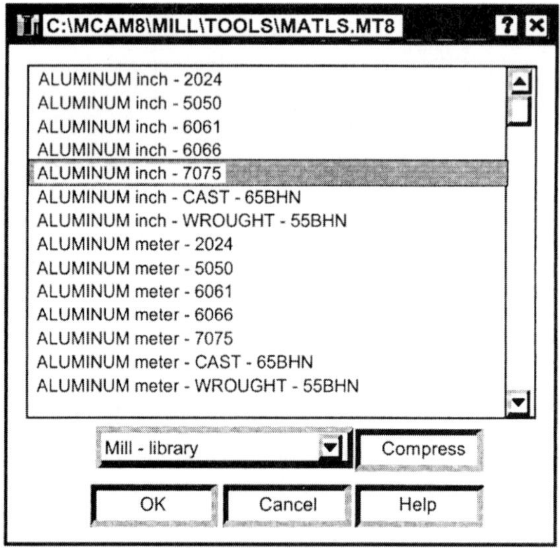

FIGURE 18–17 The material is selected.

Step 4 The operator selects the operation drill and clicks the holes to be drilled, as shown in Figure 18–18.

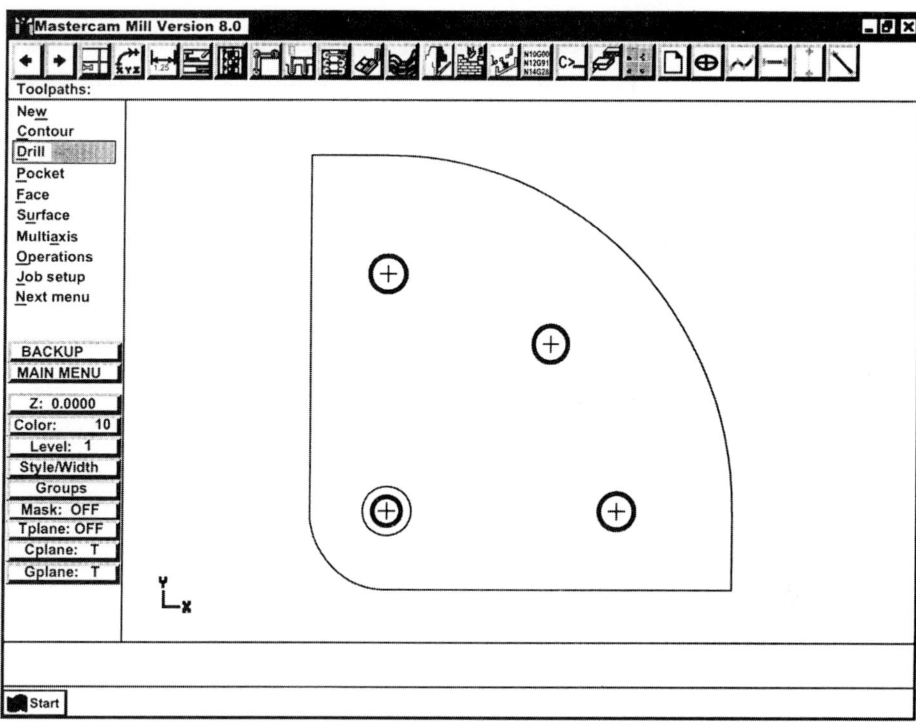

FIGURE 18–18 Clicking the holes to be drilled.

Step 5 A .201*D* drill is selected from the tool library. See Figure 18–19.

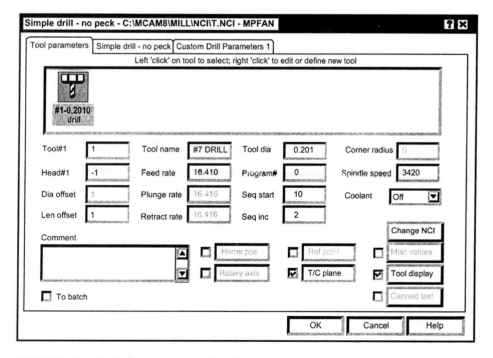

Simple drill - no peck - C:\MCAM8\MILL\NCI\T.NCI - MPFAN

Tool parameters | Simple drill - no peck | Custom Drill Parameters 1

Left 'click' on tool to select; right 'click' to edit or define new tool

#1-0.2010 drill

Tool#1	1	Tool name	#7 DRILL	Tool dia	0.201	Corner radius	0
Head#1	-1	Feed rate	16.410	Program#	0	Spindle speed	3420
Dia offset	1	Plunge rate	16.416	Seq start	10	Coolant	Off
Len offset	1	Retract rate	16.416	Seq inc	2		

Change NCI

Comment

Home pos | Ref point | Misc values

Rotary axis | T/C plane | Tool display

To batch | Canned text

OK | Cancel | Help

FIGURE 18–19 Selecting a 0.201*D* drill from the tool library.

Step 6 The operator enters the parameters for the drilling operation including the drill depth (-0.625), as shown in Figure 18–20.

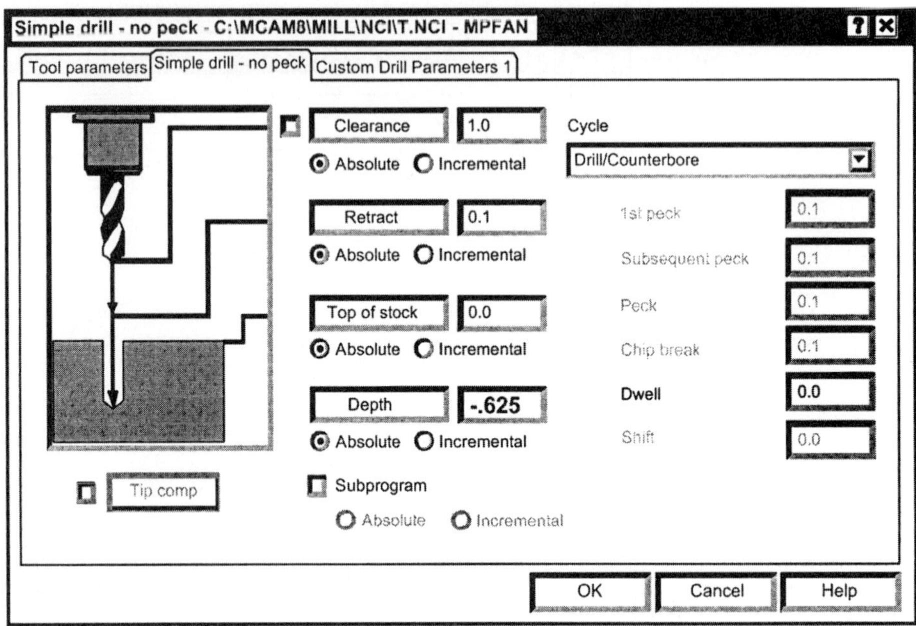

FIGURE 18–20 Entering the drilling operation parameters.

Step 7 The operator selects the operation drill and clicks the holes to be tapped. See Figure 18–21.

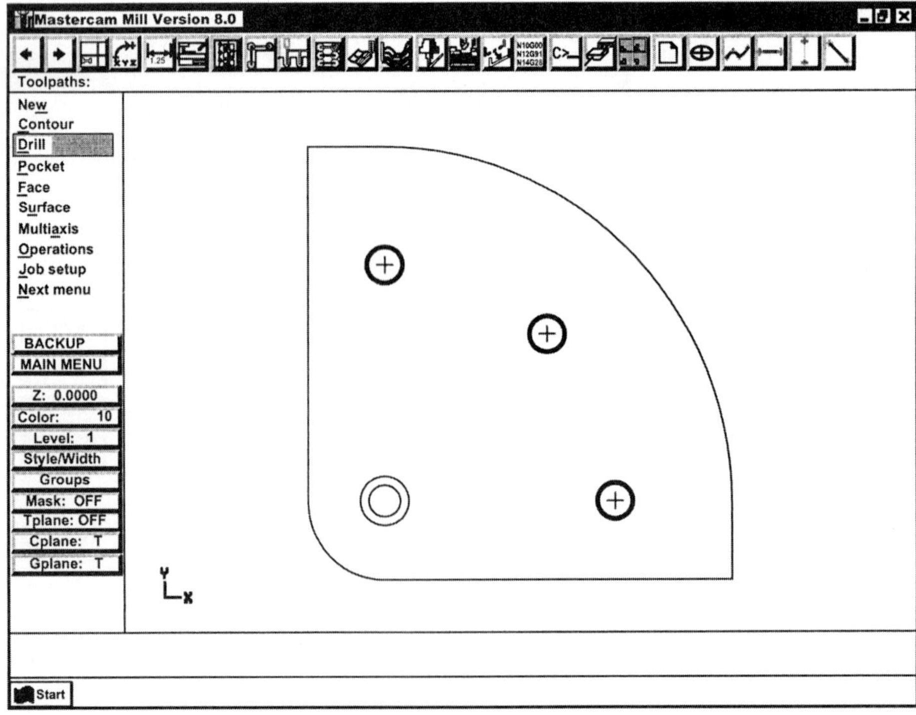

FIGURE 18–21 Clicking the holes to be tapped.

Step 8 A .2500-20in tap is selected from the tool library. See Figure 18–22.

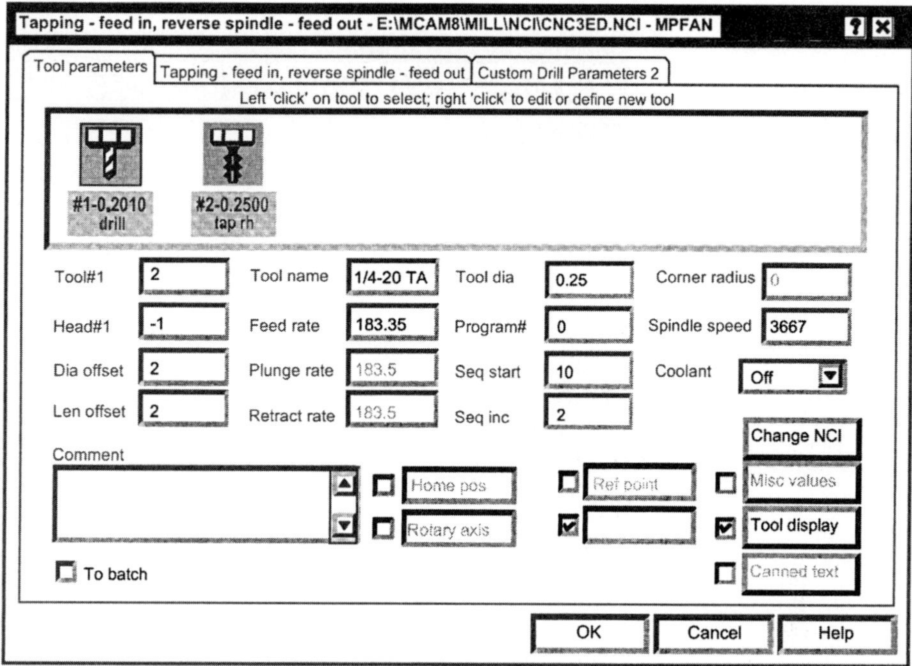

FIGURE 18–22 Selecting a 0.2500-20in tap from the tool library.

Step 9 The operator enters the parameters for the tapping operation including the tap depth (-0.625), as shown in Figure 18–23.

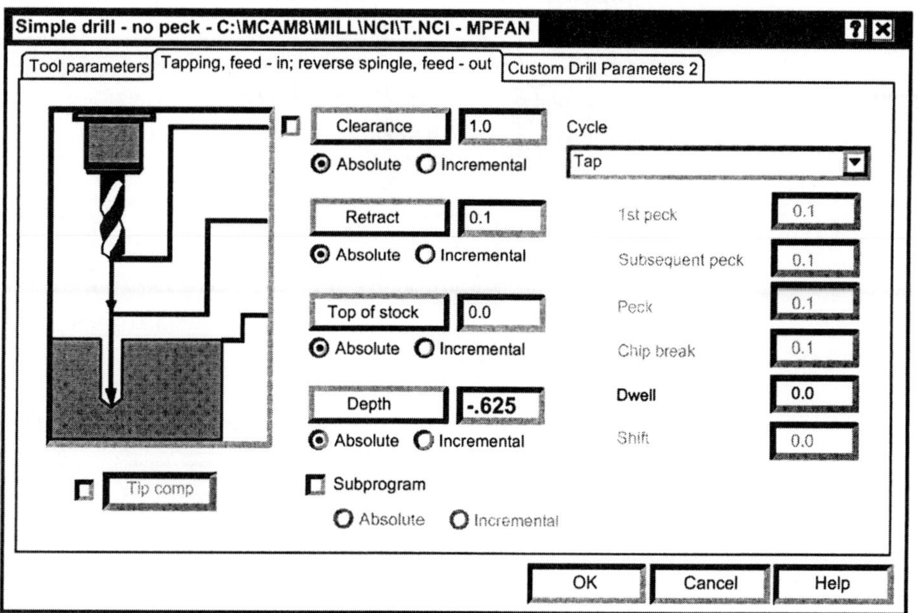

FIGURE 18–23 Entering the tapping operation parameters.

Step 10 The operator selects the operation circle mill and clicks the hole to be milled. See Figure 18–24.

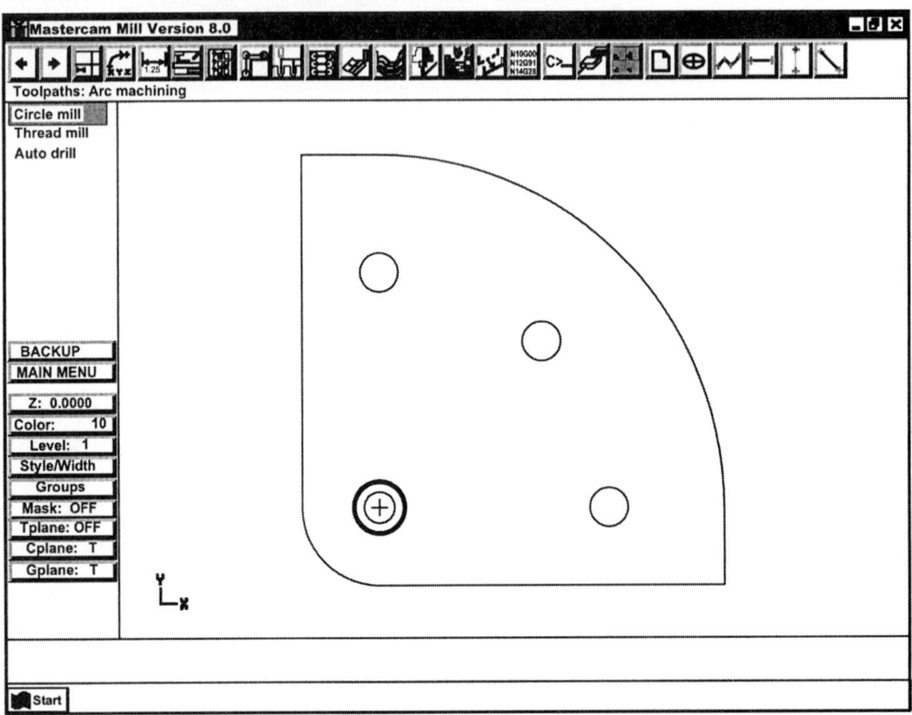

FIGURE 18–24 Clicking the hole to be circle milled.

Step 11 A 0.312-2FLT end mill is selected from the tool library. See Figure 18–25.

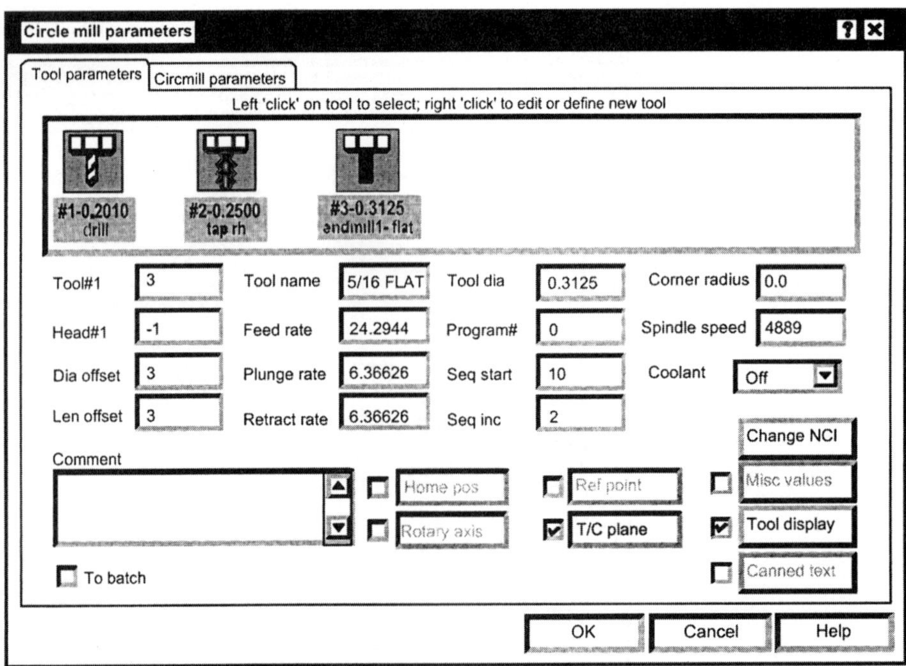

FIGURE 18–25 Selecting a 0.312-2FLT end mill from the tool library.

Step 12 The operator enters the parameters for the circle mill operation including the mill depth (-0.40) as shown in Figure 18–26.

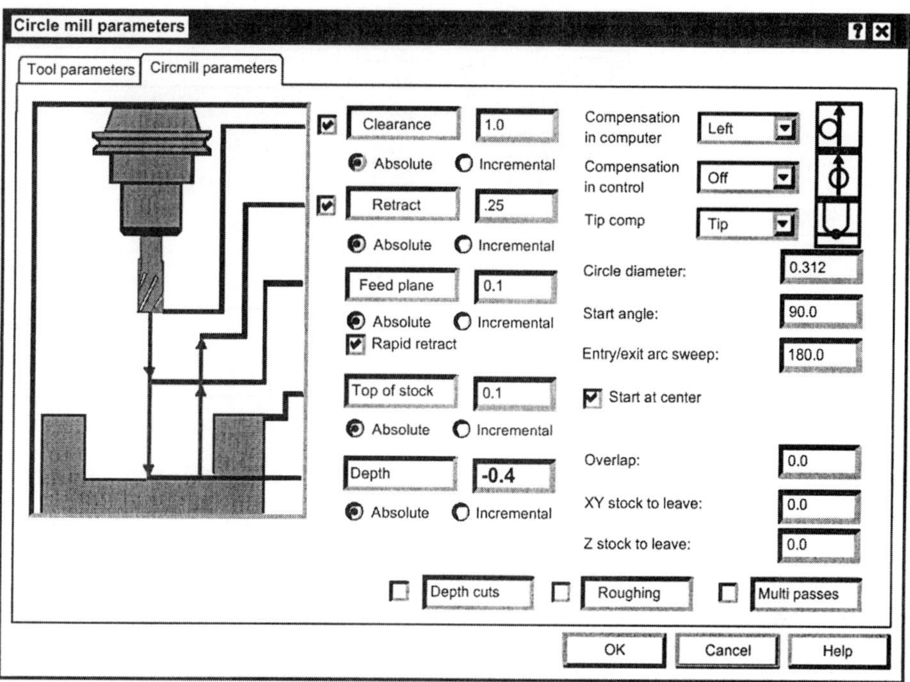

FIGURE 18–26 Entering the circle mill operation parameters.

Step 13 The operator selects the operation contour and clicks geometry to be milled. Note the system displays an arrow to indicate the direction in which milling will occur. See Figure 18–27.

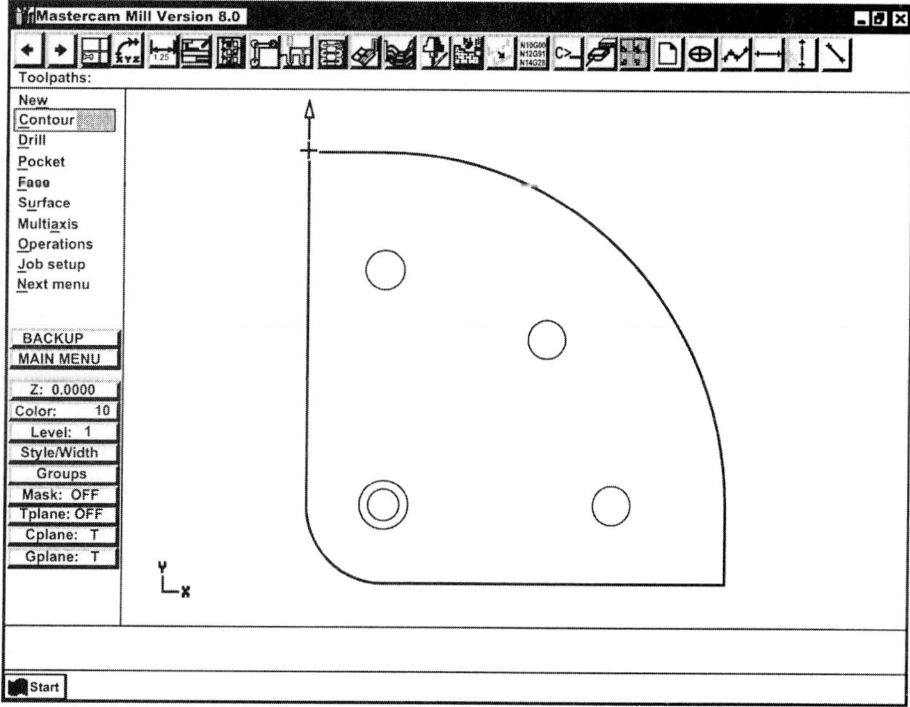

FIGURE 18–27 Clicking the geometry to be contour milled.

Step 14 A 0.5000-2FLT end mill is selected from the tool library. See Figure 18–28.

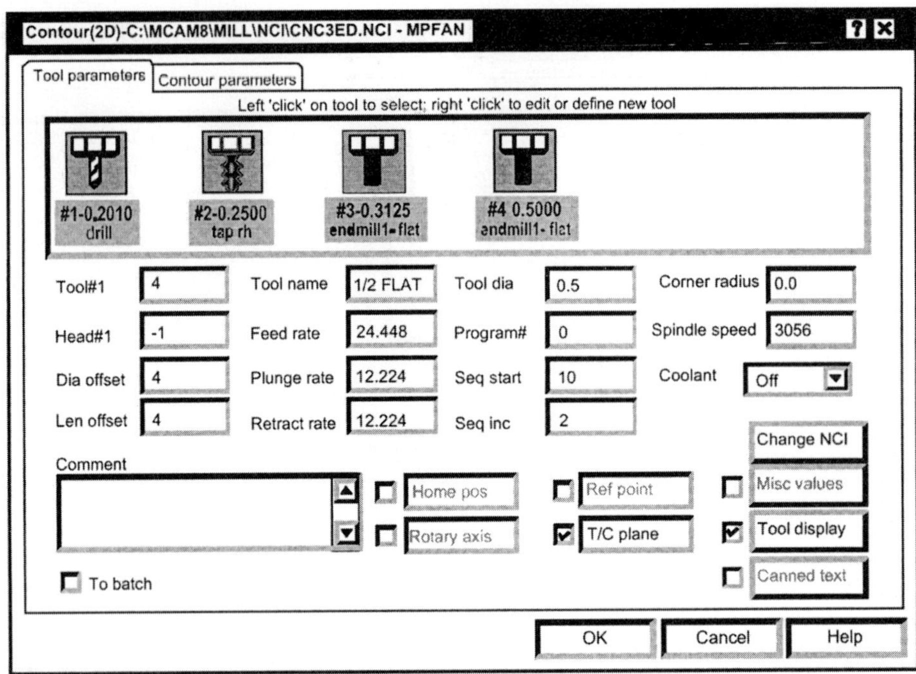

FIGURE 18–28 Selecting a 0.5000-2FLT end mill from the tool library.

Step 15 The operator enters the parameters for the contour mill operation including the mill depth (-0.625), as shown in Figure 18–29a.

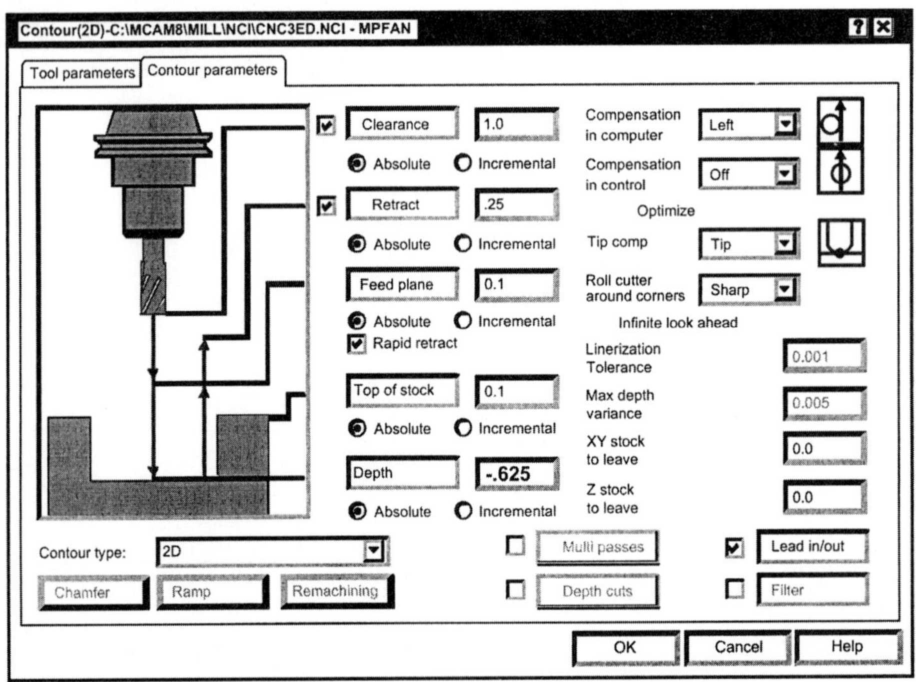

FIGURE 18–29a Entering the contour mill operation parameters.

Step 16 The operations Manager dialog box is opened. All the machining operations listed are selected for postprocessing by clicking the ***Select All*** button. The operator then clicks the ***Post*** button and selects an appropriate postprocessor (FANUC, etc). Refer to Figure 18–29b.

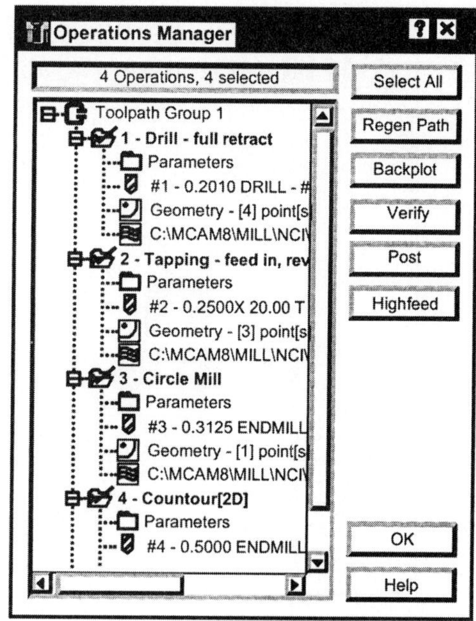

FIGURE 18–29b Specifying the machining operations to be postprocessed and selecting a postprocessor.

Step 17 The system is directed to create the word address part program. See Figure 18–30.

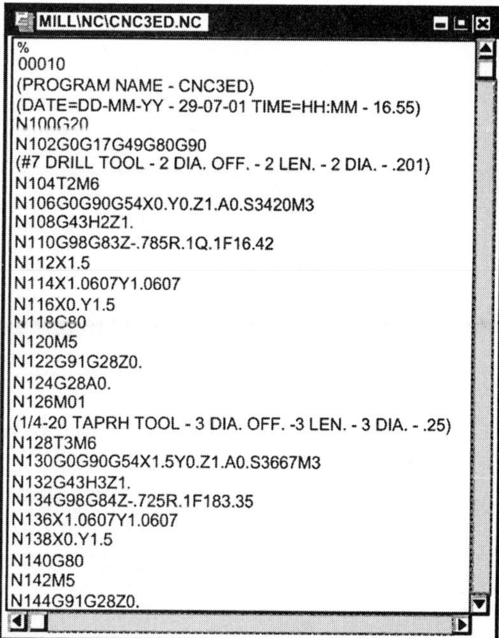


```
%
O0010
(PROGRAM NAME - CNC3ED)
(DATE=DD-MM-YY - 29-07-01 TIME=HH:MM - 16.55)
N100G20
N102G0G17G49G80G90
(#7 DRILL TOOL - 2 DIA. OFF. - 2 LEN. - 2 DIA. - .201)
N104T2M6
N106G0G90G54X0.Y0.Z1.A0.S3420M3
N108G43H2Z1.
N110G98G83Z-.785R.1Q.1F16.42
N112X1.5
N114X1.0607Y1.0607
N116X0.Y1.5
N118G80
N120M5
N122G91G28Z0.
N124G28A0.
N126M01
(1/4-20 TAPRH TOOL - 3 DIA. OFF. -3 LEN. - 3 DIA. - .25)
N128T3M6
N130G0G90G54X1.5Y0.Z1.A0.S3667M3
N132G43H3Z1.
N134G98G84Z-.725R.1F183.35
N136X1.0607Y1.0607
N138X0.Y1.5
N140G80
N142M5
N144G91G28Z0.
```

FIGURE 18–30 The word address part program created by *Mastercam*.

Step 18 The operator clicks the ***Verify*** button shown in Figure 18–29b and the ***Play*** button illustrated in Figure 18–31. *Mastercam* will then animate all the machining operations selected in ***Step 17***.

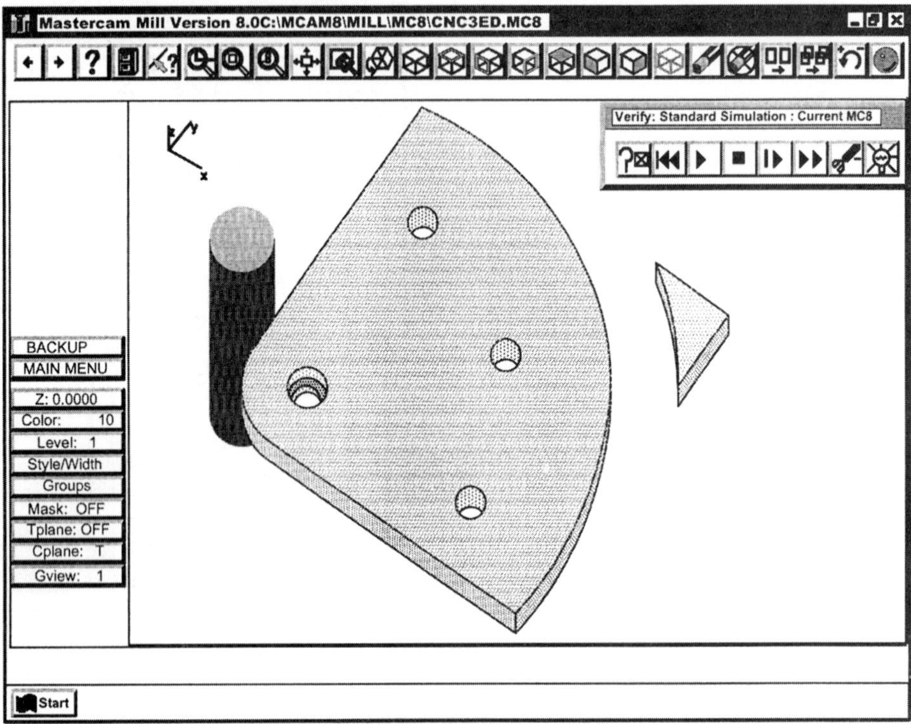

FIGURE 18–31 The machining animation displayed by *Mastercam*.

18–12 CHAPTER SUMMARY

The following key concepts were discussed in this chapter:

1. Off-line programming involves the use of a remote computer to write and store a program in word address. This technique eliminates the need for keying in the program at the machine control unit.
2. Computer-aided programming languages enhance the technique of off-line programming. A computer-aided programming language enables the programmer to define the tool path by using English-like commands. It eliminates the need for doing extensive trigonometric computations. A program written in a computer-aided programming language is universal, too. It can be automatically translated (postprocessed) into a word address program for several different types of controllers.

3. A program written in a computer-aided programming language has four main parts:
   - Setup parameter statements
   - Geometric definition statements
   - Tool motion statements
   - Miscellaneous shut-down statements
4. The most important programming language still in use today is APT (automatic programming of tools).
5. Modern off-line programming utilizes computer-aided design (CAD)/computer-aided manufacturing (CAM) software.
6. CAD allows the designer to create a part shape (model) with the aid of a computer. CAM enables the programmer to use the computer to generate a word address program for machining the CAD model. CAM involves the use of computer graphics for creating tool paths on a model.
7. CAD/CAM systems can display the tool motions resulting from a part program. This feature allows for quick and accurate off-line checking of the program. Programs can be edited at the computer as required.

REVIEW EXERCISES

18.1. How does a computer-aided programming language differ from word address (G, M code) programming?

18.2. What does APT stand for?

18.3. Identify which computer-aided programming language should be used in each case:

(a) For programming two- and three-axis machining. The computer is to enter into an interactive conversation with the user during program debugging.

(b) For programming complex surfaces.

18.4. Write APT statements to define the points shown in Figure 18–32.

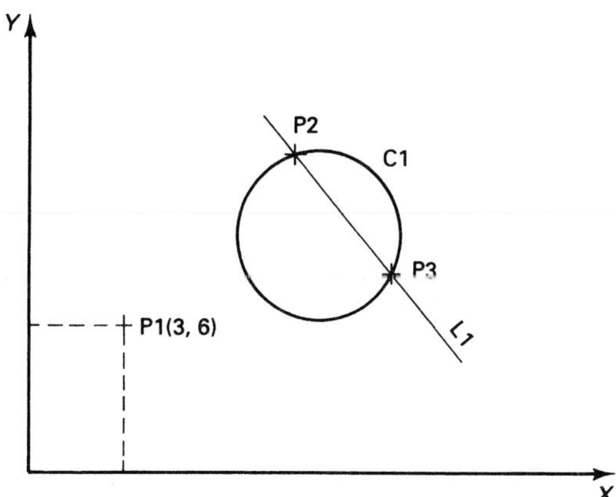

FIGURE 18–32

18.5. Write APT statements to define the lines shown in Figure 18–33.

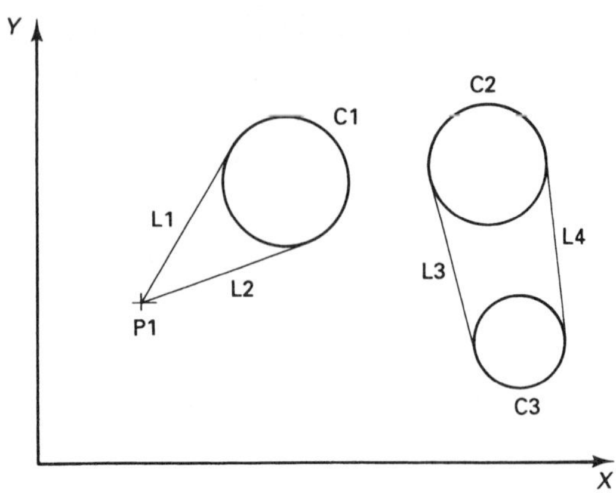

FIGURE 18–33

18.6. Write APT statements to define the circles given in Figure 18–34.

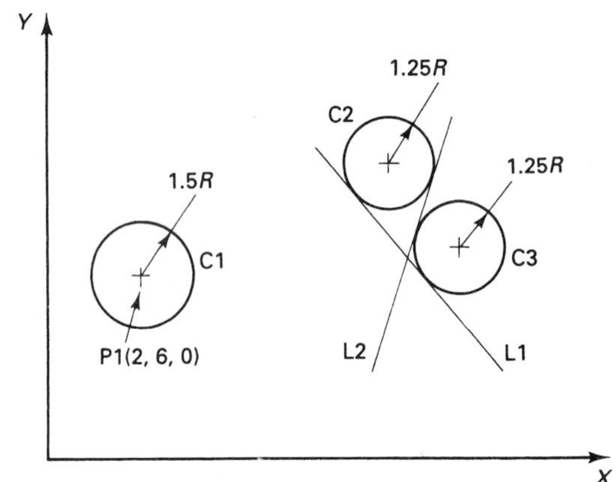

FIGURE 18–34

18.7. Name three surfaces in the APT language used to control tool motion.
18.8. What does CAD/CAM stand for?
18.9. Name six advancements offered by CAD/CAM.
18.10. Give a definition of CAD.
18.11. Name four major components of a typical CAD system.
18.12. How do the following CAD models differ?
   (a) Wire frame model
   (b) Surface model
   (c) Solid model
18.13. What systems are used to duplicate an object into a CAD model? Explain how they work.
18.14. Identify six steps used in CAM to create a word address program.
18.15. Describe the steps involved in generating a word address program using *Mastercam* software.

CHAPTER

19

Elements of the Computer-Controlled Factory

19–1 CHAPTER OBJECTIVES

At the conclusion of this chapter you will be able to

1. Explain the computer-integrated manufacturing (CIM) approach to manufacturing.
2. Identify the key components of a CIM installation.
3. State how group technology and just-in-time improve the efficiency of operations within CIM.
4. Explain what comprises a flexible manufacturing system.
5. Describe a flexible manufacturing cell.
6. Describe a local area network.
7. State the need and opportunities for training.

19–2 INTRODUCTION

CIM is becoming a key tool for competitive manufacturing in today's global market. Therefore, this chapter introduces the reader to CIM and its workings. The chapter begins with a general description of CIM and identifies the components of a typical CIM installation. The advantages and disadvantages of such a system are outlined.

Important operating strategies such as group technology and just-in-time are explained. A key element of CIM, the flexible manufacturing system (FMS), is described and illustrated. An important FMS subsystem, the flexible manufacturing cell, is also discussed.

Communication within CIM is afforded by networking. The concept of a local area network (LAN) is explained. Important LANs such as Ethernet, MAP, and TOP are explored. The chapter ends with a discussion of the need for training and the many training opportunities available today.

19–3 COMPUTER-INTEGRATED MANUFACTURING

Present and future trends point toward an increase in the use of computers and a decrease in manual methods. The driving force in this trend is the need to utilize a central computer for storing and retrieving information. Information stored in the central computer is said to reside in a common database. Computer-integrated manufacturing (CIM) is a system whereby all aspects of a company's operation are computerized and tied to a common database. An

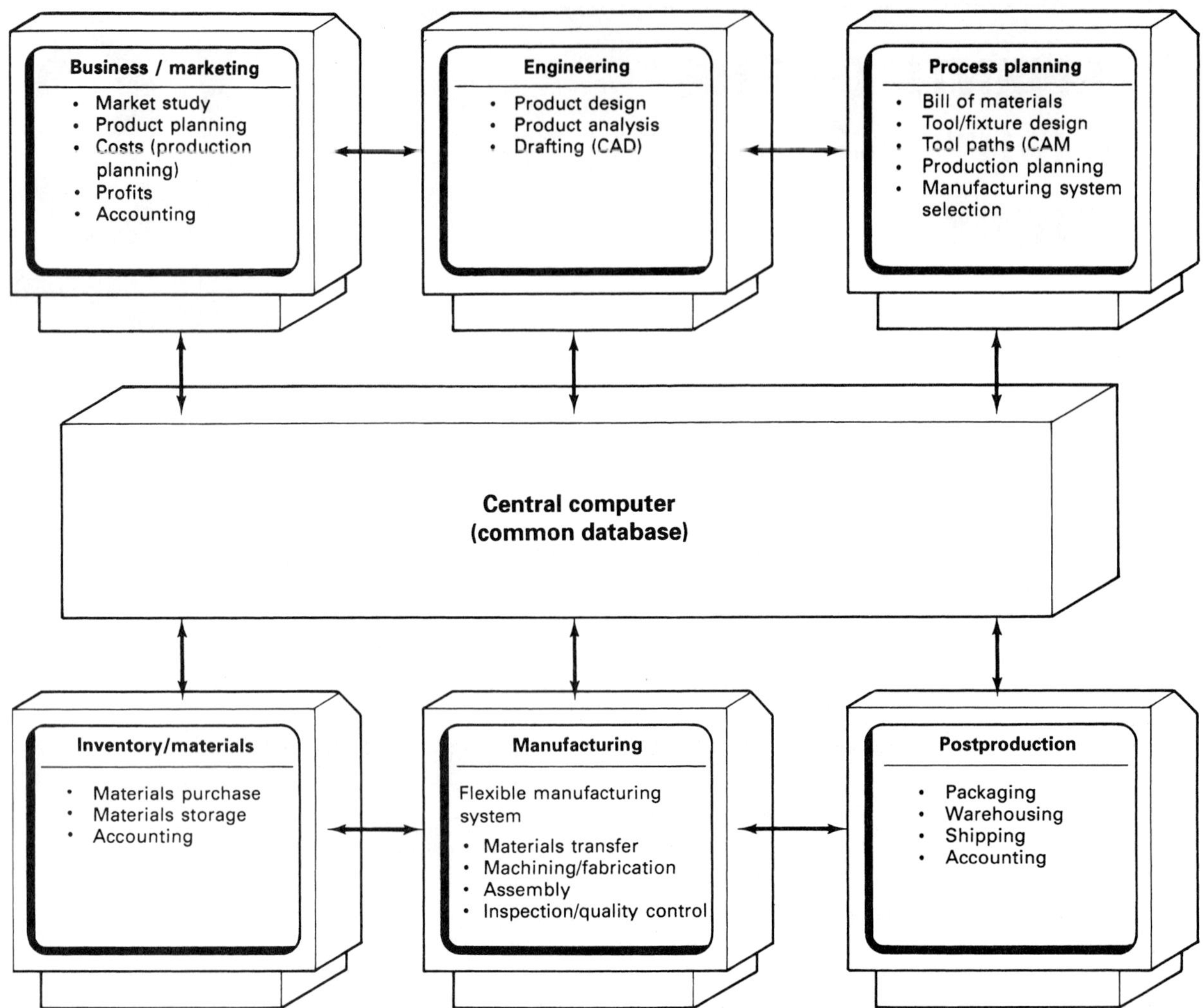

FIGURE 19–1 A configuration for a CIM system.

important element of CIM is CAD/CAM and DNC. A CIM system for unifying business, engineering, process planning, inventory, manufacturing, and post-production is shown in Figure 19–1.

When fully implemented, CIM will offer the following advantages:

- Increased integration and coordination of all company disciplines
- Increased productivity
- Decreased costs of operation
- Improved quality
- Reduced scrap
- Improved flow-through
- Reduced inventory
- Around-the-clock operation with a minimum of operator intervention

CIM also poses the following challenges:

- Substantial installation and start-up costs
- Training personnel in the use of CIM equipment (hardware and software)
- Dealing with proprietary systems and software
- Maintaining system hardware and software

- Implementing CIM fully to make it cost effective
- Operating a CIM installation for long periods of time
- Planning CIM operations

CIM should be looked upon as an ongoing development. Smaller shops are beginning to put together some of its components such as CAD/CAM and computerized accounting. Larger corporations have made substantial gains in adding many more components and some have even achieved full CIM implementation.

19–4 OPERATIONAL STRATEGIES WITHIN CIM

Many types of strategies have been formulated to increase the efficiency of operations within CIM. Chief among them are group technology (GT) and just-in-time (JIT).

Group Technology

Group technology is a system for reducing needless duplication of effort. Within this system, parts with similar characteristics are categorized into groups or families. Characteristics could include similar geometric shape (round, square, etc.) and dimensions (long, short, etc.). Additional characteristics of interest to manufacturing would be the type of manufacturing equipment and fixturing used, machining techniques, lot sizes, and manufacturing cost. A special code is used to identify the part as belonging to a specific group. Each element in the code specifies an important part characteristic. Thus, complete information describing an entire family of parts can be accessed simply by entering the proper GT code. This approach can save CAD operators substantial time in creating a design for a specific application. Previous designs for similar applications can be recalled and studied upon entry of the GT code. Similarly, previous production process solutions and manufacturing procedures for similar parts can be quickly accessed. This information can aid in speeding up these important CIM operations. Marketing and business can use the GT code to readily obtain information concerning cost versus profit for similar families of parts.

Just-in-Time

Just-in-time is a way of operating so that all unnecessary waste and product nonvalue elements are eliminated. It has been used to slash inventory waste and shape production techniques within CIM. A JIT facility has a minimum of materials stored in on-site warehouses. Instead, materials are ordered and delivered from a vendor as needed for a production run. Thus, the materials arrive at the facility "just in time" for manufacturing. The part is manufactured using a minimum number of machining operations. It is routed to as few machines as possible. Each worker on the production line acts as an inspector. Any worker can stop production if a defect is found. JIT encourages workers to form quality groups or "quality circles." In this way suggestions for increasing value and reducing waste can be discussed and implemented.

19–5 FLEXIBLE MANUFACTURING SYSTEMS

A flexible manufacturing system (FMS) is capable of automatically manufacturing complete families of related parts. FMS is designed to work closely with group technology and just-in-time.

The main distinguishing feature of a flexible manufacturing system is central computer control over its operations. These systems offer such benefits as

- Automatic production of many different types of parts in random lot sizes
- Reduction in the number of setups and routing to produce parts
- Reduction in lead time
- Production of parts having consistent high quality
- Reduction in shop floor space
- Around-the-clock unstaffed production of parts

Flexible manufacturing systems are composed of three main elements:

- A central computer control system
- An automated materials handling system
- A group of machine tools

Additionally, these systems may have a coordinate measuring machine for automated inspection of parts and a parts cleaning station. A typical flexible manufacturing system is shown in Figure 19–2.

FIGURE 19–2 A flexible manufacturing system. (Photo courtesy of Cincinnati Milacron, Inc.)

19–6 IMPORTANT ELEMENTS OF FLEXIBLE MANUFACTURING SYSTEMS

Central Computer Control System

The main CIM computer controls the operations of the flexible manufacturing system. It oversees such operations as

- Real-time routing of materials and parts into and out of the FMS
- DNC downloading of programs into the machine tool controllers
- Production scheduling
- Tool delivery and tool maintenance

See Figure 19–3.

FIGURE 19–3 Computer display of FMS operations. (Photo courtesy of Amatrol, Inc.)

Automated Materials Handling System

Two main types of systems are involved in moving materials and part elements into and out of the FMS. See Figure 19–4. The first utilizes a conveyor mechanism. The computer controls the conveyor's movements so that each element is moved at the proper time. The other employs transfer vehicles to route materials, parts, or tools from machine to machine. Transfer vehicle types include

Wire-guided vehicles: These vehicles have a sensor that follows a wire laid out on the shop floor. This provides for medium transfer flexibility.

Automatic guided vehicles (AGV): An on-board computer controls the path followed by an AGV. The path program specifies the path to be followed. The program can be changed, thus affording maximum transport flexibility.

Rail-guided vehicles: These delivery systems use a rail guide to rigidly control the vehicle's path. They provide the least amount of flexibility.

FIGURE 19–4 [(a) Photo courtesy of Amatrol, Inc. (b) Photo courtesy of Cincinnati Milacron, Inc.]

19–7 FLEXIBLE MANUFACTURING CELLS

A flexible manufacturing cell (FMC) can be considered as a flexible manufacturing subsystem. The following differences exist between the FMC and the FMS:

1. An FMC is not under the direct control of the central computer. Instead, instructions from the central computer are passed to the cell controller.
2. The cell is limited in the number of part families it can manufacture.

The following elements are normally found in an FMC:

- Cell controller
- Programmable logic controller (PLC)
- More than one machine tool
- A materials handling device (robot or pallet)

The FMC executes fixed machining operations with parts flowing sequentially between operations. An FMC is shown in Figure 19–5.

19–8 IMPORTANT ELEMENTS OF FLEXIBLE MANUFACTURING CELLS

Cell Controller

The cell controller consists of a local computer (minicomputer or microcomputer) and associated cell control software. The cell controller monitors and coordinates all the units in the cell. It stores and transmits such information as process planning programs, CNC programs, production statistics, messages, counters, and position coordinates. The cell controller passes sequencing information for a job to the programmable logic controller (PLC). The PLC in turn transmits the proper on/off signals to the cell units (robot, tool changer, CNC machine tool, etc.). In this way the PLC acts as a programmable switching mechanism or traffic controller for the cell units. See Figure 19–6.

Materials Handling Systems

It is common practice to use robots to load and unload materials, parts, and tools to the CNC machine tools in the cell. Special fixtures on the materials routing vehicles allow the robot easy access. Similarly, fixtures on the CNC machine table provide for automatic placement and clamping after loading. See Figure 19–7.

FIGURE 19–5 A flexible manufacturing cell. (Photo courtesy of KT-Swazey, Milwaukee.)

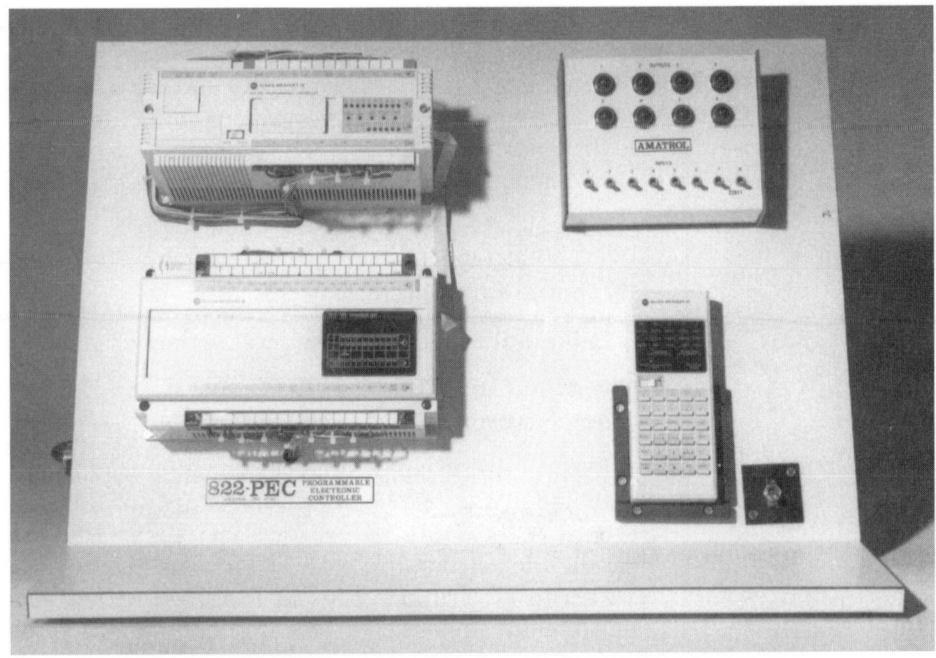

FIGURE 19–6 Programmable logic controllers. (Photo courtesy of Amatrol, Inc.)

FIGURE 19–7 A robot loading a part into the CNC machine. (Photo courtesy of Amatrol, Inc.)

19–9 NETWORKING WITHIN CIM

The equipment in each component of CIM (business, engineering, manufacturing, etc.) must be connected so that data can flow between units. A local area network (LAN) is a high-speed data communications system that operates between the units in a local area. The units can be connected in many different patterns or topologies. The most common LAN topologies are bus, star, and ring. See Figure 19–8.

For business applications the most widely used LANs include

- IBM's token ring
- Novell's Network
- Technical Office Protocol (TOP)

For industrial applications the most widely used LANs include

- Xerox and Digital Equipment Corporation's Ethernet
- Manufacturing Automation Protocol (MAP)

Each network has a protocol or set of conditions for communication between units. Many LANs have proprietary protocols. Among these are Systems Network Architecture (SNA) for IBM's token ring and Transmission Control Protocol/Interconnect Protocol (TCP/IP) for Ethernet. Communication problems have occurred when one LAN is interfaced with another. The international organization for standards (ISO) has addressed this problem by specifying a standard or open protocol that all LANs should follow. The trend today is toward the ISO seven-layer open protocol.

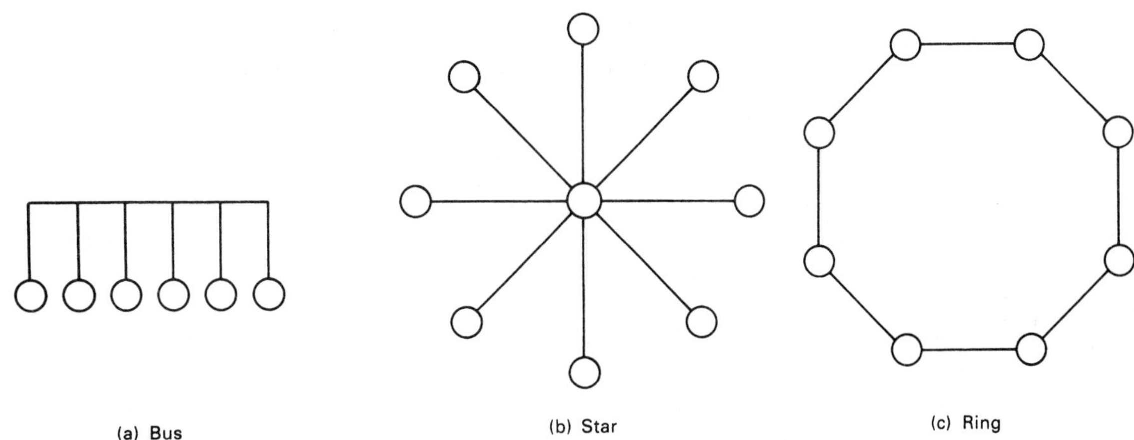

FIGURE 19–8 LAN connection topologies.

Manufacturing Automation Protocol (MAP) is a LAN specifically designed to offer the following benefits:

- Follows the ISO open protocol standards
- Networks the islands of automation (FMS, FMC) in manufacturing

MAP was developed by General Motors and is fast becoming the standard LAN for manufacturing.

A special LAN for the office environment has also been developed by the Boeing Corporation. It is called Technical Office Protocol (TOP). TOP offers the following advantages:

- Follows the ISO open protocol standards
- Networks the islands of automation within business and marketing
- Provides a data highway to all the other LANs within a CIM facility

19–10 NEED FOR TRAINING

CIM represents substantial investments in finance, time, and training. Its components (hardware and software) are often costly. A considerable amount of time must be spent in configuration planning, unit selection, and unit integration. It cannot be stressed enough that the last investment, training, may well be the most important. CIM systems are complex and require teams of highly skilled personnel for their successful operation. The payoff is that this new technology will provide the edge that companies will need in order to compete in today's and tomorrow's global market.

Many training opportunities exist for those who wish to gear up for CIM. They include

- In-house apprenticeship programs
- Co-op education programs between companies and universities
- Vocational and junior college training programs
- Community college programs (associate degree study, continuing education study, and seminars)
- Four-year technical college programs
- Four-year engineering programs
- Grant-funded training
- Self-study programs (self-paced training materials, tapes, and videos)

The reader is encouraged to aggressively pursue training on a regular basis.

19-11 CHAPTER SUMMARY

The following key concepts were discussed in this chapter:

1. CIM is a system of operating so that a company's functions are computerized and tied to a common database.
2. Group technology is a system for storing and accessing important information on similar parts.
3. Just-in-time is a way of operating so that all unnecessary waste and all product nonvalue elements are eliminated.
4. A flexible manufacturing system is an important component of CIM. It is capable of automatically manufacturing complete families of related parts. It operates under the direct control of the central CIM computer.
5. A flexible manufacturing cell is a flexible manufacturing subsystem. It is composed of a cell controller, materials handling robots, and CNC machine tools. It performs its operations in a sequential manner.
6. A local area network provides for high-speed data communication between the units in a local area.
7. MAP is an open protocol LAN for tying together the islands of automation in manufacturing.
8. TOP is an open protocol LAN for tying together the islands of automation in business and finance.
9. Training is the key to successful implementation of CIM.

REVIEW EXERCISES

19.1. What are two central ideas behind CIM?
19.2. Identify five major advantages of CIM.
19.3. Identify five challenges of CIM.
19.4. How can group technology make CIM operations more efficient?
19.5. What are two benefits of implementing just-in-time?
19.6. What is a flexible manufacturing system (FMS)?
19.7. Identify three main elements of an FMS.
19.8. List four benefits of using an FMS.
19.9. What is a flexible manufacturing cell?
19.10. Give two differences between the FMC and FMS.
19.11. Identify four elements of an FMC.
19.12. Explain the function of the following units:
  (a) Cell controller
  (b) Programmable logic controller (PLC)
19.13. What is a local area network (LAN)?
19.14. Describe three types of LAN topologies.
19.15. List two LANs for each of the applications:
  (a) Business
  (b) Industrial
19.16. Explain the significance of LAN protocol.
19.17. Identify the benefits of using the following LANs:
  (a) MAP
  (b) TOP

CHAPTER 20

VERIFYING PART PROGRAMS

20-1 CHAPTER OBJECTIVES

At the conclusion of this chapter you will be able to

1. Write, edit, and save word address CNC programs using Predator's CNC Editor.
2. Use the Predator Virtual CNC program to specify the CNC machine tool, the stock, and the cutting tools to be used for machining a part.
3. Direct the Virtual CNC to animate the machining caused by an inputted word address CNC program.
4. Inspect a machined part via the Virtual CNC program.

20-2 INTRODUCTION

The reader is instructed in the use of the Predator simulation software included with this text. The coverage includes specific off-line simulations of *drilling* and *milling* part programming examples in the text.

An overview of the entire simulation process is outlined. This is followed by a discussion of how word address part programs are created via the Predator CNC Editor. The step-by-step process of verifying part programs via the Predator Virtual CNC simulator is then presented.

The reader is encouraged to use the simulator as an aid in creating the word address programs that are required in the *drilling* and *milling* exercises.

20-3 PREDATOR SIMULATION SOFTWARE

Predator Software, Inc produces the leading state of the art verification and machining simulator available in the industry. Predator is located at 8835 SW Canyon Lane, Suite 300, Portland, OR 97225. Predator can also be reached at 503-292-7151 or at their web site *www.predator-software.com.*

This software produces a realistic machining animation as it processes an inputted word address part program. Any errors in the program are flagged. The finished part can be viewed at any angle, sectioned, and checked for dimensional correctness. The software is capable of producing machining simulations for 3-, 4-, or 5-axis milling machines and 2-axis turning machines. It supports a wide variety of CNC controllers and machine tool configurations. Simple or complex stock shapes can be created. A library of standard cutting tools is provided. The user also has the option of creating custom tools and tool holder shapes for milling as well as turning operations.

20-4 SYSTEM REQUIREMENTS

Predator Virtual CNC is a 32-bit verification and simulation package. It operates on Windows 98 with Explorer 5.0 or higher, Windows 2000, and Windows NT4.0.

The following minimum system hardware must be installed:

WINDOWS 98+Explorer 5.0 or higher; Windows 2000
- Pentium or Pentium-Pro processor
- 800 × 600 display resolution with 256 colors
- 32 Mbytes of RAM
- 40 Mbytes of hard disk space
- Parallel printer port
- Mouse

WINDOWS NT
- Pentium or Pentium-Pro processor
- 800 × 600 display resolution with 256 colors
- 64 Mbytes of RAM
- 40 Mbytes of hard disk space
- Parallel printer port
- Mouse

20-5 CONVENTIONS USED IN THIS CHAPTER

The following conventions are used throughout this chapter.

DISPLAY	MEANING	PICTORIAL
Enter	Directs the operator to *press* the Enter key	Enter
Click (N)	Means to move the mouse cursor to position (N) and press the <u>*LEFT*</u> mouse button	position (N) Mouse cursor Depress **left** mouse button after moving the cursor to position (N) Monitor Mouse
Bold	Commands to be typed at the keyboard appear in **bold**	

20–6 INSTALLATION

- Insert the CD provided at the back of this text into the CD-ROM drive and follow the instructions as displayed by the install software
- At the Windows desktop click ① on the Predator Virtual CNC icon 

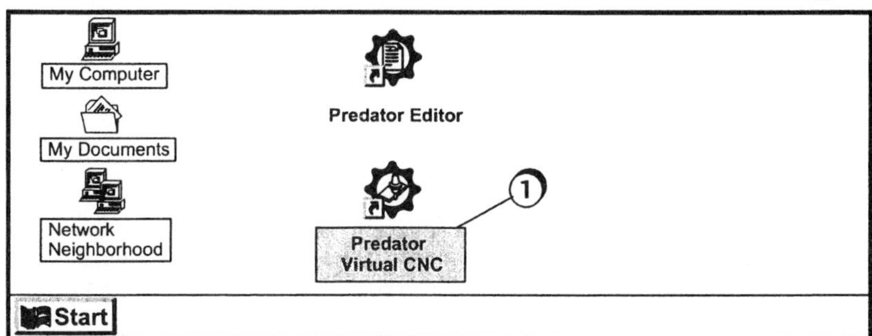

- In the Registration Information dialog box Click ② and enter **DEMO**

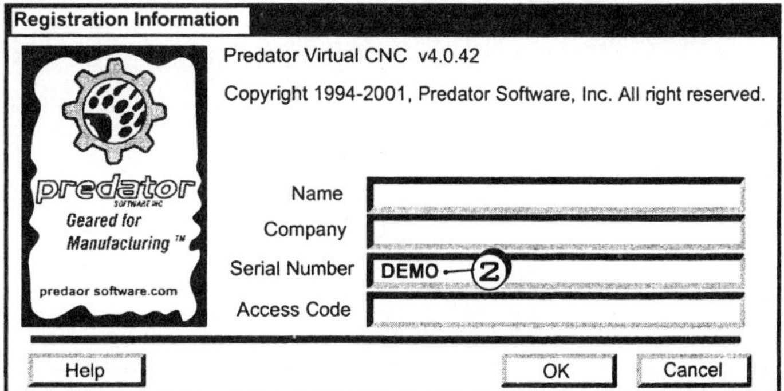

Upon launching the Virtual CNC the following message be displayed.

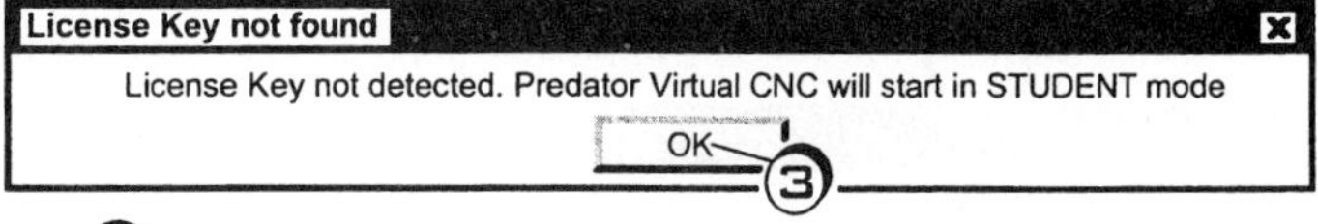

- Click ③ the OK button to run in STUDENT mode

The virtual CNC will then process up to *100 lines* of word address code.

20–7 AN OVERVIEW OF WRITING AND VERIFYING PROGRAMS

The word address program for machining a part is first written with the aid of the Predator CNC Editor. The program is downloaded to the Predator Virtual CNC simulator. The operator then inputs such job setup information as the type of CNC machine and controller used, the tooling description, the stock shape and size, and the location of the part origin. The virtual CNC simulator uses the word address program and the job setup information to produce a realistic animation of the entire machining process.

The sequence of steps to be followed in writing and simulating a program are shown in Figure 20–1.

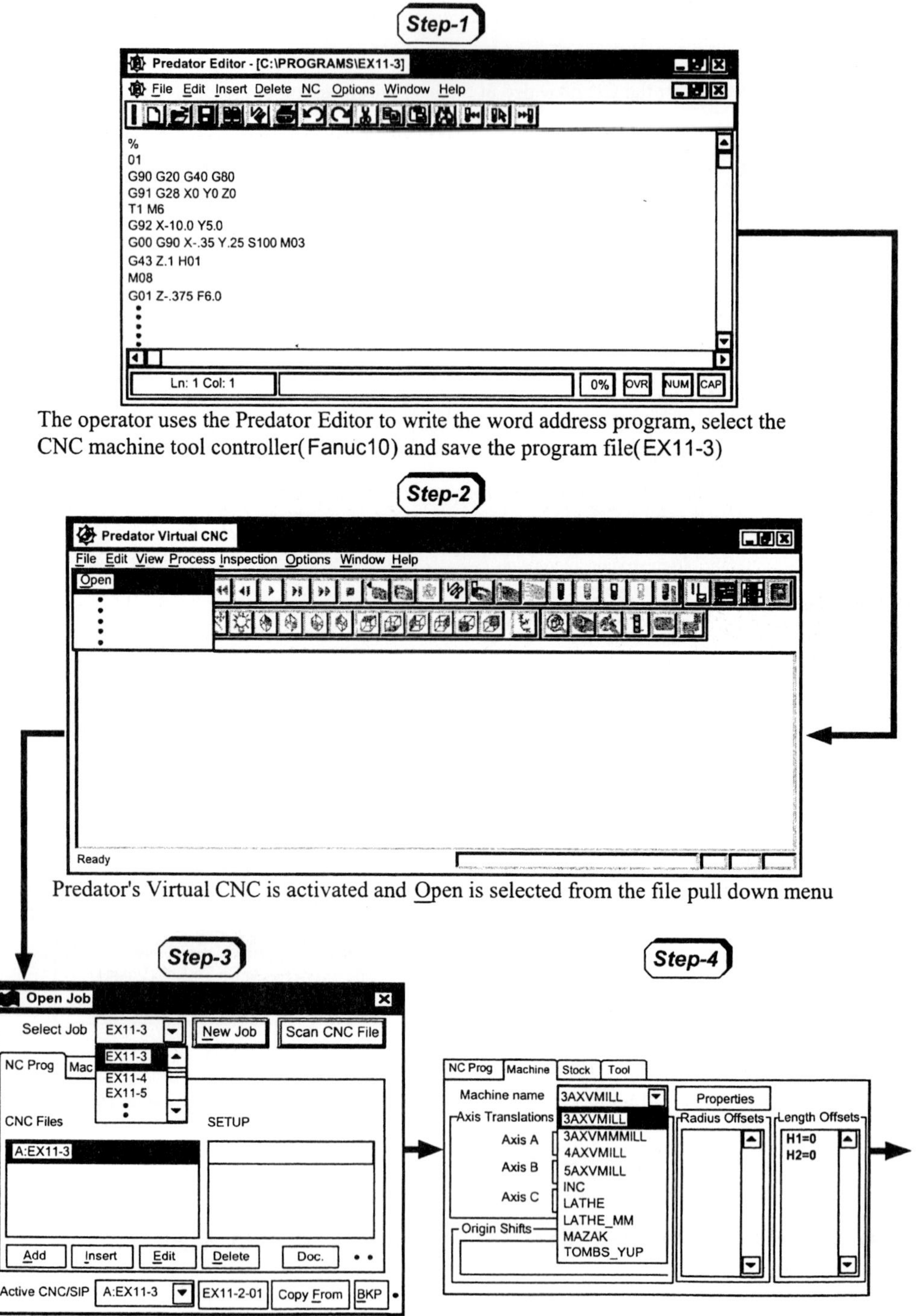

The operator uses the Predator Editor to write the word address program, select the CNC machine tool controller(Fanuc10) and save the program file(EX11-3)

Predator's Virtual CNC is activated and Open is selected from the file pull down menu

The word address job file (EX11-3) is selected

The CNC machine tool type (3AXVMILL) is selected.

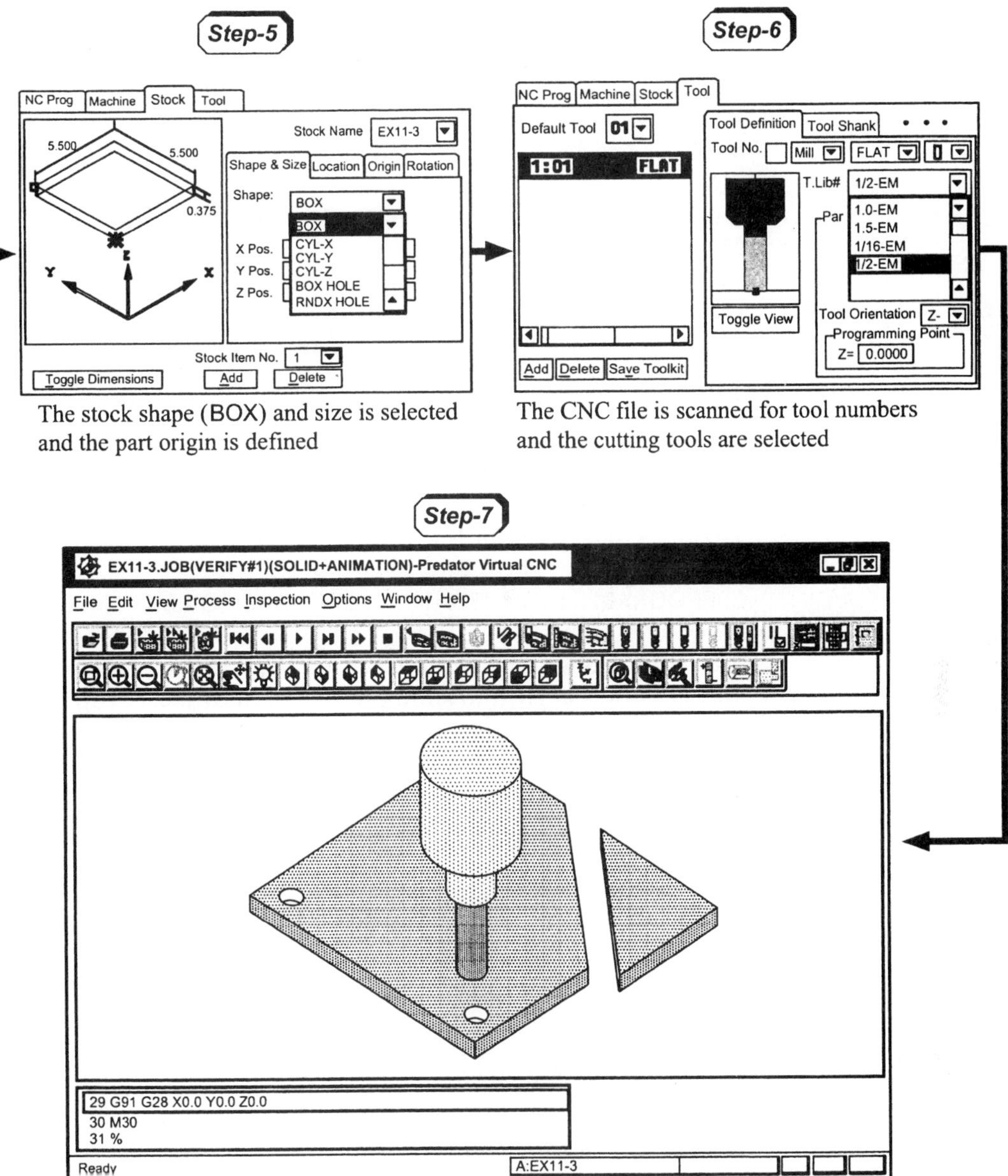

The stock shape (BOX) and size is selected and the part origin is defined

The CNC file is scanned for tool numbers and the cutting tools are selected

The operator directs Virtual CNC to display each block of the word address program as it animates the corresponding machining operation on the selected stock.

20–8 TYPES OF FILES CREATED BY PREDATOR

Program Files

The program file contains the CNC word address program created via the CNC Editor. It is stored as a general Microsoft (*.*) Windows file. The program file is saved by clicking All files (*.*) in the Save as type box.

Virtual CNC

1. Job files
   The job file contains important references to other files needed to simulate a job. Referenced files contain information concerning machine, stock, and tool settings. The job file is stored as a .JOB file in the sub-directory
   C:\PROGRAM FILES\PREDATOR-SOFTWARE\COMMON FILES\JOBS
2. Machine files
   The CNC machine type to be used for the job is stored as a .MCH file in the sub-directory
   C:\PROGRAM FILES\PREDATOR-SOFTWARE\COMMON FILES\MACHINES
3. Stock files
   The stock file to be used for the job is stored as a .STK file in the sub-directory
   C:\PROGRAM FILES\PREDATOR-SOFTWARE\COMMON FILES\STOCK
4. Tool files
   The tools to be used for the job are stored as a .TLB file in the sub-directory
   C:\PROGRAM FILES\PREDATOR-SOFTWARE\COMMON FILES\TOOLS
5. Setup files
   The setup file is used to *automatically* load the information in the .MCH and .STK files needed to simulate a particular job. Setup files have the extension .STP and are stored in the sub-directory
   C:\PROGRAM FILES\PREDATOR-SOFTWARE\COMMON FILES\JOBS

20-9 RUNNING AN OLD SIMULATION JOB STORED ON DISKETTE (MILL)

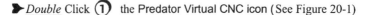

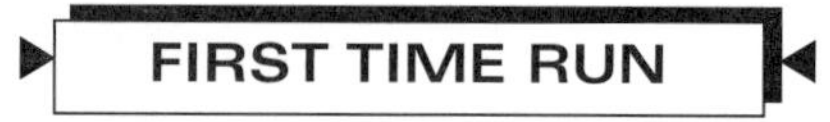

■ EXAMPLE 20-1

Run the machining simulation for the job **EX10–2** stored on the DATA diskette at the back of this text. It is assumed that this is the *first time* the job is being run. ■■

Start the Predator Virtual CNC

➤ *Double* Click ① the Predator Virtual CNC icon (See Figure 20-1)

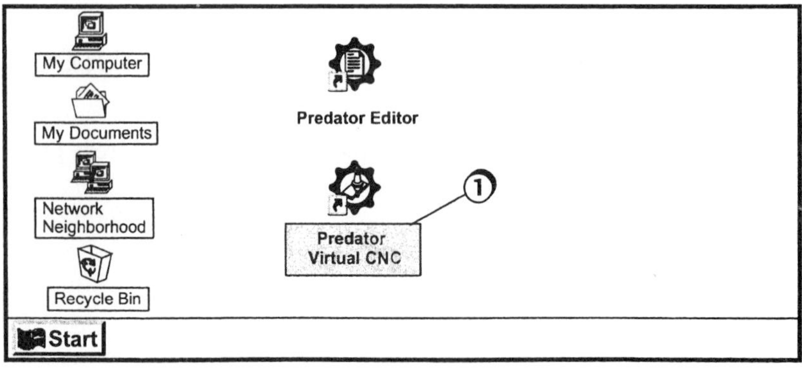

FIGURE 20-1 Starting the Predator Virtual CNC from the Windows Desktop.

The system will open a Virtual CNC session as shown below

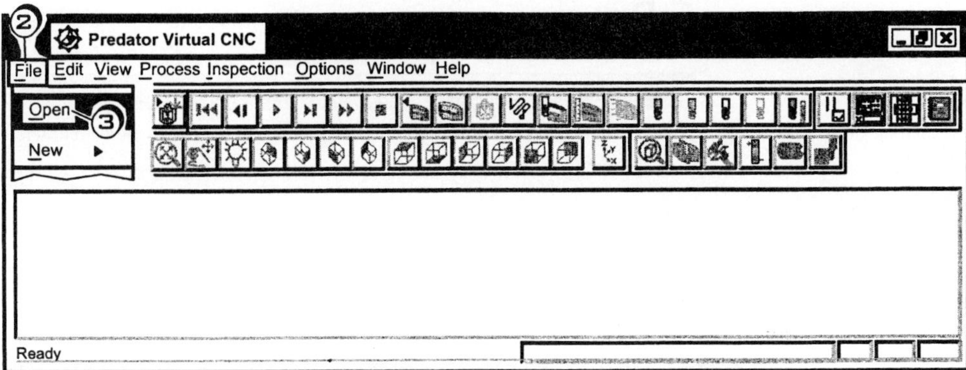

➤Remove the diskette labeled DATA DISK FOR SIMULATIONS from the back of the text and place it into the A drive.

➤Click ② the File Pull down menu

➤Click ③ the Open command

Delete Any Existing Jobs Listed

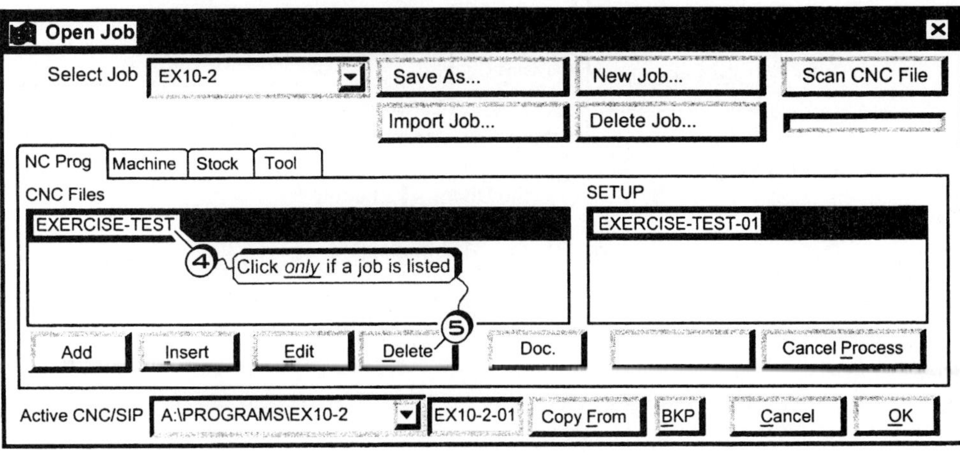

➤Click ④ on any *existing* job listed(if none are listed ignore this step)

➤Click ⑤ on the Delete key

Open the Old Job File on Diskette and Copy it to C Drive

> ***Note:*** Predator will copy all the files for running a job from the data diskette in A drive to the C drive. Predator will then work with the files on C drive when executing the machining simulation.

➤Click ⑥ the New Job button

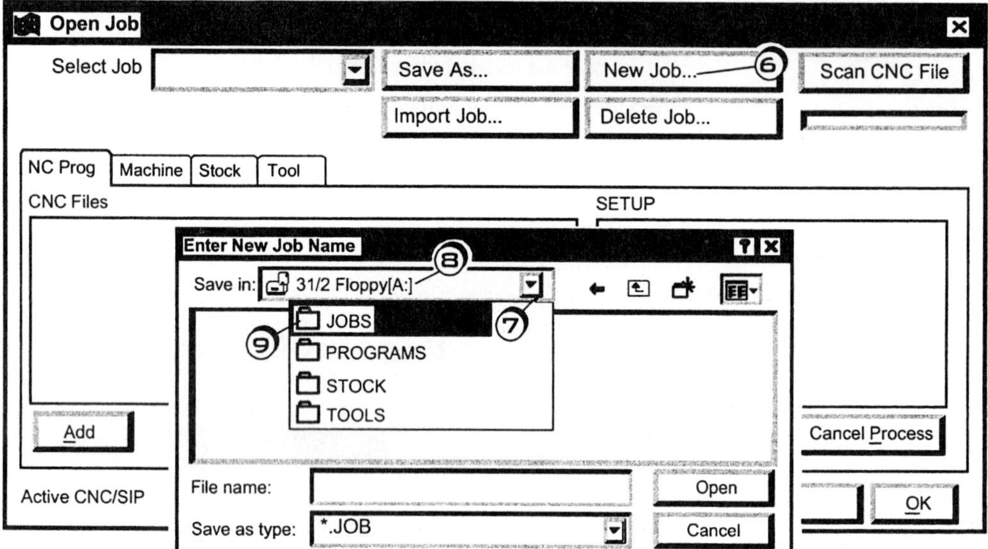

➤Click ⑦ the Down button

➤Click ⑧ the 31/2 Floppy[*A]

➤*Double* Click ⑨ on the file folder JOBS

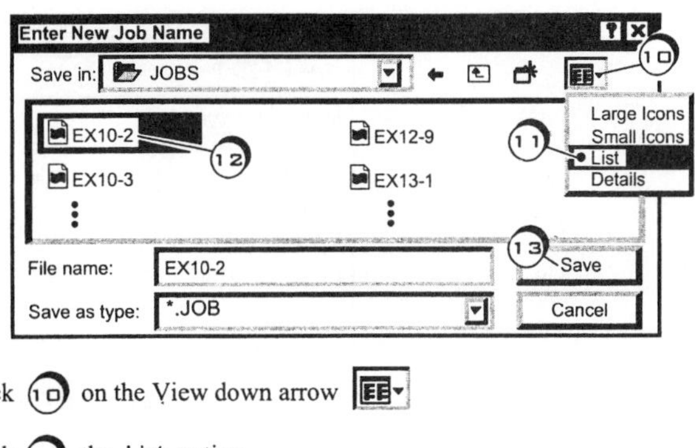

➤Click (10) on the View down arrow

➤Click (11) the List option

➤Click (12) on the job file **EX10-2**

➤Click (13) the Save button

The new job file **EX10-2** will then be created in the Predator sub-directory on C drive.

Add a Program File Reference to the Job File Created on C Drive

A reference to a particular word address program file ***must now be added*** to the new job created on C drive. When the Virtual CNC is instructed to process the job, it will display its simulations based on the ***currently referenced word address program.***

➤Click (14) the Add button

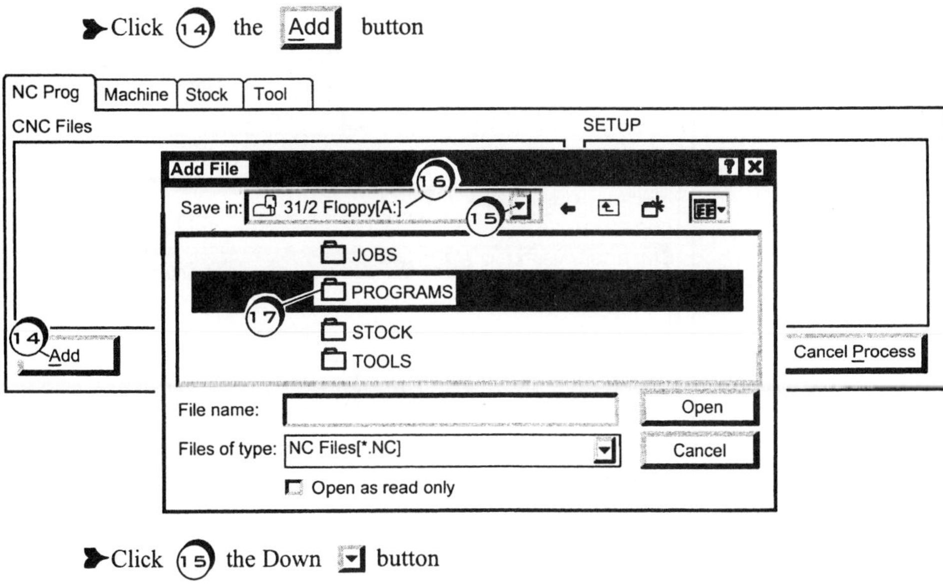

➤Click (15) the Down button

➤Click (16) the 31/2 Floppy[*A]

➤*Double* Click (17) on the file folder PROGRAMS

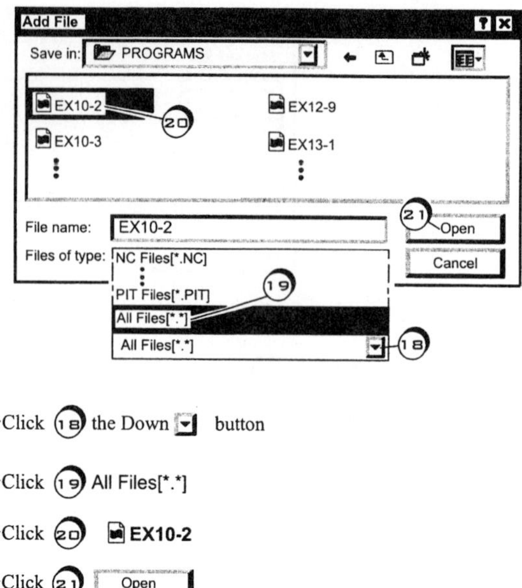

➤Click (18) the Down button

➤Click (19) All Files[*.*]

➤Click (20) **EX10-2**

➤Click (21) Open

Reference to the program file **EX10-2** will then be added to the job file on C drive.

Copy the Setup File Containing Machine and Stock Settings from A to C Drive

The setup file must now be copied to the C drive enabling Predator to *automatically* load the CNC machine and stock settings needed for the simulation.

➤Click (22) Copy From

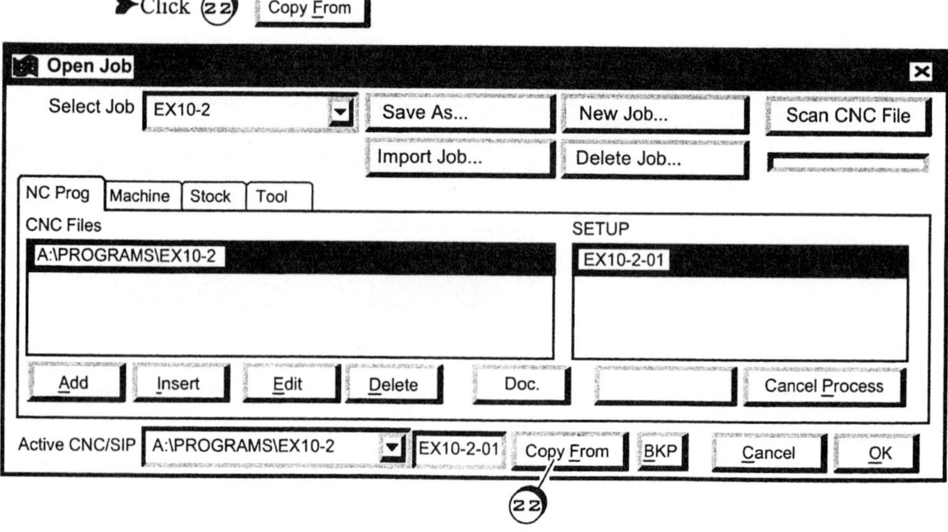

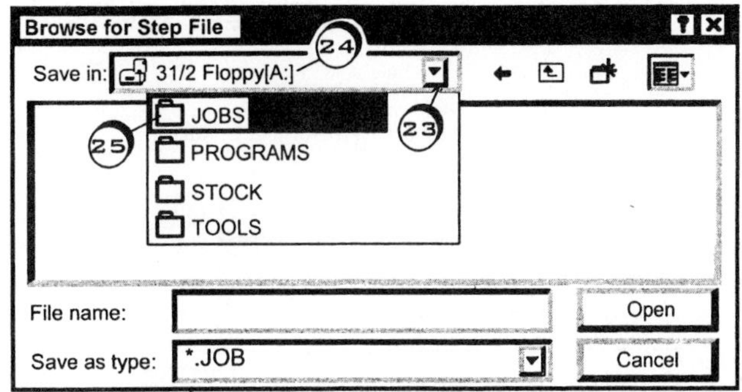

➤Click (23) the Down [▾] button

➤Click (24) the 31/2 Floppy[*A]

➤*Double* Click (25) on the file folder JOBS

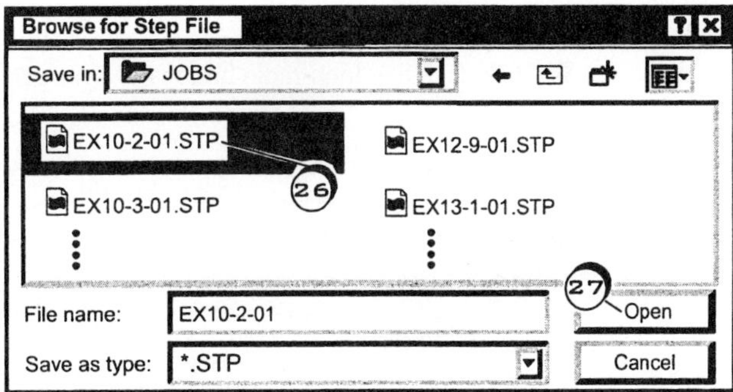

➤Click (26) on the setup file **EX10-2-01**

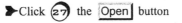

➤Click (27) the [Open] button

The setup file **EX10-2-01** will then be loaded into the Predator sub-directory on C drive. The software will consider this as the *active* setup file for the *current* simulation job.

Specify the Axis Travel Units of the CNC Machine Tool

➤Click (28) the Machine tab

➤Click (29) the Properties button

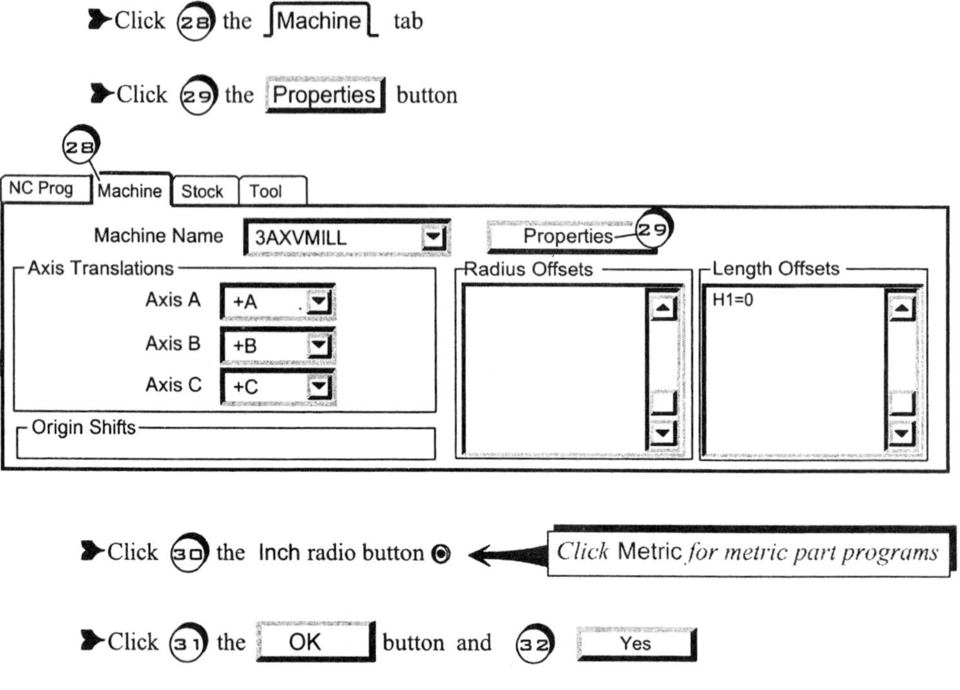

➤Click (30) the Inch radio button ◉ ← *Click* Metric *for metric part programs*

➤Click (31) the OK button and (32) Yes

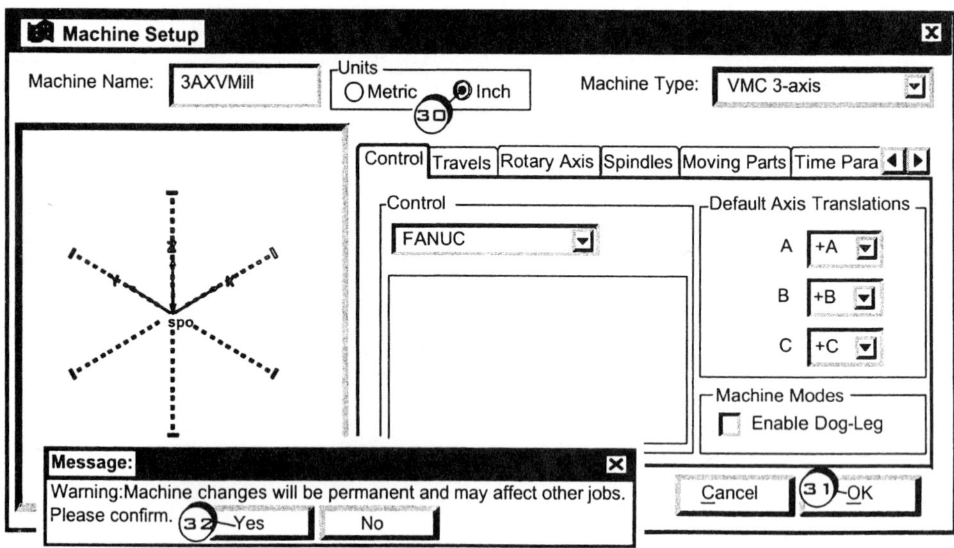

Copy the Tool File Containing the Tools Settings from A to C Drive

The tools file containing all the tools needed for the job is now copied to the C drive. ***Delete all*** the tools **NOT SET.**

➤Click (33) the Tool tab

➤Click (34) on 1:T1 NOT SET

➤Click (35) on Delete

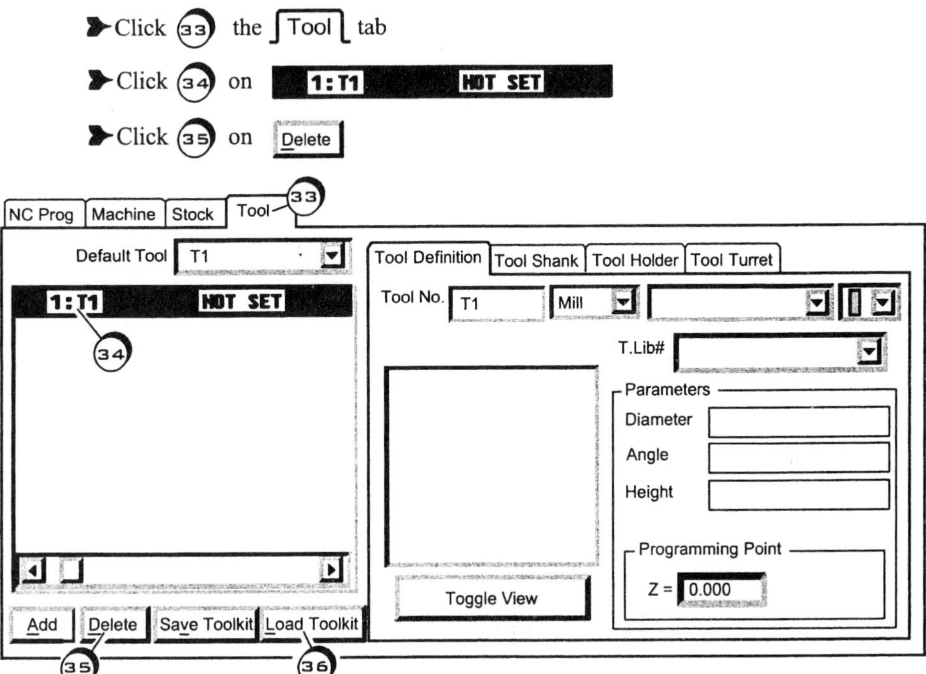

Copy the Tool File from the Disk to C Drive

➤Click (36) on Load Toolkit

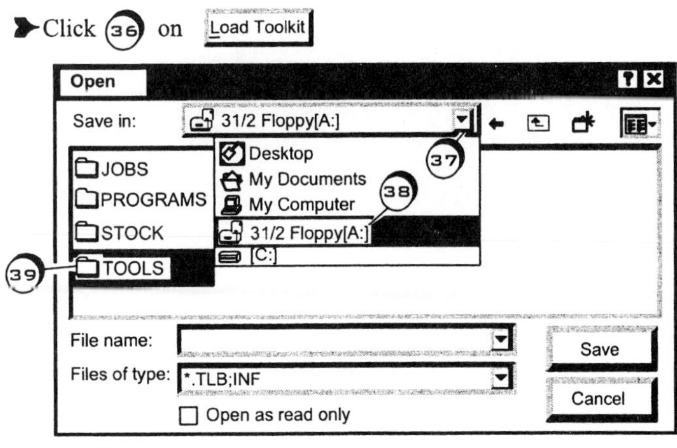

➤Click (37) the down button

➤Click (38) the desired drive : 31/2 Floppy[A:]

➤*Double* Click (39) the folder TOOLS where the file is to be stored on the diskette

➤ Click (40) the desired tool file for the job: EX10-2

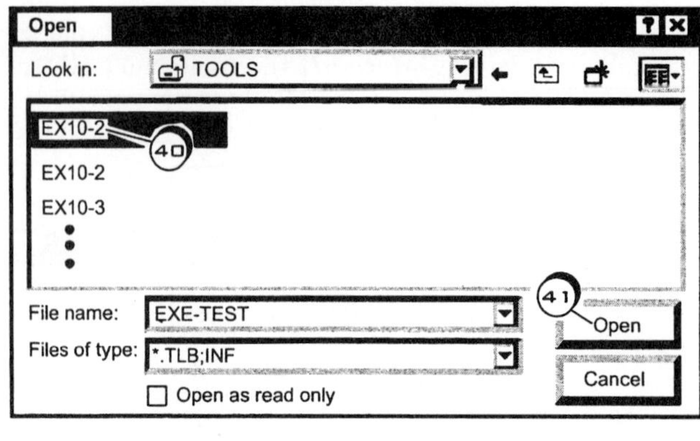

➤ Click (41) on Open

The tool page dialog box will now list the tool(s) required for the simulation

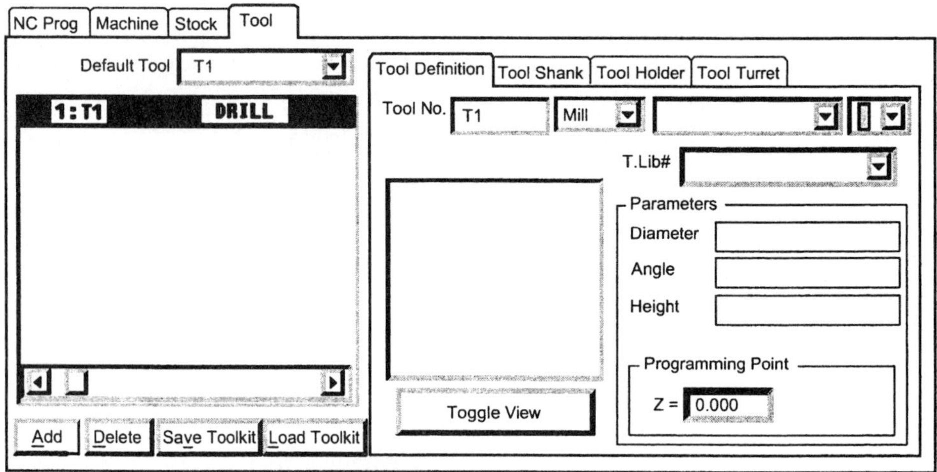

Direct the Virtual CNC to Simulate the Current Job

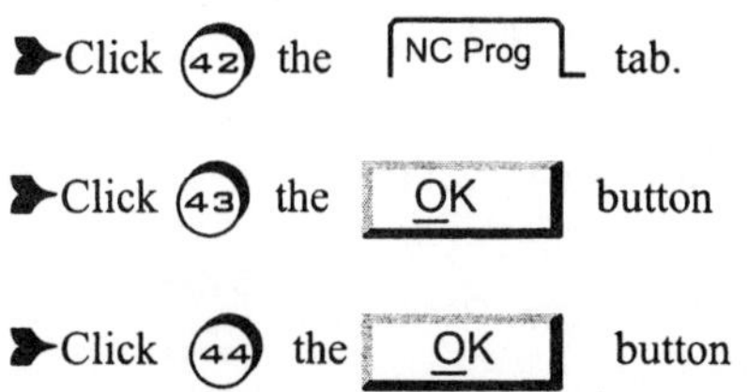

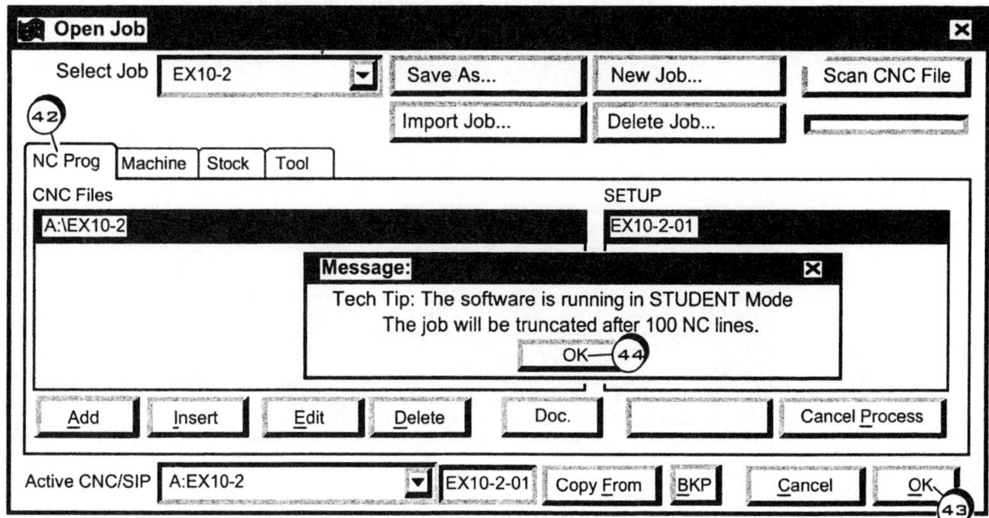

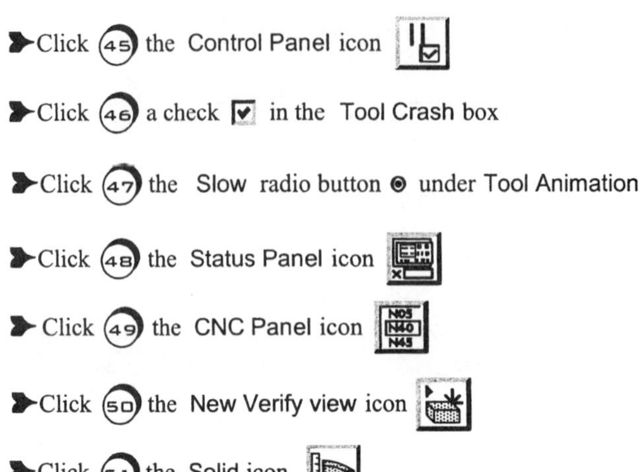

➤Click (45) the Control Panel icon

➤Click (46) a check ☑ in the Tool Crash box

➤Click (47) the Slow radio button ◉ under Tool Animation

➤Click (48) the Status Panel icon

➤Click (49) the CNC Panel icon

➤Click (50) the New Verify view icon

➤Click (51) the Solid icon

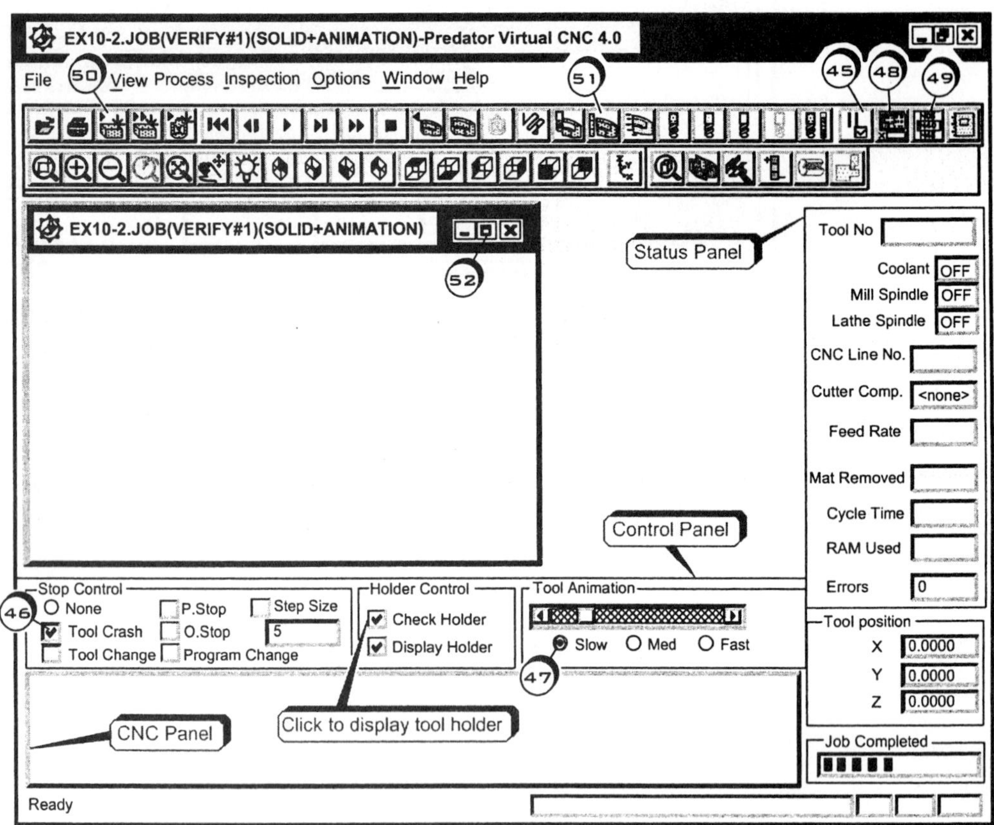

➤Click (52) the maximize button in the **SOLID+ANIMATION** window

➤Click (53) the Edit pull down menu and (54) the Background color option.

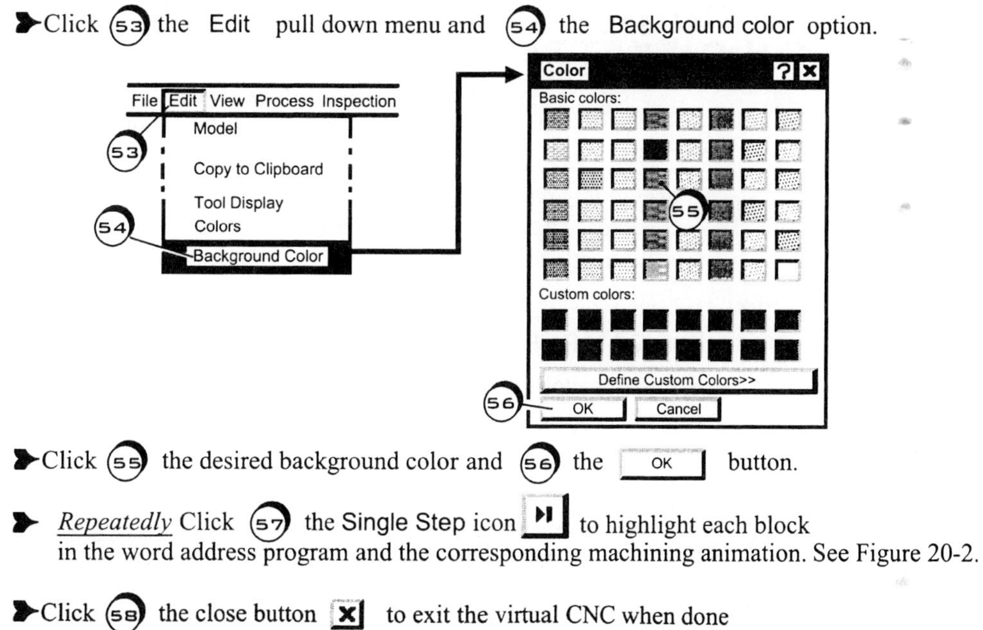

➤Click (55) the desired background color and (56) the OK button.

➤ *Repeatedly* Click (57) the Single Step icon to highlight each block in the word address program and the corresponding machining animation. See Figure 20-2.

➤Click (58) the close button to exit the virtual CNC when done

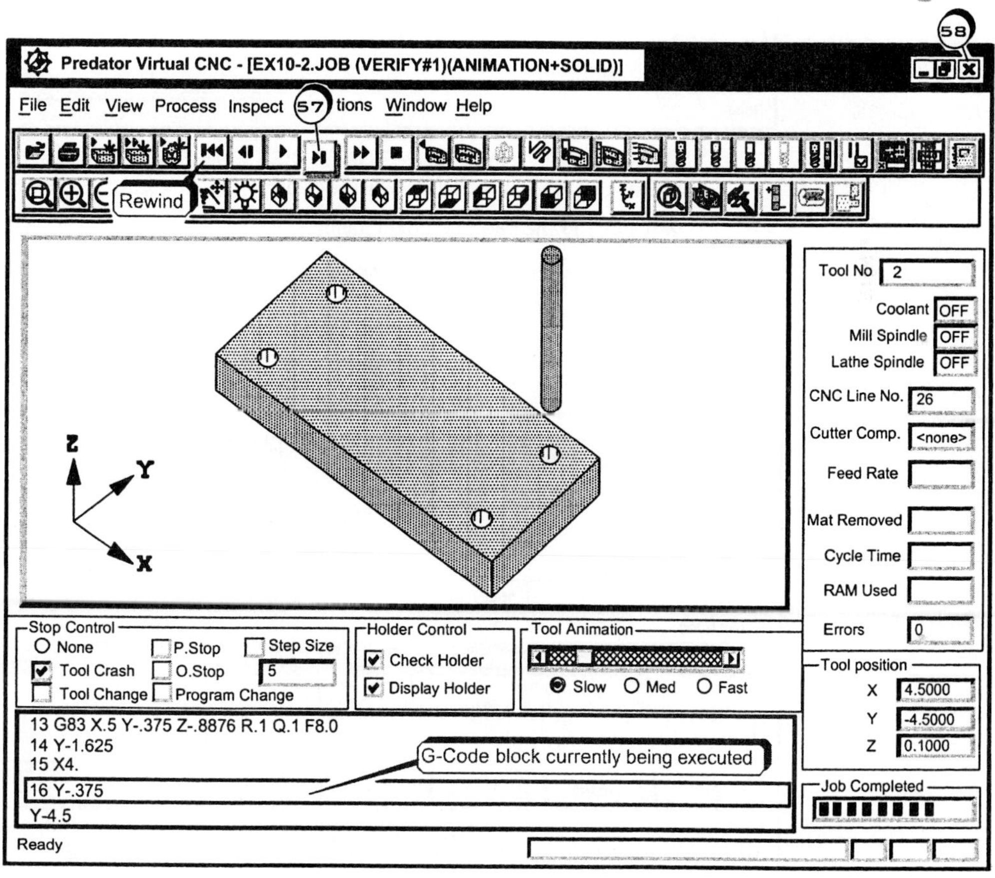

FIGURE 20–2 The machining animation produced by Predator's Virtual CNC.

20–10 RUNNING AN OLD SIMULATION JOB STORED ON DISKETTE (MILL)

JOB HAS BEEN RUN PREVIOUSLY

A previously run job has its job, machine, stock, tools, and .stp files *already* stored on *C drive*. Therefore *copying* these files to C drive is *not necessary*. The operator simply needs to *open* the appropriate job file from A and *run* the simulation.

■ EXAMPLE 20–2

Run the machining simulation for the job **EX10–2** stored on the DATA diskette at the back of this text. It is assumed the job has been run *previously*. ■■

Assume the operator is in a virtual CNC session

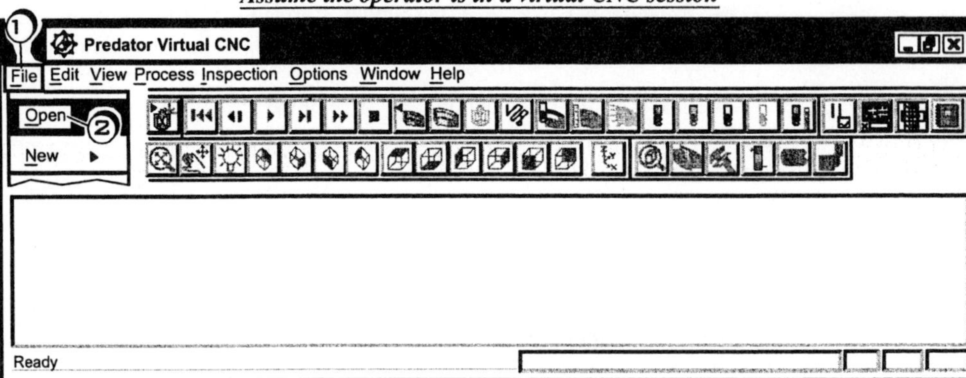

- Assume the diskette labeled DATA DISK FOR SIMULATIONS from the back of the text is currently in the A drive.
- Click ① the File Pull down menu
- Click ② the Open command

Delete Any Existing Jobs Listed

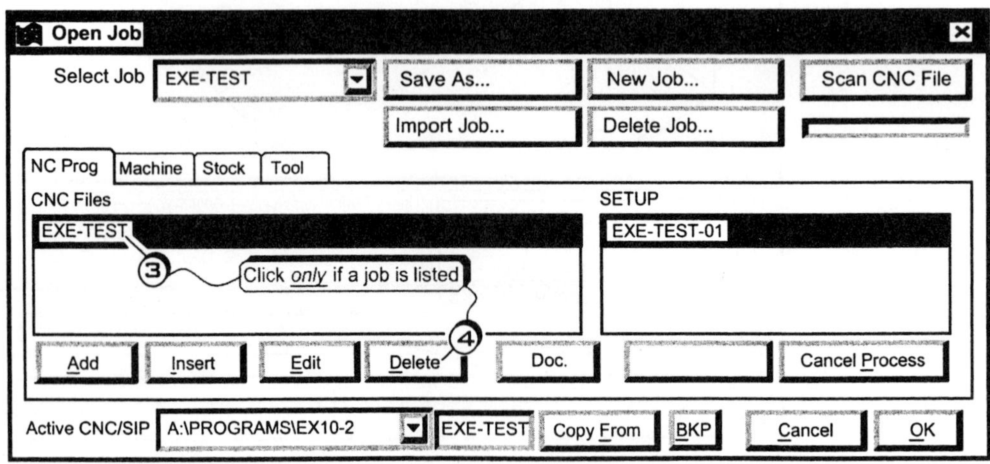

- Click ③ on any *existing* job listed(if *none* are listed *ignore* this step)
- Click ④ on the Delete key

Open the Old Job File on Diskette

➤Click ⑤ the New Job button

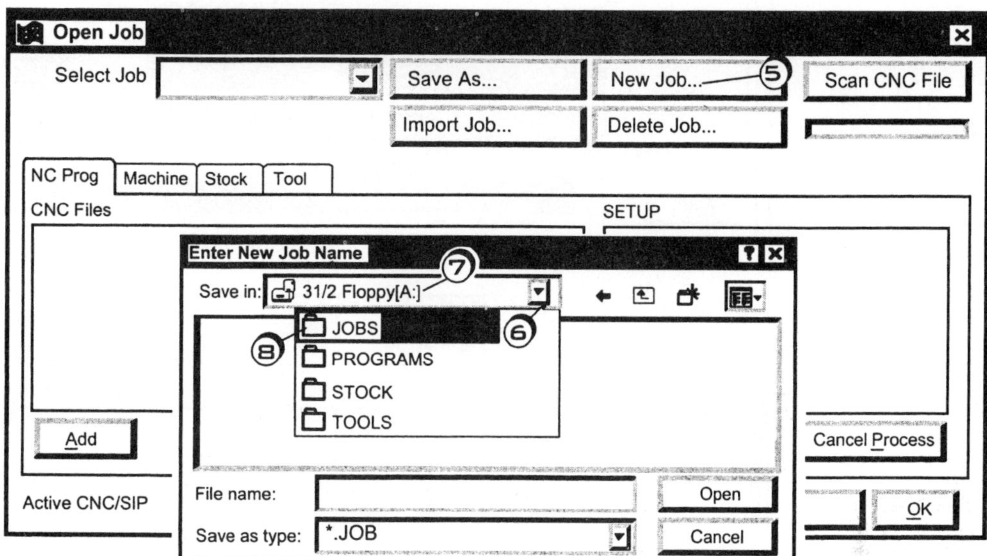

➤Click ⑥ the Down button

➤Click ⑦ the 31/2 Floppy[*A]

➤*Double* Click ⑧ on the file folder JOBS

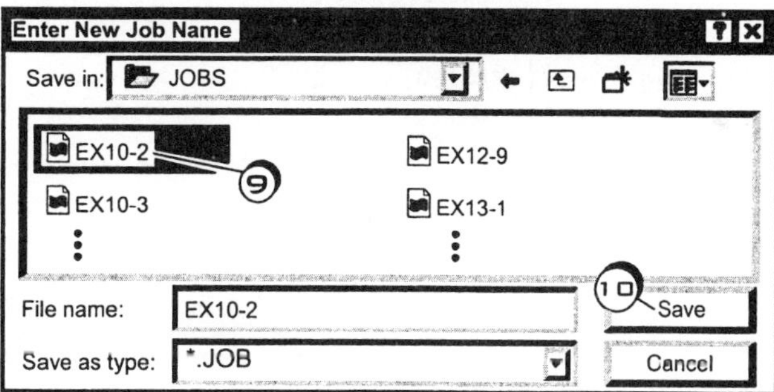

➤Click ⑨ on the job file **EX10-2**

➤Click ⑩ the Save button

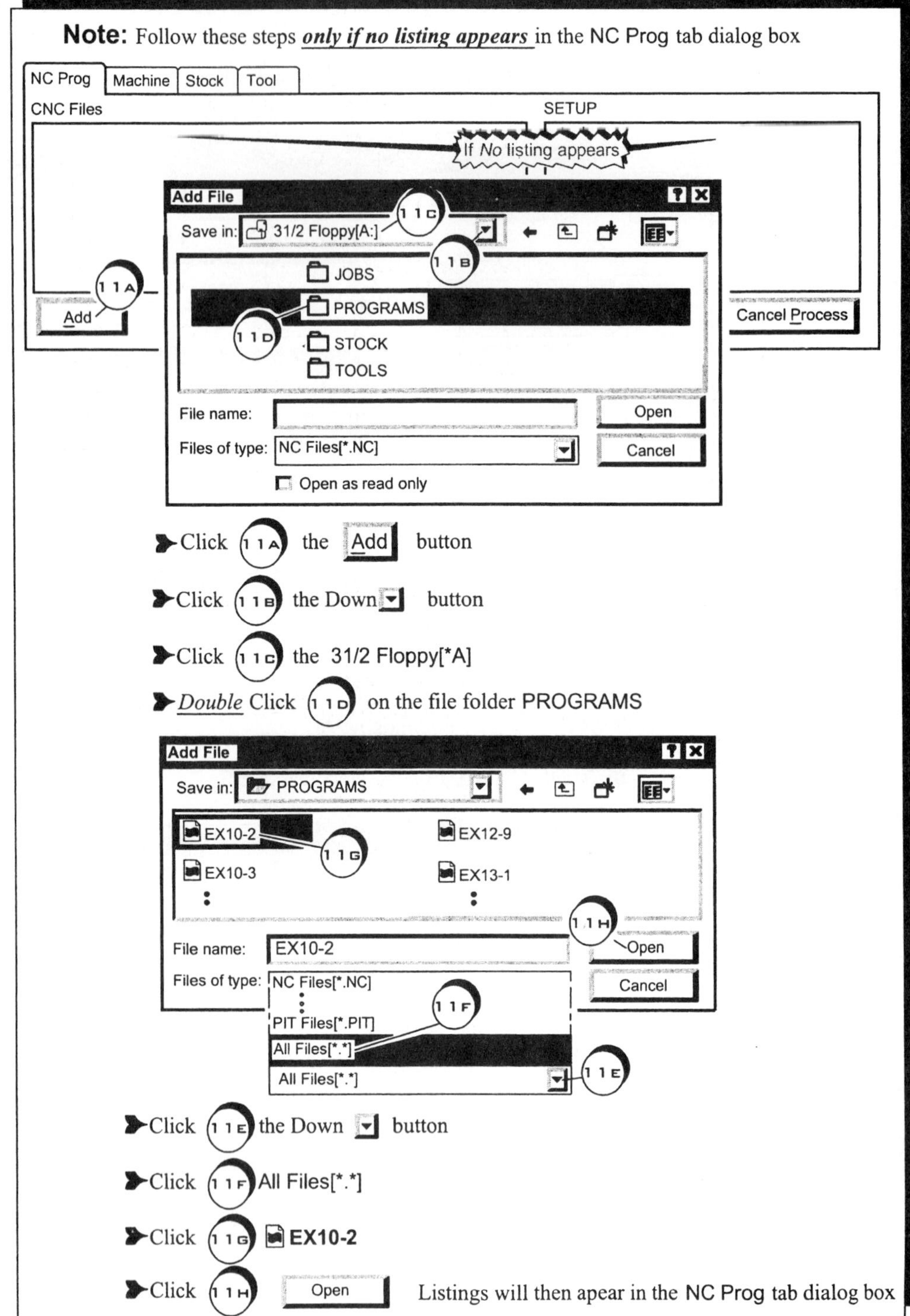
Note: Follow these steps only if no listing appears in the NC Prog tab dialog box
NC Prog
Machine
Stock
Tool
CNC Files
SETUP
If No listing appears
Add File
Save in:
31/2 Floppy[A:]
JOBS
PROGRAMS
STOCK
TOOLS
Add
Cancel Process
File name:
Open
Files of type:
NC Files[*.NC]
Cancel
Open as read only
Click 11A the Add button
Click 11B the Down button
Click 11C the 31/2 Floppy[*A]
Double Click 11D on the file folder PROGRAMS
Add File
Save in:
PROGRAMS
EX10-2
EX12-9
EX10-3
EX13-1
File name:
EX10-2
Open
Files of type:
NC Files[*.NC]
PIT Files[*.PIT]
All Files[*.*]
All Files[*.*]
Cancel
Click 11E the Down button
Click 11F All Files[*.*]
Click 11G EX10-2
Click 11H Open Listings will then apear in the NC Prog tab dialog box

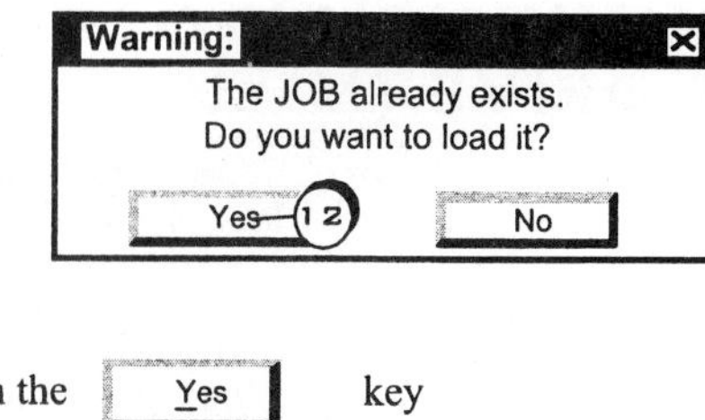

➤Click (12) on the Yes key

Run the Job

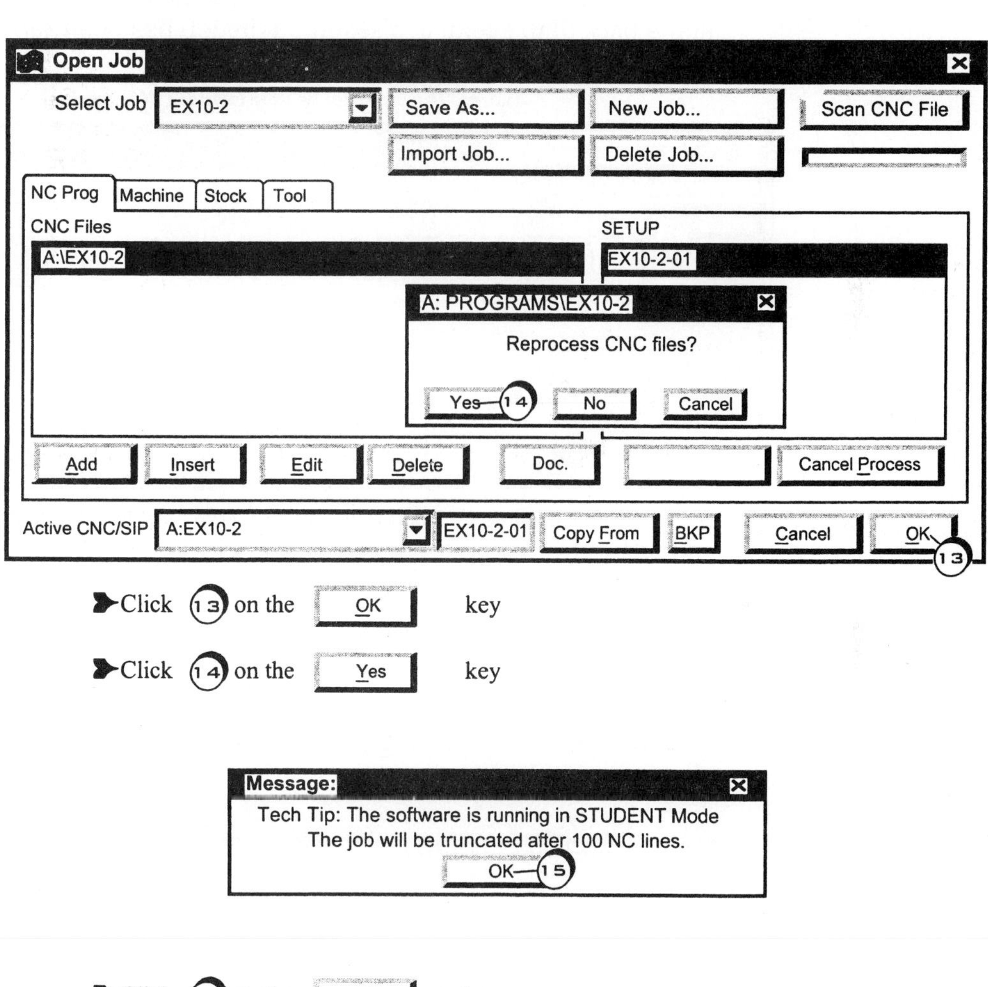

➤Click (13) on the OK key

➤Click (14) on the Yes key

Message:
Tech Tip: The software is running in STUDENT Mode
The job will be truncated after 100 NC lines.
OK (15)

➤Click (15) on the OK key

20–11 RUNNING AN OLD SIMULATION JOB FROM C DRIVE (MILL)

JOB HAS BEEN RUN PREVIOUSLY

A previously run job has its job, machine, stock, tools, and .stp files *already* stored on *C drive*. Therefore *copying* these files to C drive is *not necessary*. The operator simply needs to *open* the appropriate job file from C drive to *run* it.

■ EXAMPLE 20–3

Run the machining simulation for the job **EX10–2** stored on the DATA diskette at the back of this text. It is assumed the job has been run *previously*. ■■

Assume the operator is in a virtual CNC session

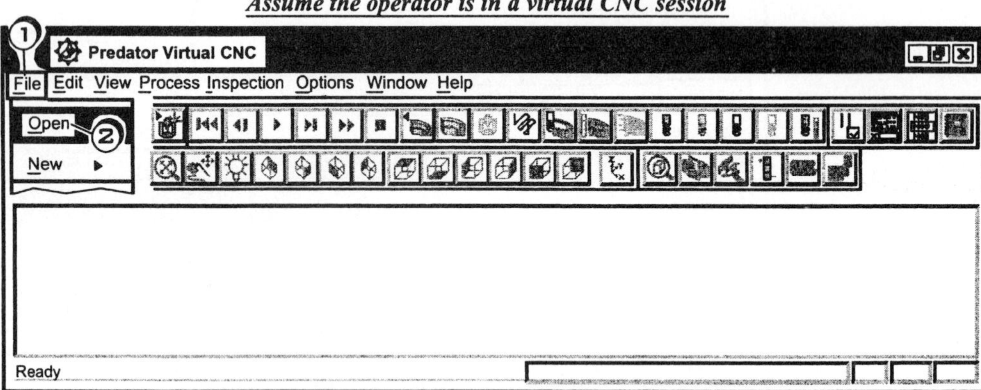

➤Assume the diskette labeled DATA DISK FOR SIMULATIONS from the back of the text is currently in the A drive.

➤Click ① the File Pull down menu

➤Click ② the Open command

Delete Any Existing Jobs Listed

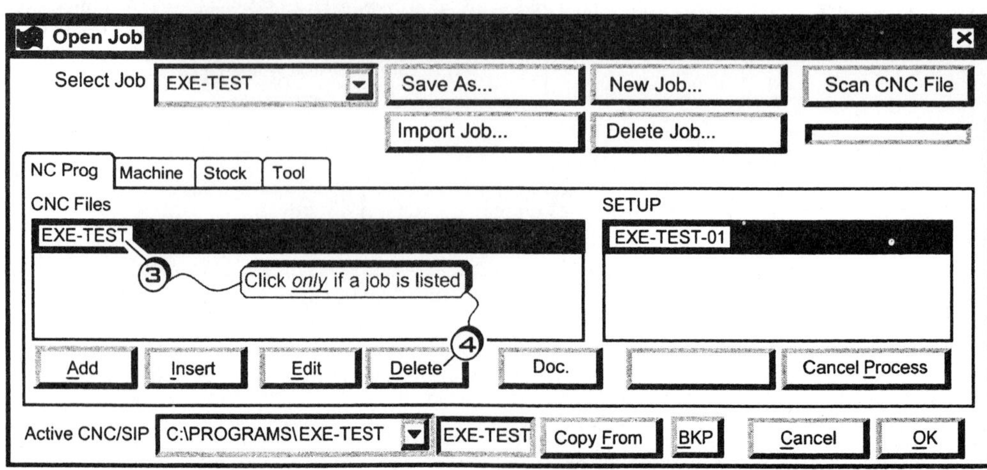

➤Click ③ on any *existing* job listed(if *none* are listed *ignore* this step)

➤Click ④ on the Delete key

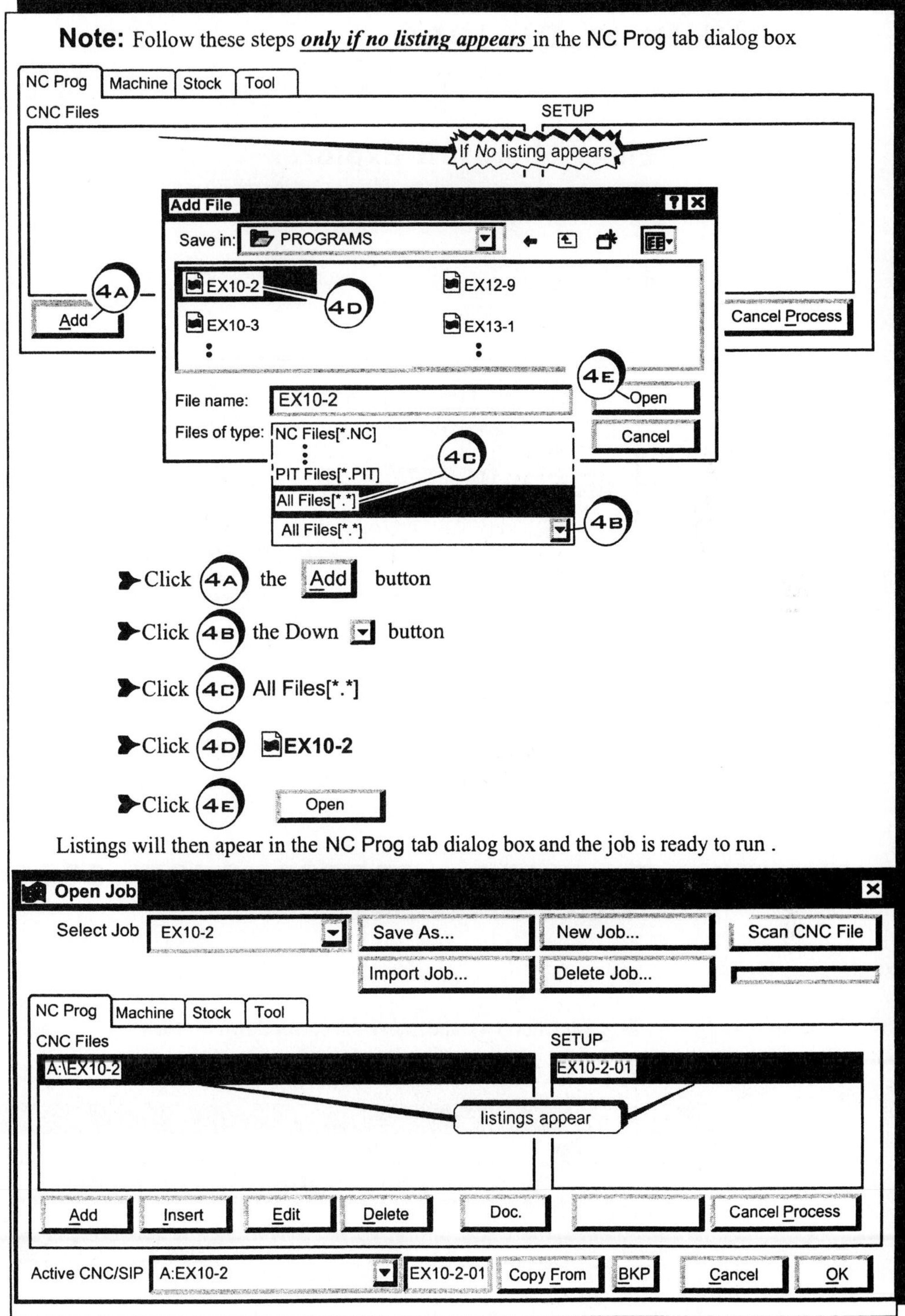

Note: Follow these steps only if no listing appears in the NC Prog tab dialog box
NC Prog
Machine
Stock
Tool
CNC Files
SETUP
If No listing appears
Add File
Save in:
PROGRAMS
EX10-2
EX12-9
EX10-3
EX13-1
Add
Cancel Process
File name:
EX10-2
Open
Files of type:
NC Files[*.NC]
Cancel
PIT Files[*.PIT]
All Files[*.*]
All Files[*.*]
Click 4A the Add button
Click 4B the Down button
Click 4C All Files[*.*]
Click 4D EX10-2
Click 4E Open
Listings will then apear in the NC Prog tab dialog box and the job is ready to run .
Open Job
Select Job
EX10-2
Save As...
New Job...
Scan CNC File
Import Job...
Delete Job...
NC Prog
Machine
Stock
Tool
CNC Files
SETUP
A:\EX10-2
EX10-2-01
listings appear
Add
Insert
Edit
Delete
Doc.
Cancel Process
Active CNC/SIP
A:EX10-2
EX10-2-01
Copy From
BKP
Cancel
OK

20-12 PREPARING A DISK FOR STORING NEW SIMULATION JOBS

Take a *new, formatted, blank* disk and insert it into the A drive.

Enter Microsoft Explorer

From the Windows desktop:

➤Click ① the [My Computer] icon and *right* click ② Explore. See Figure 20-3

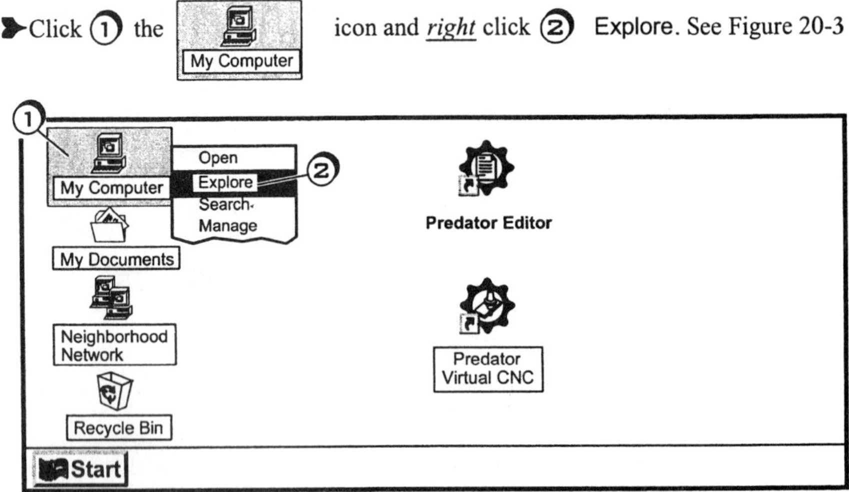

FIGURE 20-3 Entering Microsoft Explorer at the Windows Desktop.

Create folders: PROGRAMS, JOBS, STOCK and TOOLS on the Disk

➤Click ③ the 31/2 Floppy(A:) folder

➤*Right* Click ④ and move the cursor down to New

➤Click ⑤ Folder

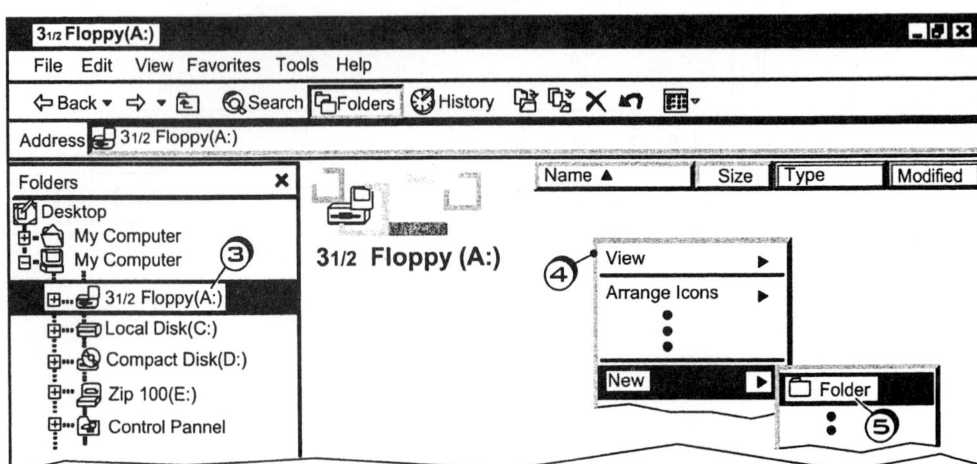

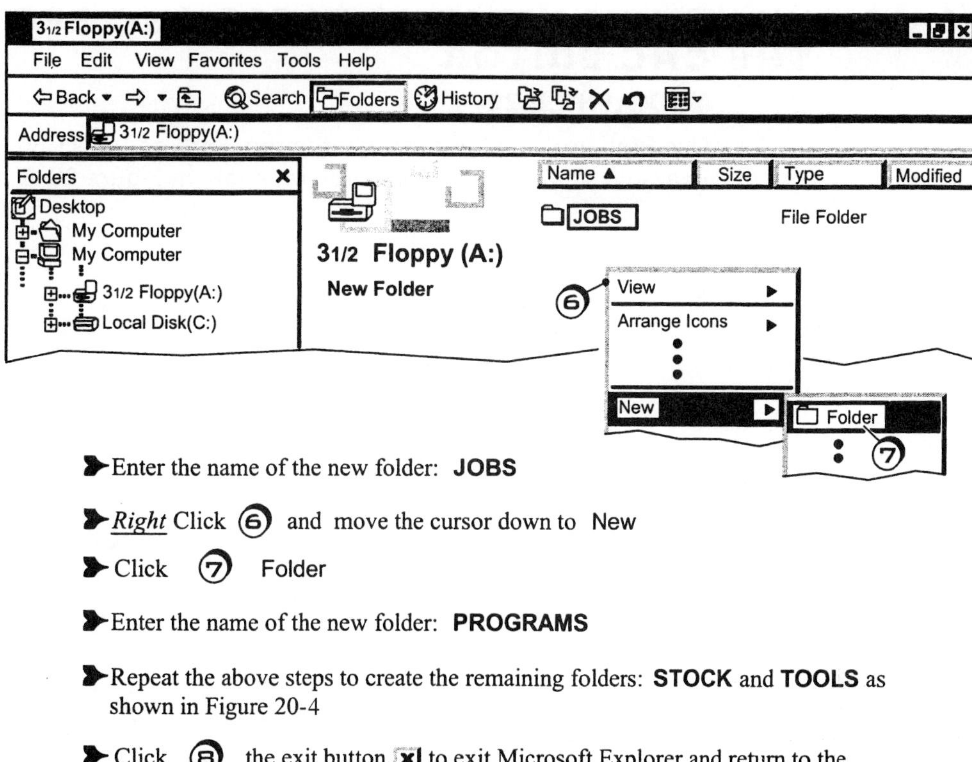

➤Enter the name of the new folder: **JOBS**

➤*Right* Click (6) and move the cursor down to New

➤Click (7) Folder

➤Enter the name of the new folder: **PROGRAMS**

➤Repeat the above steps to create the remaining folders: **STOCK** and **TOOLS** as shown in Figure 20-4

➤Click (8) the exit button **x** to exit Microsoft Explorer and return to the Windows desktop

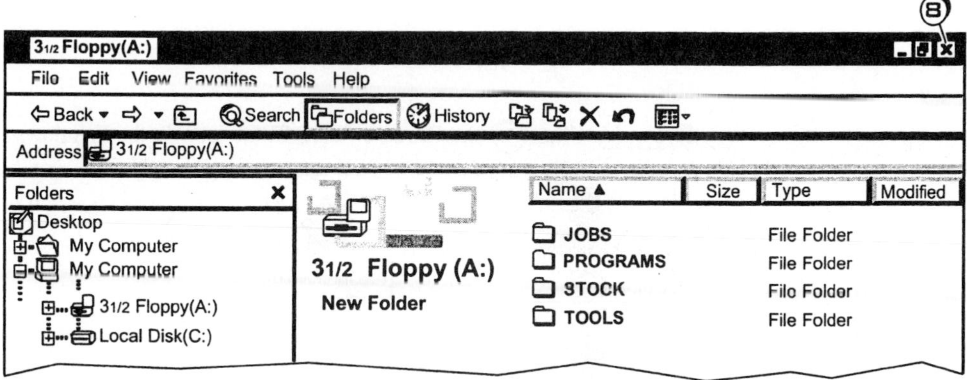

FIGURE 20–4 The new disk file folders JOBS, PROGRAMS, STOCK, and TOOLS.

20-13 WRITING A WORD ADDRESS CNC PROGRAM VIA THE CNC EDITOR

■ EXAMPLE 20-4

Use Predator's CNC Editor to write the word address CNC program given in Example 11-3, page 215 and shown again in Figure 20-5.

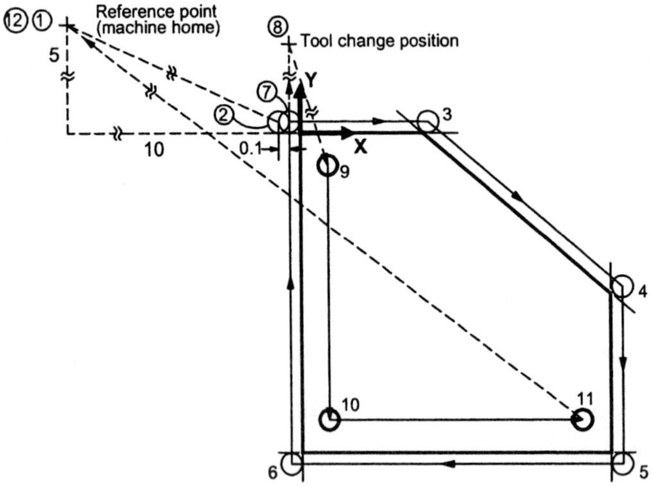

FIGURE 20-5 Part to be machined. ■■

Tool	Operation	Tooling	Speed (rpm)	Feed (ipm)
1	Profile mill contour × 0.375 deep	0.5D end mill	1500	7
2	Deep drill all holes thru	0.4375D drill	1500	7

Start the Predator CNC Editor

➤*Double* Click ① the Predator CNC Editor icon in the Windows desktop.See Figure 20-6

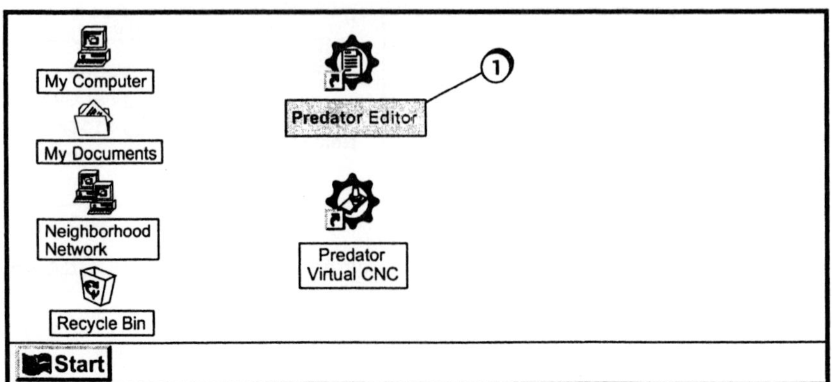

FIGURE 20-6 Starting the Predator CNC Editor from the Windows Desktop.

Create the Word Address CNC Program

The system will open a *new* program file and display the Editor as shown in Figure 20–7. Type the first block of the CNC program, hit the **Enter** key to type the next block, and so on.

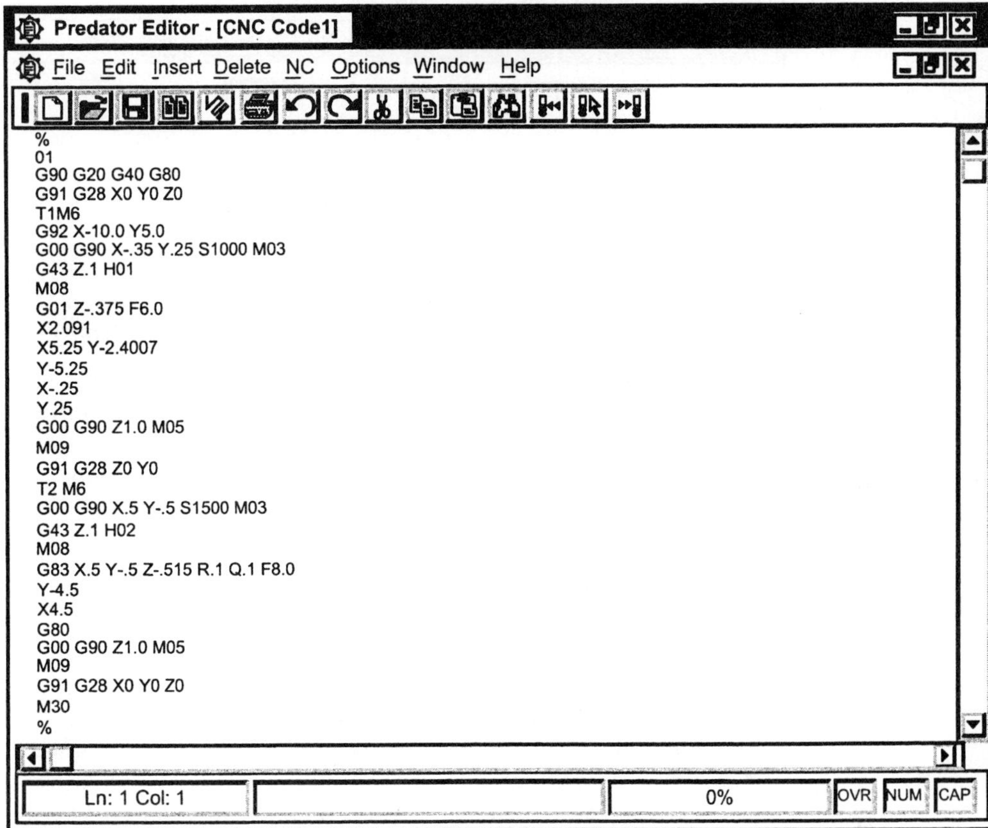

Predator Editor - [CNC Code1]
File Edit Insert Delete NC Options Window Help
%
01
G90 G20 G40 G80
G91 G28 X0 Y0 Z0
T1M6
G92 X-10.0 Y5.0
G00 G90 X-.35 Y.25 S1000 M03
G43 Z.1 H01
M08
G01 Z-.375 F6.0
X2.091
X5.25 Y-2.4007
Y-5.25
X-.25
Y.25
G00 G90 Z1.0 M05
M09
G91 G28 Z0 Y0
T2 M6
G00 G90 X.5 Y-.5 S1500 M03
G43 Z.1 H02
M08
G83 X.5 Y-.5 Z-.515 R.1 Q.1 F8.0
Y-4.5
X4.5
G80
G00 G90 Z1.0 M05
M09
G91 G28 X0 Y0 Z0
M30
%
Ln: 1 Col: 1
0%
OVR
NUM
CAP

FIGURE 20–7 The word address CNC program created via the Predator CNC Editor.

20-14 USING EDIT COMMANDS

Predator provides a number of commands for editing the word address programs that are written. Some of these are described here:

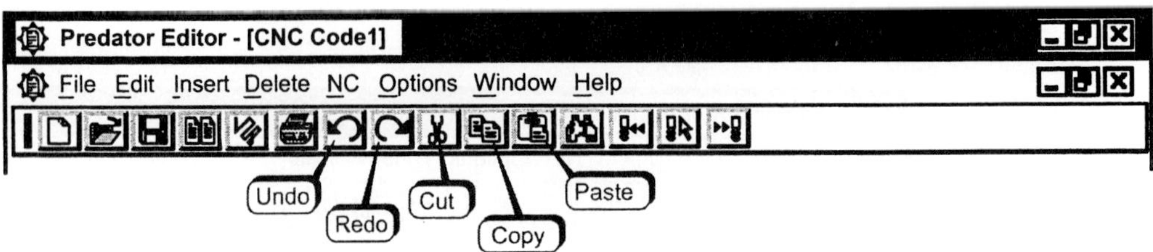

Clicking the **UNDO ICON** causes the editor to *undo* the effects of the last command entered within the active word address program.

Clicking the **REDO ICON** causes the editor to *redo* the effects of the last command entered after an Undo was issued. Redo can *only* be entered *after* an Undo command. Redo can be *repeatedly* clicked to redo the effects of the last 256 Undo commands within the current CNC program.

a. Clicking the **CUT ICON** to delete entries
   To select the entry(s) to be *deleted,* move the mouse to the beginning character. Press the *left* mouse button and, *keeping the button depressed,* move the mouse over the entry(s). Release the button.
   Click the Cut icon. The entry(s) will be cut from the active word address program and sent to the Windows clipboard.
b. Clicking the **CUT ICON** to Move Entries
   To select the entry(s) to be *moved,* position the mouse cursor on its beginning character.
   Press the *left* mouse button and, *keeping the button depressed,* move the mouse over the entry(s). Release the button.
   Click the Cut icon. The entry(s) will be cut from the active word address program and sent to the Windows clipboard. Move the cursor to the new position in the program and click the Paste icon.

Clicking the **PASTE ICON** directs the editor to *paste* the contents of the Windows clipboard into the currently active word address program. Recall that selected entry(s) from programs are sent to the Windows clipboard via the Cut or Copy icons.

To select the entry(s) to be *copied,* move the mouse to the beginning character.
Press the *left* mouse button and, *keeping the button depressed,* move the mouse over the entry(s). Release the button.
Click the **COPY ICON.** The entry(s) will be copied from the active word address program and sent to the Windows clipboard.
Recall that the Paste icon is used to *paste* a copy from the Windows clipboard into the currently active CNC program.

Adding Characters to a Block

■ EXAMPLE 20–5

Given the block of code shown in Figure 20–8 (left), add the characters as shown in Figure 20–8 (right).

Characters to be added

G00 G90 X-.35 S1000 M03 | G00 G90 X-.35 Y.25 S1000 M03

Before Adding | After Adding

FIGURE 20–8 Adding characters.

Move the mouse cursor to the location *where* the characters are to be *added*

G00 G90 X-.35 S1000 M03

Tap the Space bar [SPACE BAR] and *type* in the characters to be *added.*

G00 G90 X-.35 Y.25 S1000 M03

■■

Deleting Characters from a Block

■ EXAMPLE 20–6

Given the block of code shown in Figure 20–9 (left), delete characters as shown in Figure 20–9 (right).

Characters to be deleted

G01 Z-.375 Y.5 F6.0 | G01 Z-.375 F6.0

Before Deleting | After Deleting

FIGURE 20–9 Deleting characters.

Move the mouse cursor to the location where the characters are to be *deleted*

G01 Z-.375 Y.5 F6.0

Repeatedly tap the Del key [Del] until all the characters to be deleted are *gone*

G01 Z-.375 F6.0

■■

20-15 USING INSERT COMMANDS

Selected insert commands are discussed in this section.

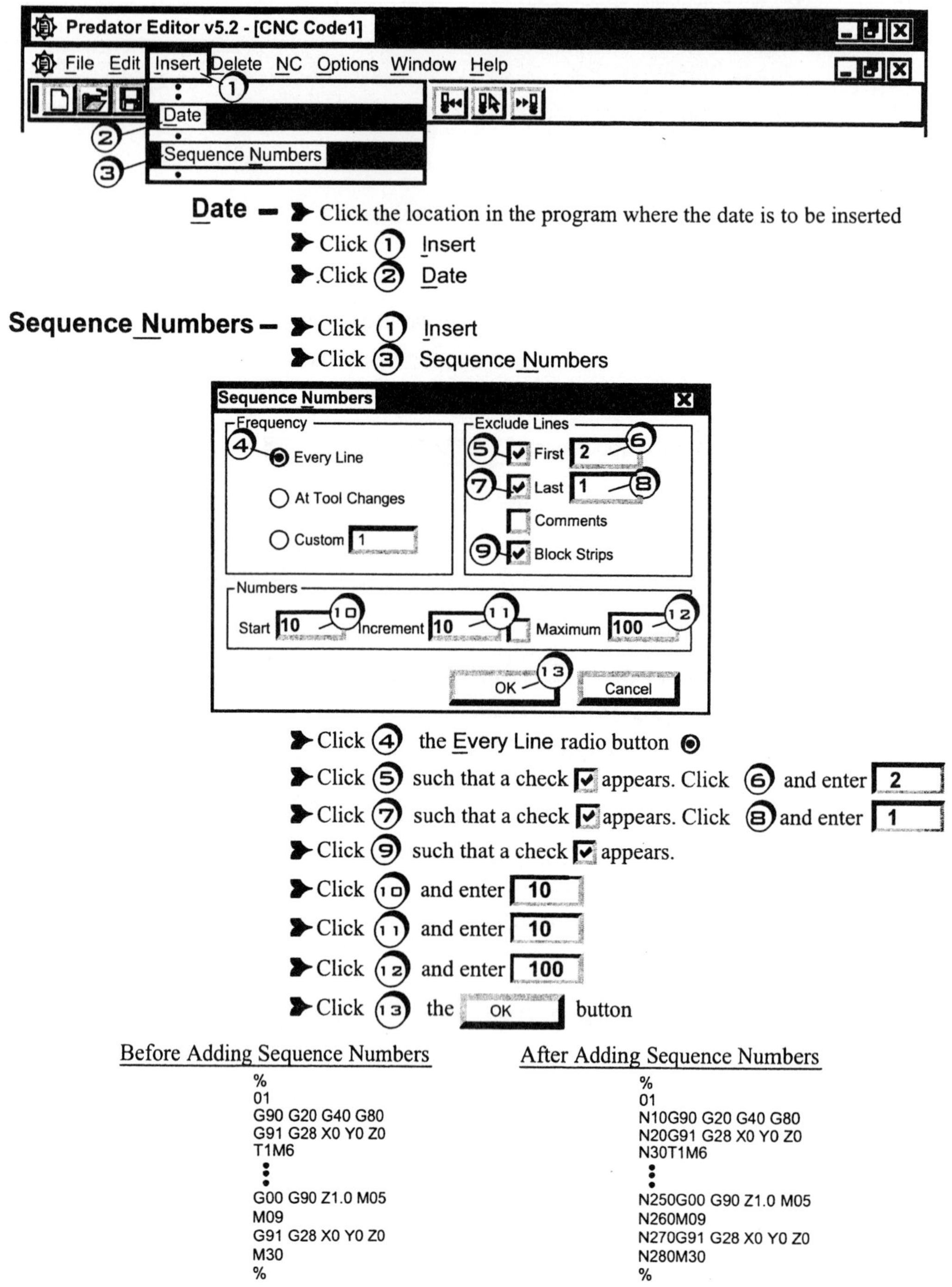

Before Adding Sequence Numbers	After Adding Sequence Numbers
%	%
01	01
G90 G20 G40 G80	N10G90 G20 G40 G80
G91 G28 X0 Y0 Z0	N20G91 G28 X0 Y0 Z0
T1M6	N30T1M6
⋮	⋮
G00 G90 Z1.0 M05	N250G00 G90 Z1.0 M05
M09	N260M09
G91 G28 X0 Y0 Z0	N270G91 G28 X0 Y0 Z0
M30	N280M30
%	%

20–16 USING FILE COMMANDS

Important file commands used within the CNC Editor shown in Figure 20–10 are presented in this section.

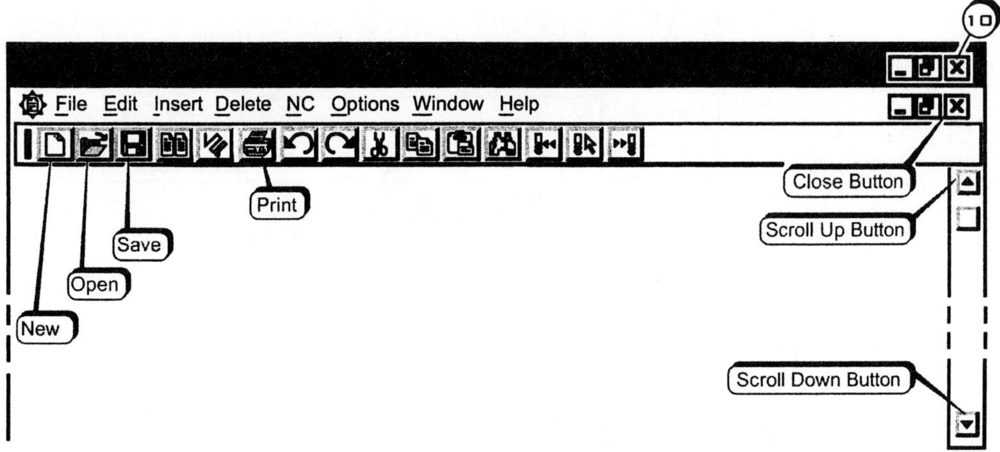

FIGURE 20–10 Important CNC Editor Pile commands.

Clicking the **NEW ICON** causes the editor to create a *new* CNC program.

Clicking the **OPEN ICON** causes the editor to open an *existing* CNC program. The editor will display the Open dialog box shown below.

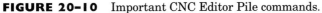

- Click ① the down button
- Click ② the drive on which the program is stored : 31/2 Floppy[A:] drive
- Click ③ the folder PROGRAMS where the CNC program is stored

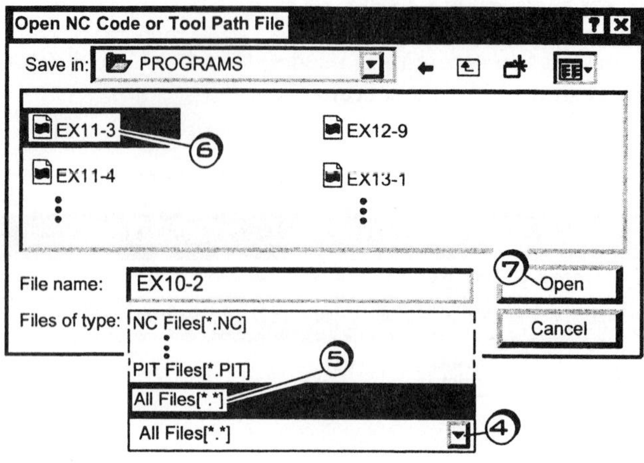

➤ Click ④ the down button

➤ Click ⑤ All Files[*.*]

➤ Click ⑥ the program file name

➤ Click ⑦ the Open button

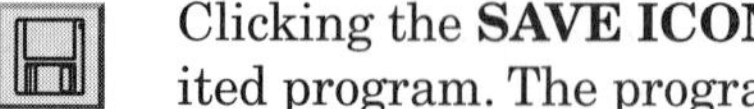

Clicking the **SAVE ICON** causes the editor to *save* the *latest version* of the edited program. The program will be saved under its *current* file name.

> ***Save As:*** The Save As command enables the operator to save a *new* program under a *desired name* in a *specified folder*. It also allows for saving an *existing open* program under a *new name* in a *specified folder*.

■ EXAMPLE 20–7

Save the program created in Example 20–4 on the disk in the file folder PROGRAMS. Use the file name **EX11–3.** ■■

➤ Click ① the File pull down menu

➤ Click ② the Save As... command

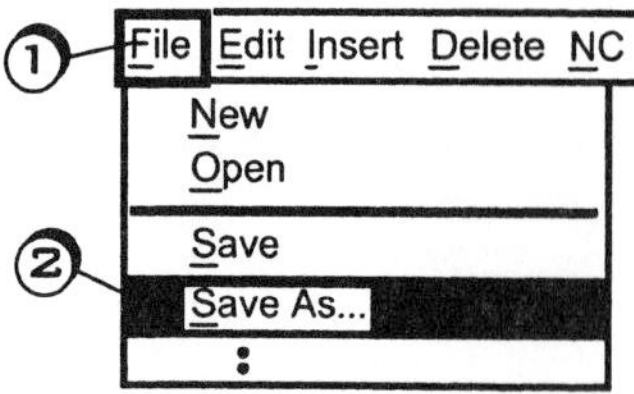

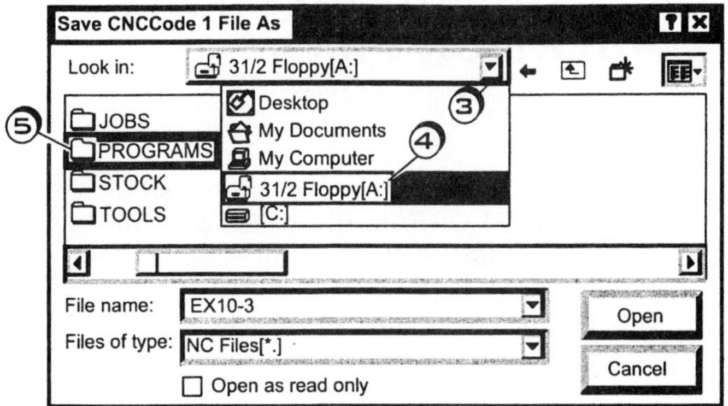

➤ Click ③ the down button

➤ Click ④ the drive on which the program is stored : 31/2 Floppy[A:] drive

➤ *Double* Click ⑤ the folder PROGRAMS where the CNC program is stored

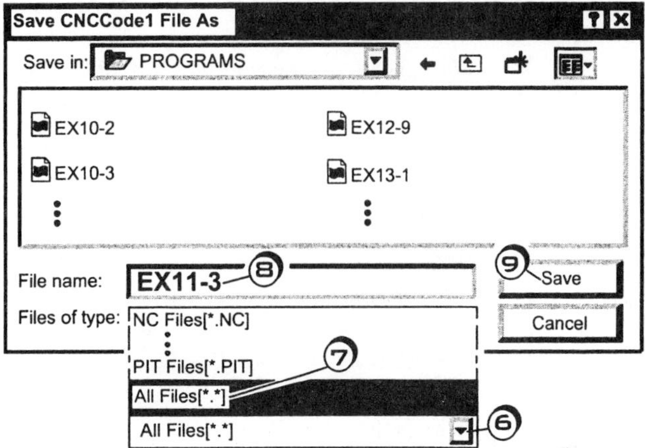

➤ Click ⑥ the down button

➤ Click ⑦ All Files[*.*]

➤ Click ⑧ in the file name box and enter : **EX11-3**

➤ Click ⑨ the Save button

➤ *Exit the CNC Editor.* Click ⑩ the close button See p475

Note: Any *valid Windows 98,2000 or NT file name* can be used as a program name

- A file name can have up to *255* characters including an optional file extension
- A file name *cannot have* the following characters: * , ; [ ] + = \ / : < > .

20-17 STARTING THE VIRTUAL CNC AND CREATING SIMULATION FILES

The Predator Virtual CNC uses the following files to generate a machining simulation:

- The JOB file, which contains references to machine, stock, and tool files
- The PROGRAM file, which contains the CNC word address program
- The MACHINE file, which contains the CNC machine settings
- The STOCK file, which contains the stock settings
- The TOOLS file, which contains the tool settings

■ EXAMPLE 20–8

Start Predator's Virtual CNC and create a new job file consisting of the CNC program file **EX11–3** saved in Example 20–7 and the setup file **EX11–3–01**. ■■

Start the Predator Virtual CNC

➤*Double* Click ① the Predator Virtual CNC icon. See Figure 20-11

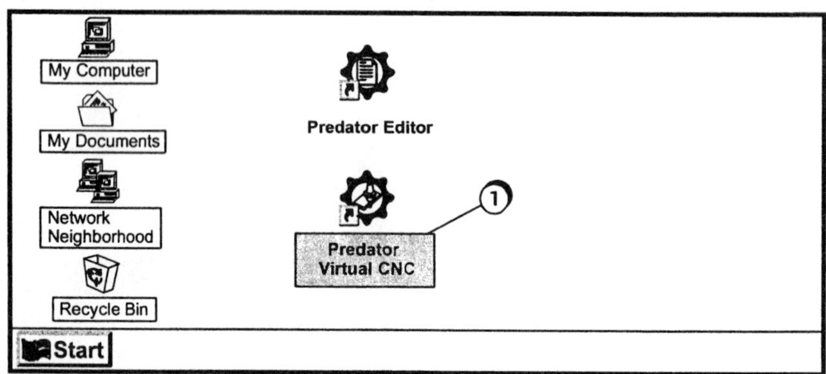

FIGURE 20–11 Starting the Predator Virtual CNC from the Windows desktop.

The system will open a Virtual CNC session as follows:

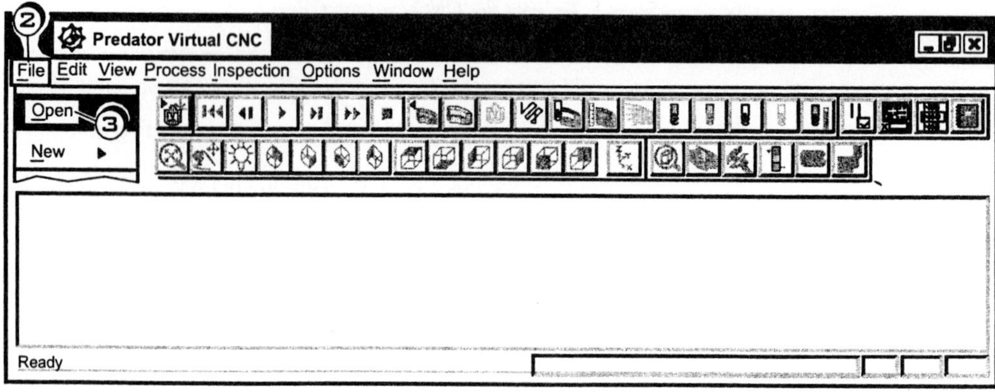

➤Click ② the File Pull down menu

➤Click ③ the Open command

Create a New Job File On C Drive

➤Click ④ the New Job button

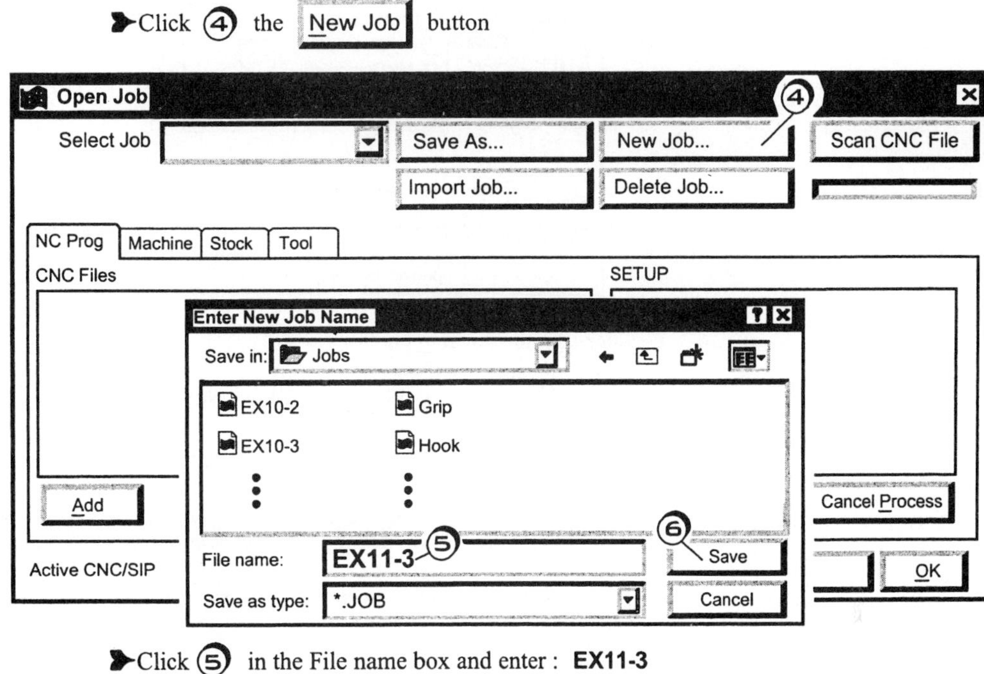

➤Click ⑤ in the File name box and enter : **EX11-3**

➤Click ⑥ Save

Copy the Job File from C drive to the Disk

The default setting for the Predator Virtual CNC is to save JOB, MACHINE, STOCK, and TOOLS files to the C drive. The Save As option ***cannot*** be used to save these files to the disk. Instead, the Copy and Paste commands ***must*** be entered to save these files to the disk.

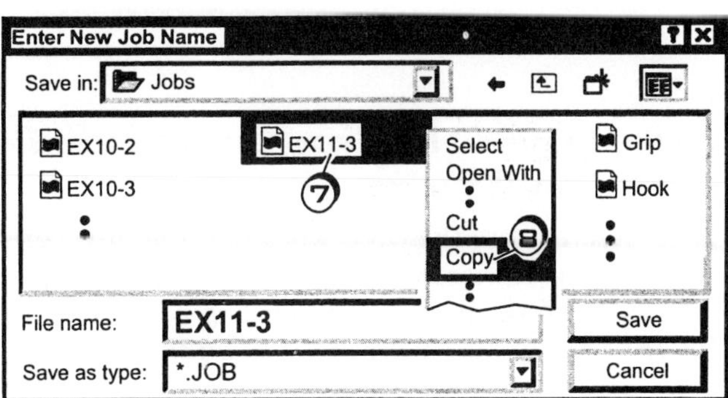

➤*Right* Click ⑦ on the job file to copy EX11-3

➤Move the cursor down and click ⑧ Copy

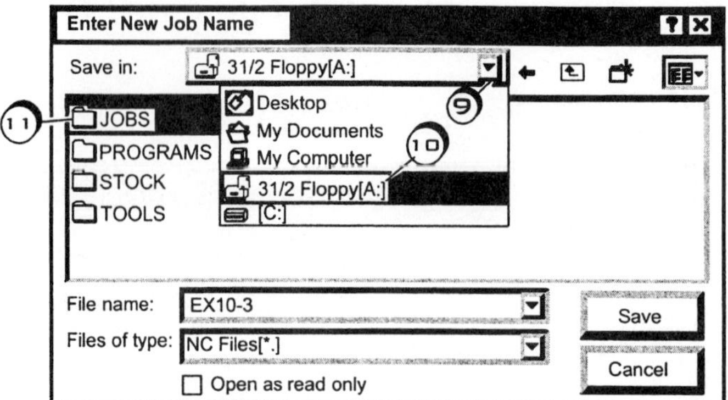

➤ Click (9) the down button

➤ Click (10) the desired drive : 31/2 Floppy[A:]

➤ *Double* Click (11) the folder JOBS where the file is to be stored on the diskette

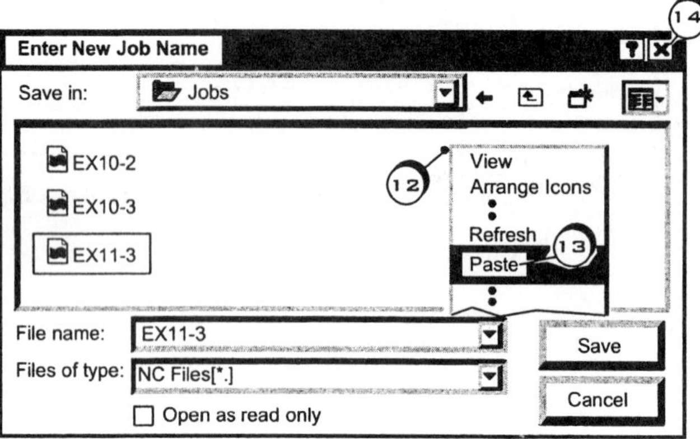

➤ *Right* Click (12) ; move the cursor down and click (13) Paste

➤ Click (14) the close button

Add the CNC Program File on Disk to the New Job File on C Drive

A reference to the CNC word address program file created in section 20–3 via the CNC Editor ***must now be added*** to the new job file **EX11–3.** The Virtual CNC will display its simulations based on the ***word address program file that is currently referred to by the the job file.***

➤Click (15) the Add button

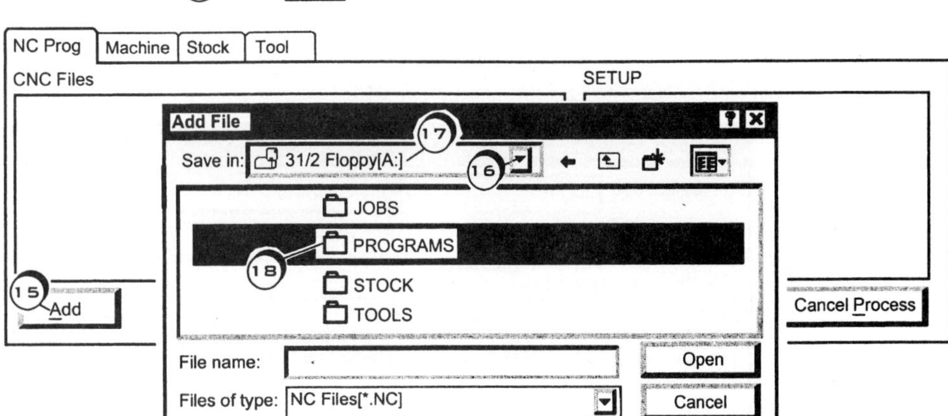

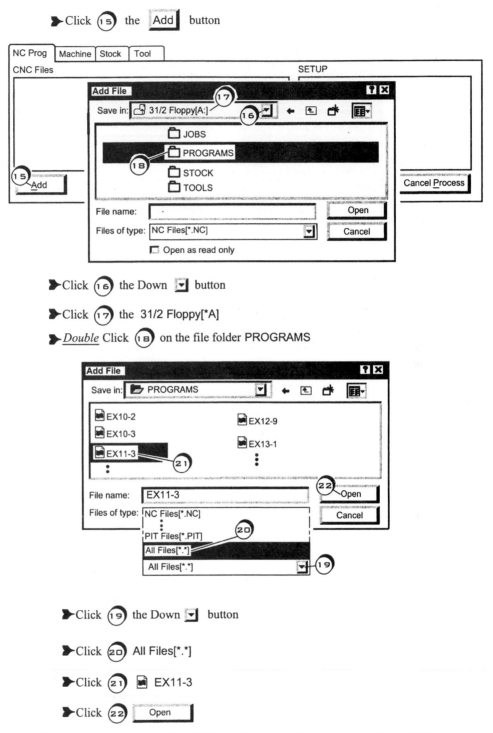

➤Click (16) the Down button

➤Click (17) the 31/2 Floppy[*A]

➤*Double* Click (18) on the file folder PROGRAMS

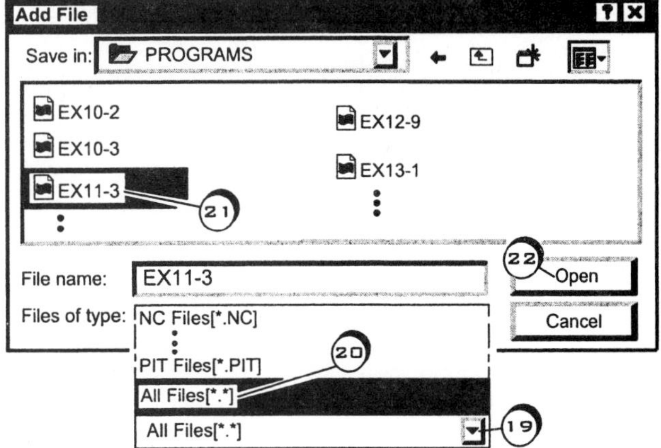

➤Click (19) the Down button

➤Click (20) All Files[*.*]

➤Click (21) EX11-3

➤Click (22) Open

Reference to the program file EX11-3 will then be added to the job file on C drive.

The JOB file EX 11-3 and the PROGRAM file EX 11-3-01 will appear listed in the Open Job dialog box as shown in Figure 20–12.

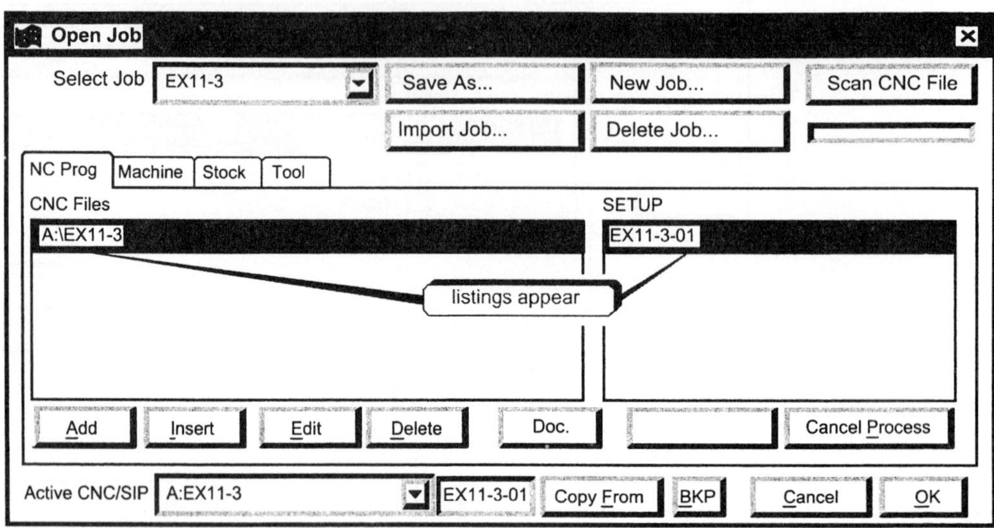

FIGURE 20–12 The new job window after creating the job and adding the program file.

20–18 ENTERING DATA INTO THE SETUP FILE ON C DRIVE

Specifying the CNC Machine Tool

■ EXAMPLE 20–9

Input the CNC machine parameters for CNC job file **EX11–3.**

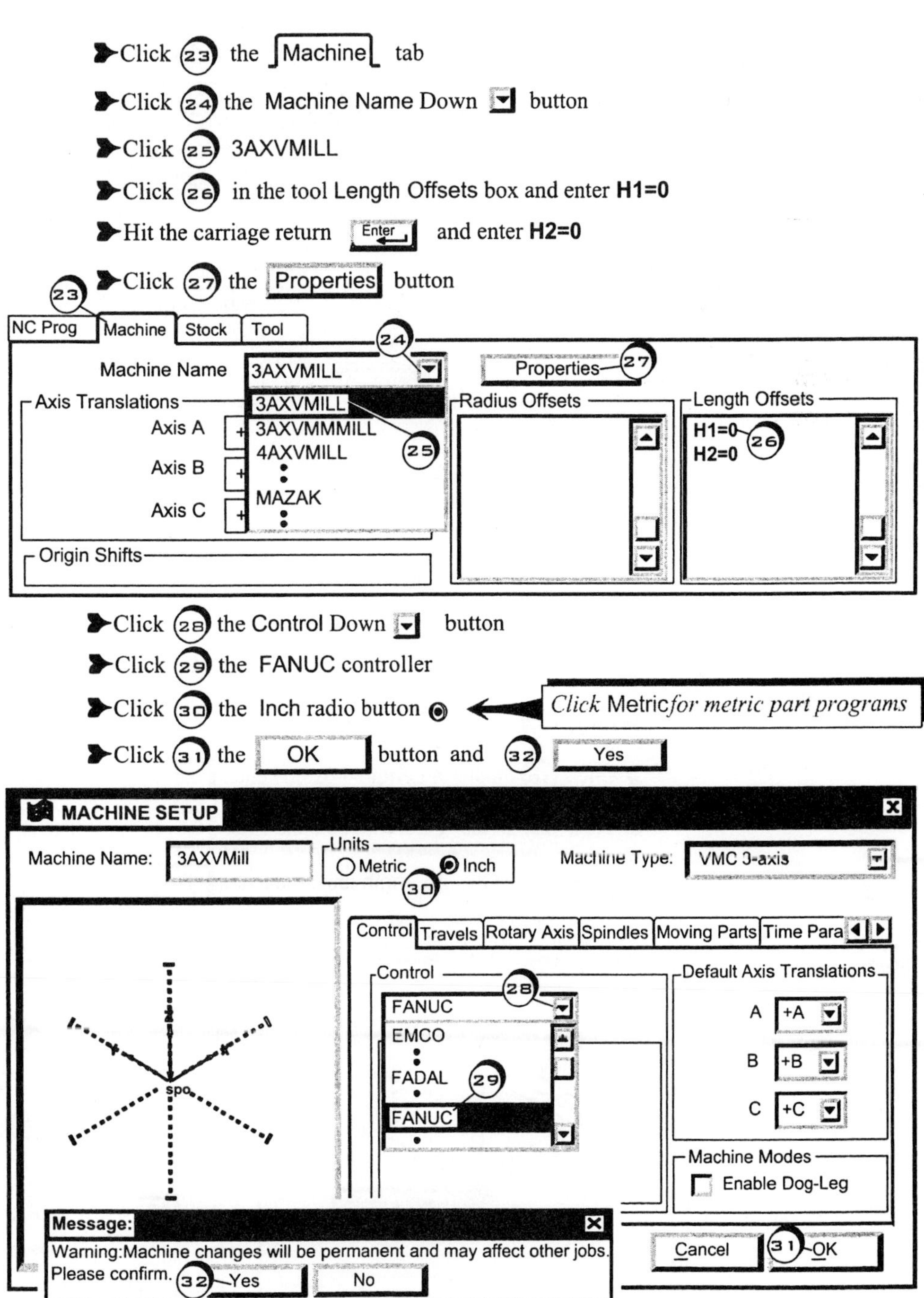

■■

Specifying the Stock

■ EXAMPLE 20–10

The stock size and part origin for CNC job file **EX11–3** is shown in Figure 20–13. Input this data into Predator's Virtual CNC.

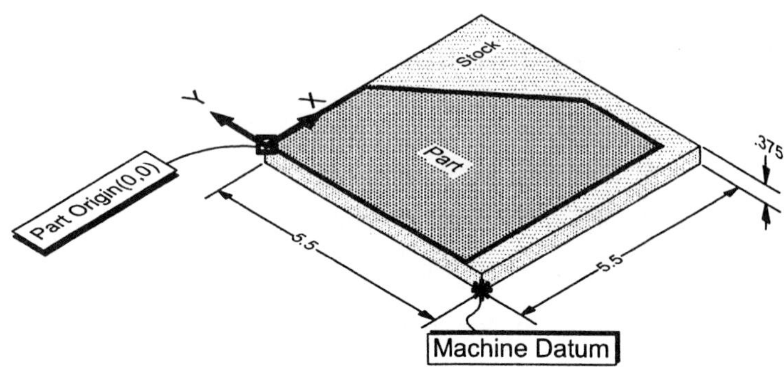

FIGURE 20–13 The part origin with respect to the machine datum.

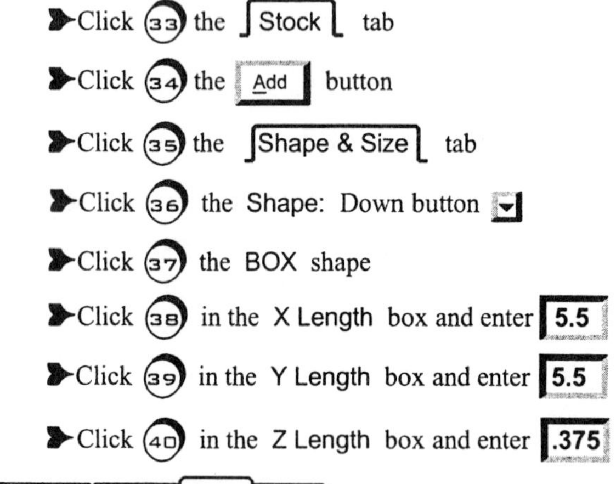

Click 33 the Stock tab
Click 34 the Add button
Click 35 the Shape & Size tab
Click 36 the Shape: Down button
Click 37 the BOX shape
Click 38 in the X Length box and enter 5.5
Click 39 in the Y Length box and enter 5.5
Click 40 in the Z Length box and enter .375

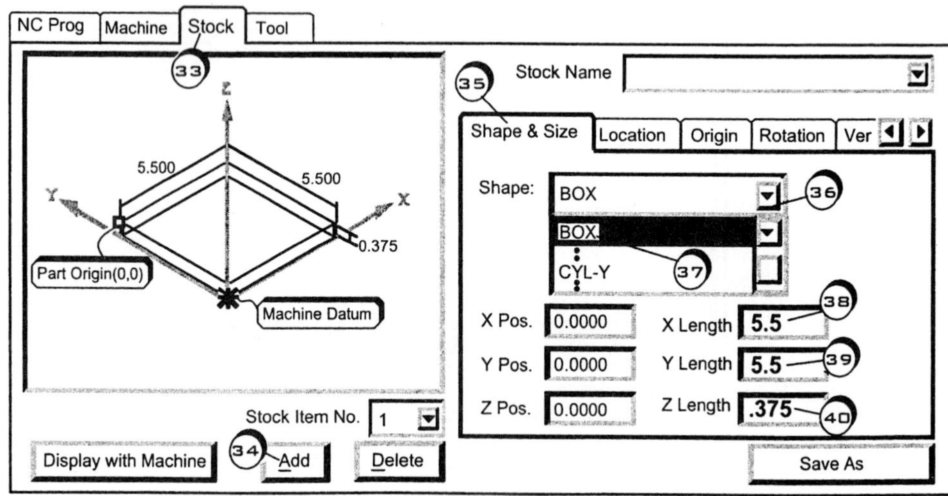

➤Click (41) the Origin tab

➤Click (42) the X-Y Origin Down ▼ button

➤Click (43) minX maxY

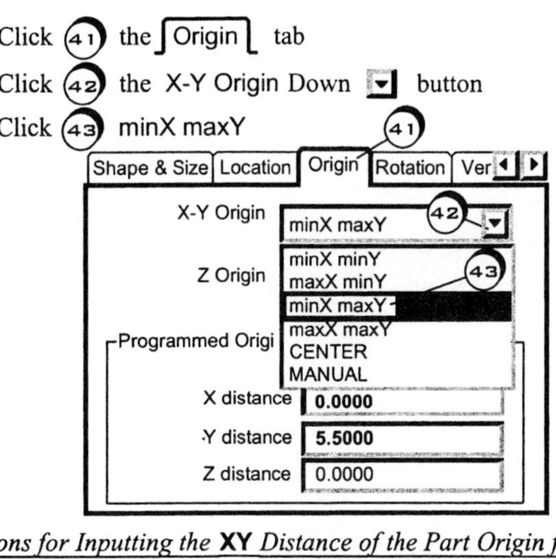

Options for Inputting the **XY** *Distance of the Part Origin from Machine Datum*

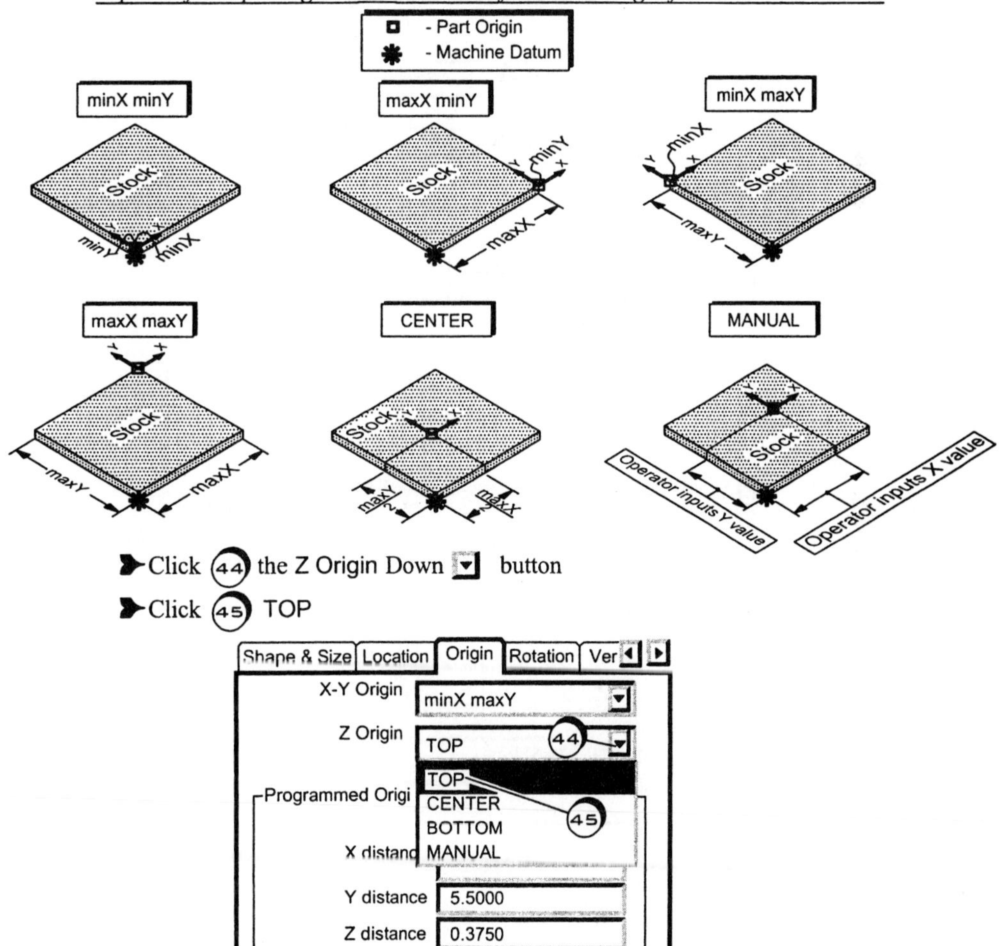

➤Click (44) the Z Origin Down ▼ button

➤Click (45) TOP

*Options for Inputting the **Z** Distance of the Part Origin from Machine Datum*

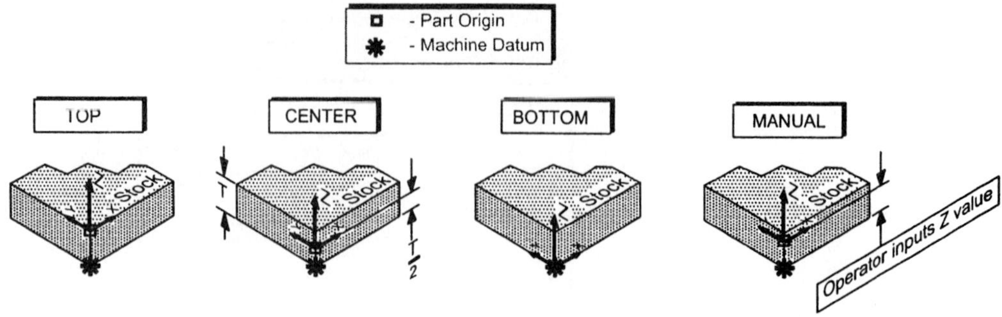

Save the Stock File for the Job on C Drive

➤ Click (46) Save As

➤ Click (47) Save

The stock name EX11-3-01 will *now appear* in the Stock Name box

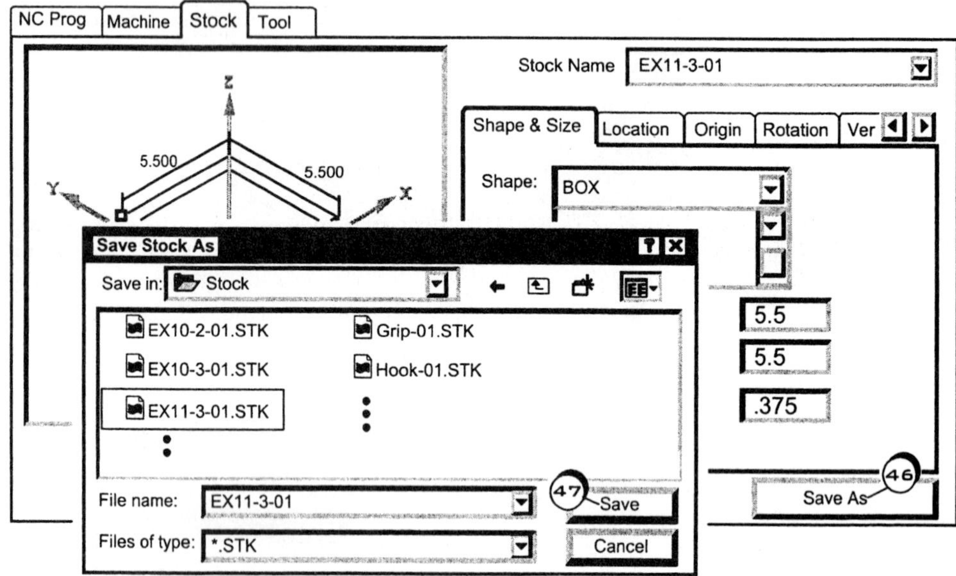

Copy the Stock File from C drive to the Disk.

➤ Click (46) Save As

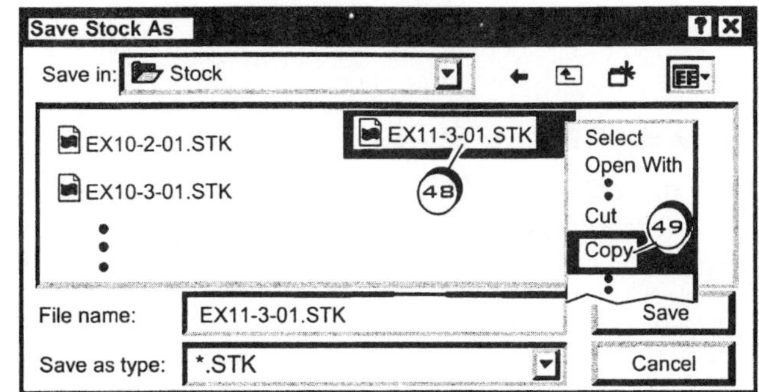

➤ *Right* Click (48) on the stock file to copy EX11-3-01.STK

➤ Move the cursor down and click (49) Copy

➤ Click (50) the down button

➤ Click (51) the desired drive : 31/2 Floppy[A:]

➤ *Double* Click (52) the folder STOCK where the file is to be stored on the diskette

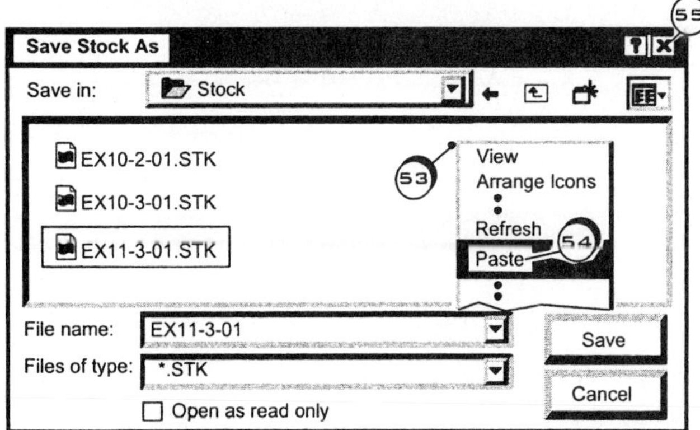

➤ *Right* Click (53) ; move the cursor down and click (54) Paste

➤ Click (55) the close button

Specify the Tooling

■ EXAMPLE 20–11

The tooling for CNC job file **EX11–3** follows. Input this data into Predator's Tools File on C drive. ■■

Tool	Tooling
1	0.5D end mill
2	0.4375D drill

> ***Important:*** The virtual CNC must first know the tool numbers of the tools in the word address program **EX11–3** contained within the job file. The **Scan CNC File** button directs the virtual CNC to scan the part program and look for tool number codes such as **T1M6, T2M6,** etc. When the virtual CNC finds a tool number code it automatically copies it into virtual CNC's tool page dialog box. The operator then opens the tool page dialog box *only after this scanning operation has been completed.*

➤ Click (56) Scan CNC File

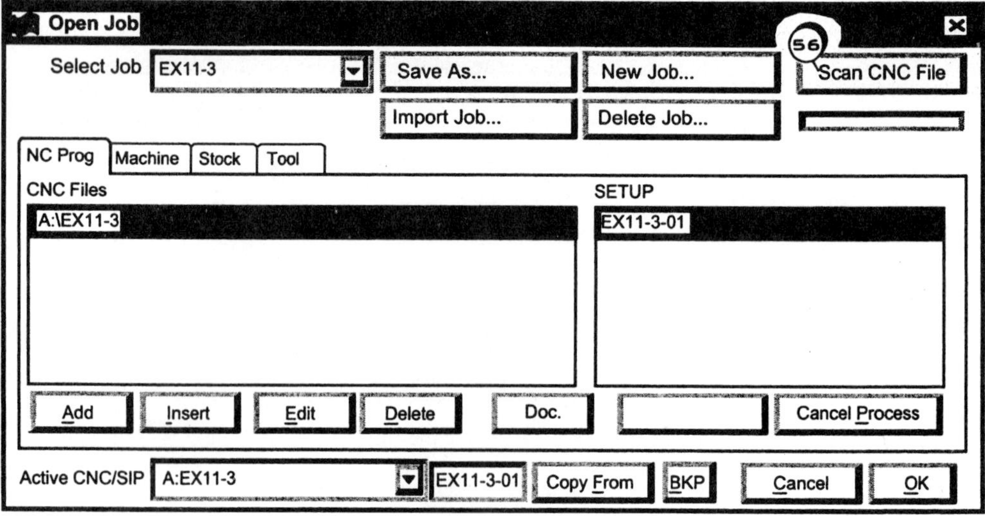

➤Click (57) the Tool tab

➤Click (58) the Tool Definition tab

➤Click (59) in the Tool type Down button

➤Click (60) the FLAT end mill tool

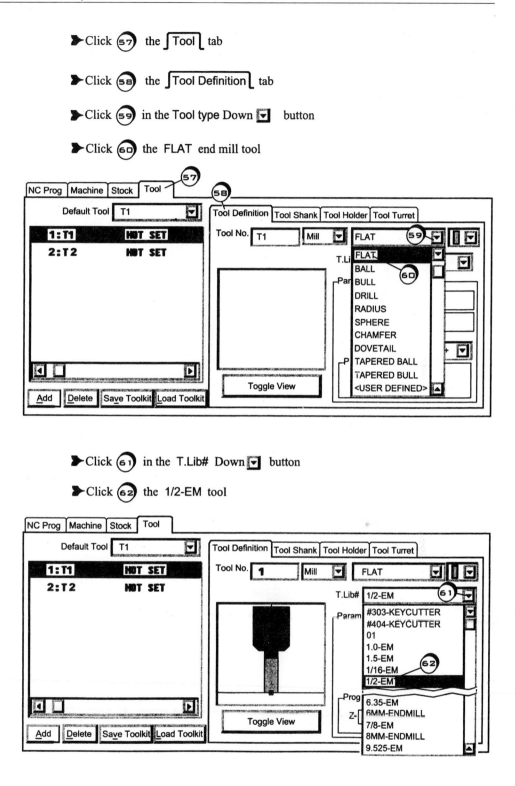

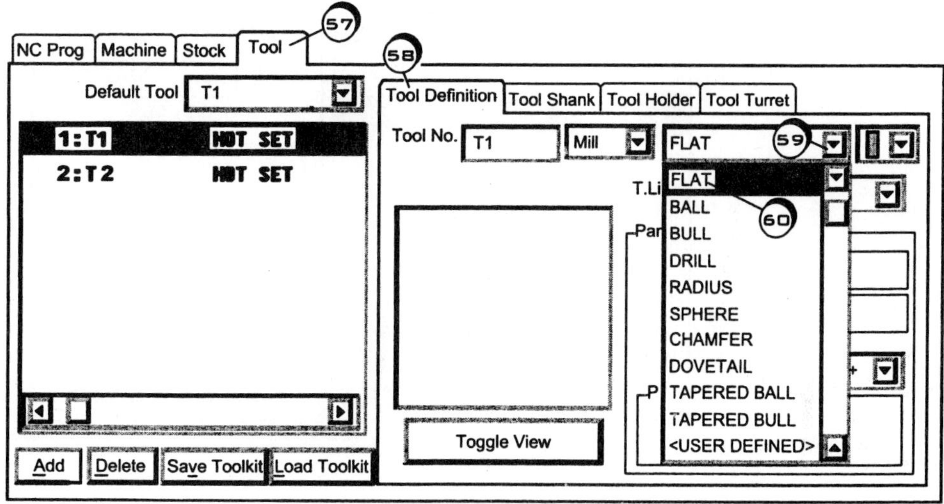

➤Click (61) in the T.Lib# Down button

➤Click (62) the 1/2-EM tool

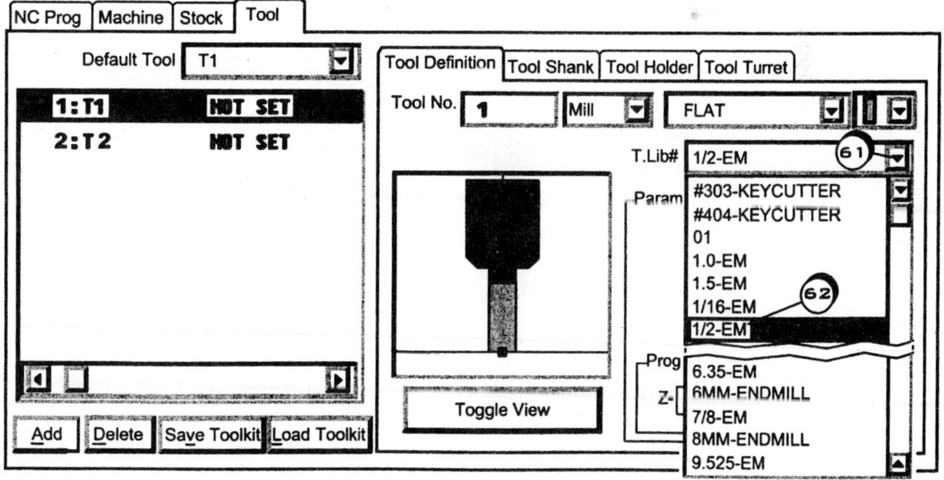

➤Click (63) 

➤Click (64) in the Tool type Down button

➤Click (65) the DRILL tool

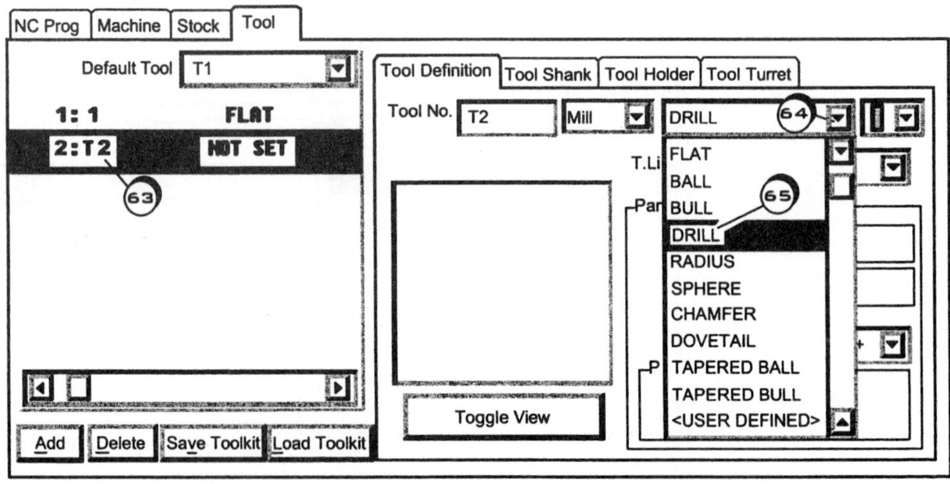

➤Click (66) the T.Lib# Down button

➤Click (67) the 7/16-DRILL tool

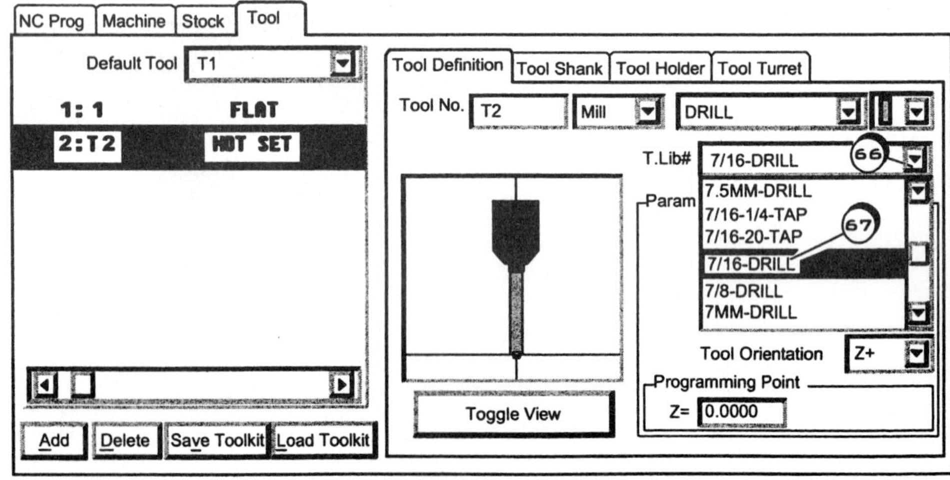

Options for Customizing the Tool Shank Dimensions

■ EXAMPLE 20–12

Input the Tool Shank dimensions in the table below into Predator's Virtual CNC.

Tool	Tooling	Tool Shank Height (in)	Tool Shank Diameter (in)
1	0.5D end mill	0.25	0.5

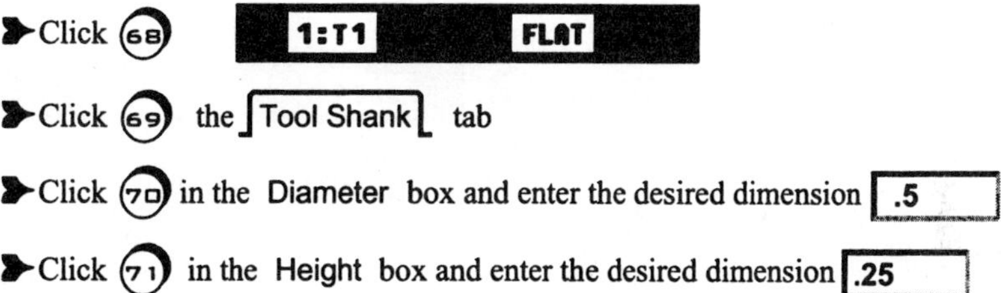

Click 68 1:T1 FLAT
Click 69 the Tool Shank tab
Click 70 in the Diameter box and enter the desired dimension .5
Click 71 in the Height box and enter the desired dimension .25

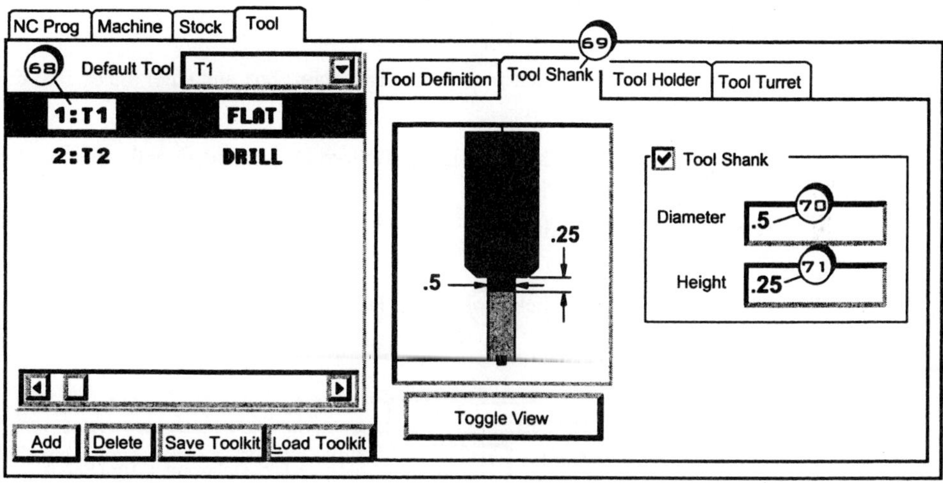

Options for Customizing the Tool Holder Dimensions

■ EXAMPLE 20–13

Set the Tool Holder shape to CHAMFER at .5 in and the Tool Shank Height at .25 in for the 1/2-EM tool listed in Example 20–12.

Tool	Tooling	Tool Holder Bottom Shape	Bottom D	Chamfer Angle (deg)	Holder Height (in)
1	0.5D end mill	CHAMFER	1.380	45	2.0

■■

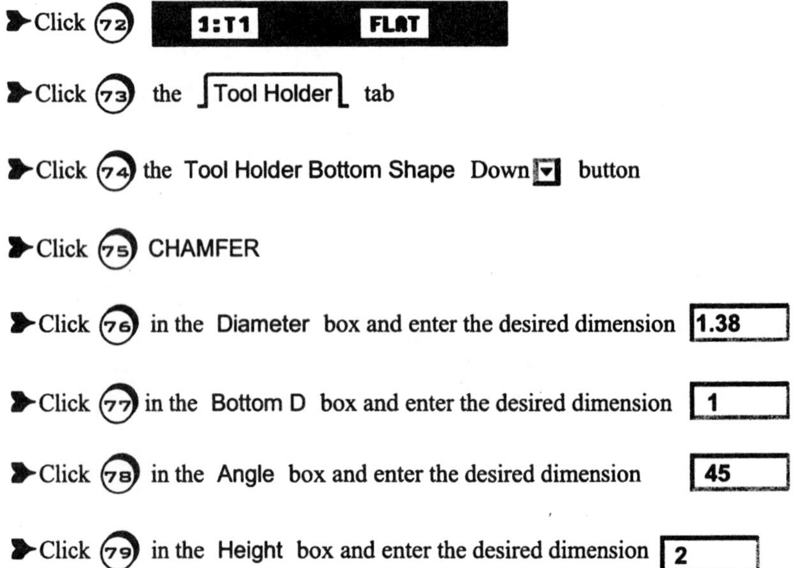

Click 72 1:T1 FLAT
Click 73 the Tool Holder tab
Click 74 the Tool Holder Bottom Shape Down button
Click 75 CHAMFER
Click 76 in the Diameter box and enter the desired dimension 1.38
Click 77 in the Bottom D box and enter the desired dimension 1
Click 78 in the Angle box and enter the desired dimension 45
Click 79 in the Height box and enter the desired dimension 2

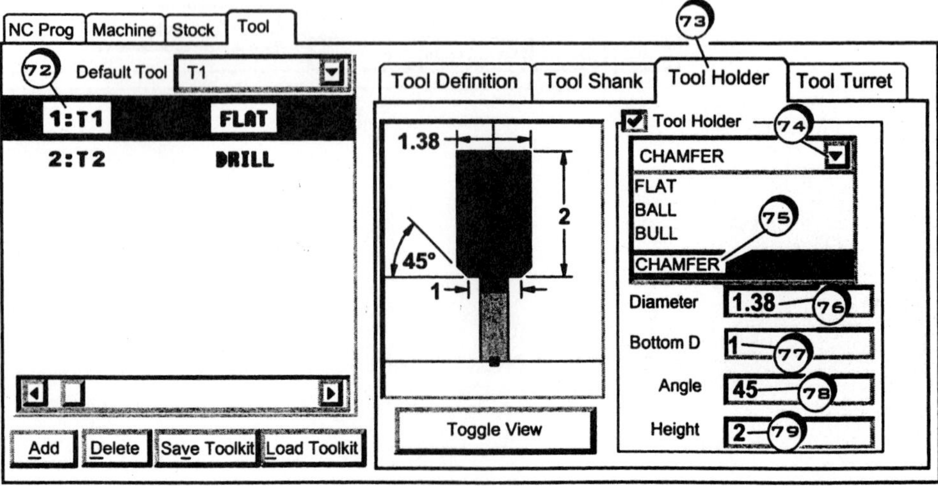

Save the Tool File for the Job on C Drive

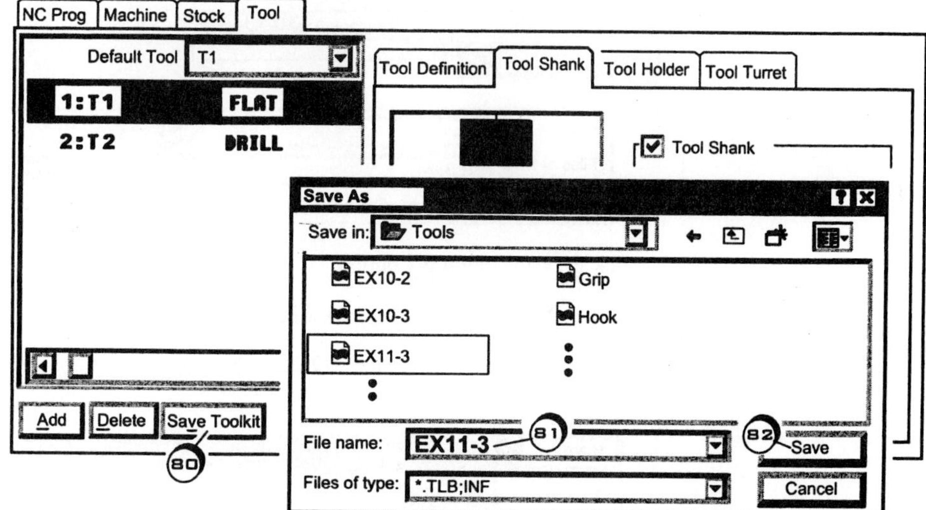

➤ Click (81) in the **File name** box and enter the name of the toolkit file to save: **EX11-3**

➤ Click (82) Save

Copy the Tool File from C drive to the Disk

➤ Click (80) Save Toolkit

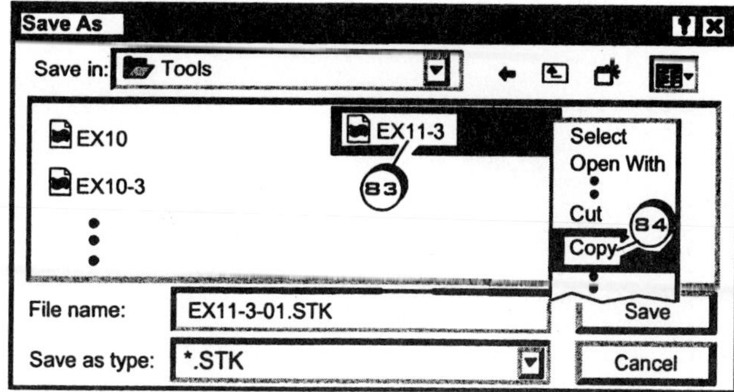

➤ ***Right*** Click (83) on the tools file to copy EX11-3

➤ Move the cursor down and click (84) Copy

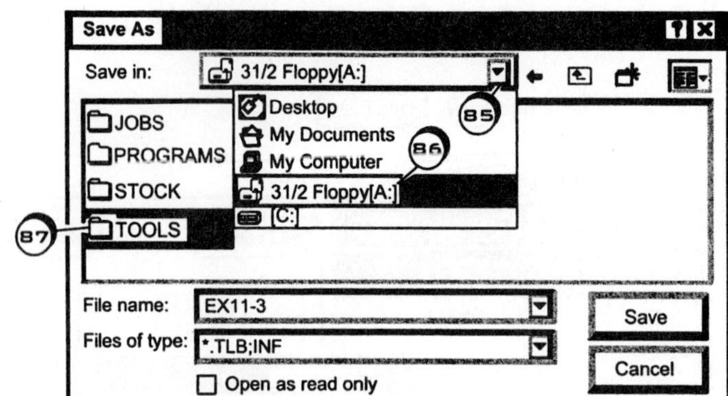

- Click (85) the down button
- Click (86) the desired drive : 31/2 Floppy[A:]
- *Double* Click (87) the folder TOOLS where the file is to be stored on the diskette

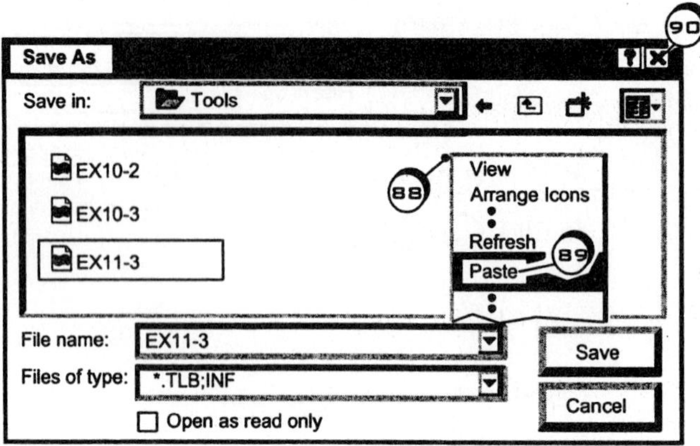

- *Right* Click (88) ; move the cursor down and click (89) Paste
- Click (90) the close button [X]

Copy the Setup File Containing Machine/Stock Settings to the Disk

➤Click (91) the NC Prog tab

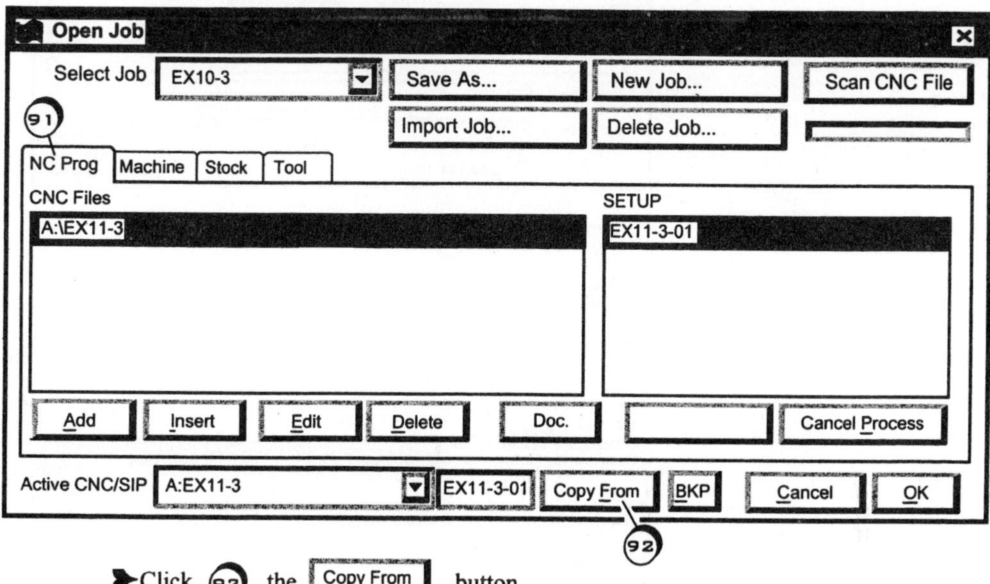

➤Click (92) the Copy From button

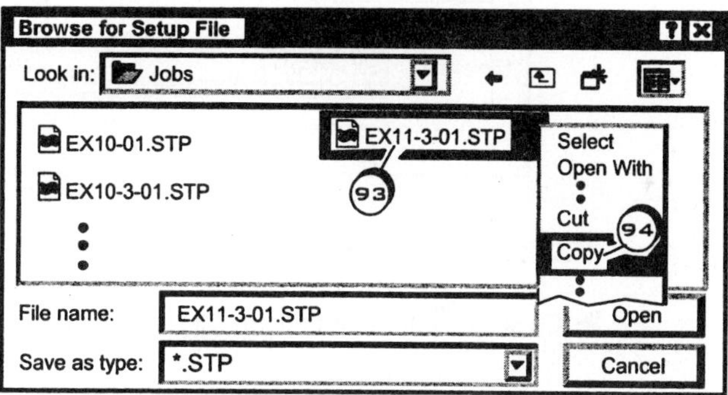

➤*Right* Click (93) on the tools file to copy EX11-3-01.STP

➤Move the cursor down and click (94) Copy

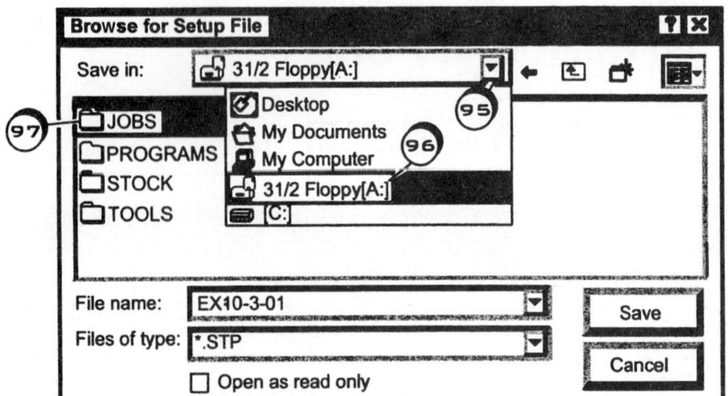

➤ Click (95) the down [▼] button

➤ Click (96) the desired drive : 31/2 Floppy[A:]

➤ *Double* Click (97) the folder JOBS where the file is to be stored on the diskette

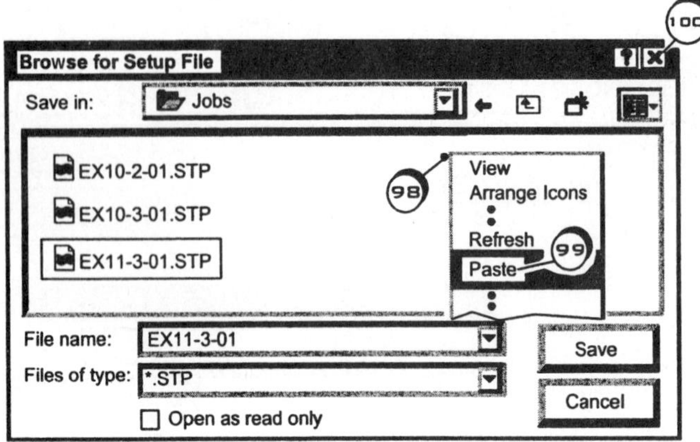

➤ *Right* Click (98) ; move the cursor down and click (99) Paste

➤ Click (100) the close button [X]

Recall that the Setup file is used to auto-load the CNC machine and stock settings into the virtual CNC simulator.

20–19 DIRECTING THE VIRTUAL CNC TO PROCESS A JOB (MILL)

■ **EXAMPLE 20–14**

Instruct Predator's Virtual CNC to process the CNC job file **EX11–3.** ■■

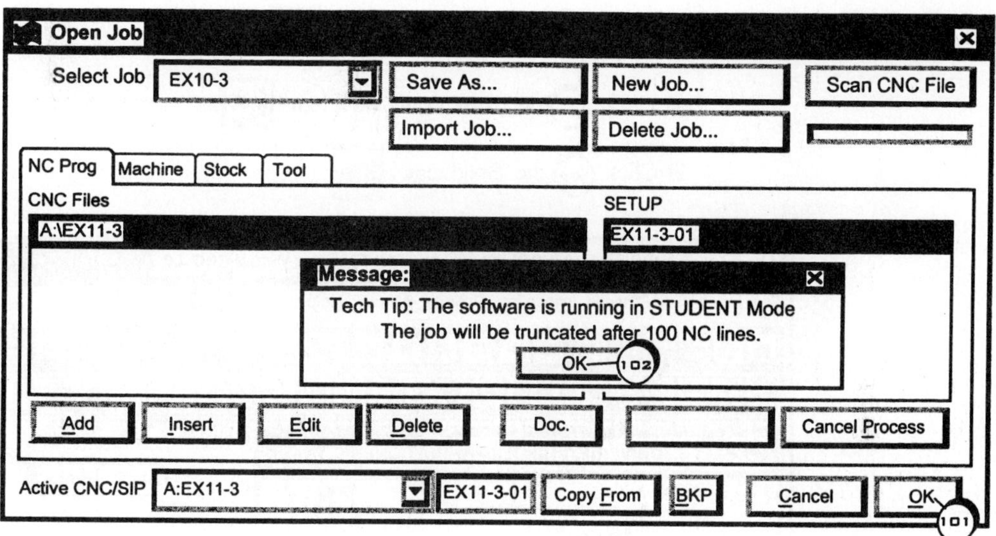

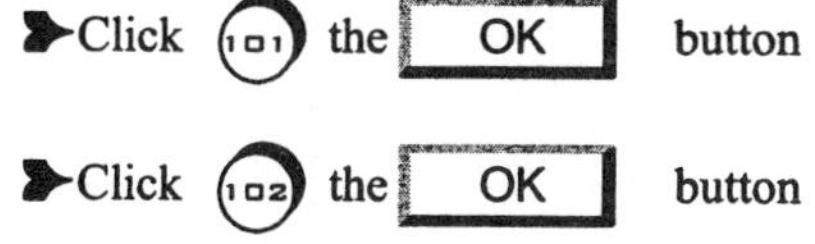

Click 101 the OK button
Click 102 the OK button

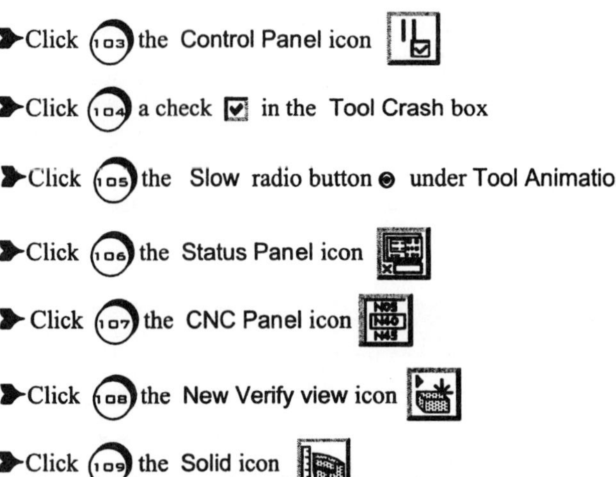

➤Click (103) the Control Panel icon

➤Click (104) a check ☑ in the Tool Crash box

➤Click (105) the Slow radio button ◉ under Tool Animation

➤Click (106) the Status Panel icon

➤Click (107) the CNC Panel icon

➤Click (108) the New Verify view icon

➤Click (109) the Solid icon

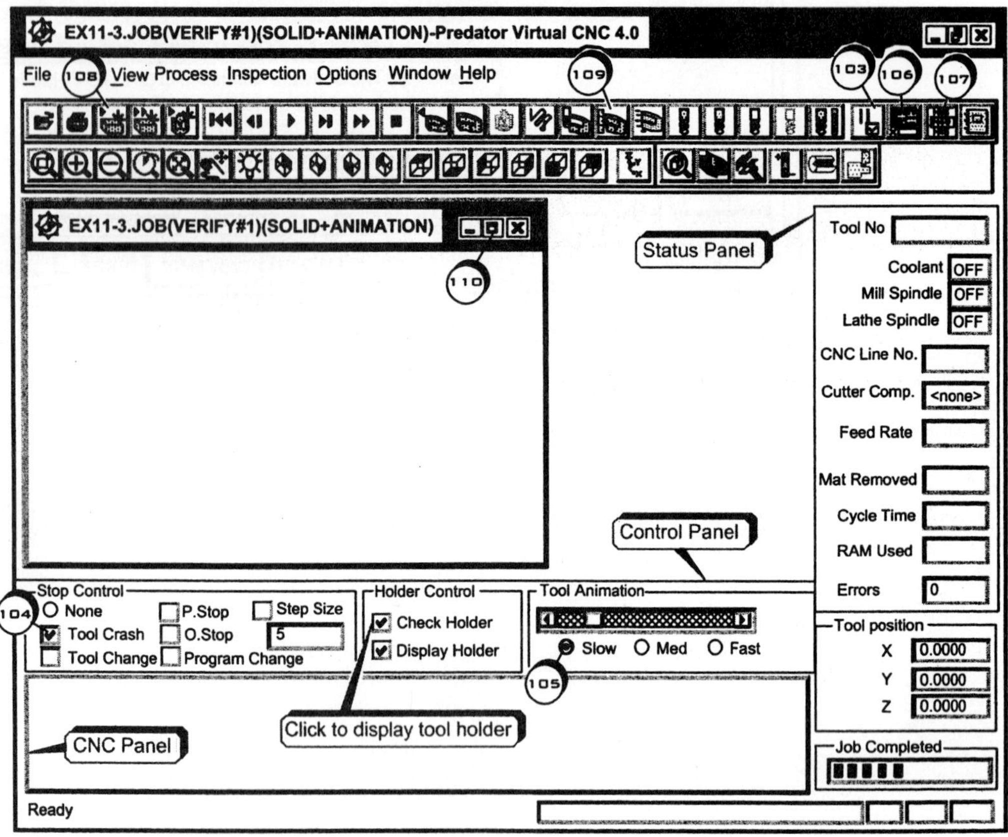

➤Click (110) the maximize button in the **SOLID+ANIMATION** window

➤Click (111) the Edit pull down menu and (112) the Background color option

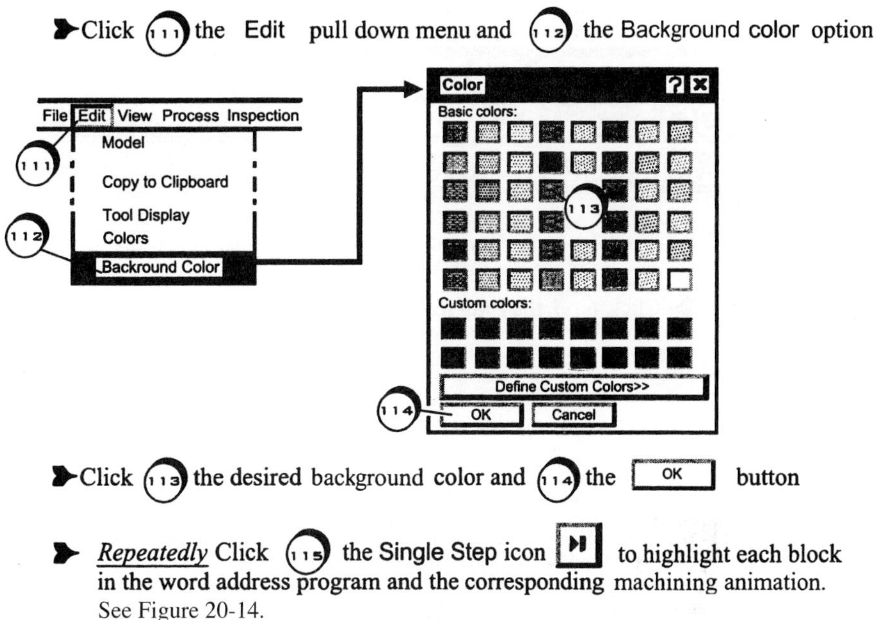

➤Click (113) the desired background color and (114) the OK button

➤ *Repeatedly* Click (115) the Single Step icon to highlight each block in the word address program and the corresponding machining animation. See Figure 20-14.

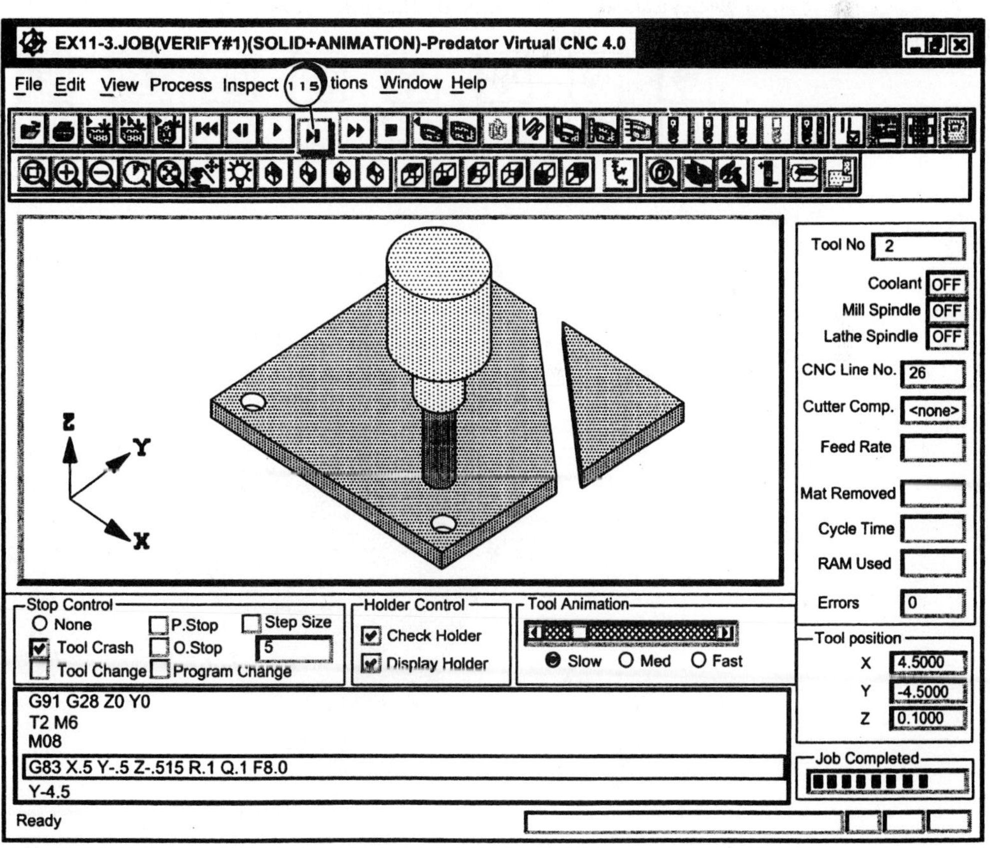

FIGURE 20–14 The machining animation produced by Predator's Virtual CNC.

- Click (116) the Rewind icon to *restart* the animation from the *beginning* of the program
- Click (117) the Play icon to see a *continuous* animation of the *entire* machining process

Viewing the Animation at Standard Orientations

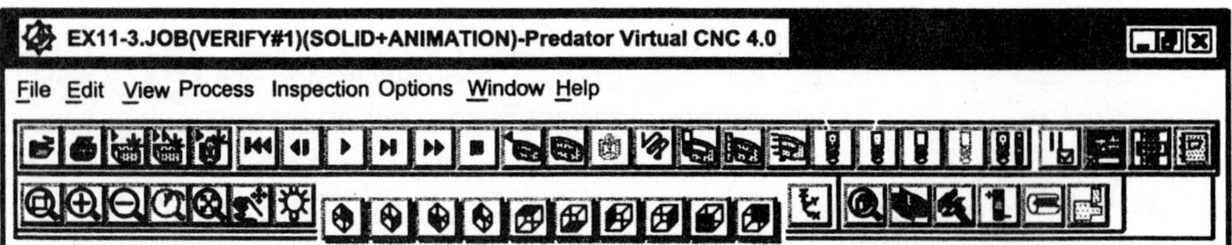

- Set a standard orientation by clicking a View icon shown below

Viewing the Animation at Any Desired Orientation

- Move the mouse cursor in the **SOLID+ANIMATION** window press the *left* mouse button and *keeping it depressed* move the mouse *up, down, left* and *right* to obtain any desired viewing orientation. *Release* the left mouse button *when done*

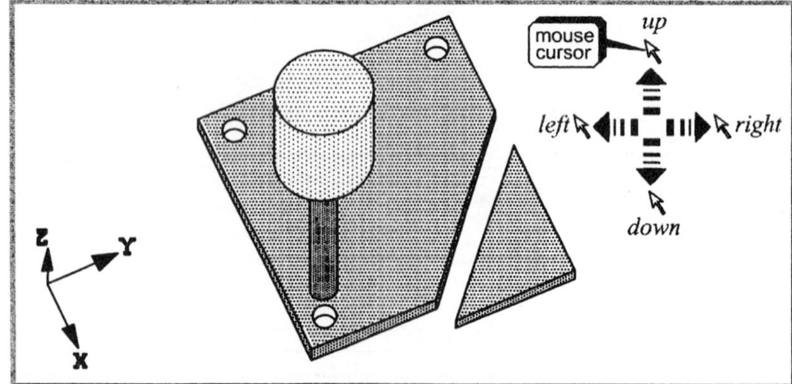

- Click (116) the Rewind icon to *restart* the animation from the *beginning* of the program
- Click (117) the Play icon to see a *continuous* animation of the *entire* machining process

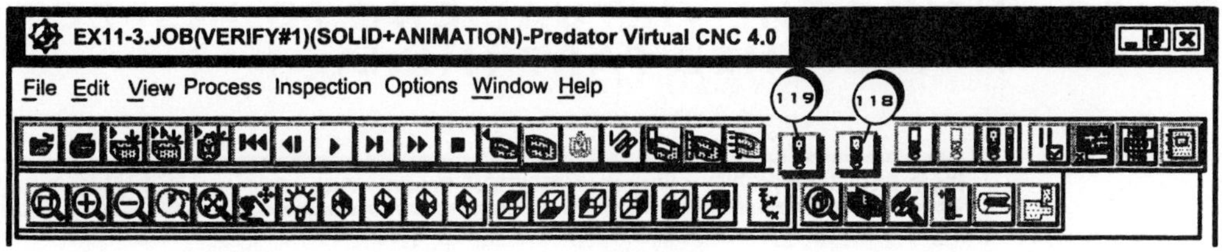

Viewing the Animation With a Translucent Tool

➤ Click (118) the Translucent Tool icon to enable a *see-through tool display*

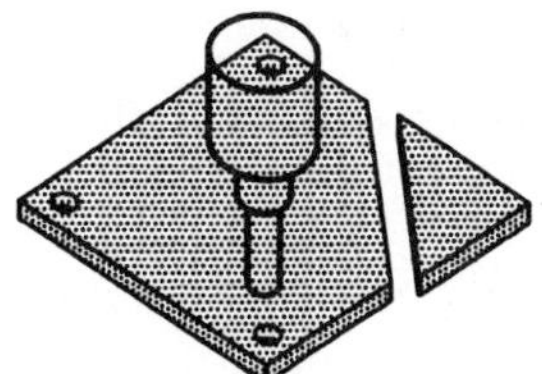

➤ Click (116) the Rewind icon to *restart* the animation from the *beginning* of the program

➤ Click (117) the Play icon to see a *continuous* animation of the *entire* machining process

Viewing the Animation With a Solid Tool

➤ Click (119) the Solid Tool icon to enable a *solid tool display*

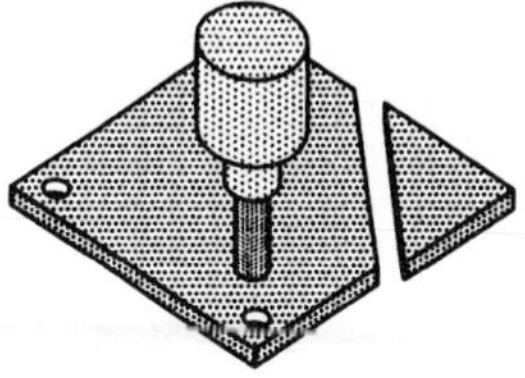

20-20 USING THE VIRTUAL CNC TO INSPECT A MACHINED PART (MILL)

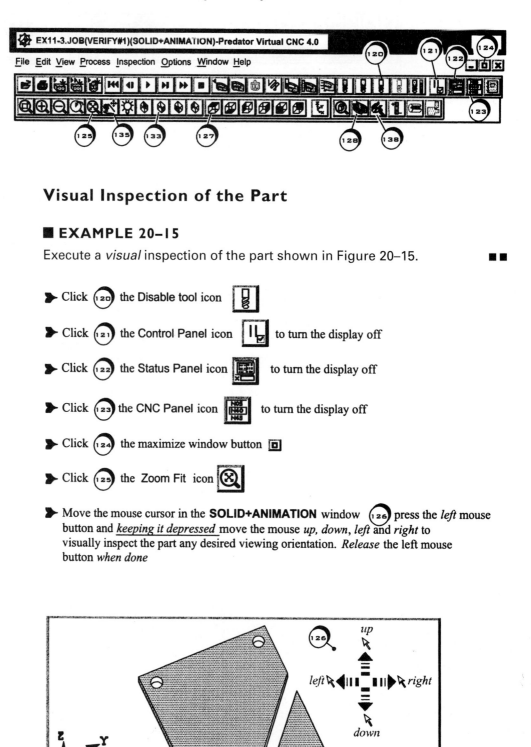

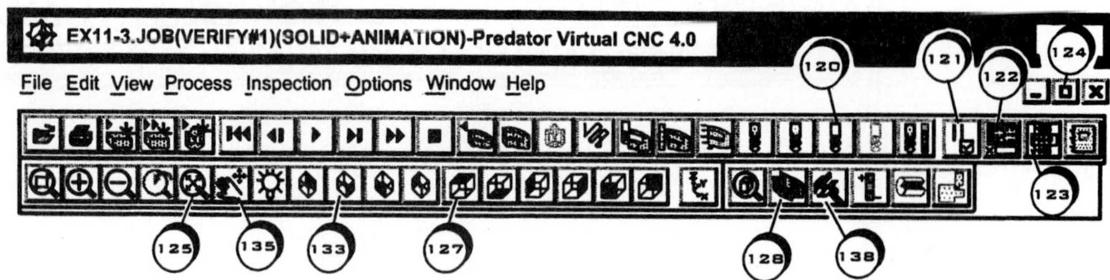

Visual Inspection of the Part

■ EXAMPLE 20-15

Execute a *visual* inspection of the part shown in Figure 20–15. ■■

- Click (120) the Disable tool icon
- Click (121) the Control Panel icon to turn the display off
- Click (122) the Status Panel icon to turn the display off
- Click (123) the CNC Panel icon to turn the display off
- Click (124) the maximize window button
- Click (125) the Zoom Fit icon
- Move the mouse cursor in the **SOLID+ANIMATION** window (126) press the *left* mouse button and keeping it depressed move the mouse *up, down, left* and *right* to visually inspect the part any desired viewing orientation. *Release* the left mouse button *when done*

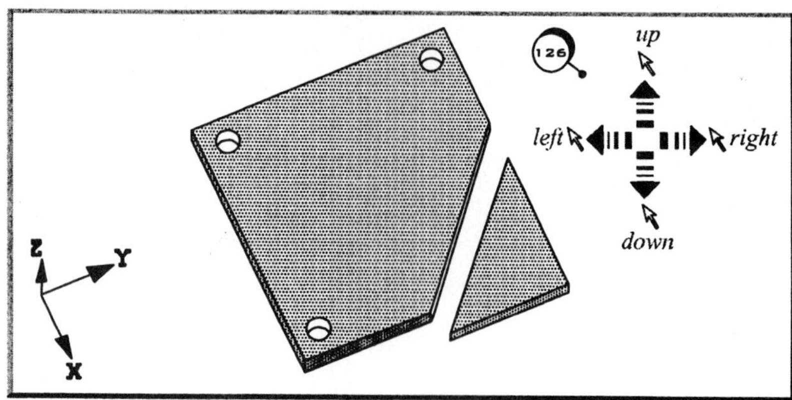

FIGURE 20-15 Executing a visual inspection of the machined part. ■■

Sectioning the Part

■ EXAMPLE 20–16

Execute inspection by *sectioning* the part shown in Figure 20–16.

- Click (127) the Top View icon . See p 502
- Click (128) the X-Section icon . See p 502

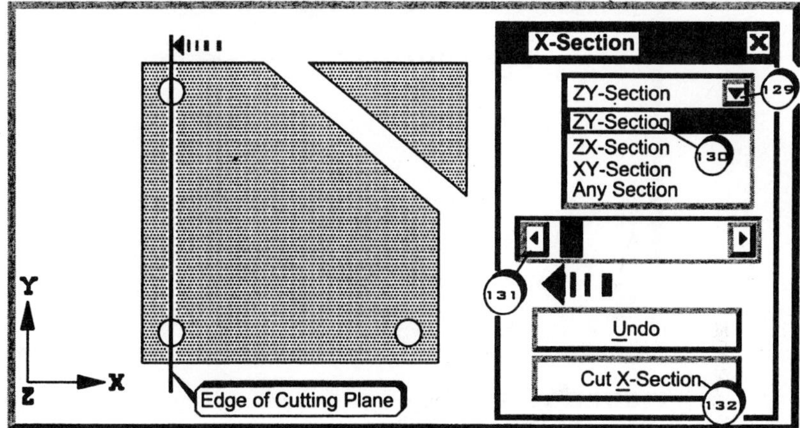

FIGURE 20–16 Specifying the cutting plane. ■■

- Click (129) the X-Section down button .
- Click (130) the ZY-Section option.
- *Repeatedly* Click (131) the Left button until the edge of the ZY cutting plane is at the desired location in the part.
- Click (132) the Cut X-Section button.
- Click (133) the Isometric View icon . See 502 .
- Click (134) the X-Section Exit button when done. See Figure 20-17.

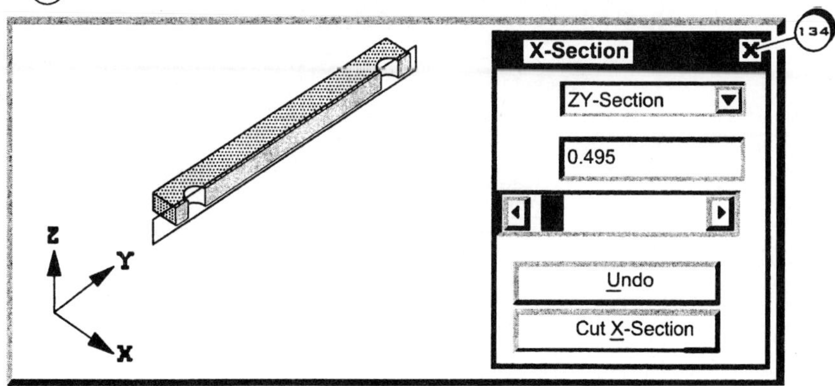

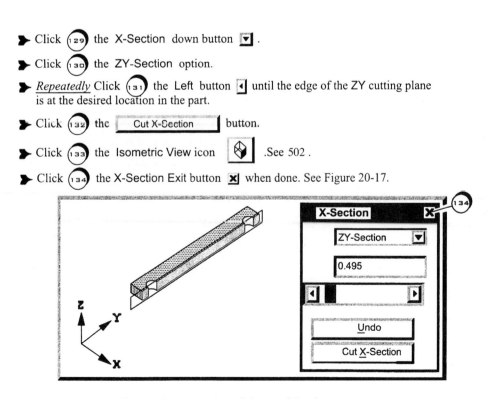

FIGURE 20–17 Inspecting a section of the machined part.

Checking the Dimensions of the Part

■ EXAMPLE 20–17

Execute inspection by checking the dimensions of the part shown in Figure 20–18.

- Click (135) the Pan icon .See p502
- Move the mouse cursor to position (136) *depress* the left mouse button and *keeping it depressed* move the cursor to position (137) and *release*

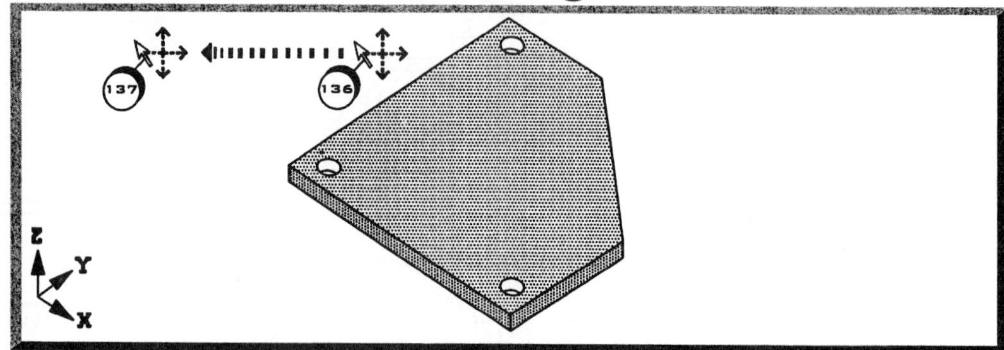

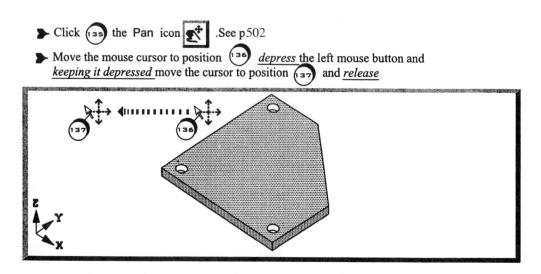

FIGURE 20–18 Panning the machined part to be inspected. ■■

Check the length of the part

- Click (138) the Inspect icon .See p 502.
- Click (139) the Features down button
- Click (140) Anything
- Click (141) the *top edge* of the part
- Click (142) the *top vertex* of the part

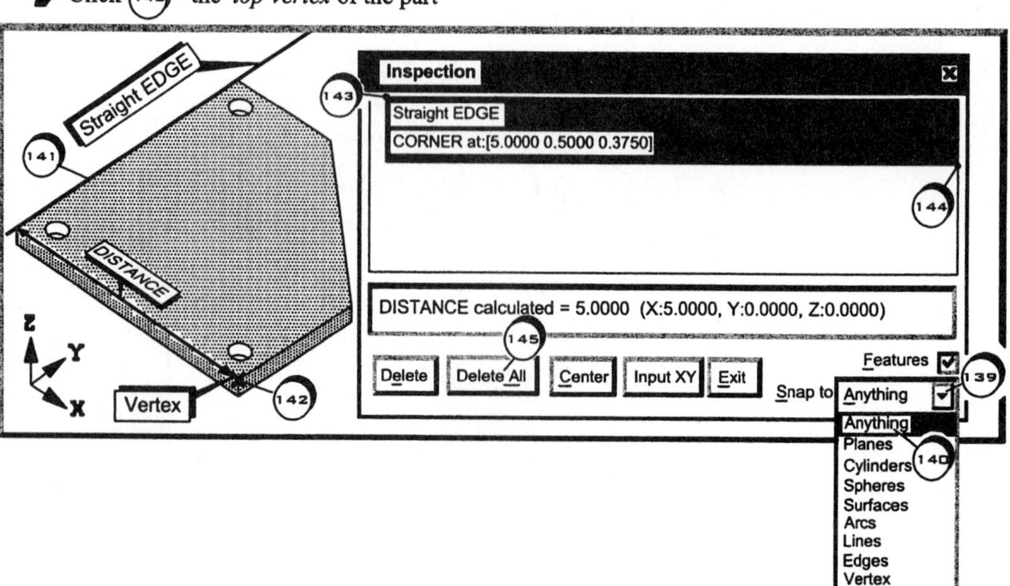

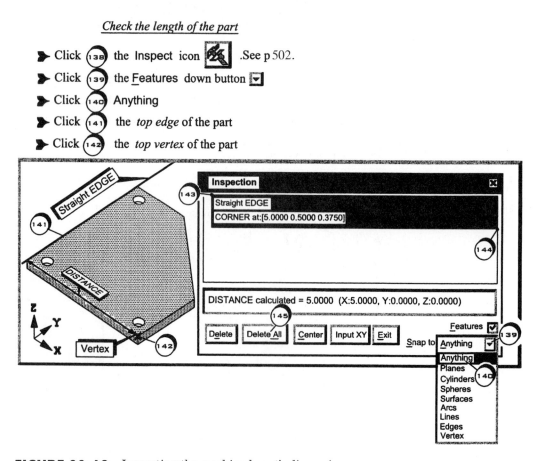

FIGURE 20–19 Inspecting the machined part's dimensions.

➤ Move the mouse cursor to position (143) *depress* the left mouse button and *keeping it depressed* move the cursor to position (144) and *release*

The system will display the ***DISTANCE*** between the two features selected

Check the location of a hole

➤ Click (145) the **Delete All** button. See p504

➤ Click (146) the *top edge* of the part

➤ Click (147) the *top edge* of the hole

➤ Click (148) the ARC feature

➤ Click (149) the **Center** button to obtain the *center point* for the hole

➤ Click (150) the **Delete** button

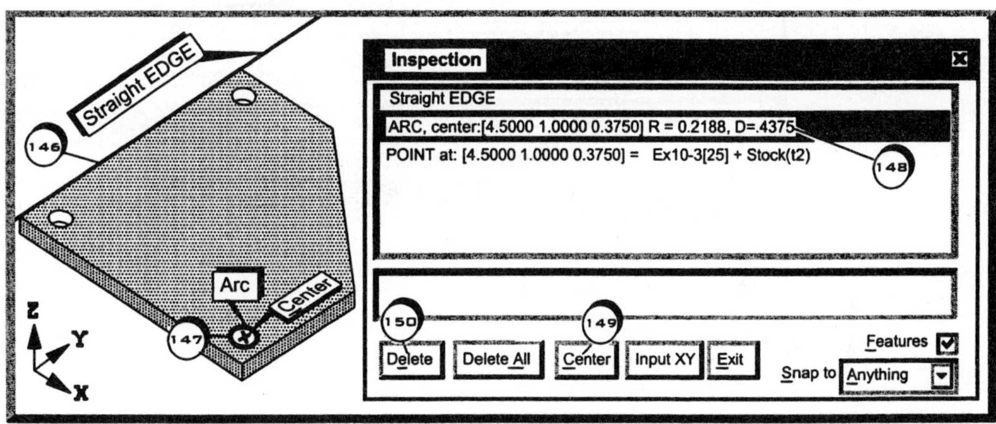

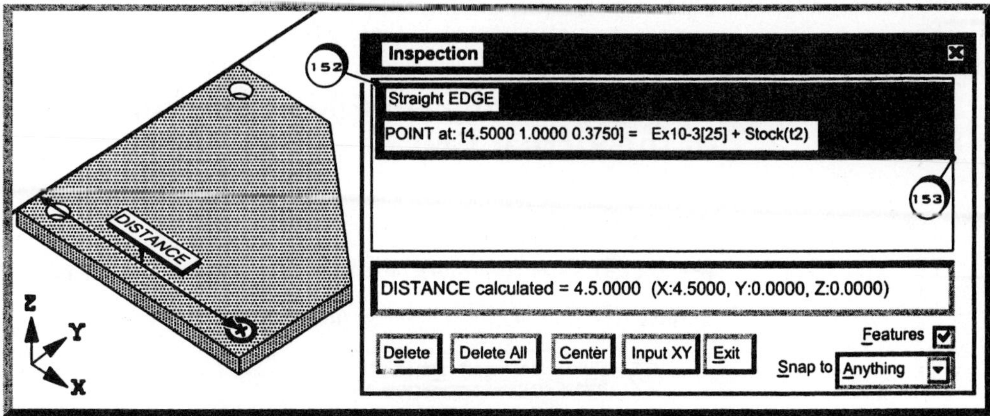

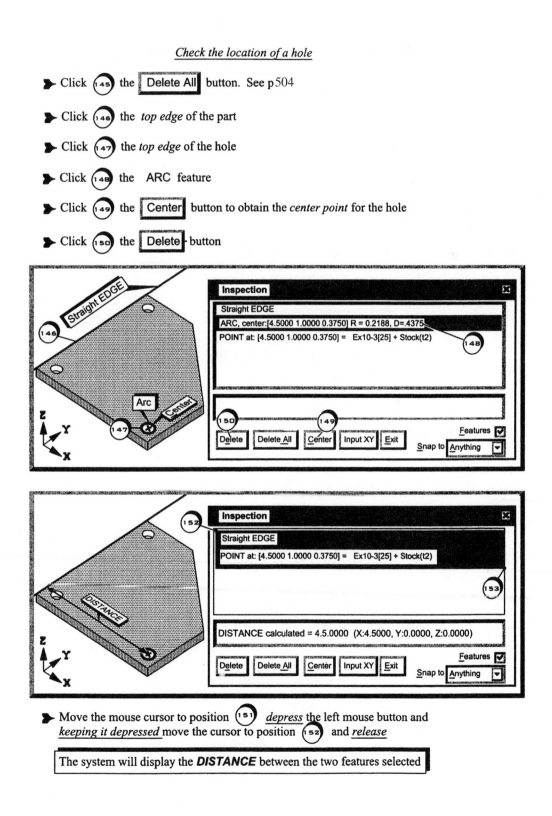

➤ Move the mouse cursor to position (151) *depress* the left mouse button and *keeping it depressed* move the cursor to position (152) and *release*

The system will display the ***DISTANCE*** between the two features selected

20–21 ENTERING THE CNC EDITOR FROM VIRTUAL CNC

The operator can enter the CNC editor and modify the current word address program directly from the virtual CNC.

Entering the CNC Editor

➤Click ① Edit

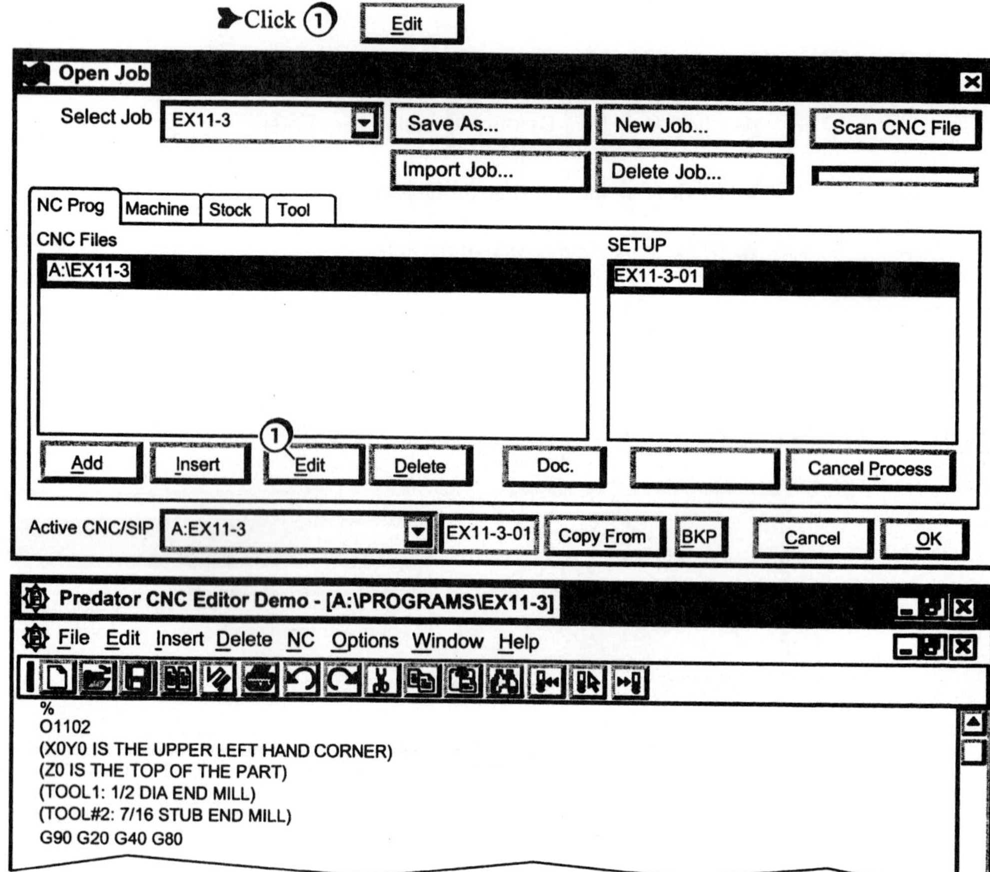

Saving the Program File and Returning to the Virtual CNC

➤Click ② File

➤Click ③ Save As *to save the edited program to the diskette in A drive*

➤Click ④ Exit *to return to the virtual CNC*

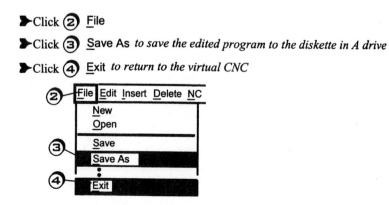

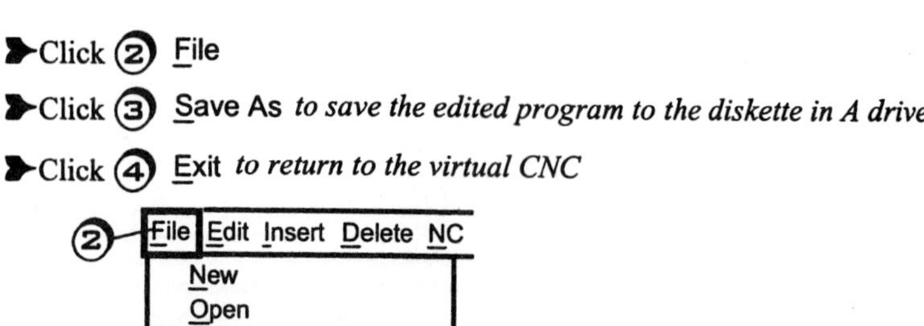

20–22 CREATING A NEW TOOL IN THE TOOL LIBRARY

If a required tool is not present in the tool library the operator can *create* a new tool. The new tool can be *saved* in the tool library for use in *other* jobs.

■ EXAMPLE 20–18

The operator currently has a job named EXE-TEST active and is ready to set up the tools required for the job. The **Scan CNC File** button has been clicked but the tool needed is not in the tool library. Create the required tool described as follows:

New Tool: .147 Dia Drill × 118° tip angle × 1.5 long.

Create the New Tool

- ➤Click ① the Tool tab
- ➤Click ② the Tool Definition tab
- ➤Click ③ the Tool Type down button
- ➤Click ④ the Tool Type DRILL
- ➤Click ⑤ in the T.Lib# box and enter the *name* of the new tool .147-DRILL
- ➤Click ⑥ in the Diameter box and enter the desired diameter .147
- ➤Click ⑦ in the Angle box and enter the desired tip angle 118
- ➤Click ⑧ in the Height box and enter the desired tool height 1.5

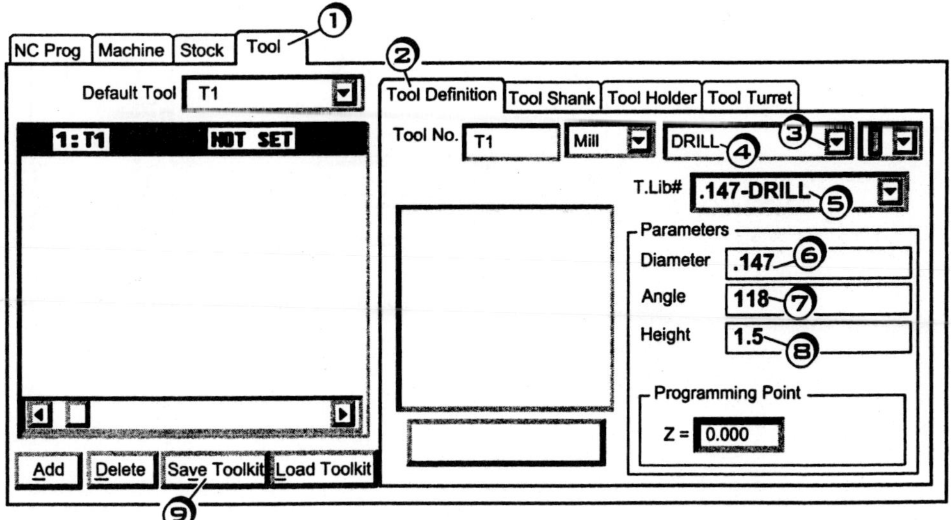

■■

Save the Tool File in the Tool Library

➤ Click (9) Save Toolkit See p507

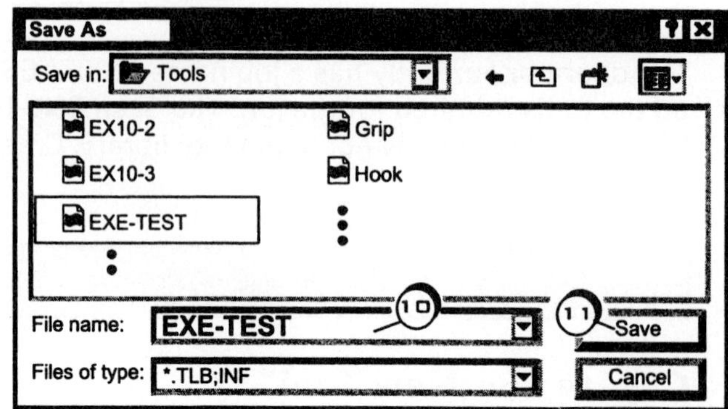

➤ Click (10) in the File name box and enter the name of the toolkit file to save: **EXE-TEST**

➤ Click (11) Save

Copy the Tool File from C drive to the Disk

➤ Click (9) Save Toolkit See p507

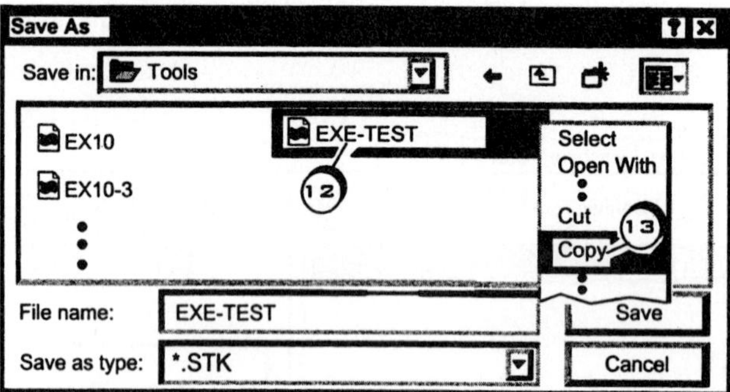

➤ *Right* Click (12) on the tools file to copy EXE-TEST

➤ Move the cursor down and click (13) Copy

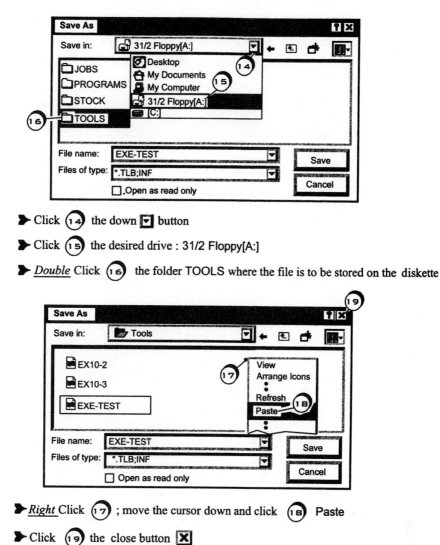

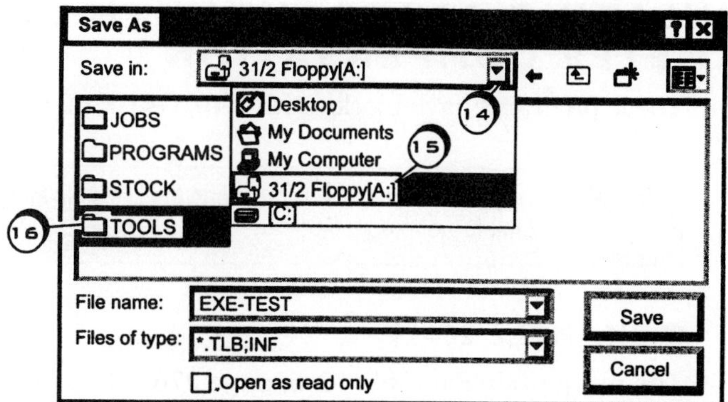

- Click (14) the down button
- Click (15) the desired drive : 31/2 Floppy[A:]
- *Double* Click (16) the folder TOOLS where the file is to be stored on the diskette

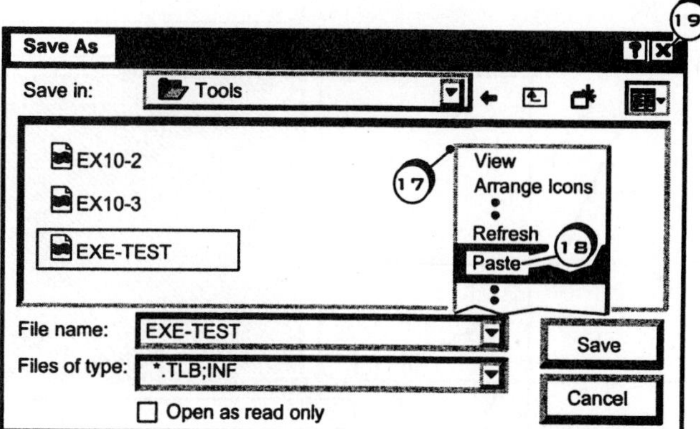

- *Right* Click (17) ; move the cursor down and click (18) Paste
- Click (19) the close button

The new .147 Dia drill tool has now been added to the tool library and is available for use in other jobs.

20–23 SIMULATING PROGRAMS WITH CUTTER DIAMETER COMPENSATION

These programs have blocks with G40, G41, G42 and D words.

> ***Note:*** The value(s) for ***D must be entered into the setup*** file for the job in order for the simulator to work properly.

■ EXAMPLE 20–19

The program in Example 13–6, page 276 is to be simulated. The machining is controlled via cutter diameter compensation as repeated here. ■■

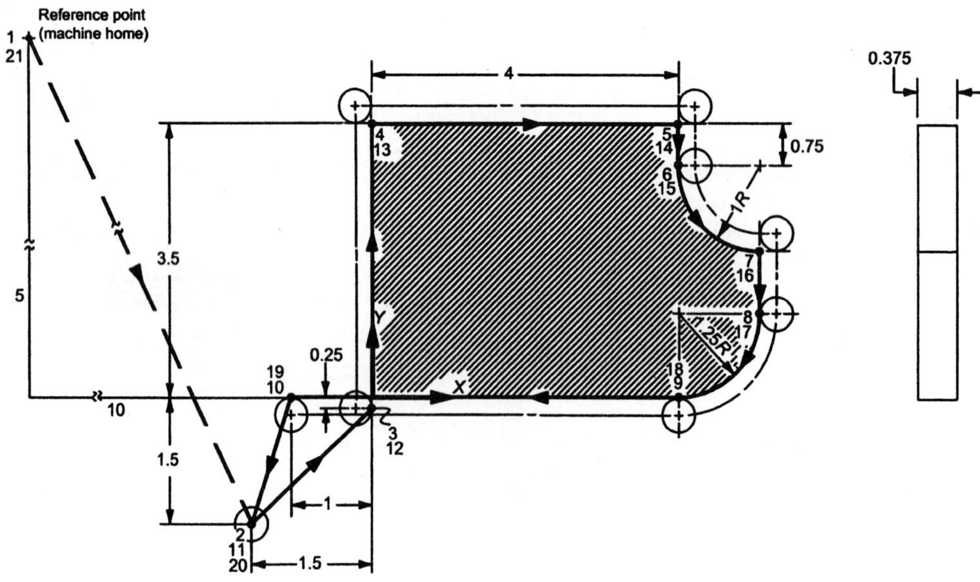

Tool	Operation	Tooling	Speed(rpm)	Feed(ipm)	D*	Effect
1	Rough profile leave .01 for finish	0.5D end mill	600	8	D21=.51	*cutter is offset by .51* CUTTER D21=.51 .01 ROUGH PROFILE FINISH PROFILE
1	Finish profile	0.5D end mill	800	6	D22=.5	*cutter is offset by .5* CUTTER D22=.5 FINISH PROFILE

* these values are entered at the MCU by the setup person

Enter the appropriate values for D21 and D22 into Predator's Virtual CNC setup file.

➤Click ① the Machine tab

➤Click ② the Machine Name Down button

➤Click ③ 3AXVMILL

➤Click ④ in the tool Length Offsets box and enter **H1=0**

➤Hit the carriage return Enter and enter **H2=0**

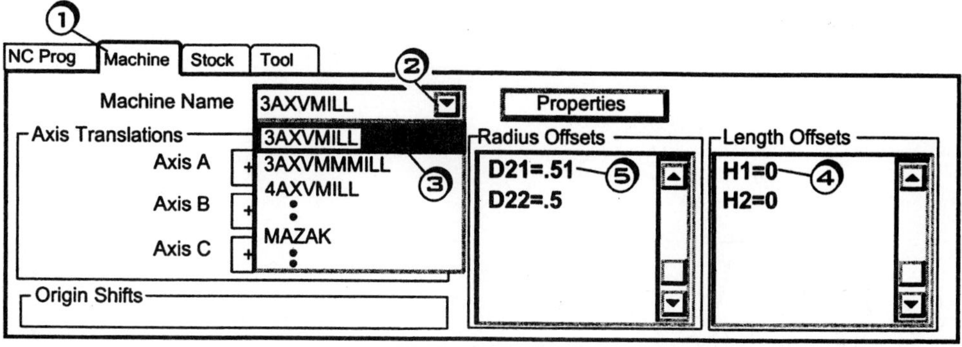

➤Click ⑤ in the tool Radius Offsets box and enter **D21= .51**

➤Hit the carriage return Enter and enter **D22=.5**

APPENDIX A

Important Safety Precautions

The reader is strongly advised to study and understand all safety precautions before entering the shop area. Special attention should be given to applying these precautions when working in the shop.

PERSONAL ATTIRE AND PERSONAL SAFEGUARDS

- Wear ANSI approved safety goggles and a protective shop apron.
- Avoid loose clothing and accessories (neckties, gloves, watches, rings, etc).
- Put long hair up under an approved shop cap.
- Use the legs, not the back, when lifting heavy loads.
- Avoid skin contact with cutting fluids.
- For cutting operation in excess of OSHA limits, wear a face mask.
- Flat non-slip safety shoes are recommended.
- Wear hearing protection for noise levels above OSHA specifications.
- Wear a face mask for dust levels above OSHA limits.
- Do not operate a CNC machine while under the influence of drugs (prescribed or otherwise).

  Regardless of how slight the injury may be, always notify the instructor. Apply first-aid treatment to any cuts or bruises.

SHOP ENVIRONMENT

- Keep the floor free from oil and grease.
- Remove chips from the floor. They can become embedded in the soles of shoes and cause dangerous slippage.
- Keep tools and materials off the floor.

TOOL SELECTION AND HANDLING

- Store tools in the tool tray.
- Make sure tools are sharp and in good condition.
- Carry sharp-edged tools with the edges or points down.
- Grind carbide or ceramic tools only in a well-ventilated area. Do not grind near any CNC machine tool.
- For carbide or ceramic insert tools, use the thickest insert possible.
- Select the thickest and shortest tool holder possible.

- Check all seating areas on holders to be sure each cutting tool rests solidly.
- Use only approved tools for the job being done.
- Do not exceed the manufacturers' recommended RPM for the tool.
- Do not force any tool.

CNC MACHINE TOOL HANDLING

- Secure all adjusting keys and wrenches before machining.
- Check compressed air equipment and any damaged parts on the CNC machine.
- Check the oil levels.
- Use only recommended accessories.
- Be sure the CNC machine is connected to a grounded permanent electrical power box and use lock-out tag-out practices.
- Keep the work area well-lighted.
- Be mindful of obstructions and sharp cutting tools when leaning into the work area.

MACHINING PRACTICES

- In case of any emergency while operating a CNC machine, hit the EMERGENCY STOP button.
- Prior to its operation, be sure to check that there are no obstacles in the machine's working area.
- Check the position height and cross-travel movements of each tool to verify noncollision with surrounding objects
  - when the tool is being lowered to a machining surface.
  - when the tool is executing a cutting operation.
  - when the tool is being moved to the tool change position.
- Use manufacturers' tables of recommended cutting speeds and feeds. Adjust these parameters based on specification of piece part accuracy, quality of surface finish, rate of tool wear, chip control, and machine capability.
- Make a dry run for safety checks.
- Check the workpiece to see if it is free of burrs and particles.
- Feed work in climb milling in the direction of spindle rotation.
- Make a deep first cut below the hard outer scale on castings, forgings, and other rough or irregular surfaces.
- Mill as close as possible to the work-holding device.
- Supply a continuous flow of cutting fluid to carbide tools when machining cast iron or steel.
- Reduce the speed and feed when drilling large-diameter holes to a depth of more than twice the drill size.
- Use slower speeds for thread cutting than for other turning operations.
- Finish cut an internal taper by moving the tool in the direction of the larger diameter.
- Do not remove debris while the CNC machine is running.

ALWAYS CONSULT WITH THE INSTRUCTOR BEFORE STARTING ANY UNFAMILIAR OPERATION.

APPENDIX

B

SUMMARY OF G CODES FOR MILLING OPERATIONS (FANUC CONTROLLERS)

G00	Function	Mode
G00	Rapid positioning (traverse tool movement)	Modal
G01	Linear interpolation (tool movement at feed rate)	Modal
G02	Circular interpolation clockwise (CW)	Modal
G03	Circular interpolation counterclockwise (CCW)	Modal
G04	Programmed dwell	Nonmodal
G09	Exact stop	Nonmodal
G10	Offset value for tool length	Nonmodal
G17	Plane selection *X, Y*	Modal
G18	Plane selection *X, Z*	Modal
G19	Plane selection *Y, Z*	Modal
G20	Input data in inches	Modal
G21	Input data in metric (mm)	Modal
G22	Programmed safety zone (no tool entry)	Modal
G23	Tool entry of programmed safety zone	Modal
G28	Return to reference point	Nonmodal
G29	Return from reference point	Nonmodal
G30	Return to second, third, and fourth reference point	Nonmodal
G33	Thread cutting autocycle	Modal
G40	Cancel cutter diameter compensation	Modal
G41	Cutter diameter compensation left	Modal
G42	Cutter diameter compensation right	Modal

G43	Tool length compensation (positive direction)	Modal
G44	Tool length compensation (negative direction)	Modal
G45	Tool offset increase	Nonmodal
G46	Tool offset decrease	Nonmodal
G47	Tool offset double increase	Nonmodal
G48	Tool offset double decrease	Nonmodal
G49	Cancel tool length compensation	Modal
G50	Scaling off	Modal
G51	Scaling on	Modal
G53	Cancels G54–G59 fixture offsets	Modal
G54-G59	Specifies fixture offset locations. G54 ″ ″ location 1 G55 ″ ″ location 2 : : G59 ″ ″ location 6	Modal
G65	Call user macro (one-time call)	Nonmodal
G66	Call user macro (repeat call)	Modal
G67	Cancel G66 function	Modal
G73	Peck drilling autocycle	Modal
G74	Counter tapping autocycle	Modal
G76	Fine boring autocycle	Modal
G80	Cancel any fixed cycles	Modal
G81	Drilling autocycle	Modal
G82	Counterboring autocycle	Modal
G83	Peck drilling autocycle	Modal
G84	Tapping autocycle	Modal
G85	Boring autocycle (return to reference level at feed)	Modal
G86	Boring autocycle (return to reference level at rapid)	Modal
G87	Back boring autocycle	Modal
G88	Boring cycle (manual return to reference level)	Modal
G89	Boring cycle (dwell, then return to reference at feed)	Modal
G90	Absolute programming mode	Modal
G91	Incremental programming mode	Modal
G92	Zero offset (programming of temporary zero point)	Nonmodal
G94	Per minute feed programming	Modal
G95	Per revolution feed programming	Modal

G98	Return to initial point in autocycle	Modal
G99	Return to *R* plane in autocycle	Modal

Note: Modal signifies a G code that remains in effect for all subsequent commands. It can only be canceled or modified by another G code that is modal. Nonmodal signifies the G code remains in effect only for the particular command line or block in which it occurs.

SUMMARY OF G CODES FOR TURNING OPERATIONS (FANUC CONTROLLERS)

G code	Function	Mode
G00	Rapid positioning (traverse tool movement)	Modal
G01	Linear interpolation (tool movement at feed rate)	Modal
G02	Circular interpolation clockwise (CW)	Modal
G03	Circular interpolation counterclockwise (CCW)	Modal
G04	Programmed dwell	Nonmodal
G09	Exact stop	Nonmodal
G10	Offset value for tool length	Nonmodal
G17	Plane selection *X, Y*	Modal
G18	Plane selection *X, Z*	Modal
G19	Plane selection *Y, Z*	Modal
G20	Input data in inches	Modal
G21	Input data in metric (mm)	Modal
G22	Programmed safety zone (no tool entry)	Modal
G23	Tool entry of programmed safety zone	Modal
G28	Return to reference point	Nonmodal
G29	Return from reference point	Nonmodal
G30	Return to second, third, and fourth reference point	Nonmodal
G33	Thread cutting autocycle	Modal
G34	Increasing lead thread cutting autocycle	Modal
G35	Decreasing lead thread cutting autocycle	Modal
G40	Cancel tool nose radius compensation	Modal
G41	Tool nose radius compensation left	Modal
G42	Tool nose radius compensation right	Modal
G50	Zero offset (programming of temporary zero point)	Nonmodal
G65	Call user macro (one-time call)	Nonmodal
G66	Call user macro (repeat call)	Modal
G67	Cancel G66 function	Modal

G68	Mirror image for double turrets on	Modal
G69	Mirror image for double turrets off	Modal
G70	Finish turning autocycle	Modal
G71	Rough turning autocycle	Modal
G72	Rough facing autocycle	Modal
G73	Repeat pattern	Modal
G74	Peck drilling autocycle (*Z* axis)	Modal
G75	Grooving autocycle (*X* axis)	Modal
G90	Absolute programming mode	Modal
G91	Incremental programming mode	Modal
G94	Per minute feed programming (G98 on some systems)	Modal
G95	Per revolution feed programming (G99 on some systems)	Modal

SUMMARY OF M CODES FOR MILLING AND TURNING OPERATIONS (FANUC CONTROLLERS)

M code	Function
M00	Program stop
M01	Optional program stop
M02	End of program (rewind tape)
M03	Spindle on (clockwise)
M04	Spindle on (counterclockwise)
M05	Spindle stop
M06	Program stop (manual tool change)
M08	Coolant on
M09	Coolant off
M13	Spindle on clockwise/coolant on
M14	Spindle on counterclockwise/coolant on
M17	Spindle off/coolant off
M19	Oriented spindle stop
M21	Mirror imaging about the *X* axis
M22	Mirror imaging about the *Y* axis
M23	Mirror imaging cancel
M30	Program end/memory reset
M41	Low-gear range for spindle
M42	High-gear range for spindle
M48	Override cancel off
M49	Override cancel on

M98	Transfer control to a subroutine
M99	Return from a subroutine

> ***Note:*** Many M codes are machine dependent and assigned by the machine tool builder. The machine tool manual should be checked for a complete list of M codes. For older controls, only one M code may be programmed in any block. If more than one is programmed, only the last code is considered effective.

SUMMARY OF CODES FOR AUXILIARY FUNCTIONS (FANUC CONTROLLERS)

Code	Function
Dn	D is the address in memory where the cutter radius offset value is stored. n(xx) is the number of the register containing the offset value.
Fn	F is the address in memory where the feedrate value is stored. n(xx.xxxx) is the numerical value of the feedrate.
Hn	H is the address in memory where the tool length offset value is stored. n(xx) is the number of the register containing the offset value.
In	Meaning varies depending upon G code. When used with G02/G03 circular interpolation (CW/CCW), I is the address in memory storing the incremental ±X distance from tool center to arc center with the tool positioned at the start of the arc cut. n(xx.xxxx) is the value of the distance. When used with G74 face grooving cycle, I is the address in memory storing the stepover/distance between grooves on the X-axis. n(xx.xxxx) is the value stepover/distance value.
Jn	J is the address in memory storing the incremental ±Y distance from tool center to arc center with the tool positioned at the start of the arc cut. n(xx.xxxx) is the value of the distance.
Kn	Meaning varies depending upon G code. When used with G02/G03 circular interpolation (CW/CCW) in the XZ or YZ plane, I is the address in memory storing the incremental ±Z distance from tool center to arc center with the tool positioned at the start of the arc cut. n(xx.xxxx) is the value of the distance. When used with G74 face grooving cycle, K is the address in memory storing the first peck distance below the clearance plane on the Z-axis. n(xx.xxxx) is the value of the distance.
Ln	L is the address in memory storing the number of times to repeat a call of a subprogram. n(xxxx) is the value of the number of times.
Nn	N is the address in memory storing a sequence number in a program. n(xxxx) is the value of the sequence number.
Pn	Meaning varies depending upon G-code. When used with a G82 counterbore/spotface cycle, P is the address in memory storing the dwell time. n(xx.xxxx) is the value of the dwell time.

	When used with G71 stock removal or G70 finishing cycle, P is the address in memory storing the sequence number of the block that starts the contour description. n(xxxx) is the value of the sequence number.
Qn	Meaning varies depending upon G-code. When used with a G83 peck drill cycle, Q is the address in memory storing the first peck distance below the R_{plane}. n(xx.xxxx) is the value of the distance. When used with G71 stock removal or G70 finishing cycle, Q is the address in memory storing the sequence number of the block that ends the contour description. n(xxxx) is the value of the sequence number.
Rn	Meaning varies depending upon G-code. When used with a G02/G03 circular profile milling, R is the address in memory storing the radius of the tool path when the arc is cut. n(xx.xxxx) is the value of the radius. When used with G81-85 hole machining cycles, R is the address in memory storing the distance to the R_{plane}. n(xx.xxxx) is the value of the distance.
Sn	S is the address in memory storing the spindle speed. When used with G97, n(xxxx) is the value of the speed in revolutions per minute (RPM). When used with G96, n(xxxx) is the value of the speed in surface feet per minute (SFM).
Tab	T is the address in memory storing tool change values. a(xx) is the number of the turret station where the new tool is located. b(xx) is the address in memory where the new tool's offset values are stored.
Xn	X is the address in memory where X-axis motion control is stored. n(xx.xxxx) is the value of the distance moved along the X-axis.
Yn	Y is the address in memory where Y-axis motion control is stored. n(xx.xxxx) is the value of the distance moved along the Y-axis.
Zn	Z is the address in memory where Z-axis motion control is stored. n(xx.xxxx) is the value of the distance moved along the Z-axis.

Note: The format n(xx.xxxx) indicates

a. the maximum range for the numerical value (the lowest range is one digit).

b. that a decimal point (.) *must always* be coded with the number.

Thus, X1 means move .0001 *in* along the X-axis

and

X1. means move 1 *in* along the X-axis.

APPENDIX C

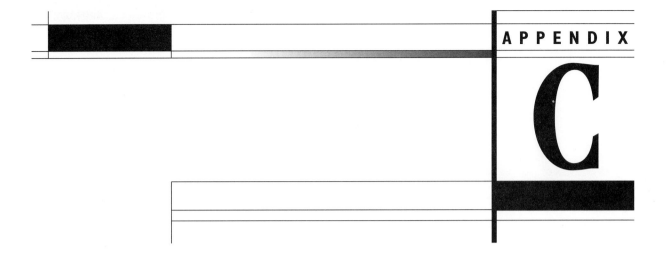

RECOMMENDED SPEEDS AND FEEDS FOR DRILLING

Drilling Speeds (High-Speed Steel Drills)

Material	Average drill speed (sfm)
Magnesium	300
Aluminum	250
Brass and bronze	200
Copper	70
Cast iron (soft)	120
Cast iron (hard)	80
Mild steel	110
Cast steel	50
Alloy steels (hard)	60
Tool steel	60
Stainless steel	30
Titanium	30
High manganese steel	15

Note: For carbide drills, double the average speeds.

Drilling Feeds	
Drill diameter (in.)	**Drill feed (ipr)**
$<\frac{1}{8}$	0.001–0.002
$\frac{1}{8}$–$\frac{1}{4}$	0.002–0.004
$\frac{1}{4}$–$\frac{1}{2}$	0.004–0.007
$\frac{1}{2}$–1	0.007–0.015
>1	0.015–0.025

Note: Use a slightly lower feed when drilling hard materials such as high carbon steel, tool steel, or stainless steel. Use a slightly higher feed when drilling softer materials such as magnesium, aluminum, or soft cast iron.

RECOMMENDED SPEEDS AND FEEDS FOR MILLING

Milling Speeds (High-Speed Steel Tools)	
Material	**Average tool speed (sfm)**
Magnesium	300
Aluminum	250
Brass and bronze	150
Copper	100
Cast iron (soft)	80
Cast iron (hard)	50
Mild steel	90
Cast steel	80
Alloy steel (hard)	40
Tool steel	50
Stainless steel	60
Titanium	50
High manganese steel	30

Note: For carbide cutting tools, double the average speeds.

Milling Feeds			
	Tool feed (inch/tooth)		
Material	Face mills	Side mills	End mills
Magnesium	0.005–0.020	0.004–0.010	0.005–0.010
Aluminum	0.005–0.020	0.004–0.010	0.005–0.010
Brass and bronze	0.004–0.020	0.004–0.010	0.005–0.010
Copper	0.004–0.010	0.004–0.007	0.004–0.008
Cast iron (soft)	0.004–0.016	0.004–0.009	0.004–0.008
Cast iron (hard)	0.004–0.010	0.002–0.006	0.002–0.006
Mild steel	0.004–0.010	0.002–0.007	0.002–0.010
Alloy steel (hard)	0.004–0.010	0.002–0.007	0.002–0.006
Tool steel	0.004–0.008	0.002–0.006	0.002–0.006
Stainless steel	0.004–0.008	0.002–0.006	0.002–0.006
Titanium	0.004–0.008	0.002–0.006	0.002–0.006
High manganese steel	0.004–0.008	0.002–0.006	0.002–0.006

RECOMMENDED SPEEDS AND FEEDS FOR TURNING

Turning Speeds (High-Speed Steel Tools)		
	Average tool speed (sfm)	
Material	Rough cuts	Finish cuts
Magnesium	400	800
Aluminum	350	700
Brass and bronze	250	500
Copper	100	250
Cast iron (soft)	100	250
Cast iron (hard)	50	150
Mild steel	100	250
Cast steel	70	150
Alloy steel (hard)	50	150
Tool steel	50	150
Stainless steel	60	180
Titanium	90	200
High manganese steel	40	100

Note: For carbide cutting tools, double the average speed.

Turning Feeds

Material	Tool feed (ipr)	
	Rough cuts	Finish cuts
Magnesium	0.015–0.025	0.005–0.010
Aluminum	0.015–0.025	0.005–0.010
Brass and bronze	0.015–0.025	0.003–0.010
Copper	0.010–0.020	0.004–0.008
Cast iron (soft)	0.015–0.025	0.005–0.010
Cast iron (hard)	0.010–0.020	0.003–0.010
Mild steel	0.010–0.020	0.003–0.010
Cast steel	0.010–0.020	0.003–0.010
Alloy steel (hard)	0.010–0.020	0.003–0.010
Tool steel	0.010–0.020	0.003–0.010
Stainless steel	0.010–0.020	0.003–0.010
Titanium	0.010–0.020	0.003–0.010
High manganese steel	0.010–0.020	0.003–0.010

APPENDIX D

Summary of Important Machining Formulas

(Courtesy of Machining Data Handbook)

Shop Formulas for Turning, Milling, and Drilling—English Units

Parameter	Turning	Milling	Drilling
Cutting speed, fpm	$V_c = 0.262 \times D_t \times \text{rpm}$	$V_c = 0.262 \times D_m \times \text{rpm}$	$V_c = 0.262 \times D_d \times \text{rpm}$
Revolutions per minute	$\text{rpm} = 3.82 \times \frac{V_c}{D_t}$	$\text{rpm} = 3.82 \times \frac{V_c}{D_m}$	$\text{rpm} = 3.82 \times \frac{V_c}{D_d}$
Feed rate, in./min	$f_m = f_r \times \text{rpm}$	$f_m = f_t \times n \times \text{rpm}$	$f_m = f_r \times \text{rpm}$
Feed per tooth, in.	—	$f_t = \frac{f_m}{n \times \text{rpm}}$	—
Cutting time, min	$t = \frac{L}{f_m}$	$t = \frac{L}{f_m}$	$t = \frac{L}{f_m}$
Rate of metal removal, in.^{3}/min	$Q = 12 \times d \times f_r \times V_c$	$Q = w \times d \times f_m$	$Q = \frac{\pi D^2}{4} \times f_m$
Horsepower required at spindle*	$\text{hp}_s = Q \times P$	$\text{hp}_s = Q \times P$	$\text{hp}_s = Q \times P$
Horsepower required at motor*	$\text{hp}_m = \frac{Q \times P}{E}$	$\text{hp}_m = \frac{Q \times P}{E}$	$\text{hp}_m = \frac{Q \times P}{E}$
Torque at spindle	$T_s = \frac{63030 \text{ hp}_s}{\text{rpm}}$	$T_s = \frac{63030 \text{ hp}_s}{\text{rpm}}$	$T_s = \frac{63030 \text{ hp}_s}{\text{rpm}}$

Symbols:
D_t = Diameter of workpiece in turning, inches
D_m = Diameter of milling cutter, inches
D_d = Diameter of drill, inches
d = Depth of cut, inches
E = Efficiency of spindle drive
f_m = Feed rate, inches per minute
f_r = Feed, inches per revolution
f_t = Feed, inches per tooth
hp_m = Horsepower at motor
hp_s = Horsepower at spindle
L = Length of cut, inches
n = Number of teeth in cutter
P = Unit power, horsepower per cubic inch per minute
Q = Rate of metal removed, cubic inches per minute
rpm = Revolutions per minute of work or cutter
T_s = Torque at spindle, inch-pounds
t = Cutting time, minutes
V_c = Cutting speed, feet per minute
w = Width of cut, inches

*Unit power data is given in the following table for turning, milling, and drilling.

Average Unit Power Requirements for Turning, Drilling, and Milling—English Units

		Unit power hp/in.³/min					
		Turning P_t HSS and carbide tools (feed 0.005–0.020 ipr)		Drilling P_d HSS drills (feed 0.002–0.008 ipr)		Milling P_m HSS and carbide tools (feed 0.005–0.012 ipr)	
Material	**Hardness Bhn**	**Sharp tool**	**Dull tool**	**Sharp tool**	**Dull tool**	**Sharp tool**	**Dull tool**
Steels, wrought and cast	85–200	1.1	1.4	1.0	1.3	1.1	1.4
Plain carbon	35–40 R_c	1.4	1.7	1.4	1.7	1.5	1.9
Alloy steels	40–50 R_c	1.5	1.9	1.7	2.1	1.8	2.2
Tool steels	50–55 R_c	2.0	2.5	2.1	2.6	2.1	2.6
	55–58 R_c	3.4	4.2	2.6	3.2	2.6	3.2
Cast irons	110–190	0.7	0.9	1.0	1.2	0.6	0.8
Gray, ductile, and malleable	190–320	1.4	1.7	1.6	2.0	1.1	1.4
Stainless steels, wrought and cast	135–275	1.3	1.6	1.1	1.4	1.4	1.7
Ferritic, austenitic, and martensitic	30–45 R_c	1.4	1.7	1.2	1.5	1.5	1.9
Precipitation hardening							
stainless steels	150–450	1.4	1.7	1.2	1.5	1.5	1.9
Titanium	250–375	1.2	1.5	1.1	1.4	1.1	1.4
High-temperature alloys							
Nickel and cobalt base	200–360	2.5	3.1	2.0	2.5	2.0	2.5
Iron base	180–320	1.6	2.0	1.2	1.5	1.6	2.0
Refractory alloys							
Tungsten	321	2.8	3.5	2.6	3.3	2.9	3.6
Molybdenum	229	2.0	2.5	1.6	2.0	1.6	2.0
Columbium	217	1.7	2.1	1.4	1.7	1.5	1.9
Tantalum	210	2.8	3.5	2.1	2.6	2.0	2.5
Nickel alloys	80–360	2.0	2.5	1.8	2.2	1.9	2.4
Aluminum alloys	30–150						
	500 kg	0.25	0.3	0.16	0.2	0.32	0.4
Magnesium alloys	40–90						
	500 kg	0.16	0.2	0.16	0.2	0.16	0.2
Copper	80 R_s	1.0	1.2	0.9	1.1	1.0	1.2
Copper alloys	10–80 R_s	0.64	0.8	0.48	0.6	0.64	0.8
	80–100 R_s	1.0	1.2	0.8	1.0	1.0	1.2

(Courtesy of Institute of Advanced Machining Sciences)

GLOSSARY

AA Stands for the Aluminum Association. This organization is responsible for issuing identification codes for wrought and cast aluminum alloys.

A-axis The axis of a circular motion of a machine tool member or slide about the *X*-axis.

Absolute dimensioning A system of specifying part dimensions whereby each new dimension is taken relative to a fixed origin.

Accuracy The difference between a programmed tool position and the actual position achieved by a CNC machine. Thus, a machine with an accuracy of ±0.001 would respond to a programmed position of $X = 1$, $Y = 1$ by moving the tool to $X =$ between 0.999 and 1.001 and $Y =$ between 0.999 and 1.001.

Actual size The actual or finished production size of a part.

Adaptive control A technique for achieving optimum cutting conditions. The speed and feed of a tool are continuously adjusted by a computer acting on sensor feedback data, which measures tool stress, heat, vibration, and so on.

Address A letter or group of letters or numbers that properly identify to the computer the type of information that follows. Thus, in the instruction X3.5 the address code X signals the computer that 3.5 is a coordinate value that refers to the *X*-axis.

Allowance The intentional minimum clearance or maximum interference assigned between mating parts.

Alphanumeric code A system of entering information using alphabetic characters (A through Z), numeric characters (0 through 9), and special characters (+, −, *, /, etc.).

Analog A system in which data is gathered continuously from a sensor that monitors some physical activity. A tachometer is an analog device that produces an output voltage signal in direct proportion to sensed speed.

Annealing A process whereby a steel part is heated to a point above its critical temperature and allowed to slow cool in a closed furnace. Annealing relieves internal stresses caused by machining.

Anodizing A process whereby an aluminum part develops a hard ceramic film of aluminum oxide when placed in an electrolyte acid bath.

ANSI Stands for the American National Standards Institute. One of the functions of this governing body is to specify drafting standards for part prints.

APT Stands for Automatically Programmed Tools. APT was the first and remains today the most powerful computer-aided part programming language. It runs on large- and medium-sized computers and can be used for four- and five-axis machining of complex part surfaces.

ASCII Stands for the American Standard Code for Information Interchange. With this system the computer represents each character, internally, as a unique series of seven bits. For example, *Y* is represented as 1011001.

ASME Stands for the American Society of Mechanical Engineers. One of the functions of this organization is to develop and issue standards for indicating dimensions tolerances in part prints.

ASTM Stands for the American Society of Testing Materials. This organization is responsible for issuing part material identification standards.

Automap Stands for automatic machining programming. This computer-aided programming language is a subset of APT. It can run on minicomputers and is used for simple profiling and point-to-point machining operations.

Autospot Stands for automatic system for positioning tools. It is a computer-aided programming language for positioning NC tools and executing straight-line cuts.

Auxiliary functions Additional programmable functions of the machine other than coordinate movement of the tool. These are referred to as M codes in word address format and are used to specify machining parameters such as speeds, feeds, and so on.

Auxiliary storage Refers to the long-time storage of information that resides temporarily in the computer's memory. Auxiliary storage devices include punched tape, magnetic tape, and magnetic disk.

Auxiliary view A view that shows the true shape of an inclined face.

Axis A line of reference for a coordinate system. In CNC programming the *X*, *Y*, and *Z* axes extend out at right angles to each other from a common origin. They are used to specify a direction for moving a tool.

Axis inhibit A feature of a CNC unit that enables the operator to withhold command information from the machine tool slide.

B-axis The axis of a circular motion of a machine tool member or slide about the *Y*-axis.

Backlash Undesirable movement that occurs between interacting mechanical parts due to looseness in the assembly.

Basic size, Basic dimension The theoretical exact size, location, profile, or orientation of a geometrical feature of an object. Tolerance variations are applied to the basic dimension.

BHN Stands for the Brinell Hardness Number. This decimal number placed to the right of BHN specifies the hardness the material is to exhibit when indented by the test's hardened steel ball.

Binary code A system for internally representing information by using binary digits (0 or 1). This is the only code the computer can understand and operate on. All other computer-assisted programming languages are automatically translated into binary prior to any computer action.

Binary-coded decimal (BCD) A system of representing numbers using a unique set of four binary bits. For example, 5 = 0101.

Bit Short for binary digit. It represents one of the two possible electronic states inside a computer's switching circuits–switch on (1) or switch off (0). It is represented as the absence or presence of a hole in punched tape or a magnetized spot on magnetic tape or disk.

Blind hole A hole drilled to a specific depth in a part.

Block A group of words representing a complete operational instruction. A block is ended by an end-of-block (EOB) character.

Block delete A slash (/) entered in the front of a block directs the CNC system to ignore the block in a program.

Body Refers to the non-threaded portion of a screw shaft.

Boss Is a circular pad extending out from the surface of a casting or forging.

Buffer storage A location for storing information in a control system or computer for subsequent use when needed. This information can be immediately transferred to active memory and acted upon. Information stored in an auxiliary storage device such as tape or disk takes longer to load than that placed in a buffer.

Bug Slang word for a mistake or malfunction in the program or the system.

Byte A sequence of adjacent binary digits that is processed as a unit. A byte is normally shorter than a word.

C-axis The axis of a circular motion of a machine tool member or slide about the *Z*-axis.

Cancel A command that will discontinue any canned cycle or offset control such as G80 or G40.

CAD Stands for computer-aided design.

CAM Stands for computer-aided manufacturing.

Canned cycle A preset sequence of events that is executed by issuing a single command. For example, a G81 code will initiate a simple drilling cycle.

Carburizing (case hardening) A process whereby the surface of a soft steel part is hardened by heating it in a carbon saturated atmosphere and allowing it to cool slowly.

Cartesian coordinates A system for defining a point in space relative to a zero position. The point is located by specifying its distance along three mutually perpendicular axes (*X, Y, Z*) that intersect at a zero point or origin.

CD-ROM Short for *Compact Disk-Read Only Memory*. This device uses optical technology to retrieve large amounts of data from a 4.75″ dia spinning plastic disk. CD disk drives are also available with read as well as write capability.

Chad Pieces of material that are removed when holes are punched in cards or tape.

Chamfer A beveled edge machined to break a sharp external corner.

Channels Paths (tracks) that run parallel to the edge of the tape along which information may be stored by the presence or absence of holes or magnetized areas. The EIA standard one-inch tape has eight channels.

Character A number, letter, or symbol that is programmed into the computer.

Chip A single piece of silicon cut from a slice by scribing and breaking. It can contain one or more circuits but is packaged as a unit.

Circular interpolation A block of entered information directing the system to cut an arc or circle.

Closed loop system A control system whereby the resulting output is measured and fed back for comparison to the input command. The system attempts to adjust itself so that the output closely tracks the corresponding input.

CNC Stands for computer numerical control. CNC is a system of manufacturing parts by using an on-board computer or machine control unit (MCU) to control an NC machine.

Code A system describing the formation of characters on a tape for representing information in a language understood by the control system.

Command A signal or group of signals for initiating a step in the execution of a program.

Command readout The display of the table slide position resulting from the control system.

Compiler A device that automatically translates an inputted program into a corresponding binary code.

Continuous path operation An operation in which rate and direction of relative movement of machine members is under continuous numerical control. There is no pause for data reading.

Contouring control system A system that can cut a curve or arc by executing simultaneous motion along two or more axes.

Counterbored hole A hole whose end has been enlarged to a specific depth. This is done to bury the head of a bolt.

Countersunk hole A hole having a conical depression at its end. This is done to bury the top of a conical head bolt.

CPU Stands for the central processing unit of a computer. This device contains the memory, logic, and arithmetic processing circuits required to execute all inputted instructions.

Crest The top of the thread teeth for external threads or the bottom of the teeth for internal threads.

CRT Stands for cathode ray tube. This device is used for displaying all programmed input and the corresponding output of text and graphics from the MCU.

Cutter diameter compensation A feature of a control system in which the inputted cutter diameter and part profile data are used to automatically place the tool on the part boundary. This feature is most desirable when compensating for tool wear.

Cutter offset The distance from the part profile to the center of the cutter.

Cutting speed The motion of the tool over the work. It is measured in surface feet per minute.

Cycle A sequence of operations that is repeated regularly.

Data A representation of information in the form of words, symbols, numbers, letters, characters, etc.

Database A collection of information stored in an auxiliary storage unit such as magnetic disk or tape.

Datum A theoretically exact point, plane, or axis from which dimensions are taken. Datums are used for CNC parts to cut down on tolerance errors.

Datum dimensioning A system of dimensioning based on a common starting point.

Debug To detect, locate, and remove mistakes from a program. To troubleshoot.

Decimal code A code that utilizes base 10 to define magnitude. Decimal code is the standard numbering system for CNC.

Delta dimensioning A system of specifying part dimensions whereby each new dimension is taken relative to the last dimension given. This is also known as incremental dimensioning.

Diagnostic test A program run to check for failure or potential failure and to determine its location.

Die Refers to the cutting tool used to cut external threads.

Digit A character in any numbering system.

Digital Refers to the discrete states of a signal (on or off). A value is created by assembling a unique pattern of these signals. Most MCU controllers are digital computers.

Display A visual representation of data.

DNC Stands for direct numerical control. A DNC system consists of several CNC machines that are connected to and receive programs from a main or host computer.

Dwell A time delay created by a command in a program.

Edit The process of modifying an inputted program.

EIA code A conventional code used for systems that execute straight-cut and contouring operations. EIA code is used in eight-track one-inch punched tape.

Electroless plating A technique for applying a hard nickel coating on a part's surface by dipping it in an aqueous solution of sodium hypophosphite and nickel salts.

Electroplating A method of applying a metallic coating to a base metal by immersing it in a solution of the plating material salts and applying a DC voltage.

End-of-block character A character placed in a program that signals the end of a command line or block.

End of program A miscellaneous (M02) function that must be placed at the end of a program to indicate its completion.

End of tape A miscellaneous (M30) function placed at the end of a program to indicate the end of the tape. The procedure is reset, coolant off, spindle off, feed off.

Executive program A set of instructions that enables an MCU to behave like a milling processor, a lathe processor, and so on. The executive program is installed by the MCU manufacturer.

Feature Refers to any surface, angle, line, hole, etc. that is to be controlled for production accuracy.

Feedback In a closed loop system this represents the return signal or response of the system to an inputted instruction.

Feedback override A manual switch that enables the operator to alter the programmed feed rate during a cutting operation.

Feed rate The rate of movement of the tool into the work as related to a particular machining operation. Feed is measured in inches per minute for milling and inches per revolution for drilling.

Feedrate override A manual switch that enables the operator to alter the programmed feedrate or spindle speed during a cutting operation.

Fillet A rounding on an interior corner formed to prevent stress concentration failures.

First angle projection A multiview orthographic projection system wherein the object is placed in the first quadrant in front of the projection planes. This technique is used for production prints in European countries.

Fixed canned cycle See canned cycle.

Flame hardening A process whereby the surface of a soft steel part is hardened by first exposing it to heat from an oxy-acetylene torch and then applying a quenching spray.

Flat taper An inclined face or bevel machined at a sharp corner.

Floating zero A feature of a machine control unit that enables the operator to establish a zero or starting point anywhere within the limits permitted by the tool travel.

Format The manner in which information must be arranged in an instruction.

FRN Stands for feed rate number.

Full Indicator Movement (FIM) Refers to open gaging and specifies that the full movement of the inspection instrument's dial indicator must not exceed the GDT tolerance(s) specified.

Full Indicator Reading (FIR) See Full Indicator Movement.

Functional gaging A complete gage fixture that is specially designed to inspect a part for compliance with a GDT tolerance specification.

Gage height A preset height above the work to which the tool retracts after an operation. This safe height, also called the *R* plane, allows for the tool to be moved parallel to the (*XY*) table plane without hitting an obstacle. Gage height is usually set at 0.1 to 0.125 in.

G code A code in word address format that signals for a preparatory function to follow. G codes initiate such operations as drilling, milling, and canned cycle execution.

GDT Stands for geometric dimensioning and tolerancing. This is a system of specifying tolerances based on how a part is to function.

Hardening A process whereby the surface and interior of a soft steel part is changed to a hard martensite structure by first heating it in a furnace and then applying controlled cooling.

Hard copy Printed output from the computer.

Hardware All physical units that make up a computer or control system. This includes the CPU, tape drive, MCU display, and so on.

Home The fixed *X, Y, Z* starting point on a machine from which all subsequent tool motion is measured.

Hot dipping A method of applying a corrosion-resistant coating to aluminum or steel parts by dipping them into certain types of molten metals.

HRA, HRB, or HRC Rockwell hardness test A,B, or C. An integer number placed to the right of the abbreviation specifies the hardness the material is to exhibit for a specific test.

Incremental dimensioning See delta dimensioning.

Induction hardening A method of hardening the surface of a soft steel part by inducing heat with the aid of an electric induction coil.

Initial level The *Z* height of the spindle at the start of a canned cycle.

Input All external information entered into the MCU control unit. Input is entered via punched tape, magnetic tape, or disk or from the MCU keyboard.

Integrated circuit A complete miniaturized electronic circuit that has been etched into a small silicon wafer or chip.

Interface A connecting unit or circuit that allows two pieces of electronic hardware to communicate information to one another.

Interpolation The process of determining an intermediate point located between two given data points. CNC systems use interpolation to generate straight-line and arc tool movements.

Jog A manually operated device used to move the tool in the *X, Y,* or *Z* direction.

Lay The direction of the predominant surface pattern caused by the method of production.

Lead The distance the thread travels along its axis during one complete revolution.

Leading zero The redundant zeros appearing to the left of a number in a program. For example, the two zeros in Y.0025 are leading zeros.

Leading zero suppression A feature of a control system that allows for the elimination of leading zeros on input. Thus, Y.0025 can be input as Y25.

Limits The largest (upper limit) and the smallest (lower limit) variations permitted from the basic size.

Linear interpolation A feature of a control system to accept two positioning points on a part profile and fit a straight line between the points. The system subsequently moves the tool along the straight line.

LMC or L Stands for least material condition. The condition at which an external feature (such as a shaft) is at the smallest size permitted and an internal feature (such as a hole) is at the largest size permitted.

Loop An instruction or series of instructions that are to be executed repeatedly to produce a desired operation.

Machining center A CNC machine that, at one setup, is capable of executing such operations as milling, drilling, boring, tapping, and reaming on one or more faces of a part.

Macro A complete set of instructions for executing a particular operation. Upon encountering the macro name in a program, the MCU executes all the instructions it contains. The parameters of a macro can be altered to fit a particular programming requirement.

Magnetic tape A plastic tape that is coated with magnetic material. Information stored on it in the form of magnetized spots can be fed into the MCU as required.

Major diameter The largest diameter on external or internal threads.

Manual data input A feature of a control system that allows the operator to manually key in a program or alter the commands of an inputted program.

Manual part programming The preparation of a manuscript in machine control language and format to define a sequence of commands for use on a CNC machine.

Manuscript A special form the part programmer uses when writing a program. The form normally contains the machining details such as processing method, tool types, feeds, and speeds.

Material hardness The resistance of the material to local penetration, scratching, machining, wear abrasion, and yielding.

Memory A unit used by a computer to store information. There are two types of memory: short term or active residing within the MCU and long term or auxiliary residing on such media as magnetic disk or tape.

Microinch (μ in) One millionth of an inch or .000001 in.

Micrometer (μ m) One millionth of a meter or .000001 m.

Microprocessor A complete miniaturized circuit capable of executing logic and control functions. The intelligence in a microcomputer is the microprocessor.

Minor diameter The smallest diameter on external or internal threads.

Mirror imaging A feature of a control system whereby the computer is capable of taking an inputted profile of part geometry from the first quadrant ($+X$, $+Y$) and automatically mirroring the profile about the Y or X axis. The system can subsequently drive a tool about the mirror images.

Miscellaneous functions Machine on/off control functions. They are designated as M codes in word address format. Thus, M03 signals for spindle on clockwise.

MMC or M Stands for maximum material condition. The condition at which an external feature (such as a shaft) is at the largest size permitted and an internal feature (such as a hole) is at the smallest size permitted.

Modal The information that stays in effect until replaced by a cancelation or newer modal information.

NC Stands for numerical control. This is a system of automatically controlling the motion of a tool by acting on numerical inputted data.

Nitriding A method of hardening the surface of a soft steel part by first placing it in a heated nitrogen atmosphere, then slow cooling it in a retort.

Nominal size Refers to the basic size of an object. Examples include stock size or thread diameter size.

Normalizing A method of improving the machinability of hardened steel parts by first heating to a point above that used for annealing then cooling in still air at room temperature.

Off-line programming A technique whereby a part program is developed away from a particular CNC machine. After it is written, manually or with the aid of a computer, the program can be transferred to the CNC machine.

Open gaging A method of determining compliance with GDT tolerance specifications by using such devices as dial indicators, height gages, surface plates, micrometers, calipers, and coordinate measuring machines (CMM's).

Open loop system A control system that has no means of comparing the output with the input for control purposes. This means feedback is absent.

Optional stop A miscellaneous (M01) command in a program that is ignored unless the operator has previously pushed a button to initiate the command.

Orthographic projection A technique for displaying an object on a projection plane that is oriented perpendicular to the parallel lines of sight.

Overshoot The amount by which a tool is moved beyond the programmed position. This is, in effect, an error caused by the mechanical and electrical inefficiencies of the control system. Overshoot is affected by part mass, feed rate, and servo motor response.

Parity check An extra hole punched in one of the track columns or channels of a tape to make an even number of existing holes odd as required by the RS-244A format. The extra hole would be punched in RS-358B tape to make the existing odd number even.

Part programmer A person who prepares the planned sequence of events for the operation of a CNC machine.

Part programs A complete set of instructions written by a part programmer in a programming language (word address, etc.). The MCU follows the part program's instructions in manufacturing the part.

Perforated tape Punched tape in which the hole pattern corresponds to the instructions of a part program.

Phosphating A method of applying a protective coating on iron and steel parts by immersing them in or spraying them with a dilute solution of phosphoric acid and other chemicals.

Pitch The distance between the adjacent crests or roots of thread teeth.

Plotter A device that will draw a plot or trace from coded CNC data input.

Polar coordinates A method of locating a point by specifying the length of a line from an origin to the point and the angle the line makes with the positive X axis.

Positioning (contouring) A control system in which the tool can remain in continuous contact with the part as it is moved from point A to point B.

Positioning (point to point) A control system in which the tool is not in continuous contact with the part as it is moved from point A to point B.

its cutting operations can only be controlled at its destination point *B*.

Postprocessor A program residing inside the MCU that takes the part program and translates it into a corresponding set of machine control instructions required to cut the part.

Preparatory function A command for changing the mode of operations of the control. In word address format, a preparatory function is given as a G plus a two-digit code appearing at the beginning of a block. Thus, G01 signals for a linear cut.

Program stop A miscellaneous (M00) function entered in a program to stop the CNC machine. Coolant, spindle, and feed activity is halted after the completion of a tool movement. The operator can restart the program again by pushing a button.

Quadrant Any of four parts into which a plane is divided by rectangular coordinate axes in that plane.

Rapid Positioning the slides at a high rate of speed, usually 150 to 400 inches per minute (IPM) before a cut is started.

Repeatability Closeness of agreement of tool movement positions from one part to another when cutting several copies of the same part.

Reset Returning a storage location in the MCU to zero or to a specified initial value.

Root Refers to the bottom of the thread teeth for external threads and the top of the thread teeth for internal threads.

Roughness Refers to the finer irregularities created on a part's surface by the production process.

Roughness average (Ra) Refers to the average roughness value over a control distance on a part's surface. Average roughness is expressed in terms of microinches (μin) or micrometers (μm).

Round A rounding at the exterior corner of two surfaces.

Row A path perpendicular to the edge of a tape along which information may be stored by the presence or absence of holes or magnetized areas. A character is represented by a combination of holes or magnetized spots.

SAE Stands for the Society of Automotive Engineers. This organization is responsible for issuing standards for identifying materials.

Sectional view A view that shows the internal features of an object.

Sequence number Consists of the letter N followed by a one- to four-digit number. A sequence number appears at the beginning of a command block to specify the block's location in the program.

Significant digit A digit that must be retained to ensure a specific accuracy in a command. Thus, in the specification Y0035620 the leading zeros are insignificant and 35620 are significant digits.

Software Part programs or instructions that are used to drive the computer's circuits in executing a particular operation.

Spherodizing A method of improving the machinability of hardened steel parts by first heating to a point below the critical temperature then cooling in still air at room temperature.

Spindle speed The rotational speed or revolutions per minute (rpm) of the machine spindle. In word address, spindle speed is specified by the letter *S* followed by the rpm value. Thus, S1200 signals for a spindle speed of 1200 rpm.

Spotface An enlargement machined at the end of a hole to a shallow depth. It is intended for seating a washer or the head of a bolt.

Storage A device used to hold information off-line until it is needed by the MCU.

Stress relieving A method of lowering internal stresses from machining operations by first heating to a point 100 to 200°F below the critical temperature and then slow cooling in a closed furnace.

Subroutine A set of instructions that executes a complete operation and is assigned a name. When the name is entered in another program, the system automatically executes all the instructions of the subroutine.

Tab A series of holes punched into a tape to separate words or groups of characters in the tab sequential format.

Tap A tapered tool with fluted cutting edges used to cut internal threads.

Tap drill A drill used to make a hole in a part to accommodate a thread cutting tap.

Tape A perforated paper or magnetized medium for storing information in the form of holes or magnetized spots.

Tempering A method of changing the hardness of steel parts by first heating to a low temperature then slow cooling. Tempering prevents cracking during storage, installation, or use.

Third angle projection A multiview orthographic projection system wherein the object is placed in the third quadrant behind the projection planes. This technique is used for production prints in the United States and Canada.

Thread depth The perpendicular distance between the crest and root of thread teeth.

Thread form The shape of the threads cut into a shaft or hole.

Thru hole A hole drilled through the entire material of the part.

Tolerance The total amount by which a dimension may vary. It may also be defined as the difference between the upper and lower limits for the dimension. Tolerances control the accuracy with which a part is made.

Tolerance zone The zone or shape in space that represents the tolerance and its position relative to the basic size. Tolerance zones may be rectangular or cylindrical.

Tool function A command that identifies a tool and calls for its selection. In word address this is programmed as the letter *T* followed by the tool register number. Thus, T01 signals for tool 1 to be selected for use.

Tool length offset The distance between the bottom of the fully retracted tool and the part Z_0.

Tool offset A correction entered for a tool's position parallel to a tool movement axis. This feature allows for compensation to be applied due to tool wear and for executing finish cuts.

Trailing zero suppression A feature of the control system that allows for the elimination of trailing zeros on input. Thus, Y2500 can be input as Y25.

UNS A numbering system developed by the SAE and ASTM for identifying materials.

Virtual condition The mating condition of the part. For external features it is the distortion that results when the part's size at M is added to the geometric tolerance at M. For internal features it is the distortion that results when the part's size at M is subtracted from the geometric tolerance at M.

Word An ordered set of characters used to execute a specific action of a machine tool.

Word address format A system of coding instructions whereby each word in a block is addressed by using one or more alphabetic characters identifying the meaning of the word.

X-axis Axis of motion that is always horizontal and parallel to the workholding surface.

Y-axis Axis of motion that is perpendicular to both the *X* and *Z* axis.

Z-axis Axis of motion that is always parallel to the principal spindle of the CNC machine tool.

Zero offset A feature of an MCU that allows the programmer to shift the zero or starting point for movements to a new position over a specified range. The system can be switched back to its old permanent origin if desired.

Zero shift Operates in a manner similar to zero offset except that the system cannot be switched back to the origin set prior to the shift.

Zero suppression Eliminating zeros either before or after the significant digits entered in an instruction.

INDEX

Entries in **Bold** indicate figure or table entry

D

R

S